NATURE versus NURTURE versus NIRVANA

An Introduction to Reality and Quantum Mechanics

Book Four of the Ultimate Model of Reality Series

By: Mark My Words

I always wanted to know how everything works; and, now I do.

The cost?

It cost me My Materialism, My Naturalism, My Nihilism, and My Atheism. It cost me my self-respect and good reputation among the Materialists, Naturalists, Darwinists, Nihilists, Behaviorists, and Atheists. They call me names and ban me from their websites. I'm no longer permitted to associate with them.

So, what did I gain after paying this cost?

I got Quantum Mechanics or Energy Mechanics, Quantum Field Theory or Energy Field Theory, Instantaneous Action at a Distance at the Quantum Level, No Speed Limits in the Quantum Realm or Spirit World, No Physical Limitations and No Entropy at the Quantum Level, the Quantum Field Model for Origins, Quantum Organization of Energy, Psychic Control of Energy at the Quantum Level, Quantum Tunneling or Teleportation of Physical Matter, the Quantum Zeno Effect or Telepathy, Quantum Processing or Quantum Waves or Thought, Instantaneous Communication at the Quantum Level, Quantum Phase-Shifting of Physical Matter, the Quantum Bubble of Protection, Quantum Transmutation, Quantum Non-Locality or Quantum Entanglement, Quantum Complementarity, Quantum Superposition or Multitasking at the Quantum Level, Quantum Consciousness, Syntropy or Conservation of Energy and Psyche, the Quantum Law of Thermodynamics, the Ultimate Law of Thermodynamics, the Quantum Law of Psyche, the Ultimate Model of Reality, Quantum Neuroscience, and the Biblical God Jesus Christ and His Atonement.

I gave away nothing, and I got everything in return. I went from being a closed-minded scientist to being an open-minded scientist. I slowly mastered critical thinking.

In 2012, I was an atheist. It took years for me to adjust; but, looking back now, I can see that the trade-off was very much worth the effort that it took for me to make these adjustments in my philosophy of life, schema, or world view. I'm now free to talk about and explore anything that interests me, where I wasn't before.

Every eastern religion defines Nirvana as some type of non-physical existence. In Hinduism's version of Nirvana, the Atman (the individual psyche or soul) unites with Brahman (God's psyche or soul); and, they become one. The psyche or soul returns to the God who gave it life, physical life. This book is about Nature (biology), Nurture (environment or society or those other psyches), and Psyche (Plato's version of soul or Hinduism's version of Nirvana). Within this book, Nirvana represents psyche or soul; and, it is my claim that all three aspects of reality are necessary to consider when trying to develop the Ultimate Model of Reality. I'm using the word "Nirvana" to represent the Quantum Realms, the Psyche Realm, the Transdimensional Realms, or the Spirit World. It works, and it has great explanatory power in the end.

This book is about including Spirit or Light into our Psychological Models and our Theoretical Models! Getting rid of Materialism and Behaviorism makes for better and more interesting Science.

Let's face REALITY. The KEY to understanding life, the universe, and everything is to eliminate Materialism or Naturalism, even if we have to find 42 different ways to do so. Once we get rid of Materialism, then suddenly everything starts to make logical sense. That has been my Scientific Observation and my Scientific Contribution to the world. Materialism and Naturalism are worthless when it comes time to explain the quantum or the non-physical. The whole quantum realm is non-local, meaning that it is non-physical. Quantum Mechanics is Energy Mechanics, or the study of the non-physical and how it works.

The Materialists never ask themselves what was there BEFORE the first physical matter was designed and created, or what was there BEFORE this physical universe was designed and created 13.8 billion years ago. The Materialists refuse to ask and refuse to consider the most interesting Scientific Questions of all.

Materialism or Naturalism is the chosen, philosophical, religious belief that the Spiritual or Non-Physical does not exist. Technically, Materialism is Creation by Physical Matter or Creation by ROCKS. The Materialists really truly believe that the ROCKS designed and created it all. But, what was there BEFORE the first rock and BEFORE the first particle of physical matter was designed and created? That's the question which the rest of us are asking.

It's obvious that Quantum Fields are non-physical and pre-physical. The Gods or the Controlling Psyches had to design, create, and make the non-physical Quantum Fields BEFORE they could create, make, and sustain physical matter. Nirvana, Psyche, or Quantum Fields are made from Energy; and, Energy is always conserved. That means that the Energy or Psyche has always existed, and it will always exist. It cannot be made, and it cannot be destroyed. That's what Conservation of Energy or Conservation of Psyche means. It's eternal and everlasting, without a beginning of days or an end of years. It's syntropic.

Nirvana is the eternal Energy Realm, or the Psyche Realm, or the Eternal Quantum Realm. It cannot be made, and it cannot be destroyed. It has always existed, and it will always exist. According to the Quantum Law of Psyche, a Psyche, Intelligence, or Consciousness can form or transform the energy under its control into anything that it wants that energy to be, including quantum fields, gravity, spirit matter or dark matter, electromagnetism, quantum waves, radio waves, or physical matter. Energy is infinitely malleable and infinitely transformable. Every psyche has a certain amount of energy that's under its direct control. The controlling psyche determines what form the energy under its control will assume. This is the way things really work. This is what has been experienced and observed.

Intelligence, or Consciousness, or Psyche, or Life Force has been experienced and observed. Has it not? The Quantum Realm, or Energy Realm, or Spirit World has been experienced and observed. Has it not? It's time for us to explain these things scientifically if we can. Whether they realize it or not, Materialism, Naturalism, Darwinism, and Atheism (Creation Ex Nihilo) have been falsified by Quantum Mechanics, Syntropy, and Quantum Field Theory. It's time for us to switch over to what's real and truly exists. Photons are obviously massless or non-physical.

My ultimate goal is to bring Science to life by infusing a generous helping of Psyche or Life into every aspect of Science.

My Author Page on Amazon:
http://amazon.com/author/science

The Associated Websites:
http://markme.website/nirvana/
https://markme.website/category/nirvana/
https://ultimate-model-of-reality.com/

The Associated Forum:
http://markme.us/forums/forum/nirvana/

The Facebook Page for Mark My Words:
https://facebook.com/MarkMyScience/

The Associated Twitter Page:
https://twitter.com/Mark_Me_Words

Table of Contents

Part 0 — PREFACE

This preface material is a collection of preparatory essays that I have written over the years, after releasing the first edition of this book. Within this book, I try to create Models of Reality that match with and explain everything that has ever been experienced and observed. Science is observation and experience after all – or it should be; and, I'm a scientist.

There seem to be an infinite number of definitions for NIRVANA, just like there seems to be an infinite number of different Interpretations of Quantum Mechanics.

Within this book and other books that I have written, I have defined NIRVANA as the Quantum Realm, Light Realm, Supernatural Realm, Spirit World, Psychic Realm, Transdimensional Realm, Non-Local Realm, the Conserved Realm, the Energy Realm, the Non-Physical Realm, or the Psyche Realm. I could have given NIRVANA many other definitions, but I chose the BEST ONE – the one that has the most explanatory power from a scientific perspective, from an experimental perspective, and from the perspective of observation and experience.

I have chosen David Bohm's (1986) Super-Implicate Order as the Interpretation of Quantum Mechanics that BEST defines and describes what is really happening both at the quantum level and the physical level – the model or interpretation of Quantum Mechanics that BEST explains what has actually been experienced and observed.

https://ultimate-model-of-reality.com/The-Super-Implicate-Order/

https://quantum-neuroscience.com/The-Essential-David-Bohm

Go with the best and get rid of all the rest. That's what I finally decided to do.

This book is part of my Ultimate Model of Reality Series. I'm a theoretician, philosopher, logician, mathematician, and scientist. I'm trying to develop models of reality that actually match with what has been experienced and observed.

For the past few years, since the summer of 2015, I have been trying to find the Ultimate Model of Reality or the True Interpretation of Quantum Mechanics. I'm getting a lot closer now that I have officially abandoned and rejected My Materialism, My Naturalism, My Nihilism, and My Atheism. Back in 2012, I was a materialist, naturalist, nihilist, and atheist. Eventually, though, Science, science experiments into Quantum Mechanics, Quantum Field Theory, and the first-hand experiences of out-of-body explorers convinced me that I was wrong. I finally chose to go with what has been experienced and observed, rather than the wishful thinking and philosophical speculation of Materialism, Naturalism, and their derivatives such as Darwinism, Creation Ex Nihilo, and Atheism.

Physical Reductionism, Scientific Naturalism, Darwinism, Classical Realism, and what David Bohm called Mechanistic Philosophy were designed to prevent us from finding and knowing the truth. They work as advertised. It's impossible to use Materialism and Naturalism and Darwinism to find the True Interpretation of

Quantum Mechanics, because Materialism and Naturalism make the claim as their primary axiom that supernatural mechanisms DO NOT EXIST, which is the same thing as saying that Quantum Fields and Quantum Mechanisms DO NOT EXIST. You can't find the True Interpretation of Quantum Mechanics by claiming that Quantum Mechanisms and Non-Physical Quantum Objects DO NOT EXIST. Quantum Fields are obviously non-physical, and prerequisites for the physical.

Something has to give. In order for me to find and KNOW the True Interpretations of Quantum Mechanics, I had to be willing to abandon My Materialism, My Naturalism, My Nihilism, and My Atheism. There was no other way. If you successfully eliminate everything that is false, or everything that has been falsified, then ONLY the truth will remain.

We have to be willing to eliminate everything that is false, if we truly want to find and know the truth. Quantum Mechanisms or Supernatural Mechanisms FALSIFY Materialism, Naturalism, the Philosophy of Mechanism, and their derivatives such as Nihilism, Atheism, and Darwinism. The false is falsified by the truth; and, the truth is repeatedly experienced and observed. Quantum Mechanisms, or Supernatural Mechanics, or Non-Physical Mechanisms are constantly experienced and observed. Space and time are obviously non-physical. Quantum Fields are obviously non-physical and prerequisites for the physical. According to Quantum Field Theory, the Gods or the Controlling Psyches had to design and make the Quantum Fields BEFORE the first particle of physical matter could be made and sustained.

The Materialists, Naturalists, Darwinists, Nihilists, Atheists, and Classical Physicists by definition, in principle, have chosen to believe that physical atoms really do exist. Most of these people have also chosen to believe that ONLY physical matter exists. Well, Quantum Mechanics is the scientific study of everything between the physical atoms, and everything within the physical atoms. Quantum Field Theory is the scientific study of the immaterial, non-physical quantum fields between the physical atoms, and the wide variety of different forces and fields and information within the physical atoms that are holding the physical atoms together as a unit. If Classical Physics is the study of physical atoms, then Quantum Mechanics is the study of everything else – everything non-physical, non-local, and non-entropic. It is obvious that at the speed-of-light, photons are massless or non-physical. Photons have to slow down to zero, in order to manifest in the physical realm.

I finally observed and accepted the fact that quantum mechanisms or supernatural mechanisms are being constantly experienced and observed. Magnetic fields, gravity waves, dark energy, quantum fields, light waves, microwaves, dark matter or spirit matter, and radio waves are obviously non-physical quantum mechanisms. Your thought processes are in fact quantum processes – they continue to exist long after your physical body and physical brain are dead and gone. This is what has actually been experienced and observed. Thoughts are quantum waves. It was finally time for me to face reality and accept the truth as it really is. I now invite you to join me. Let's go pursue the Ultimate Model of Reality or the True Model of Reality. Let's go pursue everything that has ever been experienced and observed.

Mark My Words

Quantum Mechanics Unleashed

A mechanistic or classical physics definition of Quantum Mechanics is absolutely worthless, because it deliberately excludes Quantum Mechanics. We need a better definition and a better interpretation of Quantum Mechanics than the one we are getting from the Materialists, Mechanists, Naturalists, Darwinists, Nihilists, and Atheists if we want to know how things really work at the quantum level and the psyche level.

Quantum Mechanics is instantaneous action at a distance. The word "transdimensional" means non-physical or non-local. Non-Locality means "not-located within our physical 3D space-time". Quantum Mechanics is Non-Local Physics. Quantum Mechanics is Transdimensional Physics, which means that Quantum Mechanics has NO physical limitations whatsoever. Remove all the physical limitations from classical physics, and we are left with Quantum Mechanics. Remember, Quantum Mechanics or Energy Mechanics or Psyche Mechanics at the quantum level has no physical limitations. It is instantaneous action at a distance, just as all of our science experiments have proven. Quantum Mechanics and Quantum Field Theory are constantly verified, which means that they are real and true. Photons are obviously massless or non-physical at the speed-of-light.

The proven and verified existence of Quantum Phenomena such as Action at a Distance (telekinesis), the Quantum Zeno Effect (telepathy), Quantum Tunneling (teleportation of physical matter), Quantum Superposition (omnipresence or multitasking at the quantum level), Quantum Phase-Shifting (walking through walls or between atoms), Quantum Consciousness (psyche or intelligence or choice), and Quantum Levitation (such as magnetic fields) FALSIFIES Materialism, Physicalism, Naturalism, Darwinism, Nihilism, Behaviorism, and Atheism which claim that these things DON'T EXIST. Quantum Phenomena are non-physical phenomena, and Scientific Naturalism of any kind claims that the non-physical DOES NOT EXIST. In other words, Materialism, Naturalism, Darwinism, Nihilism, and Atheism erroneously claim that Quantum Mechanics and Quantum Fields do not exist. Quantum Fields are obviously non-physical in nature and origin. In fact, Quantum Fields are pre-physical, which means that the Gods or Controlling Psyches had to design and make the quantum fields first BEFORE they could make and sustain physical matter. If there are no organized quantum fields, then physical matter is impossible to make and sustain.

A fundamental axiom of Quantum Field Theory is that particles are born, and particles die. This means that particles, including physical matter, are chosen into existence by Someone Psyche, and then later dissolved or chosen out of existence by Someone Psyche. Particles, including physical matter, are made from energy or light. Psyche or Intelligence or Consciousness is the only thing we know of that is capable of forming the energy under its control into anything that it wants that energy to be, including quantum waves, dark energy, gravity waves, microwaves, radio waves, spirit matter or dark matter, or physical matter. Energy is infinitely transformable or infinitely malleable. Energy or light can be formed or transformed into anything that its controlling psyche wants it to be, including physical matter.

13

Electron orbitals are force fields or shields that are made out of energy; and, it's these non-physical, non-local, quantum electron orbitals that make the physical atoms seem solid and real. It's a supernatural illusion, but it works.

https://en.wikipedia.org/wiki/Atomic_orbital

Here we are dealing with Nirvana, the Spirit World, the Transdimensional Realm, or the Quantum Realm. At the quantum level, a single electron can travel at an infinite velocity and literally form a solid shell or sphere around the nucleus of an atom. Electron orbitals actually have an identifiable shape. A single electron becoming a solid sphere is a quantum phenomenon and not a physical phenomenon. Such a thing is physically impossible according to Newtonian Physics or Classical Physics, but totally possible at the quantum level according to Quantum Mechanics.

What Starts Motion at the Quantum Level?

Study the vector within physics. Motion has a starting point, which means that Someone Conscious or Someone Psyche has to start that motion; otherwise, nothing would ever happen. Furthermore, vectors have direction. Someone Conscious or Someone Psyche has to choose a direction whenever it instigates motion. In the Quantum Realm or Spirit World, it has been observed that the psyche or the spirit chooses a destination before quantum tunneling or teleporting to that destination. Motion, direction, and destinations are a function of choice, which means that they are a function of psyche, consciousness, intelligence, or life force. This is what has actually been experienced and observed, both in the physical realm and in the psychic quantum realm.

Remember, physical matter is made and controlled from the quantum realm by Someone Psyche or Someone Intelligent. In contrast, Something Psyche, or Some Conscious Life Force, or Someone Non-Physical is the only thing we know of that is capable of controlling, transforming, and molding energy at the quantum level into useful things such as quantum fields, quantum waves, dark energy, forces, fields, spirit matter or dark matter, and physical matter. Remember, choice is a function or a product of Psyche or Life Force or Intelligence. Choice has no other source or cause, and this reality is especially obvious at the quantum level in the Quantum Realm or Spirit World, which is a non-physical environment to begin with.

https://en.wikipedia.org/wiki/Euclidean_vector

Study Quantum Field Theory and Feynman Diagrams. Anytime two particles interact, the psyches within each particle choose the result of their interaction from the valid choices that are possible. God or the Controlling Psyches determined the valid choices or the quantum laws; but, anytime you see a vertex within a Feynman Diagram or particles interacting, the psyches within the particles are making a choice as to how they want to respond to their interaction. This reality applies just

14

as much to physical particles as it does to the non-physical particles such as photons and virtual particles. Choice is a function or a produce of Psyche, Life Force, Intelligence, or Consciousness. Without choice, nothing valuable ever happens or could ever, and reality would be nothing but random or disorganized Chaos. Chosen obedience to quantum laws and physical laws is what brings stability, order, and organization. Quantum laws and physical laws obviously need some kind of Law Giver and Law Enforcer. Furthermore, quantum laws and physical laws obviously need some kind of intelligent and psychic Law Follower or Law Obeyer; otherwise, the Chaos or Random Disorder would continue to prevail. Remember, the Gods or the Controlling Psyches had to design and make the quantum fields and the quantum laws BEFORE they could design, make, and sustain physical matter. Quantum fields are non-physical and pre-physical. It's obvious which came first. The non-physical quantum fields were made by Someone Psyche BEFORE physical matter could be made and sustained. First things first.

https://en.wikipedia.org/wiki/Quantum_field_theory

https://en.wikipedia.org/wiki/Feynman_diagram

What Births or Starts Particles at the Quantum Level?

The primary axiom of Quantum Field Theory is that particles are born, and particles can die. This reality also includes physical particles. Physical matter is made by the different psyches who are controlling the energy or the light that goes into the construction of that physical matter. Every physical atom is compatible with every other physical atom, which means that they are being built from a common blueprint. It takes intelligence to read the blueprint and construct something from energy according to that blueprint. Physical matter is made from energy or light. The only thing we know of that can control energy or light and form that energy or light into physical matter or physical atoms is Psyche, Intelligence, Life Force, or Consciousness.

Psyche or energy or light is eternal and everlasting, which means that Psyche or Life Force is conserved. It cannot be made, and it cannot be destroyed. It has always existed, and it will always exist. In contrast, physical matter is made from energy by the psyches controlling that energy, and physical matter can be dissolved or converted back into energy by the psyches controlling that energy. In other words, physical matter is NOT conserved, as the Materialists and Naturalists claim. There's nothing fundamental about physical matter. We humans destroy or dissolve physical matter back into energy within our nuclear explosions, and we humans help to make new physical matter within our particle accelerators. Particles are born, and particles can die. ONLY the energy, light, psyche, or life force is eternal and everlasting. This is what has actually been experienced and observed. Is it not?

Most, if not all, of the mathematics for Quantum Mechanics has already been developed; and, Quantum Mechanics and Quantum Field Theory are proven and

15

verified Science. The only thing that remains is to determine what it all means. Once I used Science to falsify My Materialism, My Naturalism, My Nihilism, and My Atheism, then I realized that I didn't know how anything really works. I decided that I wanted to find out how everything works – find the correct interpretation or the correct definition for everything, including Quantum Mechanics. That's where I come in. I'm an open-minded scientist. I'm a philosopher. I'm a psychologist. And, I'm a theoretician. I notice patterns or trends that others seem to ignore or reject.

I was the perfect candidate for developing something completely new and different – like Quantum Neuroscience, Scientific Proofs of God's Existence, the Ultimate Model of Reality, Ultimate Causality, a Psyche Ontology, the Ultimate Law of Thermodynamics, the Law of Psyche, the Quantum Law of Thermodynamics, the Obviously Law of Physics, and Science 2.0. I've spent time on both sides of the fence, and I KNOW the difference between the two. I KNOW that the proven and verified existence of Quantum Field Theory, Quantum Mechanics, Conservation of Energy or Psyche, and Psyche or Intelligence or Consciousness FALSIFIES Materialism, Naturalism, Darwinism, Nihilism, Behaviorism, Determinism, Physical Reductionism, Atheism, and their derivatives. The one that has been experienced and observed FALSIFIES the one that is simply a figment of their imagination with no evidentiary support to back up their claims and wishful thinking. The non-physical or transdimensional aspects of Quantum Mechanics and Quantum Field Theory FALSIFY Scientific Naturalism, which claims that the non-physical does not exist. One has been experienced and observed; and, the other has not.

Reconciling the Theory of Relativity with Quantum Mechanics

Normally, what physicists do when it comes to the speed-of-light, they use Minkowski diagrams and Lorentz transformations to diagram everything from the perspective of two different physical objects trapped in spacetime. What interests me most is how everything seems to be from the perspective of the photon. Einstein and nobody else ever thought about it from the perspective of the photon. However, from the perspective of the photon traveling at the speed-of-light, time stops, distance goes to zero, resistance to acceleration goes to zero, and velocity goes to infinity. That's what the theory of relativity is trying to tell us. From the photon's perspective, it ceases to be a physical object and becomes a purely quantum object instead. From the photon's perspective, it simply quantum tunnels to its destination experiencing nothing during the journey because time has stopped during its journey, from its perspective.

Cool, huh?

According to the theory of relativity, a quantum object traveling at the speed-of-light in our physical dimension MAPS directly to that same quantum object traveling at an infinite velocity in the quantum realm, thanks to time dilation and length contraction. Time stops at the speed-of-light, length and distance contract to zero, mass or matter or resistance to acceleration ceases to exist, which means that velocity goes to infinity. It's calculus. Everything goes to its limits, and everything transforms into something completely different – something non-physical with NO physical limitations whatsoever.

According to quantum mechanics, while a photon is quantum, it is omnipresent and spread throughout the whole of our physical universe as a quantum field; yet, that same photon STOPS or coalesces to a single point in spacetime in order to show up in our physical realm. That explains in part how the quantum or the infinite MAPS to the physical or the limited. It basically slows down and stops. We have known this for a hundred years; but as far as I know, nobody has actually ever said it, because most of our scientists have chosen to believe that the quantum or the supernatural or the spiritual does not exist. Einstein never made the connection because he didn't believe in the quantum, or "spooky action at a distance", or infinite velocities in an infinite realm. Yet, it has to be true in order to successfully explain the observational evidence.

Remember, energy or light in its native original format is non-physical and pre-physical because it existed BEFORE the first particle of physical matter was designed and made, and because photons of light have NO mass. Mass or resistance to acceleration is what makes a particle physical. From our physical perspective, photons travel at the speed-of-light. From the photon's perspective at the quantum level, it simply quantum tunnels or teleports to its destination. You see, according to the Theory of Relativity, as an object approaches the speed-of-light, TIME STOPS, distance contracts or goes to zero, and the photon or quantum object simply quantum tunnels or teleports to its destination. From a photon's perspective, NO time passes, and it travels NO distance, even though from our perspective trapped within space-time, it took that same photon 13.2 billion years to reach us from the other side of the visible universe. That's how the Quantum Realm or the Infinite Realm MAPS to our slowed-down spacetime physical realm. There has to be a correlation, or the system wouldn't work.

Out-of-body explorers have repeatedly observed that psyches and spirit bodies can travel at an infinite velocity in the quantum realm or spirit world, if they choose to do so. They have also observed that energy or spirit matter transforms from one thing to another all the time. The energy is conserved; but, the FORM of that energy is never conserved. A spirit can take on any FORM it wants, because it is spirit matter or pure energy. Just like dark matter and dark energy, the spirit world or quantum realm is right here, right now, in your living room. It's just out of phase with our physical reality – meaning that it's non-local, transdimensional, phase-shifted, and non-physical. We are talking about multi-use space. Dozens of physical earths could be right here, right now, in the same place all out-of-phase with each other and unable to contact each other. Astrophysicists have discovered dark matter and dark energy; and, the stuff is non-physical, which means that it's quantum or non-local, which means that it is spirit matter or energy or light. It exists; it just exists in a different dimension or phase than our physical spacetime realm.

That's how Quantum Mechanics is unified with the Theory of Relativity – by studying things from our slowed-down space-time perspective, and then studying the same things from the infinite velocity perspective of the photon. From the photon's perspective, it literally travels at an infinite velocity to its destination, because it experiences NO passage of time during its journey, because TIME STOPS at the "physical limitation" which we call "the speed-of-light". At the speed-of-light, the governor is removed. From the photon's perspective, it simply quantum tunnels to its final destination traveling at an infinite velocity in order to do so. This is the notorious Instantaneous Action at a Distance or "Spooky Action at a Distance" that Einstein, the Classical Realists, and the Scientific Naturalists assure us DOES NOT EXIST.

The Theory of Relativity is the pinnacle of Classical Physics, Materialism, Naturalism, or Classical Realism. Remember, according to the Theory of Relativity itself, at the speed-of-light, time stops or time goes to zero, distance or length contracts to zero, mass or resistance to acceleration goes to zero, and the speed or velocity goes to infinity. At the speed of light, everything goes to the limit, which means that everything integrates! It becomes something different than it was before. At the speed-of-light, we switch from Classical Physics over to Quantum Mechanics or Instantaneous Action at a Distance. Mass cannot be accelerated or pushed faster than the speed-of-light; but, the Gods or Controlling Psyches can quantum tunnel it instantaneously from one end of the universe to the other, if they choose to do so.

One of Einstein's biggest blunders was to reject Quantum Mechanics, Quantum Tunneling, and Instantaneous Action at a Distance. Einstein's biggest blunder was to force himself to remain within the constraints of Classical Realism, Materialism, or Naturalism. Einstein's errors became our errors for a hundred years or more. The Theory of Relativity points us to the truth; but, Einstein refused to jump past the speed-of-light limitation of the Theory of Relativity and Classical Realism into the infinite velocity and zero distance of Quantum Mechanics. At the speed-of-light, the gloves come off and everything goes infinite; but, Einstein didn't know that because he wasn't able to accept that.

Anyone who has studied limits and calculus should be able to see with their own eyes that something very special happens at the speed-of-light. Everything goes to its limit. It integrates. It achieves unity. It becomes whole or complete. It becomes REAL and TRUE. It integrates. All physical limitations cease to exist. Time stops or goes to zero. Entropy ceases to exist. Length or distance contracts to zero; and, velocity or speed goes to infinity. The object transitions from the physical realm into the quantum realm. Physical laws no longer apply, and it switches over to quantum laws, transdimensional laws, or spiritual laws instead.

This is my best attempt to try to explain what really happens whenever a particle or an object transitions or transforms from the physical realm to the quantum realm. It's based upon the Theory of Relativity which most scientists consider to be true; and, it's also based upon Quantum Mechanics and the first-hand experiences or observations of out-of-body explorers who have observed that physical limits and speed limits cease to exist in the Quantum Realm or the Spirit World. This interpretation of Quantum Mechanics is an attempt to explain what has actually been experienced and observed. It's also an attempt to reconcile the Theory of Relativity with Quantum Mechanics.

Denying the Obvious Is a Bad Way to Do Science

The Classical Realists, Naturalists, Nihilists, Darwinists, and Atheists tell us that Instantaneous Action at a Distance does not exist. They are wrong. Instantaneous Action at a Distance has been proven and verified repeatedly through scientific experimentation. Einstein was proven wrong; Classical Realism was proven wrong; and Scientific Naturalism was proven wrong, by the verified and observed existence of Instantaneous Action at a Distance (telekinesis), Non-Locality or Non-Physicality, Quantum Tunneling (teleportation), the Instantaneous Quantum Zeno Effect (telepathy), Quantum Phase-Shifting, Quantum Forces and Fields, Quantum Waves (thoughts), Quantum Entanglement, Quantum Superposition (omnipresence or multitasking at the quantum level), Quantum Field Theory, Quantum Mechanics, the Placebo Effect (mind-over-matter), and Quantum Consciousness (psyche or intelligence).

Remember, the false is falsified by the truth; and, the truth is repeatedly experienced and observed. Intelligence, Consciousness, or Psyche is repeatedly experienced and observed. There's no sense trying to sweep it under the rug as the Materialists and Naturalists try to do, because it has actually been experienced and observed. It's time for us to try to explain this quantum phenomenon scientifically rather than trying to hide it from the world or sweep it under the rug. Psyche, energy, light, life force, or intelligence simply works differently at the quantum level than it does at the physical level where it is subjected to physical limitations. At the quantum level in the Quantum Realm, Psyche or Energy has NO physical limitations! This is what has actually been experienced and observed.

Defining and Securing the Laws of Conservation

The Classical Physicists seem to have some extremely liberal and flexible definitions for "conservation". I tried to tighten it up a bit, by introducing the Quantum Law of Thermodynamics and the Ultimate Law of Thermodynamics. If something is made, or born, or ceases to exist, or comes to an end, then it is NOT conserved. ONLY the things that are infinite and eternal are conserved, because they have always existed, and they will always exist. According to this enhanced definition of conservation, NOTHING at the physical level is conserved. The primary axiom of Quantum Field Theory states that particles are born, and particles can die. That means that physical matter and entropy haven't always existed, and there's no guarantee that they will continue to exist either, which means that they are NOT conserved. Observe that the Materialists, Naturalists, Darwinists, Atheists, and Classical Realists erroneously teach and believe that entropy, death, or heat death is conserved; and that entropic physical matter is fundamental and conserved.

19

They are wrong. Physical particles and entropy are born, and these things can also die or cease to exist by being absorbed back into the available energy from whence they came.

Anything that can end is NOT conserved. Heat Death or Entropy is the end of available energy at the physical level within a physical system, which means that Heat Death or Entropy is NOT conserved. ALL of that energy is still available at the quantum level, because energy and psyche are always conserved at the quantum level. Entropy or the Second Law of Thermodynamics is NOT a permanent situation. Entropy ends or ceases to exist whenever the passage of time (or the aging process) is removed from the physical system, such as happens at the speed-of-light or at the transition from the physical realm into the spiritual quantum realm. Entropy is not conserved, and the aging process stops whenever an object transitions into the Quantum Realm. Entropy is not conserved at the speed-of-light, when time stops, and the aging process stops. Psyche, intelligence, knowledge, or information is the ONLY thing that's needed to remove all of the entropy or disorder from any system, including a physical system. The Quantum Realm is Pure Intelligence, or Pure Psyche, or Pure Information, which means that entropy or the Second Law of Thermodynamics doesn't exist in the Quantum Realm. In fact, thermodynamics don't exist in the Quantum Realm. It's impossible for a spirit body or a psyche to freeze to death in the Quantum Realm or the Spirit World. This is what has actually been experienced and observed.

Remember, it's the Quantum Mechanics and the Conservation of Psyche and Energy that is in fact REAL, and not the classical realism and physical limitations such as entropy and the speed-of-light. Remember, entropy, physical matter, and the speed-of-light limitations are constructs. They are NOT fundamental realities. They are NOT conserved. They are the emergent properties of things that were made. They are constructs, and they are made or enforced by something non-physical – something like Psyche, Intelligence, Thought, or Consciousness. If something is designed and made, then it is NOT conserved. It can end, or it can die. Entropy, entropic physical matter, physical atoms, and the physical limitations or physical laws were designed and made by the psyches who are controlling the energy from which these things were made. Entropy and physical matter were made or had a beginning which means that they can end or cease to exist.

Entropy and entropic physical matter are NOT conserved, despite the claims of the Materialists, Naturalists, and Atheists to the contrary. The Materialists, Darwinists, Nihilists, Naturalists, and Atheists really truly believe that entropy, heat death, death, and entropic physical matter are conserved. They are wrong. Entropy is NOT a force nor a field. It is simply a measure of the lack of available energy at the physical level. ALL of that energy is still available at the quantum level, and it can be used or manipulated anytime the psyches controlling it choose to do so. The Quantum Law of Thermodynamics states that there is NO entropy or unavailable energy at the quantum level, because everything at the quantum level – or the quantum fundamentals such as psyche and energy and information – are always conserved and therefore always available at the quantum level.

This is possible because they are always conserved at the quantum level. Psyche and energy cannot be made, and they cannot be destroyed, which means that the information that enters into those things or is produced by those things is always conserved as well.

The Quantum Law of Thermodynamics is based upon what has actually been experienced and observed. NO psyche or spirit body has ever been observed freezing to death, because they can't freeze to death, because entropy or the second law of thermodynamics or heat death does not exist at the quantum level in the Quantum Realm thanks to the LAW of the Conservation of Energy and Psyche. Spirit matter is just made from energy, and energy cannot be created nor destroyed. Psyche or energy has always existed, and it will always exist, because there is NO heat death or entropy at the quantum level. There is NO thermodynamics at the quantum level in the Quantum Realm or Spirit World. This is what has actually been experienced and observed.

We also know that the Quantum Law of Thermodynamics is true from the Theory of Relativity. As an object achieves the speed-of-light, TIME STOPS, which means that the aging process stops, which means that entropy and death stops. Entropy is a product of time or a function of time. Whenever time stops, entropy or heat death ceases to exist. The Quantum Law of Thermodynamics states that there is NO heat death or entropy at the quantum level within the Spirit World or Quantum Realm. At the quantum level and the psyche level, the Conservation of Energy and Psyche reigns supreme, because there is NO second law of thermodynamics at the quantum level in the Spirit World. Psyches and energy can't suffer heat death, because they are always conserved and because there is no such thing as heat death or entropy at the quantum level. This is what has actually been experienced and observed.

I tried to upgrade the laws of physics and Science in general by tightening the definition for what it means to be conserved. My new definition for "conservation" is NO CHANGE over time. In other words, that quantum object is not subject to the passage of time, which means that the passage of time and entropy have no effect on it. There is NO aging process, which means that this object cannot die or cease to exist. It's immune to the second law of thermodynamics. In other words, it's immortal. It can't die or cease to exist, which means that it can't be born nor created either. It is eternal and everlasting. That's what it really means to be conserved. There's NO entropy or death where a conserved object is concerned. That's a different definition for conservation than what the Classical Physicists tend to give us. These people erroneously try to convince us that entropy, entropic physical matter, and death are conserved. That's just not the case. Particles are born, and particles can die, which means that entropy can emerge at the physical level, and entropy can also dissipate or be eliminated. Entropy or the second law of thermodynamics is not conserved. It's constantly changing over time; and, entropy actually ceases to exist whenever time stops, or the passage of time ceases to exist.

Under this new tightened definition for conservation, only three things seem to be truly conserved – Quantum Syntropy, Psyche or Intelligence or Information

Storage, and Energy. These things can change FORM, but they are always conserved. They don't age. They aren't subject to entropy or the passage of time. They can't be made, and they can't die. They are always conserved.

Syntropy Is the Opposite of Entropy

I wrote the book on Syntropy. Syntropy is the opposite of entropy. Syntropy has to exist, or entropy and physical matter wouldn't be possible to make and sustain. Syntropy is in fact the Conservation of Psyche and Energy.

Quantum Fields are syntropic. Quantum Fields are simply a different FORM of energy. The FORM is never conserved; but, the underlying energy and intelligence is always conserved. Whenever a physical system experiences heat death, thermal equilibrium, or maximum entropy, the Quantum Fields sustaining it continue to function and work for forever – until the controlling psyches decide to transform that energy into something else. The controlling psyches can transform the energies within entropic physical matter into anything else they want it to be, anytime they choose to do so. Whenever a particle of physical matter ceases to exist, the associated entropy ceases to exist, because entropy or the second law of thermodynamics is never conserved. Entropy is an emergent property of physical matter, and when the physical matter ceases to exist, entropy ceases to exist, and the associated energy becomes available for use once again. This reality and truth is a bit different than what your typical atheistic, naturalistic, and materialistic college teacher will tell you; but, he's never seen nor experienced the Quantum Realm or Spirit World, so he really has no idea how it truly works. At the quantum level, Psyche or Intelligence or Consciousness or Life Force or Information Storage is always conserved, because its associated energy is always conserved. In other words, energy and psyche are syntropic. They are always conserved at every level of existence. This is what has actually been experienced and observed.

Syntropy: The Answer to Life, the Universe, and Everything

https://www.amazon.com/gp/product/B07BPT3W8R/

There are many different definitions for entropy. On my physics chart, Entropy (S) is defined as J/K, where J equals Joules or Energy or Heat, and K equals Kelvin or Temperature. Therefore, as K approaches 0 (absolute zero), entropy becomes undefined. Anything divided by zero is officially undefined according to mathematics. That's unfortunate, because that really doesn't reflect what's truly happening as we approach zero.

Many of the Materialists, Naturalists, and Atheists define "maximum entropy" as "zero energy". These people erroneously tell us that Heat Death or Entropy is conserved. Some of these people also tell us that the energy has ceased to exist because it is no longer available for use. That's a faulty and falsified interpretation of entropy. Nevertheless, professional physicists like the ones at PBS Space-Time, actually produce videos where they try to demonstrate using math and theory that energy is made and energy can be destroyed. They claim that virtual particles and electrons cease to exist, which they interpret as meaning that the energy has ceased to exist. I personally know for a fact that any interpretation of math or quantum theory that violates or falsifies the Conservation of Energy is in fact a FALSE interpretation. Here is why.

Let's take Y divided by X – anything divided by X. Whenever we approach the limit of X=0 from the left or from the negative, the value of Y/X does in fact approach minus infinity. Likewise, whenever we approach the limit of X=0 from the right or from the positive side of the equation, the value of Y/X does in fact approach infinity. Therefore, a better and more realistic definition of Y/0 is in fact plus and minus infinity, because that is what is really happening.

https://www.youtube.com/watch?v=SQzjzStU1RQ

https://syntropy.site/Why-dividing-by-zero-is-undefined

Now, let's apply this to entropy.

Entropy = Energy / Kelvin.

As Kelvin or temperature approaches absolute zero, the amount of energy approaches infinity, which means that entropy approaches maximum entropy or infinity. The system achieves unity or infinity – it reaches its limits. ALL of the energy is still there. The energy is conserved. What happens is that at 0 Kelvin or "maximum entropy", the energy is no longer available for use at the physical level, because the system has achieved thermal equilibrium or heat death. It has achieved infinite entropy, or maximum thermal equilibrium, or maximum entropy. No further work can be done at the physical level. None of the energy is available for use at the physical level. But, ALL of the energy is still there at the quantum level. The quantum fields, energy, spirit bodies, dark energy, dark matter, gravity waves, and psyches are still there at the quantum level, because there is NO entropy or second law of thermodynamics at the quantum level. This is the Quantum Law of Thermodynamics. ALL of the energy is available for use ALL the time at the quantum level and the psyche level. That's what keeps the physical level in existence, even when the physical level has achieved maximum entropy, heat death, or thermal equilibrium. Someone Psyche at the quantum level continues to sustain it.

In contrast, the Materialists, Naturalists, Darwinists, and Atheists erroneously teach and truly believe that death, heat death, or entropy is always conserved. That is what they teach and believe, is it not? They teach and believe that ONLY entropic physical matter exists, which means that they teach and believe that entropy and entropic physical matter will always be conserved. They teach that heat death, death, and entropy are permanent. That is what they teach and believe, is it not? Many of them also believe that the energy goes to zero or the energy ceases to exist at K equals zero or absolute zero. They literally teach and believe that there is no energy or no heat at maximum entropy or heat death. This is what they teach and believe, is it not?

They are wrong.

At the quantum level, ALL of the energy is exergy or syntropy, which means that it is always conserved and is always available for use. The Controlling Psyches who control the energy within physical atoms can transform that energy into quantum fields, or available energy, or photons anytime they choose to do so. Anytime you see a vertex or a convergence on a Feynman diagram, you do in fact observe a couple of psyches transforming their energy from one form to a different form. The FORM is never conserved. The quantum object or quantum energies can be transformed anytime its controlling psyches choose to transform it.

Virtual Particles or Non-Physical Particles Do Not Exist, or Do They?

Materialists, Naturalists, and Atheists will tell you that virtual particles don't really exist. They are wrong again. Virtual particles are also a FORM of energy. The energy is still there – it's just in a quantum form or a non-physical form – which makes it seem like it has disappeared or ceased to exist. Virtual particles or purely quantum particles are always erroneously portrayed by the Materialists, Naturalists, and Atheists as a violation of the Conservation of Energy – they always say that the virtual particle or the electron has temporarily ceased to exist. That's what they say, is it not? They tell us that electrons pop in and out of existence all the time. They are wrong. The virtual particle or electron has simply transitioned to the Quantum Realm or the Spirit World where the Materialists and Naturalists can't currently get at it. ALL of its energy is still there, and ALL of its energy is available for use at the quantum level.

These people tell us that virtual particles don't really exist, and that electrons are jumping in and out of existence all the time. They are wrong. The energy is still there. It has simply changed form or changed its state, so that it no longer has a physical existence, but has a quantum or non-physical existence instead. These people erroneously define existence as physical existence, and then they completely ignore and reject the non-physical, syntropic, quantum side of the equation because they have erroneously chosen to believe that it does not exist. Talking about a virtual particle's non-physical existence is a much better way of explaining and defining what really happens when an object transitions or transforms into the Quantum Realm than to say that the virtual particle does not exist or that the energy associated with virtual particles ceased to exist. The energy never ceases to exist; and, any interpretation of math or science, that says that it does, is in fact a faulty and falsified interpretation of math and science. In the Quantum Realm or the Syntropy Realm, the Conservation of Psyche and Energy reigns supreme. Energy cannot be made, and it cannot be destroyed. Any theory that says that energy can be made, and that energy can be destroyed or cease to exist, is in fact a false theory.

When it comes to virtual particles – they don't cease to exist as the Materialists and Naturalists and Atheists claim. There is NO such thing as Creation Ex Nihilo or Atheism. When it comes to virtual particles, their quantum fields, quantum waves, information, intelligence, structure, and energies continue to exist, even though the virtual particles no longer exist at the physical level. Everything about a virtual particle continues to exist at the quantum level, even though its existence at the physical level has temporarily ceased. The Materialists, Naturalists, and Atheists simply have a faulty definition of reality and existence, because they claim that only entropic physical matter exists, and they are obviously wrong. The verified and proven existence of immaterial, non-physical, Quantum Fields FALSIFIES Materialism, Naturalism, and their derivatives such as Atheism or Creation Ex Nihilo, Darwinism or Creation by Entropy, Physicalism or Classical Realism, and Nihilism or the Philosophy of Mechanism.

These people are motivated to try to convince you that the non-physical DOES NOT EXIST, because these people don't want God and Psyche to exist. These people will deny their own intelligence in order to convince themselves that Psyche or Intelligence does not exist. Self-deception works, and it works every time. These people are obviously trying to trick you and deceive you, just as they have been tricked and deceived. The whole of Science (all of the observations and experiences on record) proves that the non-physical does in fact exist. Magnetism, Gravity, Radio Waves, Microwaves, Gamma Rays, X-Rays, Dark Energy, Energy, Quantum Information, Quantum Intelligence, Dark Matter or Spirit Matter, Space, Time, and Quantum Fields PROVE that the non-physical exists. Everything is non-physical, except for the stuff that we selectively define as a physical atom, or more accurately, a physical quark or a particle with mass. Mass is resistance to acceleration, which means that mass is the stuff within the physical atom that has had a time-delay programmed into it or built into it. It resists acceleration or subjects itself to velocities slower than the speed-of-light; but, mass itself is made from energy or from the non-physical.

This is obviously true. It's obvious that the non-physical exists; but, this is precisely the stuff that the Materialists, Naturalists, Darwinists, Nihilists, and Atheists refuse to accept and refuse to believe in because it's non-physical and because these people have chosen to believe that the non-physical does not exist.

In order to find and know the truth, we must identify, expose, and eliminate the lies. That's what Science is supposed to be about – the elimination of the deceptions and the lies. However, the Materialists, Naturalists, Darwinists, Nihilists, Behaviorists, Determinists, Physical Reductionists, Atomists, and Atheists ENSHRINE THE LIES or SANCTIFY THE LIES and then present those lies to the world as axiomatic truths. That's what they do, is it not? They present their claim that the non-physical does not exist as an axiomatic truth or a Universal Law. They jump to that conclusion and interpret everything accordingly.

Physical Matter and Entropy are NOT Conserved

The Materialists and Naturalists erroneously believe that entropy, heat death, death, thermal equilibrium, and the second law of thermodynamics are conserved. That's what they believe, is it not? They believe that death, entropy, and heat death are eternal and everlasting. Most of these people also erroneously believe that entropic physical matter is conserved. They are wrong. Psyche and Energy cannot die or cease to exist at the quantum level because they have always existed; and therefore, they will always be conserved. Energy and Psyche cannot be made, and they cannot be destroyed. They are immortal. They have always existed, and they will always exist. They are eternal and everlasting. Energy can change FORM. The FORM is never conserved; but, the underlying energy and the psyche who controls that energy are always conserved.

Physical matter is simply a form of energy; and, the FORM is never conserved. Only the energy is conserved. Every psyche has a certain amount of energy that's under its control, and it can transform that energy into anything that it wants that energy to be, anytime that it chooses to do so – including quantum fields, gravity, magnetic fields, microwaves, visible light, dark matter or spirit matter, dark energy, information or intelligence, quantum waves, or physical matter.

I'm simply trying to develop a better, a more observationally accurate, a more all-inclusive, and a more realistic and usable definition and explanation of physics than the limited and exclusive one that the Materialists, Naturalists, Mechanics, Darwinists, Nihilists, and Atheists try to force upon us through ridicule and intimidation. I wrote the book on Syntropy (the opposite of entropy), because it's a scientific concept that the Materialists and Naturalists completely reject and ignore. These people teach and believe that entropy, heat death, death, or the second law of thermodynamics is eternal and everlasting. They erroneously teach and believe that entropy, heat death, and death are conserved. That's what they teach and believe, is it not? Many of these people literally define entropy as "no energy", "no heat", or "the cessation of energy"; and, that's not what it means at all.

They are demonstrably wrong. Entropy, or death, or natural selection doesn't exist as a person, place, or thing. These things are a measure of the non-existence of something. Entropy is the non-existence of available energy at the physical level. Entropy is the non-existence of heat or life at the physical level. Natural selection is entropy or death – the cessation of life. The origin of species by means of natural selection means the origin of species by death and entropy. When you get selected against at the physical level, your physical body dies.

Maximum entropy means anergy or no available energy at the physical level. Maximum entropy means no available energy. At maximum entropy, ALL of the energy is still there, it's just not currently available for use at the physical level. The energy hasn't ceased to exist, when a physical system achieves maximum entropy or thermal equilibrium; and, it's an error to say that it does. Anytime the Materialists and Naturalists claim that energy has ceased to exist, they are wrong; and, they have in fact gotten the wrong interpretation of the math and the scientific evidence. Remember, the Quantum Law of Thermodynamics states that there is NO thermodynamics, NO heat death, NO death, NO entropy, and NO second law of thermodynamics at the quantum level and the psyche level. This is what has actually been experienced and observed. No psyche, quantum field, quantum wave, or spirit body has ever been observed freezing to death or ceasing to exist, because they can't. It is impossible thanks to the Conservation of Energy and Psyche.

In the process of defining and explaining what Syntropy means, I developed the Ultimate Law of Thermodynamics, which is a new law of thermodynamics that is intended to differentiate between what is truly conserved and what is not conserved. Since energy is infinitely malleable and since the controlling psyche can form the energy that's under its control into anything that it wants that energy to be, the FORM is never conserved! Physical matter is a form of energy, which means that physical matter is NOT a conserved object. The FORM is never conserved. The controlling psyches who made the physical atom in the first place can transform that physical atom into anything else they want it to be, anytime they choose to do so. Physical matter is not conserved.

Furthermore, since entropy or the second law of thermodynamics is an emergent property of physical matter, whenever the physical matter ceases to exist, the associated entropy ceases to exist. In a hydrogen bomb, for example, some physical matter ceases to exist, which means that its associated entropy ceases to exist; and, a whole bunch of energy or heat is introduced into the system as a result. This is what has actually been experienced and observed. Ultimately, entropy and physical matter are not conserved. Physical matter was made, which means that physical matter and the associated entropy can die and cease to exist anytime that its controlling psyches choose to transform that entropic physical matter into something else instead. This is what has actually been experienced and observed.

Entropy is death. The Materialists, Naturalists, and Atheists are not going to believe my claim that there is NO entropy nor thermodynamics at the quantum level and the psyche level, because to these people entropy, death, and the second law of thermodynamics are sacred, and sacrosanct, and take precedence over the First Law of Thermodynamics which states that energy and psyche and information are always conserved. These people have decided that psyche doesn't exist, and that entropy and death trump the conservation of energy every time. They are wrong, but that doesn't change what they have chosen to believe. These people truly believe that death and entropy are always conserved. These people believe that death and entropy is permanent. That's what they really believe, is it not?

The Ultimate Law of Thermodynamics states that Entropy or the Second Law of Thermodynamics is NOT conserved. Entropy is a temporary or temporal phenomenon. Entropy is a function of time – an aging process. Whenever time stops at the speed-of-light, entropy or the passage of time ceases to exist. Entropy is an emergent property of something that was born or made – entropic physical matter. According to Quantum Field Theory, particles are born, and particles can die, which means that particles are made, and particles are therefore NOT conserved. Entropy is a function of time. Entropy is produced by the passage of time. At the speed-of-light, time stops which means that entropy ends at the speed-of-light. There is NO entropy in the Quantum Realm. Psyche, Energy, and Information Storage are NOT subject to the passage of time in the quantum realm, which means that they are NOT subject to an aging process and therefore are NOT subject to entropy. Everything is conserved at the psyche level and the quantum level within the Quantum Realm or Spirit World. Quantum Objects can be transformed or changed from one form into a different form, but the energy and the psyches controlling that energy are always conserved. This is what has actually been experienced and observed.

This is how things really work at all the levels of existence that we know about. No psyche or spirit body has ever been observed freezing to death, because heat death or entropy or the second law of thermodynamics doesn't exist at the quantum level and the psyche level within the Quantum Realm or the Spirit World. Psyches, spirit matter, and spirit bodies can't freeze to death because there is NO entropy or thermodynamics in the Quantum Realm. They can change their FORM, but they can't cease to exist. They can be disassembled, but they can't cease to exist. Energy and Psyche are always conserved. They can't die or cease to exist.

A Photon Is Intelligent and Alive

A photon is the quintessential quantum, non-physical, supernatural, spiritual object. It's the prototype for all the others. We KNOW it exists by the effect that it has on our physical realm. From its massless, timeless, non-physical, quantum perspective, time has stopped, distance has gone to zero, and it literally travels at an infinite velocity to its chosen destination. A photon KNOWS where it is going to land before it launches; otherwise, it doesn't instantiate. It doesn't launch, unless it has some place to land. It creates itself for a purpose. From a photon's perspective, it travels at an infinite velocity to its chosen destination, because time has stopped and distance has contracted to zero, according to the theory of relativity. Yet, the thing has to be able to map directly onto our physical reality because it's eventually going to have to interact with our physical reality; and, the way it maps to our physical reality is through the physical phenomena that we call the speed-of-light and the collapse of the wave function. This is what the Theory of Relativity and Quantum Mechanics are trying to teach us about the quantum or the non-physical. It's smart. It's powerful, and it's fast.

A photon traveling at the speed-of-light is obviously massless or non-physical, because that's how it's able to travel at the speed-of-light. A photon has no mass or no resistance to acceleration. A photon is the very definition of a massless or non-physical object. From a photon's perspective, time has stopped, distance has gone to zero, and it literally travels to its destination at an infinite velocity; but, it also has to slow down to zero in order to show up in our physical realm on our physical detectors. A photon collapses its own wave function into a single point, when it reaches its destination and enters into our physical realm. A photon has to be able to map itself directly onto our 3D physical spacetime reality, because the thing was intelligently designed and meant to register upon our physical senses.

Even though a photon travels at an infinite velocity from its perspective, from our perspective, it can only travel at the speed-of-light until it stops. A photon is a massless, non-physical, spiritual, supernatural, quantum object that is capable of making a physical impression or capable of leaving an impression upon the physical realm. There's an intelligence there that is deeper, broader, and more eternal than anything we humans currently have access to.

A photon KNOWS when to act like a wave and when to act like a particle. It can do this because it has intelligence or psyche within it. A photon is intelligent and alive. A photon is sentient, conscious, aware, perceptive, intelligent, psychic, conscious, dynamic, and alive. A photon can sense its surroundings on a universal scale. A photon can read your mind. There's life and intelligence within the light.

Experiments with the Quantum Zeno Effect have proven that the light or the intelligence within a physical atom can also read your mind. It KNOWS when you are looking at it and when you are ignoring it. It KNOWS when it can quantum tunnel and when it's supposed to stay still. A physical atom is also intelligent and alive. In fact, it's theoretically possible to store ALL of the intelligence and knowledge in the universe within a single physical atom. A physical atom is infinitely more intelligent than we are, because we have been deliberately dumbed-down by being subjected to physical limitations.

Photons only deposit their energies in discrete bunches that we call quanta. A photon travels as a wave, yet when it reaches a barrier or a detector, it gathers all of its energy and deposits that energy at a single point of its own choosing. Again, we are describing the chosen behaviors of energy or light. Any time you observe a choice being made, you are in fact observing some type of Psyche or Intelligence at work. A change in behavior signifies the actions or the choices of some type of Intelligence or Psyche who is capable of making a choice. A photon KNOWS where it is going to land before it launches, because a photon is omnipresent or quantum before it makes itself and launches itself. From its massless, timeless perspective, it travels at an infinite velocity to its chosen destination; but, from our physical perspective, it travels at the speed-of-light to its destination. When it reaches its chosen destination, it collapses its wave function and leaves an impression on our physical realm.

Here, once again, we have the truth hiding in plain sight where nobody can see it, because they aren't looking for it. In fact, the Materialists and Naturalists don't want this to be true because it falsifies their personal beliefs.

Nevertheless, predictable physical laws and replicable quantum behavior require some type of Law-Maker, Law-Enforcer, and Law-Obeyer. The physical laws and quantum mechanisms would be absolutely worthless, without some type of Intelligence or Psyche within the atoms and photons, capable of understanding and obeying the laws that have been made. It would be nothing but random chaos without the physical laws that the Law-Maker has made, and without the Intelligent Psyches within the atoms who are capable of understanding and choosing to obey the physical laws that have been made.

I find it fascinating that they chose the letter Psi (Ψ) as the symbol for the quantum wave function. In some circles, Psi means psychic or psyche. It's as if somebody instinctively knew that electrons, quarks, atoms, and photons are psychic, intelligent, sentient, aware, perceptive, conscious, and alive. These things can CHOOSE whether to act like waves or to act like particles. Choice signifies intelligence or psyche. The quantum wave function seems to represent latitude, or freedom of choice. Choice is a function or a product of psyche or intelligence. Once again, we have the truth hiding in plain sight waiting for those who are willing to look at it and see it. It is obvious that a photon is massless or non-physical until it chooses to stop and manifest itself in the physical realm. There's no reason that a photon or a quantum wave would have to stop for a physical atom. It would be able to pass straight through a physical atom as if the atom weren't even there. A photon brakes or stops for an atom ONLY because it chooses to do so. A photon can sense the atoms and chooses to stop and interact with the atoms. A photon could pass straight through a physical atom unphased if it wanted to.

Complex behavior reveals the presence of a sophisticated intelligence. Take an electron, for example. IT KNOWS whether it is supposed to travel in a straight line through empty space, or whether it should stop and form an elliptical spherical force-field or atomic orbital shell around the nucleus of a physical atom. An electron KNOWS whether it is in empty space and is supposed to travel through a couple of slits towards a detector beyond those slits, or whether there is an atomic nucleus nearby that needs it to stop moving in order to form an atomic orbital shell around that atomic nucleus. IT KNOWS and IT CHOOSES, which means that it has some type of intelligence or psyche or consciousness within it. This reality applies to every electron that was ever made. The quantum wave function reveals the presence of latitude, which reveals the presence of intelligence and choice. It's obvious that electrons and photons are choosing whether to remain as waves, or to collapse their wave function into a single point. This choice reveals their intelligence, or the presence of some type of psyche within each and every photon and electron. It's obvious for those who are willing to look and see.

Particles are born – they make themselves. And particles can die – they choose when and how they want to cease to exist. Particles obviously collapse their own wave function. They don't require the presence of an observer to do what they need to do when they need to do it. They are smart enough to follow the laws or the standards of behavior that have been made. They can also transform from electrons to photons to virtual non-physical quantum particles back to electrons or photons. They choose what to transform into based upon the LAWS that were made and established. This signifies some type of intelligence or psyche.

I think that was Henry P. Stapp's and Paul Dirac's greatest scientific discovery – namely that Nature or Nature's Psyche collapses its own wave function. Nature's Psyche not only creates the wave in the first place, but it also chooses when and where to collapse the wave. Every vertex on a Feynman Diagram represents a point where Nature's Psyche makes a choice. It decides what it's going to transform its energy into. Every psyche has a certain amount of energy that's under its control; and, it chooses what form it wants that energy to be. The Human Psyche isn't consciously aware of any of these things because it's trapped within a physical brain. It's Nature's Psyche who collapses the wave function, transforming the quantum wave into something that shows up on our physical detectors.

There are two other associated processes that need to be recognized. The first of these is the process that selects the outcome, 'Yes' or 'No', of the probing action. Dirac calls this intervention a "choice on the part of nature", and it is subject, according to quantum theory, to statistical rules specified by the theory. I call by the name 'process 3' this statistically specified choice of the outcome of the action selected by the prior process 1 probing action. (Henry P. Stapp. *Mindful Universe*.)

Process 1 is the Human Psyche's request or the probing action. The Human Psyche decides what it wants to do with its physical body and physical surroundings. Process 2 is the collapse of the wave function. Process 3 is a choice on the part of Nature's Psyche whether or not to collapse the wave function. It's Nature's Psyche who collapses the wave function, and NOT the Human Psyche. Nature's Psyche was collapsing wave functions long before human beings and Gods came along to make the request.

The truth of this is obvious because it has been experienced and observed trillions of trillions of trillions of times. The ONLY difficulty is getting materialistic, naturalistic, and atheistic scientists to accept it as being real and true.

Quantum Consciousness or Nature's Psyche

There are many different definitions for "Nirvana", depending upon the philosophy that you have chosen to believe in. I chose to define Nirvana as the non-physical, the spiritual, the supernatural, or the quantum. Nirvana is the spirit world or the quantum realm.

When it comes to the physics of the Quantum Realm, it can be fascinating to study the experiences of out-of-body travelers such as William Buhlman. There's some type of Universal Intelligence at the quantum level that is capable of understanding and performing your commands, while you are out-of-body or dead. This Universal Intelligence seems to be something completely different than God or the Being of Light and Love, who is Jesus Christ.

https://psyche-ontology.com/buhlman/

At the quantum level within the Spirit World or Quantum Realm, there seems to be some kind of Universal Intelligence, who is capable of reading your mind and granting your requests instantaneously. It doesn't matter the direction that you choose travel, this Universal Intelligence can take your psyche and your spirit body at an infinite velocity to the location of your choosing; or, this Universal Intelligence can take you to the destination of your choosing more slowly, if it desires, so that you actually experience the journey along the way.

These out-of-body observations and experiences are important from a scientific point of view, because they actually help us to establish the Laws of Physics in the Quantum Realm. There also seem to be multiple different psyches at work in the Quantum Realm. There's the Human Psyche who issues commands or requests; and, then there seems to be some type of Universal Intelligence that I at times call "Nature's Psyche" who chooses whether or not to answer the Human Psyche's request, and how to answer it, or how quickly to answer it.

It's really important to understand that the Quantum Realm functions differently than the Physical Realm, because at the quantum level various different physical limitations such as entropy, thermodynamics, the aging process, the passage of time, death, and speed-limits don't always exist and don't always apply. Quantum Tunneling, Immortality, and Instantaneous Action at a Distance are a realistic possibility at the quantum level; but, these same things are deliberately limited at the physical level by the Universal Intelligence or the Controlling Psyches who choose to impose physical limitations on physical atoms at the physical level. In other words, Classical Physics or Physical Physics doesn't always apply and doesn't always exist at the quantum level in the Quantum Realm or Spirit World; but, Classical Physics works just fine and works reliably at the physical level, as it was designed to do and is made to do by the Universal Psyche or the Universal Intelligence. These realities and truths are an integral part of the Ultimate Model of Reality, where Psyche or Intelligence is the Ultimate Cause and the Ultimate Causal Agent. This is a Psyche Ontology where psyche is the fundamental unit of reality.

Quantum objects such as psyches and spirit matter are capable of infinite velocities, infinite speeds, instantaneous acceleration, and infinite acceleration at the quantum level, because there is NO mass or resistance to acceleration at the quantum level. It's the mass or the resistance to acceleration that makes physical objects subject to the passage of time and subject to the speed-of-light limitation. There's NO mass or resistance to acceleration at the quantum level, and the time to travel between two distant points tends towards zero at the quantum level. In other words, according to the Theory of Relativity, distance or length goes to zero at the speed-of-light or at the quantum level, time stops, and the time it takes to travel between two points can also go to zero at the quantum level. This is possible because there is no mass or resistance to acceleration at the quantum level. This is what has actually been experienced and observed by out-of-body travelers and near-death experiencers.

https://psyche-ontology.com/psyche-experienced-and-observed/

Universal Laws consistently understood, enforced, and obeyed at the quantum level signifies or reveals some type of Universal Intelligence that I have defined as "Nature's Psyche". Physical particles are identified by their mass – their resistance to acceleration. It's obvious from scientific experiments that energies at the quantum level in their native original format have NO speed limits or NO resistance to acceleration; therefore, the energies within physical matter that are showing resistance to acceleration are in fact being made or forced to show resistance to acceleration by Nature's Psyche and the psyches within the physical matter. A choice is being made to force the energies within the physical matter act that way because it doesn't do so naturally or natively; and, choice signifies or reveals Intelligence or Psyche or Consciousness at the quantum level.

The People Who Are Trying to Hide the Truth from Us

Clearly, you won't get any of this Science or Information from the Materialists, Naturalists, Darwinists, Nihilists, and Atheists because these people have chosen to believe that the Quantum Realm or the Spirit World, the Human Psyche, Nature's Psyche, and a Universal Intelligence DO NOT EXIST. In order to know the truth about these things, we have to experience them for ourselves, or choose to trust someone who has. That's just the way it is. Science is observation and experience after all. You will never find the truth about the Spirit World or the Quantum Realm from someone who has never experienced it, or from someone who has chosen to believe that it doesn't exist. You can't get blood from a stone; and, you can't get the truth about the Quantum Realm from a Materialist, Naturalist, Darwinist, or Atheist who has chosen to believe that the Quantum Realm doesn't exist. Instead, the Materialists, Naturalists, Darwinists, Physical Reductionists, Determinists, Behaviorists, and Atheists are trying to trick you and deceive you into believing that the non-physical or the quantum DOES NOT EXIST. That's their whole purpose for being or existing. They want you to believe that the quantum or the non-physical DOES NOT EXIST, because the proven and verified existence of anything non-physical FALSIFIES Materialism, Naturalism, and their derivatives which claim that the non-physical DOES NOT EXIST.

But, it is obvious that Quantum Fields are non-physical in nature and origin, and that the couple dozen different types of Quantum Fields that we have discovered so far are in fact non-physical and pre-physical. Any physics course makes it obvious that the Force of Gravity, Magnetic Force, Electrostatic Force, the Normal Force, Friction, Inertia, and the Force of Tension are non-physical in nature and origin. The different forces and fields are "non-physical action at a distance"; and, they are used to make physical matter in the first place. In other words, it's obvious that the Gods, or the Controlling Psyches, Nature's Psyche, or the Universal Intelligence – who are enforcing the physical laws – had to design and make the quantum fields BEFORE physical matter could be made and sustained, because these quantum fields are how Nature's Psyche makes and enforces physical laws. It's obvious that physical matter and the various different physical laws are the result of Some Universal Intelligence enforcing a bunch of different non-physical Quantum Fields upon the universe as a whole. Physical matter is made from energy or from a bunch of different non-physical quantum fields working in tandem, or as a whole.

It is obvious that a photon, while traveling at the speed-of-light, is massless or non-physical; yet, the Materialists and Naturalists are trying to convince us that the non-physical or the spiritual does not exist. In a very real sense, the Materialists, Naturalists, Darwinists, Nihilists, and Atheists are trying to convince us that photons or quantum waves do not exist. It's obvious that they are wrong; but, these people can't see it because they have convinced themselves that the non-physical does not exist. Self-deception works, and it works every time.

The proven and verified existence of ANY long-range force such as electrostatic force, magnetic force, gravity, dark energy, and quantum fields

FALSIFIES Materialism, Naturalism, and their derivatives which claim that the non-physical does not exist. Why? It's because long-range forces are Action at a Distance; and, the Materialists and Naturalists are constantly trying to convince us that Action at a Distance or supernatural quantum mechanisms DO NOT EXIST.

Waves are disturbances traveling through a medium. When the Materialists and Naturalists proved that the luminiferous ether does not exist, they erroneously concluded that the vacuum of space is completely empty, and that the atom is 99.999% empty space. They were wrong. All they proved is that there is NO physical matter in a vacuum of empty space, which means that NO sound waves can travel through it. However, this "luminiferous ether" does in fact exist as non-physical quantum fields; and, non-physical quantum waves, light waves, gravity waves, and dark energy waves do in fact travel through it. Unlike sound waves, light waves, gravity waves, and magnetic waves can travel through a vacuum. That's possible because these different quantum waves are non-physical; and therefore, they travel through physical matter as if the physical matter isn't even there. Quantum Mechanics falsifies Materialism, Naturalism, Darwinism, Nihilism, and Atheism trillions of different ways.

https://www.youtube.com/watch?v=h0L6YpWh0KM

Again, the Materialists and Naturalists got their science wrong simply because these people refuse to believe in the existence of the non-physical. Space is NOT empty as the Materialists and Naturalists claim. Space is chock full of a couple dozen different non-physical omnipresent quantum fields that we know about, and quantum waves travel through them. Furthermore, the quantum waves that are produced or instantiated by psyches and sent traveling through quantum fields are technically omnipresent and potentially infinite velocity as a result. If they slow down, it's because they choose to do so. Quantum waves are thoughts.

With the Michelson–Morley Experiment, they proved that there is no luminiferous ether made out of physical matter; but, they erroneously concluded that light travels through nothing at all.

https://www.khanacademy.org/science/physics/special-relativity/michelson-and-morleys-luminiferous-ether-experiment/v/michelson-morley-experiment-introduction

The flaw with this experiment is that they started with the pre-chosen conclusion that the luminiferous ether has to be made out of physical matter and therefore subject to the speed-of-light limitation and the theory of relativity.

The luminiferous ether does in fact exist; but, it's made out of non-physical quantum fields. Quantum fields are omnipresent; and, they are NOT subject to physical limitations. Quantum waves can travel at infinite velocities, if they choose to do so. It's called quantum tunneling. This infinite velocity quantum phenomenon has also been verified with the experiments in non-locality or quantum entanglement. Quantum objects can do instantaneous action at a distance, instantaneous communication, and quantum tunnel across the universe, if they choose to do so. It's a whole other dimension.

These scientists successfully proved that there is no physical matter in an empty vacuum of space; however, light will travel just fine through a vacuum. Remember, a wave is a disturbance traveling through a medium. In the case of light, light waves and quantum waves are a disturbance traveling through omnipresent quantum fields.

The luminiferous ether does exist. It just isn't physical in nature, and therefore it isn't necessarily subject to the theory of relativity or the speed-of-light limitation. The luminiferous ether is comprised of a couple dozen different quantum fields; and, the quantum waves or "particles" traveling through these quantum fields are capable of traveling at an infinite velocity if they choose to do so. If they slow down to the speed-of-light, it's because they choose to do so.

The theory of relativity is the pinnacle of classical physics; and, quantum fields have nothing to do with relativity because at the speed-of-light, time stops, distance goes to zero, resistance to acceleration ceases to exist or goes to zero, physical limitations cease to exist, physical matter ceases to exist, and the resulting quantum objects are capable of infinite velocities or quantum tunneling as well as omnipresence or omniscience. Like I said, the quantum dimension is a whole other dimension. It has no physical limitations.

The Materialists and Naturalists erroneously believe that, because they can prove the existence of the physical, they have simultaneously proven the non-existence of the non-physical. They are wrong. According to the Philosophy of Science, it is impossible to prove that something does not exist. It's impossible to prove that the non-physical does not exist. The Materialists, Darwinists, and Naturalists claim that the non-physical doesn't exist; but, they can't prove it. The truthfulness of their claim has to be taken on blind faith alone.

In fact, the whole of Science proves Materialism and Naturalism wrong. The instantaneous attraction between the earth and the sun, or the earth and the moon, is a mystical supernatural force that we call gravity. There's no physical contact between the sun and the earth; but, they are still interacting with each other at a distance. Action at a Distance as well as Instantaneous Interaction at a Distance are quantum phenomena. This is obviously true because the interaction is obvious, and it's also obvious that there is NO physical contact between the sun and the earth. The proven and verified existence of something non-physical – like gravity, space, time, dark energy, dark matter, quantum fields, quantum waves, magnetism, or photons – FALSIFIES the materialistic, naturalistic, and atheistic claim that the non-physical does not exist. Everything is non-physical, except for physical matter.

When it comes to the Materialists, Naturalists, Darwinists, Nihilists, and Atheists, their whole reason for being is to take something that has been experienced and observed, and then to try to convince you that it doesn't exist because it isn't physical. These people use a wide variety of different logic fallacies in an attempt to convince you that what you have experienced and observed didn't really happen and doesn't really exist. It works. Billions of people have fallen for the ruse. I was one of them.

It's extremely important to realize that every physical atom has BOTH instantaneous quantum components and time-delayed physical components. The time-delayed components of physical matter were built directly into the physical matter by the Psyches who are controlling the energies from which the physical matter was made. Along with the physical components that are built into a physical atom, the Psyches within those atoms are also capable of communicating and interacting with each other at a distance instantaneously to produce magnetic and gravitational effects. The non-physical components of physical matter are the parts of physical matter which the Naturalists, Atheists, Darwinists, Nihilists, and Materialists are trying to ignore, deny, eliminate, and convince you that they don't exist.

Remember, every physical atom in our physical universe is attracted to every other physical atom in our physical universe. This attraction takes place at a distance. It is mystical, supernatural, non-physical, spiritual, and quantum in nature and origin. It's capable of instantaneous action at a distance; and, since we are talking about physical atoms, this instantaneous quantum force is also capable of producing physical effects that travel throughout our universe at speeds slower than or equal to the speed-of-light. A physical atom can do BOTH. A physical atom can interact gravitationally and magnetically through instantaneous action at a distance with every other physical atom in this universe; but, that same physical atom can also produce detectable physical effects that propagate through space-time at velocities slower than or equal to the speed-of-light. A physical atom can do BOTH the quantum and the physical. Parts of a physical atom can do instantaneous action at a distance; and, other parts of it can do Newtonian Physics.

Our science experiments are constantly revealing BOTH the instantaneous action at a distance and the slower-than-light physical aspects of physical atoms and electrons.

Electrons create a force field or an atomic orbital shell because they are moving at an infinite velocity at the quantum level. The gluons are forces and fields that are interacting at an infinite velocity at the quantum level. However, electrons have a physical component that produces physical effects and that slows the electron down at the physical level; and, quarks have an obvious time component built into them that clearly slows the physical atom down to sub-light speeds and velocities. When it comes to the physical, electrons seem to be half-in and half-out. When it comes to the physical, quarks seem to be all the way in, or totally physical. When it comes to gluons, bosons, quantum fields, atomic orbital shells, and psyche or quantum consciousness, those things seem to be purely quantum and therefore seem to be instantaneous action at a distance.

When electrons are in their physical state, electrons have a bit of mass or resistance to acceleration. A lone electron will travel through space with a measurable velocity slower than the speed-of-light. That's how it manifests its physical components. However, the very same electron in orbit of a nucleus will move so fast at the quantum level that it literally forms a solid shell or force field around the atom. When electrons are in their quantum state, they seem to travel at an infinite velocity; and thereby, these electrons create solid shells around their chosen nucleus.

https://www.khanacademy.org/science/physics/quantum-physics#atoms-and-electrons

At Khan Academy, they use Bohr's Radii Model to measure or quantify an electron's physical components or physical manifestations. All the while, he is constantly saying that this isn't how it truly works in reality. A 2D orbit or circumference is NOT how an electron orbits an atom. In its quantum non-physical wave format, the electron moves at an infinite velocity and quantum tunnels everywhere so fast that it creates a SOLID force field or shell around the nucleus of the atom thereby making the atom seem solid and physical and real. This infinite velocity omnipresent sphere or shell that the electron creates is its quantum manifestation. When the electron breaks away or when the electron produces a photon, the results show up on our physical detectors according to the physics established by the Bohr Radii Model.

Every quantum object is capable of being omnipresent and capable of traveling at an infinite velocity. An electron in the presence of a nucleus uses its omnipresent and infinite velocity capabilities to form a solid shell around the nucleus of that physical atom. It KNOWS where it's at, and it KNOWS what it is doing. It is intelligent and alive. When it is in orbit of an atom, an electron chooses to form a shell or a sphere around that atom; but, when it chooses to leave the atom, it vectors straight outward in a ray of its own choosing. It changes its behavior based upon what it has chosen to do. When it reaches a detector, it knows that as well, and then it gathers its wave into a single point and makes a single impression. These behaviors reveal the presence of intelligence, consciousness, sentience, or psyche.

Remember, every electron has BOTH a physical component and a quantum component. The Materialists and Naturalists deny the existence of the quantum component, and only focus their attention on the physical manifestations. These people never once realize that it is in fact the electron's quantum component, or spiritual component, or non-physical component that produces a solid shell or wall out of a single electron, during the production and the maintenance of a physical hydrogen atom. All of these other things that they pay attention to show up ONLY AFTER the electron or photon has left the atom. While the electron is with the atom, it is producing a force-field, shell, or solid shield around the atom – it is producing physical matter or a solid wall at an infinite velocity. The electron is omnipresent and infinite velocity within that shield or shell. That's a quantum phenomenon that produces a physical effect. These people can't see it nor understand it because they don't want to. They don't want it to be true because it falsifies their pre-chosen beliefs. These people have chosen to believe that the spiritual, or the non-physical, or the quantum does not exist. They are wrong. These people don't understand the science, because they don't want to. Quantum mechanics is spiritual mechanics or supernatural mechanics. The best, most powerful, and most explanatory parts of it are non-physical in nature and origin.

There is an Intelligence or Psyche within each electron who KNOWS whether the electron is supposed to act like a physical particle and travel in a straight line through space, or whether it is supposed to act like a quantum object and travel at an infinite elliptical velocity and form a specific type of shell around the nucleus of an atom.

Every particle seems to have BOTH quantum components and physical components; whereas, quantum fields seem to be purely non-physical, omnipresent, and instantaneous in their interaction. The truth of this becomes obvious once we start to study physics for real with an open mind. Yet, the quantum, or the spiritual, or the non-physical components go completely unnoticed by the Materialists and Naturalists who have formally and officially rejected these things. These people are in a perpetual state of denial. These people are the creators of the denialistic philosophies, the fictional philosophies, or the falsified philosophies.

There's a whole collection of math that has been developed around Quantum Mechanics and Quantum Field Theory. Chances are high that some of it is incomplete or even incorrect because we scientists have yet to develop the True Interpretation of Quantum Mechanics. We still use the Bohr Radii Classical Physics Model to calculate electron volts and other things, even though we know that it has been falsified or replaced by the Atomic Orbital Model. It's possible that our interpretation of the Atomic Orbital Model is inaccurate or incomplete as well. This is because our mathematicians and scientists deny the existence of the non-physical and deny the existence of anything that travels faster than the speed-of-light. You can't find the truth through denial or through the pursuit of non-existence, which is exactly what the Materialists and Naturalists try to do.

The Long-Range Forces Are Instantaneous Action at a Distance

Many different people have stated:

The force of gravity exerted by the Earth is infinite in range (i.e. pulls on things infinitely far away), but the force gets smaller and smaller the farther away an object is separated from the Earth.

Infinite range or long-range forces are Action at a Distance; and, these are quantum phenomena, and NOT physical forces or contact forces as described by classical physics. The quantum or the non-physical was used to FORM the physical – not the other way around as the Materialists and Naturalists claim.
Instantaneous Action at a Distance is purely a quantum phenomenon or a non-physical phenomenon. It works just as well on physical objects as it does on non-physical objects. In fact, it is non-physical quantum fields and quantum waves that are used to make physical matter in the first place. The only reason that a physical atom seems solid is because it has one or more electrons traveling at an infinite velocity forming an omnipresent sphere or shell around the nucleus of that atom. A single electron transforming itself into a sphere or a shell or a force-field is a

quantum phenomenon, and NOT a physical phenomenon. Such a thing is physically impossible according to classical physics, which means that it must be a non-physical quantum phenomenon. The electron is NOT in direct contact with the nucleus, which means that once again we are dealing with action at a distance and a non-physical long-range force even when we are dealing with atomic orbital shells that are being used to make the atom seem solid or physical. Omnipresent, infinite velocity quantum fields permeate quarks, electrons, and atoms as if these things weren't even there. Remember, the Obvious Law of Physics states that the smaller dwells within and controls the larger. Ask yourself what dwells within and controls the quarks, electrons, and physical atoms. It obviously has to be something quantum and non-physical because quarks and electrons are used to make physical atoms.

The long-range forces are quantum or spiritual in nature and origin; whereas, the contact forces are obviously physical in nature but quantum in origin. The verified and proven existence of Quantum Mechanics and Action at a Distance FALSIFIES Materialism and Naturalism which claim that these things DO NOT EXIST. The Materialists and Naturalists are constantly getting it wrong because they are constantly making the claim that the truth does not exist. They are the ones who originated the idea that science and the scientific method cannot be used to find and know the truth. In contrast, the best and the fastest way to find and now the truth is to live it and experience it for yourself, or to choose to trust someone who has.

I KNOW that substance-induced psychosis is a living hell because I have lived it and experienced it. Likewise, WE KNOW that non-physical quantum mechanisms, quantum waves (thoughts), quantum fields, and action at a distance are real and true because WE have lived them and experienced them in our daily lives.

Quantum Field Theory Playlist

https://www.youtube.com/playlist?list=PLsPUh22kYmNBpDZPejCHGzxyfgitj26w9

It's obvious from Quantum Field Theory that quantum fields are non-physical and prerequisites for the physical; and, the proven and verified existence of non-physical quantum fields and non-physical quantum mechanisms FALSIFIES the primary axiom of Materialism, Naturalism, Darwinism, Nihilism, and Atheism which states that the non-physical does not exist. The false is falsified by the truth; and, the truth is repeatedly experienced and observed. Non-physical quantum fields, and non-physical quantum mechanisms are repeatedly verified, experienced, and observed; therefore, Materialism and Naturalism of any kind are obviously false.

I have observed that ALL of the different forces and fields are invisible, intangible, and non-physical. There's physical matter; and, then there's everything else. Everything else is non-physical, which means that everything else is spiritual, quantum, intangible, action at a distance. There are the contact forces that depend upon contact between particles of physical matter; and, then there is everything else – the long-range forces. The long-range forces appear to be instantaneous and omnipresent, which means that they are quantum in nature and origin.

Watch the following video and count how many different times he says that Forces are instantaneous. There is NO delay. They have NO speed limit. Particles have speed-limits, NOT forces and fields.

https://www.youtube.com/watch?v=VfpKzwrhmqQ

Any time a physicist talks about instantaneous partner forces or instantaneous long-range forces that exhibit no delay, he or she is in fact talking about non-physical forces or quantum forces – and NOT physical matter. Delay or the passage of time is introduced into the system through particles – particularly through physical atoms. Physical particles were designed to experience delay or the passage of time. Interaction between forces is instantaneous and omnipresent, which means that these are quantum phenomena and NOT physical phenomena. It's obvious, for anyone who is willing to look and see.

Instantaneous Action at a Distance is the SIGN of Quantum Mechanics, Spiritual Mechanisms, Psyche Mechanisms, Non-Physical Mechanisms, Transdimensional Mechanisms, Non-Local Mechanisms, or Supernatural Mechanisms. Forces and fields have NO physical limitations; therefore, they are not physical. Yet, we typically use the physical in order to detect the presence or the influence of the non-physical. Magnetism and gravity and dark energy are the first to come to mind. These things are intangible. We can't get our hands on them; and, they are instantaneous and omnipresent. The physical matter has speed limits; but, the forces and fields do not. The contact forces require interaction between adjacent physical particles; but, the long-range forces do not. It's obvious that the long-range forces are quantum in nature and origin – NOT physical in nature and origin.

Particles are slowed down or can be slowed down so that they can interact with the physical realm; but, the omnipresent forces and fields have no speed-limits or range-limits, unless their controlling psyche deliberately gives them a limit.

Everything besides Physical Matter IS Non-Physical

Mass or resistance to acceleration is the is the component that makes a quantum object physical in nature, because it slows the object down to sub-light velocities. If an object has mass or resistance to acceleration, then it can't travel at the speed-of-light or be accelerated to the speed-of-light. It has to be massless, or have no resistance to acceleration, in order to travel at the speed-of-light. Photons are obviously massless or non-physical. All types of light are massless or nonphysical, until they slow down to zero and show up in our physical realm on our physical detectors. Physical objects have to quantum tunnel or teleport in order to travel faster than the speed-of-light because they can't be accelerated to the speed-of-light due to their resistance to acceleration or mass.

In contrast, photons can inherently or natively travel at infinite velocities because they have no resistance to acceleration, so if they travel slower than that, it is because they have chosen to do so. Photons have to slow down to zero in order to show up in our physical realm on our physical detectors. If the photons kept going at the speed-of-light or faster, they would simply pass through the atoms as if the physical atoms weren't even there. The photons we have observed obviously CHOSE to slow down to zero when they reached their chosen destination. Photons will brake for physical atoms, which means that photons can sense the presence of physical atoms. This obvious truth reveals the presence of intelligence or psyche within every photon that has chosen to slow down and manifest itself in our physical realm. Photons can change their behavior based upon what they sense around them. They travel at the speed-of-light until they reach their chosen destination, at which point, they slow down to zero and show up on our physical detectors. A change in behavior based upon sensory input signifies some type of intelligence or psyche or consciousness. Cool, huh? The truth is hiding in plain sight where nobody seems to be able to see it or find it. Quantum waves are waves traveling through the spirit realm, or the supernatural realm, or the non-physical realm. They will never slow down unless they choose to do so; and, they are capable of traveling at infinite velocities if they choose to do so. The quantum has no physical limitations.

Quarks are the ONLY thing in this universe that seem to be physical all-the-time; yet, even quarks have a quantum component. Quarks are made from energy by psyches who are also energy or light. Psyches are living light or conscious light. The energies and psyches are always conserved which means that they are always quantum in nature and origin. The psyches or intelligences are the things that impose mass or physical limitations or the aging process on the quarks and the other particles that have mass, or speed-limits, or resistance to acceleration.

When it comes to any particle in this physical universe, its quantum components have to be identified and separated from its physical components. In contrast, its internal forces and fields seem to be purely quantum in nature, origin, and function. Instantaneous Action at a Distance, or Long-Range Forces and Omnipresent Quantum Fields, are obviously non-physical phenomena; and, they are therefore purely quantum and psychic in nature and origin. You'll never find the truth by rejecting the quantum, the supernatural, the spiritual, the transdimensional, or the non-physical. There will always be something missing from your theories and ideas once you choose to believe that the non-physical or the immaterial does not exist. I KNOW because I used to be a materialist, naturalist, nihilist, and atheist back in 2012. When I was in that frame of mind, the quantum didn't make any sense to me at all, because I didn't have what I needed in order to figure it out and understand it. You can't find and know the truth by rejecting the truth. It's obvious that non-physical, immaterial, quantum fields truly exist; therefore, you will never find and know the truth by denying their existence as the Nihilists and Atheists do. It's obvious that non-physical magnetic fields, gravity, dark matter, radio waves, and dark energy truly exist; therefore, you will never find the truth by denying their existence as the Materialists and Naturalists do. Remember, at the speed-of-light, photons are massless or non-physical.

Whenever I study physics, and I love to study physics, it's obvious that everything besides physical matter is intangible, invisible, quantum, spiritual,

41

supernatural, and non-physical. Others have noticed it too. I'm not alone. Out-of-body travelers and near-death experiencers have observed that we continue to think and remember long after our physical body and physical brain are dead and gone. Thoughts are quantum waves; and, psyche, the spark of life, quantum consciousness, or intelligence is a pinpoint of light that is capable of producing quantum waves, transmitting quantum waves, receiving quantum waves, processing quantum waves, and storing quantum waves. Quantum waves or thoughts are Wi-Fi at the quantum level; and, Psyche is the transceiver, processor, and storage unit for quantum waves. It's obvious for anyone who is willing to look and see. Photons at the speed-of-light are obviously massless or non-physical.

These truths show up subtly everywhere, once we start looking for them. For example, we have this following quote from the internet:

The coefficient of kinetic friction is a number with no units. We call these types of quantities dimensionless quantities since they have no units (i.e. no physical dimensions).

Anytime physicists start talking about things that have NO physical dimensions, they are in fact talking about quantum phenomena – invisible and intangible forces and fields. "Dimensionless Quantities" with no physical dimensions actually exist in the quantum dimension, or the transdimensional realm, or the supernatural realm, or the Spirit World. They exist and exert an influence – they just don't have any physical dimensions. Space and time are non-physical. We can't get our hands on them or manipulate them. The same thing can be said for psyche, information, consciousness, or intelligence. Start looking for dimensionless quantities, and you end up finding the non-physical. Purely quantum phenomena exist is the quantum dimension or supernatural dimension, rather than the physical space-time dimension; yet, the non-physical or the quantum clearly influences the physical. Anytime you find and observe something non-physical, you have in fact used Science (observation and experience) to falsify Materialism, Naturalism, Darwinism, Nihilism, and Atheism which claim that the non-physical does not exist.

Sal Khan stated that he finds the gravitational attraction between two physical objects almost mystical. It is mystical because it is quantum – it is instantaneous action at a distance. The gravitational attraction between the earth and the moon is a mystical, supernatural, quantum, spiritual phenomenon. The thing that the Materialists and Naturalists don't realize is that each physical atom has BOTH an instantaneous action-at-a-distance component and a delayed physical component. When two neutron stars merge, the gravitational fields and forces acting between them is instantaneous. At the point of merger, the merger creates ripples in space-time that are physical in nature and that ripple throughout space at velocities less than or equal to the speed-of-light. Physical objects have BOTH the instantaneous action-at-a-distance quantum phenomenon built into them and the delayed physical space-time phenomenon built into them. A physical atom has capabilities of doing both. Einstein was only interested in the delayed speed-of-light limitation of physical matter because he didn't believe in the instantaneous action-at-a-distance quantum component. Most of our scientists are like Einstein, and they choose to completely ignore the quantum or spiritual or non-physical aspects of physical matter. But, you can't do that if you want to find and know the truth, and figure out how things really work.

We scientists, the real scientists, use experiments with the physical to detect the presence or the influence of the non-physical. Play around with a magnet and some iron filings. When I was a kid, I would run a magnet through the sand in the sandbox and collect pill bottles full of iron. The physical iron was used to detect the presence or the influence of the non-physical and invisible quantum fields or magnetic fields. You can see with your own eyes the effects of the non-physical magnetic waves upon the physical iron simply by running a magnet under a piece of paper with iron filings on top of it. That's action at a distance, a quantum phenomenon. There's no direct physical contact between the magnet and the iron filings thanks to the piece of paper in between. Invisible, instantaneous, non-physical "action at a distance" acting on physical matter reveals the presence of quantum forces and fields. Long-range forces reveal the presence of the quantum or the spiritual, because long-range forces are action at a distance; and, instantaneous action at a distance is quantum mechanics. Studying physics constantly reveals the presence and the influence of the non-physical. It's hidden all throughout physics just waiting for someone to find it. Only the Materialists and Naturalists deny its existence because they are biased or motivated to do so.

The non-physical obviously exists. The presence and the influence of the non-physical has been experienced and observed. The only people who are trying to prevent you from discovering these truths are the Materialists, Naturalists, Darwinists, Nihilists, Behaviorists, Determinists, Physical Reductionists, and Atheists. They each have a specific agenda. The Materialists and Physicalists and Mechanists are trying to convince you that the non-physical does not exist. The Naturalists are trying to convince you that the supernatural, the spiritual, or the quantum does not exist. The Darwinists and Atheists are trying to convince you that Intelligence, Psyche, Quantum Consciousness, Nature's Psyche, Universal Intelligence, and God's Psyche do not exist. They preach spontaneous generation, abiogenesis, chemical evolution, and "creation from nothing by nothing" as a replacement for God, or Universal Intelligence, or Nature's Psyche. The Behaviorists and Determinists are trying to convince you that choice, free will, and volition do not exist because "choosing between two competing options" reveals the presence of Psyche, or Intelligence, or Consciousness. Psyche is the ultimate chooser; and therefore, Psyche is the Ultimate Cause and the Ultimate Causal Agent. Furthermore, out-of-body travelers and near-death experiencers have observed that we continue to think and choose long after our physical brain and physical body are dead and gone. These Materialists and Naturalists are trying to hide the truth from you. It's what they do. It's their reason for being.

I have studied physics for years looking for signs of the non-physical; and over time, I have found precisely what I was looking for. No seeking, then no finding. I couldn't see it until I started looking for it. Self-deception works, and it works every time. But nowadays, it's obvious to me that photons traveling at the speed-of-light are completely massless or non-physical. It's also obvious to me that quantum fields are non-physical.

By studying the non-physical aspects of physics, you can see all the different things that the Materialists, Naturalists, Darwinists, Nihilists, and Atheists are trying to hide from you and are trying to convince you that they don't exist. You can see the truth of it, unless you are a Materialist, Naturalist, Darwinist, Nihilist, Behaviorist, Determinist, or Atheist. I KNOW, because I have lived it and experienced it. Back in 2012, when I was a materialist, naturalist, nihilist, and atheist, I was completely oblivious to all of these different non-physical forces and fields, because I only wanted physical matter to exist. At the time, I was in a bad way, and I personally wanted to die and cease to exist. I came close to dying; but, I continued to exist. I lost most of my knowledge and most of my memories; but, they continued to exist and were eventually restored. I needed a scientific explanation for what happened to me, and I wasn't getting it through Materialism, Naturalism, Nihilism, Atheism, and their derivatives. I had to look someplace else for the truth. I eventually found what I was looking for within Quantum Consciousness, Quantum Field Theory, Quantum Forces and Fields, Quantum Mechanics, and Instantaneous Action at a Distance – all of which are non-physical or supernatural in nature and origin.

In order to find and know the truth, we must be willing to identify and eliminate everything that is false. The materialistic, naturalistic, and atheistic claim that the non-physical does not exist is obviously false; so, it must be eliminated from science and from our scientific explanations if we really want to find and know the truth. The purpose of this and my other books is to unleash Quantum Mechanics and then use it to find and know the truth.

Mark My Words

The True Interpretation of Quantum Mechanics

Without realizing it, I have been pursuing or looking for the True Interpretation of Quantum Mechanics for half a decade now (May 2019). I'm closer to finding the True Interpretation of Quantum Mechanics than I was back in 2012, when I wasn't looking for it, when I didn't know anything about it, and when I was a materialist, naturalist, nihilist, and atheist.

As I see it, Quantum Theory or the Meaning of Quantum Mechanics is the most interesting part of Quantum Mechanics and Quantum Field Theory because it's still up for grabs or is still yet to be determined and agreed upon.

The math for Quantum Field Theory and Quantum Mechanics has been developed. More importantly, Quantum Mechanics has been repeatedly and robustly verified through scientific experimentation. The only thing left to do with Quantum Mechanics and Quantum Field Theory is to figure out what they truly mean. In other words, we need an interpretation of Quantum Mechanics that is real and true; and, this is where we continue to fall down and fail, because there is a "school of thought" or a group of philosophers within the scientific community who do in fact deny the existence of Quantum Mechanics or Supernatural Mechanisms.

A person will NEVER find the True Interpretation of Quantum Mechanics if he or she continues to insist that quantum mechanisms or supernatural mechanisms DO NOT EXIST. That's just the way it is. Seek and ye shall find. Knock, and it shall be opened unto you. NO seeking, then NO finding. That's just the way it is.

From the "shut up and calculate" school of Quantum Mechanics, these people don't want you thinking about what Quantum Mechanics means. These people are typically Materialists, Naturalists, Nihilists, Mechanists, and Atheists. They are only interested in having you use Quantum Mechanics to make them some money. They aren't interested in thinking about what consciousness is, or where the quantum realm is, or what it takes to collapse a wave function, or what is capable of collapsing a wave function. These people formally and deliberately reject the non-physical or supernatural interpretations of Quantum Mechanics because whenever Quantum Mechanics is interpreted in that manner, it FALSIFIES Materialism, Naturalism, Darwinism, Nihilism, Determinism, Behaviorism, and Atheism (Creation Ex Nihilo).

I no longer fit well with this group, because I'm no longer a materialist, naturalist, nihilist, and atheist. One day I decided that I wanted to figure out what everything truly is and how it really works. That change in emphasis doesn't fit well with a philosophy that is based upon the denial of the existence of things that obviously exist. Though they don't realize it, the Materialists and the Naturalists effectively state that Quantum Mechanics and Quantum Fields DO NOT EXIST because these people claim that the non-physical does not exist and that the supernatural does not exist. Quantum Mechanics by definition in principle is obviously supernatural; and, quantum fields are obviously non-physical and pre-physical. In other words, the proven and verified existence of Quantum Mechanics and Quantum Fields

FALSIFIES Materialism and Naturalism which claim that quantum fields and quantum mechanisms DO NOT EXIST.

You see, when it comes to Quantum Mechanics, the math has been developed and the experiments have been run; and, Quantum Mechanics has passed the test and has become LAW or proven Science. Quantum Mechanics and Quantum Field Theory have been repeatedly verified and proven true. The math for it all has been developed as well.

When it comes to Quantum Mechanics, the ONLY thing left to do is to figure out what it all means. That's where I come in. I'm an open-minded theoretician, logician, philosopher, and scientist. I sample or examine the different interpretations of Quantum Mechanics in an attempt to find the ones that have the MOST explanatory power or to find the ones that match MOST closely with what has actually been experienced and observed. I then choose to go with a preponderance of the experiential evidence. A theory or philosophy is absolutely worthless if it has NEVER been experienced nor observed by anyone!

For example, the Multiple Worlds interpretation of Quantum Mechanics is worthless, because it has NEVER been experienced by anyone!

https://en.wikipedia.org/wiki/Interpretations_of_quantum_mechanics

Likewise, many of the interpretations of Quantum Mechanics on the Wikipedia are completely worthless because they have NO explanatory power. The interpretations of Quantum Mechanics which attempt to reduce Quantum Mechanics to Classical Realism, Classical Physics, Materialism, Naturalism, or Atheism (Creation Ex Nihilo) are absolutely worthless because they effectively deny the existence of Quantum Mechanics or Supernatural Non-Physical Mechanisms.

I've been trying for years to discover, find, and/or develop interpretations of Quantum Mechanics that actually match with reality, have been experienced and observed either in the flesh or in the spirit world (quantum realm), that explain everything that has ever been experienced and observed, and that make logical sense when I'm done.

I also, of course, need an interpretation of Quantum Mechanics that allows for the existence of the physical level or the physical aspect of our universe, because it obviously exists as well. To be worthwhile, our interpretation of Quantum Mechanics has to be able to explain both the physical and the non-physical, the natural and the supernatural. It needs to be able to work at both the quantum level and the physical level. Our interpretation of Quantum Mechanics needs to be able to explain what the Human Psyche and Nature's Psyche are doing at the quantum level in order to get things done for us at the physical level. So far, I have found only three interpretations of Quantum Mechanics that fulfill these requirements or needs. The rest of them have proven to be pretty much inadequate or even totally worthless.

Henry P. Stapp's Orthodox Interpretation of Quantum Mechanics divides quantum theory up into three or four different processes. It actually tells us mathematically what the Human Psyche is doing, what Nature's Psyche is doing, and which one is in fact collapsing the wave function. According to the Orthodox Interpretation of Quantum Mechanics, the Human Psyche chooses what it wants to do with its physical body and physical brain, then Nature's Psyche decides whether that is permitted or not, and then it is in fact Nature's Psyche (the psyches within the atoms themselves) who collapses the wave function and/or fires the neurons within our physical brain. Nature's Psyche collapses the wave function, not the Human Psyche; and, the math from von Neumann and Dirac actually distinguishes between what the Human Psyche and Nature's Psyche are doing. It's there in the math. The Human Psyche isn't consciously aware of any of this, because it's all being handled by Nature's Psyche. This reality and truth explains everything that has ever been experienced and observed.

Since Materialism and Naturalism have been FALSIFIED by Quantum Mechanics and Quantum Field Theory, materialistic and naturalistic interpretations of Quantum Mechanics that attempt to reduce Quantum Mechanics to classical physics or classical realism or scientific naturalism end up being completely worthless. Quantum Mechanics, Psyche Mechanics, Energy Mechanics, or Transdimensional Physics is very much a supernatural phenomenon. Quantum fields are obviously non-physical; and, quantum fields of necessity MUST EXIST. The non-physical MUST EXIST or the physical wouldn't be able to exist, because the non-physical or the quantum is the substrate or the foundation of the physical. The Gods or the Controlling Psyches had to design and make the quantum fields BEFORE they could make and sustain physical matter. That's just the way it is. Physical matter is impossible without quantum fields; and, quantum fields are obviously non-physical and pre-physical.

We have to learn to detach from the physical, if we truly want to find and understand the quantum, or the non-local, or the non-physical. Most scientists don't want to do this; but, I'm no longer 'most scientists'. In my theorizing, I've deliberately let go of physical limitations, in order to pursue a true and realistic understanding of the quantum or the non-physical or the supernatural. Other scientists have started to do so as well. It's the only way to find and know the truth about the quantum or the non-physical. Let's face reality here. Quantum Fields are obviously non-physical, immaterial, pre-physical, and supernatural. It's finally time for us to face up to this reality and to try to explain it scientifically, by examining what has actually been experienced and observed.

The following is a selection of what I have encountered so far while trying to find and develop a True Interpretation of Quantum Mechanics. You will have to decide for yourself how much of it you agree with it, and how much of it you either reject or are able to falsify with experiential evidence.

Scientific LAW: The Obvious Law of Physics states that the smaller dwells within and controls the larger. It's obvious to most people that the smaller dwells within and controls the larger.

Scientific Observation: It is obvious that at the speed-of-light, photons are massless or non-physical. Quantum waves or photons are purely quantum objects or massless non-physical objects while traveling at the speed-of-

light. Photons have to slow down to zero or splash-down, in order to manifest in the physical realm. This reality is obviously true because it has been experienced and observed. It's obvious that the massless or the non-physical exists. At the speed-of-light, photons are quantum phenomena, supernatural phenomena, spiritual phenomena, or non-physical phenomena. At the speed-of-light, we can't get our hands on them.

Scientific or Experiential Observation: A psyche or intelligence is observed at the quantum level or the psyche level or the thought level as being a photon, or a pinpoint of light, or a spark, or a point particle. This is what has actually been experienced and observed by out-of-body travelers and near-death experiencers. Remember, photons or pinpoints of light have been experienced and observed both at the quantum level and the physical level. Psyche is observed as being a pinpoint of light or a photon at the quantum level; and, psyche is experienced first-hand as an immaterial viewpoint in space. This is what has actually been experienced and observed. Observation and experience should trump philosophical speculation and wishful thinking every time; but, they don't where the Materialists and Naturalists are concerned.

https://psyche-ontology.com/psyche-experienced-and-observed/

Scientific Definition: Psyche or Life Force or Consciousness or Intelligence is a point particle, an infinite singularity, or a photon at both the quantum level and the psyche level. Every Psyche or Life Force has a certain amount of energy that's under its control. Photons or point-particles are intelligent, conscious, self-aware, psychic, and alive.

Scientific Definition: A Psyche Ontology states that Psyche, Photons, or Point-Particles are the fundamental unit of reality and existence. Psyche is the Ultimate Cause or the Ultimate Causal Agent. Some type of Psyche obviously designed and made the first physical atom or the first particle of physical matter. First things first.

Scientific LAW: Energy, or Psyche, or Consciousness, or Intelligence, or Life Force is always conserved. This means that it has always existed, and it will always exist. It cannot be made, and it cannot be destroyed. This is the LAW of the Conservation of Energy and Psyche. This is the fundamental law of physics both at the quantum level and the physical level. The energy is always there waiting to be reused, even when it seems that entropy or physical laws have taken it temporarily out of play.

Scientific Corollary: The Law of Psyche states that each psyche has a certain amount of energy that's under its control. The Controlling Psyche can transform the energy under its control into anything that it wants that energy to be, including quantum waves, quantum fields, dark energy, dark matter or spirit matter, gravity, radio waves, or physical matter. Energy is infinitely malleable, which means that its Controlling Psyche can transform that energy into anything that it wants that energy to be, any time that it chooses to do so. The Controlling Psyche chooses what it wants to do with the energy or the light that's under its control. A Psyche makes quantum waves, transmits quantum waves, manages or controls quantum waves, receives quantum waves, processes quantum waves, and stores quantum waves. Quantum waves are thoughts. Thoughts are transmitted by Psyches through quantum waves. Quantum waves or thoughts are Wi-Fi at the quantum level. Psyche is the transmitter and the receiver of these quantum waves or thought waves. Psyche is a quantum transceiver.

Scientific Corollary: The Quantum Law of Thermodynamics states that there is NO thermodynamics or entropy at the quantum level and the psyche level. This is because the Conservation of Energy and the Conservation of Psyche reign supreme at the quantum level and the psyche level. ALL of the energy is available ALL of the time at the quantum level and the psyche level. In contrast, thanks to Entropy or the Second Law of Thermodynamics, not all of the energy is available all of the time at the physical level because it was actually designed to be that way by the Gods or the Controlling Psyches. There are physical limitations being imposed at the physical level by the Controlling Psyches who are controlling the energies that make up the physical level. The Quantum Law of Thermodynamics is a corollary to the LAW of the Conservation of Energy and Psyche. The Quantum Law of Thermodynamics or the Conservation of Energy and Psyche trumps all the other LAWS including the Second Law of Thermodynamics, because it has always been true and it will always be true. There is NO thermodynamics or entropy at the quantum level and the psyche level.

Scientific Corollary: The Ultimate Law of Thermodynamics differentiates between what is being conserved and what is NOT conserved. It is a corollary to the Law of the Conservation of Energy and Psyche. The materialists, naturalists, nihilists, and atheists erroneously teach and believe that entropic physical matter and the resulting entropy, heat death, or second law of thermodynamics are conserved. Something obviously has to give. Entropy and death cannot be eternal if energy or psyche is always conserved. Either the Second Law of Thermodynamics is true, or the Conservation of Energy is true. They both can't be true at the same time, because they contradict each other or falsify each other. The materialists, naturalists, and atheists always use the Second Law of Thermodynamics to falsify or eliminate the First Law of Thermodynamics or the Conservation of Energy. They are wrong to do so; and, here is why.

Making a Case for the Ultimate Law of Thermodynamics: First of all, the Second Law of Thermodynamics is temporary or temporal, which means that it isn't conserved. The Second Law of Thermodynamics hasn't always existed; and, that means that it won't of necessity always exist. Second of all, based upon Quantum Field Theory, which states that particles are born and particles can die, it is obvious that particles (including physical particles) are NOT conserved. Particles are NOT eternal and everlasting, which means that they are not conserved. Particles are born, and particles die, which means that they are NOT conserved. When an entropic particle of physical matter is dissolved back into the energy from whence it came, the associated entropy or temporality goes with it. The clock is reset. The energy is made available for use once again, and the entropy ceases to exist. The controlling psyches who control the energy from which physical atoms are assembled or made can choose at any time to transform that energy into something else like a quantum field, dark energy, available energy, or anything else they want it to be. The Second Law of Thermodynamics remains true only as long as the controlling psyches decide that it is going to remain true. Remember, the Second Law of Thermodynamics is temporary or temporal. When the controlling psyches get tired of the entropic physical matter, they can transform it into anything else that they want it to be, any time they choose to do so. Psyches make choices at the quantum level. Psyches control quantum objects and quantum waves. Energy or Psyche is always conserved; whereas, entropic physical matter is NOT conserved. This is the Ultimate Law of Thermodynamics which states that entropic physical matter is NOT conserved. We human beings destroy or dissolve physical matter in our nuclear bombs. We convert the physical matter back into the energy from whence it came. We human beings effectively make new physical particles from energy or light within our particle accelerators. Heat death, or entropy, or death in general is NOT conserved thanks to the Conservation of Psyche and Energy. Entropic physical matter and entropy are NOT conserved, especially at the quantum level or the psyche level. This is the Ultimate Law of Thermodynamics. It states that the Second Law of Thermodynamics is temporary, which means that entropy and entropic physical matter are NOT conserved. Heat death, or entropy, or the Second Law of Thermodynamics will cease to exist at the very moment that the controlling psyches choose for it to end; and suddenly, ALL of that energy will become available for use once again.

Scientific Observation and Scientific Definition: Instantaneous Action at a Distance – Non-Local Quantum Mechanisms – have been experienced and observed and verified experimentally. There are NO speed limits at the quantum level. There are NO physical limitations at the quantum level. Yes, things can slow down and stand still at the quantum level, but they can also travel at an infinite velocity, quantum tunnel, or teleport to their chosen destination. Instantaneous Action at a Distance is the same thing as an infinite velocity, teleportation, or quantum tunneling. There are indications that the distance is indeed traveled; but, it's done instantly at the quantum level, meaning at an infinite velocity. Quantum Mechanics, Spiritual Mechanics, Energy Mechanics, Psyche Mechanics, or Psychic Mechanics is Instantaneous Action at a Distance or Infinite Velocity. Remember, Quantum Mechanics is Instantaneous Action at a Distance, or what Einstein called, "spooky action at a distance." We know from these observations that the speed-of-light limitation at the physical level is chosen into existence by the Psyches who are controlling the physical level, because the speed-of-light limitation doesn't exist at the quantum level. The speed-of-light limitation at the physical level is the result of choice, and NOT the result of necessity. The Controlling Psyche chooses what it wants to do with the energy that's under its control. Remember, Instantaneous Action at a Distance has been repeatedly proven or verified through scientific experiments. It's proven and verified Science, which means that the speed-of-light limitation at the physical level is a construct or something that was chosen into existence or made to exist.

Scientific Observation: Quantum fields, photons, and quantum waves (thoughts) are non-physical. They have no mass or resistance to acceleration, which means that they are theoretically capable of traveling at an infinite velocity at the quantum level. They have no physical limitations or speed limits, unless their controlling psyche imposes such a thing upon them.

Scientific Observation: Quantum Fields fill the immensity of space and permeate the smallest particles such as electrons, quarks, and atoms. Space isn't empty. It's filled with a couple dozen different Quantum Fields. Dark Energy is one of these quantum fields. It has its own set of rules or laws, and it fills the immensity of space and permeates every elementary particle. The same can be said for gravity. Gravity and Dark Energy and Quantum Fields are obviously non-physical. They don't have a physical component. They are obviously non-physical, which means that they were obviously designed and made by the non-physical Psyches or Intelligences who are controlling the energy from which they were made. Remember, the Obvious Law of Physics states that the smaller dwells within and controls the larger. This reality applies at every level of physics or existence, including the Quantum Fields.

Scientific Logic and Scientific Conclusion: Quantum Fields are obviously non-physical and pre-physical as well. Something very small and non-physical at the quantum level or psyche level had to design and make and implement the non-physical quantum fields and the first particle of physical matter or the first physical atom. It's obvious from Quantum Field Theory that the Gods or the Controlling Psyches who made our physical universe had to design and make the Quantum Fields first, before they could make and sustain physical matter.

Scientific Logic or Scientific Conclusion: Physical matter and physical atoms are made from energy or light by the Psyches or Intelligences who are controlling that energy or light. Remember, when it comes to physics, the smaller dwells within and controls the larger. This is the Obvious Law of Physics.

Scientific Conclusion: Some type of non-physical Psyche, or Intelligence, or Consciousness, or Life Force obviously designed and made the first quantum field and the first particle of physical matter. First things first. Psyches have always existed and will always exist, because they are always conserved. In contrast, quantum fields and physical particles are made, which means that they can be disassembled and turned into something else instead.

Conclusion: You, your psyche, is going to have to decide for itself whether you believe that any of this is true, or not. It works for me; but, it's left up to you to decide for yourself if any of it works for you. That's what psyches do. They choose what they want to believe and what they want to reject. Something like love, justice, friendship, charity, grace, forgiveness, or compassion does not exist until after some psyche has believed it into existence or chosen it into existence. That's the power of psyche. It believes things into existence and chooses things into existence – including quantum waves, quantum fields, and physical matter. Psyches control the energies from which physical matter is made.

This huge syllogism represents my personal attempts to derive a true and comprehensive interpretation of Quantum Mechanics and Quantum Field Theory. It's theoretical or philosophical rather than mathematical. The math has already been developed, but Quantum Theory or the Meaning of Quantum Mechanics remains unresolved. I'm trying to bring about some resolution based upon what has actually been experienced and observed. Within this syllogism, I tried to reconcile everything that has ever been experienced and observed. I may or may not have gotten it 100% correct; but, it's the best I have to go with so far. It fits, and it's self-contained or self-consistent at least.

My interpretations of Quantum Mechanics and Quantum Field Theory are infinitely more expansive and all-inclusive than anything I have encountered so far. I couldn't find what I needed from the self-proclaimed experts in these fields, so I had to create my own. I kept working on it year after year, until it finally started to make some logical sense to me. I'm an open-minded scientist, theoretician, logician, philosopher, mathematician, and generalist. I'm good at everything and master of nothing. One day, after I got tired of being lied to and deceived by the Materialists, Naturalists, Darwinists, and Atheists, I finally decided that I wanted to figure out how everything really works and how everything truly fits together – before I die.

I've spent a lot of time sampling different interpretations of Quantum Mechanics trying to find or develop The True Interpretation of Quantum Mechanics. I sense that I still have a ways to go; but, I'm infinitely closer to the True Interpretation of Quantum Mechanics than I was a few years ago when I was a materialist, naturalist, nihilist, and atheist. Quantum Mechanics only makes sense from a spiritual perspective or a supernatural perspective. It never made any sense from a materialistic, classical physics, or classical realism perspective.

From our perspective trapped in space-time and subject to a physical mortal body, Quantum Mechanics is indistinguishable from magic. In fact, we deliberately go out of our way to avoid it and prevent it, especially when it comes to something like our computer chips where quantum tunneling or teleportation at the quantum level actually ruins the functionality and reliability of the computer chips – a situation where the electrons seem to develop a mind of their own. With computer chips, we have to build in barriers wide enough to prevent the electrons from quantum tunneling or jumping because we mere mortals can't control Quantum Mechanics from the physical level. ONLY Psyche, Intelligence, Consciousness, or Life Force can control quantum mechanics – both at the quantum level and at the physical level.

Every Psyche, or Spark of Life, or Intelligent Consciousness has a certain amount of energy that's under its control. It decides what form that energy will assume. Quantum Mechanics cannot be controlled at the physical level with physical matter such as our physical genes.

In our comic books, we have super heroes with superpowers; and, these are almost always presented to us as being caused and controlled by the person's superior DNA – X-Men and Mutant X and the Inhumans – or you have Kilgrave in the Jessica Jones series who uses viruses to control the minds of others. That's physically impossible. DNA cannot be used to give us superpowers, or quantum powers, or spiritual powers. It doesn't work that way. Viruses cannot be used to control our minds or psyches. Remember, according to the Obvious Law of Physics, the smaller dwells within and controls the larger. The Psyches within the viruses and DNA can indeed control them, make them move, and make them do things; but, viruses and DNA can't dwell within and control Psyches or Minds. DNA simply codes for proteins at the physical level. DNA has nothing to do with Quantum Mechanics, or Superpowers, or Magic, or Miracles.

Likewise, it's physically impossible for Kilgrave to use viruses to control Psyches or Minds. It doesn't work that way. In order for someone like Kilgrave to exist, it would have to be Mind-over-Matter (the Placebo Effect) or it would have to be transpersonal telepathic Mind-over-Mind. It would have nothing to do with his physical matter nor his physical brain.

Whenever we find ourselves in the situation where we are experiencing physical-matter-over-mind, such as when we are addicted or going through withdrawal, then we are really in trouble, and only one's Psyche can break the pattern by choosing to go cold turkey and suffering the results of withdrawal. When I was addicted to prescription drugs, my physical body wanted me to keep taking the drugs. My medical doctors, psychiatrists, and pharmacists wanted me to keep taking the drugs. BioPsychoSocial! My biology wanted me taking the drugs, and my society wanted me taking the drugs. I was suicidal. They had me on benzos, an anti-psychotic, an SSRI, a pain killer, an anti-depressant, an emergency anti-psychotic, and an anti-anxiety drug; and, they had me cycling through four different sleeping pills. Any time that I would complain about the side-effects, they would give me another drug in an attempt to cover up the side effects. Any time that I started to go through withdrawal, I would go psychotic, suicidal, delusional, and insane. The people on the radio knew my name and were telling me to kill myself. I had to get out of there. I tried to end it all by taking most of the sleeping pills. I just had to get out of there. I was at the pinnacle of matter-over-mind. They were trying to manage my psychology or my mind through my physical matter or physical body, and they literally drove me insane in the process. It doesn't work!

After my trip to the looney bin, they had gotten me addicted to benzos again, and they had prescribed everything for me all over again, except for the sleeping pills that I had overdosed on. I still had a couple of bottles of sleeping pills in the cabinet, and along with the new prescription for Valium, I could have tried it all over again. My wife was headed out to the pharmacy to buy all the junque all over again, and I told her not to bother because I wasn't going to be taking anything anymore. One day, that day, I finally had enough.

One day, I finally realized that I was never going to get better unless I quit taking their drugs. I, my Psyche, chose to go cold turkey, and I went psychotic, delusional, hallucinatory, and insane for six months as a result during the withdrawal period. It was only after I got sober that I first realized how insane, psychotic, and delusional I had been during the withdrawal periods and while I was on their drugs. It was Mind-over-Matter that saved me, not my biology, and not my society! I, my Psyche, chose to quit taking the drugs and chose to get sober. Mind-over-matter, or the placebo effect, is a superpower!

Superpowers are spiritual, or quantum, or supernatural in nature and origin – mind-over-matter or mind-over-energy. A physical enhancement of your DNA is NEVER magically going to grant you superpowers. Drugs or toxic goo are never going to grant you superpowers. A spider bite is never going to grant you superpowers. Superpowers have to be mind-over-matter in order for them to work. A physical improvement cannot produce spiritual gifts, quantum capabilities, psychic gifts, or super-hero supernatural superpowers. Natural adjustments do NOT produce supernatural results. This is what has actually been experienced and observed.

Remember, adjustments to nature, or to one's physical body, or to one's genome do NOT produce supernatural, psychic, spiritual, or quantum results. In fact, it's the other way around. Psyche, Intelligence, Mind, Soul, Consciousness, or Life Force is the thing that decides and chooses what form the energy under its control will take. The Obvious Law of Physics states that the smaller dwells within and controls the larger. Every Psyche has a certain amount of energy that's under its control. Psyche can form that energy into quarks, electrons, photons, quantum fields, chemical bonds, quantum waves or thoughts, spirit matter or dark matter, entropic physical atoms, or anything else it desires. Psyches are the things who design and make physical matter, from the quantum level up. Psyches are the thing that control Quantum Mechanics, or Energy Mechanics, or Transdimensional Physics. There is nothing that we can do at the physical level to control Quantum Mechanics except to build bigger and wider barriers or walls to prevent it from happening.

It is obvious that at the speed-of-light, photons are massless or non-physical. Quantum waves or photons are purely quantum objects or massless non-physical objects while traveling at the speed-of-light. Photons have to slow down to zero or splash-down, in order to manifest in the physical realm. This reality is obviously true because it has been experienced and observed. It's obvious that the massless or the non-physical exists. At the speed-of-light, photons are quantum phenomena, supernatural phenomena, spiritual phenomena, or non-physical phenomena. At the speed-of-light, we can't get our hands on them.

This is the way things really work in the real world. This is what has actually been experienced and observed.

Mark My Words

It Is Time to Adjust Our Mathematical and Theoretical Models

Within all the different Models of Reality that I have developed during the past few years, I have repeatedly observed that photons and psyches are massless and therefore non-physical. In fact, they seem to be the same thing, from what I can tell. Whenever out-of-body travelers have observed a psyche at the quantum level while in the spirit world or quantum realm, a psyche is always observed as being a pinpoint of light or a photon, even at the quantum level and the psyche level.

A photon is defined as a massless particle, which means that at the speed-of-light, a photon is obviously non-physical, spiritual, supernatural, or quantum in nature. A photon has no mass which means that it has NO resistance to acceleration. A photon is theoretically capable of infinite velocities. If it travels slower than that, it's because it chooses to do so. Furthermore, a photon has to slow down to zero in order to manifest or show up in our physical realm; otherwise, it remains as a non-physical quantum object or a quantum wave. Quantum Mechanics in its purest form is Non-Physical Mechanics.

Clearly, photons and massless particles exist; therefore, it is obvious that the non-physical exists. The proven and verified existence of anything non-physical FALSIFIES Materialism, Naturalism, Darwinism, Nihilism, Atheism and their derivatives. It's time for us to adjust our mathematical and theoretical models accordingly. Quantum Fields are obviously non-physical and pre-physical. The Gods or Controlling Psyches had to design and make the quantum fields BEFORE they could make and sustain the existence of physical matter. Anything that has mass or resistance to acceleration ends up being the physical aspect of a quantum object.

Many mathematicians who specialize in Quantum Mechanics and Quantum Field Theory have noticed that the math works better and makes more sense when they use complex numbers or imaginary numbers. What's happening here? Well, these mathematicians have discovered that the Quantum Realm is a completely different and completely separate dimension from our "real number system", or "real realm", or physical realm. A switch to imaginary numbers or complex numbers represents a switch to a completely different dimension. The Spirit World or Quantum Realm is a completely different dimension. It's out-of-phase with our physical dimension. It's here in the very same space as our physical dimension; but, it's out-of-phase with our physical dimension.

Quantum Superposition is one of the things that makes this reality possible. The Quantum Realm is the Transdimensional Realm, or the Non-Physical Realm. Transdimensional means non-physical. Pure Quantum Mechanics is in fact Transdimensional Physics. It's a whole other dimension.

It's time for scientists and mathematicians and theoreticians to introduce Nirvana, the Spirit World, the Supernatural Dimension, or the Quantum Realm into our theoretical models and mathematics. It's time for the human race to be introduced to Reality and Truth. It is time for us to adjust our mathematical models and theoretical models – and actually go out of our way to make them match with what has actually been experienced and observed by out-of-body travelers and near-death experiencers. It's finally time for us to use Quantum Mechanics and Quantum Field Theory to FALSIFY Materialism, Naturalism, and their derivatives. The Materialists, Naturalists, Darwinists, and Atheists won't let it happen in their classrooms; but, there's nothing to stop us from introducing the world to the truth elsewhere besides our public schools.

The ironic thing about the complex and confusing math behind Quantum Mechanics and Quantum Field Theory is the possibility that there may in fact be a simpler, better, more efficient, and more effective way to do the math, to explain the quantum phenomena, and to explore the quantum phenomena; but, we will never look for it and never find it if we don't allow for the possibility. We have to be willing to adjust, if we are ever going to find and know the truth or find the ultimate best way to mathematically and theoretically model quantum phenomena. As I see, the best that we can do for now is to find math and theoretical models for Quantum Mechanics that actually match with what has actually been experienced and observed.

https://psyche-ontology.com/psyche-experienced-and-observed/

The Ultimate Model of Reality and the True Interpretation of Quantum Mechanics that I personally have chosen to conform my mathematical models and theoretical models with is David Bohm's (1986) Super-Implicate Order, as contained within his book, "The Essential David Bohm". For Bohm, his explicate order is our physical realm. The Super-Implicate Order is the Supernatural Realm, the Non-Physical Realm, Command and Control, the Psyche Realm, or the Quantum Realm. There are literally an infinite number of different interpretations of Quantum Mechanics that I could have gone with; but, I chose to go with the ONE that has actually been experienced and observed – the ONE that actually matches with reality and the truth as we know it to be.

I'm a scientist, and each scientist chooses the Model of Reality and the Interpretation of Quantum Mechanics that he or she personally wants to believe in and support. I have chosen mine. I chose Bohm's Super-Implicate Order because it matches with and explains everything that has ever been experienced and observed. It also matches with the Theory of Relativity and explains the Theory of Relativity in quantum terms. You can't get better than that. Let me try to demonstrate.

Bohm: If you want to relate it to modern physics (light and more generally anything moving at the speed of light, which is called the null-velocity, meaning null distance), the connection might be as follows. As an object approaches the speed of light, according to relativity, its internal space and time change so that the clocks slow down relative to other speeds, and the distance is shortened. You would find that the two ends of the light ray would have no time between them and no distance, so they would represent immediate contact. (This was pointed out by G. N. Lewis, a physical chemist, in the 1920s.) You could also say that from the point of view of present field theory, the fundamental fields are those of very high energy in which mass can be neglected, which would be essentially moving at the speed of light. Mass is a phenomenon of connecting light rays which go back and forth, sort of freezing them into a pattern. So, matter, as it were, is condensed or frozen light. Light is not merely electromagnetic waves but in a sense other kinds of waves that go at that speed. Therefore, all matter is a condensation of light into patterns moving back and forth at average speeds which are less than the speed of light. Even Einstein had some hint of that idea. You could say that when we come to light we are coming to the fundamental activity in which existence has its ground, or at least coming close to it.

https://ultimate-model-of-reality.com/The-Super-Implicate-Order/

This is it. This reconciles both the Theory of Relativity and Quantum Mechanics with EVERYTHING that has ever been experienced and observed by out-of-body travelers, scientists, and near-death experiencers.

It's really simple to understand. The truth always is. They are both made from energy; but, a photon acts differently than a physical atom because a photon is massless or non-physical. As an object approaches the speed-of-light, time slows down and distance contracts or is shortened. What's the NET effect of all of this as this object approaches the limit that we call the speed-of-light? Well, everything about this object is approaching a limit, and the object literally integrates and transforms into the Quantum Realm or Spirit World when it achieves the speed-of-light.

What's the NET effect when an object is traveling at the speed-of-light? Well, TIME STOPS. DISTANCE GOES TO ZERO, which means that velocity goes to infinity. We switch from the physical realm into the realm of Instantaneous Action at a Distance. We literally switch from physical limitations over to Quantum Tunneling, Quantum Field Theory, Quantum Mechanics, and the Quantum Zeno Effect.

TIME STOPS, which means that entropy or the passage of time ceases to exist at the quantum level within the Spirit World or the Quantum Realm. There is NO such thing as a spirit body or a psyche freezing to death. That has never been observed, because energy or psyche is always conserved at the quantum level. Within the Quantum Realm or the Spirit World, the Conservation of Energy and Psyche reigns supreme, because entropy or the second law of thermodynamics ceases to exist at the quantum level. There is NO thermodynamics, death, or heat death at the quantum level.

Think about it logically. What has actually been experienced and observed. Take any physical system that you can possibly imagine, move that system to maxim entropy or complete thermal equilibrium, and what do we observe? Well, at the physical level, we observe what the scientists call Heat Death. However, at the quantum level, the photons and other quantum objects like psyches, quantum fields, and spirit bodies continue to work and live unimpeded by the Heat Death that has taken place at the physical level. This is possible because there is NO heat death at the quantum level in the psyche realm or the quantum realm. Entropy ceases to exist at the quantum level. This is what has actually been experienced and observed, both from our physical level and from the quantum level by out-of-body explorers and near-death experiencers. It's time for a new law of thermodynamics called the Quantum Law of Thermodynamics which states that there is NO entropy, NO second law of thermodynamics, and NO thermodynamics at the quantum level in the Psyche Realm, Spirit World, or Quantum Realm.

The proof of concept comes from the observed and proven FACT that quantum fields continue to exist, and photons continue to travel and exist, even within physical systems that have achieved maximum entropy, heat death, or thermal equilibrium. The Quantum Law of Thermodynamics truthfully and accurately states that there is NO heat death, thermodynamics, death, or second law of thermodynamics within the Quantum Realm. Instead, at the psyche level and the quantum level – Bohm's Super-Implicate Order – the Conservation of Energy and Psyche reign supreme. This is what has actually been experienced and observed – both at the physical level and within the Quantum Realm or Spirit World. It's time for us to adjust our Science and Mathematical Models to make them match with what has actually been experienced and observed.

So, at the speed-of-light, EVERYTHING reaches some type of limit, and the object transforms from a physical object into a quantum object. TIME STOPS. Entropy ceases to exist. Death and heat death cease to exist. Conservation of Energy and Psyche takes over as the dominant LAW of the Quantum Realm and Quantum Field Theory. Heat Death may take place at the physical level, but the Quantum Fields, Quantum Laws, and Quantum Objects continue to function normally and eternally even though the explicate order or physical order above them has suffered heat death.

As an object approaches the speed of light, according to relativity, its internal space and time change so that the clocks slow down relative to other speeds, and the distance is shortened. You would find that the two ends of the light ray would have no time between them and no distance, so they would represent immediate contact.

At the speed-of-light, from the perspective of the photon, TIME HAS STOPPED, distance has gone to zero; and, that photon from its perspective experiences Instantaneous Action at a Distance, Immediate Contact, or what we call Quantum Tunneling. From the perspective of the photon traveling at the speed-of-light, it simply quantum tunnels or teleports to its destination instantaneously experiencing NO TIME and NO DISTANCE while it travels. You could say that from its perspective it literally travels at an Infinite Velocity in the direction that it has chosen to travel.

From our perspective at the physical level, where we are trapped within space-time, it seems as if it took that photon 13.2 billion years to reach us. But, from the photon's timeless perspective at the quantum level, where everything has integrated or has reached the limit, NO TIME PASSES, distance shortens or contracts to zero, and the quantum object simply quantum tunnels or teleports instantaneously to its chosen destination. When that vector is first established, that photon literally chooses a direction and a destination. In fact, it has been said that a photon won't launch or instantiate if it has no place to land. A photon literally KNOWS where it is going to land before it launches, otherwise, it doesn't launch. How is this possible? It's because there is a little bit of psyche or intelligence or energy or living light within every photon, quantum field, and quantum object. Every psyche has a certain amount of energy that's under its control; and, that controlling psyche can literally choose to do anything with the energy that's under its control. It can transform that energy into quantum fields, photons, spirit matter or dark matter, dark energy, mass or resistance to acceleration, quantum waves or thoughts, or even entropic physical matter. Energy is infinitely malleable and infinitely conserved. It has always existed, and it will always exist. Energy and psyche cannot be made, and they cannot be destroyed, because they are always conserved. This is the way things really work at the quantum level within the Quantum Realm; and ironically, this is precisely what has been experienced and observed by out-of-body explorers and near-death experiencers.

You could also say that from the point of view of present field theory, the fundamental fields are those of very high energy in which mass can be neglected, which would be essentially moving at the speed of light. Mass is a phenomenon of connecting light rays which go back and forth, sort of freezing them into a pattern. So matter, as it were, is condensed or frozen light.

I was taught in my physical science classes that as a physical object approaches the speed-of-light, its mass or "resistance to acceleration" approaches infinity. Once again, we see calculus taking place here and the physical object is approaching a limit. So, what happens when that physical object reaches the speed-of-light or reaches infinite mass? It transforms. It integrates. It reaches the limit.

So, what is the limit of mass or "resistance to acceleration" when an object achieves the speed-of-light? Well, David Bohm actually said that mass can be neglected at the speed-of-light. In other words, mass or "resistance to acceleration" literally goes to ZERO at the speed-of-light. In other words, mass ceases to exist at the speed-of-light in the quantum realm or spirit world. NO resistance to acceleration in the quantum realm literally means INFINITE VELOCIETIES in the quantum realm. What has actually been experienced and observed by out-of-body travelers and near-death experiencers? Well, they have observed spirit bodies and other quantum objects jumping, or teleporting, or quantum tunneling at the quantum level within the Spirit World or the Quantum Realm. Yes, objects such as spirit bodies can stand still, or walk, or interact if they choose to do so; but, they can also travel at an Infinite Velocity or quantum tunnel or teleport if they choose to do so. Just as David Bohm said, there is NO mass or resistance to acceleration at the quantum level within the spirit world or the quantum realm.

In one single paragraph, while trying to explain his Super-Implicate Order, David Bohm has literally explained EVERYTHING that has ever been experienced and observed both at the physical level and at the quantum level. That's the kind of Interpretation of Quantum Mechanics that we need – the type that explains everything that has ever been experienced and observed.

In contrast, the Mechanists, Physical Reductionists, Materialists, Naturalists, Nihilists, Darwinists, and Atheists tell us and assure us that spirit matter doesn't exist, that psyche doesn't exist, that instantaneous action at a distance doesn't exist, and that the quantum realm or spirit world does not exist. How does telling us that quantum mechanics, quantum realms, quantum objects, and quantum fields DO NOT EXIST explain what they are and how they work? It doesn't! Instead, the verified and proven existence of Quantum Tunneling, Instantaneous Action at a Distance, Non-Locality, the Quantum Zeno Effect, the Quantum Realm or Spirit World, and Psyche or Intelligence or Consciousness or Living Light FALSIFIES Materialism, Naturalism, Classical Realism, and their derivatives such as Darwinism, Nihilism, and Atheism. If the ONE can be demonstrated to be real and true, then the other has been falsified. So, which one has been experienced, observed, and verified; and, which one has been falsified? Quantum Mechanics or Materialism? Quantum Field Theory or Scientific Naturalism? Creation by Intelligence, or Spontaneous Generation and Creation Ex Nihilo?

Quantum Fields are supernatural phenomena – they are non-physical phenomena. They are based upon instantaneous action at a distance, which means that they have NO mass or NO resistance to acceleration. Quantum Fields have actually been experienced and observed and verified. They really do exist. Magnetism, radio waves, microwaves, thoughts or quantum waves, gravity, and dark energy are non-physical quantum fields that have actually been experienced and observed – both at the physical level and at the quantum level. Their very existence FALSIFIES Materialism and Naturalism which claim that the non-physical or the immaterial DOES NOT EXIST.

The false is falsified by the truth; and the truth is repeatedly experienced and observed. Quantum Fields are repeatedly experienced and observed; whereas, Materialism, Naturalism, and their derivatives are constantly FALSIFIED by quantum objects, quantum fields, and Quantum Mechanics, or Spiritual Mechanics, or Supernatural Mechanics, or Energy Mechanics, or Psyche Mechanics, or Psychic Mechanics.

Remember, ALL of the memories that survive the death of your physical brain and show up in your after-death Life Review have to be stored someplace else besides your physical brain. After your physical body and physical brain die, your Psyche and Spirit Body continue to live and exist in the quantum realm, because psyche and energy (and the information that they contain) are always conserved at the quantum level in the Quantum Realm or Spirit World.

Some of the physicists have proposed the Quantum Law of Information Conservation. It's a thing that they have created. Well, where's that information going to be stored at the quantum level, especially before the first particle of physical matter was designed and created? Well, that information had to be stored as quantum waves or thoughts within some kind of non-physical, immaterial, quantum storage device. The only thing that fits the requirements is a photon or a spark of light that we call Psyche, Intelligence, Consciousness, or Life Force. It would be some type of point-particle or an infinite singularity. These are things as well that have been proposed, and that we sometimes hear about in physics. The only ones who have any problem with this are the Materialists, Darwinists, Naturalist, Physicalists, and Atheists – the people who claim that psyche, quantum waves (thoughts), supernatural mechanisms or quantum mechanisms, and non-physical quantum fields do not exist.

The Quantum Law of Information Conservation is ONLY possible if there is in fact something at the quantum level capable of storing and processing information. That something is what we call Psyche, Intelligence, Consciousness, or Living Light. Whenever a psyche has been seen at the quantum level within the Quantum Realm, it is observed as being a pinpoint of light, a true point-particle, a spark, or a photon. Psyche is a conscious, intelligent, sentient, and living photon, even at the quantum level within the quantum realm. Psyche is the ultimate point-particle – an infinite singularity. This is what has actually been experienced and observed in the Quantum Realm or the Spirit World. It's time for us to start modeling and explaining it, because it has actually been experienced and observed.

A few years ago, while pursuing this matter from the philosophical direction, I introduced a Psyche Ontology to the world. A Psyche Ontology is the Ultimate Model of Reality and Existence. According to a Psyche Ontology, Psyche (or that pinpoint of light that we call "intelligence" and "consciousness") is the fundamental unit of existence and reality. Before the Gods gave you a spirit body, you were an intelligence or a psyche. If your spirit body were ever to cease to exist, you would still be a psyche or an intelligence, because energy and psyche are always conserved at the quantum level in the Quantum Realm. Your psyche or your spark has always existed, and it will always exist. It cannot be made, and it cannot be destroyed. It is eternal and everlasting. It is always conserved. This is what has actually been experienced and observed.

Remember, it's the psyche or the intelligence within physical atoms that in fact gives those atoms the ability to do the quantum things which the natural physical laws prevent them from doing – things like hydrogen molecules coalescing into planets and stars and galaxies, or things like atoms organizing into functional proteins and genes, or things like ions quantum tunneling to the nearest ion receptor when a specific motor neuron needs to instantiate or start an action potential immediately. Such things are based upon intelligence and choice, and NOT happenstance or chance.

Bohm: So matter, as it were, is condensed or frozen light. Light is not merely electromagnetic waves but in a sense other kinds of waves that go at that speed. Therefore, all matter is a condensation of light into patterns moving back and forth at average speeds which are less than the speed of light. Even Einstein had some hint of that idea. You could say that when we come to light we are coming to the fundamental activity in which existence has its ground, or at least coming close to it.

Physical matter is light that has been deliberately slowed down. It has had mass or "resistance to acceleration" introduced into it. Physical matter is light that has been woven into solidity. Physical matter is a bunch of different quantum objects that have been transformed into a physical object that we call a physical atom. Entropy or the passage of time has been introduced into it.

Once again, mass or "the resistance to acceleration" has been programmed into it. According to Quantum Field Theory, particles are born, and particles can die. Physical matter is made. It hasn't always existed, and there's no guarantee that it will continue to exist. The controlling psyches, who made the first particle of physical matter or the first physical atom, can choose to transform that physical atom into anything else that they want it to be including photons, quantum waves, quantum fields, spirit matter or dark matter, radio waves, magnetic waves, available energy, gravity, dark energy, or anything else they want it to be. Energy is infinitely malleable, and just because the controlling psyches have chosen to form some of that energy into physical matter doesn't mean that that energy has to remain as physical matter for the rest of eternity. The controlling psyches can transform that energy into anything they want it to be, anytime they choose to do so. The second law of thermodynamics doesn't exist and doesn't apply at the quantum level, because entropy and the passage of time or the aging process doesn't exist at the quantum level.

Through the introduction of physical matter, the Gods or the Controlling Psyches introduced the ultimate consensus reality. They slowed everything down so that we can live it, experience it, and remember having done so. You see, at the speed-of-light, at an infinite velocity, during quantum tunneling, there is NO passage of time, which means that there is NOTHING to experience and remember. From the photon's perspective traveling at the speed-of-light, time has stopped, distance has contracted to zero, and it literally experiences nothing while it travels from one side of our physical universe to the other. From a photon's perspective, it's instantly at its destination the very moment I launches. There's nothing for it to experience. A photon doesn't experience anything because time has stopped at the speed-of-light.

I experienced quantum tunneling, when God teleported me and the car I was driving to safety. I had no scientific explanation for it at the time, so I tended to dismiss it and forget about it. I was traveling 40mph to the north, a car pulled out in front of me, time stopped as I was ready to merge with the side of that car, everything froze, and then instantly I was in the middle of a lawn, with the car completely stopped and the engine turned off. I have no memory or knowledge of how I got there. From my perspective, it was as if I had passed out. From my perspective, time stopped, and then I was just someplace else instead. If I would have kept going at that speed in that new direction, I would have gone over the edge down onto the freeway into oncoming traffic. Instead, the car was turned off, and the car was stopped.

A man came running out of the store and said, "That's the coolest thing I've ever seen. There's no way you could have missed hitting that car, but you did." It was nothing that I had done. I wasn't conscious nor aware of the event. From my perspective, time froze, and then I was instantly somewhere else instead. I assume that I traveled that distance and passed through that other car unphased and unscathed; but, I have no way to know for sure because I have no conscious memory of any of it. Time stopped, and then I was simply somewhere else instead. That's precisely what happens at the speed-of-light – TIME STOPS and distance contracts to ZERO, and the effects of mass or "resistance to acceleration" temporarily cease to exist; and then, an infinite velocity or instantaneous action at a distance becomes possible.

This happened to me 40 years ago when I was 18. It has taken me 40 years to find a scientific explanation for what I experienced during that event. It was a quantum event, a supernatural event. Forty years ago, we didn't have a scientific explanation for those kinds of things; but, now we do.

At the speed-of-light, there is NOTHING to experience because TIME STOPS. From your perspective, everything is instantaneous, and distance goes to ZERO. From your perspective at the speed-of-light or at the limit of the physical realm, you simply quantum tunnel or teleport to your destination, experiencing nothing in between. You are one place one second, and then someplace completely different the next second – having experienced nothing during the journey.

The Gods or Controlling Psyches introduced mass, or "resistance to acceleration", or the passage of time, or space-time into physical matter and this physical realm so that we actually have something to experience while moving from point A to point B. At the physical level, because of mass or resistance to acceleration, it actually takes some time to get from point A to point B, and we experience events during the interim due to the passage of time that takes place at the physical level within this physical realm.

You can see the truth of it because it has actually been experienced and observed. Our theoretical and mathematical models for Quantum Mechanics and Quantum Field Theory need to be adjusted so that they match with what has actually been experienced and observed.

So, what have out-of-body travelers and near-death experiencers experienced and observed while in the Quantum Realm? Well, no psyche or spirit body has ever been observed freezing to death or suffering heat death, because they can't die. There is no entropy or thermodynamics or heat death in the

63

Quantum Realm. Therefore, our math and theories have to be adjusted accordingly to match with the observational evidence.

Furthermore, out-of-body explorers and near-death experiencers have observed that spirit bodies, and their psyche or consciousness, can quantum tunnel or teleport long distances instantaneously. Therefore, our math and theories need to be adjusted so as to allow for the possibility of quantum tunneling or teleportation or instantaneous action at a distance at the quantum level in the Quantum Realm, or Spirit World, or Non-Physical Transdimensional Realm. Not only can spirit bodies travel at huge velocities faster than the speed-of-light within the Quantum Realm, they can also quantum tunnel or teleport instantaneously in the Quantum Realm. If we want to find and know the truth, our mathematical and theoretical models have to be adjusted a bit so as to make them reflect what has actually been experienced and observed.

Science is observation and experience and experimentation after all, so our Science needs to be adjusted from time to time in order to make it match with what has actually been experienced and observed. Psyches, quantum fields, and spirit bodies can't die, or freeze to death, or cease to exist within the Quantum Realm, because they are pure energy or light; and, energy, psyche, intelligence, consciousness, light, syntropy, and information storage are always conserved at the quantum level within the Quantum Realm.

Take any physical system that you want that has achieved maximum entropy or thermal equilibrium or heat death; and, what do we observe? Its quantum fields continue to function normally because the energies within those quantum fields are always conserved, which means that they are infinite and eternal and immortal. They can't suffer heat death or cease to exist. The structure and composition of those physical objects continue to exist. Something at the quantum level continues to sustain them. The energies within the quantum fields can be transformed into something else like physical matter; but, they cannot be destroyed or cease to exist. The second law of thermodynamics or entropy doesn't exist at the quantum level. This is what has actually been experienced and observed. We need to adjust our mathematical and theoretical models accordingly.

It's time for us to introduce the purely quantum or the purely transdimensional or the purely non-physical into our theoretical and mathematical models, because it's real and truly exists. Non-Physical Quantum Fields are proven and verified science. It's time to take notice of them and start to use them in our mathematical and theoretical models. We have the truth, and then there is everything else. It's time for us to adjust our science and math, and make them match with what has actually been experienced and observed.

Mark My Words

The Scientific Methods

Due to the wide variety of different logic fallacies which are built directly into the Scientific Methods, it is technically impossible to find the truth and to KNOW the truth using the Scientific Methods.

The original meaning for the word "Science" was knowledge; however, it is technically impossible to prove truth or to find knowledge using Science and the Scientific Methods.

Since this is the case, doesn't that technically mean that Science and the Scientific Method are absolutely worthless? It would seem that way, wouldn't it? What good is it if it can't be used to prove the truth?

Well, all is not lost. Science and the Scientific Methods do have their uses, as the Lived Experiences of the human race testify.

We can use Scientific Methods to falsify theories or to prove theories false!

Science is very good at falsifying theories or proving theories false if we let it do so. Most scientists, though, don't let it do so, because of their prejudices or confirmation biases. But, I personally no longer suffer from those kinds of illusions and delusions, because I am no longer a Materialist or an Atheist. I'm now free to let science tell me whatever it wants to tell me.

We can use Science and the Scientific Methods to falsify theories because *negating the consequent* is philosophically and logically sound!

The argument structure for falsification, or *negating the consequent*, looks like this:

Scientific Hypothesis: If Theory X is true, then we will observe Y.

Scientific Observation: We don't observe Y.

Scientific Conclusion: Therefore, Theory X is false, and Theory X has been falsified.

That's the Scientific Method in action, doing what it was meant to do, falsifying a theory!

Let me provide an example to show how this works in practice.

Scientific Definition of the following Theories: Materialism, Naturalism, Darwinism, and the Theory of Evolution are defined as "Design and Creation by Physical Matter". The Materialists, Naturalists, and Darwinists contend that raw inert physical matter designed, created, and manufactured the first genomes and the first living cells, because according to these people physical matter is the only thing that exists and therefore the only thing that could have designed, programmed, engineered, created, and manufactured the first genomes and the first life forms on this planet. We choose to take these people at their word whenever they claim to believe that physical matter is the only thing that exists. They really believe that! These people truly believe that the rocks designed and created the first genomes and life forms. These people call it Science.

Premise or Axiom or Law: The Scientific Methods can be used to falsify theories and to prove them false.

Scientific Hypothesis: If Materialism, Naturalism, Darwinism, and the Theory of Evolution are true, then we will observe the rocks and physical matter designing, creating, and manufacturing genomes and life forms from scratch.

Scientific Observations: We NEVER observe the rocks designing and creating anything. The rocks can't design and create anything at all.

Scientific Conclusion: Therefore, Materialism, Naturalism, Darwinism, and the Theory of Evolution are false, and have been successfully falsified by the Scientific Method.

That's the Scientific Method in action, doing what it was meant to do, falsifying a theory or two!

I just falsified Materialism, Naturalism, Darwinism, and the Theory of Evolution! I love doing that! I wish I would have known how to do that forty years ago!

Furthermore, Atheism is "Creation by Nothing"; and, we all know that nothing cannot design and create anything.

Scientific Hypothesis: If Atheism is true, then we will observe nothing designing and creating something.

Scientific Observations: We NEVER observe nothing designing and creating anything.

Scientific Conclusion: Therefore, Atheism is false, and has been successfully falsified.

I just falsified Atheism, using the Scientific Method.

Best of all, using the Scientific Methods to falsify theories or to prove theories false IS philosophically and logically sound! In other words, I HAVE IN FACT FALSIFIED Materialism, Naturalism, Darwinism, Atheism, and the Theory of Evolution. The debate has ended. I have just PROVEN Materialism, Naturalism, Darwinism, Atheism, and the Theory of Evolution FALSE, using the scientific method. We now KNOW that these theories are false, because we KNOW why they are false. They are FALSE because the rocks and blind luck have never been caught in the act of design and creation; and, they NEVER will be. Case closed!

Since we now KNOW that Materialism, Naturalism, Darwinism, Atheism, and the Theory of Evolution are FALSE, we must look someplace else if we want to find the truth. So, where else are we going to look for the truth?

Well, it's my contention that THE TRUTH cannot be falsified by the Scientific Methods. When we are dealing with THE TRUTH, the Scientific Methods will continuously verify the truth for us, as long as we have our definitions for everything correct so that we are getting the correct interpretation of the Scientific Evidence. Using the Scientific Methods, we can center in on THE TRUTH through a process of elimination by using the Scientific Methods to falsify the falsehoods such as Materialism, Naturalism, Atheism, and Darwinism.

So, what's the opposite of Materialism and Naturalism? Since these have been falsified by the Scientific Method, then their opposite is most likely true – through a process of elimination. If you knock down all the falsehoods, THE TRUTH will be the only thing left standing. It's elementary my dear friend!

Psyche or Non-Local Consciousness is the opposite of Materialism. Spirit, spirituality, non-locality, the supernatural, and God are the opposite of Naturalism and Atheism.

I recently discovered that Lived Experience or Psyche Experience is in fact the BEST and the most efficient way of finding and KNOWING the TRUTH. Lived Experience or Direct Observation is vastly superior to philosophical guesswork, scientific hypotheses, and philosophical interpretations of scientific data. In fact, the BEST definition for Science is "Observation". We observe things, so that we can come to understand things and know things.

For example, we KNOW from the Lived Experiences of the human race that Psyche or Non-Local Consciousness exists, because people have lived it and experienced it firsthand for themselves during their Out-of-Body Experiences and Near-Death Experiences. We don't have to run any science experiment on Psyche, because we already KNOW that it exists. KNOWING is vastly superior to hypothesizing, theorizing, experimenting, and philosophizing.

We KNOW from observation and lived experience that Psyche or Intelligence can design, program, engineer, manufacture, and create anything that it sets its mind to. Psyche can do science! Psyche has been caught in the act! We KNOW this is the TRUTH, because it has NEVER been falsified! It's impossible to use the Scientific Methods to falsify THE TRUTH!

Intelligent Design, Creation by Psyche, and Manufacturing by Human Beings or Human Psyches have been observed and verified trillions of trillions of times with an infinite more times to go. THE TRUTH is always verified by our Scientific Observations and NEVER once falsified, because THE TRUTH can never be falsified by our Scientific Methods or Observations! God is KNOWN by living Him and experiencing Him. God is KNOWN by observing Him! THE TRUTH IS KNOWN by living it and experiencing it for yourself, or by choosing to trust someone who has. This is the way Science should be and should be done!

We KNOW from Observation and Lived Experience that Psyche exists and that Psyche or Intelligent Beings can design and create anything that they set their minds to. This Reality has been verified trillions of trillions of times and has NEVER been falsified. THE TRUTH cannot be falsified by the Scientific Methods, because it will always be true, stay true, and hold true. In contrast, Materialism, Naturalism, Atheism, and Darwinism have been falsified thousands of different ways and trillions of different times.

That's the way you do Science; and, that's the way Science should be done!

But, guess what's going to happen now. It happened before, and it's going to happen some more. The Materialists, Naturalists, Darwinists, and Atheists are going to chime in and sound off and assure you authoritatively as scientists and PhDs that I don't understand Science, that I don't understand the Scientific Method, and that I don't understand the power of Evolution. They are going to tell you that I'm an idiot, because that's what these people do. I KNOW because I have lived it and experienced it. I KNOW, because I used to be a Materialist, Nihilist, and Atheist and was in complete denial of Reality at the time. I KNOW how it goes, because I have lived it and experienced it. These people have to DENY it, or they will no longer be Materialists, Naturalists, Darwinists, and Atheists. But, I have observed that the Materialists, Naturalists, Darwinists, and Atheists are always wrong.

Consequently, I finally realized that if I wanted to KNOW the TRUTH, then I had to look someplace else besides Materialism, Naturalism, Darwinism, and Atheism because these philosophies have in fact been falsified by Science and the Scientific Methods thousands of different ways and trillions of different times. In fact, the major premises or the primary assumptions of Materialism, Naturalism, Darwinism, and Atheism are impossible to verify! They have NEVER been verified, and will NEVER be verified, because they can't be. They have to be taken on blind faith as being true, because they are in fact FALSE and have been falsified by the Scientific Method.

That's the way Science and the Scientific Methods work, or should work!

Cool, huh?

I bet you'll never get anything like this from your Science teacher or college professor, because they don't have anything like this to give you.

Mark My Words

Some Flaws in the Scientific Method

Psychology, or the study of the human psyche, is one of my specialties. In a number of my Psychology courses, they have had us take the Implicit Association Test to demonstrate that we are all prejudiced against women. I object to or reject that particular interpretation of the Implicit Association Test or use of the Implicit Association Test. I don't believe that it detects prejudice as the scientific researchers claim.

Where the IAT self-assessment is concerned, I can say that I really haven't bought into their prechosen conclusions where the Implicit Association Test is concerned. When you are just dealing with colors – their names not matching with their actual color – you do just as poorly on an Implicit Association Test as you do when the test is designed to test for prejudice against women. There's no prejudice against women involved when you are trying to match the name of the color with the actual color; but, the Implicit Association Test can be interpreted to mean that you are prejudiced against colors, which is technically a faulty or invalid interpretation of the test. Therefore, whenever the Implicit Association Test is designed to detect prejudice against women, I don't believe that the test is completely valid – it's not really testing what it was designed to test. The Implicit Association Test accurately tests how counterintuitive a match is, not necessarily the amount of prejudice that is built into a person. Their interpretation of the test may in fact be faulty or counterintuitive, and the IAT may not have anything to do with prejudice at all.

As designed, whenever they make the Implicit Association Test counterintuitive, our time to completion slows down and expands, whether they are testing for colors, bath products, kitchen aids, power tools, or prejudice against women. On an Implicit Association Test, I got the expected Gender Bias of +33. That's supposed to mean that I'm prejudiced against women. The thing is that women will end up with the very same score when taking the same test, but they might not necessarily be prejudiced against women. They would get the same score if they were being testing to see if they are prejudiced against men, and so would a man.

I already know that I have prejudices and biases against women, and I have had them all my life. We just naturally want our women to be women. What my psyche or mind tells me is in fact more believable than what the Implicit Association Tests are telling me. I just don't find their conclusion convincing, because I also know through past experience that with sufficient practice on the different Implicit Association Tests, I could in fact make it seem as if I have no prejudice against women or no prejudice against colors and power tools. In other words, I really don't believe that the Implicit Association Test is in fact accurately testing our prejudice as much as it is testing our test-taking skills and training. I personally don't believe that the Implicit Association Test is a valid test – it really isn't testing what it was supposedly designed to test, especially when it is used to measure prejudice. Other researchers and scientists have chosen to believe differently. To each their own!

When it comes to the Scientific Method, the personal interpretation that comes at the end of the science experiment is based upon the Affirming the Consequent logic fallacy, and this is where the Scientific Method always fails to match with reality and truth. Each piece of scientific evidence can literally be interpreted in an infinite number of different ways. There are as many different interpretations of Quantum Mechanics as there are physicists on this planet. They can't all be right, because they are contradicting each other. Human interpretation is the flaw in the Scientific Method, and it's also the flaw in many of our science experiments including the Implicit Association Test. It has been a hundred years since they discovered Quantum Mechanics, and they are still trying to figure out what it truly means. They are still trying to figure out how it should be interpreted.

https://en.wikipedia.org/wiki/Interpretations_of_quantum_mechanics

When it comes to Science and the Scientific Method, I find it helpful to study the various different interpretations of Quantum Mechanics that have been proposed. It sort of reminds me of the Implicit Association Test and faulty interpretations in general. Ironically, the most useful and most plausible and defendable interpretations of Quantum Mechanics that I have found so far aren't even on the Wikipedia list, because the Wikipedia was made by Naturalists, Nihilists, Darwinists, Mechanists, Materialists, and Atheists for the Scientific Naturalists and Darwinists. They don't allow the interpretations of Quantum Mechanics that falsify Naturalism, Materialism, Atheism, and Physicalism. They call them pseudo-science in an attempt to discount them and dismiss them from consideration. The Darwinists, Naturalists, Nihilists, and Atheists will never allow into evidence anything that successfully falsifies Materialism, Naturalism, Darwinism, Behaviorism, Physicalism, Determinism, or Atheism. They won't allow Quantum Mechanics, or Non-Physical Mechanisms, into evidence because Quantum Mechanics or Transdimensional Non-Physical Physics falsifies Materialism, Naturalism, Darwinism, Nihilism, and Atheism which claim that these things DO NOT EXIST.

The flaw in Science and the Scientific Method comes into play every time that a person decides how he or she wants the evidence to be interpreted. When it comes to personal interpretation, there are literally an infinite number of possibilities for each piece of scientific evidence, and they can't all be right, especially when they contradict or falsify each other. Anyway, that's what I got from my Philosophy of Science course. I was introduced to the fatal flaw within the Scientific Method itself – the Affirming the Consequent logic fallacy. They are also employing the Affirming the Consequent logic fallacy within the Implicit Association Tests whenever they are using such tests to demonstrate prejudice against colors, power tools, toiletries, women, or men. There are literally an infinite number of ways that an Implicit Association Test can be interpreted, and they can't all be right because many of them contradict each other and falsify each other. Anyway, I have given this some thought, and chose to put it into writing.

Likewise, the most effective, most verified, and most believable definitions and interpretations of Quantum Field Theory and Quantum Mechanics involve Instantaneous Action at a Distance, Non-Locality or Non-Physicality, Quantum Entanglement, Quantum Fields, and Quantum Waves which are ALL non-physical and pre-physical phenomena. Remember, the Gods or Controlling Psyches had to design and make the quantum fields from raw chaotic disorganized energy BEFORE they could make and sustain physical matter. This reality is one of the most convincing Scientific Proofs of God's Existence that I have encountered so far. Someone Psyche, or some type of non-physical Life Force, had to design, form, and make the quantum fields BEFORE physical matter could be made and sustained. That's just the way it is. It's obvious that the first physical atom was designed and made by something non-physical, because physical matter didn't exist yet.

Even when it comes to the Big Bang Theory, it is obvious that the physical matter is made from the non-physical energy because physical matter didn't exist until 380,000 years after the Big Bang according to the Big Bang Theory. Physical matter is always made from the non-physical, and the proven and verified existence of the non-physical FALSIFIES Materialism, Naturalism, Darwinism, Nihilism, and even Atheism or Creation Ex Nihilo.

The Materialists, Naturalists, Darwinists, Nihilists, and Atheists portray physical matter as making itself. This is both true and false, depending upon how one chooses to interpret the evidence or the observations.

Physical matter is obviously made out of energy or light. Physical matter is made. According to Quantum Field Theory, particles are born, and particles can die. Where do the particles go when they die or cease to exist? They go back into the energy reserve from whence they came. Particles don't come from nothing when they are born. They are made from energy or light. Particles don't cease to exist when they die – instead, they are converted back into usable energy or light. The energy, or the psyche, or the light is always conserved! It cannot be made, and it cannot be destroyed. It has always existed, and it will always exist. Physical matter and particles of any kind are made from energy or light. Whenever they cease to exist or "die", they are dissolved back into the energy or the light from whence they came. Physical matter is made from energy; and, physical matter becomes available energy whenever a physical particle dies or "ceases to exist". This is true because the LAW of the Conservation of Energy reigns supreme, both at the quantum level and the physical level. Energy or light or life force cannot be made, and it cannot be destroyed. It has always existed, and it will always exist. This is what we have actually experienced and observed. The false is falsified by the truth; and, the truth is repeatedly experienced and observed.

Scientific Axiom: Anything that is obviously made obviously has a Maker who made it.

Scientific Observation: Physical matter and particles in general are obviously made from energy or light.

Scientific Conclusion: Therefore, it is obvious that photons, electrons, quarks, quantum fields, quantum waves, and physical matter have a Maker who makes them or organizes them from raw unorganized chaotic energy or light.

What do we know of, or what has been experienced and observed, that actually has the power and the ability to organize or form raw chaotic energy into quantum waves, photons, electrons, quarks, quantum forces, quantum fields, and physical matter? Well, it's NOT physical matter as the Materialists and Naturalists claim. Physical matter is the product, NOT the cause or the source of the product.

Psyche, Life Force, Intelligence, or Consciousness is the thing that is capable of organizing the energy under its control into things like quantum waves (thoughts), quantum forces and fields, electrons, quarks, dark matter or spirit matter, dark energy, gravity, microwaves, radio waves, and physical matter.

Psyche has been experienced and observed. Even at the quantum level and the psyche level, it looks like and acts like a photon or a pinpoint of light. Psyche is a living, conscious, intelligent, and self-aware photon or pinpoint of light. Psyche is living light, or Psyche is conscious self-aware light. It's really easy to understand once a person chooses to do so.

https://psyche-ontology.com/psyche-experienced-and-observed/

Psyche, Intelligence, Life Force, or Consciousness is NOT theory, philosophy, or speculation. It has actually been experienced and observed. Science is observation and experience and knowledge, after all – or it should be.

Quantum Fields are obviously non-physical and obviously pre-physical. Quantum Fields are prerequisites for physical matter, including entropic physical matter. Something Non-Physical or Something Psyche has to design and make the quantum fields BEFORE it can design, make, and sustain physical matter.

Electrons and quarks are obviously non-physical BEFORE they are given mass and made physical. Mass is resistance to acceleration. Gluons, or force fields, or electron orbitals are obviously made out of energy. Physical matter is the result of the organization of these things, NOT the cause of the organization of these things. Physical matter doesn't exist yet while the Controlling Psyches or Nature's Psyche are making the quantum fields, quantum laws for the electron orbitals, the electrons, and the quarks. It's only AFTER these things are made and implemented by the Controlling Psyches or Nature's Psyche that it becomes possible to organize them into physical atoms thereby making physical matter.

You have got to get the right interpretation for all of this, or it will never make any sense. I know, because Quantum Mechanics didn't make any logical sense while I was a materialist, naturalist, nihilist, and atheist because Quantum Mechanics and Quantum Field Theory are at their very core non-physical or transdimensional phenomena. Quantum Field Theory and Quantum Mechanics explain to us how the non-physical or the spiritual or the transdimensional works. In contrast, the Materialists and Naturalists and Atheists simply state that the non-physical or the psychic DOES NOT EXIST. How does claiming that Quantum Mechanics or Spiritual Mechanisms do not exist explain what they are and how they work? It' doesn't! It's a cop out! It's a dodge. It's lazy and ineffective. Materialism and Naturalism have NO explanatory power when it comes to the non-physical such as Psyche Mechanics, Energy Mechanics, Spiritual Mechanics, or Quantum Mechanics, which are the same thing in the end.

The fundamental axiom of Quantum Field Theory states that particles are born, and particles can die. According to the Materialists, Naturalists, Darwinists, Nihilists, and Atheists, particles magically cease to exist when they die. Furthermore, these people erroneously claim that particles, including physical matter, literally come from nothing ex nihilo when they are born. These people claim that electrons are blinking into existence and out of existence all the time. These people are wrong.

Where do the particles go when they die? They go the very same place that our Psyche or Life Force goes when our physical body dies – back into the pool of raw unorganized energy. The LAW of the Conservation of Energy reigns supreme both at the quantum level and the physical level. When the electrons die or "cease to exist" or "blink out of existence", they in fact switch back into virtual particles, or quantum waves, or spirit waves, or energy, or light from whence they came. Electrons are half-in and half-out, which means that they are half in the physical world and half in the quantum realm (or spirit world). Electrons don't cease to exist as the Materialists and Naturalists claim. They simply phase-shift temporarily into a different dimension – the quantum dimension. Phase-Shifting is yet another proven, verified, and observed Quantum Phenomenon. When electrons "cease to exist" here on the physical plane, their energy shifts into a different dimension that we call the Quantum Realm or the Spirit World. It's a temporary phase-shift or dimension shift. And, since electrons are effectively quantum tunneling or moving faster than the speed of light at the quantum level, they literally form a shell or a force field around the nucleus of an atom thereby bringing the solidity of physical matter into existence in the process here at the physical level.

The explanatory power of Quantum Mechanics is through the roof to infinity and beyond. In contrast, the Materialists and Naturalists and Atheists simply state that Quantum Mechanics, Spiritual Mechanisms, or Psychic Mechanisms DO NOT EXIST. There's NO explanatory there when it comes to Scientific Naturalism. Claiming that Quantum Mechanics, or Energy Mechanics, or Non-Physical Mechanics DOES NOT EXIST doesn't explain what it is and how it works. Transdimensional means non-physical; and, claiming that Transdimensional Physics DOES NOT EXIST (as the Materialists and Naturalists do) DOES NOT EXPLAIN what Transdimensional Physics is and how it works.

Science is supposed to have explanatory power. Materialism, Naturalism, Darwinism, Nihilism, and Atheism were instead designed to explain things away or to deny the existence of the non-physical, quantum, psychic, or spiritual. Scientific Naturalism or Classical Realism is worthless because it has NO explanatory power, especially when it comes to the non-physical, pre-physical, transdimensional, and the quantum.

There are literally an infinite number of different interpretations of Quantum Mechanics. I have observed that the interpretations of Quantum Mechanics that are based upon Classical Realism, Naturalism, Atheism (Creation Ex Nihilo), and Physicalism or Materialism are in fact worthless and false because they actually deny the existence of Instantaneous Action at a Distance or Quantum Mechanisms. The Denialistic Philosophies (such as Materialism, Naturalism, and their derivatives) have NO explanatory power when it comes to the transdimensional, spiritual, quantum, or non-physical. Remember, Quantum Mechanics or Spiritual Mechanics is proven and verified Science; whereas, Physical Reductionism and Scientific Naturalism have been successfully FALSIFIED by Quantum Mechanics, Psyche Mechanics, Energy Mechanics, Non-Physical Mechanics, or Spiritual Mechanics. Quantum Mechanics is constantly verified; and, both Materialism and Naturalism are constantly falsified by Quantum Mechanics.

Since Quantum Mechanics and Scientific Naturalism are mutually exclusive, if we can verify or prove one of them to be true, then we have in fact falsified or proven that the other one is false. Ask yourself which one has actually been experienced and observed, and which one is based upon nothing more than wishful thinking in the complete absence of both observations and experience. Has the non-existence of energy, light, consciousness, life, and intelligence been experienced and observed? OR, has the existence of these things been experienced and observed? The one that has been experienced and observed is in fact the one that is real and truly exists. Quantum Mechanics, or Spiritual Mechanics, or Psyche Mechanics, or Energy Mechanics is the thing that has actually been experienced, observed, and verified – NOT Materialism, Naturalism, and Atheism (Creation Ex Nihilo).

Mark My Words

How the SELF Controls Its Brain

In 1994, neuroscientist John C. Eccles wrote a book entitled, "How the SELF Controls Its Brain", in which he tries to explain to the world how the SELF, Psyche, Mind, Intelligence, or Consciousness controls its physical brain.

https://www.scribd.com/document/334415538/eccles-john-how-the-self-controls-its-brain-doc

His conclusion is the ONLY conclusion that makes logical sense. The Psyche or SELF is obviously a non-physical quantum object. Psyche is experienced first-hand as being an immaterial viewpoint in space, and it is seen or observed at the quantum level as being a pinpoint of light, a spark of light, a photon, or a point-particle. There's ONLY one way that such a non-physical, immaterial, quantum photon could ever possibly control a physical brain and that's through Quantum Mechanics, Quantum Fields, and Quantum Waves. Ironically, Quantum Mechanisms, Quantum Fields, and Quantum Waves have actually been discovered and verified and proven to exist through scientific experimentation – just as non-physical radio waves, x-rays, magnetic waves, light rays, and gamma rays have been experienced, observed, and proven to exist for real despite the claims of the Materialists and Naturalists who assure us with certainty that the non-physical DOES NOT EXIST.

Neuroscientist John C. Eccles drew the very same conclusion that I did. The verified and proven existence of Intelligence, Consciousness, Instantaneous Action at a Distance, Quantum Mechanics, Quantum Field Theory, Non-Local Non-Physical Quantum Objects, the Conservation of Energy and Psyche, Dark Energy, Dark Matter or Spirit Matter, Radio Waves, Microwaves, Magnetism, Quantum Waves, and Psyche FALSIFIES Materialism, Naturalism, Behaviorism, Hard Determinism, Classical Realism, Darwinism, Atheism, and their derivatives which claim with surety that the non-physical DOES NOT EXIST. The proven and verified existence of the non-physical FALSIFIES Scientific Naturalism, Physical Reductionism,

Mechanism, and their derivatives such as Darwinism and Atheism (Creation Ex Nihilo or Magic).

Within his book, John Eccles states that the Materialists are deathly afraid of consciousness (or psyche). It's true. The proven and verified existence of Intelligence, Consciousness, Psyche, or Life Force FALSIFIES Materialism and Scientific Naturalism which state that these things DO NOT EXIST. There's the truth, and then there's everything else. Materialism, Mechanism, Darwinism, and Scientific Naturalism were designed to trick us and deceive us; and, they work as advertised.

Quantum Mechanics and Quantum Field Theory are NOT magic – or creation from nothing by nothing – but from our limited perspective trapped within our physical 3D space-time, they seem like magic because we have no way to control them or manipulate them with our physical instruments. The best we can do to "control" Quantum Mechanics is to build barriers wide enough within our computer chips in the hope of preventing electrons from quantum tunneling out-of-bounds. The Gods or Controlling Psyches purposefully designed our physical reality so that electrons and atoms can still quantum tunnel – just not very far. Physical atoms, electrons, and quarks are given a very SHORT De Broglie Wavelength, so that they can only quantum tunnel a very short distance – maybe the length of a synapse. Physical objects can no longer quantum tunnel across the length of our universe, because physical objects no longer have a long or infinite De Broglie Wavelength. This is a good thing; otherwise, the atoms within your physical body would eventually quantum tunnel away from you at will, and your physical body would literally dissolve into thin air, if the physical atoms still had their original long or infinite De Broglie Wavelength.

Who or what would be capable of assigning a SHORT De Broglie Wavelength to a physical atom, an electron, or a quark? Who or what could enforce such a thing at a universal scale? Well, it's NOT going to be physical matter as the Materialists and Naturalists and Darwinists claim. It's going to be some type of conscious, sentient, intelligence, and self-aware Psyche, Intelligence, Life Force, Consciousness, Spark, Photon, or Quantum Object. The Obvious Law of Physics states that the smaller dwells within and controls the larger. The Psyche or Psyches obviously dwell within and control the electrons, quarks, and physical atoms. It's also obvious that we are unable to control psyches, electrons, quarks, quantum fields, and quantum waves with our physical instruments and physical brains.

Quantum Tunneling Is Faster than Light

https://www.youtube.com/watch?v=-IfmgyXs7z8

Within his book, John C. Eccles uses Henry P. Stapp's Orthodox Interpretation of Quantum Mechanics to explain how the SELF or the Psyche controls its physical brain. Quantum Mechanics is the ONLY way that it could possibly be done, and we have already discovered Quantum Mechanics and have already proven that Quantum Mechanics and Quantum Fields are REAL and truly exist.

Neuroscientist John C. Eccles states that he spent his whole life fighting against and falsifying Materialism, Naturalism, and the Dominant Materialist Philosophers.

Materialism or Physicalism states that the non-physical, the immaterial, or the quantum DOES NOT EXIST.

Scientific Naturalism states that the supernatural, the quantum mechanical, the spiritual, and the quantum fields DO NOT EXIST. Scientific Naturalism states that dark matter (spirit matter), quantum fields, and dark energy DO NOT EXIST.

Scientific Naturalism IS the dominant philosophy within Science. Ironically, it's philosophy and NOT Science. There's NO way to demonstrate, prove, or verify through scientific experimentation that quantum mechanisms, quantum fields, intelligence, consciousness, and psyche DO NOT EXIST. You can't prove a negative. Furthermore, the proven and verified existence of quantum mechanisms, action at a distance, quantum waves, quantum fields, intelligence, life force, and consciousness FALSIFIES Materialism, Naturalism, and their derivatives such as Darwinism and Atheism.

These two philosophies – psychic mechanisms and scientific naturalism – are mutually exclusive. If one can be demonstrated to be real and true, then the other one is automatically falsified.

Well, there's NO way to demonstrate the truthfulness of Materialism and Scientific Naturalism; and, Psychic Mechanics or Quantum Mechanics has already been demonstrated and proven true; so when it comes to these two philosophies or worldviews, WE KNOW which one is true, and which one has been FALSIFIED by Science and scientific experimentation. Materialism and Naturalism have lost the war, but we are still waiting for the Atheists and the Naturalists to get the memo.

A few other people besides me have caught onto this reality and truth. See for example:

Eccles's Psychons Could be Zero-Energy Tachyons

https://neuroquantology.com/index.php/journal/article/viewFile/169/169

Abstract

This paper suggests that mental units called psychons by Eccles could be tachyons defined theoretically by physicists some time ago. Although experiments to detect faster-than-light particles have not been successful so far, recently, there has been renewed interest in tachyon theories in various branches of physics. We suggest that tachyon theories may be applicable to brain physics. Eccles proposed an association between psychons and what he called dendrons which are dendrite bundles and basic anatomical units of the neocortex for reception. We show that a zero-energy tachyon could act as a trigger for exocytosis (modeled by Friedrich Beck as a quantum tunneling process), not merely at a single presynaptic terminal but at all selected terminals in the interacting dendron by momentarily transferring momentum to vesicles, thereby decreasing the effective barrier potential and increasing the probability of exocytosis at all boutons at the same time. This is consistent with the view of tachyons, which treats them as strictly non-local phenomenon produced and absorbed instantaneously and non-locally by detectors acting in a coherent and cooperative way.

A tachyon is spirit matter, a quantum object. This means that tachyons are NOT entropic physical matter, because they can quantum tunnel or travel faster than the speed of light.

The Psyche or the SELF controls its physical brain through Quantum Mechanisms. I prefer quantum waves as the controlling mechanism rather than spirit matter or tachyons, because out-of-body explorers have observed that spirit matter or tachyons have a different function than quantum waves or thoughts. In the Quantum Realm or the Spirit World, tachyons, dark matter, or spirit matter can be used to make things that seem physical and real at the quantum level; whereas, quantum waves and quantum fields seem to be the way that Psyches communicate with each other transpersonally at the quantum level.

But, from the perspective of a physical atom, whether the atom uses spirit matter to communicate with other atoms or whether it uses quantum waves to communicate with other atoms, it will work either way. Physical cells such as neurons can use quantum waves to communicate with each other. Quantum waves are thoughts. Quantum waves are WiFi at the quantum level. Psyches communicate transpersonally through quantum waves. Quantum waves have been verified and proven to exist. A Psyche or Photon at the quantum level transmits quantum waves, receives quantum waves, processes quantum waves, and stores quantum waves. Quantum waves are thoughts or WiFi at the quantum level. A Psyche is a Quantum Transceiver. A Psyche is also a Quantum Computer.

According to the LAW of the Conservation of Psyche and Energy, a Psyche cannot be made, and it cannot be destroyed. It has always existed, and it will always exist, because it is always conserved. I sometimes wonder how long my Psyche or Intelligence dwelt within the quantum fields and then dwelt within the rocks before finally being called up and given a spirit body, as a SON of Heavenly Father and Heavenly Mother. God knows, but I don't. According to the Bible, we human beings are the children of the Gods, which explains in part why we are so intelligent compared to the other animals around us. Our surname is God. We are descended from the Gods, both physically and spiritually, which means that our physical bodies and spirit bodies descended from the Gods.

Of course, the atoms or the rocks are infinitely more intelligent than we are because they don't have a veil of forgetfulness placed upon them like we do. They aren't being tested like we are. A Psyche is an infinite singularity, and there are a bunch of psyches within a physical atom. Each psyche, and therefore each atom, is theoretically capable of storing ALL of the information and knowledge in the universe within them, because they have NO physical limitations where information storage capacity is concerned. This is the holographic principle – another quantum phenomenon that our scientists are just beginning to observe and verify. Each part

77

of a hologram contains the whole hologram within it. Each psyche or intelligence, and therefore each physical atom, can theoretically contain ALL of the knowledge and information in the universe within it. Whether it does or doesn't is another matter; but, it's theoretically possible because a Psyche is an infinite singularity or the ultimate point-particle. It has NO physical limitations and is theoretically capable of storing an infinite amount of knowledge and information.

The explanatory power of Quantum Mechanics and Quantum Field Theory is through the roof to infinity and beyond. It's infinitely better and a lot more explanatory than the materialistic and naturalistic claim that the quantum, the quantum tachyons (spirit matter), and quantum psyches (intelligence) DO NOT EXIST. Quantum mechanisms are supernatural mechanisms. They are obviously non-physical. They are also pre-physical. The Gods or Controlling Psyches had to design and make the quantum fields BEFORE physical matter could be made and sustained. First things first!

Ask yourself which model or philosophy or worldview has the most explanatory power from a scientific perspective – Scientific Naturalism which claims that psyche and quantum (supernatural) mechanisms DO NOT EXIST – or the proven and verified existence of Instantaneous Action at a Distance, Intelligence, Consciousness, Quantum Mechanisms, Quantum Waves, and Quantum Fields.

Ask yourself which way that the SELF is most likely to control its physical brain – through Scientific Naturalism which DOES NOT EXIST – or through Quantum Mechanisms or Supernatural Mechanisms which have been proven to exist through scientific experimentation.

We have KNOWN for over twenty-five years how the SELF or Psyche controls its physical brain; but, the Materialists, Darwinists, and Naturalists have yet to get the memo. They don't receive the memo, because they don't want it and don't want it to be true. I KNOW because I used to be a materialist, naturalist, nihilist, and atheist back in 2012; and, I didn't want my eternal, everlasting, indestructible Psyche to exist at the time. My chosen beliefs back in 2012 were demonstrably wrong; but, I didn't know that at the time, because I had been successfully deceived by the Materialists, Naturalists, Nihilists, and Atheists who had convinced me that my Psyche or Soul or Spirit DOES NOT EXIST. I believed them because I wanted to believe them. I believed them on blind faith because I wanted their Materialism, Naturalism, Nihilism, and Atheism to be true. There's no evidence to support their claims. In fact, ALL of the evidence falsifies their claims. But, that doesn't matter when one is operating on blind faith and truly wants Materialism and Naturalism to be true. I desperately wanted Scientific Naturalism to be true at that point in my life.

We have Quantum Field Theory or Quantum Mechanics on one side, and Materialism or Scientific Naturalism on the other side. They were designed to falsify each other. If one can be demonstrated to be true, then the other has been falsified. They are mutually exclusive. The false is falsified by the truth; and, the truth is repeatedly experienced and observed. So, which one has been experienced and observed, and which one has been falsified?

Except for the Materialists, Darwinists, and Naturalists, it's obvious to everyone which philosophical worldview is true, and which one is false and has been falsified. Darwinism, Materialism, and Naturalism were designed to trick us and deceive us into believing that the non-physical, or the supernatural, or the quantum DOES NOT EXIST. They work as advertised. I KNOW because I used to be a materialist, naturalist, nihilist, and atheist; and, I had successfully convinced myself that the psyche, spirit, mind, or soul does not exist. I was wrong; but, I didn't know that at the time. Self-deception works, and it works every time. Nobody is immune.

The Placebo Effect is also a quantum phenomenon. It is mind-over-matter, or psyche-over-matter. It, too, is a proven and verified phenomenon. Observe how the truth is constantly and repeatedly verified, experienced, and observed; whereas, the false has never been experienced and observed by anyone.

The non-existence of quantum mechanisms – or Scientific Naturalism – has NEVER been experienced nor observed by anyone, because it's false. Design and creation by Natural Selection, or Spontaneous Generation, or Chemical Evolution, or Abiogenesis, or Macro-Evolution, or Creation Ex Nihilo has NEVER been experienced nor observed by anyone, because it is false. Louis Pasteur falsified these things back in 1859, when he falsified spontaneous generation; but, the Atheists, Naturalists, Darwinists, and Materialists NEVER got the memo because they refused to receive it. Macro-evolution, chemical evolution, abiogenesis, spontaneous generation, or the origin of species by means of natural selection was FALSIFIED back in 1859 BEFORE the theory of evolution was introduced to the world. We've had the truth all the way along, but have chosen to reject it, preferring the lie or the Darwinism instead. It's a choice that one's psyche makes. Choice reveals to us our true nature and what we want most.

Remember, genetics and genomes were designed to prevent macro-evolution from happening; and, they work as they were designed to do. Remember, Louis Pasteur falsified spontaneous generation, abiogenesis, and macro-evolution in 1859.

It's obvious to everyone, except for the Darwinists and Naturalists, that genetics and genomes were designed to prevent macro-evolution from happening. It's physically impossible for "natural selection" to design and create a compatible Mr. and Mrs. Mutant at the same place and at the same time for millions of years on end, so that some chimp-like ancestor can evolve naturally into chimpanzees and human beings over millions of years of time. Genetics and genomes prevent that type of macro-evolution from happening in the real world. Macro-evolution is physically impossible thanks to genetics and genomes.

Remember, genetics and genomes were designed to prevent macro-evolution from happening; and, they work as they were designed to do. Remember, Louis Pasteur falsified spontaneous generation, abiogenesis, and macro-evolution in 1859.

79

Our beliefs are chosen into existence by our Human Psyche. We believe what we choose to believe. I KNOW because I have lived on both sides of the fence, and I have watched my beliefs change over the years as further information and knowledge slowly came my way. Things like justice, compassion, friendship, forgiveness, charity, and love DO NOT EXIST until some Psyche chooses them into existence or believes them into existence. That's the power of Psyche. Your psyche is the ONLY thing that you are guaranteed to have for the rest of your existence, so make it a good one. Your Psyche or Intelligence is you at your very core. Remember, your SELF uses its Psyche to control its physical brain through quantum waves or quantum mechanisms. This is what has actually been experienced and observed and verified to be true.

Mark My Words

Quantum Neuroscience

As neuroscientists in the 21st century, we basically KNOW what is happening at the physical level within the physical brain. Now, it's time for us to complete our science and training by trying to figure out what's happening at the quantum level within the physical brain.

That's why I introduced Quantum Neuroscience to the world. It was a science whose time had come. Quantum Mechanics and Quantum Field Theory are proven and verified sciences. It's time for us to start using Quantum Mechanics within Neuroscience in order to explain Neuroscience; and, the explanatory power of Quantum Neuroscience is through the roof to infinity and beyond. That's what makes it so valuable to us. When it comes to Quantum Mechanics and Quantum Field Theory, there doesn't seem to be a whole lot more to discover, because it has ALL been proven and verified as being real and true. We simply need to start using it, since we have already discovered it and proven that it is true.

Remember, Quantum Mechanics and Quantum Field Theory are used to explain the physically impossible. In other words, from our limited perspective trapped within space-time, Quantum Mechanics and Quantum Field Theory are indistinguishable from magic, because we can't control the quantum from the physical level. All of these different quantum mechanical functions, that we have discovered, happen behind the scenes whether we believe in their existence or not. There is a whole quantum world happening at the quantum level completely outside the awareness of us physical beings, who are trapped within and live within space-time at the physical level. Our physical brain has NO control over any of it! In fact, the opposite is true. Something or someone (or a lot of somethings and someones) at the quantum level has complete control over our physical brain.

Remember, your physical brain doesn't have control over anything at the quantum level. Your physical brain is simply used to control your physical body at the physical level. Quantum Mechanics and Quantum Field Theory turn everything upside down and rearrange our priorities. Suddenly, we find ourselves dealing with

NEW science that nobody has ever thought of before. It's a brave new world – a world that the Materialists and Naturalists can't understand nor accept. These people are literally left behind in the dust once we step into the realm of Syntropy, Psyche, Consciousness, Quantum Mechanics, and Quantum Field Theory. Everything physical is designed, made, and controlled at the quantum level by a bunch of different Psyches, Intelligences, or Life Forces that we can't see nor detect with our physical instruments at the physical level. The Quantum World or the Spirit World is a whole other world just waiting for us to discover and explore.

The question is, is whether you will explore it, or simply allow yourself to be left behind? It's happening with you or without you; so, I figured that it's time that I finally get involved and get with the program. Quantum Mechanics, Energy Mechanics, or Spiritual Mechanics is here to stay, whether we like it or not.

It Is All about the Quantum Aspects of Neurotransmitters

Neurotransmitters are how brain functions are handled at the physical level. Neurotransmitters are physical keys that open physical ion gates, letting the atoms or ions flow into the neuron and change its functioning at the physical level.

It has been observed that Dopamine and Serotonin have to be maintained at a certain level within the cerebral spinal fluid for proper brain function.

Insufficient levels of Dopamine within the neurons and cerebral spinal fluid results in Parkinson's Disease. Parkinson's disease is an inability of the mind or psyche to use its physical brain to control and regulate the functioning of one's physical body. In fact, we ALL would have Parkinson's disease if Materialism, Naturalism, Darwinism, Physical Determinism, and Behaviorism were 100% true because everything would be firing randomly and happening randomly at the physical level within the physical brain if Materialism and Naturalism were in fact true. In other words, our spirit, psyche, consciousness, or mind would have NO control whatsoever over our physical brain and physical body if Materialism and Naturalism were real and true.

The very fact that we consciously choose to control our physical body through our physical brain when our physical brain is functioning properly is Proof of Psyche and proof that Quantum Mechanics is real and true. We would ALL exhibit symptoms of Parkinson's Disease and have no control over our physical body if Materialism and Naturalism were actually real and 100% true. The whole thing would be random with nobody in the driver's seat, just as the Naturalists and Atheists believe it to be. It's obvious when you are looking at a physical body where nobody is in control – the thing is either dead or epileptic. A physical body and physical brain NEED Something Psyche to control them, or they do nothing but sleep and randomly fire. In the movie *Serenity*, the people of the planet Miranda just sat there and starved to death because they had been pacified, drugged, or lulled so much that they no longer had any incentive or desire to do anything.

That's what would happen to all of us if Materialism and Naturalism were actually true.

People with Parkinson's Disease, due to insufficient levels or amounts of dopamine within their physical brain, cannot fully and efficiently control their physical body and physical movements. Their mind or psyche tells their body to do one thing; but, because the physical machinery isn't working right or isn't fully there, their body doesn't always do what their psyche or their spirit wants their body to do. You also end up with thrashing, because there are times when their body finally starts to do what these people want, long after they no longer want it to do that and instead want it to do something else. There are FALSE starts and stops all along the way because the system isn't being properly regulated. All fluidity of motion is lost, due to all of these FAILED starts and actual FALSE starts that are happening within their physical brain simultaneously. These people's psyche or consciousness is no longer in full control of their physical brain and physical body. For them, it's as if Materialism, Naturalism, Determinism, and Behaviorism are actually true; and, they don't have full control over their physical body through their physical brain. Their brain and body seem to develop a mind of their own. These people can no longer fully control their body and brain.

Serotonin seems to be the neurotransmitter that was designed to give us a sense of well-being – a sense of being in control of the physical aspects of our emotions and feelings and lives. I find it interesting that physical brains, neurotransmitters, and synapses can be designed and mapped at the quantum level in order to produce or perform certain emotions or feelings or sensations at the physical level. It's all just energy; but, the pain and pleasure make it seem real.

Anyway, a certain level of Serotonin within the cerebral spinal fluid is necessary in order to maintain a constant sense of well-being, or what we sometimes call happiness and joy. On my mother's side, they are 100% Scandinavian. If you notice carefully, there are high levels of depression, anxiety, and mental illness within the Scandinavians due to the fact that their bodies never learned how to properly regulate themselves, due to the fact that these people never sleep during the summer and constantly sleep or hibernate during the winter.

Scandinavians are easily addicted. They can't hold their liquor; and, alcoholism is rampant among the Scandinavians.

I inherited all of these weaknesses from the Scandinavians on my mother's side. I'm easily addicted. I also seem to have naturally low levels or insufficient levels of Serotonin within my system. I have benefited noticeably from Zoloft, which is an SSRI.

For a sense of well-being and a sense of control, certain neurons within our physical brain need to be able to reach out telepathically and telekinetically into the cerebral spinal fluid nearby and "grab" or quantum tunnel a Serotonin molecule into an ion-gate in order to properly regulate the necessary flow of ions into the neuron.

Remember, at the physical level, atoms and molecules can still quantum tunnel or teleport – they just can't go very far. The Gods are preventing your atoms and molecules from quantum tunneling away from you to the other side of the universe, so that your brain and physical body don't dissolve into thin air on you.

There has to be Serotonin molecules there within the cerebral spinal fluid somewhere nearby; otherwise, there would be nothing to grab or quantum tunnel to the Serotonin receptor when the post-synaptic neuron needs a Serotonin molecule to open the requisite ion gate. NO Serotonin, then NO sense of well-being or NO sense of control. Once again, your brain seems to develop a mind of its own because you no longer have full control over it, if you are lacking in Serotonin.

You see, when it comes to Dopamine and Serotonin, there has to be a certain level of these neurotransmitters maintained within the cerebral spinal fluid for proper brain functioning. Because when the post-synaptic neuron or the "first neuron to fire" needs one of these molecules to start an action potential going, the molecule actually has to be there nearby to quantum tunnel into position, because the post-synaptic neuron can't quantum tunnel in the requisite Serotonin from the opposite side of the brain or from another brain. The Serotonin, or Dopamine, has to be there within the synaptic cleft, because that's pretty much the greatest distance that something can quantum tunnel or teleport at the physical level within our physical universe, without direct intervention from the Gods who have complete control over Quantum Mechanics at all levels of existence.

Our materialistic, atheistic, and naturalistic Neuroscientists erroneously portray neurotransmission as a random physical process. It's NOTHING of that sort! If it were truly random as portrayed, then there would be tons of misfiring, random misfiring, or NO firing whatsoever. Your psyche or mind would tell your body to move, and at times NOTHING would happen whatsoever, if Materialism and Naturalism were true and everything was simply happening randomly as the Materialists and Naturalists claim. It would be constant spasticity or constant cerebral palsy for ALL of us if Materialism and Naturalism were actually true! The fact that most of us can control ourselves is Proof Positive that Psyche and Quantum Mechanics are real and true, and solid evidence falsifying the claims of Materialism and Naturalism which claim that everything within the physical brain happens randomly under natural physical processes. If Materialism and Naturalism were actually true, then NONE of us would be able to feed ourselves!

You see, if the whole neurotransmission process were completely random as the Atheists and Naturalists claim that it is, then theoretically it could take a million years for the right and necessary neurotransmitter to cross the synaptic cleft and finally attach with the proper neuroreceptor on the post-synaptic neuron. Random means random, which means NO reliability or dependability whatsoever. You tell your arm to move, and if it were simply random as the Materialist and Naturalists claim, the neurotransmitters would dump into the synaptic cleft; but then, it could in fact take a million years for the connection to be randomly achieved according to the requirements of Materialism and Naturalism, which means that it could potentially take trillions of years before your arm finally moved. Randomness does NOT produce movement or progress! Randomness doesn't produce anything but chaos at the physical level.

When it comes to neurotransmitters, someone or something is driving these things. It's as if they are alive. OR, someone or something is reaching out telepathically or telekinetically into the synaptic cleft and literally grapping or teleporting the specific neurotransmitter that it needs immediately to the ion gate that needs to be opened in order to make a specific neuron fire at a specific time.

In other words, when it comes to neurotransmitters, brain functioning, and quantum mechanics, there is NOTHING random about it. If it were in fact random at times, then your arm would move of its own accord, when in fact you wanted your arm to stay still. The fact that you can actually decide and choose when to move your arm and when to hold your arm still – the fact that it isn't random at all – is Proof of Psyche and Proof of Quantum Mechanics. Your arms and legs would be firing on their own all the time – randomly – if Materialism and Naturalism were 100% true; and, we all in fact would be nothing but a spastic mess, if Naturalism, Darwinism, Atheism, Nihilism, Physicalism, and Mechanism were 100% true. We wouldn't and shouldn't have any control over our physical bodies whatsoever if Scientific Naturalism were in fact true. The whole thing would be random, just as the Determinists and Behaviorists claim that it is. These people really would have NO control over themselves, just as B. F. Skinner claimed. It's obvious to everyone but B. F. Skinner that his Radical Behaviorism or Atheism was WRONG. Self-deception works; and, it works every time, even on geniuses like B. F. Skinner, Einstein, Stephen Hawking, and Richard Dawkins. If these people were actually right, then we ALL would be constantly spastic or have constant cerebral palsy, and everything would in fact be random just as these people claim that it is.

BUT, the Behaviorists and Determinists are obviously wrong. There's NOTHING random about neurotransmission. The whole thing is deliberately done at the quantum level by Someone Psyche. Neurotransmission is chosen into existence at the quantum level by Someone Psyche or Something Psyche. Neurotransmitters, like atoms, proteins, and other molecules are Quantum Objects, which means that even though they are physical, they can quantum tunnel or teleport short distances at will. Quantum Mechanics works because it is telepathically controlled at the quantum level by some kind of Psyche or Consciousness. There's nothing random about it. Randomness comes into play at the physical level, NOT the quantum level.

I KNOW from personal experience that all of this is true, and so do you, because we have lived it and experienced it.

I have also experienced brain zaps, when the neurons within my brain fire randomly, without my telling them to fire. I think we have all experienced a brain zap or a random discharge within our brain – that we didn't request and that we didn't have any control over. That's the way it would be ALL the time if Materialism and Naturalism were true. Your brain and body would be firing or going off all the time, and you wouldn't be able to control it or stop it.

Not only does Quantum Neuroscience have massive explanatory power; but, Quantum Neuroscience also has practical applications. Let me demonstrate with a personal example.

Like I said, I have Scandinavian ancestry on my mother's side, which means that I'm prone to depression, anxiety, OCD, and a bunch of other brain malfunctions because my brain doesn't naturally produce enough Serotonin. I benefitted noticeably from Zoloft, an SSRI (Selective Serotonin Reuptake Inhibitor). The purpose of an SSRI is to increase the levels of Serotonin in the synaptic cleft and cerebral spinal fluid to normal levels for proper functioning.

Most Neuroscientists and Medical Doctors don't really understand how all of this works, because they can't visualize how it works at the quantum level because they have chosen to believe that it's all happening randomly at the physical level. The problem with their "random hypothesis" is that it isn't dependable, replicable, nor reliable. Therefore, they can't understand how SSRI's really work, nor how they should be properly administered. I understand, because I have lived it and experienced it for a decade, so I KNOW.

Now let me explain what I KNOW from my past experiences with Zoloft and SSRI's.

When I was first put onto Zoloft, the idiots put me on the full dosage, which is 200 mg per day. They don't understand how any of this works, so they just automatically put you on the full dosage. The full dosage is determined by their observation that if they give you more than the full dosage, then it becomes toxic and starts to kill you. So, they put you on the full dosage because that's how they make their money, by feeding you pills and getting you dependent upon their pills.

Well, at the full dosage, I was catatonic all day, just sitting there staring at the wall or lying there staring at the ceiling. You could call it maximum mellow. I was non-functional; and, I was that way for about six months while I was also going through withdrawal from my addiction to benzos or Valium. It took me six months of delusions, hallucinations, and withdrawal from substance-induced psychosis before I started thinking rationally again – after I chose to go cold-turkey on the benzos and the Seroquel anti-psychotic. Whenever I stopped taking the benzos and the anti-psychotics, I would automatically go psychotic and start having delusions and the occasional hallucination.

I finally went to a neuropsychiatrist who actually understood how all of this really works and how brain-altering medications should be administered. He told me that my medical history and genetics were telling him that I'm overly sensitive to medications and that from then on I was to tell every psychiatrist to NEVER put me onto benzos or anti-psychotics or sleeping pills or any of the other junk that they had me on. He said, "PICK ONE, AND GET RID OF ALL THE REST". You should only be on ONE brain-altering drug – not a dozen like they had me on. By then, I had gone cold-turkey, was nearly six months into my withdrawal, starting to think and understand again, and I was on only ONE thing – the Zoloft. They wouldn't let me quit the Zoloft; but, I had managed to get off of everything else. Six months is a long time for withdrawal; but, that's what it took to get my brain to normalize after I went cold-turkey from all the rest.

Then he said that my ancestry and medical history made it clear that I needed the Zoloft or something like it. HOWEVER, he told me that I should NEVER let the psychiatrists put me on the full dosage of anything ever again. I'm overly sensitive to medications, and I should NEVER be on the full dosage of anything. He told me that he wanted me to immediately cut the dose in half to 100 mg, and then try it for a year and see how it goes. Then he wanted me to cut the dose in half again to 50 mg per day and see how that goes. He told me to keep cutting the dose in half until I got down to a maintenance dose, where I was getting all of the benefits with NONE of the side-effects. So, that's what I did.

When I cut the dose in half, suddenly, I could sleep at night, the racing thoughts would stop, and I could breathe well because the Zoloft had an antihistamine and decongestant effect. Yet, I would wake up in the morning, was alert and wake, able to drive, and able to go back to school and go to work. I normalized on 100 mg of Zoloft per day – half the normal adult dosage. It was working, just as he had suggested.

Somewhere along the way, I began to notice that at 200 mg, the brain zaps during the night seemed relatively constant. I would be zapped or jerked awake every hour or so, while trying to sleep; and, I would have periodic brain zaps during the day too – when my brain would just randomly misfire and send an electric pulse through my body.

At 100 mg, the brain zaps were there and relatively constant as well. I was on 100 mg for a year, and then I cut the dose in half to 50 mg. Again, the results were even better at 50 mg per day. I was getting all of the benefits with noticeably fewer side effects. It was then that I started to notice that the brain zaps seemed to be decreasing in regularity. At 100 mg, the brain zaps seemed to be a daily occurrence, especially at night. At 50 mg, I would go days without brain zaps, only having two or three per week on average. I tried going lower, but I felt the depression and anxiety and OCD coming back; so, 50 mg ended up being my maintenance dose for six or seven years.

It was only recently, when I started to study Neuroscience and started to develop Quantum Neuroscience, that I finally began to figure out that SSRI's actually cause brain zaps – random misfiring of the brain or random connections within the brain – because you have too much Serotonin in the synaptic cleft and contact is actually randomly achieved at times and misfires are the result of a random docking of a neurotransmitter with a post-synaptic neuroreceptor. Remember, when it comes to neuroscience and brain science RANDOM CONNECTIONS cause misfires or brain zaps within the brain circuitry. In other words, the neuron actually fires randomly without Someone Psyche telling it to do so.

Suddenly, I understood how everything works – seven years later. Brain zaps were a sign that I have too much Serotonin in my system. More than what's needed. Seven years later, I was still having the occasional brain zap two or three times a week, at 50 mg per day of Zoloft. It was time to cut the dose in half again. I'm still experimenting with that; but so far, six months later, I have noticed that my brain zaps have decreased to about once per week, but I'm tapering down so I do 25 mg one day and 50 mg the next day. I suspect that when I get down to 25 mg every day, I may not have brain zaps any more at all, or maybe once a month. In other words, by lowering the dose and lowering the amount of Serotonin in my system, I'm in fact lowering the RANDOMNESS or the misfiring brain zaps that come from having too much Serotonin in the system.

You need to have a certain level of Serotonin in the system, because when the post-synaptic neuron needs a Serotonin molecule to trigger an action potential, there has to be one nearby that it can quantum tunnel into position to open the gate. If there is no molecule nearby, then the post-synaptic neuron can't open the gate, and thus the post-synaptic neuron can't fire and can't do its job.

However, if you have too much Serotonin in the system, then things start connecting randomly, and then you end up with lots of random brain zaps or random misfiring that you have no control over because it's happening randomly. It's all about balance. You need enough Serotonin and Dopamine in the cerebral spinal fluid and synaptic cleft so that when a neuron needs to fire, it can reach out telepathically and quantum tunnel in the necessary neurotransmitters so that the post-synaptic neuron can fire at will. Yet, you can't have the synaptic cleft chock full of neurotransmitters because if you do, then random connections will start to happen with greater and greater frequency as the levels of neurotransmitters continue to rise. Random connections and random firing of the neurons is worthless, because they are unwanted. We would be nothing but a spastic mess if that were going on all the time.

So, with the medication or the Zoloft, I have to find the right dose so that I have enough Serotonin in the system that I can maintain that sense of well-being that Serotonin provides; but, I need to decrease the amount of Zoloft I'm taking until I get to the point where I'm no longer having all the random brain zaps that I used to have. Too much Serotonin in the system causes random misfiring or random brain zaps. The brain zaps where the physical sign that I had too much Serotonin, and therefore too much Zoloft in my system. As I decrease the amount of Zoloft, I decrease the amount of Serotonin, and therefore I decrease the average number of brain zaps for misfires that happen randomly. The goal with any brain-altering drug is to keep cutting the dose in half until you reach a maintenance dose where you are getting all of the benefits with NONE of the side effects. Brain zaps or random misfiring is a side effect of Zoloft and SSRI's. Cutting the dose in half decreases the average number of misfires or brain zaps. Hopefully, at 25 mg per day, all of the brain zaps will be gone, but the sense of well-being will remain.

It's an experiment, a science experiment; and, Quantum Mechanics and quantum tunneling were the only way that I could explain all of this to myself logically and rationally in a way that actually made sense.

When it comes to Dopamine and Serotonin, certain levels have to be maintained, so that the post-synaptic neuron can reach out telepathically and quantum tunnel one to the neuroreceptor when a Serotonin molecule is needed. If there's nothing in the synaptic cleft, and nothing in a vesicle to dump into the synaptic cleft, then the neuron can't fire when it needs to fire; and, a certain sense of uneasiness starts to arise within the psyche, spirit, or soul because you can simply feel that your brain isn't working right for some unexplained reason.

At the other extreme, if you get too much Serotonin in the system, then random misfiring or brain zaps start to increase – again, you start to lose control of the system. You have got to have the right balance when it comes to Serotonin and Dopamine, for maximum function. Too little is not good, and too much is not good. Random misfiring represents a loss of control, and NO firing when you want it to fire also represents a loss of control. The goal is for the psyche, spirit, or soul to be able to maintain control all the time at optimum levels, when it comes to the spirit's physical body and physical brain. You want control, not random misfiring and not NO-firing at all.

We ALL automatically and immediately KNOW whenever our brain zaps or misfires. We know that we didn't do it; and, we KNOW that there is nothing that we could have done to stop it either. It just simply happens randomly – out-of-the-blue – as if Materialism and Naturalism were actually true. We would have NO control over our physical body whatsoever, if Materialism and Naturalism were actually true. The whole thing would be firing randomly, and neurotransmitters would be connecting randomly, and it could at times take a million years for your arm to move once you decided that you wanted it to move – if Materialism, Naturalism, Darwinism, and Atheism were actually true. You see, Someone Psyche, or Someone Intelligent, or Someone Alive at the quantum level actually has to choose to open a vesicle of neurotransmitter and dump those neurotransmitters into the synaptic cleft for the FIRST TIME, in order to get the first action potential or the first choice going in the first place. Choosing to move your arm doesn't just happen randomly. Someone Psyche makes that choice and sets the whole process into motion.

Remember, when it comes to neurotransmitters and action potentials, there's NOTHING random about it at all. The whole thing is chosen into existence by Someone Psyche at the quantum level. If it weren't so, then your arms and legs would be firing and misfiring randomly all the time, and there's no way physically possible that you could ever control yourself and your physical body. Randomness at the physical level is highly unreliable and very undesirable. You don't want your arm randomly steering you into oncoming traffic while you are driving a car. You want to be in complete control of your physical body while you are driving a car or operating heavy machinery. That's just the way it is. While driving a car, you definitely don't want Materialism, Naturalism, Darwinism, Nihilism, and Atheism to be true because you want complete control over what you are choosing to do with your physical body and physical car.

By letting Psyche and Quantum Mechanics in to play, we can literally explain everything that comes our way. Suddenly, WE KNOW how neurotransmitters and neurons really work, because we KNOW how things work at the quantum level. When a neuron needs to fire, so that your arms or legs will move, it needs to be able to reach out telepathically and grab the nearest neurotransmitters, and quantum tunnel them into the neuroreceptors, so that the neuron can fire immediately; and, your arm will move immediately. It needs to happen immediately, when your psyche or consciousness decides that it needs to happen now. You can't afford to sit around for a couple of minutes or a million years waiting for it to happen randomly. That neurotransmitter is needed right now; and, it is quantum tunneled into position immediately, so that your arms and legs will move right now, not a random number of hours from now. Behaviorism, Materialism, Naturalism, and Darwinian Randomness are absolutely worthless when you need to move out of the way of the oncoming truck right now. You NEED Psyche and Quantum Mechanics right now to get the job done right now – not the randomness and unreliability of Classical Physics, Materialism, and Naturalism just so that the Atheists can feel comfortable with themselves and their unbelief.

The good thing about Quantum Mechanics is that it all happens automatically at the quantum level under the control of Nature's Psyche or the Atom's Psyche whether you believe in it or not. The bad thing about Quantum Mechanics is that we fallen, physical, mortal, entropic beings don't seem to have any control over any of it whatsoever. We can't trigger it, we can't control it, and we can't do anything with it. I can't quantum tunnel my physical body to work at will; and, we are completely unaware of all the quantum tunneling that's taking place within the synaptic clefts of our own physical brain, because we aren't controlling any of that either. We make NO choices as to what synapses to build, where to build them, what types they will be, or how they are to be operated or controlled. Nature's Psyche or the Psyche within the atoms is controlling all of that for us. The only thing we have control over is what we choose to do with the physical body and physical brain that we have been assigned. Everything else is controlled at the quantum level by Nature's Psyche or God's Psyche instead, completely outside our awareness and control.

WE KNOW that this is true, because we have proven the Placebo Effect or Mind-Over-Matter to be true. Some kind of mind or psyche is controlling all of this at the quantum level; and for the most part, it isn't the Human Psyche but rather Nature's Psyche instead. Remember, the quantum can't be controlled from the physical level. The quantum has to be controlled from the quantum level or the psyche level instead. WE KNOW that this is true because the Obvious Law of Physics states that the smaller dwells within and controls the larger. This Obvious Law of Physics is obviously true at all levels of existence. The smaller always dwells within and controls the larger. Something dwells within and controls the atoms, molecules, neurotransmitters, and neurons – Something Psyche.

The Materialists, Naturalists, Darwinists, Nihilists, Behaviorists, Determinists, and Atheists tell us and assure us that the non-physical DOES NOT EXIST. These people are obviously wrong. Quantum Fields are obviously non-physical in nature and origin. In fact, to my surprise, I eventually discovered and observed that Quantum Fields are "pre-physical". In other words, the Gods or Controlling Psyches had to design and make the Quantum Fields BEFORE they could design, make, and sustain physical matter. They had to make the Quantum Fields first, or physical matter could NEVER have been made. That's just the way it is. That's what the Science is trying to tell us. The quantum or the spiritual had to be organized FIRST before it became possible to make and sustain physical matter. First things first. Even the Gods or Controlling Psyches have to obey physics. They had to organize and make the Quantum Fields BEFORE they could make physical matter. The physical matter didn't design and make itself. Physics doesn't work that way. Someone Psyche or Someone Intelligent designed and made everything that we see and experience around us here today. This book didn't just spontaneously spring into existence from nothing. It didn't just make itself. Someone Psyche or Someone Intelligent designed it and made it. That's just the way things really work. This is what we have actually experienced and observed.

Quantum waves are obviously non-physical. That's how the physical atoms communicate with each other – through quantum waves or thoughts. The Quantum Zeno Effect has repeatedly been proven to be real and true – physical atoms freeze or stop moving while you are looking at them. They are telepathic, psychic, and can read your mind. They KNOW when you are looking at them and when you are not looking at them. They don't quantum tunnel and they don't move while you are looking at them. They only quantum tunnel or jump the barriers while no human being is looking at them or paying attention to them. This Quantum Zeno Effect is a psychic phenomenon and a quantum phenomenon; and, it has NO explanation whatsoever in terms of Classical Physics, Materialism, and Naturalism.

Physical atoms have a mind or a psyche of their own. They KNOW when you are looking at them, and they KNOW when you are not. In fact, a case can be made that the atoms, molecules, proteins, transcription enzymes, and neurotransmitters are infinitely more intelligent than we are. We human beings or human psyches have been dumbed-down to the extreme, so that our test her in mortality will be a valid and effective test. We are completely oblivious to the psychic and the quantum, unless we actually go looking for it. Typically, we are completely oblivious to all these things, and we have to approach Quantum Mechanics, Energy Mechanics, Psyche Mechanics, or Spiritual Mechanics on faith if we are to discover and learn anything about them. They are NOT a part of our normal physical experience.

At the quantum level and the psyche level, it's all about Psyche or Energy. It has nothing to do with the physical, except for the fact that it can serve as the foundation for the physical. We can't control Quantum Mechanics from the physical level. Here at the physical level, the quantum, spiritual, and psychic happen completely outside of our conscious awareness and conscious control; and, we actually have to choose to go looking for them before we start seeing signs of them. Something as simple as magnetism or magnetic waves is completely invisible and non-physical; and, we only know that it exists by the effects that it has on physical matter – which we can see with our physical eyes. Unless we are paying attention, though, we are normally completely oblivious to all the different magnetic waves and quantum waves that surround us and pass through us, because these things are non-physical in nature and origin.

The proven and verified existence of magnetism and magnetic waves FALSIFIES the claims of Materialism and Naturalism which state that the non-physical DOES NOT EXIST. These people are demonstrably wrong whenever they start claiming that the non-physical or the immaterial does not exist. All you have to do is study the electromagnetic waves in order to see for yourself that these people are wrong. Psyche is just concentrated energy or concentrated light; and, we have already proven beyond a shadow of a doubt that energy and light do in fact exist, and that energy and photons and light are in fact massless and non-physical. The proven existence of light FASLIFIES Materialism, Naturalism, and their derivatives such as Atheism, Darwinism, Behaviorism, Determinism, and Nihilism – which are trying to tell us that the Psyche or the Light DOES NOT EXIST.

According to the Law of Psyche, every Psyche has a certain amount of energy that's under its control; and, the Controlling Psyche decides what form the energy under its control will assume or take. Energy is infinitely malleable. Energy can be formed into anything, including quantum waves, quantum fields, quarks, electrons, photons, spirit matter or dark matter, and the different types of physical matter. It's all made from energy; and, it's the Psyche who decides what form the energy under its control will become and be, and for how long it will be so. NONE of this happens at the physical level. It all happens at the quantum level and the psyche level, where the Psyches or Intelligences have complete control over the energy that has become theirs to control. This is the way things really work. It's a Psyche Ontology, wherein Psyche or Consciousness is the fundamental unit of reality. This is the Ultimate Model of Reality. It's falsified by Materialism, Naturalism, Darwinism, Nihilism, and Atheism; and, IT FALSIFIES Materialism, Naturalism, Darwinism, Nihilism, and Atheism. These two philosophies or worldviews are mutually exclusive. If one of them is proven true, then the other has automatically been proven false. So, which one has been proven true, and which one has been falsified?

Remember, the false is always falsified by the truth; and, the truth is repeatedly and constantly experienced and observed. So, have we experienced and observed the non-existence of Psyche and Consciousness; OR, have we experienced Psyche and Consciousness? Which one have you experienced? Have we scientists experienced and observed Quantum Tunneling, the Quantum Zeno Effect, Quantum Waves, Quantum Fields, the Conservation of Energy or Psyche, Quantum Superposition, Quantum Entanglement, Non-Locality, and Action at a Distance? OR, have we experienced the non-existence of these things. Materialism and Naturalism tell us that these things DO NOT EXIST. Are they right? Or have these people been proven wrong by Science itself? Which has been experienced and observed? The existence of these things or the non-existence of these things? Have you experienced moments of Intelligence and Consciousness, or have you gone throughout your life completely oblivious to these things? Have you made your arms and legs move, or have they simply moved randomly and purposelessly throughout your life? Which is it?

The truth is obvious, because the truth is constantly experienced and observed. Only the Materialists, Naturalists, Darwinists, Nihilists, and Atheists are capable of denying the truth and telling us that the sun doesn't exist while they are looking at the sun. Self-deception works, and it works every time. These people are Sophists. They are trying to trick you and deceive you as they have been tricked and deceived. It's obvious, now that I'm no longer a Materialist, Naturalist, Nihilist, and Atheist; but, it was not the least bit obvious while I was trapped within it. That's the way psychology works. It's obvious to me now that I was totally psychotic and insane while I was going through withdrawal and trying to overcome my substance-induced psychosis; but, while I was trapped within it and going through it, I didn't have a clue.

While you are insane, delusional, and hallucinating – while you are psychotic and going through withdrawal – you often have no clue whatsoever how insane and how far gone you truly are. It's only when you get past it, and get outside of it, that you can truly see how insane and irrational and illogical you truly were. The same reality ended up applying to My Materialism, My Naturalism, My Nihilism, and My Atheism. I had no idea how stupid those things were making or how much they were holding me back, until I finally got past them and overcame them. Looking back now, though, I can see that I was completely crippled and completely insane. That's why I decided to study Psychology and Quantum Neuroscience. I'm now trying to grasp the things that My Materialism and My Naturalism were preventing from understanding and accepting before I saw the light – or saw the Quantum Mechanics.

The grass really is greener and a whole lot more vibrant and alive on the Quantum Mechanics side of the fence than it ever was on the Atheistic and Naturalistic side of the fence. I KNOW from personal experience that this is true.

Mark My Words

Einstein's Biggest Blunders

I personally believe that Einstein's biggest blunder was his ongoing rejection of Quantum Mechanics, along with his attempt to promote Classical Realism, Materialism, and Naturalism instead. Materialism and Naturalism have been falsified by Quantum Mechanics and Quantum Field Theory; whereas, Quantum Mechanics and Quantum Field Theory have been repeatedly verified and proven true, over and over again in a multitude of different ways. It's a mistake to choose Classical Realism, Materialism, Naturalism, and their derivatives over the immaterial and supernatural Quantum Field Theory. Materialism and Naturalism have lost every time.

https://markme.website/einsteins-biggest-blunders/

My purpose is to set everything straight; and, it began with the long and painful process of setting myself straight, first. Back in 2012, I was a Materialist, Naturalist, Nihilist, and Atheist. I had a long way to go before I finally saw the light and figured out how things really work.

Einstein's biggest blunder or greatest error came when he rejected Action at a Distance and concluded that the speed-of-light limitation applies at the quantum level in the Quantum Realm, Energy Realm, Psychic Realm, or Spirit World. This error is based upon the conclusion that physical matter is the only thing that exists; and, it has been proven false.

https://ultimate-model-of-reality.com/falsifying-einstein-and-classical-realism/

Instantaneous Action at a Distance is the rule or the law in the Quantum Realm. My best friend (and thousands of other people) has been there in the Quantum Realm or the Spirit World, and he told me that he can testify as an eyewitness that the speed-of-light limitation does NOT apply in the Spirit World or the Quantum Realm. He was participating and communicating instantaneously with humans who were on opposite ends of the universe. Even though these people reside on different planets in vastly separated parts of this universe, their consciousness was there witnessing and experiencing the same event that my friend was witnessing and experiencing. There are NO physical limitations in the Quantum Realm! In the Spirit World, they communicate telepathically, at a distance, instantaneously.

It's obvious that it has to be this way. There is no entropy and can be no entropy in the Quantum Realm; otherwise, something like the Big Bang would never have been possible. If the Big Bang were simply a random physical event as the Materialists and Naturalists claim, then there is nothing to stop it from happening randomly once again in your back yard, wiping us out in the process.

Einstein's other really great blunder was to assume that entropy applies to the Quantum Realm or Spirit World. It doesn't. It can't. The Quantum Law of Thermodynamics states that there is no entropy and no thermodynamics at the quantum level in the Light Realm or the Quantum Realm. It has to be this way, or we wouldn't exist. There could never have been something like a Big Bang (or the creation of a physical universe) if entropy were in fact building up in the Quantum Realm. This is a serious blunder that we all have made. We deny the Conservation of Energy (or the Conservation of Psyche) at the most fundamental level by insisting that entropy or the second law of thermodynamics applies at the quantum level in the Psychic Realm or Quantum Realm. It's a mistake to do so, because it isn't true. It can't be true.

There has to be some type of Syntropy or Organizing Force or Life Force at the quantum level, or the Big Bang could never have happened, the quantum fields could never have been made, and physical matter would have been impossible to build and sustain. You see, the Gods or the Controlling Psyches had to design and make the quantum fields BEFORE it became possible for them to make and sustain physical matter. Quantum fields are obviously made from energy by Someone Psyche, which means that the same has to be said of physical matter. There has to be Syntropy or Conservation of Energy at the quantum level, or our physical level would not have been possible to make.

Syntropy: The Answer to Life, the Universe, and Everything

https://www.amazon.com/gp/product/B07BPT3W8R/

Einstein didn't know about any of this, because he was stuck on Classical Realism, and he couldn't accept Quantum Mechanics and Spook Action at a Distance as being real and true. He didn't accept any of this Action at a Distance stuff, which means that he would have never been able to develop the Law of Syntropy. At the quantum level and the psyche level, Syntropy trumps entropy every time, because entropy or the second law of thermodynamics doesn't exist at the quantum level and the psyche level in the Quantum Realm or the Spirit World.

Einstein's biggest blunder came when he assumed that physical laws, physical limitations, and physical restrictions apply at the quantum level in the Quantum Realm; and, our biggest blunder comes whenever we choose to believe him. Whenever we choose to give physical limitations and physical laws priority over Syntropy or the Conservation of Energy and Psyche, then we make the most serious blunder that we can possibly make in physics and effectively destroy everything that we are trying to explain, sustain, and understand. Materialism and Naturalism destroy science, especially Quantum Field Theory and Quantum Mechanics. They both can't be true simultaneously. They are mutually exclusive. Something has to give.

During the past few years, I have repeatedly observed that Quantum Mechanics and Quantum Field Theory are constantly verified; whereas, Materialism and Naturalism are constantly falsified. Therefore, we KNOW which one of them is true, and which one of them is false. When it comes to Science, the false is falsified by the truth, and the truth is repeatedly experienced and observed. Materialism and Naturalism state that the non-physical or the immaterial – quantum fields, psyche or light or energy, and supernatural mechanisms or quantum mechanisms – DOES NOT EXIST. Materialism and Naturalism are demonstrably false. The proven and verified existence of quantum fields and quantum mechanisms FALSIFIES Materialism, Naturalism, and their derivatives such as Nihilism, Darwinism, and Atheism.

We have repeatedly observed that physical limitations and physical laws DO NOT APPLY at the quantum level in the Psyche Realm, Syntropy Realm, or Quantum Realm. The Science – the Real Science – changed the way I look at things; and, it helped me to overcome some of Einstein's greatest blunders.

One of my greatest conceptual breakthroughs came when I first understood what really happens when an object or particle reaches the speed-of-light. According to the theory of relativity, at the speed-of-light, time stops, length contracts to zero, resistance to acceleration or mass goes to zero, and distance goes to zero. Do you understand what this really means? It means that at the speed-of-light, in the quantum realm, velocity or speed goes to infinity. Speed is instantaneous or infinite in the Quantum Realm at the quantum level. Objects in the quantum realm can be stationary, of course; but, whenever they choose to move, they can quantum tunnel or teleport instantly to their destination at an infinite velocity. They have NO speed limit in the quantum realm.

An interesting note: My physics teachers taught me that at the speed-of-light, mass or resistance to acceleration goes to infinity – in other words, the object becomes omnipresent. It's the same difference. Whether the mass goes to infinity or the mass goes to zero, the result is the same – infinite velocities in the quantum realm, or instantaneous action at a distance and instantaneous quantum tunneling to anywhere in the universe, in the Quantum Realm or Spirit World. The Gods or the Controlling Psyches introduced mass or physical matter or space-time into the equation, in order to slow things down for us, so that we can experience them, learn from them, and actually remember having done so. Space-time and physical matter are used to create the ultimate consensus reality – a reality that is highly dependable and predictable. There are real consequences to our choices at the physical level. At the quantum level, everything is uncertain and tends towards the infinite or the eternal. You don't have to worry about dying at the quantum level. In contrast, at the physical level, everything is predictable, and death is certain.

One of my greatest conceptual breakthroughs came when I finally reconciled the Theory of Relativity with Quantum Mechanics. It came with the realization of what happens to a photon traveling at the speed-of-light.

You see, from our perspective trapped in space-time and this physical reality, that photon took 13.2 billion years to reach us as it traveled through space. However, at the speed-of-light from the perspective of the photon, time stopped, entropy stopped, there was no passage of time, distance collapsed to zero, length contracted to zero, mass or resistance to acceleration went to zero, and that photon from its perspective literally quantum tunneled or teleported instantaneously to its destination. In other words, that photon from its perspective at the speed-of-light experienced nothing from start to finish. It simply teleported or quantum tunneled instantaneously to its destination. This leads to the quantum realization that at the quantum level in the Quantum Realm from the perspective of the photon, a photon knows where it is going to land before it launches, or it doesn't launch.

This "infinite velocity" or "quantum tunneling" at the quantum level also lead to my realization that the Gods or Controlling Psyches created this physical realm for us, in order to slow things down for us, so that we can actually experience them, learn from them, experiment with them, and remember having done so. You see, at the speed-of-light, space-time goes to zero, distance goes to zero, time stops or goes to zero, mass or resistance to acceleration goes to zero, and length contracts to zero. When time stops, there's nothing to experience. When time stops, there is no entropy and no aging process. Isn't that so? Space-time as we know it ceases to exist at the quantum level in the Quantum Realm or the Spirit World. They have to switch over to some other time dimension or timeless dimension in the Spirit World because the space-time dimension that we are familiar with here in this physical realm ceases to exist at the quantum level in the Energy Realm or the Quantum Realm. Everything integrates at the speed-of-light – it either goes to infinity or it goes to zero – it goes to the limit; it transmutes; it becomes something completely different. At the speed-of-light, the object transfers into the Quantum Realm, becomes subject to Quantum Laws; and, all physical limitations including entropy cease to exist for that particular quantum object.

Just like when two gases combine to produce water, a huge transmutation takes place whenever a particle achieves the speed-of-light. It transmutes from a physical object into a quantum object; and, all physical limitations cease to exist. Its physicality ceases to exist. In order to regain its physicality, or its presence in space-time, it has to slow down to below the speed-of-light.

This perspective-shift successfully reconciles the Theory of Relativity with Quantum Mechanics. From the perspective of the photon, it simply quantum tunnels to its destination experiencing NO passage of time while it is "traveling" to its destination. From our perspective mired in space-time, it took that same photon 13.2 billion years to reach us. This perspective-shift from the quantum realm to the physical realm successfully reconciles quantum mechanics with the theory of relativity. Does it not? We are talking about the same photon but from different perspectives. From its timeless perspective, a photon quantum tunnels or teleports to its destination. From our space-time perspective, it is limited by the speed-of-light. It's the same photon but different perspectives; and, it works! It matches with reality. This observation and reality reconciles Quantum Mechanics with the Theory of Relativity. It's all relative depending upon the perspective of the human being or the perspective of the photon. They see and experience the same things differently.

The Theory of Relativity is the pinnacle of Classical Physics, Materialism, Naturalism, or Classical Realism. Remember, according to the Theory of Relativity itself, at the speed-of-light, time stops or time goes to zero, distance or length contracts to zero, mass or resistance to acceleration goes to zero, and the speed or velocity goes to infinity. At the speed-of-light, everything goes to the limit, which means that everything integrates! It integrates, or transmutes, or transforms. It becomes something different than it was before. At the speed-of-light, we switch from Classical Physics over to Quantum Mechanics or Instantaneous Action at a Distance. Mass cannot be accelerated or pushed faster than the speed-of-light; but, the Gods or Controlling Psyches can quantum tunnel it instantaneously from one end of the universe to the other, if they choose to do so. One of Einstein's biggest blunders was to reject Quantum Mechanics, Quantum Tunneling, and Instantaneous Action at a Distance. Einstein's biggest blunder was to force himself to remain within the constraints of Classical Realism, Materialism, or Naturalism. Einstein's errors became our errors for a hundred years or more. The Theory of Relativity points us to the truth; but, Einstein refused to jump past the speed-of-light limitation of the Theory of Relativity into the infinite velocity and zero distance of Quantum Mechanics.

These truths and observations led to my discovery of the Quantum Law of Thermodynamics. So, what is the Quantum Law of Thermodynamics? Well, at the speed-of-light, time stops according to the Theory of Relativity which means that any aging process stops, which also means that entropy stops. In other words, there is NO entropy, or thermodynamics, or aging process at the quantum level in the Quantum Realm. Instead, energy or psyche is syntropic in the Quantum Realm, which means that energy or psyche is always conserved at the quantum level in the Quantum Realm. Energy or psyche is eternal and everlasting in the Quantum Realm or Psyche Realm – without a beginning of days or an end of years.

https://quantum-neuroscience.com/the-quantum-law-of-thermodynamics/

https://syntropy.site/syntropy-must-exist/

The greatest blunder of any Materialist, Naturalist, Nihilist, Classical Physicist, Darwinist, or Atheist is to assume that the Second Law of Thermodynamics applies at the quantum level in the Energy Realm, Spirit World, or Quantum Realm. They are wrong. There is NO entropy at the quantum level in the Transdimensional Quantum Realm or Spirit World. No psyche or spirit body has ever been observed freezing to death, because it can't. Psyche and the energies within spirit matter are eternal and everlasting. They are syntropic. They cannot be made, and they cannot be destroyed. Entropy ceases to exist at the speed-of-light, because time stops at the speed-of-light. Entropy or the second law of thermodynamics is exclusively a physical phenomenon. It does not apply and does not exist in the Quantum Realm or the Spirit World. For me personally, this was a huge and monumental scientific discovery.

The Quantum Law of Thermodynamics states that ALL of the energy is available all of the time at the quantum level and the psyche level. In other words, anergy, entropy, unavailable energy, thermodynamics, heat death, death, and the second law of thermodynamics DO NOT EXIST in the Quantum Realm. This is what has been experienced and observed by Out-of-Body Travelers and Near-Death Experiencers. The BEST and most convincing Proof of Heaven is to go there and see it for yourself, or to choose to trust someone who has.

Well, that's a major discovery, isn't it? But, you'll never find out about it from the Materialists, Naturalists, Darwinists, Nihilists, Behaviorists, and Atheists because they have rejected it. According to these people, the non-physical Psychic

97

Realm or Quantum Realm does not exist. According to these people, Psyche, Energy, Life Force, or Light does not exist. According to these people, only physical matter exists, which means that only entropy exists. This also means that these people really truly believe that entropy and entropic physical matter are conserved. The Materialists and Naturalists erroneously teach and believe that entropy is conserved or that entropy is eternal. These people teach and believe that only entropic physical matter exists; therefore, these people erroneously teach and believe that physical matter and entropy are conserved. Entropy is death. These people literally teach and believe that entropy or death is conserved. Do they not? That's what they really believe!

They are wrong.

The fact that these people have been proven wrong by Science led me to the Ultimate Law of Thermodynamics. So, what is the Ultimate Law of Thermodynamics?

The Ultimate Law of Thermodynamics differentiates between what is being conserved and what is not conserved. Psyche, Life Force, Energy, Intelligence, Information, and Syntropy are conserved at the quantum level; whereas, physical matter, death, and entropy are not conserved. The First Law of Thermodynamics tells us that the energy is constant, and that the energy is conserved – not the FORM of that energy. Energy can be transformed from one form to another, which means that the FORM is never conserved. The Ultimate Law of Thermodynamics teaches that the FORM of the energy is never conserved. Only the energy or the psyche is conserved. Consequently, physical matter, death, and entropy are NOT conserved at the quantum level. In fact, they don't really exist at the quantum level. There is no entropy or death in the Quantum Realm or Energy Realm because everything is conserved in the Syntropic Realm. Everything is syntropy in the Quantum Realm, meaning that it's always conserved. It can change FORM at will, but its energy and psyche are always conserved.

http://ultimate-law-of-thermodynamics.com/defined/

I used to be a Physicalist, Naturalist, Nihilist, and Atheist; but, I'm no longer satisfied being dumb and blind without a soul or a mind. Nowadays, I have to know what things are and how they work. Like I said, my purpose is to set everything straight; and, that process began by setting myself straight.

Psyche is the innate intelligence within all the different forms of energy or quanta. Quanta are organized packets of energy. Quanta or "particles" are waves of energy moving through a quantum field. Someone Psyche has to start the quanta moving in the first place or nothing would ever happen. Syntropy is the Conservation of Energy. Syntropy is the First Law of Thermodynamics. Syntropy is the opposite of entropy. Psyche or Energy is syntropic, which means that Psyche or Energy is conserved. There is NO entropy and NO resistance to acceleration at the quantum level in the Quantum Realm.

In contrast, entropic physical matter is made from different forms of energy. The form is NEVER conserved. The Ultimate Law of Thermodynamics states that the form is never conserved. Energy is infinitely malleable or infinitely transformable. Each psyche has a certain amount of energy that's under its control; and, that controlling psyche can form the energy under its control into anything that it wants that energy to be. At some point in the future – if not before – when all of the physical matter in our physical universe has reached thermal equilibrium or heat death, the Controlling Psyches who made the physical matter in the first place can simply convert that physical matter back into raw reusable energy, and then use that energy again to make quantum fields and another physical universe. There's no such thing as heat death in the Quantum Realm. Heat or thermal disequilibrium is irrelevant at the quantum level, where everything is made and sustained. Again, NO psyche, intelligence, or spirit body has ever been observed freezing to death, because it can't freeze to death. There is NO death or entropy in the Quantum Realm. This is what has actually been experienced and observed.

Physical matter, entropy, and time are made or caused to begin, which means that they are NOT conserved. Anything that has a beginning is something that cannot be conserved. The Gods or Controlling Psyches or Controlling Intelligences can transform entropic physical matter into a different form of energy anytime they choose to do so. This is the Ultimate Law of Thermodynamics. It states that Psyche, Energy, Life Force, and Syntropy are conserved; whereas, entropic physical matter is NOT conserved. The Ultimate Law of Thermodynamics simply differentiates between what is being conserved and what is NOT conserved. Physical matter is subject to the Second Law of Thermodynamics; whereas, spirit matter, energy, and psyche ARE NOT. Spirit matter, energy, and psyche really can't die and cease to exist. They can transform themselves or reshape themselves; but, they can't die and cease to exist. The First Law of Thermodynamics, Syntropy, or the Conservation of Energy and Psyche reigns supreme in the Quantum Realm whereas the Second Law of Thermodynamics ceases to exist in the Quantum Realm. This is what has actually been experienced and observed; and, it's time for us to finally put it into some kind of Law or Axiom.

The Ultimate Law of Thermodynamics is a necessary adjustment or refinement to the classical laws of thermodynamics because the Materialists, Naturalists, Darwinists, Nihilists, and Atheists erroneously teach and believe that physical matter, death, and entropy are conserved because these people erroneously teach and believe that entropic physical matter is the ONLY thing that exists. The Ultimate Law of Thermodynamics corrects that flaw by stating that entropy, death, and physical matter are NOT conserved. Therefore, if there is any conflict between the first law of thermodynamics and the second law of thermodynamics, the Ultimate Law of Thermodynamics declares the first law of thermodynamics the winner every time.

The Ultimate Law of Thermodynamics states that Psyche, Energy, Life Force, and Syntropy are always conserved at the quantum level; whereas, entropy, death, and physical matter are NOT conserved and actually cease to exist at the quantum level. This is what has actually been experienced and observed by Out-of-Body Travelers and Near-Death Experiencers. Science is observation and experience, not wishful thinking and philosophical speculation. The Ultimate Law of

Thermodynamics gives precedence to whatever has been experienced and observed. The Ultimate Law of Thermodynamics gives precedence to the Conservation of Energy and Psyche, both at the quantum level and at the physical level. In harmony with the Quantum Law of Thermodynamics, the Ultimate Law of Thermodynamics states that entropy or the second law of thermodynamics ceases to exist at the quantum level in the Quantum Realm or Spirit World.

Think about it logically. What have we observed? Take any physical system that has achieved thermal equilibrium or maximum entropy. Its underlying quantum fields continue to exist and continue to be conserved. The quantum fields don't die simply because the physical system has died. The quantum fields can't die or cease to exist, because they are pure energy and pure intelligence. The second law of thermodynamics has absolutely NO effect on the quantum fields. The quantum fields can be transformed into something else; but, they can't die and cease to exist. This is what has actually been experienced and observed.

https://syntropy.website/the-law-of-psyche/

https://psyche-ontology.com/psyche-and-energy-are-essentially-synonymous/

Truth sustains truth. Within Science, the false is falsified by the truth; and, the truth is repeatedly experienced and observed.

According to the Law of Psyche, every Psyche, Intelligence, Life Force, or Consciousness has a certain amount of energy that's under its control. The controlling psyche can form the energy under its control into anything that it wants that energy to be. Energy is infinitely malleable or infinitely transformable, because energy or psyche is always conserved. Energy is the ultimate perpetual motion machine. It never wears out because it is always conserved.

The Law of Psyche states that Psyche, Consciousness, Life Force, or Intelligence has been experienced and observed; therefore, it is real, truly exists, and must be explained and explainable by Science, when our Science finally has enough depth to do so.

Psyche or Consciousness is the innate intelligence within all the different forms of energy. The Ultimate Law of Thermodynamics states that Psyche, Intelligence, Life Force, Energy, the Quantum Realm, and Syntropy are conserved; whereas, physical matter, death, and entropy are NOT conserved. The Ultimate Law of Thermodynamics simply differentiates between what is being conserved and what is NOT conserved. It all fits together perfectly because it is true. It is true because it has actually been experienced and observed.

https://ultimate-law-of-thermodynamics.com/

The Quantum Law of Thermodynamics states that heat death is impossible at the quantum level because the second law of thermodynamics or entropy doesn't exist at the quantum level. The Quantum Realm is an Isothermal Realm which means that in the Quantum Realm or Spirit World all of the energy is available for use all of the time, and that energy is always conserved. There is NO anergy or entropy in the Quantum Realm. Heat death is impossible at the quantum level. Consequently, the Quantum Law of Thermodynamics states that death is impossible at the psyche level. Your psyche, life force, or intelligence was not made, and it cannot be destroyed. Your Psyche cannot die. It is always conserved. This is what has been experienced and observed.

https://quantum-law-of-thermodynamics.com/

You see, the fermions comprise matter and the bosons communicate or transfer the fundamental forces. Alas, both fermions and bosons are made from energy by some controlling psyche. The controlling psyche or controlling intelligence chooses whether the energy under its control will be fermions or bosons. David Bohm stated that physical matter is frozen light or frozen energy. It's energy that's slowed down or brought to a standstill within locality or space-time. In contrast, bosons are quantum waves or thoughts. They travel from one psyche to the next instantaneously communicating information or forces. Massless bosons have NO speed limit, which means that they are capable of Instantaneous Action at a Distance at the quantum level. It's the fermions that have been slowed down and made subject to physical limitations, by having some type of clock or entropy or aging process placed within them.

Energy is the ultimate perpetual motion machine – you get the same amount out as what you put in – because energy or psyche is always conserved. But these "perpetual motion machines" are only realistically functional at the quantum level. They are impossible at the physical level thanks to the Laws of Thermodynamics that exist at the physical level. The physical has physical limitations. The quantum doesn't. That's something that Einstein and the other materialistic and naturalistic geniuses never realized. These people truly believed that the speed-of-light limitation applied just as much to the quantum realm as it does to the physical realm; and, they were wrong, and have subsequently been proven wrong.

PBS NOVA "Einstein's Quantum Riddle – Instantaneous Action at a Distance Proven True Once Again" - 1/9/19

https://www.youtube.com/watch?v=ZEhoR-LCDlo
https://www.amazon.com/gp/video/detail/B07MJK3GMV/

Einstein's biggest blunders became our biggest blunders.

Every time that Instantaneous Action at a Distance, Non-Locality, Non-Physicality, or Quantum Entanglement is proven true or verified during our science experiments, Materialism, Naturalism, Darwinism, Nihilism, Behaviorism, Determinism, and Atheism are FALSIFIED. The false is falsified by the truth; and, the truth is repeatedly experienced, observed, witnessed, and verified. Instantaneous Action at a Distance, Telepathy, Quantum Tunneling, Quantum Entanglement, Non-Locality, Non-Physicality, Psychic Interaction, Instantaneous Transpersonal Communication, or Infinite Velocity at the Quantum Level is repeatedly verified during our science experiments; and, its verification falsifies the claims of Materialism, Naturalism, and their derivatives which state that these things do not exist. The repeated verification of Quantum Phenomena or Psychic

Phenomena or Action at a Distance FALSIFIES Materialism, Naturalism, Darwinism, Nihilism, Behaviorism, Determinism, and Atheism which claim that these things do not exist.

Einstein's biggest blunder, and the one that most of our scientists continue to make, is the chosen belief that physical limitations such as entropy and the speed-of-light apply at the quantum level in the Quantum Realm or Spirit World. They don't! There is NO entropy or thermodynamics at the quantum level where energy or psyche is concerned. Instead, psyche or energy is always conserved at the quantum level.

Furthermore, there are no speed limits enforced at the quantum level. At the quantum level in the Spirit World, it has been observed that spirit bodies and quantum objects are capable of quantum tunneling or teleporting across the universe instantaneously if they choose to do so. There are no speed limits in the Quantum Realm, which means that physical limitations do not apply in the Spirit World. Things don't age, and things don't wear out and die in the Spirit World. Psyche or energy is conserved in the Quantum Realm, which means that it doesn't age or wear out. Entropy or death, as we know it, doesn't exist at the quantum level in the Quantum Realm.

This is it. This is what I have been searching for, for all of my life.

As a scientist, one of the biggest blunders I ever made was to assume that Natural Selection or Evolution can design and create. I was wrong. Natural Selection or Evolution is entropy or death. Entropy and death cannot design and create. They can only destroy. Natural Selection doesn't do anything except sit around and wait for you to die. In the case of Natural Selection, "selection" is death. It is natural for us to die or get selected against. Natural Selection doesn't touch our genes. It can't, because it isn't a physical objective being. It also doesn't have a soul, intelligence, consciousness, life force, or a mind. Only psyche or intelligence can design and create. This is what we have actually experienced and observed. Natural Selection doesn't have the power nor the ability to do anything. It's simply a philosophical concept and nothing more. Go figure! I don't know if Einstein fell for this particular blunder, but most of our modern-day scientists have.

https://evolution-is-entropy.com

Remember, genetics and genomes were designed to prevent macro-evolution from happening; and, they work as they were designed to do. Remember, Louis Pasteur falsified spontaneous generation, abiogenesis, and macro-evolution in 1859.

The original creative powers – the ones that existed BEFORE the first particle of physical matter was designed and made – were found at the quantum level within Consciousness, Intelligence, Life Force, or Psyche. This is obviously true. In fact, my research into Consciousness, Quantum Field Theory, and Quantum Mechanics eventually led to my discover of what I call The Obvious Law of Physics, which states that the smaller dwells within and controls the larger. Energy is infinitely malleable. You can form energy into anything that you want it to be. Who or what within the energy itself decides whether to form that energy into an electron or a quark? Someone conscious and intelligent is making that choice. It doesn't just happen randomly. There is a choice being made. Someone intelligent and psychic within the energy actually knows how to form the energy under its control into either an electron or a quark. It knows how an electron or a quark is supposed to act, and how they are supposed to interact. There's intelligence there and consciousness there. Energy carries within it intelligence, consciousness, life force, light, intention, perception, attention, and information.

According to the Law of Psyche, every psyche has a certain amount of energy that's under its control; and, that controlling psyche decides what form the energy under its control will assume. That controlling psyche KNOWS what an electron is and what a quark is, and that controlling psyche DECIDES whether the energy under its control will be an electron, or a quark, or part of a quantum field, or a quantum wave, or a photon. Remember, according to the Obvious Law of Physics, the smaller dwells within and controls the larger. There is some type of intelligence, life force, consciousness, or psyche within each electron, photon, field, force, and photon who actually decided that the energy under its control would function as one of these things and be one of these things.

There are different levels of reality, or different phases or dimensions in the Spirit World. It's layered, and it has depth. Objects out of phase with each other can occupy the same space at the same time without colliding, interacting, or interfering with each other. This is what has actually been experienced and observed. The Quantum Realm, Non-Physical Energy Realm, Psychic Realm, or Spirit World is right here within us; and, it is completely different than our physical realm with its obvious physical limitations and speed-limits. Instantaneous Action at a Distance, Conservation of Psyche or Energy, a complete lack of Entropy or Thermodynamics, and Potential Infinite Velocities or Quantum Tunneling are an integral part of the Quantum Realm. Einstein's biggest blunder was to assume that physical limitations such as entropy, the aging process, death, the theory of relativity, and his speed-of-light limitation apply to the Quantum Realm or the Spirit World. It's a blunder or a mistake that our modern-day physicists continue to make because they don't want Action at a Distance and the Immortality of their Psyche to be true.

I personally believe that Einstein's biggest mistake was his ongoing rejection of Quantum Mechanics, along with his attempt to promote Classical Realism, Materialism, and Naturalism instead. Materialism and Naturalism have been falsified by Quantum Mechanics and Quantum Field Theory; whereas, Quantum Mechanics and Quantum Field Theory have been repeatedly verified and proven true, over and over again in a multitude of different ways. It's a mistake to choose Classical Realism, Materialism, Naturalism, and their derivatives over the immaterial and supernatural Quantum Field Theory. Materialism and Naturalism have lost every time.

Remember, Einstein's biggest blunders became our biggest blunders. I find all of this fascinating – figuring out what I and Einstein got wrong, and why we got it wrong. After a hundred years of verifying Quantum Mechanics and falsifying Naturalism or Classical Physics, we are now in a much better position to see what Einstein got wrong than they were back then. I like Quantum Mechanics and Quantum Field Theory because they make me think and learn.

Space Is Not Empty

https://www.youtube.com/playlist?list=PLsPUh22kYmNAHB1W2_Ka2F83sOb dczwKr

Quantum Mechanics

https://www.youtube.com/playlist?list=PLsPUh22kYmNCGaVGuGfKfJI-6RdHiCjo1

Quantum Field Theory

https://www.youtube.com/playlist?list=PLsPUh22kYmNBpDZPejCHGzxyfgitj2 6w9

Quantum Tunneling Is Faster than Light

https://www.youtube.com/watch?v=-IfmgyXs7z8

Quantum tunneling is faster than light, except at the physical level, where everything was designed to be slower than the speed-of-light. It's a good thing that God gave physical matter a short De Broglie Wavelength; otherwise, the atoms within your physical body could quantum tunnel away from you at will, and you would literally dissolve into thin air. It's a good thing that the atoms in your physical body can only quantum tunnel short distances and that they are subject to physical laws and physical restrictions. If the atoms within your physical body had a long De Broglie Wavelength or an infinite De Broglie Wavelength, then they could quantum tunnel to the other side of the universe instantaneously at will, leaving your spirit with no physical body or physical world to live within. God put the physical laws and physical restrictions into place for a good reason.

The Impossibility of Perpetual Motion Machines

https://www.youtube.com/watch?v=rckrnYw5sOA

Perpetual motion machines are physically impossible at the physical level because the conservation of energy or the first law of thermodynamics states that the best we can do is to get out precisely what we put in. In other words, it's impossible to get more energy out than what we put in at the physical level, without tapping into some kind of hidden resource at the quantum level – think of a nuclear explosion which is a quantum phenomenon. When it comes to a nuclear explosion, we tend to get more energy out than what we put in, and that's thanks to quantum mechanics or energy mechanics. $E = mc^2$. Typically, though, when it comes to machines at the physical level, the combined influence of friction, gravity, and entropy interfere, which means that we always get out less energy than what we put in when it comes to our physical machines.

The only way to get true perpetual motion is at the quantum level where entropy and thermodynamics don't exist and where energy or psyche is always conserved. If we could tap into the Spirit World or the Quantum Realm, we would never run out of energy because energy or psyche is always conserved at the quantum level and there are no thermodynamics to have to deal with at the quantum level.

It's pretty obvious that this is the way things really work; but, it wasn't the least bit obvious when I was a Materialist, Naturalist, Nihilist, and Atheist. I couldn't see it, until I saw it. There are a lot of brilliant physicists who still can't see it because they don't want it to be true. They are banking on the heat death of the universe instead.

Back in 2012, I was a Materialist, Naturalist, Nihilist, and Atheist. Then I started interacting with these people online; and as the months passed by, I started to notice that these people were lying to us and many of them were deliberately trying to trick us and deceive us. It was obvious to me; and, it was also obvious that if their philosophies, theories, and ideas were in fact correct, true, and right as they claimed they were, then there should be no need for them to lie to us, trick us, or deceive us in order to get us to believe what they wanted us to believe.

When I realized that the Scientific Naturalists, Darwinists, Nihilists, Behaviorists, Determinists, and Atheists are lying to us and trying to trick us and deceive us, that's when they began to lose me. Eventually, I decided to go in search of the truth instead. The truth ended up being noticeably different than what the professional Darwinists and Naturalists were claiming that it is. The more I studied, the more obvious it became that these people were lying to us and were deliberately trying to trick us and deceive us. I was more interested in the truth than in having these people like me. I wanted to know how everything really works.

Eventually I realized that the Materialists, Naturalists, Darwinists, Nihilists, and Atheists have no observations or experiences to support their claims that the non-physical does not exist, the supernatural or transdimensional does not exist, intelligence or psyche does not exist, the quantum or the supernatural or the spiritual does not exist, a pre-mortal life and an after-life do not exist, and God or the Being of Light and Love do not exist.

Instead, ALL of the observations and experiences were on the side of Quantum Mechanics, Energy Mechanics, Intelligence or Psyche, Choice or Psyche Determinism, Action at a Distance, Non-Locality or Non-Physicality, Quantum Field Theory, and the observed and experienced existence of the resurrected Biblical God Jesus Christ. It is obvious to me that Quantum Fields are non-physical and pre-physical in nature and origin, which means that God or the Controlling Psyches had to design and make the Quantum Fields BEFORE they could make and sustain physical matter.

It also became obvious to me that our resurrected Lord Jesus Christ, or that Being of Light and Love experienced while out-of-body during Near-Death Experiences, has in fact been experienced and observed. Furthermore, it is obvious that Intelligence, Consciousness, Life Force, Psyche, or Soul has in fact been experienced and observed. If you can read and understand this, then it is obvious that you are Intelligent, Conscious, Alive, and have some type of Psyche or Soul within you somewhere at some level.

It was then that I developed a new Philosophy of Science that is based directly upon observation and experience and verification. I called it Science 2.0. Science 2.0 is based upon everything that has been experienced and observed. If it has been experienced, observed, or witnessed, then it qualifies as a part of Science 2.0. Soon I noticed that the false is always falsified by the truth; and, the truth is repeatedly experienced and observed.

Science 2.0: I Upgraded My Science

https://www.amazon.com/gp/product/B0771K6WTX/

Then, I was well on my way to developing a new model of reality, the Ultimate Model of Reality, which is a model of reality that is based upon what has actually been experienced and observed. If it hasn't been repeatedly witnessed, experienced, and observed, then it is worthless and isn't a part of reality as we know it and have experienced it. Materialism, Naturalism, Darwinism, Chemical Evolution, Design and Creation by Natural Selection, Spontaneous Generation, Abiogenesis, Macro-Evolution or Cats Giving Birth to Dogs, Non-Existence, and Creation Ex Nihilo or Atheism are NOT a part of reality as we have experienced it to be. These falsified philosophies are wishful thinking or a denial of reality rather than actual observations and experiences. That's what I experienced and observed. It's time for a new model of reality that actually matches with reality.

Remember, genetics and genomes were designed to prevent macro-evolution from happening; and, they work as they were designed to do. Remember, Louis Pasteur falsified spontaneous generation, abiogenesis, and macro-evolution in 1859.

David Bohm is probably my most favorite theoretical physicist; but, it takes a bit of effort to translate what he is talking about into ordinary everyday English.

https://epdf.tips/the-essential-david-bohm.html

https://philosophy-of-science.com/The-Essential-David-Bohm

With his implicate order, David Bohm opened the door to the Quantum Realm and started to explain how it really works.

What we are suggesting here is that the implicate order, the non-physical order, the energy order, the psyche order, the spiritual order, or the quantum order be taken as fundamental. What Bohm is repeatedly suggesting is that we start with the truth; and, the truth is that the physical is made from the non-physical by Consciousness or by Someone Psyche. In contrast, all the Materialists, Naturalists, Nihilists, Darwinists, Behaviorists, and Atheists can tell us is that the non-physical does not exist. That's worthless! Claiming that the non-physical doesn't exist doesn't tell us what it is or how it works.

According to Bohm, the essence of consciousness resides within the Holomovement or the Implicate Order, which is the quantum level or the spirit world. It has to reside someplace, because consciousness has been experienced and observed. It's real. It truly exists. Furthermore, the Explicate Order or our physical reality unfolds or explicates from this much deeper Holomovement or Implicate Order. The physical is built from the quantum, or the spiritual, or the non-physical, just like Quantum Field Theory states and just like the Biblical God Jesus Christ tries to teach us. It would take quite a bit of Intelligence, Information, or Psyche to design and create a quantum field or a holomovement and then convince the energy within it to act that way.

Bohm's Holomovement is synonymous with his Implicate Order, which are synonymous with quantum fields and quantum mechanisms. Quantum Mechanics explains how energy, or psyche, or light acts at the quantum level or the implicate level.

Scientific Axiom: Energy or light can take on any form that it desires to assume, including spirit matter and physical matter. Energy or psyche is always conserved, but it can change form at will. This is what has been experienced and observed.

Scientific Observation: Anything that is obviously organized obviously has an Organizer who organized it.

Scientific Observation: Quantum fields were obviously organized from raw chaotic energy.

Scientific Conclusion: Therefore, quantum fields obviously have an Organizer who organized them and taught the energy within them to act that way or to form that way.

David Bohm was teaching these very same principles back in 1980. He wrote:

> **Indeed, if one applies the rules of quantum theory to the currently accepted general theory of relativity, one finds that the gravitational field is also constituted of such "wave-particle" modes, each having a minimum "zero-point" energy. As a result, the gravitational field, and therefore the definition of what is to be meant by distance, cease to be completely defined. As we keep on adding excitations corresponding to shorter and shorter wavelengths to the gravitational field, we come to a certain length at which the measurement of space and time becomes totally undefinable. Beyond this, the whole notion of space and time as we know it would fade out, into something that is at present unspecifiable. So, it would be reasonable to suppose, at least provisionally, that this is the shortest wavelength that should be considered as contributing to the "zero-point" energy of space.**

> **When this length is estimated it turns out to be about 10^{-33} cm. This is much shorter than anything thus far probed in physical**

experiments (which have got down to about 10^{-17} cm or so). If one computes the amount of energy that would be in one cubic centimeter of space, with this shortest possible wavelength, it turns out to be very far beyond the total energy of all the matter in the known universe.

What is implied by this proposal is that what we call empty space contains an immense background of energy, and that matter as we know it is a small, "quantized" wavelike excitation on top of this background, rather like a tiny ripple on a vast sea. (David Bohm, *The Essential David Bohm*, p. 98.)

Imagine it! According to Zero-Point Theory, ALL of the energy needed to make ALL of the physical matter in our observable physical universe can be found or contained within a single cubic centimeter of space at the quantum level.

We have the truth on the one hand and the lie on the other hand. Most scientists have chosen to believe in the lie. However, the truth is that it's all made from energy or light – what the new agers call the Quantum Sea of Light. We'll never run out energy that can be converted into physical matter. And, whenever physical matter reaches thermal equilibrium or maximum entropy, the psyches controlling it can turn it back into raw usable energy anytime they choose to do so. In other words, entropy is a non-starter. Entropy is an illusion. There is more energy in one cubic centimeter of space at the quantum level than in all of the physical matter in the whole universe combined. All it takes is the energy of one cubic centimeter of space at the quantum level to make the whole physical universe that we see around us today. Are you starting to see it yet?

There is no such thing as heat death, or entropy, or running out of energy where the Zero-Point Field of Light is concerned. At the quantum level, there's enough energy in one cubic centimeter of space to make a whole other physical universe just like this one, at will. In other words, if you are a Psyche with a cubic centimeter of space under your complete control, you have enough energy under your control to make a whole other physical universe like this one out of the energy that is under your control. That's completely different than what the Materialists, Naturalists, Nihilists, and Atheists are teaching us when they tell us that we and the universe are going to end in entropy or heat death, and that we cease to exist when we die.

So, who is right and who is wrong? Which model has actually been seen, observed, experienced, and experimentally verified? Do we observe the existence of Psyche or Intelligence; or, do we observe the non-existence of Psyche, Consciousness, or Intelligence? Answer that, and you figure out how everything really works! As I see it, in order to be an Observer or a Scientist, you actually have to be conscious or intelligent, which means that you have to have some kind of Psyche or Life Force in there somewhere, or you couldn't be an Observer nor a Scientist.

Back in 2012, I was a Materialist, Naturalist, Nihilist, and Atheist; but, I was also an open-minded scientist who was looking for the truth. All I wanted was the truth, and I eventually got tired of these people lying to me. After interacting with these people online for two or three years, I started to realize that they were lying to us and trying to trick us and deceive us. I started to see through the charade. Eventually, I got to the point where I wanted to find out what they were getting wrong. So, what is the thing that Einstein, the Materialists, Naturalists, Atheists, Darwinists, Nihilists, Classical Physicists, and Mechanists are getting wrong?

As I see it at this point in time (2019), the main problem that these people have is that they don't realize that space is Multi-Phasic. The Multiphasic nature of space is what they are constantly getting wrong. Only a few of the Quantum Physicists like David Bohm, Henry P. Stapp, and Pim Van Lommel are in fact getting it right. Everyone else has been getting it wrong.

Again, the answer and the truth are hidden in what David Bohm states here:

When this length is estimated it turns out to be about 10−33 cm. This is much shorter than anything thus far probed in physical experiments (which have got down to about 10−17 cm or so). If one computes the amount of energy that would be in one cubic centimeter of space, with this shortest possible wavelength, it turns out to be very far beyond the total energy of all the matter in the known universe.

What is implied by this proposal is that what we call empty space contains an immense background of energy, and that matter as we know it is a small, "quantized" wavelike excitation on top of this background, rather like a tiny ripple on a vast sea. (David Bohm, *The Essential David Bohm*, p. 98.)

How is it possible for one cubic centimeter of space to contain enough energy to make ALL of the physical matter in our known universe? Isn't that an oxymoron or a contradiction in terms? It is, if physical matter and our physical universe are the only thing that exists, as the Materialists, Naturalists, Darwinists, Nihilists, Mechanists, and Atheists claim. However, these people are always wrong, and always proven wrong.

So, what gives? What are they always getting wrong?

The answer is that these people don't realize and accept the fact that Space is multi-use or multi-phasic. Space has depth, or levels, or dimensions, or phases. There's enough energy in a centimeter of space at the quantum level or the zero-point level to make all of the physical matter that we see around us here at the physical level. Space is comprised of different levels of reality, different phases, or different dimensions. Space is not all one thing, and it's not completely empty as the Materialists and Naturalists claim. At the quantum level in the Spirit World, space is chock full of energy. The Spirit World is right here within us. The dark matter and dark energy are right here within us. The gravity waves and the quantum waves and the psychic telepathic waves are right here within us. It's all taking place within the same space at the same time, just at different levels of existence. There's a psyche level, there's a quantum level, and there's a physical level.

The FLAW and WEAKNESS of the Materialists, Naturalists, Darwinists, Nihilists, and Atheists is their claim that physical matter or the physical level is the only thing that exists. These people really truly teach and believe that entropy, death, and therefore entropic physical matter are conserved. They are demonstrably wrong. Space is multiphasic. It contains multiple levels within it. Einstein's biggest blunders became our biggest blunders. Remember, Quantum Field Theory and Quantum Mechanics FALSIFY Materialism, Naturalism, and their derivatives such as Atheism (Creation Ex Nihilo By Nihilo), Nihilism (Space Is Empty), and Darwinism (Spontaneous Generation or Abiogenesis).

Within his theoretical physics, David Bohm talks about an explicate order or a physical level of existence, an implicate order or a quantum level of existence, and a super-implicate order or an organizing principle which would in fact be a psyche level of existence. These are different levels or phases of reality that exist within the same space at the same time. The physical level or explicate order appears to be, from our perspective at the physical level, rarefied or spread-out or limited in energy. Almost all of the energy is hidden at the lower levels where we can't get at it. In contrast, Psyche and Spirits can get at that hidden energy all the time, because at their level the energy is available all the time. Energy is exergy (fully available) at the quantum level and the psyche level. There is NO entropy or anergy at the quantum level. This is the thing that the Materialists and Naturalists don't realize. There are NO physical limitations at the quantum level within Bohm's implicate order.

At the quantum level and the psyche level, energy is isothermal and non-thermal and exergic, which means that all of the energy is available all the time and also means that the energy is always conserved or preserved. At the quantum level there is no thermodynamics, which means that the second law of thermodynamics or entropy does not exist at the quantum level. Only the conservation of energy or the conservation of psyche exists at the quantum level. At the quantum level or the psyche level, energy is exergy, which means that there is NO entropy at the quantum level or the psyche level. This is the thing that Einstein, and the Materialists and Naturalists are getting wrong. They have erroneously chosen to believe that the physical limitations such as the speed-of-light and entropy apply at the quantum level and the psyche level, and they don't.

Thus, the super-quantum potential expresses the activity of a new kind of implicate order. This implicate order is immensely more subtle than that of the original field, as well as more inclusive, in the sense that not only is the actual activity of the whole field enfolded in it, but also all its potentialities, along with the principles determining which of these shall become actual.

I was in this way led to call the original field the first implicate order, while the super-quantum potential was called the second implicate order (or the super-implicate order). In principle, of

course, there could be a third, fourth, fifth implicate order, going on to infinity, and these would correspond to extensions of the laws of physics going beyond those of the current quantum theory, in a fundamental way. But for the present I want to consider only the second implicate order, and to emphasize that this stands in relationship to the first as a source of formative, organizing, and creative activity.

It should be clear that this notion now incorporates both of my earlier perceptions – the implicate order as a movement of outgoing and incoming waves, and of the causal interpretation of the quantum theory. So, although these two ideas seemed initially very different, they proved to be two aspects of one more comprehensive notion. This can be described as an overall implicate order, which may extend to an infinite number of levels and which objectively and self-actively differentiates and organizes itself into independent sub-wholes, while determining how these are interrelated to make up the whole.

Moreover, the principles of organization of such an implicate order can even define a unique explicate order, as a particular and distinguished sub-order, in which all the elements are relatively independent and externally related. To put it differently, the explicate order itself may be obtainable from the implicate order as a special and determinate sub-order that is contained within it.

All that has been discussed here opens up the possibility of considering the cosmos as an unbroken whole through an overall implicate order. Of course, this possibility has been studied thus far in only a preliminary way, and a great deal more work is required to clarify and extend the notions that have been discussed in this paper. (David Bohm, *The Essential David Bohm*, pp. 196-197.)

Space had depth. Space has layers. Space is Multi-Phasic. Space at the physical level contains tons of non-physical stuff within it! Space at the physical level contains lots of non-physical layers within it. Energy is mostly non-physical. The only time energy is physical is when the Gods or the Controlling Psyches have formed that energy into entropic physical matter. The rest of the time, energy is non-physical. The verified and proven existence of the non-physical FALSIFIES Materialism, Mechanism, Classic Physics, Newtonian Physics, Naturalism, Physicalism, Atomism, Darwinism, Atheism, and their derivatives such as Behaviorism and Determinism.

https://psyche-ontology.com/buhlman/

https://psyche-ontology.com/psyche-experienced-and-observed/

https://psyche-ontology.com/psyche-observed/

This is it. This explains what the Materialists, Naturalists, Darwinists, Nihilists, Behaviorists, Determinists, and Atheists are getting wrong. Physical space contains within it some type of quantum space or spiritual space, which contains within it some kind of psyche space or organizing space or conscious space, which contains within it all of the energy or exergy. It's nested, or layered, or multiphasic; and for all we know, there could be an infinite number of levels at the same time within the same cubic centimeter of space. That's how it's possible for one cubic centimeter of space at the quantum level or zero-point level to contain within it enough energy to make all of the physical matter that we see around us here at the physical level within our physical universe.

This means that there is NO limit to the number of physical universes that the Gods or the Controlling Psyches can make from all of the exergy or energy that exists at the quantum level, and the psyche level, and the infinite number of levels that exist at the zero-point level or the exergy level – where entropy, thermodynamics, and anergy do not exist and cannot exist thanks to the Conservation of Energy that reigns supreme at the super-implicate level or lowest levels of existence.

The destroyed exergy, or expended energy, or used exergy has been called anergy or entropy. Exergy is the energy that is available to be used. We see and experience a lot of entropy or anergy at the physical level that simply doesn't exist at the quantum level. Entropy, or thermodynamics, or the aging process is exclusively a physical phenomenon. It doesn't exist at the quantum level. It's ALL exergy or syntropy at the quantum level. It has to be, or the quantum level would have filled up with entropy eons ago, and we wouldn't and couldn't exist today. Someone Psyche at the quantum level has to KNOW how to do syntropy, or the reversal of entropy, or the conservation of energy; otherwise, quantum fields and physical matter would be impossible to make and sustain today.

Whenever I choose to let Psyche and Quantum Mechanics in to play, I can literally explain everything that comes my way. The explanatory power of Intelligence and Quantum Mechanics is vastly superior to anything that we can get from the Materialists, Naturalists, Darwinists, Nihilists, and Atheists who assure us that Psyche, Intelligence, Consciousness, Quantum Fields, Life Force, Syntropy, and Quantum Mechanics DO NOT EXIST. How does claiming that something does not exist explain what it is and how it works? It doesn't! And, that's why Materialism and Naturalism fail in the end. They lack explanatory power!

Remember, exergy is the energy that is available to be used; and at the quantum level, ALL of the energy is exergy thanks to the Conservation of Psyche or the Conservation of Energy that predominates at the quantum level and the psyche level. There is NO thermodynamics, anergy, or entropy at the quantum level. It doesn't exist at the quantum level. Only the First Law, or the Conservation of Energy, or Syntropy, or Exergy exists at the quantum level. This is the Quantum Law of Thermodynamics.

This is what the Materialists, Naturalists, Darwinists, Nihilists, Mechanists, and Atheists are constantly getting wrong, and this is why their theories, philosophies, and beliefs are constantly being FALISFIED by Quantum Mechanics, Action at a Distance, the Quantum Zeno Effect, Quantum Tunneling, and Quantum

Field Theory. The proven and verified existence of Quantum Field Theory, Action at a Distance, Quantum Mechanics, the Quantum Zeno Effect or Telepathy, Quantum Tunneling or Teleportation, Quantum Phase-Shifting, and the Conservation of Psyche or Energy FALSIFIES Materialism, Naturalism, and their derivatives such as Darwinism and Atheism.

The false is falsified by the truth; and, the truth is repeatedly experienced and observed. Psyche, Intelligence, or Consciousness is constantly being experienced and observed; yet, the Materialists and Naturalists are constantly telling us that it doesn't exist. It's obvious that they are wrong. I'm seeing or observing Intelligence, Consciousness, Self-Awareness, Life Force, or Psyche all around me all the time. It's unavoidable.

My dog is just smart enough to get into trouble all the time, but not smart enough to get out of it every time. The same reality applies to the Materialists, Naturalists, Darwinists, Classical Physicists, Mechanists, and Atheists. They are just smart enough to create a whole host of unsolvable problems, but they aren't smart enough to solve them. Psyche, Intelligence, Life Force, Consciousness, Quantum Mechanics, Quantum Field Theory, the Quantum Zeno Effect, Quantum Tunneling, Quantum Multi-Phasing, Quantum Superposition, and Quantum Field Theory SOLVE everything that the Materialists and Naturalists can't solve with their falsified philosophies and beliefs. This is what has been experienced and observed.

You are going to have to discover these truths for yourself, or go without. The best I can do for you is to show you where to look; but, you are going to have to do the work of studying and learning for yourself. You won't find these truths in our public school system, though, because in the Western World our public education systems are run by the Materialists and Naturalists and the purpose of a public education is to indoctrinate you in the religious dogma of Materialism, Naturalism, Darwinism, Nihilism, and Atheism. If you are not an Atheist by the time you graduate from college, then your college professors have failed in their primary mission where you are concerned. Their ultimate goal is to have you dumb and blind without thoughts or a mind. It's what they do.

Ultimately, the Materialists, Naturalists, Darwinists, Nihilists, Behaviorists, Mechanists, and Atheists weren't able to explain to me what I needed to know and wanted to know, so it was time for me to upgrade my Science to something better. And, I did. With Psyche and Quantum Mechanics under my sway, I can literally explain everything that comes my way. It looks like to me that Quantum Mechanics, Action at a Distance, Psyche, and Quantum Field Theory are here to stay. Materialism and Naturalism are dead. Long reign the new paradigm – Quantum Field Theory and Quantum Mechanics!

It's good to finally have the truth and to know how everything really works.

Mark My Words

Falsifying Einstein's Classical Realism with Non-Locality, Human Choice, Action at a Distance, and Quantum Entanglement

https://ultimate-model-of-reality.com/falsifying-einstein-and-classical-realism/

Everything that Einstein proposed is true at the physical level; but, it doesn't exist at the quantum level in the Spirit World or the Quantum Realm. The Theory of Relativity and Classical Realism is the pinnacle of Classical Physics; but, the theory of relativity, entropy, the aging process, and speed-limits don't really exist at the quantum level and the psyche level, once we have left the physical behind and switched completely over to the Quantum Realm instead. There's an obvious marriage between the quantum and the physical; but, at the truly quantum or the truly supernatural, physical limitations no longer exist and no longer apply. Once the physical has been eliminated from a system, ONLY the quantum remains; and, the rules of Quantum Mechanics are vastly different than the rules of physical mechanics, because there are NO physical limitations or speed-limits in the Quantum Realm or Spirit World. At the quantum level, everything integrates or reaches a limit, transmutes, transforms, and then changes into something completely different than it was at the physical level. At the quantum level, the barriers are broken.

If you want to make a quantum leap in your Science, Logic, Knowledge, and Understanding of Reality, then search for and study ALL of the different things which the Materialists, Naturalists, Darwinists, Nihilists, and Atheists are trying to hide from you. There are thousands of such things, waiting for you to find them and understand them. I encourage you to take that quantum leap with me.

As of 2019, the last loophole for Materialism, Naturalism, Darwinism, and Einstein's Classical Realism has been FALSIFIED; and, Quantum Entanglement or Spooky Action at a Distance has once again been verified. That's officially the end of Materialism and Naturalism. They have been falsified by Quantum Mechanics or Energy Mechanics. They have been falsified by Non-Locality, Action at a Distance, and Quantum Entanglement.

PBS NOVA "Einstein's Quantum Riddle - Entanglement Theory Proved" - 1/9/19

https://www.youtube.com/watch?v=ZEhoR-LCDlo
https://www.amazon.com/gp/video/detail/B07MJK3GMV/

Here are the paraphrased highlights. As scientists, we are officially done with Materialism and Naturalism. Instantaneous Action at a Distance reigns supreme in Science. Distance, Locality, Time, Space, and Entropy cease to exist at the quantum level or the psyche level (energy level) of existence and reality. Space and time are created by entangled particles. Our three-dimensional reality is an illusion or an emergent property of the timeless and spaceless quantum realm or spiritual realm. Entanglement or Action at a Distance is what forms the true fabric of the universe. It's non-physical or non-local. It's based upon Quantum

Mechanics or Energy Mechanics. It has NO physical limitations and NO physical restrictions. The physical laws and physical limitations ONLY exist at the physical level. They cease to exist at the quantum level or the psyche level. The Theory of Relativity has been FALSIFIED or has been limited exclusively to the physical realm. The Theory of Relativity is the ultimate theory of Classical Physics, Materialism, and Naturalism; but, it doesn't exist at the quantum level or the psyche level.

Space, space-time, or distance literally disappears and ceases to exist at the quantum level in the quantum world or the spirit world. In the quantum mechanical world, there is no space, no time, and no distance. There are NO physical limitations.

Understanding Quantum Mechanics will only happen when we put ourselves on the entanglement side, and we stop privileging the world that we can see and touch, and start thinking about the world as it actually is. We must use Quantum Mechanics, Quantum Field Theory, Quantum Tunneling, Psyche or Intelligence or Life Force, and Action at a Distance to FALSIFY Classical Physics, Materialism, Naturalism, Darwinism, Chemical Evolution, Spontaneous Generation, and Creation Ex Nihilo or Atheism; and then, we need to start thinking about the world and the universe as it actually is. We need to stop breaking things down into parts or particles, and start thinking of the universe as one united whole or one universal Matrix of Quantum Fields. We are done with Materialism and Naturalism.

The basic motivation is just to learn how Nature and Reality truly work. We need to know what's really going on, when it comes to the quantum level and Action at a Distance. Time, Space, Distance, and Entropy literally disappear and cease to exist at the quantum level. Einstein said it nicely when he stated that he's not interested in this little detail or that little detail. Einstein just wanted to know, what were God's thoughts when He created the world and our universe? This from NOVA and PBS itself!

Scientific Proof of Non-Locality or Quantum Entanglement

https://www.iflscience.com/physics/quantum-entanglement-proved-to-be-correct-even-billions-of-lightyears-away/

https://journals.aps.org/prl/pdf/10.1103/PhysRevLett.121.080403

https://news.mit.edu/2018/light-ancient-quasars-helps-confirm-quantum-entanglement-0820

https://news.mit.edu/2017/loophole-bells-inequality-starlight-0207

https://www.iflscience.com/physics/physicists-prove-quantum-spookiness-and-start-chasing-schr-dinger-s-cat/

https://www.iflscience.com/physics/one-of-einsteins-major-theories-just-got-disproved-by-a-bunch-of-amateurs/

https://arxiv.org/pdf/1805.04431.pdf

http://liu.diva-portal.org/smash/get/diva2:1209417/FULLTEXT01.pdf

https://news.mit.edu/2017/loophole-bells-inequality-starlight-0207

https://www.iflscience.com/physics/play-this-game-today-to-help-test-the-laws-of-quantum-mechanics/

https://resonance.is/big-bell-test-human-free-will-verifies-quantum-violation-local-realism/

https://physicsworld.com/a/computer-gamers-close-freedom-of-choice-loophole-of-quantum-entanglement/

https://www.nature.com/articles/s41586-018-0085-3

https://thebigbelltest.org/

http://www.mtnmath.com/willbe/rv.pdf

https://theconversation.com/to-test-the-effect-of-gravity-on-quantum-entanglement-we-need-to-go-to-space-27614

https://news.nationalgeographic.com/news/2013/08/130812-physics-schrodinger-erwin-google-doodle-cat-paradox-science/

https://qutech.nl/wp-content/uploads/2017/03/Loophole-free-Bell-inequality-violation-using-electron-spins-separated-by-1.3-kilometres.pdf

https://arxiv.org/pdf/1509.03763.pdf

https://theconversation.com/explainer-quantum-physics-570

https://www.nature.com/news/2001/011129/full/news011129-15.html

https://www.bbc.com/news/uk-northern-ireland-29904682

https://quantumfrontiers.com/2014/11/23/bells-inequality-50-years-later/

http://www.mtnmath.com/whatrh/node81.html

https://theconversation.com/schrodingers-cat-gets-a-reality-check-37278

https://theconversation.com/physicists-prove-quantum-spookiness-and-start-chasing-schrodingers-cat-48190

Einstein's Quantum Riddle

https://www.youtube.com/watch?v=ZEhoR-LCDlo

https://www.amazon.com/gp/video/detail/B07MJK3GMV/

https://www.google.com/search?q=proof+of+quantum+entanglement

https://www.space.com/41569-ancient-quasars-evidence-quantum-entanglement.html

https://curiosity.com/topics/the-worlds-biggest-quantum-entanglement-experiment-proved-einstein-wrong-curiosity/

https://www.cnet.com/news/physicists-prove-einsteins-spooky-quantum-entanglement/

https://www.iflscience.com/physics/quantum-entanglement-proved-to-be-correct-even-billions-of-lightyears-away/

http://www.astronomy.com/news/2018/08/distant-quasars-confirm-quantum-entanglement

https://phys.org/news/2017-08-bell-prize-scientists-spooky-quantum.html

http://news.mit.edu/2018/light-ancient-quasars-helps-confirm-quantum-entanglement-0820

https://gizmodo.com/scientists-finally-prove-strange-quantum-physics-idea-e-1798433666

https://futurism.com/physicists-take-one-large-step-towards-proving-quantum-entanglement

In Science, It's Time to Replace Falsehoods with Truths

This NOVA PBS episode, "Einstein's Quantum Riddle", is officially the end of Materialism, Naturalism, Physicalism, Atomism, and their derivatives.

Now all we need is a similar episode or series from National Geographic providing all the different things that FALSIFY the Theory of Evolution or Darwinism, and we will finally have the truth in our hands from which we can start to build something new and better. They can start with the following:

https://www.youtube.com/watch?v=W1_KEVaCyaA

https://www.amazon.com/dp/B06X9DVRQW/

https://www.youtube.com/results?search_query=Falsifying+Evolution

The Probability of a Protein Forming by Chance

https://www.youtube.com/watch?v=W1_KEVaCyaA
https://evolution-is-entropy.com/wp-content/uploads/2018/05/Probability-of-a-Protein-Forming-by-Chance.zip

The Probability of Making a Protein

Evidence for Creation by Outside Intervention

These videos are worth owning, watching, and keeping. I archived them so that they can't disappear on me while I'm alive.

Falsifying Classical Realism

Anytime that instantaneous Action at a Distance is verified, Materialism, Naturalism, Classical Physics, Classical Realism, the Faster-than-Light Speed Limit, and Einstein's Theory of Relativity are falsified. It's Einstein's Theory of Relativity that is false or inadequate, NOT Quantum Mechanics, Quantum Field Theory, Action at a Distance, and Conservation of Energy. Whenever a choice has to be made between Einstein's Theory of Relativity (Classical Physics) and Quantum Mechanics (Instantaneous Action at a Distance), Quantum Mechanics should be chosen every time because Action at a Distance or Quantum Entanglement has been experimentally verified every time.

Quantum objects do not have physical limitations. They are unlimited. When objects achieve the speed of light, time stops, and distance goes to zero. From the perspective of photons traveling at the speed of light, time stops, distance goes to zero, and they simply quantum tunnel to their destination. Photons experience nothing from start to finish because, from their perspective, there is no passage of time, there is no distance from start to finish, and therefore there is nothing for them to experience.

From our perspective trapped in physical space-time, some of those photons take 13.2 billion years to reach us; but, from their perspective, they simply quantum tunnel to their destination and experience no passage of time during their journey. That's the way the theory of relativity really works. As photons and other quantum objects achieve the speed-of-light, time stops, and distance goes to zero.

Everything is instantaneous at the quantum level. There is no quantum speed-limit because there is no distance and there is no space-time or entropy at the quantum level. Everything is omnipresent at the quantum level. In the quantum realm, physical space-time and entropy cease to exist. These are physical phenomena or physical limitations – God's attempt to slow everything down for us so that we can live it, experience it, learn from it, and remember having done so. At the quantum level, there are NO physical limitations.

Whether they realize it or not, this is officially the end of Classical Physics, Classical Realism, Materialism, Naturalism, Darwinism, Nihilism, the Theory of Relativity, and even Atheism or Creation Ex Nihilo. Physical matter is NOT the fundamental stuff from which our universe was made. Energy is. Quantum Mechanics or Spiritual Mechanics is in fact Energy Mechanics – the way that energy or psyche really works at the quantum level.

What is Energy?

https://www.youtube.com/watch?v=PUn2izowBkw

In this PBS Space-Time video, Matt repeatedly states that energy is NOT fundamental. So, what could be more fundamental than energy? I want to know. Don't you?

He never really answers that question except to suggest that the Laws that the energy is using to form itself into physical matter would have to be more fundamental or more essential than the energy itself. Said another way, the "potential" has to be more fundamental than the energy which contains that potential.

At the beginning of the video, Matt defines energy as "a mathematical relationship between more fundamental quantities". Later, he also says that energy is a hint of something more fundamental. I'm curious to know what could be more fundamental than energy. Eventually, Matt makes the claim that the Theory of Relativity, the Expanding Universe, and Physical Laws in General are more fundamental than the conservation of energy and therefore more fundamental than energy.

In this video, Matt treats energy as a property or a function or a product of Classical Physics or Newtonian Mechanics. Matt limits his thought experiment to space-time or physical reality. Matt basically defines energy's quantum aspects out-of-existence from the start, so that he can then demonstrate violation or falsification of Conservation of Energy, which seems to be his ultimate goal. Matt binds energy to space-time so as to make energy compatible with Einstein's special relativity, general relativity, and the speed-of-light limitation on physical matter.

As a dedicated Materialist, Naturalist, and Atheist, Matt treats energy as a physical phenomenon or from the perspective of classical physics, and he then teaches that Conservation of Energy and "Living Force" are falsified or broken by friction, general relativity, and an expanding physical universe.

Matt erroneously states that energy is NOT conserved on the scale of an expanding universe, which he then says leads to effects like dark energy and the accelerating expansion of the universe. This error derives from his determination to define energy as a physical phenomenon. Physical phenomena are NOT conserved. At the physical level, energy seems to be unconserved at the scale of an expanding universe because that energy is being converted into dark energy and being used

119

to expand the universe – a quantum phenomenon and NOT a physical phenomenon. The energy is always conserved, just not at the physical level. Matt makes this error because he deliberately gives priority to Classical Physics and the Theory of Relativity over Quantum Mechanics and the Conservation of Energy. It's an error that everyone continues to make because they don't believe in Spooky Action at a Distance or Universal Conservation of Energy.

Energy is always conserved at the quantum level. When we convert physical matter into energy during a nuclear explosion, the physical matter is NOT conserved; but, all of the energy is conserved at the quantum level.

Matt also states that, "Conservation of energy is invalid in the context of Einstein's theory of general relativity due to the potential time evolution of space. Energy is not fundamental". In other words, General Relativity and Space-Time falsify Conservation of Energy, thereby making Einstein's Theory of Relativity and Classical Physics more fundamental than energy. He seems to be saying that Einstein's Theory of Relativity, Classical Physics, Materialism, Creation Ex Nihilo, and Naturalism are more fundamental than energy or more fundamental than the conservation of energy. Since the physical laws are more fundamental than energy, according to Matt, my question becomes, "Who made the physical laws in the first place, and what within the energy is making that energy obey these physical laws?"

Matt deliberately gives priority to Classical Physics, Materialism, Naturalism, and Atheism thereby giving us a model of reality that seems to violate the conservation of energy. In contrast, thanks to recent scientific experimentation, we have the proven and verified existence of Non-Locality, Quantum Entanglement, Universal Conservation of Energy, Quantum Fields, Quantum Mechanics, Quantum Tunneling, the Quantum Zeno Effect, and Action at a Distance FALSIFYING Einstein's Theory of Relativity, Classical Physics, Newtonian Mechanics, Materialism, Naturalism, Mechanism, and even Darwinism or Atheism.

They both can't be right.

If Action at a Distance is true, then energy is conserved at all levels of existence, and both Matt and Einstein have been proven wrong. If Matt, Einstein, Newtonian Mechanics, Classical Physics, and the Speed-Limits of the Theory of Relativity are true, then Action at a Distance or Quantum Entanglement or Non-Locality can't possibly be true. So, which is it? ALL of the science experiments are telling us that Matt and Einstein are wrong, and that Action at a Distance, Conservation of Energy, and Quantum Mechanics are true.

Whenever a choice has to be made between Einstein's Theory of Relativity (Classical Physics) and Quantum Mechanics (Instantaneous Action at a Distance), Quantum Mechanics should be chosen every time because Action at a Distance or Quantum Entanglement has been experimentally verified every time.

Matt repeatedly violates this proven truth in his PBS Videos by constantly giving Einstein's Theory of Relativity and Classical Physics priority over Quantum

Mechanics and Instantaneous Action at a Distance. It's what they all do, even though the experiments prove them wrong.

Energy or light seems to be the fundamental stuff from which everything is made, including psyche, life force, intelligence, thought, quantum waves, quantum fields, spirit matter or dark matter, gravity, magnetism, and physical matter.

But, there does indeed have to be something or someone even more fundamental who designed and made the universal blueprints or the physical laws from which physical matter is made. All the atoms across our universe seem to be compatible with each other, which means that they are being made from the same universal blueprint or the same set of universal laws. This Law-Maker and Law-Enforcer, who designed and made the quantum fields and the universal blueprints upon which physical matter depends, would in fact have to be more fundamental and more essential than the energy itself. This organizing principle, or organizing consciousness, or syntropy, or organizing force would have to be more fundamental and primal than the energy that's being organized.

The universal blueprints or knowledge or information, that the energy within physical atoms is choosing to conform with, has to be more fundamental than the energy itself. Now we are starting to move down into David Bohm's third and fourth implicate orders.

https://epdf.tips/the-essential-david-bohm.html

https://philosophy-of-science.com/The-Essential-David-Bohm

There is a Law of Physics that I developed which states that the smaller always dwells within and controls the larger. I also observed that Quantum Mechanics is in fact Energy Mechanics, or the way that energy and psyche really work at the quantum level in the non-physical realm.

In other articles, it became obvious that Psyche or the Spark-of-Life is in fact Bohm's "organizing principle", second implicate order, or super-implicate order. Psyche or that pin-prick of light seems to have size, which means that there are probably quantum computers and quantum ram within that psyche or spark-of-light, which means that the quantum machinery that is producing, transmitting, receiving, processing, and storing quantum waves or thoughts would have to be smaller than that Psyche or Spark-of-Life; and therefore, that quantum machinery within the Psyche would be something like Bohm's third implicate order. Then the different quantum waves that are being transmitted instantaneously and telepathically between Psyches would be some type of fourth implicate order. Then whoever is making all of that quantum machinery and producing all of those quantum thoughts would have to be a fifth implicate order, or some type of super-super implicate order, or a Super Psyche. Bohm says that the implicate order could go all the way down to infinity. Or, somewhere along the way, there is going to be a point-particle, an infinite singularity, or an intelligence that has NO size and takes up NO space; and, that thing will end up being the fundamental unit of reality and existence.

The proven and verified existence of Quantum Entanglement, Non-Locality, Action at a Distance, or Non-Physicality proves that Einstein's Classical Realism, Einstein's Special Relativity (light-speed limit), Classical Physicists, and Atheistic Naturalists like Matt are wrong. Quantum Information, or Energy, or Thought is being conserved at the quantum level, even though it seems to be getting lost at the physical level into entropy and black holes.

Psyche, Energy, Life Force, Information, and Thought are always conserved. In contrast, particles are born, and particles can die, according to Quantum Field Theory. In other words, physical limitations such as the Theory of Relativity and Entropy are NOT conserved. Physical matter is NOT conserved. Entropy is NOT conserved at the quantum level. Space-time and entropy do NOT exist at the quantum level. They are physical phenomena. They are not conserved. It's the underlying energy, intelligence, psyche, or life force that's always being conserved – NOT the physical limitations, and NOT the physical laws.

Time Dilation and Length Contraction

Normally, what physicists do when it comes to the speed-of-light, they use Minkowski diagrams and Lorentz transformations to diagram everything from the perspective of two or three different physical objects trapped in spacetime. What interests me most is how everything seems to be from the perspective of the photon. Einstein and nobody else ever thought about it from the perspective of the photon. However, from the perspective of the photon traveling at the speed-of-light, time stops, distance goes to zero, resistance to acceleration goes to zero, mass or physicality goes to zero, entropy or the passage of time goes to zero, and velocity goes to infinity. From the photon's perspective, it ceases to be a physical object and becomes a purely quantum object instead. From the photon's perspective, it simply quantum tunnels to its destination experiencing nothing during the journey because time has stopped during its journey, from its perspective. Cool, huh? This is what you get when everything is taken to its limits. Mass goes to zero and the photon ceases to be a physical object at the speed-of-light. At the speed of light, everything goes to its limits; and, the object literally transforms into something completely different – something quantum, or spiritual, or transdimensional, or non-physical. Transformation is what energy does. Energy is always conserved; but, it can be transformed into anything that its controlling psyche wants it to be, including something non-physical or something capable of infinite velocities. You can't find the truth by denying its existence.

They have worked things out from the physical perspective; but, they have yet to work things out from the quantum perspective or the non-physical perspective. They don't think about things from the non-physical perspective because they don't believe that the non-physical (or the spiritual or the supernatural) exists.

Einstein's greatest omission came when he rejected quantum mechanics and failed to look at things from the photon's perspective or from the quantum perspective.

"Einstein's Relativity is WRONG" by Run Ze Cao:

https://www.youtube.com/watch?v=5_tig3NaTjI

In this video, I was hoping that Run Ze Cao would have used quantum mechanics to falsify Einstein's claim that the speed-of-light is constant or Einstein's claim that nothing can move faster than the speed-of-light. Run Ze Cao failed to do so.

Instead, within this video, Run Ze Cao converts Special Relativity back into Newtonian Mechanics and Classical Physics, and then uses Newtonian Mechanics to falsify Einstein's Theory of Special Relativity. **Cao's conclusion is that there is NO time dilation and NO length contraction.** Essentially, Cao states that there is NO relativity. It's all just Newtonian Physics or Classical Physics. That would definitely be the case at the physical level where everything is nice and slow and it's technically impossible to accelerate mass faster than the speed-of-light; but, what about the quantum level?

At the quantum level things are completely different. According to Einstein's Theory of Relativity, when a quantum object travels at the speed-of-light, time stops and distance contracts to zero. This is exactly the way things actually work at the quantum level. Time stops, length contracts to zero, and distance goes to zero. Mass or resistance to acceleration goes to zero; and, velocity goes to infinity at the speed-of-light. Everything is instantaneous at the quantum level. Everything functions at an infinite velocity at the quantum level. When the passage of time is zero, distance is zero, and velocity is infinite or instantaneous, Classical Physics and Newtonian Mechanics literally cease to exist, and so does Einstein's claim that nothing can move faster than the speed-of-light. At the quantum level, there are NO physical limitations, and both Run Ze Cao's Newtonian Physics and Einstein's Special Relativity cease to have meaning. It's impossible to run math on zero time, zero length, zero distance, and infinite velocities. We have to switch over to something completely different once an object achieves the speed-of-light and transfers into the Quantum Realm or the Spirit World, because the quantum object has become massless or non-physical at that point.

While attempting to falsify Einstein's Special Relativity, Run Ze Cao concludes that **there is no time dilation nor length contraction**.

Time dilation has been verified by scientific experimentation which means that Run Ze Cao's conclusions are wrong, which means that Run Ze Cao got something wrong that Einstein managed to get right.

https://www.physlink.com/education/askexperts/ae433.cfm

https://www.scientificamerican.com/article/einsteins-time-dilation-prediction-verified/

http://www.alternativephysics.org/book/TimeDilationExperiments.htm

https://futurism.com/the-most-accurate-clocks-in-the-world-just-confirmed-that-time-is-not-absolute

http://math.ucr.edu/home/baez/physics/Relativity/SR/experiments.html

https://www.physicsforums.com/threads/evidence-for-length-contraction.6842/

Time dilation has been experienced and observed. Ultimately, the purpose of Science is to explain what has been experienced and observed. When Run Ze Cao states that there is NO time dilation, Run Ze Cao "explains" it by ignoring it or deleting it. Claiming that time dilation does not exist doesn't explain what it is or how it works. Time dilation has been experienced and observed. Einstein actually attempts to explain what time dilation is and how it works. Run Ze Cao does not. In materialistic and naturalistic fashion, Run Ze Cao sweeps time dilation and length contraction under the rug and pretends that they don't exist. Technically, that's dogmatism, and NOT Science. Run Ze Cao is dogmatic about his claim that time dilation and length contraction do not exist. He's also wrong.

In order to falsify Einstein's Special Relativity, Run Ze Cao employs Newtonian Mechanics and Classical Physics to "prove" that there is NO time dilation and NO length contraction or distance contraction. Run Ze Cao essentially falsifies Quantum Mechanics as well, by converting everything back into Newtonian Mechanics. I don't buy it.

The science experiments prove that Cao is wrong, and that Einstein was right. I choose to go with what has been experienced and observed.

According to the Theory of Relativity, as an object achieves the speed of light, time stops, distance goes to zero, length contracts to zero, and velocity goes to infinity. The results match with Quantum Tunneling and Quantum Mechanics, which have been experienced and observed. The purpose of Science is to explain what has been experienced and observed. In the Quantum Realm, everything is capable of quantum tunneling, or infinite velocities, or instantaneous action at a distance. This is what has actually been experienced and observed. Therefore, we need to find a way to explain it.

At the quantum level, at the speed-of-light, time literally goes to zero, which means that distance either goes to zero or quantum objects simply teleport or quantum tunnel to their destination every time. Quantum tunneling or teleportation has been experienced and observed. Therefore, as scientists, we need to find a way to explain it.

Einstein and Cao don't understand the implications of Quantum Mechanics and instantaneous Action at a Distance, because they insist on remaining within the realm of Classical Physics. Both Classical Physics and Cao's Newtonian Physics can't explain instantaneous Action at a Distance, which has been experienced and observed.

Remember, at the speed-of-light at the quantum level from a quantum perspective, time goes to zero, distance goes to zero, length contracts to zero, and velocity goes to infinity. There is NO speed limit at the quantum level! At the speed-of-light, we switch from Locality, Classical Physics, the Theory of Relativity, Physical Matter, Thermodynamics, Classical Realism, and Newtonian Mechanics OVER TO Quantum Mechanics, Quantum Field Theory, Non-Locality, Quantum Entanglement, Quantum Tunneling, Instantaneous Psychic Telepathy, Non-Physicality, and Instantaneous Action at a Distance. The speed-of-light is the turning point from Classical Physics to Quantum Mechanics. This is what has actually been experienced and observed. In the Quantum Realm, entropy, physical space-time, locality, the theory of relativity, and physical limitations cease to exist. At the speed-of-light, we switch from the Physical Realm to the Quantum Realm. Quantum Mechanics is Energy Mechanics, not Physical Mechanics. At the speed-of-light, physical matter or mass ceases to exist, energy is always conserved, psyche or consciousness reigns supreme, time stops, distance goes to zero, length contracts to zero, and velocities can be infinite.

Therefore, all of Cao's and Einstein's thought experiments and train experiments with internal observers and external observers go out the window and cease to be valid at the quantum level because time stops, distance goes to zero, length contracts to zero, and velocity goes to infinity at the quantum level. Entropy, the Theory of Relativity, Locality, Distance, Space-Time, and the passage of time no longer exist and no longer apply at the quantum level. Everything is omnipresent at the quantum level. Energy or Psyche is always conserved. The Psyche Realm or the Quantum Realm is a whole other reality that has nothing to do with Einstein's Relativity nor Cao's Newtonian Mechanics. There's a completely different set of physical laws in the Quantum Realm at the speed-of-light or faster. At the physical level, though, mass or physical matter cannot be accelerated faster than the speed-of-light because once it achieves the speed-of-light, then it no longer has mass and it is no longer physical matter. It is energy or spirit matter instead.

Run Ze Cao uses Newtonian Physics, or common sense, to falsify Einstein's Special Relativity; and, I believe that he fails to do so, especially since he comes to the conclusion that there is no time dilation and that there is no length contraction.

I found and developed a better and more convincing way to falsify Einstein's Special Relativity by using Quantum Mechanics, Quantum Tunneling, Action at a Distance, and Quantum Entanglement to falsify Einstein's claim that nothing can travel faster than the speed-of-light. Demonstrating the existence of faster-than-light interactions is a better, more convincing, and more verified way of falsifying Special Relativity than Cao's use of Newtonian Mechanics to falsify the Theory of Relativity.

The title of Cao's other video, "Surpassing the Speed of Light", hints at this reality. If we can get anything going faster than the speed-of-light, then we have in fact falsified Einstein's Special Relativity or Einstein's claim that nothing can move faster than the speed-of-light.

Surpassing the Speed of Light

https://www.youtube.com/watch?v=FYtKqyjxVug

I believe that Cao is wrong and that it is in fact impossible to accelerate mass or physical matter faster than the speed-of-light at the physical level. The speed-of-light was designed and implemented as a legitimate or lawful speed-limit for mass or physical matter at the physical level. Once we achieve the speed-of-light, then we are no longer in the physical realm but are in the quantum realm instead.

At the physical level, time dilation has been verified which means that Cao's conclusions have been falsified. However, at the quantum level, there are no physical limitations; and therefore, at the quantum level there is no entropy, no passage of time, no space-time, no distance, no length, and no locality. Everything at the quantum level can quantum tunnel or teleport at will. At the quantum level, Action at a Distance reigns supreme; and, the proven and verified existence of Action at a Distance is the thing that successfully FALSIFIES Einstein's Special Relativity and Einstein's claim that nothing can move faster than the speed-of-light.

Einstein was wrong, but not for the reason that Cao says that Einstein was wrong. Einstein was wrong for the reason that both Cao and Einstein were wrong – Action at a Distance or Faster-than-Light Interactions. Remember, according to the Theory of Relativity, at the speed-of-light, time stops, length contracts to zero, and distance goes to zero. The whole thing integrates or goes to unity, like in calculus. At the speed-of-light, Classical Physics or Newtonian Mechanics ceases to exist, and we switch over to Quantum Mechanics instead. At the quantum level, everything is capable of infinite velocity. There is no distance and no passage of time because there is NO space-time at the quantum level and NO entropy at the quantum level.

It's impossible to do math with zero time or stopped time, zero distance, zero length, and infinite velocity, which explains why the Gods created physical matter, physical limitations, and physical space-time in the first place, so that we could measure it and understand it. A physical reality is the ultimate consensus reality. In a physical reality, the Gods slow things down for us and stretch things out for us so that we can experience them, learn from them, experiment with them, and actually remember having done so. When time goes to zero, distance goes to zero, and velocity goes to infinity in the Quantum Realm, we are dealing with a whole other reality that has nothing to do with Entropy, Thermodynamics, Classical Physics, Classical Realism, Newtonian Mechanics, the Theory of Relativity, Space-Time, Distance, Length, or Physical Matter. The Quantum Realm has its own set of rules.

ALL of Cao's and Einstein's math and thought experiments become completely meaningless at the quantum level when space-time and distance cease to exist, and everything can teleport instantaneously to any destination of its choosing. We need a whole other conception, a whole other theory, and a whole other set of mathematics when time stops, distances goes to zero, length contracts to zero, velocity becomes infinite, and everything is omnipresent as it is in the Quantum Realm or Spirit World. This is the thing that nobody seems to get because they don't want it to be true. Yet, it all makes perfect logical sense once we choose to let go of Classical Physics or Newtonian Mechanics at speeds faster than the speed-of-light. Instantaneous Action at a Distance, Quantum Tunneling, and Teleportation are easy to understand and explain once a person chooses to do so. Infinite velocity or zero distance is really easy to understand, once we realize that they actually exist in the Quantum Realm.

Remember, whenever a particle achieves the speed-of-light, we should switch from the Theory of Relativity and Classical Physics OVER TO Quantum Mechanics, Action at a Distance, and Quantum Field Theory in order to explain it. We don't explain it and we can't explain it by claiming that it does not exist! Instantaneous Action at a Distance has been experienced and observed. It's time for scientists to start to explain it rather than stating that it does not exist. Stating that it doesn't exist doesn't explain what it is and how it works. Likewise, stating that time dilation and length contraction do not exist, as Run Ze Cao does, doesn't explain what they are and how they work. Time dilation has been experienced and observed. As scientists, we must explain what it is and how it works. Einstein attempts to do so; and, Run Ze Cao does not.

Searching for the True Interpretation

The first thing you should learn from an effective and true course in the Philosophy of Science is that the Scientific Method is flawed. The flaw in the Scientific Method comes in the last step of the process – it comes while interpreting the science experiment or interpreting the scientific data. Human interpretation is the FLAW in the whole of Science – getting the wrong interpretation.

You see, for each piece of scientific data, there are theoretically an infinite number of possible interpretations that can be given to that specific piece of scientific data. These interpretations are contradictory and mutually exclusive, which means that they can't all be true. This reality and truth has severe ramifications when it comes to Science.

Observe reality!

Quantum Mechanics and Quantum Field Theory are proven and verified Science. Energy Mechanics, Non-Local Mechanics, Psyche Mechanics, or Quantum Mechanics has been verified every time that it has been sufficiently tested. Quantum Mechanics or Psyche Mechanics is the TRUTH. So, where's the FLAW?

The fatal FLAW comes in the fact that there are literally an infinite number of different contradictory interpretations of Quantum Mechanics, and they all can't be true because they are mutually exclusive and falsify each other. There is NO universal agreement among scientists and mathematicians as to what Quantum Mechanics means and as to what Quantum Mechanics is. Each scientist has his or her own personal agenda, which means that each scientist has his or her own personal interpretation of Quantum Mechanics, an interpretation that is most likely wrong.

Here's another materialistic, naturalistic, and mechanistic interpretation of Physics and Quantum Mechanics that came to my attention.

"Einstein's Relativity is WRONG" by Run Ze Cao:

While attempting to falsify Einstein's Special Relativity, Run Ze Cao concludes that **there is no time dilation nor length contraction**. Run Ze Cao is demonstrably wrong, at least where time dilation is concerned. Since time dilation and length contraction go together as one united whole, if Cao is wrong about time dilation, then he is also automatically wrong about length contraction. There is something going on here that Cao is unable to perceive because he has limited himself to Newtonian Mechanics or Classical Physics.

Run Ze Cao has an interpretation of Physics and Quantum Mechanics that states that there is NO time dilation and NO distance contraction. Where the time dilation aspect is concerned, Run Ze Cao's interpretation has been falsified by scientific experimentation. Time dilation has been experienced and observed, which means that we need a scientific explanation for it, as well as an interpretation of Quantum Mechanics that takes it into account. We won't get that from Run Ze Cao. Instead, we will have to turn to scientists who have interpretations of Quantum Mechanics and Physics that actually take time dilation and length contraction into consideration, because Run Ze Cao is wrong and says that these things don't exist, when in fact they do.

During my research, I eventually decided that an interpretation of Quantum Mechanics is worthless, unless it actually attempts to explain what Psyche, Intelligence, Consciousness, or Life Force is doing at the quantum level in order to get things done for us at the physical level. I want my science to have explanatory power. I want my science to explain what's actually been experienced and observed. Intelligence or Consciousness has actually been experienced and observed. It's time to explain it scientifically.

Most, if not all, of the "official" interpretations of Quantum Mechanics on the Wikipedia are worthless, because they are based upon Materialism and Naturalism which state that Psyche or Intelligence or Consciousness does not exist. A materialistic, naturalistic, and atheistic interpretation of Quantum Mechanics is powerless to explain what Psyche, or Consciousness, or Life Force is doing at the quantum level, because these people really don't believe in a quantum level, or a spirit world, or some type of non-physical non-local consciousness. Materialism, Naturalism, Mechanism, Newtonian Physics, and Classical Physics completely lack explanatory power when it comes to the spiritual, the non-physical, or the quantum. They were designed to be that way. Materialism, Naturalism, and their derivatives were designed to falsify Non-Local Mechanisms, Spiritual Mechanisms, Psyche Mechanisms, or Quantum Mechanisms.

Since I started doing research into Quantum Mechanics a few years ago, I have come across ONLY three different interpretations of Quantum Mechanics that actually attempt to explain what Psyche is doing at the quantum level in order to get things done for us at the physical level. Even then, two of these three interpreters didn't really believe in Psyche, but their interpretations of Quantum Mechanics at least made allowances for the possibility of Non-Local Consciousness.

The first is Pim Van Lommel's Non-Local Consciousness interpretation of Quantum Mechanics. My first book on Quantum Mechanics was based upon his interpretation of Quantum Mechanics.

Quantum Mechanics: From a Non-Physical Spiritual Perspective

https://www.amazon.com/gp/product/B01J023TGU/

I relied on this interpretation of Quantum Mechanics for a couple of years, because it was the ONLY interpretation of Quantum Mechanics that I knew about that actually took Non-Local Consciousness, or Psyche, or Intelligence into consideration.

However, I found it to be inadequate when it came time for me to develop and present Quantum Neuroscience.

Quantum Neuroscience: The Answer to Life, the Universe, and Everything

https://www.amazon.com/gp/product/B079Z6QQQB/

In Quantum Neuroscience, I switched over to Henry P. Stapp's Orthodox Interpretation of Quantum Mechanics because it actually explains what the Human Psyche and Nature's Psyche are doing at the quantum level in order to get things done for us at the physical level. It all shows up in the math. Stapp's Orthodox Interpretation of Quantum Mechanics provided the complete package that I needed in order to successfully develop, present, and defend Quantum Neuroscience.

I thought that was it. I thought that I was done with Quantum Mechanics, until I found *The Essential David Bohm*. Within that book, I found a third interpretation of Quantum Mechanics that actually makes allowances for Psyche, Intelligence, Life Force, an Organizing Principle, or Non-Local Consciousness.

The Bible teaches us that God is in the light. God is in the energy or the light. This means that God's Psyche, or God's Intelligence, or God's Influence is in the energy or the light. Quantum Mechanics is Energy Mechanics, which means that Quantum Mechanics explains how psyche controls energy at the quantum level in order to get things done for us at the physical level.

Run Ze Cao concludes that **there is no time dilation nor length contraction**. I know that Run Ze Cao is wrong, because time dilation has actually been experienced and observed. I needed a better interpretation of Quantum Physics than the Newtonian Physics, Materialism, and Naturalism that Run Ze Cao is relying upon to prove his point. I NEED an interpretation of Quantum Mechanics that actually takes time dilation, distance contraction, and length contraction into consideration; and, I found exactly that, though *The Essential David Bohm*.

129

Over 200 pages of my book, *God Is in the Light*, is in fact a book review of David Bohm's book, *The Essential David Bohm*. I go through David Bohm's book and try to demonstrate how his unorthodox interpretation of Quantum Mechanics has been confirmed and verified and vindicated throughout the years.

God Is in the Light: God is light, and in Him is no darkness at all.

https://www.amazon.com/gp/product/B07168S37N/

I now have three different interpretations of Quantum Mechanics and three different books of my own that explain what the Human Psyche and Nature's Psyche and God's Psyche are doing at the quantum level in order to get things done for us at the physical level. That's pretty good actually. I think that it's more than anyone else has managed to do.

Obviously, I much prefer David Bohm's interpretation of Physics and Quantum Mechanics over Run Ze Cao's interpretation, because Bohm's ideas have been verified and vindicated; whereas, Run Ze Cao's claims have been falsified.

I'm searching for the True Interpretation of Quantum Mechanics, the one that matches with observation and experience and experiments, rather than simply trying to falsify Psyche, or Non-Local Consciousness, or Intelligence by pretending that they don't exist.

David Bohm said and wrote:

Bohm: If you want to relate it to modern physics (light and more generally anything moving at the speed of light, which is called the null-velocity, meaning null distance), the connection might be as follows. As an object approaches the speed of light, according to relativity, its internal space and time change so that the clocks slow down relative to other speeds, and the distance is shortened. You would find that the two ends of the light ray would have no time between them and no distance, so they would represent immediate contact. (This was pointed out by G. N. Lewis, a physical chemist, in the 1920s.) You could also say that from the point of view of present field theory, the fundamental fields are those of very high energy in which mass can be neglected, which would be essentially moving at the speed of light. Mass is a phenomenon of connecting light rays which go back and forth, sort of freezing them into a pattern. So, matter, as it were, is condensed or frozen light. Light is not merely electromagnetic waves but in a sense other kinds of waves that go at that speed. Therefore, all matter is a condensation of light into patterns moving back and forth at average speeds which are less than the speed of light. Even Einstein had some hint of that idea. You could say that when we come to light we are coming to the fundamental activity in which existence has its ground, or at least coming close to it.

This is the way things really work.

Bohm said the exact opposite of what Run Ze Cao said. They both can't be right, because they contradict each other. So, who is right, and who is wrong?

Run Ze Cao said that there is no time dilation nor length contraction.

In contrast, Bohm said that at the speed-of-light a transition or transformation takes place, and physics switches over to null distance. In other words, distance or length literally contract to zero at the speed-of-light. Null distance is zero distance – at the speed-of-light. There is NO distance at the quantum level because there is no space-time at the quantum level. There is NO distance at the quantum level because length contracts to zero and time goes to zero at the quantum level. As an object approaches the speed-of-light, according to the theory of relativity, time slows down, distance is shortened, and we switch over to Quantum Mechanics and Quantum Field Theory. That's precisely what the Theory of Relativity and Quantum Mechanics claim!

Rather than falsifying or denying time dilation and length contraction, David Bohm explains what it is and how it works. That's what science is supposed to do, since time dilation has in fact been experienced, observed, documented, and verified.

Within his theorizing, David Bohm reconciles and unites the Theory of Relativity and Quantum Mechanics by building a bridge between them. I think that's significant. At the speed-of-light, particles switch into the quantum realm, they transmute into something completely different, and everything integrates (like in calculus with an infinite sampling) into one united whole. It's a whole different set of rules and a completely different type of physics at the quantum level within the Spirit World.

Remember, at the speed-of-light, time stops, distance goes to zero, length contracts to zero, and velocity goes to infinity. At the speed-of-light, there is NO time and NO distance between start and finish, because at the quantum level everything is in immediate contact. In other words, everything can move at an infinite velocity at the quantum level because there is NO time and NO distance at the quantum level. That's precisely what Bohm said, and that's precisely how things work at the quantum level. This is precisely what has been experienced and observed. It also matches perfectly with the Theory of Relativity. In other words, the Theory of Relativity predicts no time, no length, no distance, and infinite velocity at the quantum level; and, it actually works that way.

Go figure!

Not only does Bohm's unorthodox interpretation match perfectly with Einstein's Theory of Relativity; but also, Bohm's unorthodox interpretation of Quantum Mechanics matches perfectly with what has been experienced and observed by out-of-body travelers and near-death experiencers. At the quantum level in the spirit world, people can quantum tunnel or teleport instantaneously to their chosen destination. It takes NO time to travel huge distances, because there are technically NO distances to travel at the quantum level. Everything is omnipresent at the quantum level. At the speed-of-light, time stops, distance goes to zero, length contracts to zero, and velocity goes to infinity. That's the way it really works at the quantum level, and it matches perfectly with Einstein's Theory of Relativity too.

Relativity explains how things work at the physical level, below the speed-of-light; and, Quantum Mechanics explains how things work at the psyche level or the quantum level at the speed-of-light, which is an infinite velocity at the quantum level. Remember, the speed-of-light is the ultimate speed limit at the physical level; but, the speed-of-light is infinite or instantaneous at the quantum level because there are NO speed limits and NO physical limitations at the quantum level. There is NO distance and NO time between the two ends of a light ray at the quantum level. Everything is in immediate contact at the quantum level. Bohm is right. This is what has actually been experienced and observed.

It all makes perfect sense. From our perspective at the physical level, it took that photon 13.2 billion years to reach us, because we are trapped in space-time which has distance and time within it. But, from the photon's perspective at the speed-of-light, it experienced NO passage of time and NO entropy from start to finish. From the perspective of the photon, it simply teleported or quantum tunneled to its destination instantaneously. The photon literally experienced nothing between start and finish because it moved at infinite velocity from its perspective at the quantum level. Bohm is right. From the perspective of the photon, we find that the two ends of the light ray have no time between them and no distance, because they represent immediate contact, teleportation, or quantum tunneling at the quantum level.

This is the way things really work! This is what has been experienced and observed. This also meshes Quantum Mechanics and the Theory of Relativity together seamlessly. Bohm successfully reconciles the Theory of Relativity with Quantum Mechanics. Mass cannot be accelerated faster than the speed-of-light. The speed-of-light was designed by the Gods to be a physical limitation for mass and physical matter. However, once something achieves the speed-of-light or NO mass, then it in fact achieves an infinite velocity at the quantum level; and, from its perspective it simply quantum tunnels to its destination instantaneously without experiencing anything from start to finish, because at the quantum level distance goes to zero, time goes to zero, length goes to zero, and velocity goes to infinity. Here we have the Theory of Relativity and Quantum Mechanics united into one great whole; and, it all makes perfect sense. It explains things from a sub-light velocity and sub-light perspective at the physical level; and, it explains the same things at an infinite velocity or timeless perspective at the quantum level. I restate this a number of different ways because it's extremely important to understand, because it explains everything that we have ever experienced or observed. It's good science.

Run Ze Cao falsifies the Theory of Relativity and Quantum Mechanics with Newtonian Physics; whereas, David Bohm explains and unifies the Theory of Relativity and Quantum Mechanics by explaining how things really work both at the quantum level and the physical level. So, which theory is the better theory? Which theory is more likely to be true? Which theory has the most explanatory power? Which theory has the most observation, experience, and evidence supporting it?

I know what I know and believe; but, you are going to have to decide for yourself what it is that you want to believe and know.

My ultimate goal has been to find and know the truth. Theories are worthless to me if they are false and have been falsified. If I can falsify philosophical theories or I have falsified philosophical theories, then they are worthless because they lack explanatory power. The false is falsified by the truth; and, the truth is repeatedly experienced and observed.

Bohm and many others repeatedly state that Quantum Field Theory is the bridge between the Theory of Relativity and Quantum Mechanics. If we finally find the truth – finally find the True Interpretation of Quantum Mechanics and the Theory of Relativity – it is all going to match together perfectly and seamlessly and come together into one united whole. It's all going to make sense.

Bohm's unorthodox interpretation of Quantum Mechanics made perfect sense to me because it explains everything precisely as it has actually been experienced and observed. With David Bohm, Henry P. Stapp, and Pim Van Lommel, I have finally found the truth. The explanatory power goes through the roof to infinity and beyond. Good enough! I found what I was searching for.

Falsifying Special Relativity

Einstein's theory of Special Relativity states that nothing can travel faster than the speed-of-light. This claim is demonstrably false; and therefore, the Theory of Relativity is demonstrably false.

Think about it logically. What have we observed?

A majority of our scientists claim to believe in the Big Bang Theory and truly believe that the Big Bang happened. If they are right, then we had to have a period of faster-than-light inflation just after the Big Bang in order to eliminate the horizon problem. The way that this was possible according to the theorists is due to the fact that nothing was physical yet, and space-time didn't exist yet. In other words, just after the Big Bang, everything was still quantum in nature and origin, which meant that it was able to travel faster than the speed-of-light, and which means that our early physical universe was able to expand faster than the speed-of-light thereby falsifying or overcoming Einstein's Theory of Relativity. If you believe in the Big Bang Theory, then you absolutely must believe in inflation, which means that you must believe in non-physical quantum phenomena that travel or expand faster than the speed-of-light. There is no other way to explain inflation logically. Inflation had to be quantum and faster-than-light in order to have worked. It could never have been a physical phenomenon.

Modern-day scientists have concluded that thanks to dark energy, our current physical universe is now expanding faster than the speed-of-light. The only way that's possible is if Einstein's Special Relativity is false. Anytime the scientists discover Action at a Distance or Faster-than-Light Expansion, we are in fact looking at something that falsifies Einstein's Special Relativity and his claim that nothing can travel faster than the speed-of-light. Einstein was wrong. Remember, the repeated experimentally proven and verified existence of Quantum Tunneling, Action at a Distance, Non-Locality, FTL Expansion, and Quantum Entanglement FALSIFY Einstein's Theory of Relativity, Classical Realism, Classical Physics, Materialism, Naturalism, and Einstein's claim that nothing can travel faster than the speed-of-light. The false is falsified by the truth; and, the truth is repeatedly experienced and observed.

The Quantum Tunneling or Teleportation of physical matter has been experienced and observed. It's real. It's a proven phenomenon that has to be taken into consideration while making CPU's, RAM, and other computer chips. By definition, in principle or actuality, Quantum Tunneling is faster-than-light. The fact of Quantum Tunneling – the fact that electrons and physical atoms can quantum tunnel or teleport under certain conditions – FALSIFIES Einstein's Theory of Relativity and Einstein's claim that nothing can move faster than the speed-of-light. In this case, it's not movement that we are talking about. It's teleportation. Remember, Quantum Tunneling is always faster-than-light, which means that Quantum Tunneling whenever it is observed falsifies Einstein's Special Relativity and Einstein's claim that nothing can move faster than the speed-of-light.

The proven and verified Quantum Zeno Effect FALSIFIES Materialism, Naturalism, Classical Physics, Classical Realism, and even Atheism or Creation Ex Nihilo. Atoms freeze or hold still whenever we look at them. Atoms stop quantum tunneling whenever we look at them. Atoms can read your mind at a distance – psyche-to-psyche. They know when you are looking at them, and they know when you are not looking at them. Atoms are psychic, telepathic, intelligent, conscious, and alive. This is telepathy or Action at a Distance in action. Anytime that we prove or verify the existence of Action at a Distance, Non-Locality, Non-Physicality, or Quantum Entanglement of any kind, we in fact falsify Einstein's Theory of Relativity, Classical Physics, Materialism, Naturalism, and Einstein's claim that nothing can move faster or function faster than the speed-of-light. Both the Quantum Tunneling of physical matter and the Quantum Zeno Effect have been observed. These quantum phenomena happen faster-than-light and are in fact examples of Action at a Distance. Every time that the scientists observe and verify Action at a Distance, they do in fact falsify the part of Einstein's Special Relativity which states that nothing can travel faster than the speed-of-light. The false is falsified by the truth; and, the truth is repeatedly experienced and observed. Quantum Tunneling, Action at a Distance, Quantum Entanglement, and the Quantum Zeno Effect have been repeatedly experienced and observed.

According to the Theory of Relativity, when a photon or any other quantum object reaches the speed-of-light, time stops, and distance goes to zero. Distance goes to zero! At the speed-of-light, quantum objects become omnipresent at the quantum level. In other words, they seem to travel at an infinite velocity. From the perspective of the photon traveling at the speed-of-light, time has stopped, entropy has stopped or ceased to exist, the photon experiences NO passage of time, and distance literally goes to zero. The photon literally quantum tunnels to its destination. From its perspective, the photon travels at an infinite velocity. In fact, a photon doesn't launch if it knows that it has nowhere to go. A photon knows where it is going to land before it launches. From our perspective a photon can take 13.2 billion years to reach us; but, from the photon's perspective at the speed-of-light, time stops, and that photon experiences no passage of time, and it simply quantum tunnels instantaneously to its destination experiencing nothing during the interim. In this situation, the truths associated with the Theory of Relativity actually falsify Einstein's claim that nothing can move faster than the speed-of-light and therefore these truths falsify Special Relativity. The truth is that at the quantum level, entropy, the passage of time, space-time, time, locality, physical limitations, speed limits, and distance cease to exist or go to zero. This is what has actually been experienced and observed. It ALL falsifies the part of Einstein's Special Relativity which states that nothing can travel faster than the speed-of-light.

Mass is resistance to acceleration. That means that objects that are given mass end up manifesting at the physical level in the physical realm; and, they are then subjected to all the physical restrictions or physical limitations of Classical Physics or the Theory of Relativity, including entropy, the passage of time, distance, locality, and a sub-light speed limit. It's physically impossible to accelerate mass or physical matter faster than the speed-of-light because mass is resistance to acceleration. However, all the different non-physical quantum objects that have NO mass such as photons, quantum waves, and quantum fields have NO resistance to acceleration which means that they have NO speed limit at the quantum level. Quantum Fields and Consciousness are omnipresent at the quantum level, and both quantum waves and photons have NO speed limit at the quantum level because space-time, entropy, the passage of time, time, distance, and locality cease to exist at the quantum level from their perspective. The whole thing is relative to one's perspective. Mass and physical limitations literally cease to exist at the quantum level in the spirit world or the psyche realm. There is no resistance to acceleration at the quantum level, which means that non-local quantum objects travel at an infinite velocity in the spirit world or the quantum realm. In the spirit world, quantum objects move in a different time stream than the one that exists at the physical level, because all the physical limitations cease to exist at the quantum level or the psyche level including entropy or the passage of time. Space-time is a physical phenomenon. Space-time, or distance and time, go to zero at the speed-of-light, which means that there are NO speed limits at the quantum level in the Psyche Realm.

At the speed-of-light, physical space-time literally goes to zero. Time stops, and distance goes to zero. The physical realm effectively ceases to exist at the speed-of-light. All physical limitations, including entropy or the passage of time, cease to exist at the quantum level. At the quantum level, everything is capable of infinite velocities or omnipresence. In other words, everything can quantum tunnel at the quantum level. There is no such thing as thermodynamics at the quantum level. At the quantum level, psyche or energy is always conserved.

According to Quantum Field Theory, particles are born, and particles can die; but, the energy is always conserved. The fundamental axiom of Quantum Field Theory states that particles are born, and particles can die. That means that particles are NOT conserved. Only the energy within the particles is conserved. Physical matter and entropy are NEVER conserved. We destroy physical matter in nuclear explosions; and, we make new physical matter from energy in our particle accelerators. Physical matter is made from energy and then destroyed and converted back into energy all the time. Physical matter and entropy are NEVER conserved. Only energy or psyche or the quantum is conserved.

Energy is infinitely malleable. Every psyche or intelligence or life force has a certain amount of energy that's under its control. The controlling psyche can form the energy under its control into anything that it wants that energy to be, including quantum waves, quantum fields, spirit matter or dark matter, or even physical matter. This is what has been experienced and observed. Quantum fields are non-physical and pre-physical. The Gods or Controlling Psyches had to design and make the Matrix of Quantum Fields BEFORE they could then make and sustain physical matter and other types of particles. The quantum fields had to be made before mass and physical matter could be made.

Here's the rub. The Theory of Relativity is a part of Classical Physics, and NOT a part of Quantum Mechanics. Physical limits do not exist at the quantum level or the psyche level. There is NO speed-of-light limitation at the quantum level! It has been observed that the speed-of-light is a physical limit and only applies at the physical level within Bohm's explicate order. Speed limits are non-existent at the quantum level within Bohm's implicate order. There are NO physical limitations at the quantum level or the psyche level. The speed-of-light limit, entropy, time, distance, and space-time cease to exist or go to zero at the quantum level or the psyche level. That's the way things really work. Even the other parts of the Theory of Relativity that are true tell us that that's how things really work at the quantum level and the psyche level. At the speed-of-light, time stops, distance goes to zero, entropy or the passage of time ceases to exist, and the particle functions according to quantum laws instead of functioning according to physical laws and physical limitations.

The whole of science and quantum physics as we know them to be and have observed them and experienced them to be FALSIFY Special Relativity and Einstein's claim that nothing can move faster than the speed-of-light. The false is falsified by the truth; and, the truth is repeatedly experienced and observed.

People correctly state that the Scientific Method cannot be used to prove anything to be true. However, there is a loophole that most scientists choose to ignore. The Scientific Methods can be used to prove theories false. Observation and experience can be used to prove theories false. Science experiments can be used to prove theories false. All of these things have been used to prove Einstein's speed-limits false, meaning that all of these have been used to prove parts of the Theory of Relativity to be false.

You see, if you successfully eliminate everything that is false, then ONLY the truth will remain. This process of elimination starts by FALSIFYING Classical Realism, Einstein's Light-Speed Limit, the parts of the Theory of Relativity that are false, Classical Physics, Materialism, Naturalism, Darwinism, Nihilism, Behaviorism, Determinism, Physical Reductionism, Atomism, Spontaneous Generation, Chemical Evolution, Macro-Evolution, Design and Creation by Natural Selection, and Creation

Ex Nihilo or Atheism. If you successfully eliminate everything that is false and everything that has been falsified, the ONLY the truth will remain.

Remember, genetics and genomes were designed to prevent macro-evolution from happening; and, they work as they were designed to do. Remember, Louis Pasteur falsified spontaneous generation, abiogenesis, and macro-evolution in 1859.

After you have eliminated everything that is false and everything that has been falsified – everything that has NEVER been experienced NOR observed – then you end up with the truth which is comprised of Psyche or Intelligence or Consciousness, Intelligent Design and Creation at all levels of reality and existence, Quantum Mechanics, Quantum Field Theory, Quantum Entanglement, Quantum Tunneling, Action at a Distance, Quantum Entanglement, Non-Locality, the Quantum Zeno Effect, the Conservation of Energy and Psyche, a Quantum Realm or Spirit World, some type of After-Life, and the resurrected Biblical God Jesus Christ. All of these things have been experienced and observed zillions of times by millions of different people. The false is falsified by the truth; and, the truth is repeatedly experienced and observed.

Mark My Words

--

Reference Material

https://ultimate-model-of-reality.com/falsifying-einstein-and-classical-realism/

https://psyche-ontology.com/

https://psyche-ontology.com/buhlman/

https://psyche-ontology.com/psyche-observed/

https://psyche-ontology.com/psyche-experienced-and-observed/

https://syntropy.site/Quantum-Zeno-Effect-Verified/

https://www.youtube.com/results?search_query=NDE+Jesus

https://www.youtube.com/watch?v=ZEhoR-LCDlo

https://www.amazon.com/gp/video/detail/B07MJK3GMV/

https://www.google.com/search?q=proof+of+quantum+entanglement

https://www.amazon.com/gp/video/detail/B078JN83BP/

https://www.youtube.com/watch?v=W1_KEVaCyaA

https://www.amazon.com/dp/B06X9DVRQW/

https://ultimate-model-of-reality.com/

Mastering the Philosophy of Science

Ever since I started pursuing the True Model of Reality or the Ultimate Model of Reality back in 2016, I have found it extremely important to learn and master the Philosophy of Science.

The Scientific Method is based upon the Logic Fallacy that we philosophers call, *Affirming the Consequent* or *Jumping to Conclusions*. Consequently, the Scientific Method is inherently flawed and often misused to prove the truth of something that has in fact been falsified by observational, logical, and experiential evidence.

How's that possible? It's because most of our scientists have actually learned how to lie with statistics. In fact, my brother defined Statistics as the art of lying with mathematics. Let me try to demonstrate.

As a way of an introduction to the Philosophy of Science, this time around, let's try to identify the ultimate purpose of Statistics. For that, I will turn to the text of my recent college statistics course.

Pagano, R.R. (2010). *Understanding Statistics in the Behavioral Sciences* (9th ed.). Belmont, CA: Wadsworth.

> **Statistics uses probability, logic, and mathematics as ways of determining whether or not observations made in the real world or laboratory are due to random happenstance or perhaps due to an orderly effect one variable has on another. Separating happenstance, or chance, from cause and effect is the task of science, and statistics is a tool to accomplish that end. Occasionally, data will be so clear that the use of statistical analysis isn't necessary. Occasionally, data will be so garbled that no statistics can meaningfully be applied to it to answer any reasonable question. But I will demonstrate that most often statistics is useful in determining whether it is legitimate to conclude that an orderly effect has occurred. If so, statistical analysis can also provide an estimate of the size of the effect.** (*Understanding Statistics*, p. xxvii.)

Within his book, Pagano includes statistics as an integral part of the Scientific Method. Why? In order to mitigate the inherent flaws within the Scientific Method.

What is the main flaw of the Scientific Method?

The primary flaw of the Scientific Method is the faulty and never-ending personal interpretations that are given to the scientific evidence and scientific data – a logic fallacy that we call *Jumping to Conclusions* or *Affirming the Consequent*. According to the Philosophy of Science, it's theoretically possible for there to be an infinite number of different and competing and contradictory interpretations for each and every piece of scientific data and scientific evidence that we encounter.

There are as many different and contradictory interpretations of Quantum Mechanics as there are physicists on this planet. That is the FLAW inherent within the Scientific Method – the final step of the Scientific Method, which we call "interpreting the data". *Jumping to Conclusions* or *Affirming the Consequent* is a logic fallacy – the fundamental logic fallacy upon which the Scientific Method is based. The truth can never be established based upon logic fallacies, because logic fallacies are actually used by scientists and statisticians to deceive us, trick us, and lie to us; and, it requires critical thinking to detect them and expose them.

Statistics were allegedly designed as an attempt to REMOVE happenstance or chance as a 'scientific explanation' or a 'scientific interpretation'? Can you think of any "science" that makes CHANCE its causal agent? I can think of a few. Can you? The goal of statistics, according to Pagano, is to remove CHANCE from science as a causal agent, because CHANCE can never be a legitimate and reliable causal agent. CHANCE isn't dependable, replicable, reproducible, demonstrable, nor reliable; and, these things are the hallmarks of Real Science and True Science. Using CHANCE as a scientific explanation is a fatal flaw that automatically falsifies any conclusion that we are trying to make or achieve.

As Pagano stated, the purpose of Science is to separate chance from "cause and effect". Instead, we actually have a few "sciences" that USE chance or happenstance as their causal agent. Using chance as a causal agent IS a logic fallacy – a logic fallacy that most of our biological sciences are currently based upon.

What is the causal agent behind the Theory of Evolution?

CHANCE!

Within the physical realm, chance is synonymous with magic. It doesn't really exist as a causal agent. You can't replicate it, which means that you can't rely upon it. Chance isn't a person, place, or thing. Yet, the Darwinists rely exclusively on chance. Do they not?

According to the Darwinists, Materialists, Naturalists, Behaviorists, Determinists, Classical Realists, and Atheists, CHANCE designed, programmed, engineered, made, field-tested, and mass-produced your genome. That's physically impossible.

Natural Selection is based exclusively on chance. When you are "selected against", you die. Your physical body dies. That's what Natural Selection means. It means that you die. The date and time of your death is based almost exclusively on CHANCE, because the rest of you is actually working against it if you have any measure of sanity or normalcy within you. Natural Selection is entropy or death. Therefore, the "origin of species by means of natural selection" is in fact "the origin of species by means of chance and death". Death obviously can't design and create. Chance cannot design and create. Therefore, Natural Selection, Chemical Evolution, Macro-Evolution, and the Theory of Evolution are in fact FAULTY INTERPRETATIONS of scientific data, because they are based exclusively on chance.

The whole purpose of science and statistics is to eliminate CHANCE as a causal agent; yet, the Darwinists and Naturalists enshrine it and worship it instead.

Chance is statistically unreliable, and as Pagano said, the whole purpose of science and statistics is to REMOVE chance as a causal explanation. In other words, the whole purpose of statistics is to remove Natural Selection, Abiogenesis, Spontaneous Generation, Chemical Evolution, Macro-Evolution, the Theory of Evolution, Creation Ex Nihilo, and any other type of CHANCE from science as a scientific explanation or a causal explanation.

This is the Philosophy of Science at its very best – using the Philosophy of Science to identify and eliminate from science everything that has already been falsified by Science. That starts by eliminating chance as a causal agent. Chance can NEVER function as a causal agent! That's what statistics is trying to teach us.

The Theory of Evolution and its derivatives are inherently flawed and faulty because they use chance exclusively as their "scientific explanation" or their "causal agent". According to the Laws of Statistics, it is a logic fallacy to use CHANCE as a causal agent or a scientific explanation within the science of your choice. With chance, there is NO identifiable "size of effect" because there is no reliable cause and effect. That's how statistics identifies chance, by identifying the things that have no sizable effect. By definition, chance has no sizable effect; therefore, chance cannot be used as a causal agent within science if we want to have a True Science or a Realistic Science when we are done. That's what statistics is trying to teach us, and that's precisely the part of statistics that the Darwinists and Naturalists reject and try to hide from us. Their ultimate goal is to hide anything from us that successfully falsifies Materialism, Naturalism, Darwinism, and Atheism; and, practically everything falsifies Materialism, Naturalism, and their derivatives such as Darwinism and Atheism.

Our atheistic, naturalistic, and materialistic college professors NEVER teach us nor tell us about any of this, because the Philosophy of Science or the "purpose of science and statistics" actually FALSIFIES the philosophies of Materialism, Naturalism, Mechanism, Darwinism, Nihilism, Behaviorism, Determinism, Physical Reductionism, and Atheism be eliminating chance as a scientific explanation or a causal agent.

Remember, using chance as a causal agent or a scientific explanation is a logic fallacy. Anything that uses chance as a causal agent or a scientific explanation is automatically false and has automatically been falsified; and if you observe carefully, the Darwinists, Naturalists, and Atheist refuse to allow into evidence anything that FALSIFIES Materialism, Naturalism, Darwinism, and Atheism. They cheat, in an attempt to trick us and deceive us.

In other words, these people *Affirm the Consequent* or *Jump to Conclusions* rather than using the Rules of Science and the Laws of Statistics to falsify Materialism, Naturalism, Darwinism, Nihilism, and Atheism as they should. These people desperately need some Philosophy of Science education in order to identify and correct their mistakes; but, they refuse to look at these kinds of things because the Philosophy of Science actually falsifies their personal theories, interpretations, and beliefs. The Philosophy of Science actually identifies and exposes the flaws and the mistakes that the Darwinists are making while they are trying to do science.

An atheist friend told me not to tell him the statistics. He preferred to remain in the dark. Materialism, Naturalism, Atheism, Spontaneous Generation, Abiogenesis, Macro-Evolution, Natural Selection, Chemical Evolution, Atheism or Creation Ex Nihilo, and Darwinism are based exclusively on chance. They are based upon wishful thinking, with NO evidence whatsoever to support them. Why do I say that there is NO evidence to support them? It's because according to the Laws of Statistics, chance cannot be used as scientific evidence; and, that's precisely what these people do – they deliberately use chance as scientific evidence and as a causal agent.

Using chance as a causal agent is a logic fallacy. There's NO CHANCE that chance could ever act reliably as a causal agent. The people who use chance as a causal agent and as a scientific explanation end up developing theories like Boltzmann Brains and the Theory of Evolution – where entropy, or random quantum fluctuations, or chance becomes their creator and their god.

In physics thought experiments, a Boltzmann brain is a self-aware entity that arises due to extremely rare random fluctuations out of a state of thermodynamic equilibrium. For example, in a homogeneous Newtonian soup, theoretically by sheer chance all the atoms could bounce off and stick to one another in such a way as to assemble a functioning human brain (though this would, on average, take vastly longer than the current lifetime of the universe).

https://en.wikipedia.org/wiki/Boltzmann_brain

Using entropy or chance to design and create a human brain or the human genome IS precisely the types of logic fallacies that Statistics was designed to identify and eliminate from science. It's physically impossible for chance or entropy to make a human brain or a functional genome. Instead, what these people do is use 'the average' or use statistics to make it seem theoretically possible, and then since it's obvious that our universe does indeed have physical brains and physical genomes, these people jump straight to the conclusion that entropy or random chance or quantum fluctuations designed and made our physical universe (the Big Bang), our earth, our galaxy, our sun, our genomes, and our brains. In other words, they use statistics to lie to us and to trick us and deceive us into believing that chance, entropy, death, and random fluctuations can in fact design and create if given enough time to do so. The idea that chance can design and create is faulty logic and therefore falsified science. There's NO CHANCE that chance can design and create anything complex or anything mechanical such as a physical universe, a gene, a protein, a genome, or a brain.

Of course, if we were to use statistics and logic as they were designed and intended to be used, then we would automatically see that according to the laws of probability there is NO CHANCE that chance, or random fluctuations in atoms, or entropy, or death (natural selection) could ever design and create functional proteins and their matching genes, nor could they ever design and create a functional thinking human brain.

Why?

It's because it has actually been observed that some atoms naturally repel each other, and they actually have to be forced together by enzymes in order to become functional molecules such as proteins and DNA. Hydrogen molecules and helium atoms naturally repel each other. That's why hydrogen and helium balloons float in the air, because hydrogen molecules and helium atoms naturally repel each other. There's no way in the universe that an expanding cloud of hydrogen and helium gas will ever coalesce naturally into a planet, star, galaxy, genome, or brain. It's physically impossible for it to do so. Likewise, it's physically impossible for "random fluctuations in entropy" to produce functional genes, proteins, cells, and brains out of expanding clouds of hydrogen and helium gas.

Our Scientific Given: Assume the Big Bang happened for real; then, the eventual result of the Big Bang was expanding clouds of hydrogen and helium gas.

Scientific Observation: Hydrogen atoms initially repel each other. Hydrogen molecules continue to repel each other. Helium atoms repel everything. This is what we have actually experienced and observed.

Scientific Conclusion: There's no way in the universe and no chance whatsoever that expanding clouds of hydrogen and helium gas will ever coalesce into planets, stars, and galaxies. There's NO CHANCE whatsoever that expanding clouds of helium and hydrogen will ever coalesce into Boltzmann Brains, living cells, proteins, and genomes. It's physically impossible.

What do we observe whenever we start pumping hydrogen or helium into a metal canister? Does it ever coalesce into a planet, star, or galaxy? NEVER! It doesn't matter how strong that canister is or how much pressure you apply, it never coalesces into anything; but instead, the hydrogen molecules and helium atoms continue to repel each other to the very end. The metal canister will break before the hydrogen and helium coalesces. It takes a focused concentrated nuclear explosion to make hydrogen atoms coalesce and fuse. That's NEVER going to happen naturally within an expanding cloud of hydrogen and helium gas out there in the vacuum of space. The hydrogen molecules and helium atoms will always repel each other instead.

There's NO CHANCE that entropy, death, or random quantum fluctuations could ever design and create a functional protein, gene, genome, or brain; therefore, we have now used Statistics to effectively eliminate chance from science as a causal agent or a creative agent – just as we should have done at the very beginning when we first started to create and use science.

There's NO CHANCE that chance or entropy or random fluctuations could ever design and create anything complex or mechanical, whether we are talking about the quantum level or the physical level. There's NO CHANCE that chance could have designed and made our physical universe and the physical constants or the physical laws that are now being enforced throughout our physical universe. There's NO CHANCE that Creation Ex Nihilo, Abiogenesis, Chemical Evolution, Macro-Evolution, Natural Selection, or any other type of chance could ever design and create anything complex or mechanical from nothing. There's NO CHANCE that random quantum fluctuations could have ever designed and created our physical universe. The whole purpose of statistics is to remove chance from science as a causal explanation, because chance is never a realistic or reliable causal explanation for anything.

It is obvious to many of us that it's physically impossible for clouds of hydrogen or helium gas to coalesce into planets, stars, and galaxies naturally or by chance because hydrogen molecules repel each other, and helium atoms repel everything. It's obvious that the original hydrogen and helium atoms that were used to make our sun were in fact quantum tunneled or forced into that position, because they would never have coalesced naturally into planets, stars, and galaxies.

It's obvious that stars and galaxies are made-in-place whole all at once or are born all at once, especially where hydrogen and helium are concerned, because expanding clouds of hydrogen and helium gas are never going to coalesce into stars naturally or by chance because hydrogen molecules repel each other, and helium gas repels everything. In fact, it's a miracle that two atoms hydrogen atoms can merge together into a single hydrogen molecule, because they each vigorously repel each other initially, until they transform their orbitals and decide to cooperate instead. It takes a huge amount of external pressure or energy to make hydrogen molecules coalesce and then later to fuse. The only way to make a star would be to quantum tunnel all of the hydrogen molecules and helium atoms into place all at once, and then let gravity take over from there.

Remember, the ultimate goal of Science and Statistics is to remove chance as a causal explanation. There's NO CHANCE that expanding clouds of hydrogen gas are ever going to coalesce into something like a planet, star, or galaxy. The hydrogen atoms within all the stars had to be forced into that position or quantum tunneled into that position in order for them to be there, because the hydrogen atoms in a gas cloud in space would have repelled each other rather than coalescing naturally into a planet, star, or galaxy. Here on this earth, even under pressure and the effects of gravity, hydrogen molecules never coalesce but instead fight to get away from each other. It takes a focused atomic explosion to make hydrogen atoms fuse or coalesce; and, it takes intelligence to make an atomic explosion. We start to get closer to the truth whenever we eliminate chance as a causal explanation and a causal force. That's why statistics was designed to identify and eliminate chance from Science, scientific explanations, and scientific interpretations because chance can never be an effective and reliable causal agent. In fact, where clouds of hydrogen and helium are concerned, chance and natural law actually work together to keep the hydrogen molecules and helium atoms separated from each other as much as possible. It's obvious to many of us that physical stars are made; and, anything that is obviously made obviously has a Maker who made it.

Mira is a variable red giant (Mira A) along with a white dwarf companion (Mira B). The variable star, Mira, is a bat out of hell. During its estimated 30,000-year life span, it has been traveling against the rotation of our galaxy – against the galactic flow – currently at 130 kilometers per second, or one-2300th the speed of light; and so far, it has left behind it a tail 13 light years long. That's fast for a physical object vectoring in the wrong direction against the flow! Since Mira was born or made traveling at a fraction of the speed-of-light against the galactic flow, we can assume (according to the laws of physics and from the huge bow shock in front of it) that it has been slowing ever since it was made 30,000 years ago. This is about as close to the Signature of God that cosmologists and astrophysicists will ever get. There's no way in this universe that a coalescing cloud of expanding hydrogen and helium gas is ever going to merge together naturally or by chance into a star that is traveling against the natural galactic gas flow at near the speed-of-light when that star is born or made. Such a thing is physically impossible; yet, there it is.

Mira was obviously made; and, Mira was obviously made to travel against the direction of the rotation of our physical galaxy at nearly the speed-of-light when it was made. There's no way in this universe that entropy, chance, or random quantum fluctuations could have ever made such a thing. Mira defies all the physical laws that we know about. Its existence is physically impossible. Anything that was obviously made obviously has a Maker who made it and started it moving. Mira is one such thing. It's obvious that Mira was made and then vectored in the wrong direction at nearly the speed-of-light. Only a God could do such a thing. It's obvious that it didn't happen naturally or by chance. There are no physical laws that would allow such a thing to happen naturally or by chance.

https://www.nasa.gov/mission_pages/galex/20070815/c.html

https://www.youtube.com/watch?v=XyuXBYWZegY

The same thing can be said about your genome. Your genome is complex hardware and a complex computer program all rolled into one. We KNOW for a fact that computers and computer programs never assemble themselves from nothing. They are designed and made. There's no way in the universe that random quantum fluctuations could have ever designed and made your genome – or your brain. In fact, entropy slowly destroys genomes and brains. Entropy, death, or natural selection doesn't make genomes – it slowly destroys them. Your genome was obviously made, and anything that is obviously made obviously has a Maker who made it. Your genome is God's Signature, and that signature is written into every cell within your physical body, except for your mature red blood cells which don't have a nucleus within them.

Once again, there's NO CHANCE that entropy, death, or random quantum fluctuations could ever design and create a functional protein, gene, genome, or brain; therefore, we have now successfully used Statistics to effectively eliminate chance from science as a causal agent or a creative agent – just as we should have done at the very beginning when we first started to create and use science.

Remember, the purpose of Statistics is to eliminate happenstance or chance from science as a scientific explanation or a causal agent. Any time we manage to eliminate chance from our science as a causal source, we are in fact coming closer to the truth of what really happened.

What are we left with when we eliminate CHANCE from Darwinism, Chemical Evolution, Macro-Evolution, Natural Selection, Abiogenesis, Creation Ex Nihilo or Atheism, Spontaneous Generation, and the Theory of Evolution? We are left with nothing. These "sciences" or philosophies or religions cease to exist, the very moment that we remove chance from the equation. Then we are left looking for a true and realistic causal agent instead.

Materialism, Naturalism, and their derivatives are based exclusively on chance; and because of that, they are faulty and have no real foundation, because chance is NOT a legitimate cause, especially where science is concerned. Instead of providing us with a plausible truth, what these people do is turn everything upside-down by claiming that anything that is based upon intelligent causes is in fact pseudo-science and should be rejected. The Materialists, Naturalists, and Atheists have created a long list of pseudo-sciences on the Wikipedia, many of which are in fact based upon legitimate intelligent causes capable of producing real effects. In other words, these people are actively and deliberately trying to lie to us, trick us, and deceive us in order to make their case. I used to be a materialist, naturalist, nihilist, and atheist; but, I caught these people in one too many lies, and they lost me as a result. I don't like being lied to. If your case is legitimate, true, and just, then you shouldn't need to lie in order to promote it.

This is Philosophy of Science at its very best – using philosophy to identify, expose, and eliminate the false philosophies, or the deceptions and the lies.

Nevertheless, this is knowledge and information that nobody wants. I can't even give it away because nobody wants it. Materialists, Naturalists, Nihilists, Darwinists, and Atheists will pay millions of dollars to be told that God does not exist; but, nobody is going to pay anything to be told why Materialism and Naturalism are false. That's just the way it is. Most people don't want to know the truth. They would rather be lied to instead.

You'll never see anything like this in your public schools, because our public schools have become temples and shrines to Materialism, Naturalism, Nihilism, Darwinism, and Atheism; and, these people will NEVER allow anything into evidence that successfully falsifies Materialism, Naturalism, and their derivatives such as Darwinism, Behaviorism, and Atheism. If you are not an atheist by the time you graduate from your public university or college, then your teachers have failed in their purpose for being there. There's no chance that these people will ever allow into evidence anything that falsifies Materialism, Naturalism, and their derivatives. If you want to find and know the truth, then you are going to have to find it on your own, because your typical college professor never will.

Mark My Words

Genetic Entropy

Web Page: https://evolution-is-entropy.com/genetic-entropy/

Genetic Entropy is one of my most favorite scientific discoveries. This scientific discovery doesn't belong to me, though. It originates with John Sanford. After reading his book, *Genetic Entropy*, I simply KNEW that the theory of evolution is false because I now KNOW why it is false.

The theory of evolution is typically defined as Creation by Mutation/Selection – particularly, 'the origin of species by means of natural selection'. The first part of the title of Darwin's book is, "On the Origin of Species by Means of Natural Selection".

Creation by Natural Selection IS science fiction. Natural selection doesn't touch our genes! It can't. It's physically impossible for natural selection to get at our genes and change them. Natural selection cannot design and create anything, let alone a genome.

In truth, natural selection is NOT the mechanism of change behind the theory of evolution. It's the random mutations that produce genetic change, NOT natural selection! Natural selection or survival of the fittest doesn't do anything. It just waits for you to die. Natural selection is entropy or death. They built a whole "science" on a fictional, immaterial, invisible process that doesn't even touch our genes – natural selection! And, they literally give natural selection ALL the credit for designing, programming, creating, and producing our genomes and our physical bodies. For these people, natural selection is their god. They worship it with a passion. Natural selection is a man-made god, an idol.

Natural Selection: The evolutionary process by which heritable traits that best enable organisms to survive and reproduce in particular environments are passed to ensuing generations.

Everyone who has taken introductory psychology has learned that nature and nurture together form who we are. As the area of a rectangle is determined by both its length and its width, so do biology and experience together create us.

As *evolutionary psychologists* remind us, our inherited human nature predisposes us to behave in ways that helped our ancestors survive and reproduce. We carry the genes of those whose traits enabled them and their children to survive and reproduce. Thus, evolutionary psychologists ask how natural selection might predispose our actions and reactions when dating and mating, hating and hurting, caring and sharing. Nature also endows us with an enormous capacity to learn and to adapt to varied environments. We are sensitive and responsive to our social context.

To explain the traits of our species, and all species, the British naturalist Charles Darwin (1859) proposed an evolutionary process.

146

Follow the genes, he advised. Darwin's idea, to which philosopher Daniel Dennett (2005) would give "the gold medal for the best idea anybody ever had," was that natural selection enables evolution.

Natural selection implies that certain genes — those that predisposed traits that increased the odds of surviving long enough to reproduce and nurture descendants — became more abundant.

Natural selection, long an organizing principle of biology, has recently become an important principle for psychology as well. *Evolutionary psychology* studies how natural selection predisposes not just physical traits suited to particular contexts — polar bears' coats, bats' sonar, humans' color vision — but also psychological traits and social behaviors that enhance the preservation and spread of one's genes. We humans are the way we are, say evolutionary psychologists, because nature selected those who had our traits — those who, for example, preferred the sweet taste of nutritious, energy-providing foods and who disliked the bitter or sour flavors of foods that are toxic. Those lacking such preferences were less likely to survive to contribute their genes to posterity.

As mobile gene machines, we carry not only the physical legacy but also the psychological legacy of our ancestors' adaptive preferences. We long for whatever helped them survive, reproduce, and nurture their offspring to survive and reproduce.

"The purpose of the heart is to pump blood," notes evolutionary psychologist David Barash. "The brain's purpose," he adds, is to direct our organs and our behavior "in a way that maximizes our evolutionary success. That's it." (*Social Psychology*, p. 8, 159.)

Everything they wrote here is false or incomplete.

It's NOT a rectangle, it's a triangle! It's not just nature and nurture that form us. There's an essential third component!

Do you know what it is?

NATURE vs. NURTURE vs. NIRVANA: An Introduction to Reality

https://www.amazon.com/dp/B01JWRCSVA

https://www.amazon.com/dp/1521132615

The third component is deliberately eliminated from science by the Materialists, Naturalists, Darwinists, Nihilists, and Atheists. These people state that it does not exist. The BioPsychoSocial Model tells us that it does exist. Somebody is right, and somebody is wrong. They both can't be right.

These people have been teaching for over 150 years that Natural Selection made you, that evolution made you; but, that's physically impossible. Natural selection can't make anything. Natural selection and the theory of evolution are based exclusively on entropy. Natural selection results in entropy or death. Evolution is entropy. Entropy is death. Entropy cannot make anything at all. Entropy or death can only destroy. The different types of evolution can only destroy. The different types of evolution or entropy can only produce death and extinction.

Natural selection doesn't predispose anything! Natural selection doesn't organize anything! It can't. Natural selection doesn't touch our genes! Natural selection doesn't endow us with anything! Natural selection has NO ability to learn anything. Natural selection doesn't enable anything. Natural selection doesn't do anything. Natural selection is supposed to be dumb and blind without a soul or a mind, according to the Darwinists. Creation by Natural Selection wins the rotten tomato for the most stupid, illogical, irrational, and ineffective idea ever created.

The theory of evolution is correlational, NOT observational. NO type of evolution has ever been caught in the act of design and creation. It's physically impossible for natural selection and random mutations to design and create something. Entropy prevents them from doing so. Chemical evolution or macro-evolution is prevented from happening by random diffusion or entropy. Macro-evolution is also prevented from happening by genetics. The genes are there to prevent macro-evolution from happening. Evolution of any type is entropy and death. Death cannot create life!

In fact, evolution (genetic change), random mutations, and natural selection didn't even exist until AFTER God designed, programmed, engineered, field-tested, fine-tuned, manufactured, created, and produced the proteins and their matching genes in the first place.

Remember, genetics and genomes were designed to prevent macro-evolution from happening; and, they work as they were designed to do. Remember, Louis Pasteur falsified spontaneous generation, abiogenesis, and macro-evolution in 1859.

The theory of evolution is a fictional story that they made up out of thin air after-the-fact to fit the facts. NO part of it can actually design and create. It's a fictional story, not science.

We are sensitive and responsive to our social context, NOT our genes!

Who is this **WE** that they keep talking about in our Social Psychology textbooks? **WE** can't be our society, environment, or social context that **WE** are sensitive to and responsive to! And, it's definitely NOT our genes. Our genes aren't sensitive and responsive to anything according to the Evolutionists! The genes, natural selection, and random mutations are supposed to be dumb and blind without a soul or a mind. Our genes can't be sensitive nor responsive to anything.

Personal pronouns imply a person or a psyche – NOT our genes (nature) and NOT our environment (nurture).

Natural selection and your genes DO NOT and CANNOT pass your Psyche Legacy (psychological legacy) from one generation to the next! Natural selection doesn't touch your genes, and it definitely doesn't touch nor change your Psyche either. It's science fiction to imply that it does. The theory of evolution is science fiction. Design and creation by natural selection is science fiction. The idea that your genes carry your "longings" or "desires" from one generation to the next is science fiction. There's no such thing as genetic memory, at least not at the physical level. All of our thoughts and memories are carried as quantum waves from one generation to the next through our Psyche or Quantum Non-Local Consciousness.

Your genes don't care whether you live or die. Only YOU care whether you live or die. Your Psyche cares whether you live or die; but, your genes do not. In order for your genes to care, they would have to have some sort of Psyche, or Intelligence, or Consciousness, or Awareness. But, if your genes have a Psyche, then the very existence of that Psyche falsifies Materialism, Naturalism, and Darwinism which claim that Psyche does not exist. Physicalism, Naturalism, and the Theory of Evolution are self-defeating. They don't work as advertised because they can't work as advertised.

Natural selection results in entropy and death, NOT X-Men and new unique life forms. Natural selection cannot design and create and program genomes. Natural selection doesn't touch our genes.

Random mutations are also entropy; but, at least random mutations by definition in principle actually change our genes. However, random mutations cannot design and create anything either.

Entropy is death. Death cannot design and create new unique genomes and life forms. That's physically impossible! Mutation and Selection can only produce entropy or death. They are based exclusively on entropy or death. Death cannot design and create life. Death can only end life.

Remember, the theory of evolution is Creation by Entropy or Creation by Death. That's NEVER going to work because it's physically impossible!

Isn't it refreshing to finally have access to the truth, rather than all the science fiction that the Evolutionists have been feeding us throughout our lives?

Well, I think it is.

I used to be a Materialist, Naturalist, Nihilist, and Atheist until I finally started to study the evidence. The evidence and the truth set me free! The Science and the Scientific Evidence convinced me that God must exist in order to have done all the Science and Fine-Tuning which natural selection and evolution could NEVER have done.

Mark my Words

—

Can Natural Selection Create?

The Physicalists, Naturalists, Darwinists, Nihilists, and Atheists teach that natural selection can create anything that it sets its mind to. These people teach that natural selection made you.

Are they right?

They are not!

They are deceiving themselves and trying to trick us and deceive us as well.

Creation by Natural Selection is demonstrably false, which means that it has been falsified by Scientific Evidence. Natural selection doesn't do anything. Natural selection doesn't touch our genes, nor does it pre-determine our future. Natural selection doesn't have a mind. There is NO intelligence or psyche within natural selection. There may (or may not) be some type of intelligence or psyche within our genes; but, there is NOTHING there when it comes to Natural Selection or Evolution. Natural selection is a fictional concept that they made up out of thin air. It doesn't really exist as a person or an entity. The same can be said of evolution. I'm not the only scientist to have figured this out by now. Natural selection is worthless as a creative agent and can't function as a creative agent.

[Editorial Note: I have written permission from John C. Sanford to use all of the quotes from John C. Sanford which I use in my books, so long as I cite the sources which I have done.]

START OF THE QUOTE FROM "GENETIC ENTROPY" BY JOHN SANFORD — USED BY PERMISSION FROM THE AUTHOR JOHN SANFORD.

Chapter 9: Can Natural Selection Create?

Newsflash — Mutation/Selection cannot even create a single gene.

We have been examining the problem of genomic degeneration and have found that deleterious mutations occur at a very high rate. Natural selection can only eliminate the worst of these, while all the rest accumulate — like rust on a car. Might beneficial mutations at other sites in the genome compensate for this continuous and systematic erosion of genetic information? The answer is that beneficial mutations are much too rare, and are much too subtle to keep up with such relentless and systematic erosion of information. This is carefully documented by Sanford et al. (2013), and Montañez et al. (2013). It is very easy to systematically destroy information, but apart from the operation of intelligence it is very hard (arguably impossible) to create information.

This problem overrides all hope for the forward evolution of the whole genome. However, some limited traits might still be improved via Mutation/Selection. Just how limited is such progressive ("creative") Mutation/Selection? By now it should be clear that random spelling errors in an instruction manual could never give rise to an airplane component (say a molded aluminum part), which then resulted in a significantly improved overall performance of a jet plane. Not even with an unlimited number of flight trials/crashes and an unlimited budget. So, it is certainly reasonable to ask the parallel biological question, "Could Mutation/Selection create a single functional gene from scratch?"

A gene is like a book, book chapter, or an executable program — and minimally consists of a text string with 1,000 characters. Mutation/Selection could not create a single gene because of the enormous preponderance of deleterious mutations, even within the context of a single gene. The net information must always still be declining, even within a single gene or linkage block. Even if a gene was 50% established, deleterious mutations would degrade the completed half of the gene much faster than beneficials could create the missing half of the gene. However, to better understand the limits of forward selection, let us for the moment discount all deleterious mutations and only consider beneficial mutations. Could Mutation/Selection then create a new and functional gene?

1. Defining our first desirable mutation. The first problem we encounter in trying to create a new gene via Mutation/Selection is defining our first beneficial mutation. By itself, no particular nucleotide (A, T, C or G) has more value than any other, just as no letter in the alphabet has any particular meaning outside of the context of other letters. So, selection for any single nucleotide can never occur except in the context of the surrounding nucleotides (and in fact, within the context of the whole genome). A change of a single letter within a word or chapter can only be evaluated in the context of the surrounding block of text. This brings us to an excellent example of the principle of "irreducible complexity" within the genetic realm. In fact, it is irreducible complexity at its most fundamental level. We immediately find we have a paradox. To create a new function, we will need to select for our first beneficial mutation, but we can only define that new nucleotide's value in relation to its neighbors — and we are going to have to be changing most of those neighbors also. We create a circular path for ourselves. We will keep destroying the "context" we are trying to build upon. This problem of the fundamental inter-relationship of nucleotides is called epistasis. True epistasis is almost infinitely complex, and virtually impossible to analyze, which is why geneticists have always conveniently ignored it. Such bewildering complexity is exactly why language and information (including genetic language and genetic information) can never be the product of chance, but always requires intelligent design. The genome is literally a book, written literally in a language, and short sequences are literally sentences. Having random letters fall into place to make a single meaningful sentence, by accident, would require more tries (more time), than earth history can provide (i.e., "methinks it is like a weasel" would take $27 \wedge 28$ tries — that is 10 followed by 40 zeros). The same is true for any functional string of nucleotides. If there are more than a dozen nucleotides in a functional string, we know that realistically they will never just "fall into place". This has been mathematically demonstrated repeatedly. But as we

will soon see, neither can such a sequence arise by selecting one nucleotide at a time. A pre-existing "concept" is required as a framework upon which a sentence or a functional sequence must be built. Such a concept can only pre-exist within the mind of the author. Starting from the very first mutation, we have a fundamental problem even in trying to define what our first desired beneficial mutation should be.

2. Waiting for the first mutation. Let's assume we can know the first desired mutation. How long do we have to wait for it to happen? Human evolution is generally assumed to have occurred in a small population of about 10,000 individuals. The mutation rate for any given nucleotide, per person per generation is exceedingly small (very roughly about one mutation per 30 million individuals, for a given nucleotide site). Within a population of 10,000, one would have to wait 3,000 generations (at least 60,000 years) to expect a specific nucleotide to mutate. But two out of three times, it will mutate into the "wrong" nucleotide. So, to get a specific desired mutation at a specific site just in one individual will take three times as long, or at least 180,000 years. Once the mutation arises in one individual, it has to become "fixed" (such that each individual in the population will eventually have a double dose of that mutation). Because a newly arisen mutation arrives in a population as just a single copy, it arrives on the brink of extinction. The vast majority of new mutations soon drift back out of the population, even the ones that are beneficial. So, any specific desired mutation must arise many times before it "catches hold" in the population. Only if the mutation is dominant and has a very distinct benefit does selection have any reasonable chance to rescue it from random elimination via drift. According to population geneticists, apart from effective selection, in a population of 10,000, our given new mutant has only one chance in 20,000 (the total number of non-mutant nucleotides present in the population) of NOT being lost via drift. Even with some modest level of selection operating, there is a very high probability of random loss, especially if the mutant is recessive or is weakly expressed (we actually know that most mutations will be both recessive and nearly neutral). Therefore, even a beneficial mutation will be randomly lost due to genetic drift most of the time. Our numerical simulations suggest a weakly beneficial mutant will be lost about 99 out of 100 times. So, a typical mildly-beneficial mutation must happen about 100 times before it is likely to "catch hold" within the population. So, on average, in a population of 10,000 we would have to wait 180,000 × 100 = 18 million years to stabilize our first desired beneficial mutation, to begin building our hypothetical new gene. So, in the time since we supposedly evolved from chimp-like creatures (6 million years), there would not be enough time to realistically expect our first desired mutation to go to fixation in the genomic location where our required gene is hopefully going to arise. A vast amount of mutations would arise during 18 million years, but only once would that specific nucleotide mutate to that specific new nucleotide — such that it's not lost due to genetic drift and is fixed.

3. Waiting for the other mutations. After our first desired mutation has been found and fixed, we need to repeat this process for all the other nucleotides encoding our hoped-for gene. A gene is minimally 1,000 nucleotides long. More realistically, a human gene is on average about 50,000 nucleotides long, when regulatory elements and introns are included. To be extremely generous we will only consider a gene of 1,000 nucleotides (and we assume each nucleotide is by itself selectable). If this process was a straight, linear, and sequential process, it would require about 18 million years × 1,000 = 18 billion years to create the smallest possible gene. This is more than the time since the reputed Big Bang! So, it is a gross understatement to say that the rarity of desired mutations limits the rate of evolution. Furthermore, single nucleotides do not carry any information by themselves, and cannot be selectively favored. Specified information requires many characters (minimally, a sentence or similar text string is needed). Like any message, a genetic message which specifies some life function requires many nucleotides to reach its "functional threshold". Functional threshold is the minimal number of characters (or nucleotides) needed to convey a meaningful message. Below the functional threshold, individual letters or nucleotides have no benefit and cannot be favored by selection. This means that realistically, waiting time will be much, much longer — because no selection can happen until the minimum string of nucleotides falls into place by chance. If the functional threshold for selection is 12 (no selection until all 12 letters are in place), the waiting time in our hypothetical human population becomes trillions of years.

Sanford, John (2015-02-23). Genetic Entropy (Kindle Locations 1684-1755). FMS Publications. Kindle Edition. USED BY PERMISSION.

END QUOTE.

—

Trillions of years!

Well, that's the END of the Theory of Evolution, isn't it?

The Darwinists NEVER use their God-given brains to stop and think about these kinds of things. At the best possible average pace, with God making sure that there are NO deleterious mutations and NO devolution taking place, it would take on average 18 million years to fixate and stabilize a SINGLE beneficial mutation through "Natural Means" into a population of 10,000 apes which God has already designed and created in the first place and kept alive and functional during those 18 million years, just so that population of 10,000 God-created apes can achieve their first beneficial mutation through "Natural" Hands-off Mutation and Selection. 18 million years on average per beneficial mutation! Think about it!

If those apes need 1,000 such beneficial mutations in order to become men, then you are looking at 18 billion years on average to produce those targeted 1,000 beneficial mutations; and, that's with a population of 10,000 apes that God has already designed and created in the first place and that God is making sure receive ONLY beneficial mutations and NO devolution or deleterious mutations. And, that's also with God keeping that population of 10,000 apes alive during those 18 billion years so that they can indeed "evolve" their necessary 1,000 beneficial mutations and become men all on their own through "Natural Means".

Furthermore, it has been estimated that it would in fact take at least 20 million such beneficial mutations to convert chimpanzees into humans through "Natural Means". With that targeted goal in mind and assuming NO deleterious mutations or extinctions along the way, how long would it take on average to convert 10,000 chimpanzees into 10,000 humans using Natural Selection and Random Mutations to do the job? So, what do you get if you multiply 20 million beneficial mutations with 18 million years per beneficial mutation? At the BEST possible pace, with God keeping those 10,000 chimpanzees alive all along the way, and with God making sure that there is NO devolution, NO extinction, and NO deleterious mutations taking place, the quickest on average that Mutation and Selection could convert a chimpanzee into a human through "Natural Means" is 360 trillion years.

John Sanford isn't exaggerating whenever he says that it could take trillions of years for Mutation/ Selection to design and create something useful "naturally". And, it really isn't Natural Evolution if God has to design and create the 10,000 chimpanzees in the first place, and then keep them alive for 360 trillion years by blocking ALL deleterious mutations and preventing ALL extinctions that might take place during that period of time, just so He can convert 10,000 chimpanzees into 10,000 humans "naturally" or through "Natural Means".

Think about it! At the BEST possible average pace, it would take at least 360 trillion years for Mutation and Selection to convert a population of 10,000 chimpanzees into 10,000 humans through "Natural Means", with God keeping those 10,000 mutants alive and preventing deleterious mutations during the whole time. And, that's with 10,000 chimpanzees that God has already designed and created in the first place! How old did they say our universe is? How long would it take to convert a bacterium into a human through "Natural Means" when billions of beneficial mutations are needed? Wouldn't it be easier and faster to just let God design and create those 10,000 humans in the first place?

YES, it would be!

The Darwinists and Materialists NEVER stop and use their God-given brains to think about and calculate these kinds of things. You will NEVER get these kinds of calculations and truths from the Darwinists because they don't DO this kind of science. It's too difficult and painful for them. I'm willing to wrap my mind around these kinds of things. Your typical Darwinist isn't. Your typical Darwinist is afraid of it because they don't want to be proven wrong. For the Materialists and Darwinists, ignorance is bliss! But, ignorance is the reason why the Darwinists and Materialists truly believe that Mutation/Selection can design and create anything that it sets its mind to. The rest of us KNOW BETTER!

If you think about it, this is radically advanced science — the best that humans are able to come up with! Can you see and understand now why the 9th Chapter of "Genetic Entropy" put an END to the Theory of Evolution for me? It's because I understood what John Sanford was talking about and chose to believe that it is true. Now the onus is on you.

What do the Darwinists typically do when presented with these kinds of Statistical Models of the Mutation/Selection Process?

Assuming that they don't go head-in-the-sand and actually study them instead, the Darwinists try to shave the figures in half or by one-tenth, which is exactly what they do when designing Mutation/Selection Models of their own. They cheat. They choose parameters that are scientifically inaccurate and don't match with Reality in order to shave those estimates down to something that they might be willing to accept. They keep shaving and cheating until they get the numbers that they want, and then they call the results "Science".

If they really take John Sanford's Models seriously, the Darwinists will demand that God artificially accelerate the Mutation/Selection Process so as to make it possible for the Theory of Evolution to be true. But, even if God were to speed up the process a thousand-fold, it's still going to take 360 billion years for chimpanzees to evolve into humans through "Natural Means"; and, the Darwinists are still going to complain, even though we are starting with Chimpanzees that God has designed and created in the first place.

It can be fascinating and entertaining to watch a Darwinist try to shave trillions of years off a Statistical Estimate that he doesn't like, all in an attempt to increase the possibility that the Theory of Evolution might be true.

And, that's just the beginning of the problems for the Theory of Evolution. It only gets worse from there on forward, because John Sanford actually has TEN points in chapter 9 of "Genetic Entropy", each of which decreases the likelihood of Mutation/Selection creating anything at all, even if it has an infinite number of years to do so. Evolution by random mutations and evolution by natural selection CANNOT design, create, and deploy anything! It has been conclusively and finally demonstrated that it is so. It has been empirically and logically observed to be so. The Theory of Evolution doesn't work and can't create new unique genomes from scratch, so we have no choice but to declare the whole thing to be FALSE.

Many different scientists taught me that Random Mutations and Natural Selection (Evolution) cannot design and create genomes and life forms. They met their burden of proof and demonstrated to me that it must be so. (See the partial list of Reference Materials below for a selection of some of the best scientific evidence that falsifies the Theory of Evolution.)

I chose to believe the scientific evidence rather than the claims of the Materialists, Naturalists, and Darwinists who teach that Psyche, Intelligence, and Syntropy do not exist.

Evolution Is Creation by Death

I did make a scientific discovery in recent days (May 2018) that I think has merit and value to the Scientific Community.

I do seem to be the first person on the planet to realize that NO type of evolution can do selection. Selection of any kind involves choice; and, choice is exclusively the product of a Psyche or a Mind. Evolution of any type by definition, in principle, is dumb and blind without a soul or a mind. The very definition of Evolution eliminates its ability to do selection, a priori. Evolution can't do selection because evolution can't do Psyche or Choice. Evolution doesn't exist as a Psyche, Person, Intelligence, Mind, or Soul capable of doing choice or selection.

Furthermore, any attempt to imbue evolution with a soul, psyche, or mind automatically FALSIFIES Materialism, Naturalism, Darwinism, Nihilism, and the Theory of Evolution which claim that Psyche or Syntropy does not exist. The theory of evolution is a non-starter because evolution of any kind cannot do selection or choice.

Always remember, creation by natural selection is science fiction and wishful thinking because evolution of any kind cannot do selection or choice.

Materialism, Naturalism, Darwinism, Nihilism, Behaviorism, Determinism, Physical Reductionism, Atheism, Classical Physics, and the Theory of Evolution are based exclusively on entropy. Entropy is death. Death cannot design, create, and produce life. Such a thing has never been experienced nor observed. Darwinism or the theory of evolution is Creation by Death, or Creation by Entropy. Creation by Death or the theory of evolution is impossible. It can't happen, which means that it didn't happen.

Therefore, the theory of evolution cannot be used to explain the origin of life. The most that the theory of evolution can explain is death, extinction, devolution, and genetic entropy. Evolution of any kind is a function of death or entropy. Evolution is entropy. Evolution is death. Evolution or death, of any kind, can't be used to explain the origin of life. The truthfulness of this reality becomes obvious, once a person realizes and accepts the fact that the Theory of Evolution, Physicalism, Naturalism, Atheism, Nihilism, Classical Physics, and Darwinism are based exclusively on entropy. Entropy or death cannot produce life. The different types of evolution or entropy cannot produce life. It's impossible for entropy or death to produce life.

At times I've wondered why nobody else has been able to see and understand these obvious truths; but then, I used to be a Materialist, Naturalist, Nihilist, and Atheism, and there I find my answer. At the time, I wasn't able to see nor understand these obvious scientific truths because I didn't want to see them, understand them, nor accept them. No seeking, then no finding. I wasn't looking for any of this, so I never found it. I only found it after I started looking for it. I had convinced myself that this type of information doesn't exist. Self-deception works, and it works every time, especially when it comes to scientists like me.

The axiom stating that evolution is dumb and blind without a soul or a mind, if taken as being true, prevents evolution of any type from being able to do selection. In other words, the very definition of evolution as being dumb and blind FALSIFIES the Theory of Evolution by preventing evolution of any kind from being able to do selection or choice. Evolution is entropy; and, entropy is death. Death cannot do selection or choice. Death cannot do life. Entropy or death can only destroy. That is what has been experienced and observed.

Meanwhile, the Materialists, Naturalists, Darwinists, Nihilists, and Atheists DEMAND that you accept on blind faith that their claims – that death, entropy, or evolution can produce life – are true.

We scientists have FALSIFIED the Theory of Evolution trillions of times in thousands of different ways, but we choose to ignore the evidence because it isn't telling us what we want to hear. That's the way we do science in this world – by ignoring the evidence, discounting the evidence, banning the evidence, and destroying the evidence. We do our science this way so that we can prove to ourselves and to others that the Theory of Evolution is true.

There is another way to do science, though – a better way of doing science. I call it Science 2.0; and, it involves allowing ALL of the evidence into evidence. Once we choose to do so, then ALL of the evidence that we have on hand as a race FALSIFIES the claims of Materialism, Naturalism, Darwinism, Nihilism, Behaviorism, Determinism, Physical Reductionism, and Atheism which claim that this evidence does not exist.

Once we have eliminated all of the falsehoods such as Materialism, Naturalism, Atheism, Nihilism, and Darwinism, then we are left staring at THE TRUTH, which is that ONLY Psyche can design, program, engineer, field-test, fine-tune, manufacture, create, and do science.

By eliminating all of the falsehoods or pseudo-sciences, it becomes obvious that God's Psyche must of necessity exist in order to have DONE all of the Science that needed to be done, which evolution and the rocks could NEVER have done. Remember, evolution and the rocks can't do science because they can't do selection or choice.

The observation and realization, that evolution of any type cannot do selection, just might be one of my greatest scientific discoveries even if I don't end up being the first person on the planet to have made this discovery.

Obviously, everyone across the world is now starting to make these kinds of scientific discoveries right and left because we have finally started to take our blinders off. More and more of us are willing to see, which makes us able to see. Nowadays, it's obvious that the Theory of Evolution is false; whereas, we couldn't see it before because we didn't want to see it, and we didn't know where to look.

All you want is the truth. Everything else is worthless in the end.

Mark My Words

—

Source

Science 2.0: I Upgraded My Science

https://www.amazon.com/dp/B0771K6WTX

The Scientific Method Proves That the Theory of Evolution Is False

https://www.amazon.com/dp/B01IAAIRT2

https://www.amazon.com/dp/1521133611

NATURE vs. NURTURE vs. NIRVANA: An Introduction to Reality

https://www.amazon.com/dp/B01JWRCSVA

https://www.amazon.com/dp/1521132615

Myers, D. G. (2010). *Social Psychology* (10th ed.). New York: McGraw-Hill.

Reference Materials

Wells, J. (2000). *Icons of Evolution: Science or Myth? Why Much of What We Teach About Evolution Is Wrong*. Washington, DC. Regnary.

Sanford, J. (2014). *Genetic Entropy* (4th ed.). Cornell University: FMS Foundation.

Sanford, J. C., Marks, R. J., Behe, M. J., Dembski, W. A., & Gordon, B. L. (Eds.). (2013). *Biological Information: New Perspectives*. Hackensack, NJ: World Scientific.

Meyer, S. C. (2010). *Signature in the Cell: DNA and the Evidence for Intelligent Design*. New York: HarperCollins.

Meyer, S. C. (2013). *Darwin's Doubt: The Explosive Origin of Animal Life and the Case for Intelligent Design*. New York: HarperCollins.

Mark My Words. (2016). *The Scientific Method: Proves That the Theory of Evolution Is False*. Kindle. Retrieve from: https://www.amazon.com/dp/B01IAAIRT2

Mark My Words. (2016). *The Theory of Evolution Proved to Me that God Exists: Why I Am No Longer an Atheist and Why I No Longer Believe in the Theory of Evolution*. Kindle. Retrieve from: https://www.amazon.com/dp/B01HZYBZ7K

Kin Selection

Web Page: https://evolution-is-entropy.com/2018/05/18/kin-selection/

Natural Selection is science fiction. By definition, in principle, Natural Selection can't do CHOICE or selection. Natural selection can't do what they say it does. By definition, in principle, natural selection or survival of the fittest is supposed to be dumb and blind without a soul, psyche, or mind. Therefore, natural selection can't do choice or selection. Instead, natural selection results in entropy or death. Natural selection, or entropy and death, cannot design and create and produce anything. Design and creation by Natural Selection is prevented from happening by entropy, random diffusion, or the second law of thermodynamics. Random mutations are also entropy. Entropy is death. The theory of evolution is Creation by Entropy, or Creation by Death. That's not going to work because it's physically impossible. Death or entropy cannot design and create life. This reality is obviously true.

Do you want the truth, or do you want the science fiction?

Can you handle the truth? Most of our scientists can't.

Kin Selection is another smoking gun where Psyche is concerned.

Kin Selection: The idea that evolution has selected altruism toward one's close relatives to enhance the survival of mutually shared genes.

Our genes dispose us to care for relatives. Thus, one form of self-sacrifice that *would* increase gene survival is devotion to one's children. Compared with neglectful parents, parents who put their children's welfare ahead of their own are more likely to pass their genes on. As evolutionary psychologist David Barash wrote, "Genes help themselves by being nice to themselves, even if they are enclosed in different bodies." Genetic egoism (at the biological level) fosters parental altruism (at the psychological level). Although evolution favors self-sacrifice for one's children, children have less at stake in the survival of their parents' genes. Thus, according to the theory, parents will generally be more devoted to their children than their children are to them. (*Social Psychology*, p. 452.)

By definition, in principle, evolution is dumb and blind without a soul or a mind. This means that evolution of any type cannot DO choice or selection. It's a deceptive lie to say that evolution can do choice or selection. It's science fiction.

Without realizing it, these people have painted themselves into a corner with this one. The idea of genes helping each other even if they are enclosed in different bodies is physically impossible. There's NO physical mechanism in place whereby the genes can communicate with each other, especially the genes in different bodies! Such an idea is ludicrous. If the genes are communicating with each other and recognizing each other, then they are doing so at the quantum level or the psyche level because they can't do so at the physical level.

The ONLY way to make Kin Selection true is if we all axiomatically agree in advance that the genes (and evolution) are psychic and are therefore capable of determining telepathically which genes are related to them and which genes are not. In order for the genes to be nice to each other, they have to know each other, perceive each other, recognize each other, and show favoritism to each other. They have to be psychic and have some kind of psyche. The genes can't be dumb and blind if we want Kin Selection to work as advertised. The genes have to communicate with each other, know each other, and recognize each other in order for them to be able to help themselves and be nice to themselves. The genes also have to be subliminally communicating their desires to the Human Psyche in order to make the Human Psyche be nice to the genes too. The only way to make Kin Selection work as advertised is if we agree in advance that the genes have some sort of Psyche or Mind and are telepathically connected with each other.

However, if we agree in advance axiomatically that genes have a psyche, that genes perceive each other telepathically, that genes know each other and recognized each other, and that genes are therefore psychic, then we have in fact FALSIFIED Materialism, Naturalism, Darwinism, Nihilism, Atheism, even the Theory of Evolution in the process.

These people personify the genes – imbue them with Psyche and Intelligence – and in the process falsify the major premises of Materialism, Naturalism, Darwinism, Nihilism, and Atheism which state that Psyche or Syntropy does not exist. Thereby, these falsified philosophies or falsified religions end up being self-defeating. The truth cannot be built upon falsehoods

When it comes to Science, we observe that Psyche, Quantum Mechanics, or Syntropy always ends up being the best possible explanation that can be given to ALL of the evidence that we are observing and experiencing, including the physical evidence. Remember, entropy and physical matter would not exist without a massive initial infusion of Syntropy, Intelligence, Power, Quantum Mechanics, or Psychic Intervention somewhere sometime along the way.

Remember, perception is a function and a product of Psyche. At the physical level, the genes have no way of knowing or perceiving which genes are related to them and which genes are not; and at the physical level, the genes have NO way to pass that information on to the physical body or physical brain even if the genes were to know who is related to them and who is not. At the physical level, the genes have no way to perceive each other, thereby falsifying the claims of Kin Selection. By restricting and limiting everything to the physical level, the Materialists and Naturalists automatically falsify anything and everything that needs Psyche or Intelligence or Perception in order to become true – things such as Kin Selection and Creation by Mutation/Selection.

Remember, evolution can't do selection! By definition, in principle, evolution cannot do CHOICE! Selection requires some type of choice, and choice requires some type of Psyche or Mind! By definition, in principle, evolution is dumb and blind without a soul or a mind. The theory of evolution is self-defeating because evolution of any kind can't do selection or choice. Evolution doesn't exist as some type of Psyche or Person who is capable of making choices or doing selection. The false is falsified by the truth; and, the truth is repeatedly experienced and observed. The theory of evolution is obviously false because evolution of any type by definition in principle cannot do selection, psyche, or choice.

Evolution of any type cannot do science, but the Human Psyche and God's Psyche certainly can. They've been caught in the act of doing so.

With just a bit of scientific observation, it's easy to see that Kin Selection is a *fictional ad hoc just-so story* that they made up out of thin air after-the-fact to match with what we have experienced and observed from Intelligent Beings or the Human Psyche. They took Kin Selection, a philosophical idea or religious idea, and they personified it, humanized it, anthropomorphized it, and deified it. Kin Selection and the Theory of Evolution are man-made idols or man-made gods. The Theory of Evolution is our modern-day form of idolatry, which is a belief in false gods that are incapable of delivering the goods.

The theory of evolution is self-defeating because it can't do what they say it does. The genes can't be nice to each other without a psyche or a soul; and, if the genes have a psyche or a soul, then the very existence of Psyche or Syntropy or Soul falsifies Materialism, Naturalism, Darwinism, Nihilism, Behaviorism, Determinism, Physical Reductionism, Atheism, and the Theory of Evolution. The false is falsified by the truth; and, the truth is repeatedly experienced and observed.

Mark My Words

—

Source

God Is in the Light: God is light, and in Him is no darkness at all.

https://www.amazon.com/dp/B07168S37N

Reference

Myers, D. G. (2010). *Social Psychology* (10th ed.). New York: McGraw-Hill.

A Philosophical Justification for Psyche

Web Page: https://philosophy-of-science.com/philosophical-justification-for-psyche/

I own and have read from many different books and articles that use philosophical sophistry to prove that psyche, mind, consciousness, choice, or syntropy does not exist. The best they can do is to convince you that psyche, spirit, or soul does not exist. They really can't prove that it doesn't exist. It's impossible to prove that something doesn't exist; but, the Materialists, Naturalists, and Atheists try to do so anyway.

I also own and have read from many different books and philosophers who use philosophical sophistry to provide philosophical justification or a philosophical foundation for psyche, mind, consciousness, awareness, choice, or syntropy. Although interesting to study and think about at times, it's really not convincing either if you choose not to believe in it. Our beliefs are chosen into existence by our Human Psyche; and, the Human Psyche can choose to believe that the Human Psyche doesn't exist.

There are times when I don't like philosophy and philosophers. They can say a lot without actually saying anything useful or interesting. It ends up being lots of obtuse, unintelligible, circular, tautological wordiness that goes nowhere when they are done. I have learned to despise philosophical sophistry that never actually gets to the point and never gets to the point of making sense. An essay and book called "Reconsidering Psychology" by Faulconer and Williams is a case in point – there are sixty pages of words within the first essay of the book that never seem to come to a point and never seem to make a point. They say a lot without actually saying anything useful or interesting; and, when you are done, you still have no idea what point they were trying to make. Science needs a philosophical and metaphysical foundation; but, it also needs to make sense when they are done.

Instead of philosophical sophistry, I like getting to the point, revisiting the point over and over again from many different angles, and hammering on the point to the point of exhaustion until it becomes solid, obvious, and clear to me how things really work. I learn best by trying to explain my scientific theories, hypotheses, and ideas to someone else. I want everything to be plain and simple to understand. I want everything to be obvious, clear, and to-the-point when I'm done with it. I like power points. I like scoring points. It's what I choose to do.

It all comes down to choice. Choice is a product of Psyche, Intelligence, Consciousness, or Awareness. Psyche is synonymous with choice, and Psyche or Intelligence is cornered, pin-pointed, and identified by the choices that it makes.

The Gods can form energy into anything and can control everything, except for our choices. The Gods cannot create Psyche or Intelligence, and the Gods can't destroy it either, which means that the Gods cannot control our choices. The Gods can determine consequences for our choices; but, they cannot make our choices for us because then they would no longer be choices. It all comes down to choice. Choice is a function of psyche or intelligence. Choices are produced by psyche or intelligence. Psyche or Intelligence is identified by the choices that it makes. Choice is the surest sign of Psyche's existence.

As I see it, the only philosophical justification that psyche or mind really needs is the empirical fact that it has been experienced and observed. Science is observation and experience; and, the Psyche has been experienced and observed, therefore we KNOW that it is real and truly exists. That's all the philosophical justification that it needs. Observation and experience trump philosophical sophistry every time. Once it has been experienced and observed, the philosophical debate should be over. Of course, that doesn't stop the Materialists, Naturalists, Darwinists, Nihilists, and Atheists from trying to convince you otherwise; but, these people are wrong to do so.

Psyche or Intelligence is experienced first-hand by out-of-body explorers as an immaterial viewpoint in space. Whenever they go looking for their own psyche, there's nothing there to be seen or found. Psyche or Intelligence is seen as a pinprick of light, a point particle of light, or a spark of light.

If there is such a thing as a point particle or an infinite singularity, Psyche is it. Psyche, Intelligence, or Thought is the fundamental unit of reality. Psyche is the elementary particle that the physicists are looking for. A psyche ontology is the ultimate model of reality and existence. Psyche is the ultimate cause, which means that psyche is the ultimate causal agent or the ultimate causal force. Psyche gets straight to the point.

Out-of-body explorers have noticed that it is their immaterial intangible Psyche or Consciousness who has experiences and forms memories, not their spirit body and not their physical body. Your spirit body and physical body don't have experiences and don't form memories while your psyche is separated from them. It's the Psyche or the Intelligence who has experiences and forms memories of those experiences. That is what has been experienced and observed.

Psyche or Intelligence has been experienced and observed by out-of-body explorers and during near-death experiences; therefore, we KNOW that it is real and truly exists. That's all the philosophical justification and philosophical foundation that it needs.

Mark My Words

—

Reference

Buhlman, William L. (1996). *Adventures Beyond the Body: How to Experience Out-of-Body Travel*. New York: HarperCollins.

Experiential and Observational Proof of Psyche

Website: https://psyche-ontology.com/psyche-experienced-and-observed/

The Materialists, Naturalists, Darwinists, Nihilists, Behaviorists, Determinists, Physical Reductionists, and Atheists teach and believe that there is NOTHING fundamental about energy or consciousness. I have even heard them say the words, "There is nothing fundamental about energy." These people teach us and tell us that Psyche or Non-Local Consciousness DOES NOT EXIST. These people erroneously teach and believe that ONLY entropic physical matter exists; consequently, these people teach and believe that energy or consciousness is an emergent property of entropic physical matter. Whether they realize it or not, these people also teach and believe that entropic physical matter and entropy are conserved. They got their priorities and science upside-down.

Psyche is the innate intelligence within all the different FORMS of energy which gives that energy the inherent ability to understand, follow, and obey God's Laws and God's Commands. Energy is psychic and intelligent. Energy is sentient, conscious, aware, and alive. Psyche is Energy, and Energy is Psychic. Energy or Psyche is the fundamental unit of reality because Psyche, Intelligence, Life Force, or Energy is always conserved. In contrast, physical matter, particles, and entropy are simply different FORMS of energy. The FORM is never conserved. This is what has been experienced and observed. Particles or quanta are popping in and out of existence all the time. However, they don't really cease to exist as the Materialists and Naturalists claim. They simply change state, form, phase, or dimension while their underlying Intelligence or Energy is always conserved. Particles or quanta are made and then destroyed. They are NOT conserved. Only the underlying energy, psyche, or intelligence is conserved. This is what has been experienced and observed. Entropic physical matter is NOT the fundamental unit of reality after all. Entropic physical matter is MADE from many different FORMS of energy; and, the FORM is never conserved. This is the Ultimate Law of Thermodynamics.

I learn best through comparison and contrast by comparing what has been experienced and observed with the philosophical speculation and wishful thinking of the Materialists, Naturalists, Darwinists, Nihilists, and Atheists.

Psyche has been experienced and observed. Psyche is experienced as an immaterial, non-physical, intangible viewpoint in space. Psyche is seen or observed by out-of-body travelers as a pinpoint of light, a point particle of light, or a spark of light.

As scientists, we have to expand, enhance, and upgrade Theoretical Physics and make it explain what has already been experienced and observed. Psyche is like a photon, or a particle of light, or a quantum of light, except for the fact that Psyche actually emits light, or emits quanta, or emits and stores quantum waves. Thoughts and memories are quantum waves. Psyche is like a permanent, infinitely stable, perfectly conserved particle of energy or quantum of energy. Psyche is NOT ephemeral – coming and going – transforming and being transformed – like the other elementary particles or quanta that have been discovered and observed. Psyche, and the energy within psyche, is conserved. It's stable. It's eternal and everlasting. Because Psyche is conserved, its FORM never changes. Psyche is always a pinprick of light, or a point particle of light, or a spark of light. Psyche is like a Starbase that instantiates, or starts, or makes the other quanta or quantum waves. The traveling quanta have to come from someplace, and they come from the Starbases that we call Psyche. Psyche makes and stores quantum waves. That's how Psyche is able to communicate with other Psyches – through quantum waves. Psyche ends up being the fundamental unit of reality, or the conserved unit of reality. Psyche is Energy, and Energy is Psychic.

The vacuum of space is NOT empty. It is comprised of many different fields of energy which are omnipresent in time and space. The various different Quantum Fields were MADE from energy. "Particles" or quanta are excitations or "packets of energy" within Quantum Fields. Psyche is the innate intelligence within ALL the different FORMS of energy which gives that energy the inherent ability to understand, follow, and obey God's Laws and God's Commands.

The **gluons** binding quarks together comprise 99% of the energy or mass in a proton and neutron. The quarks themselves comprise only 1% of the energy or mass in a proton and neutron. The massive energy of the **gluons** within a proton, as compared to the rest energy of the quarks alone in the QCD vacuum, accounts for almost 99% of the mass within a proton. It's ALL made from energy, not entropic physical matter! The gluons or energy are what give the neutrons and protons stability and make them last essentially for forever. Proton decay has NOT been observed. Why? It's because the underlying energy is always conserved. Gluons and quarks are just different FORMS of energy. **Gluons** are viciously strong attractive forces within the Quantum Chromodynamic Field. Who is making these quarks and gluons ACT this way rather than some other way? It has to be Someone Psyche. Therefore, Psyche or Consciousness ends up being the fundamental unit of reality – NOT quarks and gluons.

Protons and neutrons can be transformed into each other which means that they are NOT conserved; but, the underlying energy is always conserved. Remember, the FORM of the energy is NEVER conserved; but, the underlying energy is always conserved. If the Gods were to pull all of the **gluons** out of this physical universe into a different dimension or universe, the whole thing would dissolve back into the chaos or anarchy from whence it came. The gluons, quarks, forces, fields, laws, restrictions, order, and organization were MADE, which means that entropy and physical matter were also made. Anything that is made or organized is NOT being conserved. In contrast, the underlying Energy or Psyche or Intelligence was NOT made and it cannot be destroyed because it is always conserved. This is what has been experienced and observed.

I define Science as observation and experience; and, I have observed that the observations and experiences of the human race as a whole FALSIFY Materialism, Naturalism, Darwinism, Nihilism, Atheism, and their derivatives. Contrary to what these people claim, entropic physical matter is NOT the only thing that exists. Entropic physical matter is NOT the fundamental unit of reality. Materialism, Naturalism, Darwinism, Nihilism, and even Atheism are FALSIFIED by Quantum Field Theory and the Ultimate Law of Thermodynamics. Psyche or Quantum Non-Local Consciousness ends up being the fundamental unit of reality. A Psyche Ontology ends up being the Ultimate Model of Reality.

We have to expand, enhance, and upgrade our Theories of Physics to include Psyche or the Conserved; otherwise, we will NEVER be able to explain what has been experienced and observed.

Quantum Field Theory supports and predicts the Ultimate Law of Thermodynamics which states that physical matter and the other FORMS of energy such as entropy are NOT conserved. It's all made from energy, and entropic physical matter consists of different FORMS of energy. Quanta or "particles", including physical matter, are packets of energy. Quanta or particles are changing FORM all the time, which means that they are NOT conserved. Quantum Field Theory supports these ideas. The energy in the quantum fields is conserved; whereas, the "particles", or physical matter, or quanta come and go which means that they are NEVER conserved. You can see the truthfulness of these assertions with your own eyes by studying the Feynman Diagrams. There is NOTHING fundamental about physical matter. Everything is made from different FORMS of energy, including physical matter. Made by whom? Organized by whom? By Someone Psyche! Who else would be able to organize energy into its different FORMS at the quantum level or the psyche level?

Packets of energy, fields of energy, quarks made from energy, gluons made from energy, and the Psyche or Intelligence within all that energy ends up being the fundamental basis of reality. Psyche or Intelligence ends up being the fundamental unit of reality or the causal unit of reality – the conserved unit of reality. In contrast, entropic physical matter is MADE from many different quanta or packets of energy that we often call "particles"; and, when it comes to the electrons and the virtual photons in the Feynman Diagrams, it is obvious that they are NOT conserved. They are being transformed all the time! ONLY the underlying Energy, or Psyche, or Intelligence is conserved. This is what has been experienced and observed!

The Secrets of Feynman Diagrams

https://www.youtube.com/watch?v=fG52mXN-uWI

https://syntropy.website/wp-content/uploads/2018/08/Feynman-Diagrams.zip

http://hyperphysics.phy-astr.gsu.edu/hbase/Forces/feyns.html

Remember, Psyche is the person who makes the particles or the quanta and starts them moving in the first place. Psyche or Intelligence is the causal force or organizing force behind the quantum fields, quantum packets of energy, "particles", and the choices being made during the particle interactions that are diagrammed on the Feynman Diagrams. Psyche or Intelligence is the reason why the quarks, gluons, bosons, and electrons ACT the way they ACT rather than acting some other way or not acting at all. Psyche or Intelligence explains why there is something rather than nothing. Psyche is the thing that instantiates quanta, makes quanta, or causes the "particles" to begin in the first place. Psyche is the thing that chooses or decides what type of quantum a newly generated energy packet will be. Psyche is the Organizing Force behind everything else. Psyche is like a permanent Starbase – it generates, transmits, receives, and stores quantum waves or packets of energy. Psyche is conserved. Psyche is the fundamental unit of reality and existence. By allowing Psyche and Quantum Mechanics in to play, we can literally explain everything that comes our way. Try it, you might like it!

Due to the observations and the experiences of the human race as a whole, I upgraded my science to Science 2.0. According to Science 2.0, the BEST and FASTEST way to find and know the truth is to live it and experience it first-hand for yourself, or to choose to trust someone who has. The Human Psyche is KNOWN by living it and experiencing it, especially while your physical brain and physical body are dead and gone. Science 2.0 allows all of the evidence into evidence and then pursues a preponderance of that evidence. That's how I was able to identify and discover the Ultimate Law of Thermodynamics. That's how I was able to FALSIFY Materialism, Naturalism, Darwinism, Nihilism, and Atheism.

Phenomenology or the Lived Experiences of the human race had a very powerful influence in changing and altering my life, my science, my philosophy of life, and my worldview. The Phenomenologists, Near-Death Experiencers, and Out-of-Body Travelers changed the way that I look at the world. Thanks to these people, I'm no longer the same individual that I used to be. I'm no longer a Materialist, Naturalist, Nihilist, and Atheist.

In order to fully comprehend Quantum Mechanics and Syntropy, you must understand and accept the impact and influence of Quantum Non-Local Consciousness (or Psyche) on quantum mechanisms. Without that knowledge and understanding, your comprehension of Quantum Mechanics will always be incomplete. You must find an interpretation of Quantum Mechanics and Syntropy that explains what the Human Psyche and Nature's Psyche are doing at the quantum level or psyche level in order to get things done for us at the physical level. Without this bridge between the quantum and the physical, your knowledge and understanding of science will always be incomplete and ineffective. You must know how Psyche or CHOICE fits into the picture, if you want to understand how science really works at every level of existence.

Here's a list of some of the observational experiences of Psyche or Intelligence that changed my life.

Psyche or Consciousness without a Spirit Body

Nowadays, I define Science as observation and experience; and, I treat experience as scientific evidence. After reading William Buhlman's account of how he left his physical body, asked to see his spirit body, and found that he was an immaterial viewpoint in space while looking at his spirit body, I started looking for corroborating evidence to verify what William Buhlman reported. I began looking for the people who had the experience of being an immaterial pinpoint of consciousness, a pinprick of light, or a spark on the ceiling instead of looking for themselves and seeing parts of their spirit body. I have been looking for the experience of Psyche and the observation of Psyche.

William Buhlman wrote the following:

Journal Entry, October 2, 1982

I hear the buzzing, engine-like sounds and will myself out-of-body. [He left his physical body behind.] **I step to the bedroom door and automatically request "Clarity now!" My vision improves, and I step through the door, into the living room. Still feeling a little out of sync, I verbally repeat my request with more emphasis, "Clarity now!" I feel my awareness and vision snap into place.**

My thoughts are clear, and I make a verbal demand, "I need to see the form I'm in now!" Instantly I feel an intense sensation of being drawn within myself. I'm suddenly different, weightless as though I'm floating in space. As I look forward I see a sparkling, bluish white form. For some reason, I seem to know that I'm looking at my nonphysical body from a different perspective. I stare in amazement at this form before me that shines and flows with energy and light. It looks like an energy mold created from a million tiny points of light; it radiates a bluish glow but appears to have a defined outer structure. The body of light before me is naked and is identical to my physical form. Even though my body looks firm, there is a noticeable energy motion and radiation present. I can see what appears to be an ocean of blue stars throughout my body. It's difficult to describe because the stars are stable yet moving at the same time; the light and energy of my [spirit] **body appear to change and flow almost like the waves of an ocean.**

As I stare at the body of light, it hits me that I must be in another body. Yet I can't perceive any form or substance; I'm like a viewpoint in space without shape or form of any kind. [He, his immaterial viewpoint in space, is pure psyche or intelligence or consciousness.] **As I reflect upon my new state of being, I feel a sensation of rapid motion and I'm instantly back within my physical body.**

Lying still and reviewing my experience, I'm struck by an inescapable conclusion: I must possess multiple energy-bodies. The form I just experienced was noticeably lighter (less dense) than even my second nonphysical body. I realize that the traditional view of our possessing two bodies — a physical body and a spiritual body — is far too simplistic; we are much more complex than this. Just as there are multiple nonphysical energy dimensions within the universe, each of us must consist of multiple energy-bodies or vehicles of expression.

Now I seriously wonder just how many nonphysical bodies or forms this involves. I suspect that there must be one within each dimension of

the universe and that all of these are interrelated and connected, just as the physical body is connected to its first nonphysical (spiritual) body.

https://psyche-ontology.com/buhlman/

[See: Buhlman, W. L. (1996). *Adventures Beyond the Body: How to Experience Out-of-Body Travel*. New York: HarperCollins.]

For me personally, this was the turning point. This was the straw that broke the camel's back. After reading this book, and this journal entry in particular, I've KNOWN that Psyche or Intelligence is something completely different than one's spirit body. Since that moment, I've KNOWN that Psyche or Intelligence is an immaterial viewpoint in space – the part of us that has experiences and forms memories of those experiences. Ever since then, I've KNOWN that our spirit body and physical body don't have experiences and don't form memories while our Psyche is separated from them.

I treated this account as Scientific Evidence; and, I went searching for confirming evidence. Seek and ye shall find. I found what I was looking for, which tells me that it is true. Psyche is experienced as an immaterial viewpoint in space. Psyche is seen or observed as a point particle of light, a pinprick of light, or a spark of light. Psyche has been experienced and observed! That REALITY makes it Science because Science is observation and experience.

Science 2.0 allows ALL of the evidence into evidence, and then it pursues a preponderance of the evidence. The BEST and FASTEST way to find and know the truth is to live and experience it for yourself, or to choose to trust someone who has. If a phenomenon is real, then you should be able to find corroborating evidence; and I do, when it comes to Psyche or Quantum Non-Local Consciousness.

When people are separate from their spirit body and physical body, and they are viewing their physical body while they are an immaterial spark on the ceiling, they are experiencing their Psyche first-hand rather than their spirit body. It happens to some out-of-body travelers, but not all of them.

A more common out-of-body experience is to see parts of their spirit body – hands, arms, legs, and feet – but not their skull or eyes. If you close your eyes right now and try to sense where you are when your physical eyes are closed, you will sense that you (your psyche) is in the third eye position just behind your skull. That's the position from which you perceive things as well while inside your spirit body exploring the astral plane.

Psyche has been experienced and observed. Our job as scientists is to figure out what it is and how it works rather than trying to explain it away, dismissing it, and pretending that it doesn't exist. Denial of evidence and rejection of evidence makes for bad science. Feynman Diagrams map or portray the interaction of "particles" or quanta within Quantum Fields. Psyche is the thing that makes those particles and starts those particles moving in the first place.

One of my most interesting and useful Scientific Discoveries came to me when I first realized that Intelligences or Psyches, whenever they are

169

seen or observed, manifest as a Pinpoint of Light, a Spark of Light, a Pinprick of Light, or a Photon of Light BOTH at the quantum spiritual level and at the macro physical level.

Direct experiential evidence of Psyche is rather rare; but, I have found a few experiential accounts of the nature and functionality of Psyche or Non-Local Consciousness while it is separated from its spirit body. Most of them were found on YouTube. Here are a few of them:

A Point of Consciousness in a Wonderful Black Void and Pinpricks of Light

https://www.youtube.com/watch?v=quU1xPeOtWs

She went looking for her body, hands, and feet and couldn't find them. "I was a point of consciousness," she said. She also saw other pinpricks of light that she recognized as souls that had passed on.

This NDE and OBE is probably the closest to what I experienced when I died. I remember floating in a crystal sharp, bright, black, obsidian void; and, for the first time in memory, I was completely and totally at peace. A part of me died. My five decades of fear and anxiety died. I experienced some kind of spiritual rebirth.

I didn't see my spirit body, didn't look for my spirit body, and didn't see anything else either that I remember. It was bright, black, and I was totally at peace. I had finally found rest and peace for my troubled soul.

Based upon my experience, the Atheists are right, there's nothing there when we die; however, the Atheists are wrong, because I was there when I died.

Each person who dies has a different experience and comes back with a different perspective, which is why it's important to allow ALL of the evidence into evidence, and then pursue a preponderance of the evidence.

I seemed to experience many of the benefits of a Near-Death Experience and none of the side-effects or bad-effects – without having any kind of astounding or amazing experience worth reporting. I was finally at peace; and, it stayed with me to this very day. The fear and anxiety were gone. I'm finally at peace. I see it as a gift from God.

A Unit of Consciousness

She talks about not being in a spirit body anymore but being a unit of consciousness. She went on a tour of the universe with the Being of Light at the speed of thought. The Human Psyche or Human Intelligence can leave its spirit body behind and separate from its spirit body.

https://www.youtube.com/watch?v=WUjP9kKoXfU

Lightning Strike Near-Death Experience – Dr. Anthony Cicoria

When Dr. Anthony Cicoria was struck by lightning, he died. He, his spirit body, was outside his physical body. As he walked up the stairs, his spirit body dissolved beneath him and transformed into an orb of light. He, his Psyche, was looking at his spirit body and later that orb of light from an immaterial third-person perspective or that Third Eye Perspective. This NDE is Scientific Evidence or Empirical Proof that Psyche, or Intelligence, is a different entity than our spirit body and our physical body.

https://www.youtube.com/watch?v=WUXzj0Tczz4

Consciousness without a Body Attached to It

https://www.youtube.com/watch?v=DmBnCTuQUOc

She Described Herself as Pure Consciousness While Out-of-Body

https://www.youtube.com/watch?v=V7xWffB2nH0

No Body – Just Pure Consciousness

https://www.youtube.com/watch?v=1kql9eD9qO4

Near Death Experience of Barbara Wilcox

In the next NDE, Barbara Wilcox explains not having a body, or spirit, or a soul but being Total Intelligence.

https://www.youtube.com/watch?v=JQzap_jFXT8

The Human Psyche or Human Intelligence is something completely different than our spirit body; and, our spirit body is something completely different than our physical body.

Intelligence or Psyche pre-dated the birth and organization of our Spirit Body. Our Psyche is described as an immaterial viewpoint in space, a point of consciousness, a spark on the ceiling, a pinprick of light, a speck of dust, a mote, or total intelligence.

Our spirit body seems to be malleable and can take on different shapes at will. In non-consensus realities, your spirit body can take on any shape that you want it to be. An orb of light is a common shape used for describing a spirit body. For the spirit children of God, our human shape is the most common shape that we see whenever our spirit body is viewed; however, Satan, the devils, and the evil spirits have denied and rejected their heritage as the spirit children of God, and these evil spirits can and do take on any shape that they want – serpents, dragons, alien grays, spacecraft, Nordic aliens, purple monstrosities, forked tail, pitchfork, red skin, and horns.

Our physical body is a part of the Ultimate Consensus Reality, which means that there's not a lot that we can do to change our physical body at will, because our physical body was designed to stay the same.

Both our physical body and our spirit body have NO memorable experiences and form NO new memories while the Human Psyche is separated from them. It's the Human Psyche who forms and stores new memories and has experiences worth remembering. Our spirit body is simply a convenient way to interface with the spirit realm; and, our physical body is an essential necessity for interfacing with this physical reality, which is the Ultimate Consensus Reality.

Susan Volt Talks about the Divine Spark within Each of Us

https://www.youtube.com/watch?v=-Gig4lwKNfQ

The word "spark" is used quite often by NDErs and others to describe the Human Psyche or the Human Intelligence. It's the spark of life. Psyche is life. Psyche is the Life Force.

Even in science fiction – something like the *Transformer* movies – they talk about the Spark or the AllSpark, which is the Life Force within us and within them.

The series *Earth: Final Conflict* did a great deal of discussion about Psyches and our spirit or ghost, often portraying them as two different things. There's a lot of talk about Quantum Mechanics within that series, because Psyche and Quantum Mechanics are integrated together and inseparable. In that series, Psyche or Consciousness has NO physical limitations, which means that it isn't limited by the speed-of-light. Entropy and the speed-of-light limitation applies only to physical matter. They don't apply to quantum mechanisms, spirit matter, and psyche. Syntropy is the natural state in the quantum realm, or the psyche realm, or the spirit world. There are NO physical limitations at the quantum level or the psyche level.

The Materialists and Naturalists have a hard time accepting and understanding Quantum Mechanics and Psyche, because Quantum Mechanics or Transdimensional Physics is supernatural in nature and origin. Quantum Tunneling or Teleportation is a supernatural phenomenon that happens automatically at the quantum level or the psyche level but is greatly limited and restricted at the physical level by the physical limitations that God has imposed on physical matter. You don't want your physical atoms within your physical body quantum tunneling away on you one at a time until you dissolve into thin air. God has imposed physical limitations on physical matter to prevent that from happening to you.

It's interesting to observe that whenever the Human Psyche is separated from its spirit body, it's the Human Psyche or Human Intelligence who is having the experiences and forming the memories, and NOT the spirit body. The Life or the Spark is something completely different than the spirit body and the physical body. It's always the Psyche or the Spark who has the experiences and forms the memories, not the spirit body and not the physical body.

Nicole Swann Is an Awareness Looking Out but No Body

https://www.youtube.com/watch?v=Bo9IAGeYG7A

It's our psyche, or intelligence, or non-local consciousness who has and experiences awareness. It's not our spirit body, and it's definitely not our physical body. Our physical body has no experiences and forms no memories while our psyche is separated from our physical body. The same can be said of our spirit body. It has no experiences and forms no memories while our psyche is separated from our spirit body.

<u>Jessica Near-Death Experience of a Different Kind</u>

In the following NDE, Jessica describes being a Point of Light, a Small Spec, and a Pinpoint of Light during her near-death experience. She, too, is describing Psyche or Non-Local Consciousness. Jessica also described the fact that we don't see everything and don't understand everything which is there to be seen and understood while out-of-body. Instead, there is ever-growing awareness and understanding, with still more to go that is never reached. Jessica could sense the presence of others, but she didn't engage with them during her NDE as much as others seem to do. Hers is the most "self-centered" NDE that I have encountered so far. Jessica's Psyche is very much the center of her universe.

https://www.youtube.com/watch?v=Ve6RG9K3qrA

https://www.youtube.com/watch?v=k_dKVndThVg

https://www.youtube.com/watch?v=VD0Pd7twFBY

There are many things about Jessica's NDE that are atypical, abnormal, weird, unusual, and strange – making some people believe that she's faking the whole thing. I include her in this short list, because she talks about being that immaterial spark during her NDE. I've never heard Jessica talk about her spirit body. She always describes herself as a spark; and because of my research, I define that "immaterial spark" or "viewpoint in space" as Psyche, Intelligence, or Non-Local Consciousness within my books.

It's interesting how they each describe something different, yet something similar as well.

Due to all the empirical evidence or observational evidence from NDEs and OBEs, I believe that I have finally found the Correct Model of Reality or the Ultimate Model of Reality.

"The Ultimate Model of Reality: Psyche Is the Ultimate Cause"

https://www.amazon.com/dp/B071NC9JK6

Psyche or Intelligence or Non-Local Consciousness is something completely different than a spirit body. The psyche and spirit body exist on different frequencies in different dimensions. The Psyche or Intelligence is eternal and everlasting without beginning of days or an end of years – PURE SYNTROPY.

In contrast, our spirit body had a point in time when it was organized by the Gods – a birth date. If you can read this, then your spirit body is a child of God, a child of your Heavenly Father and Heavenly Mother. Raw spirit matter is also PURE

SYNTROPY, without beginning of days or an end of years. But, the birthdate of our spirit body or the organization of our spirit body did indeed have a beginning.

It's ALL matter and energy, all the way down to the most elementary of particles. If there is such a thing as a point particle, then Psyche or Intelligence is it. Psyche is some kind of smart dust or intelligent energy. Psyche or Intelligence is smaller than a string. Psyche is the thing that sets the frequency at which the string vibrates. Thoughts and memories are like quantum waves or vibrations within a string. The smaller can dwell within and control the larger. Psyche produces and controls the quantum waves or the vibrations within the strings. Psyche is like an infinite singularity.

Two particles of matter can occupy the same space at the same time if they are out of phase with each other, existing a different frequency or dimension from each other. Consequently, a spirit body and its assigned physical body as well as their Psyche can all occupy the same space at the same time because they are out of phase with each other. The quantum functionality known as phase-shifting allows the Psyche, the Spirit Body, and the Physical Body to occupy the same location or the same space at the same time. It works due to a feature known as Quantum Superposition. Different quantum waves can occupy the same space at the same time in a cumulative or additive fashion, forming a united whole, but also at the same time remaining completely separate from each other as well.

Due to the same Quantum Superposition feature, it's possible for your Psyche and your Spirit Body to be in two different places on opposite sides of our earth at the same time. Due to the fact that our physical body is part of the Ultimate Consensus Reality, this dual-presence feature doesn't seem to apply to our physical body. Although dual-presence is theoretically possible for a physical body thanks to Quantum Superposition, God seems to prevent that from happening where the physical body is concerned, because physical matter and physical bodies have been subjected to physical limitations and entropy by God in order to make our physical reality dependable, reliable, predictable, and controllable. There are NO physical limitations at the quantum level or the psyche level, so infinitely more is possible at the quantum level or the syntropy level.

Quantum Mechanics is infinitely more versatile and infinitely more powerful than Materialism, Naturalism, Nihilism, and Classical Physics. Quantum Mechanics has infinitely more explanatory power than Materialism and Naturalism do. Quantum Mechanics makes for better science and better philosophy than Materialism, Naturalism, and Classical Physics do.

WE KNOW that Psyche or Intelligence exists and that it is REAL because it has been observed as pinpricks of light and experienced first-hand from a first-person perspective. The existence of Psyche or Intelligence or Non-Local Consciousness has been VERIFIED by observation and experience.

In contrast, something like the Multi-Me Theory or the Parallel Worlds Theory has been FALSIFIED due to a complete lack of observation. Nobody has observed this phenomenon nor experienced it, outside of science fiction. Likewise, chemical evolution and the associated theory of evolution have been FALSIFIED due to a complete lack of observational evidence because chemical evolution of genes and proteins from atoms is physically impossible thanks to entropy or the second law of thermodynamics.

Do you see how that works?

Truths are repeatedly VERIFIED by being observed and experienced. In contrast, falsehoods and lies and fiction – like chemical evolution or the evolution of completely new and different life forms from random mutations and natural selection – are repeatedly FALSIFIED by a complete lack of verification or a complete lack of observational evidence.

According to Science 2.0, the BEST and FASTEST way to find and know the truth is to observe it, live it, experience it, and VERIFY it. The second-best way to find and know the truth is to use the *negating the consequent* version of the Scientific Methods to falsify and eliminate everything that is false so that only the truth remains. If you successfully eliminate everything that is false, then only the truth will remain. This is Logic 101.

Science 2.0: I Upgraded My Science

https://www.amazon.com/dp/B0771K6WTX

Materialism, Naturalism, Darwinism, Nihilism, and Atheism are FALSIFIED due to a complete lack of observational evidence, or a complete lack of supporting evidence, or a complete lack of verification. That's the same way that we have falsified the Multi-Me Theory or the Parallel Worlds Theory. Chemical evolution, spontaneous generation, abiogenesis, and macro-evolution have ALL been FALSIFIED due to a complete lack of observation. These things are physically impossible and prevented from happening by entropy or the second law of thermodynamics.

Every aspect of the Theory of Evolution has been FALSIFIED due to a complete lack of observation, or a complete lack of empirical evidence, or a complete lack of verification. Random mutations and natural selection have NEVER been caught in the act of designing and creating completely new and different life forms from the pre-existing life forms that God designed and created; and, evolution (genetic drift), random mutations, and natural selection did NOT exist until AFTER God designed and created the genes, genomes, proteins, eyes, brains, and life forms in the first place. These truths are so obvious that I sometimes wonder how I was able to overlook them for fifty years of my life.

Science is all about observation and experience. Go with the experienced and the observed; and then, get rid of all the rest – the philosophical speculation and wishful thinking.

Remember, Psyche or Quantum Non-Local Consciousness has been experienced and observed. Psyche or Intelligence is a massless quantum or a massless elementary particle; and, Psyche looks like and acts like a Photon whenever it is seen or observed, both at the quantum spiritual level and at the macro physical level. At every level of existence, Psyche or

Intelligence uses mass, energy, or matter in order to get things done. Photons and other massless quantum particles have been experienced and observed which means that different types of Psyches or Intelligences have also been experienced and observed. The massless elementary particles within fermions, leptons, quarks, electrons, and neutrinos are Controlling Psyches or Controlling Particles who command and control the mass, energy, forces, and fields that surround them. This is what has been experienced and observed.

Mark My Words

My Time in the Void

I don't consider my experience to be an NDE. I don't think it happened when I was near-death during my suicide attempt, but sometime later during the withdrawal phase of my addiction when I was near death or wishing I could die. For me, withdrawal lasted six months after I decided to go cold turkey from all the different prescription drugs that they had me on and had gotten me addicted to. During this six-month withdrawal period, my sense of time was gone, my sense of reality was distorted, I was psychotic and delusional, my sense of right and wrong was gone, I was apathetic, I wanted to die, I couldn't remember how to turn on a computer or television, and I was in a lot of pain and confusion.

I was tripping the light fantastic. I found out later that one of the psychiatrists had diagnosed me as being schizophrenic. That was a misdiagnosis, because I was actually suffering from substance-induced psychosis and withdrawal symptoms; but, I didn't figure out any of this until years later. I didn't know what the hell was going on, but it was hell.

I didn't believe in God anymore. I was an Atheist. But, at a couple of the worst times during the withdrawal, I remember crying out, "God help me. I can't take this anymore." And apparently, He did. When the pain is severe enough, you will say and do anything to get it to stop.

At some point along the way, a part of me did die; and, for the first time in my existence, I finally achieved peace. I don't think it was a gift from my suicide attempt, but more of a gift from finally getting sober. Like I said, I lived for over six months without any real sense of time; and, I was very much in an atheistic frame of mind. I was delusional and hallucinating the whole time. I'm one of those who convinced himself that my NDE didn't happen – that I simply imagined it all – but the profound change in psyche, personality, psychology, and perspective was definitely real and life-changing.

After dreaming of the void or floating in the void completely at peace, my psychology completely changed from anxiety, dread, fear, paranoia, and anger to one of peace and compassion. It was a spiritual rebirth of some sort, and the peace has continued to last ever since.

176

I consider October 31, 2012, to be my birthday. By then I was back to the land of the living, and I fully realized how insane, psychotic, delusional, and paranoid had really been. By that day, I had finally started thinking rationally once again. All the addictive medications were out of my system. My brain had rebalanced or normalized, and the withdrawal period was coming to an end.

They wouldn't let me quit the Zoloft, but by October 31, 2012, they had had me tapering down the Zoloft to a maintenance dose for a month or more; and, I felt an infusion of happiness, peace, and joy. I was back, and better than before. Good enough.

The Blind Can See While Out-of-Body

Spiritual consciousness is like the internet. It continues to exist when the computer or the brain is turned off. Spiritual vision continues to exist, even though the physical brain was never able to see.

https://www.youtube.com/watch?v=gKyQJDZuMHE

https://www.youtube.com/watch?v=azIh8gsXVRg

Powerful Life Reviews

https://www.youtube.com/watch?v=gF_Dj6EduLY

Merging and Becoming One or Omniscient

Renee Pasarow - Near Death Experience

https://www.youtube.com/watch?v=qlFTanblpvg

https://www.youtube.com/watch?v=rSrHE8zkwYg

https://www.youtube.com/watch?v=xB-T78qgfHM

Eben Alexander: A Neurosurgeon's Journey through the Afterlife

https://www.youtube.com/watch?v=qbkgj5J91hE

Confirmed Out-of-Body Experiences

Saw and Described the Operating Theater

https://www.youtube.com/watch?v=J5_x8U7SR0I

Maria Sees a Tennis Shoe on a Ledge While Dead and Out-of-Body

https://www.youtube.com/watch?v=3gGqpxa32oq

The Dead Person Describes the Operating Theater

https://www.youtube.com/watch?v=JL1oDuvQR08

A Shared-Death Experience (SDE)

https://www.youtube.com/watch?v=a1qGJfSZ_LQ

The Bubble of Protection, Quantum Tunneling, and Teleportation

https://www.youtube.com/watch?v=DmBnCTuQUOc

Near-Death Experience: Conversations with God

https://www.youtube.com/watch?v=Zrx8C2lxhJI

Rudolf Smit - The Self Does Not Die: 104 Cases of Verified NDEs

https://www.youtube.com/watch?v=2qJ5UGbBGhq

NDE Research Proves Afterlife Exists

https://www.youtube.com/watch?v=mcb2cQMTPRg

Dr. Jeffrey Long - God and the Afterlife - Science & Spirituality Collide

https://www.youtube.com/watch?v=SyhZV-LGtJ8

Dr. Gary Habermas - Near Death Experiences

https://www.youtube.com/watch?v=ac9pF32gRxU

Penny Satori – Documented Cases Near-Death Experiences

https://www.youtube.com/watch?v=yS2ITQzSPLk

https://www.youtube.com/watch?v=F6TRsTUj8WM

https://www.youtube.com/watch?v=MwXQaRwqUpk

https://www.youtube.com/watch?v=T_KJNQPPoZk

https://www.youtube.com/watch?v=n1JAPxxFkkA

https://www.youtube.com/watch?v=YkW0ikd8i7U

Shared-Death Experiences

https://www.youtube.com/watch?v=Z31cI73DI7M

https://www.youtube.com/watch?v=-0-R3nz0cdg

https://www.youtube.com/watch?v=lWjYjsh8i0w

https://www.youtube.com/watch?v=N8_2P8s77lc

William Buhlman Out-of-Body Experiences

William Buhlman has turned out-of-body experiences into a business, a science, and a way of life. There is no end to what he has available for free on the internet. What you won't get from William Buhlman is any contact with God or religion because he seems to have had no experience with either.

https://psyche-ontology.com/buhlman/

I used to be a Materialist, Naturalism, Nihilist, and Atheist; so, William Buhlman was the first person to get through to me and convince me that there's more to existence than this physical reality, because he treated his OBEs as a science and actually experimented with the phenomenon.

https://www.youtube.com/results?search_query=William+Buhlman

https://www.youtube.com/watch?v=v-Oa2lWrKOg

https://www.youtube.com/watch?v=FjbwXI2-0n8

https://www.youtube.com/watch?v=HtHEtWntLiw

https://www.youtube.com/watch?v=6HWGPSRLBTo

https://www.youtube.com/watch?v=4OR6Kiwlohw

https://www.youtube.com/watch?v=ZlZNmwCD1pA

https://www.youtube.com/watch?v=JEimkt3Wl98

https://www.youtube.com/watch?v=NaaUkJF2JMc

https://www.youtube.com/watch?v=Apw2WpuW60o

https://www.youtube.com/watch?v=Zy1-0wbRJpY

https://www.youtube.com/watch?v=IoQ9T7H4OrE

https://www.youtube.com/watch?v=gewT3DtsRWM

After Effects of NDEs

Alicia Fagan on some of the after-effects of an NDE.

https://www.youtube.com/watch?v=71zXemHbHaE

Nancy Rynes on NDE after-effects and adjustments.

https://www.youtube.com/watch?v=ii1UDGWi6Gk

The After Effects of My Personal Experiences

After my brush with death, spiritual death, and spiritual rebirth, I have experienced some of the positive effects of NDEs. I had fifty years of anxiety and fear. The fear is gone. For the first time in my remembered existence, I'm at peace. I'm not afraid of death.

I worked out a system with God to get answers to my questions. I want to know how things really work; and, whenever I ask God a question, He either tells me in my mind what book to read next where I find my answer, or He reveals to me the next morning while waking up the answer to my question if the answer is not easily available in one of the books that I own.

Sometimes, I'll spend a month or two writing a book based upon a hypnopompic flash of insight that came to me in answer to prayer while waking up the next morning. It can take months to put into words that single flash of insight, which I had months before in answer to my prayer.

I don't see dead people, and don't have visions or any of that. That doesn't seem to be my calling or gift. But, I do seem to get answers to my prayers whenever it comes to science, reality, and how things really work at all levels of existence or in every dimension of existence. Sometimes the answers have been quite surprising, and far outside the box of the natural sciences or the physical science.

Once I got rid of My Materialism, My Nihilism, and My Atheism, it opened me up to infinitely more. Materialism, Naturalism, and their derivatives are the greatest bane and hindrance to scientific discovery that mankind has ever created. Quantum Mechanics and Psyche are supernatural in nature and origin. The very existence of Psyche and Quantum Mechanics as well as verified proof of quantum mechanisms FALSIFIES Materialism, Naturalism, Darwinism, Nihilism, Behaviorism, Scientism, Determinism, and Atheism.

Quantum Mechanics and Naturalism are mutually exclusive because Quantum Mechanics is supernatural and Naturalism states that the supernatural does not exist. They both can't be true at the same time. If one of them is true, then the other has to be false. Quantum Mechanics has observational evidence, experimental evidence, and verified evidence supporting it; whereas, there is NO evidence and can NEVER be any evidence to support the claim of Naturalism which states that the supernatural does not exist. It's logically impossible to provide evidence of something's non-existence. It can't be done. It's physically impossible.

Quantum Mechanics is supernatural. Its very existence falsifies Naturalism, Materialism, and their derivatives such as Darwinism and Atheism. Quantum Mechanics and Psyche are true and verified; whereas, Materialism, Naturalism, and their derivatives are false and have been falsified. If you choose to follow the evidence and choose to allow ALL of the evidence into evidence, it is immediately clear that Psyche and Quantum Mechanics are true because they have been observed and experienced, and it's equally clear that Materialism and Naturalism are false because their claims are impossible to observe and experience. Observation trumps philosophical speculation every time, or at least it should.

Science is observation. Materialism and Naturalism are unscientific, because their hidden assumptions or major premises cannot be observed nor experienced. Psyche and Quantum Mechanics have been observed and experienced. Materialism and Naturalism have not been observed nor experienced because they can't be. You can't experience nor observe the non-existence of something, such as the non-existence of the supernatural or the non-existence of Quantum Mechanics. You cannot observe the non-existence of God and Psyche, either; but, you can definitely experience these things and observe these things for yourself or choose to trust someone who has.

Can you see the difference between the observed and the verified in comparison to the unobservable, unverified, and falsified? The one is true and the other is false. Quantum Mechanics and Naturalism are mutually exclusive. If one of them is true, then the other one is automatically false. Psyche, Intelligence, Life, Syntropy, the Supernatural, and Quantum Mechanics are true, which means that Materialism, Naturalism, Darwinism, and their derivatives are false. It's really simple to understand once a person chooses to do so.

How Quantum Physics and Psychology Affirm NDEs

https://www.youtube.com/watch?v=rBshmwf-iaw

https://www.youtube.com/watch?v=V9KnrVlpqoM

Since I'm a scientist by nature, I also turn to Quantum Mechanics in order to explain Quantum Non-Local Consciousness or Psyche, Out-of-Body Experiences, and Near-Death Experiences. Quantum Mechanics is vastly superior to classical physics when it comes time to try to explain what's happening to us at the quantum level or the psyche level. Quantum Mechanics is a proven and verified science, and it's time that we as a race start using Quantum Mechanics to explain our observational experiences while we are out-of-body exploring the Astral Plane.

Your consciousness, the Human Psyche, is what makes your reality in any plane of existence that we can possibly imagine. Wherever you go, there you are. You don't cease to exist when your physical body and physical brain die or go offline for an extended period of time.

Scientists Who Have Had NDEs

Life-long scientists have interesting and unique perspectives on NDEs, especially after they have had one. By best friend is a patent holding scientist. He is also a seer and has had NDEs and visions. He has been in the presence of Jesus Christ our Savior, and he has had a lot of other amazing and interesting experiences while out-of-body or in-the-spirit.

I have gone out of my way to collect NDEs from scientists who have had them; and, TED talks have proven to be a good source.

https://www.youtube.com/watch?v=EtbiUsX1klk

https://www.youtube.com/watch?v=Y8WIdDz4RxI

https://www.youtube.com/watch?v=MyaBeHeRK6M

https://www.youtube.com/watch?v=rbnBe-vXGQM

https://www.youtube.com/watch?v=mMYhgTgE6MU

There are others:

https://www.youtube.com/watch?v=PMICW2aaplA

https://www.youtube.com/results?search_query=Eben+Alexander

I'm a scientist. I define Science as observation and experience. As I see it, we scientists have to take these observations and experiences seriously if we want to find out what things really are and how they truly work. The BEST and FASTEST way to find and know the truth is to live it, experienced it, and observe it for yourself, or to choose to trust someone who has.

Encounters with God or Jesus Christ

Whenever the being of light and love identifies himself, it's always Jesus Christ whom NDErs encounter and experience when they have separated from their physical body. These various NDEs convinced me that Jesus Christ does in fact exist. Should you find yourself in hell, Jesus Christ will come to you and get you out of there just for the asking. That's probably the most valuable lesson that I have learned from all of this.

Howard Storm Saved from Hell by Jesus

https://www.youtube.com/watch?v=UPj4wci_bcI

https://www.youtube.com/watch?v=Y9AjcfM75gI

Taught by Jesus Christ: Ralph Jensen Shares NDE (Near Death Experience)

https://www.youtube.com/watch?v=uWshfNnyEQA

Lee Stoneking Addresses UN General Assembly

https://www.youtube.com/watch?v=FYt8sv4vzQs

George Ritchie

https://www.youtube.com/watch?v=DsQIU2dNY44

Richard Met God

https://www.youtube.com/watch?v=HAR5MpYNnRE

Erica Had Two Different Types of Life Review during Her Encounter with God

https://www.youtube.com/watch?v=xG_hEi8E4U8

Ian McCormack - NDE - former atheist - near death experience

https://www.youtube.com/watch?v=sTU7MfOgDKM

Dr. Mary Neal - Raised from the Dead

https://www.youtube.com/watch?v=DX473dF7ChY

https://www.youtube.com/watch?v=ULsl92H-Noc

https://www.youtube.com/watch?v=63wY2fylJD0

A Wide Variety of NDE Encounters with Jesus Christ

https://www.youtube.com/results?search_query=nde+jesus

A preponderance of the evidence tells us that Jesus Christ is the being of light and the being of love whom we encounter after we die. A preponderance of the evidence tells us that Jesus Christ exists and that He truly rose from the dead. Science is supposed to be about observation and experience, not philosophical speculation. Science is supposed to be phenomenological, not just hypothetical wishful thinking. The Materialists, Naturalists, Darwinists, Behaviorists, Determinists, Nihilists, and Atheists don't want God to exist, so these people deliberately reject, censor, block, ban, and destroy any evidence to the contrary. That's NOT science. That's the actions of religious dogmatism, fanatical extremism, wishful thinking, and blind faith. It's a perversion of rationality, logic, and science to block, ban, censor, ridicule, and destroy evidence. Materialism, Naturalism, Atheism, and their derivatives are based upon a refusal to look at evidence. *Refusing to look at evidence* is a logic fallacy.

Truth Is Known through a Preponderance of the Observational Evidence

According to Science 2.0, the BEST and FASTEST way to find and know the truth is to observe it, live it, and experience it for yourself, or to choose to trust someone who has. Through phenomenology or lived experiences, we can go directly to knowing the truth without having to run expensive and time-consuming science experiments. Furthermore, our physical science experiments definitely infer that the quantum level, quantum realm, and quantum mechanisms are real and truly exist; but, our quantum experiments cannot provide direct evidence that the quantum level or psyche level is there nor explain what these things are truly like. When it comes to the quantum level, psyche level, or syntropy level, the BEST and FASTEST way to know that it is real and truly exists is to observe it, live it, and experience it for yourself or to choose to trust someone who has.

The preponderance of the evidence tells me that the NDE and OBE phenomenon is real, and that there really is an afterlife and a spirit world.

In contrast, we have the Materialists, Naturalists, Nihilists, and Atheists telling us that there is no such thing as psyche, God, non-locality, spirit, non-local consciousness, an afterlife, NDEs, OBEs, SDEs, Life Reviews, quantum mechanisms, supernatural mechanisms, and spiritual experiences.

How do they know?

They don't know. They can't know. It's impossible to know that something does not exist.

How do they prove that they are right?

The Materialists and Naturalists can't prove that they are right. There's NO way to prove that something doesn't exist. It's physically impossible, philosophically impossible, and logically impossible.

The major premises or primary assumptions of the Materialists, Naturalists, Nihilists, and Atheists are FALSIFIED by a complete lack of observational evidence or a complete lack of verification. You can't observe or verify that something doesn't exist.

ALL of the observational evidence, experiential evidence, empirical evidence, and eye-witness evidence tells us that the Materialist, Naturalists, Darwinists, Nihilists, Behaviorists, Determinists, and Atheists are wrong. Materialism, Naturalism, and their derivatives are FALSIFIED by observational evidence and a preponderance of that evidence.

People choose to trust and believe the Materialists, Naturalists, Darwinists, Nihilists, Behaviorists, Determinists, and Atheists; BUT, these people haven't even observed, nor experienced, nor seen anything at all. It's impossible to observe that God does not exist. These people are flying on blind-faith and wishful thinking, and nothing more. When it comes to science, truth, and reality, you are much better served by choosing to trust and believe someone who has actually observed and experienced something.

Science 2.0 elevates observation and experience over the philosophical speculation of the Materialists, Naturalists, and Atheists. Science 2.0 allows all of the evidence into evidence and then pursues a preponderance of the evidence. Science 2.0 is the way that science should have always been done but wasn't. We should NEVER have let science be hijacked by the Materialists, Naturalists, Darwinists, Nihilists, and Atheists, because these people don't KNOW anything because they have never experienced anything and never observed anything.

Science 2.0: I Upgraded My Science

https://www.amazon.com/dp/B0771K6WTX

Remember, the BEST and FASTEST way to find and know the truth is to observe it, live it, and experience it for yourself, or to choose to trust someone who has. In contrast, know that Materialism, Naturalism, Nihilism, and Atheism are FALSIFIED due to a lack of observational evidence supporting their primary assumptions or major premises.

The truth is repeatedly observed, repeatedly experienced, and therefore repeatedly VERIFIED. In contrast, falsehoods like Materialism, Naturalism, and their derivatives are repeatedly FALSIFIED by a complete lack of observational evidence or a complete lack of verification. All you want is the truth, unless of course you are a Materialist, Naturalist, Darwinist, Nihilist, Behaviorist, or Atheist – then any old lie will do.

Mark My Words

—

Source

Putting Psyche Back into Psychology: Restoring Science to Consciousness

https://www.amazon.com/dp/B071NC987S

References

The Ultimate Model of Reality: Psyche Is the Ultimate Cause

https://www.amazon.com/dp/B071NC9JK6

Science 2.0: I Upgraded My Science

https://www.amazon.com/dp/B0771K6WTX

Quantum Neuroscience: The Answer to Life, the Universe, and Everything

https://www.amazon.com/dp/B079Z6QQQB

The Science Experiment We All Want to Experience and Observe

Physical objects such as atoms and electrons – objects with mass or "resistance to acceleration" – have been caught in the act of teleportation or quantum tunneling.

If the barriers are not wide enough within a computer chip, electrons will quantum tunnel or bleed out of the circuit into a neighboring circuit and ruin the calculations. I've done stress testing of CPUs; and, whenever the CPU gets hot enough, or the electrons are agitated enough, the electrons begin to quantum tunnel past their barriers, and faulty calculations in the Prime95 stress test are the result. I'm a scientist. I have experienced it. It's real. It truly happens.

https://www.mersenne.org/download/

Physical objects such as atoms and electrons can quantum tunnel at will if they choose to do so. Theoretically, they could quantum tunnel to the other side of our observable universe, if they chose to do so.

What's interesting, though, is that something external to the atoms and electrons is making their De Broglie Wavelength very short so that they can't quantum tunnel very far. However, if we heat up the electrons, agitate them, or pour pure energy into them, it increases the distance that they can quantum tunnel; and within a computer chip, if the thing gets hot enough and the electrons get agitated enough, they can jump their barriers or quantum tunnel out of their circuits, thereby producing errors in the Prime95 computations. Heat or energy increases an electron's De Broglie Wavelength so that they can quantum tunnel farther than they can when they are cold. At the quantum level, quantum objects such as psyche and spirit matter seem to have an infinite De Broglie Wavelength and no physical limitations; but, at the physical level, very short De Broglie Wavelengths and very short quantum tunneling distances seem to be the rule. The Science explains it all, if we are willing to allow it to do so.

In this next science experiment, the scientists used Bose Enhancement to increase the range at which particles will tunnel by making particles quantum tunnel through five barriers instead of one.

https://quantum-neuroscience.com/Long-Range-Tunneling-of-Quantum-Particles

The quantum tunneling of physical atoms and electrons is a proven and verified phenomenon.

I want to meet and observe the individual who can quantum tunnel or teleport large physical objects from one location to another at will. This is the science experiment that I want to experience and observe.

I want to take a permanent marker and write a "QT" on the head's side of penny of my choosing; and then, I want you to teleport or quantum tunnel that penny from the other side of the room, while Professor Observer watches it disappear, and while I watch that same penny appear on the table in front of me. I want to observe the quantum tunneling of something larger than a single atom or a bunch of electrons. Don't you?

Quantum Tunneling or Teleportation of physical matter is an observed and a proven quantum phenomenon. It happens. It can be done. So, why isn't it being done?

I wanted a scientific explanation as to why it isn't being done, and now I have a few, thanks to my ongoing studies of Quantum Mechanics and Quantum Field Theory.

First of all, we need to understand the Quantum Zeno Effect – what it is and how it works.

https://quantum-neuroscience.com/Quantum-Zeno-Effect-Verified

Atoms won't move while you watch them. When the imaging laser was off, or turned on only dimly, the atoms tunneled freely. But as the imaging beam was made brighter and measurements made more frequently, the tunneling reduced dramatically. We now have the unique ability to control quantum dynamics purely by observation.

What do we learn from this science experiment and others like it? First of all, we learn that physical atoms can quantum tunnel at will – just not very far. We also learn that physical atoms are sentient, psychic, telepathic, and intelligent. Physical atoms can read your mind. They KNOW when you are looking at them and when you are not. This is possible because each physical atom has within it a controlling psyche or a controlling intelligence who can read your mind at a distance and thereby KNOWS whether you are looking at it or not.

Physical atoms choose to freeze or choose to hold still whenever they KNOW that you are looking at them – whenever they KNOW that you might need them. There's some type of intelligence or psyche within each physical atom and within each physical electron. They KNOW when they are being observed or when they are needed. This is Action at a Distance or telepathy that we have observed here with the Quantum Zeno Effect. Physical atoms and physical electrons seem to have assigned themselves the task of servicing us. Why? What could persuade them to do so? We human beings aren't consciously aware of any of these quantum or non-physical phenomena. So, who is?

Physical Atoms can and do quantum tunnel freely, just not while you are observing them, looking at them, or recording their positions. They don't quantum tunnel when they think that you might need them.

When atoms quantum tunnel, they don't seem to quantum tunnel very far. Theoretically, according to Quantum Mechanics, physical atoms should be able to quantum tunnel instantaneously across the length of our observable physical universe at will; but, they don't. Why don't they?

It's because the God's or the Controlling Psyches have deliberately given physical atoms a very short De Broglie Wavelength so that the atoms in your physical body can't quantum tunnel away from you at will. Physical atoms can quantum tunnel just enough to interact with each other; but, not enough so that your physical body will dissolve into thin air and cease to exist.

Quantum Tunneling Is Faster than Light

https://www.youtube.com/watch?v=-IfmgyXs7z8

Matt starts this Space Time Video by asking:

Wouldn't it be nice to be everywhere at once? According to quantum mechanics you are, at least a little bit.

Omnipresence and Quantum Superposition combine to explain this particular quantum phenomenon. Quantum Objects are omnipresent at the quantum level, and they are additive and multitasking. They can be more than one place at once. So, why aren't we omnipresent? Why do physical atoms seem to hold still? Why isn't quantum tunneling observed at the physical level, with large physical objects on a daily basis?

Any material object is really a matter wave. It can be described as a wave packet, and that wave packet has a wavelength. This De Broglie Wavelength defines how well determined an object's position is. A large wavelength means a highly uncertain position or omnipresence. A small wavelength means a well-defined or limited position. That's true of subatomic particles, and that's true of anything, including your physical body, physical house, or physical car.

The Gods or the Controlling Psyches have given the physical atoms within your physical body a very short De Broglie Wavelength so that the atoms within your physical body cannot quantum tunnel very far. The best that atoms and ions can do is to quantum tunnel across the length of a synapse. They can't go much further than that, because some Psyche or Intelligence is preventing them from doing so.

A LAW or a limitation of any type, including a short De Broglie Wavelength, requires some type of Law Giver and Law Enforcer. This Law Giver and Law Enforcer has to be enforcing this short De Broglie Wavelength onto the physical atoms within your physical body; otherwise, you would literally dissolve into thin air as the atoms within your physical body quantum tunneled away from you to the other sides of the universe as quickly as they chose to do so.

There are non-physical laws or telepathic laws at the quantum level that are forcing the atoms within your physical body to remain in place. Someone had to make these laws, and more importantly, someone has to be enforcing them. Laws are no good, unless they are being enforced. Furthermore, there has to be something psyche or something intelligent within the atoms and electrons capable of understanding and choosing to obey the laws that have been made. Laws are no good, if they are never followed and obeyed. It's obvious that there has to be some type of psyche or intelligence within the atoms who is capable of understanding and obeying the physical laws and physical restrictions that have been made. Otherwise, chaos would reign supreme. Law is what produces order. There has to be a Law-Maker, a Law-Enforcer, and a Law-Obeyer; otherwise, laws are absolutely worthless. So, what can function as a Law-Maker, Law-Enforcer, and Law-Obeyer at the quantum level and the psyche level, and therefore at the physical level as well? Psyche, Intelligence, Life Force, or Consciousness is the only thing we know about.

Something within the physical atoms has to be making the physical atoms obey and follow the physical laws or physical limitations that were designed and made, before the first particle of physical matter was designed and made. Something non-physical and pre-physical had to design and make the first particle of physical matter, because according to Quantum Field Theory, particles are born, and particles can die or be unmade. The only thing that we have discovered so far that is capable of controlling physical matter from the quantum level is Psyche, Intelligence, Living Light, or Consciousness.

We only know of one individual who was capable of controlling physical matter at the physical level through the power of his mind or his word of command.

That was Jesus Christ.

Luke 24: 31:

31 And their eyes were opened, and they knew him; and he **vanished** out of their sight.

Jesus Christ quantum tunneled or teleported out of their sight.

Whenever Jesus told the Jews that he is Jehovah, the God of the Old Testament, the creator of the heavens and our earth, they always took ahold of him and tried to kill him. They considered his claim to be blasphemy – the worst crime that they could possibly imagine.

At Nazareth, his home town, when he effective told his neighbors that he is Jehovah, the God of the Old Testament, and the creator of the heavens and this earth, they grabbed him and proceeded to toss him off the cliff at the edge of their city, so as to kill him for committing blasphemy.

Luke 4: 16 -31:

16 And he came to Nazareth, where he had been brought up: and, as his custom was, he went into the synagogue on the sabbath day, and stood up for to read.

28 And all they in the synagogue, when they heard these things, were filled with wrath,

29 And rose up, and thrust him out of the city, and led him unto the brow of the hill whereon their city was built, that they might cast him down headlong.

189

30 But he **passing through the midst of them** went his way,

31 And came down to Capernaum, a city of Galilee, and taught them on the sabbath days.

Jesus Christ disappeared. He either phase-shifted, disappeared, and walked away from them; or, he quantum tunneled to a different location. In either case, these were quantum phenomena that Jesus used to get away from them. Quantum mechanics is supernatural mechanics or non-physical mechanics.

Matthew 14: 29:

29 And he said, Come. And when Peter was come down out of the ship, he **walked** on the **water**, to go to Jesus.

Walking on water and standing in the air is a quantum phenomenon. Faith in Christ gave Peter the temporary ability to walk on water.

Repeatedly, over and over again, Jesus Christ used quantum mechanics to demonstrate and prove to the Jews that He is the Son of God, the God of the Old Testament, and the creator of our heavens and earth.

Matthew 15: 34-38:

34 And Jesus saith unto them, How many loaves have ye? And they said, Seven, and a few little fishes.

35 And he commanded the multitude to sit down on the ground.

36 And he took the seven loaves and the fishes, and gave thanks, and brake them, and gave to his disciples, and the disciples to the multitude.

37 And they did all eat, and were filled: and they took up of the broken meat that was left seven baskets full.

38 And they that did eat were four thousand men, beside women and children.

This time around Jesus Christ did quantum transmutation by converting the air or the quantum fields into loaves and fishes – enough to feed thousands of men, women, and children.

Mark 4: 39:

37 And there arose a great storm of wind, and the waves beat into the ship, so that it was now full.

38 And he was in the hinder part of the ship, asleep on a pillow: and they awake him, and say unto him, Master, carest thou not that we perish?

39 And he arose, and rebuked the wind, and said unto the sea, Peace, be still. And the wind ceased, and there was a great calm.

40 And he said unto them, Why are ye so fearful? How is it that ye have no faith?

41 And they feared exceedingly, and said one to another, What manner of man is this, that even **the wind and the sea obey him**?

Mind-over-matter, or the placebo effect, is a quantum phenomenon. Commanding the elements and having them obey you is a quantum phenomenon. The elements or physical matter have covenanted to obey God's command. He is the Law Giver or the Creator of the Physical Laws, and the elements obey whenever He commands.

If He had a reason for doing so, Jesus Christ could quantum tunnel that penny from one side of the room to the other. He could also quantum tunnel that penny to the end of our observable universe instantaneously if He chose to do so.

Jesus Christ employed Quantum Mechanics at the physical level to prove to the rest of us that He is the Son of God, the God of the Old Testament, and the creator of our heavens and our earth.

Of course, most people didn't believe him, and most people still don't believe him; but, that doesn't change the fact of what was actually experienced and observed by the eyewitnesses who were present at these events.

As scientists we KNOW and we have proven beyond a shadow of a doubt that Quantum Mechanics, Quantum Field Theory, and the various different Quantum Mechanisms are real and truly exist. Through Jesus Christ, we got to observe what can happen when a physical being has the innate ability to control Quantum Mechanisms through the power of His mind or through His Word of Command. It's precisely what we observe in the various different Avenger Movies whenever physical beings have somehow gained the powers of a God. It's mind-over-matter or psyche-over-energy. It's Quantum Mechanics.

We allow it to exist in our science fiction and fantasy; but, we won't allow it to exist within the Living God, Jesus Christ, because we don't want to believe in Jesus Christ and don't want Him to be our God.

We human beings are an interesting set of contradictions. It's okay for Thanos, Thor, Captain Marvel, Loki, or Dr. Strange to have superpowers; but, there's no way in the universe that we are ever going to allow Jesus Christ to have superpowers, because when it comes to Jesus Christ, we all unitedly claim that superpowers DO NOT EXIST and there's no such thing as psyche and quantum mechanics.

I KNOW how it goes, because I used to be a materialist, naturalist, nihilist, and atheist; and at the time, I didn't want psyche, quantum mechanics, and God to exist. From my perspective, they didn't exist because I didn't want them to exist, and I basically went out of my way to make sure that they didn't exist. I was without God in the world. We each get what we choose to accept.

Mark My Words

Part I — INTRODUCTION TO NIRVANA

The word "Spirit", with a capital letter, is defined as a Psyche and Spirit Body combination, and signifies a person or a personality or living individual. The word "spirit" in lower case is often defined as spirit matter and sometimes defined as a spirit body. But, I and others don't always follow this convention in our writing, so every time we encounter the word "spirit", we have to stop and try to determine what type of spirit the author might be talking about.

I had to modify and enhance the Nature vs. Nurture debate in order to get it to make logical sense to me. I had to add in the NIRVANA or the PSYCHE!

NIRVANA is a spiritual concept, and not a physical location. In some religious traditions, NIRVANA is a place of bliss where our spirit goes after we die. NIRVANA is often associated with liberation, peace of mind, and release. Some people believe that we can achieve NIRVANA, enlightenment, self-actualization, and peace while we are still mortal and still living here in this physical realm.

I have observed that the human Psyche's pursuit of NIRVANA, happiness, freedom, and peace is typically a much more powerful motivator and modifier of behavior than our NATURE and our NURTURE. Therefore, I have modified and even solved the Nature vs. Nurture debate by including the NIRVANA, or the human Psyche's pursuit of life, liberty, and happiness.

In this "Nature vs. Nurture vs. Nirvana Model of Reality", I make NIRVANA synonymous with PSYCHE. I needed to come up with something that started with the letter, "N", that could be defined as Psyche; and, Nirvana was the closest thing that I could come up with.

I hold these truths to be self-evident, that all men are created equal, that they are endowed by their Creator with certain unalienable Rights, that among these are Life, Liberty and the pursuit of Happiness. I hold this truth to be self-evident, that NIRVANA should be added to NATURE and NURTURE, if we truly want to know what's driving and motivating human beings and what's really going on in this physical universe. The NIRVANA, or the human psyche's pursuit of happiness and peace, is the missing ingredient which caused the Nature vs. Nurture debate in the first place.

When trying to determine the cause of a specific behavior, we should in fact examine NATURE versus NURTURE versus NIRVANA. Nature is genes and heredity. Nurture is the physical environment in which the physical body finds itself. And NIRVANA, or the Spirit, or the Psyche, or the Consciousness is the "WHO" that decides what it is going to do with the NATURE and the NURTURE which it has received or inherited.

Put it all together, and suddenly there is no Nature vs. Nurture debate. Instead, we find ourselves looking at the True Reality of all things, and realize that the NIRVANA can trump NATURE and NURTURE any time that the NIRVANA wants to do so.

I developed this model of reality BEFORE I learned anything about the BioPsychoSocial Model of therapy. They are similar. In fact, I used this book as the starting point for my Ultimate Model of Reality or Psyche Ontology which I present in my much larger books.

1. *The Ultimate Model of Reality: Psyche Is the Ultimate Cause.*

2. *Putting Psyche Back into Psychology: Restoring Science to Consciousness.*

3. *BioPsychoSocial: Including Psyche or Light into Our Theoretical Models.*

I retain this *NATURE vs. NURTURE vs. NIRVANA* book as a short introduction to those much larger books.

In the second edition of this book, I added a Part III, which is my standard introduction to *THE ULTIMATE MODEL OF REALITY*.   In the third edition, I added Part IV in which I define and explain Psyche or Light.

Determining the Most Appealing Theoretical Approach

The purpose of this college assignment was to determine for myself which Theoretical Approach to Psychology that I found most appealing — choosing among the Psychoanalytic, Behavioral, and Humanistic schools of thought.

These are the tree main schools of thought in Psychology, and I was only permitted to choose among these three. Unfortunately, they all are based upon Materialism and Naturalism to one extent or another; but, I didn't know that at the time. I instinctively gravitated towards the one that seemed to permit the most psychology or the most spirituality.

There is no denying it. I left Freud and Behaviorism far behind back in the 1980's after reading and actively applying David Burns' book, "Feeling Good". Cognitive Therapy was what worked to me when I needed help the most at that point in time, even though it wasn't the established paradigm.

After getting addicted to prescription drugs a few years back (2010 – 2012) and having everything go weird on me for a while, I realize now how important it is to be sober and to be in complete control of one's own thoughts and feelings and emotions. The drugs had made me psychotic, suicidal, and paranoid — that was the official diagnosis which I received from my psychiatrist, while I was going through the six-month long withdrawal process. A person can harbor some really strange ideas and beliefs when the mind is no longer in control of the matter. It's a really bizarre experience to have your brain chemistry in control of you, instead of the other way around and being in control of your brain, as anyone who gets addicted to something knows. The loss of freedom and control, along with the withdrawal process, is very frustrating and confusing.

It's good to be sober again! I pray to God every day that I never have to go back to my addictions. It was hell. I am four years old now (2016). I have been sober and semi-sane for nearly four years now.

When it comes to Psychology and my most favorite School of Thought, I have always been interested in the Recent Trends. While I was trying to get sober, my Therapist had me working on Mindfulness — learning to be present in the here-and-now rather than living in depression in the past, or fearing the future. I was always interested in Cognitive Therapy; and, I am most interested in the attempts that are currently being made among cutting-edge Psychologists to merge Consciousness with Neuroscience and Philosophy of Mind into one science — the study of the psyche, or the mind, or the human spirit. I'm interested in what our textbook and other psychologists call the Mind-Brain Problem. I have also realized that the Mind-Brain Problem is immediately solved once we get rid of Materialism.

I took the little survey of personal interests that was required for this college assignment, in an attempt to determine my preferred theoretical approach to Psychology.

I got a 10 on the Psychoanalytic theoretical approach or school of thought. I also got a 10 on the Behavioral school of thought or theoretical approach. However, I got a huge 27 on what my teacher called the "Humanistic" school of thought, in that little introductory survey test that I took as part of my Psychology course.

I have always equated humanism with atheism or secularism, so I thought it strange that that "aptitude test" or proclivity test was calling me a humanist, because atheism is not my current frame of mind. Furthermore, the phrase "humanistic school of thought" or "humanistic theoretical approach" had no meaning to me at the time; and, our college textbook doesn't use that phrase either. I found myself wondering what I had been labeled as being.

I later found out that the Author of our textbook, James Kalat, is a behaviorist, materialist, and evolutionist, which explains why he buried Humanistic Psychology at the back of his book and gives it only a brief one-paragraph mention and makes absolutely NO mention of Transpersonal Psychology or Parapsychology. The Materialists by definition don't believe in Spirituality or Parapsychology or Transpersonal Psychology, which means that they don't believe in Humanistic Psychology.

The author of our college textbook IS a Behaviorist, Evolutionist, and Materialist; consequently, I was forced to go to Wikipedia and look up "Humanistic Psychology" for the definition that I desired to have, because the term wasn't even mentioned in our text until the very end of the book.

> **Humanistic Psychology** is a psychological perspective which rose to prominence in the mid-20th century in response to the limitations of Sigmund Freud's psychoanalytic theory and B. F. Skinner's behaviorism. With its roots running from Socrates through the Renaissance, this approach emphasizes individuals' inherent drive towards self-

actualization, the process of realizing and expressing one's own capabilities and creativity.

It helps the client gain the belief that all people are inherently good. It adopts a holistic approach to human existence and pays special attention to such phenomena as creativity, free will, and positive human potential. It encourages viewing ourselves as a "whole person" greater than the sum of our parts and encourages self-exploration rather than the study of behavior in other people. Humanistic psychology acknowledges spiritual aspiration as an integral part of the human psyche. It is linked to the emerging field of transpersonal psychology.

Primarily, this type of therapy encourages a self-awareness and mindfulness that helps the client change their state of mind and behavior from one of reactions to a healthier one with more productive self-awareness and thoughtful actions. Essentially, this approach allows the merging of mindfulness and behavioral therapy, with positive social support.

After getting a definition for Humanistic Psychology from the Wikipedia, I could easily see why this is my preferred theoretical approach and the school of thought which I allegedly find most appealing, when it comes to the study of Psychology.

I know for a fact that it is the human Consciousness or human Spirit that has the innate ability to ACT. In contrast, Physical Matter was designed to REACT and to be ACTED UPON.

The Behaviorists and Materialists treat human beings as if they are nothing more than physical matter and thus capable of nothing more than reacting. Humanistic Psychology recognizes the fact that we humans have a Consciousness or a Spirit within us that is in fact capable of self-awareness and thoughtful actions, as well as the ability to design and create and manufacture the things that we set our minds to. I have learned that once we get rid of Materialism, then everything starts to make logical sense.

There are now Psychologists, real Scientists, who are actively trying to solve the Mind-Brain Problem; and, they are doing so by deliberately studying the Scientific Evidence instead of suppressing the Scientific Evidence, as the Behaviorists and Naturalists and Materialists and Freudians tend to do. Psychology is starting to get back to its roots, the study of Psyche, or the study of the Human Spirit.

> The rejection of any source of evidence is always treason to that ultimate rationalism which urges forward science and philosophy alike. — A. N. Whitehead, "The Function of Reason".

The Materialists and Behaviorists deliberately eliminate and exclude any Scientific Evidence which contradicts their chosen worldview or religion. The Materialists refuse to look at any evidence that supports a Spiritual Perspective, including any Scientific Evidence that supports a Spiritual or Non-Physical Worldview, because the existence of the Spiritual or the existence of Light proves Materialism false. The mantra of the Materialists is, "I ain't gonna look at it!"

I tired of dealing with atheistic, naturalistic, materialistic Scientists who deliberately reject and suppress volumes of Scientific Evidence in order to maintain their chosen worldview. These self-proclaimed scientists, who label the Scientific Evidence that they don't like as "Not Science", are the worst of the bunch in my humble opinion. They are traitors to science and have held back scientific discovery for decades with their Materialism and Naturalism. We would be decades ahead of where we are now, if we would have abandoned Materialism and Naturalism centuries ago. I learned to hate Materialism, because Materialism IS the bane to Scientific Exploration, Scientific Discovery, and Scientific Understanding.

As I studied "Humanistic Psychology" and tried to observe its ramifications on my own psychology or worldview, a month later I finally came across something that is called "Transpersonal Psychology", which is a sub-domain or an off-shoot from "Humanistic Psychology". I also eventually encountered "Positive Psychology".

It was interesting that the term "Transpersonal Psychology" first achieved meaning to me while I was reading the book, "The Holographic Universe", by Michael Talbot — on page 70 of the 2011 Reissued Edition. I hadn't remembered hearing the term, so I looked it up on the Wikipedia and Amazon only to discover that Transpersonal Psychology had broken off from Humanistic Psychology, and to discover that I had indeed been introduced to the term a month before and hadn't realized it at the time. It took me a month after I had been introduced to "Humanistic Psychology" to learn and realize that the area of psychology that clearly interests me the most is the study of "Transpersonal Psychology" and "Parapsychology".

It takes time to learn the meaning of these things, and we each have to start where we currently are. It doesn't help, either, when the Authors of our College Textbooks are Materialists and Evolutionists and deliberately hide this kind of information from us.

Therefore, a month after I wrote most of the preceding, I came in here and provided a definition for Transpersonal Psychology from the Wikipedia, because it's the area of psychology that interests me the most:

> **Transpersonal psychology** is a sub-field or "school" of psychology that integrates the spiritual and transcendent aspects of the human experience with the framework of modern psychology. It is also possible to define it as a "Spiritual Psychology". The transpersonal is defined as "experiences in which the sense of identity or self extends beyond (trans) the individual or personal to encompass wider aspects of humankind, life, psyche, or cosmos". It has also been defined as "development beyond conventional, personal, or individual levels".
>
> Issues considered in transpersonal psychology include spiritual self-development, self beyond the ego, peak experiences, mystical experiences, systemic trance, spiritual crises, spiritual evolution [or learning or growth],

religious conversion, altered states of consciousness, spiritual practices, and other sublime and/or unusually expanded experiences of living. The discipline attempts to describe and integrate spiritual experience within modern psychological theory and to formulate new theory to encompass such experience.

Transpersonal psychology has made several contributions to the academic field, and the studies of human development, consciousness, and spirituality. Transpersonal psychology has also made contributions to the fields of psychotherapy and psychiatry.

Since my discovery of Transpersonal Psychology, I have been buying different books about Transpersonal Psychology and Humanistic Psychology. These two "schools of psychology" are often treated together in the same book as if they are one. I have also acquired a few books about Parapsychology and Positive Psychology. Good enough!

Before my discovery of Transpersonal Psychology, I had developed a keen interest in the Mind-Brain Problem and Consciousness, and had already acquired many different books about Consciousness and the Mind-Brain Problem.

Here are listed a few books that I purchased, cutting-edge science, which deal directly with the Mind-Brain problem. Some of the Scientists are now trying to merge the study of Consciousness, Neuroscience, Spirituality, and Philosophy of Mind into one science. It's cool stuff! This is the area of psychology that has always interested me the most — the study of the psyche or the human spirit — the kind of psychology that the Materialists and Behaviorists refuse to look at.

Irreducible Mind: Toward a Psychology for the 21st Century:
http://www.amazon.com/dp/1442202068/

Beyond Physicalism: Toward Reconciliation of Science and Spirituality:
http://www.amazon.com/dp/1442232382/

Consciousness Beyond Life: The Science of the Near-Death Experience:
http://www.amazon.com/gp/product/0061777250

Some of the other books that I found and acquired regarding the Mind-Brain Problem:
http://bookrev.allthings.computer/consciousness-and-the-mind-brain-problem/

Inspired by the complete lack of coverage of Transpersonal Psychology in James Kalat's "Introduction to Psychology, 9th Edition" as well as Kalat's worship and adoration of Evolution, I went ahead, critically evaluated the evidence in great detail as he suggested, and wrote a bunch of different books debunking

Materialism, Physical Monism, and the Theory of Evolution. Some of those books are now available on Kindle ready for you to purchase them and read them.

https://www.amazon.com/author/science

https://www.facebook.com/MarkMyScience/

My disagreement with Kalat's Materialism and Behaviorism, along with his worship of Evolution, motivated me to write a dozen different books debunking Materialism, Atheism, Behaviorism, Darwinism, and the Theory of Evolution; and, half of them are finished and online now, with more to come. I am thorough when I finally set my mind to something.

I learned that EVERYTHING makes perfect logical sense once we get rid of Materialism or Physicalism, and embrace the Quantum Non-Local or the Non-Physical, or what is often called "The Spiritual". Even the Mind-Brain Problem is immediately solved once we get rid of Materialism.

The Materialists and Darwinists will shame you and ridicule you for discovering that Materialism is false and why it is false. Mocking and ridicule and compulsion are the ONLY weapons that the Materialists have to use against us, because there is NO evidence and can be NO evidence whatsoever that actually supports Materialism, or the claim that the Spiritual does not exist.

The Materialists, Darwinists, Naturalists, and Atheists can mock me all they want; but, I am not ashamed of discovering the truth. In fact, I have found The Truth extremely liberating. I have discovered hundreds of different reasons why Materialism is false. Those discoveries have set me free. I have also observed that there is No Scientific Evidence and can be No Scientific Evidence to support Materialism, which is the claim that the Spiritual or the Non-Physical does not exist. In fact, the most interesting Scientific Disciplines and Scientific Discoveries ARE Non-Physical or Spiritual in nature.

The KEY to understanding life, the universe, and everything is to eliminate Materialism, even if we have to find 42 different ways to do so. Once we get rid of Materialism, then suddenly everything starts to make sense.

While critically evaluating the evidence, I have observed that EVERYTHING makes perfect logical sense once we have gotten rid of Materialism. That Scientific Observation was a major epiphany in my life, and it changed everything for the better.

When it comes to Materialism, they can have it. I don't want it. All Materialism does is create problems for us that the Materialists don't know how to solve.

Mark My Words!

The Mind-Brain Problem

The word "Spirit", with a capital letter, is defined as a Psyche and Spirit Body combination, and signifies a person or a personality or living individual. The word "spirit" in lower case is often defined as spirit matter and sometimes defined as a spirit body. But, I and others don't always follow this convention in our writing, so every time we encounter the word "spirit", we have to stop and try to determine what type of spirit the author might be talking about.

The four major philosophical issues are:

1) Free Will vs. Determinism

2) Dualism vs. Monism (mind-brain problem)

3) Nature vs. Nurture

4) Spiritual vs. Physical

ALL of these problems cease to exist and are immediately solved once Materialism is completely removed from the equation and replaced by some kind of Spiritualism or Spirituality or Non-Local Quantum Mechanics. Materialism is the thing that creates these problems for us in the first place; and, once we have jettisoned Materialism for good, then all of these problems are immediately solved. Materialism IS the ultimate bane to Scientific Understanding and Scientific Discovery, especially when it comes to the Spiritual Sciences and the Spiritual Aspects of any Scientific Discipline. Materialism is a scam, a form of job security, wherein the Materialists create unsolvable problems that can't be answered nor solved in terms of Materialism alone. The Materialists paint themselves into a corner and leave us chasing our tails in a never-ending pursuit of nothing.

I have discovered that every Scientific Discipline and every Scientific Concept has a Non-Physical or Spiritual Dimension, to one degree or another. In fact, the Physical derives from, is comprised of, and is composed of the Spiritual.

According to Quantum Mechanics, EVERY physical particle started out as a Spiritual, Non-Physical, Trans-Dimensional Wave — what some Scientists call a Quantum Probability Wave. "Trans-Dimensional" by definition means Non-Physical or not located in our Physical 3D Space-Time. The various Quantum Probability Waves or Spiritual Waves do not reside nor exist in our Physical Space-Time Dimension, until after they are Consciously Observed and then localized and particalized into our Physical 3D Space-Time. Quantum Probability Waves or Spiritual Waves are Trans-Dimensional, meaning that they are not located in our 3D Physical Space-Time. The Quantum or the Spiritual is by definition, Extra-Dimensional and Trans-Dimensional. That's what Quantum Mechanics is trying to tell us. Quantum Objects or Spiritual Waves become Physical in nature only after they are Consciously Observed by a Living Mind, particalized, and then located or given location in our 3D Physical Space-Time. Materialism is a DENIAL of all of this! Materialism is a denial of REALITY.

Mr. Materialist wrote: "Just a reminder. . . Theories don't provide proof of anything. They're theories."

My reply: "Yes, but if you are cunning enough, you can indeed derive a bunch of proofs out of theories and theorems, and such things do in fact become axiomatic laws. The message of my books is not to cripple ourselves with semantics."

Some people are satisfied with a theory and actually hide behind their theories. I, myself, want THE TRUTH.

Materialism is the philosophical claim or theory that the Spiritual or the Non-Physical does not exist. There is NO way possible to provide Scientific Evidence to support such a claim. Instead, the existence of Light proves to us that Materialism is false. Light is massless, non-physical, and immaterial in nature. Spirit is Light; and, Light is Spirit.

Our physical instruments can only capture and study the effects of Light. We can't use our physical instruments to capture a Quantum Object or a "Photon" while it is in motion and while it is in the Trans-Dimensional Realm or Spirit Realm. That Quantum Object has to localize or physicalize into our 3D Physical Space-Time before we can get our physical instruments onto it to study it. Even Quantum Mechanics makes sense once we have successfully removed Materialism from consideration.

Quantum Nonlocality is the KEY to understanding Quantum Mechanics. What is Nonlocality? Something is said to be Non-Local if it doesn't currently have a location in our Physical 3D Space-Time. In other words, Non-Local means Non-Physical! Non-Local means Spiritual. A Quantum Object is Non-Local if it currently IS spiritual or non-physical or wave-like in nature. When a Non-Local or Spiritual Quantum Object is consciously observed by a Living Consciousness, then its wave function collapses into a physical particle; and then, that Quantum Object becomes localized and particlizes into 3D Space-Time and thus becomes physical in nature. All of this is really easy to understand once Materialism has been eliminated from the equation.

The existence of Quantum Nonlocality proves to us that Materialism is false. Non-Local by definition means Non-Physical and Trans-Dimensional. Quantum Nonlocality by definition, in principle, proves to us beyond a shadow of a doubt that Materialism is FALSE. Materialism violates Quantum Mechanics, which is why the Materialists can't understand Quantum Mechanics. See the Light!

Light is Non-Physical. Spirit is Non-Physical. Light is Spirit. Spirit is Light.

A Non-Local Non-Physical Consciousness can function as a CAUSAL AGENT. Physical Matter can't. In fact, it is the Non-Local Consciousness that CAUSES Physical Matter to come into existence in the first place. Physical Matter doesn't cause anything. Physical Matter simply reacts and does whatever Consciousness tells it to do. Physical Matter was designed to react and to be acted upon by Living Consciousness. Physical Matter cannot function as an ACTOR or a CAUSAL AGENT. Consciousness IS Life; and thus, Consciousness or Spirit or Life is the thing that has the innate ability to design, create, manufacture, act, and cause things to happen.

This truth is really easy to understand once we have gotten rid of Materialism. Everything becomes obvious and clear once we have removed Materialism from the equation.

It is important to get a brutally honest and correct definition for these various philosophical concepts. Materialism is Creation by Physical Matter or Creation by ROCKS. Darwinism is Creation by Evolution, or Creation by Chance, or Creation by Physical Matter. Atheism is Creation by Chance or Creation by NOTHING. Naturalism is Creation by Natural Reactions, or Creation by Physical Matter, or Creation by the Weather, or Creation by ROCKS. Once we have the correct definition for these things, the truth of the matter becomes obvious and clear for most of us.

The Atheistic Scientists and Materialistic Darwinists want me to talk about the Scientific Evidence. I would love to do so, but what they don't realize is that there is NO SCIENTIFIC EVIDENCE and can be NO SCIENTIFIC EVIDENCE to support Creation by Rocks, Creation by Chance, Creation by Evolution, or Creation by NOTHING, because these things cannot design and create! How can we in all honesty discuss the Scientific Evidence for Creation by ROCKS when there is NO SCIENTIFIC EVIDENCE for Creation by ROCKS? The ROCKS have never been caught in the act of designing and creating anything from scratch.

A Materialist is spiritually blind; and thus, a Materialist is severely handicapped when it comes to Science, Scientific Research, Scientific Discovery, Scientific Evidence, and Scientific Understanding. What the Materialists don't realize is that every Scientific Discipline has a Spiritual Aspect to one degree or another, which becomes obvious and clear once a person has permitted himself or herself to start looking for it. The Spiritual forms everything and informs everything.

According to Quantum Mechanics, the Physical arises from or comes from the Spiritual or the Non-Local or the Non-Physical. The Materialists fail to recognize all of this because they aren't looking for it and don't want to find it. It's not that the Materialists are right, it's that the Materialists refuse to look and see. The Materialists make for the worst kind of scientists because they are so one-sided and closed-minded. There are reams of Scientific Evidence, Observational Evidence, Logical Evidence, and Experiential Evidence that the Materialists and Atheists simply refuse to look at and consider. It's their problem, not ours. But, it can get very annoying whenever we encounter an arrogant, prideful, condescending, and boastful PhD Materialist who knows less about spirituality than a three-year-old does.

Most PhD Materialists and Atheists are proud of their ignorance and wear their ignorance as a badge of honor. They don't realize how shallow they are or how uneducated they come across as being. I know this for a fact, because I used to be an Atheist and a Materialist for a while — you don't see how stupid Materialism is until long after you have successfully overcome it. Materialists and Atheists are blind to the true realities of our existence, because they want to be. No seeking, then no finding!

We can't get our physical instruments to record our thoughts and our dreams, because our thoughts and our dreams reside in the realm of the Spiritual or the Non-Physical or the Quantum Non-Local. Our physical instruments can indeed detect brain activity and thus detect the presence of thoughts and dreams, but our physical instruments will never be able to tell us what a person is thinking about and dreaming about. Physical Instruments cannot be used to detect and record Spiritual Realities or Non-Local Non-Physical Realities.

Materialism causes and creates an uncountable number of problems for Science; and, ALL of these problems are efficiently and immediately resolved once we get rid of Materialism and shine a little Light or Spirit onto the subject at hand. Materialism IS the problem! Materialism is the philosophy or religion that causes these various philosophical issues and "unsolvable" problems. Get rid of Materialism, then all of these problems are easily resolved.

There is NO Mind-Brain Problem once Materialism has been removed from the equation. Mind is Non-Physical Non-Local Consciousness; and, the Physical Brain is the way that our Consciousness or Non-Local Mind interfaces with this physical reality. If the brain is damaged, the physical interface is weakened or broken, but the mind or spirit or consciousness is still there and alive. The Mind-Brain Problem ceases to exist once Materialism has been removed from consideration. There is NO Mind-Brain Problem in Transpersonal Psychology or Parapsychology where the Non-Local Non-Physical Consciousness or Living Light is taken into consideration as a matter of fact. If Psychology is taken back to its roots as the "study of the human psyche" or "the study of the human spirit", then the Mind-Brain Problem ceases to exist, because we just know that both the Spiritual and the Physical co-exist in harmony and that all is right with that world. The Mind-Brain Problem is no problem once we have successfully rid ourselves of Materialism.

In fact, one of the reasons why we experience a single stream of experiences or a single stream of consciousness is because all of the various different physical mechanisms in the body and the brain register upon a single Non-Local Consciousness. Everything in our physical body and Physical Brain is communicated to a single Non-Physical Non-Local Consciousness that exists separate from our Physical Body and separate from our Physical Universe.

Scientists have observed that there are some parts of our Physical Brain that are completely separate from other parts of our Physical Brain, which makes it impossible for the Materialists to explain to us how we each experience Unity or experience Reality as a single Conscious Event. The Materialists can't figure out how the different parts of our Physical Brain are communicating with each other and providing a united and unified picture of Reality to each one of us. This is called the Binding Problem. All of these problems and questions disappear instantly once we have rid ourselves of Materialism. There is no Mind-Brain Problem and no Binding Problem once we have thrown Materialism into the trash-heap where it belongs.

There is NO reason to be merciful to Materialism, because Materialism provides us with most of our falsehoods and lies. Materialism is THE LIE that the most people in this world want to believe in and desperately want to be true. Get rid of Materialism, then suddenly everything starts to make logical sense.

The Spiritual vs. Physical debate shows up in all of these different philosophical issues; but, it is immediately resolved once we remove Materialism from the equation. Then we just know and understand that the Spirit or Consciousness resides in harmony with the Physical, and coexists in a fully functional duality. Once we get rid of Materialism, even Quantum Mechanics makes perfect sense! Quantum Mechanics is Spiritual Mechanics — the science of how spirit matter really works.

There is no Nature vs. Nurture debate once we bring NIRVANA or the spiritual aspect of human existence into consideration. In this usage, I define NIRVANA as PSYCHE. The human spirit's pursuit of NIRVANA, or happiness, or bliss can override NATURE and NURTURE at will. The introduction of NIRVANA, or the human spirit's pursuit of happiness, into the equation actually resolves all of the contradictions and controversies associated with the Nature vs. Nurture debate. The NIRVANA, or the PSYCHE, or the human spirit in pursuit of its happiness, is the missing ingredient or the secret ingredient that resolves all of the Materialists' problems; and, they don't even know it.

Light is Non-Physical. Spirit is Non-Physical. Light is Spirit. Spirit is Light. Psyche or Intelligence is Truth and Light.

Whenever I bring up the NIRVANA, or our spirit's pursuit of freedom and peace, I am in fact playing with Light. The NIRVANA or the human spirit or human consciousness resolves ALL of the problems that the Materialists have created for themselves.

It's Materialism that causes these various unresolvable problems; and, getting rid of Materialism solves all of these "unsolvable" problems. Isn't it ironic that Nature vs. Nurture ceases to be an issue once we reject Materialism and include NIRVANA, or the human spirit's pursuit of happiness and freedom and peace, into the mix? But, ALL of these problems or philosophical issues cease to be a problem once Materialism is removed from consideration. Obviously, it's Materialism that causes all of these unsolvable problems; and thus, eliminating Materialism resolves them all.

All you really want is THE TRUTH, unless of course you are a Materialist or an Atheist.

There is NO free-will versus determinism issue, once we realize and accept the fact that it is the human spirit or human consciousness who determines what it will do with its genetic inheritance and the environment in which it finds itself. The human spirit or human psyche in pursuit of its NIRVANA, or happiness, trumps NATURE (genetics) and NURTURE (environment) every time! The human spirit can override genetics and environment at will. While pursuing happiness and peace, the NIRVANA or PSYCHE determines what he or she wants to do with its NATURE and NURTURE. Get rid of Materialism, and there is NO free-will versus determinism issue. The free-willed Consciousness, or "NIRVANA", or Psyche, or human spirit determines and decides what it's going to do with its NATURE and NURTURE.

Genetics establishes our physical potential and our physical limitations, and sets up physical possibilities. Environment establishes our physical needs and our physical opportunities. But, it is the human spirit who chooses and decides what it is going to do with its genetics and its environment. The NATURE versus NURTURE debate will never be complete and never be resolved without including the NIRVANA or the PSYCHE, or the human spirit's pursuit of its happiness and bliss and peace. This is so obvious that the Materialists and Atheists completely miss it, because they don't want to look at it, nor consider it, nor understand it.

NONE of these things can be explained by Materialism, because Materialism completely lacks explanatory power when it comes to the Non-Physical, or the Non-Local, or the Spiritual. This is why these various philosophical issues become such a serious problem for the Materialists and why the Materialists can't provide an answer or solution to these various problems. The major philosophical problems cannot be solved in materialistic terms, because they are Spiritual in nature. This little piece of information is extremely useful to us, whenever we try to pursue the truth, or the true reality of our existence. I'm not going to apologize to the Materialists for choosing to go searching for the True Reality of all things. Science is supposed to be the Pursuit of Truth, after all. It's the Materialists who need to apologize to us for trying to deceive us.

According to the Darwinists and Materialists, we are simply dancing to our DNA — we are robots under the control of our DNA, or genetics, or heredity, or NATURE. According to them, NOBODY is driving the bus. The bus is simply driving itself. Many of these people believe that they have no control over themselves and simply have to do whatever their NATURE or DNA or genes tell them to do. Their mantra is that "NATURE made them do" all of the horrible things that they have chosen to do. According to them, they have no choice but to stab their neighbor in the back, because it's in their NATURE to do so.

The Behaviorists love to believe that we are robots that are completely under the control of their conditioning techniques, thus completely controlled by our NURTURE or our environment. If the Behaviorists were in fact 100% correct, then there would be NO extinction associated with any kind of conditioning. In other words, there would be no way for any of us to break our addictions and bad habits and conditioning. Thankfully, they are wrong.

Contrary to what the Behaviorists have chosen to believe and are trying to force us to believe, the human spirit or human consciousness causes the behavior, because our spirit or consciousness or psyche is the only part of us that has the innate ability to deliberate and ACT. However, the various physical mechanisms relay information to the Consciousness so that our spirit or consciousness can make an informed decision or choice. Reflexes are an automated response designed to keep us out of danger; but, a spirit or Non-Local Consciousness can actually override the physical body's reflexes if it wants to do so. With enough determination and concentration, your spirit or consciousness can force your hand to stay on that fiery hot stove until your hand is burned to a crisp. Such a thing would not be possible if the Behaviorists were right.

If we add in the NIRVANA, or the human spirit's pursuit of happiness and bliss and comfort and peace, then suddenly everything becomes possible and everything makes logical sense. The NIRVANA, or PSYCHE, or human spirit can override NATURE and NURTURE at will. Before you know it, with NIRVANA or spirit added into the picture, Materialists and Darwinists instantly gain the power to control themselves, to repent of their sins, and to strive to overcome their physical limitations. Before you know it, with the NIRVANA or human spirit added into the equation, people are going cold-turkey from their addictions and succeeding; and, any kind of conditioning can be broken or go extinct in a matter of hours.

Those of us, who fully understand the human spirit's pursuit of NIRVANA or freedom or peace, just know that Alex in a "Clockwork Orange" could in fact break his conditioning in a matter of hours if he wanted to do so, once he has gotten away from his conditioners. He could choose to go cold-turkey from all of his conditioning, because he has a human spirit within him. I was addicted to a handful of different prescription drugs, chose to go cold-turkey, and succeeded. Yes, there were six months of confusion, psychosis, paranoia, and hell during the withdrawal process, but I succeeded in staying off the crap; and, I have been in NIRVANA or complete and total peace ever since I got sober. Ironically, the Materialists or Psychiatrists caused my mental illness in the first place. Getting rid of their Medication, their Materialism, solved most of my psychological issues.

A human spirit or human consciousness, in pursuit of its NIRVANA or happiness and peace, can make a LIAR out of the Materialists, Darwinists, and Behaviorists any time that it wants to do so! I can stop dancing to my DNA or stop succumbing to my NATURE any time I want to do so. I can start the extinction process, break my habits and conditioning, and override the influence of my NURTURE or environment any time that I choose to do so. This is because I have a human spirit or human psyche within me, who can choose to go in pursuit of NIRVANA or freedom or peace, any time that I want to do so.

This REALITY and TRUTH is what I got out of the evidence from the Materialists and Behaviorists, when I chose to critically evaluate their evidence. I quickly realized that the Nature vs. Nurture debate is immediately trumped and solved, when we choose to add in the NIRVANA or the human spirit's pursuit of happiness, peace, and bliss. NIRVANA can trump and override NATURE and NURTURE at will, instantly making liars out of the Materialists, Darwinists, Determinists, and Behaviorists any time that the NIRVANA wants to do so.

Once I KNEW that Materialism is FALSE and more importantly why it is false, then I found myself actively searching for the Spiritual, the Non-Physical, and the Non-Local within EVERY Scientific Discipline. Once I started looking for it, I discovered that the Spiritual is there to be found within EVERY Scientific Discipline that we can possibly imagine.

I also discovered that Materialism or Naturalism is the greatest bane to Scientific Discovery that humankind has created so far. Materialism cannot provide an adequate and believable explanation for Non-Local Consciousness or Quantum Trans-Dimensional Consciousness. Materialism can't explain the Trans-Dimensional or the Non-Local. Materialism cannot explain to us why Consciousness continues after the Physical Brain dies or why Consciousness can exist and function separately outside of our Physical Body during an Out-of-Body Experience or a Near-Death Experience. Materialism is completely worthless when it comes to the Non-Local or the Non-Physical.

Consciousness IS Life. Without Consciousness of some kind, physical matter cannot be alive. PSYCHE or NIRVANA is truth, light, and life.

According to the LAWS of Quantum Mechanics, without Non-Local Consciousness of some kind, physical matter cannot come into existence. Jesus Christ says that He is the Light and the Life of this universe, which is another way of saying that He is the Primal Consciousness that brought this physical universe into existence and brought life or consciousness into this physical universe.

The final frontier of Psychology and Science is to try to get some kind of handle on the non-physical or the spiritual.

In Psychology, the non-physical or the spiritual is classified under "The Mind-Brain Problem", the question between physical monism and dualism. Monism represents Materialism. Dualism is the combination of both the Spiritual and the Physical working in harmony. This problem is immediately solved once Materialism is removed from consideration. In Quantum Mechanics, Quantum Nonlocality is synonymous with Spirituality or Non-Physicality. This reality is obvious and clear once Materialism has been removed from consideration.

Synesthesia is most likely a manifestation of the human spirit in action, because the Psychologists can't explain synesthesia in physical terms. In Philosophy the spiritual is called "The Philosophy of Mind", or "Metaphysics", or "Ontology". Many scientists, including some of the Neuro-Scientists and Neuro-Surgeons, call it the study of "Consciousness" or "Awareness". There are also Psychiatrists and Neuroscientists who are studying Near-Death Experiences, which is an indirect way of observing the human spirit and the human psyche in action. Medical Doctors and Psychiatrists indirectly refer to "Spirit" as the Placebo Effect — a type of mind over matter. The mystics call it "Mysticism" or "Transcendental Meditation" or some other type of meditation — an attempt to get in touch with one's own spirit, or inner light, or consciousness. The religious in general often call interaction with the non-physical a "Spiritual Experience".

In "Introduction to Psychology 9th Edition" by James W. Kalat, he mentions the fact that brilliant PhD scientists can't determine the source of "Resilience" or "Determination". The reason they can't do so is because they are Materialists and are in fact looking in the wrong place for the source of Resilience and Determination and Free Will. The human spirit IS the source of these things, NOT physical matter! You will never find a single molecule of mercy, friendship, resilience, determination, or love within physical matter! Such things do NOT reside in physical matter. I even wrote an essay entitled, "NATURE versus NURTURE versus NIRVANA", about this subject and this truth; and, I now include that essay in a number of my books that I have available for sale online under my pen name, "Mark My Words".

The Materialists will NEVER find the source for one's NIRVANA, or happiness, or bliss because the Materialists refuse to look at and consider the true source for these things — the human spirit or the human psyche. And, it is in fact our spirit's pursuit of NIRVANA, or happiness, or freedom, or bliss that gives us the Resilience and Determination which mystify the Materialists to no end.

NIRVANA, or the human spirit's pursuit of happiness and bliss, has to be added to NATURE and NURTURE if we want a true understanding and representation of all the different sources of behavior and motivation which we witness within a human being. It should in fact be the "NATURE versus NURTURE versus NIRVANA" debate, and not just the Nature vs. Nurture debate. If we desire to face REALITY for real, then the NIRVANA or the human psyche must be added into the mix.

We have the Materialists and Behaviorists to thank for the removal of the NIRVANA, or the human spirit's pursuit of its happiness and bliss and peace, from the equation. As more and more Scientists abandon the restrictions of Materialism and Naturalism, the NIRVANA is getting reintroduced into the Nature vs. Nurture debate and actually solves all the problems related to that debate. The NIRVANA or PSYCHE is the missing ingredient or the secret ingredient. NIRVANA is a spiritual concept, not a physical place or physical location. It's NATURE (genetics) and NURTURE (physical environment) that are in fact physical in nature, although it can be successfully argued that NURTURE can have a spiritual component as well.

I discovered that there are thousands of different ways to get at Spirit and Spirituality, once we have successfully rejected Materialism and eliminated Materialism. All we have to do is to open the eyes of our understanding and see.

The New Age proponents call the Non-Physical by dozens of different names. The Paranormal, the Astral Plane, Parapsychology, Transpersonal Psychology, Humanistic Psychology, the Supernatural, UFO Phenomena, Remote Viewing, Spiritism, Channeling, Metapsychology, Ontology, and Metaphysics come to mind here, just to name a few. The Occult and Mediums and Meta-Physicists will talk about contacting the spirits of the dead, avoiding evil spirits, clairvoyance, telepathy, telekinesis, channeling the extraterrestrials, or channeling the spirit of a dead loved-one. The Parapsychologists are starting to come up with convincing Scientific Evidence for some of these kinds of things, by observing and measuring the physical effects of these things.

In Biology and many of the other Physical Sciences, it's called learning how to abandon or overcome the Materialistic Worldview or the Naturalistic Worldview — for some scientists that requires seeing and understanding the multitude of different falsehoods and lies that the Theory of Macro-Evolution is based upon. Some of the Christians call Spirituality "The Light of Christ" or "The Mind of God".

In Physics, the Non-Physical or Spiritual IS called "Quantum Mechanics". A Photon of light is massless, which means that it is non-physical or immaterial; and yet, there are ways that we can detect its effects. God tells us in His scriptures that

Spirit is Light, and that Light is Spirit. The existence of Light proves to us that Materialism is false. Materialism is the philosophical belief that Spirit and Spirituality do not exist. There is no way possible to provide any Scientific Evidence to support that belief. Materialism is unscientific and requires an infinite amount of blind-faith in order to believe that it is true. The fact that Light is Spirit and that spirit matter is made of light is in reality evidentiary proof that Materialism is false.

This is just a small sample of all the different Science and Religious disciplines that are trying to get a handle on the non-physical or the spiritual. There is something going on here, and the evidence cannot be denied, although the Materialists and Atheists have always been in a state of denial.

As anyone who has been addicted or has had a handicap knows, it can be very frustrating and confusing when your mind or spirit wants your body to do something that your brain and body are incapable of performing. People in this position quickly realize that "I am not my physical body." Often the real you wants your brain or your physical body to do things that you just simply can't get your brain or your physical body to do. The greater the disconnect between the spirit and the brain or the greater the brokenness of the brain, the more noticeable this mind-brain problem becomes. The ONLY mind-brain problem that we truly have is when the Mind or Consciousness or Spirit cannot control the physical matter, for one reason or another.

Our physical body is the way that our spirit or psyche or consciousness interfaces with this physical reality. If there is something wrong with the physical machinery, then the spirit or the mind can't get that physical machinery to do what the psyche wants it to do. Things get really weird when the physical brain starts feeding in false sensations, false perceptions, false feelings, false ideas, false beliefs, false desires, and false realities to the human spirit or the human psyche. When the physical machinery breaks down, the human spirit can have a difficult time figuring out what's really going on, out there in the physical universe. I KNOW, because the medications to which they got me addicted created some really strange mental illnesses that I had never experienced before. Addiction is Materialism and Naturalism at their VERY WORST! Although the Atheism doesn't do you much good either, from a psychological or spiritual standpoint.

Naturalism and/or Materialism has been a bane to science and has held back and prevented scientific discovery for the past couple of centuries. It's time for Scientists like us to band together and get rid of Naturalism and Materialism so that we can each start to make some progress in our respective scientific disciplines. If you notice carefully, the Scientists who have deliberately abandoned the curse of Naturalism and Materialism have noticeably more depth and are a lot more interesting and knowledgeable than the Scientists who have chained themselves to Naturalism, Materialism, and Atheism. The Atheistic Worldview strangles and suffocates Scientific Research and Scientific Discovery. There are dozens of recent scientific discoveries that we would have discovered decades sooner if we would have abandoned Naturalism, Darwinism, and the Theory of Evolution decades sooner.

In "Introduction to Psychology 9th Edition" by James W. Kalat, we have this quote from page 145:

> "The difference between having color vision and lacking it depends almost entirely on genetics; whereas the difference between speaking English and speaking some other language depends on where you were reared. Most behavioral differences depend upon differences in both heredity and environment."

Then you have to add in the spirit's individual desires and wants!

Behaviorists, Naturalists, and Materialists deliberately overlook the third dependency, which is the human consciousness or spirit or mind — our free will. They deliberately overlook the human spirit's pursuit of NIRVANA, or happiness, or bliss, or freedom.

Genes establish possibilities, or potential and limitations. Environment establishes our needs and our opportunities. Spirit or mind or consciousness establishes our desires, wants, resilience, determination, decisions, and free will. Our spirit or psyche has the power to override our genetics and our environment at will or willfully. Thanks to our spirit or psyche, someone is at the helm when it comes to our environment and our inherited genetics. Someone is actually driving the bus.

Remember, it is the human spirit that decides what it wants to do with its genes and its environment. When it comes to human behavior, we cannot fully separate the effects of heredity, environment, and free will. Don't forget! Humans can do more than just react. Humans can choose to act, or choose not to act. Behaviorists and Materialists say that we can only react, because physical matter can only react. They completely ignore the Actor, or the human spirit in its pursuit of NIRVANA, happiness, peace, and freedom of choice.

We have all read the example in the materialistic and atheistic college textbooks where the boy is born with the tall gene, and then that boy finds himself in a basketball environment, and because of the multiplier effect inevitably becomes a basketball star. All of these examples invariably overlook an essential and critical factor — what if the boy doesn't want to play ball? Desire plays a key foundational role in all of our actions and behavioral choices! If the boy doesn't want to play ball, then all of his genetic and environmental advantages will never come into play in the first place. That's the power of spirit, or desire, or free will. Some people are stunned speechless whenever they encounter a tall black man from the ghetto who doesn't know how to play basketball. Clearly, the dude didn't like playing ball. His Nature and Nurture never came into play. They were trumped by his NIRVANA, or PSYCHE, or free will.

Our NIRVANA, or human spirit, has the ability to trump its NATURE and NURTURE every time that it decides to do so! The NIRVANA or human spirit can override the NATURE and the NURTURE at will. The Behaviorists and Materialists don't realize this and don't have a clue when it comes to the human spirit and its pursuit of NIRVANA, happiness, freedom, and peace.

You will never learn anything about Spirit, Psyche, and Spirituality from a Materialist or an Atheist. The way that the Materialists deal with Spirit and Spirituality is to try to convince themselves that it does not exist. The Materialists simply stop doing Scientific Research into the subject, and instead spend all of their time mocking and ridiculing the Scientists who are studying and researching Spirit, Psyche, Non-Local Consciousness, and Spirituality. Materialism is the lazy and boring way of doing science; and, many of these people choose to become rude and unpleasant when it comes to their science.

In "Introduction to Psychology 9th Edition" by James W. Kalat, we have this quote from page 153:

> "It is remarkable that an occasional "high risk" child — small at birth, exposed to alcohol and other drugs before birth, from an impoverished or turbulent family, a victim of prejudice, and so forth — overcomes all odds to become healthy and successful. Resilience (the ability to overcome obstacles) is poorly understood."

Yes, resilience is poorly understood by the physical sciences because you guys are looking in the wrong place for resilience. Resilience is indeed poorly understood in terms of bad genetics and/or a bad environment. There isn't a physical or materialistic explanation for resilience. You should be looking at the human spirit and its pursuit of NIRVANA for the source of resilience. It's so obvious and clear, that the Materialists and Naturalists can't even see it, because they have deliberately and purposefully blinded themselves to it. In his book, Kalat concentrates on the physical explanations, or physiological explanations, and completely ignores the spiritual or non-local quantum explanations. As a Materialist, he has no other choice but to do so, or he would no longer be a Materialist.

In psychology, we have the Mind-Brain Problem, the Nature-Nurture Problem, the Binding Problem, and the Free-Will vs. Determinism Problem. ALL of these problems simply disappear and are immediately solved when Materialism is eliminated from the equation. In other words, it's Materialism that is causing the various problems. NONE of these problems exist when Materialism has been successfully eradicated from a person's chosen worldview or chosen religion.

The obvious problem that every discipline encounters is the fact that "spirit" or the "non-physical" is not directly empirically observable with our physical instruments — that's the very definition of being "non-physical". New, interesting, and unique Scientific Methodologies have to be developed in order to detect or sense these things indirectly, which is easy enough to do once a person starts looking for the spiritual or the non-physical. Seek, and ye shall find. Knock, and it shall be opened unto you.

We can't observe a massless, immaterial, wave-like, non-physical Photon of light directly and get it under a microscope, but we can definitely experience and observe its effects. This is not difficult to understand, but many people refuse to accept it as being true. That is their right and choice. If they want to cripple their understanding, God lets them do so. We see the effects and the results of the human spirit in its pursuit of NIRVANA or freedom or peace; but, we will never be able to get our physical instruments onto the human spirit nor trap the human spirit in some kind of physical cage for examination. Only God knows how to bind a specific Spirit or Consciousness to a physical body. We're not there yet.

Physics has probably come the closest to identifying, documenting, and revealing the non-physical or the spiritual to us. In most of the other sciences, the human spirit or the human psyche is inferred as a possible explanation. In Physics and Quantum Mechanics they have the observations and the numbers to prove the existence of the mystical realm or the spiritual realm or the non-physical quantum realm. String Theory is also revealing the mystical realm or spiritual realm and the extra-dimensions of reality to us. These extra dimensions or realms of existence are not visible to us, yet String Theory says that they must exist. Likewise, extra time dimensions or other timelines and other universes are not detectable to us who are stuck in this particular "box", but the Sciences and Mathematics infer that other universes and other timelines for those universes must exist as well. String Theory is an attempt to marry the micro "invisible" Quantum Mechanical world with the macro physical world described by the Law of Relativity. Physics seems to be touching upon the immaterial, the invisible, and the empirically undetectable all the time.

Many of the founding fathers of Quantum Mechanics and modern-day Professors of Quantum Mechanics have turned to mysticism for an explanation of what they are seeing in their mathematical equations and science labs. The Science of Physics is finding a multitude of different ways of accessing the non-physical or documenting the effects of the non-physical. It's a most fascinating subject to study. The books out there that discuss and document this issue have been multiplying exponentially in the last decade or two. Science is finally starting to catch up with Religion where this subject is concerned, and many believe that the gulf between Science and Religion will continue to narrow until they become One and converge upon The Truth.

I have even published a book on this subject entitled, "Quantum Mechanics from a Non-Physical Spiritual Perspective".

https://www.amazon.com/dp/B01J023TGU

CONSCIOUSNESS is the other part of the Non-Physical, Non-Local, and Spiritual which we must take into consideration. Conscious Observation is what collapses the Quantum Wave Function or the Spiritual Wave Function. Conscious Observation is what converts a Spiritual Probability Wave into a Physical Particle. A Quantum Object is wave-like and spiritual and non-local in nature before Conscious Observation collapses that wave function into a physical particle with locality in

Physical 3D Space-Time. CLEARLY, Consciousness or Psyche is something completely different and completely separate from Spirit Matter or Quantum Matter or the Quantum Probability Wave. Consciousness IS the Actor, and it is the Spirit Matter or the Quantum Object that reacts.

A Quantum Object or Particle of Matter has two modes of existence that we know of — a Non-Physical, Non-Local, Trans-Dimensional, Spiritual, Wave-like nature; and, a Physical, Localized, Particalized atom or particle in 3D Space-Time. Consciousness or Psyche is NOT a Quantum Object. Consciousness is a third type of thing and/or a third mode of existence or reality. Consciousness or Living Light seems to follow a completely different set of rules and seems to be everywhere present in everything.

The physicists and mystics have documented and observed that those immaterial or non-physical photons traveling at the speed of light are situationally aware. They are conscious of their environment. They can "see" whether there are two slits open or only one slit open, and they can act or react accordingly. They seem to be able to coordinate or communicate with other photons in the vicinity past, present, and future. How do they do that? When it comes to particles and atoms, it's as if there is a little mind or consciousness hidden inside the atoms and other particles, somewhere between the empty space of the electrons and protons. There seems to be an invisible driver or a nonphysical driver in there somewhere telling these particles and photons how to act, what to do, and what physical laws to obey. That's what Quantum Mechanics is telling us.

Non-physical or immaterial entities like photons "act" like their physical counterparts — atoms and molecules and such. They all seem to have some kind of spirit or mind that is guiding them. They are all made up of Light or energy. They seem to be able to choose whether to obey or rebel. They are also aware of the times when another "consciousness" or "spirit" or "overlord" is looking at them or observing them. These little particles can see you or sense you looking at them! How do they do that? The Scientists, the Real Scientists, are trying to figure out how the different particles are doing that; and, some of them have succeeded in doing so.

Scientific experimentation revealed to us that a photon knows where it is going to land before it launches; otherwise, it won't launch. How is this possible? It's possible because of Non-Local Consciousness or Psyche. You see, in Trans-Dimensional Non-Locality, there is NO space, NO time, and NO distance. Everything is simultaneously and instantaneously present in the Non-Local Realm or the Non-Physical Realm or the Spiritual Realm. From our physical perspective, a photon can take millions of years to travel from one galaxy to another; but, from its massless and timeless perspective, no time and no distance whatsoever passes from the moment a photon launches to the moment it reaches its destination. From its perspective, a photon knows its destination and is at its destination the very moment it launches, or it doesn't launch. That's the power and advantage of Non-Local, Non-Physical Consciousness. From its perspective, there is no space, no time, and no distance whatsoever.

There are NO speed limits in Non-Local "Space". Nonlocality, or the Quantum Realm, or the Trans-Dimensional Realm takes up less physical space than the period at the end of this sentence. The Non-Local Realm or Spirit Realm appears to be an infinite singularity. Think about it! That's kind of cool, isn't it?

This is a fascinating area of study that overlaps and impacts every Scientific Discipline including Naturalism, Materialism, and the Theory of Evolution, which so many people seem to admire and worship. I'm surprised at times how much the Atheistic Scientists and Darwinists seem to worship the Theory of Evolution. Yet, Non-Local Consciousness or Quantum Consciousness is radically advanced Science in comparison to something like the Theory of Evolution. Evolution can't even begin to design, create, and manufacture something from scratch. Living Consciousness or Psyche can and does.

The Theory of Evolution IS Creation by Evolution. The more critically that I evaluated the evidence, the clearer it became to me that evolution of any kind cannot design, create, manufacture, nor produce anything at all from scratch. Creation by Evolution cannot be OBSERVED, because Creation by Evolution of any kind does not in reality exist. Evolution cannot design and create. That's what The Scientific Method is trying to tell us.

When it comes to the Origin of the first genome and the first life form, there can be NO evolutionary explanation, because evolution cannot design and create. Furthermore, random mutations and natural selection cannot come into play until AFTER God has designed and created and manufactured the first genomes and the first life forms in the first place. The Darwinists and Materialists and Atheists don't realize this because they have deliberately blinded themselves to it.

I have watched the Atheistic Darwinists online and on DVD worshipping the Theory of Evolution and extolling its many virtues. Call me picky or choosy if you want; but, if I'm going to worship something, I prefer that it be Conscious and Alive, and not deader than a rock like the Theory of Evolution is. Darwinism or the Theory of Evolution is just a modern-day form of idol worship, where the Materialists and Atheists are found falling on their faces and worshipping their man-made gods. Evolution cannot hear and answer our prayers; but, the Biblical God certainly can, because the Biblical God is Conscious and Alive. That's the power of Non-Local Consciousness. Consciousness or Psyche can think, and live, and answer prayers and requests for help.

The pursuit of "Consciousness" affects everyone, whether they consider themselves scientists or not. There is one caveat, though. Once you discover the human spirit and what it can really do, what it can overcome and what it can override, you can make the mistake of completely discounting the effects of genetics and environment and delude yourself into believing than an individual should be able to overcome or conquer all obstacles in his way. Those born with Downs Syndrome or Fetal Alcohol Syndrome have genetic and developmental disabilities that sometimes no amount of spiritual desire or environmental support can overcome. Genetics have a definite influence on what's possible, and genetics often do impose physical and mental limitations and handicaps on an individual that he or she cannot overcome with any amount of spiritual desire or environmental support. We humans weren't born with wings, so we can't fly like the birds fly, no matter how much we might want to do so. Our genetics or NATURE or heredity opens up possibilities for us, but also provides limitations to us.

The Mind-Brain Problem is an integral part of Psychology — the Study of the Psyche, or the Study of the Mind, or the Study of the Human Spirit. Psychology itself is coming home to its original definition because of this new awakening to the non-physical that's happening among the Scientific Community at-large. It's a great time to be alive if you are a Scientist. Some of us are starting to abandon Materialism, Naturalism, and Macro-Evolution; and consequently, we are finally starting to make some real progress towards understanding how things really work.

Getting rid of Materialism and Behaviorism makes for better and more interesting Science.

Let's face REALITY. The KEY to understanding life, the universe, and everything is to eliminate Materialism, even if we have to find 42 different ways to do so. Once we get rid of Materialism, then suddenly everything starts to make sense. That has been my Scientific Observation and my Scientific Contribution to the world.

The Materialists never ask themselves what was there BEFORE the first particle of physical matter was designed and created, or what was there BEFORE this physical universe was designed and created 13.8 billion years ago. The Materialists refuse to ask and refuse to consider the most interesting Scientific Questions of all.

Materialism is the chosen philosophical religious belief that the Spiritual or Non-Physical does not exist. Technically, Materialism is Creation by Physical Matter or Creation by ROCKS. The Materialists really truly believe that the ROCKS designed and created it all. But, what was there BEFORE the first rock and BEFORE the first physical matter were designed and created? That's the question which the rest of us are asking.

Many Scientists from a wide variety of different Science backgrounds are now starting to converge upon the True Reality or the True Construct of the Universe, and the cutting-edge Scientists of today are discovering that the Prophets in the Bible got there first two thousand to three thousand years before they did. We are watching Science and the Bible merge in real-time right before our eyes, assuming of course that we have our eyes open and are actually looking at what's happening with an open mind.

Science and Religion are starting to converge upon the same Truth or the same Reality, and behind it all is some kind of Psyche or Conscious Mind. And then we realize that the Mind-Brain Problem is no problem at all, once we have removed Materialism and Atheism from the equation. Everything makes perfect logical sense once we finally get rid of Materialism.

Mark My Words!

Reference Materials for the Mind-Brain Problem

Once I got rid of Materialism, then everything made perfect logical sense, including Quantum Mechanics and the Mind-Brain Problem. These things ceased to be a problem for me, once I eliminated Materialism from my worldview.

The books that I found and acquired regarding the Mind-Brain Problem:
http://bookrev.allthings.computer/consciousness-and-the-mind-brain-problem/

Favorite books about Consciousness and the Mind-Brain Problem:

Beyond the Cosmos: The Extra-Dimensionality of God: What Recent Discoveries in Astronomy and Physics Reveal about the Nature of God:
http://www.amazon.com/Beyond-Cosmos-Extra-Dimensionality-Discoveries-Astronomy/dp/0891099646/

Irreducible Mind: Toward a Psychology for the 21st Century:
http://www.amazon.com/dp/1442202068/

Beyond Physicalism: Toward Reconciliation of Science and Spirituality:
http://www.amazon.com/dp/1442232382/

Quantum Enigma: Physics Encounters Consciousness 2nd Edition:
http://www.amazon.com/Quantum-Enigma-Physics-Encounters-Consciousness/dp/0199753814/

The Conscious Mind: In Search of a Fundamental Theory (Philosophy of Mind) 1st Edition:
http://www.amazon.com/gp/product/0195105532

Consciousness Beyond Life: The Science of the Near-Death Experience:
http://www.amazon.com/gp/product/0061777250

The Holographic Universe: The Revolutionary Theory of Reality
http://www.amazon.com/dp/0062014102/

The Character of Consciousness (Philosophy of Mind):
http://www.amazon.com/gp/product/0195311116

Biocentrism: How Life and Consciousness are the Keys to Understanding the True Nature of the Universe:

The End of Materialism:

There are many other books with similar content from a multitude of different authors. Wherever you might find one of these books online, just look at the suggestions offered down below each book's main review for additional books on the same topic. Each book is a rabbit hole that goes down far and deep. This whole subject is starting to achieve a great deal of depth. In comparison, Materialism is boring and has no depth. Once Mr. PhD Materialist has said that spirit does not exist, then he is cooked and done; and then, he has no more to contribute to the subject. That's very limited and boring!

There are dozens of different books from different authors entitled, "**Philosophy of Mind**", which also touch upon this subject of Spirituality or Non-Physicality in many different ways.

In "Introduction to Psychology 9th Edition" by James W. Kalat, his Materialism and Behaviorism combined with his worship of Evolution spurred me into action researching, studying, and evaluating the evidence. I soon discovered that there is No Scientific Evidence and can be No Scientific Evidence to support Materialism or Creation by Evolution, because these things cannot design and create. ONLY a living, conscious, being can actually design and create and manufacture things like Genomes and Life Forms. Evolution could never have done the job because Evolution of any kind cannot design and create anything at all, because Evolution is not conscious and not alive. Therefore, Kalat's claim in his book that Evolution designed and created your brain is false. Evaluating the evidence critically and logically will reveal to you why it is false, because it is obvious and clear to most of us that Evolution cannot design and create anything at all. Evolution is dumb and blind without hands and a mind.

I have written a dozen books of my own on this subject; and, some of those books are now available on Kindle, ready for you to purchase them and read them.

I have been busy, just not directly with this psychology course. This psychology course was the stimulus and inspiration that got me studying Materialism, the Theory of Evolution, Consciousness, and the Mind-Brain Problem. We take our inspiration and motivation wherever we might find them. I'm not sure that James Kalat would be pleased with where the evidence led me, but I am pleased with what my critical evaluation of the evidence has brought to me. For me personally, it finally brought an end to Materialism. I'm free at last! Thank God Almighty, I'm free at last!

Mark My Words!

--

Helpful Informative Reviews from Amazon

I have personally observed that Quantum Mechanics makes perfect logical sense once we get rid of Materialism. Quantum Mechanics is BEST described and explained from a Non-Physical Spiritual Perspective.

I have even written a book about this realization or epiphany entitled, "Quantum Mechanics from a Non-Physical Spiritual Perspective", which can be found at this link.

https://www.amazon.com/dp/B01J023TGU

In my essays and books, I now debunk Materialism and Darwinism first thing, because I have observed that EVERYTHING makes logical sense once we have eliminated Materialism. Materialism IS the problem and the bane that is holding back Scientific Discovery, Scientific Exploration, and Scientific Understanding. It has been that way all along.

—

—

A Helpful Review from the Amazon Website regarding the "Quantum Enigma" book:

By kaon2009 on August 9, 2013:

I am a theoretical physicist, but I must admit I did not fully appreciate the Quantum Enigma until I read the first edition of this book a few years ago. I first learned quantum mechanics over 40 years ago and have actively practiced it. That is, I used it to calculate theoretical predictions. It was only in the last 10 years or so that I asked myself, "What is the electron actually doing when light is emitted from a hydrogen atom?" After reading this book I realized the answer is, "Nobody has the slightest idea!" Fully appreciating the vast gap between the "classical" world we live in and the "quantum world" took some time for me. That kind of profound ignorance takes time to appreciate. I now better understand what I have read in biographical books about Bohr, Einstein, Heisenberg, and Schrodinger. As the realization slowly set in as to what quantum mechanics was saying, these men and other physicists struggled with each other in an almost religious battle. Now over 80 years later we know no more than we did then. In the end, everyone has to come to appreciate the profound ignorance we have at this point in history. For any interested layman or scientist, the Quantum Enigma is a must-read item.

—

—

Another Helpful Review from the Amazon Website regarding the "Holographic Universe" book:

Are individual experiences valid scientific data?

By Damian Nash on January 8, 2001:

This is one of the most provocative books I have read in years. In the first few chapters Mr. Talbot describes the emerging holographic paradigm in science, drawing on David Bohm's work in quantum physics and Karl Pribam's work in neuroscience. I found both descriptions to be fascinating, and especially enjoyed the historical context for the work of these two seminal thinkers. As a person with a master's degree in neuroscience and chaos/complexity theory, I found a couple of his simplifications misleading, but would give him high marks for his overall comprehension of the conclusions of Pribam and his followers.

The remaining 2/3 of the book is a discussion of how the holographic paradigm may provide a rational basis for interpreting a wide variety of phenomenon located around the fringes of established science. He looks at everything from strange historical "miracles" like stigmata and appearances of the Virgin Mary to modern psychic abilities and LSD experiences, from out-of-body and near-death-experiences to UFO abductions. In addition, he compares language used in the modern scientific discussion of holography with the language used by ancient mystical traditions.

Mr. Talbot's writing style is unusually clear and lucid. All of this makes for a highly engaging book. It kept me up late every night for more than a week. I am a person who has had an OBE/NDE (out-of-body, near-death-experience), and can tell you that his description of such events is an astoundingly accurate portrayal of what I experienced.

I am also a scientist, and know that most of my highly rational, empirical colleagues would have trouble accepting a majority of Mr. Talbot's conclusions. This work addresses something so completely out of the realm of everyday experience for most people, and probes a world that is normally invisible to the five senses. Hence, objective, empirical science — as defined by a conventional theorist or practicing technician — simply cannot address these experiences. They are outside the range of focus of the tool that Western minds currently rely on.

The service that Mr. Talbot provides is a challenge to rethink the conventional definition of science so that it can take into account a much wider range of human experience. What he argues for is the acceptance, as valid scientific data, of the experiences of individual humans, across cultures and throughout history that are remarkably consistent with one another. These experiences address aspects of reality that are invisible to the skeptical eye, but become obvious to the person who chooses to develop other forms of perception.

As a person who was unwittingly thrown into an OBE/NDE experience, I am naturally inclined to read a book like this one with an open mind, and felt immensely rewarded for doing so. However, if I had reviewed the same book before having my own personal experience of some of the phenomena it describes, I would have reviewed it as a new-age excursion into a realm of fantasy. I am completely sympathetic to some of the reviewers who see it that way, and respectfully disagree.

I believe there is an extraordinary synthesis happening among the realms of human experience, one that can validate each individual's story, however unusual, and also one that honors all the different ways of knowing. I see Mr. Talbot's work as one of the more important bridges yet constructed between traditional science and spirituality, between rational discourse about repeatable, empirically verifiable phenomenon and the quirky, esoteric or mythological elements of personal experience that actually define most people's experience of reality. This book is a "must read" for any passionate seeker of truth.

—

—

Review from page 153 of "Proof of Heaven: A Neurosurgeon's Journey into the Afterlife", Simon & Schuster, 2012:

For those still stuck in the trap of scientific skepticism, I recommend the book Irreducible Mind: Toward a Psychology for the 21st Century, published in 2007. The evidence for out-of-body consciousness is well presented in this rigorous scientific analysis. Irreducible Mind is a landmark opus from a highly reputable group, the Division of Perceptual Studies, based at the University of Virginia. The authors provide an exhaustive review of the relevant data, and the conclusion is inescapable: these phenomena are real, and we must try to understand their nature if we want to comprehend the reality of our existence.

(Eben Alexander III, MD, Neurosurgeon and author of *Proof of Heaven* and *The Map of Heaven*).

—

—

Getting rid of Materialism and Behaviorism makes for better and more interesting Science.

Let's face REALITY. The KEY to understanding life, the universe, and everything is to eliminate Materialism, even if we have to find 42 different ways to do so. Once we get rid of Materialism, then suddenly everything starts to make sense. That has been my Scientific Observation and my Scientific Contribution to the world.

The Materialists never ask themselves what was there BEFORE the first physical matter was designed and created, or what was there BEFORE this physical universe was designed and created 13.8 billion years ago. The Materialists refuse to ask and refuse to consider the most interesting Scientific Questions of all.

Materialism is the chosen philosophical religious belief that the Spiritual or Non-Physical does not exist. Technically, Materialism is Creation by Physical Matter or Creation by ROCKS. The Materialists really truly believe that the ROCKS designed and created it all. But, what was there BEFORE the first rock and BEFORE the first physical matter were designed and created? That's the question which the rest of us are asking.

Mark My Words!

NATURE versus NURTURE versus NIRVANA

The word "Spirit", with a capital letter, is defined as a Psyche and Spirit Body combination, and signifies a person or a personality or living individual. The word "spirit" in lower case is often defined as spirit matter and sometimes defined as a spirit body. But, I and others don't always follow this convention in our writing, so every time we encounter the word "spirit", we have to stop and try to determine what type of spirit the author might be talking about.

In the NATURE-NURTURE debate the all-knowing Materialistic Scientists completely eliminate the NIRVANA or the PSYCHE, or the spiritual and personal desires, or the Consciousness and Intelligence, or the light and the life.

When it comes time for the Scientists to determine the source or the cause of a specific behavior, not only do they have to consider NATURE (heredity) and NURTURE (environment), but they also have to consider NIRVANA (the human spirit, or human desires, or the human psyche's pursuit of Bliss and Nirvana) — if they want to find The Truth of the situation under examination.

If you are trying to explain behavioral differences you do indeed have to take into consideration heredity and environment, but you also have to add-in the spirit's individual desires and wants. This is something that the Behaviorists refuse to do; and consequently, the Behaviorists and Materialists can't define Consciousness nor explain Determination and Resilience in any way that makes logical sense to them. The Behaviorists and Materialists are clueless when it comes to these kinds of things, because they are missing an essential ingredient — the NIRVANA or the spirit's desires and wants.

The human Spirit actually has the power and the ability to override and veto NATURE and NURTURE in the pursuit of its NIRVANA, or Happiness, Freedom, and Bliss!

It doesn't matter if the kid is seven feet tall by the time he is a teenager and has been playing basketball all of his life, if that kid decides that he's never going to pick up a basketball ever again, then his spirit or personality has actually trumped his NATURE and his NURTURE. The Spirit's pursuit of NIRVANA or Bliss can override NATURE and NURTURE any time that it wants to do so! It doesn't matter how tall the kid is, or how skilled he is, if he has no desire to play ball, then he won't play ball. The Behaviorists completely overlook this reality, because they want to overlook it! The Behaviorists are determined to overlook the NIRVANA aspect, or the spiritual aspect, in order to make their case.

Genes establish possibilities or potential, and some limitations. Environment establishes needs and opportunities. Spirit or mind or consciousness establishes desires, wants, decisions, choices, determination, resilience, and free-will. The human spirit or human psyche decides what it wants to do with its genes and its environment.

In other words, the NIRVANA or the PSYCHE decides what it wants to do with its NATURE and its NURTURE. It is in fact the NIRVANA, or Spirit, or Consciousness who is in control of the situation. It is in fact the NIRVANA or the Psyche who is actually conscious and alive and aware of its NATURE and its NURTURE. Consciousness IS Life, and Life IS Consciousness! Consciousness is known to pursue its NIRVANA or Bliss or Happiness.

Remember, humans can do more than react. They can choose to act! And, many of their choices will actually override their genetics and their environment. Free will, or choice, will always be the 800-pound gorilla in the room which the Behaviorists, Materialists, and Darwinists are trying to ignore. But, as human beings, we often override our NATURE and our NURTURE in pursuit of our NIRVANA.

It's the Spirit or the Living Light or the Consciousness that chooses and acts; and, it's the physical matter that was designed to react and be acted upon. The Life is in the Psyche or the Light or the Consciousness. Spirit is Light. Light IS Consciousness.

We hold these truths to be self-evident, that all men are created equal, that they are endowed by their Creator with certain unalienable Rights, that among these are Life, Liberty and the pursuit of NIRVANA. Eternal life, liberty or peace of mind, and NIRVANA or bliss or happiness are spiritual qualities — NOT physical qualities. The most important things in our lives are spiritual and not physical! You could tear the whole universe down and never find a single atom of charity or love or happiness, because unconditional love is not to be found in physical matter.

In the psychology textbooks, they sometimes talk about "Resilience," and the PhD scientists can't explain it! The reason that these brilliant Materialists and Darwinists and Behaviorists can't explain Resilience is because they are looking in the wrong place. They should be looking at the human spirit or the psyche's pursuit of NIRVANA for the source of Resilience, not physical matter! Raw physical matter has NO conscious desire or will; therefore, the Materialistic Scientists are not going to get any Resilience or Gumption or Initiative out of that ROCK sitting on their desk.

The human Spirit in its pursuit of NIRVANA can override any genetic programming and genetic defects, contrary to what the Darwinists and Behaviorists and Materialists try to tell us. The human Spirit can break its behavioral conditioning at will. The human Spirit can push past its genetic handicaps and developmental handicaps and environmental handicaps in the pursuit of its NIRVANA or Freedom or Bliss. The human Spirit can choose to go cold-turkey and end years-long habits and addictions if it decides to do so.

I have some friends who have Downs Syndrome, Alcohol Fetal Syndrome, or Autism — their spirits are capable of accomplishing and doing many amazing things despite their physical, developmental, and mental handicaps. They are NOT just their brain or their physical body! They are NOT just dancing to their DNA! There is a person in there who has wants, desires, interests, and needs. In contrast, evolution has no wants, interests, plans, desires, or needs because evolution is not alive, and Creation by Evolution does not exist for real.

I learn best by comparison and contrast. When it comes to something like evolution or natural selection, there really is nobody at home upstairs. The same thing can't be said about my Autistic and mentally handicapped friends.

When it comes to Behaviorists, Materialists, and Darwinists and their Materialistic Worldview, a human being with a human spirit can decide to prove them wrong and succeed every time he or she wants to do so. In my college textbook, the PhD author told us not to dismiss Behaviorism, implying that it is difficult to think of an example that proves it wrong. Two examples jumped immediately into my mind.

We know that Alex in "A Clockwork Orange" would immediately start experiencing the extinction of his conditioning once he was no longer in the hands of his conditioners. He could break his conditioning in a matter of hours if he really wanted to do so. In the pursuit of our NIRVANA or Bliss or Happiness, we can override our crappy Environment, and our horrible Genetics, and our Behavioral Conditioning any time that we want to do so.

Furthermore, I have dogs and they provide an even better example. Every time that I am presented with a dog, I have a choice to make. Radical Behaviorism says that every time I see a dog, I have to kick the dog. I have no choice in the matter — I have to kick the dog whenever I encounter one, because that's the natural pre-programmed response that evolution has given me. In reality, though, because I have a human Spirit and free will, whenever I encounter a dog, I can choose to love and pet the dog, or I can choose to feed the dog, or I can choose to walk the dog, or I can choose to ignore the dog. I can choose to hug the dog, even if I have already kicked the dog a thousand times before. Or, I can choose to kick the dog, even if I have never kicked the dog before. The human Spirit with its free will and ability to choose can make a Behaviorist and a Materialist a liar any time that it wants to do so. The human Psyche in the pursuit of NIRVANA can override NATURE and NURTURE any time that it wants to do so. Powerful, isn't it!

Consciousness by definition records significant stimuli and perceptions. Many psychologists (the Materialists) have despaired ever finding any scientific approach to Consciousness, because Consciousness is a non-physical spiritual phenomenon.

In order to study Consciousness or psyche or spirit, the Real Scientists have had to develop some new and interesting and unique Scientific Methodologies in order to get the job done! In contrast, the Materialists gave up millennia ago and simply state that there is no such thing as Consciousness or Spirit — "we are just dancing to our DNA." The Behaviorists and Materialists consider us to be pre-programmed robots and nothing more. The Materialists do indeed compare us to robots, but they never stop to ask who did all of the programming that needed to be done to create those robots in the first place! The Materialists don't even try to find signs of Consciousness, or Psyche, or Spirit, or Life. But, there is more to life than just our genes and our physical matter!

The Real Scientists have noticed that the spirit or psyche or subconscious decides to move before the physical body and the brain are even aware of it. Brain awareness lags behind the unconscious or spiritual decision. It takes time for the body and the brain to respond to a decision that the Spirit has already made. When it comes to science and everything else, the Materialists ALWAYS lag behind the Spiritualists.

Evolution and Natural Selection are powerless to touch the human Spirit. Spirituality is infinitely superior to Materialism, and vastly more interesting too!

When it comes to Spirit and spirituality, the only thing that you have to be careful about is that, once you discover the human Spirit, you can make the mistake of completely discounting the effects of genetics and environment. If there are NO basketballs and no basketball standards in your country, then you will never be playing basketball in your country no matter how much your Spirit might want to do so. If your genetics top you out at three-feet tall or a meter tall, you will never become an NBA basketball star no matter how much your Spirit might want to become a basketball star.

Genes establish possibilities or potential, as well as limitations. Environment establishes needs and opportunities. Spirit or mind or consciousness establishes desires, wants, decisions, choices, determination, resilience, and free will. The human spirit or human psyche decides what it wants to do with its genes and its environment.

The Behaviorists, Materialists, and Darwinists limit themselves to NATURE and NURTURE; and therefore, they can't explain various psychological phenomenon in scientific terms or Materialistic terms, because these things can only be explained in terms of NIRVANA or the Spirit's pursuit of happiness.

The Theory of Evolution doesn't figure into any of this, because evolution of any kind cannot design, program, engineer, create, manufacture, implement, and deploy even a single GENE. The Materialists deliberately cripple themselves with their theories.

The Behavioral Scientists, Materialists, and Darwinists ALWAYS overlook the spiritual possibilities — desires or the NIRVANA aspect of life — because they want to. But, Behaviorism is just another form of Materialism, and just as useless and FALSE.

Think about it! That's why God gave you a Spirit or a Mind! You can still think about these things even if your physical brain has been damaged. You can't interface with the physical world as well with a damaged brain, but you can still think about things and have dreams and desires even with a damaged or addicted physical brain!

You feel the frustration, but choose the response. Our responses aren't programmed or conditioned into us. Yes, we do indeed feel the fear, or the frustration, or the anger; but, we always choose the response! Any kind of conditioning or habituation or addiction is subject to extinction and can be broken by the human Spirit pursuing its bliss or freedom or peace or NIRVANA. I broke my addictions because I wanted to, and I'm a lot happier as a result. This is really cool stuff, if you take the time to think about it.

When it comes to the sun, moon, earth, stars, and rocks, we assume that these things are not alive or self-aware, but we could in fact be wrong. However, we KNOW that evolution is not alive, because the Theory of Evolution has no substance, in which a spirit, psyche, intelligence, mind, desires, or personality could actually reside. Evolution has NO desires, because it has no Spirit or Soul.

In contrast, the reason we know that Jesus Christ is alive and has risen from the dead is because thousands of different people have seen Him and touched Him after He rose from the dead, as documented in the Bible, Book of Mormon: Another Testament of Jesus Christ, Doctrine and Covenants, and Pearl of Great Price. There is NO such evidence for the Theory of Evolution, because Evolution is not alive! Unlike goo-to-you Darwinian evolution, our resurrected Lord Jesus Christ has been observed in the wild.

The existence of these revelations from God and revelations of God are bulletproof and irrefutable evidence that Materialism and Naturalism are FALSE. Any time that our Resurrected Lord Jesus Christ makes an appearance in the flesh to mere mortals like us, that reality and experience PROVES to us that Naturalism and Materialism are FALSE! Try getting Evolution to make an appearance to you in the flesh. Common sense logic tells you that it will never happen. Evolution isn't alive and doesn't really even exist.

Darwinism is Creation by Evolution. Darwinian molecules-to-man evolution could NEVER have done any of the things which the Darwinists say that evolution did because evolution is NOT conscious and alive! Therefore, common sense tells us that the Theory of Evolution is FALSE, and that God must of necessity exist in order to do all of the different things that evolution could NEVER have done.

Getting rid of Materialism and Behaviorism makes for better and more interesting Science.

Let's face REALITY. The KEY to understanding life, the universe, and everything is to eliminate Materialism, even if we have to find 42 different ways to do so. Once we get rid of Materialism, then suddenly everything starts to make sense. That has been my Scientific Observation and my Scientific Contribution to the world.

The Materialists never ask themselves what was there BEFORE the first physical matter was designed and created, or what was there BEFORE this physical universe was designed and created 13.8 billion years ago. The Materialists refuse to ask and refuse to consider the most interesting Scientific Questions of all.

Materialism is the chosen philosophical religious belief that the Spiritual or Non-Physical does not exist. Technically, Materialism is Creation by Physical Matter or

Creation by ROCKS. The Materialists really truly believe that the ROCKS designed and created it all. But, what was there BEFORE the first rock and BEFORE the first particle of physical matter were designed and created? That's the question which the rest of us are asking.

—

In this essay, I tended to define NIRVANA as the bliss, happiness, will, and blessedness of our own internal Light or Spirit. Spirit IS Light. Light IS Spirit. NIRVANA is PSYCHE.

It is the Spirit or Living Light or NIRVANA within each one of us that has the inherent ability to design and create — NOT our physical matter! Our physical matter can't do anything except to react to whatever our Spirits tell it to do.

A rock has physical matter, but a rock does NOT have the kind of spirit within it that has the ability to design and create new things. Even though a rock has a lot of physical matter, you will NEVER catch a rock in the act of designing and creating new things from scratch. It isn't the physical matter that designs and creates! Physical matter was designed to be molded and acted upon by Spirits, or by Living Lights, or by Living Consciousness. Physical matter was designed to be occupied and enlivened by Spirits, or by Living Lights, or by Consciousness. Inert physical matter cannot design and create! And, Materialism is Creation by Physical Matter, which makes Materialism a nonsensical oxymoron.

There are different types of Spirit or Light. While in Spirit, people have seen the consciousness or awareness of rocks and nails. But, there isn't a designing and creating type of Spirit within a rock; and, rocks don't have the physical makeup necessary to design, create, and manufacture. There isn't a loving and charitable Spirit within a rock either. A rock is a spiritual recording device, nothing more.

It's our Human Spirit, or Human Psyche, or Living Light, or Consciousness which gives us our desires, intelligence, will, and ability to choose — NOT our physical matter. Living Intelligent Spirits like ours are Actors, Designers, and Creators. In contrast, physical matter was designed to be acted upon. Spirits or Living Lights act. Physical Matter reacts and is acted upon. This is the True Reality of the situation.

The reason we humans can design and create is that Our Spirits inherited that particular ability from the Parents of Our Spirits — God the Father and Heavenly Mother. The Spirit or Living Light within human beings is unique among God's creations, because we are the Children of God. We humans can have Lived Experiences and Spiritual Experiences, write them down on paper, and share them with other human psyches.

Our Living Light, or Psyche, or Spirit or NIRVANA enlivens our physical matter, giving that formerly inert physical matter purpose, intelligence, desires, goals, and in the case of humans the ability to design, plan, repent, and create. The pursuit of love, friendship, life, sanity, sobriety, freedom, peace, bliss, happiness, and NIRVANA is the most powerful motivator that we know of; and, ONLY Psyche can

225

do these kinds of things for real. Our genetic and environmental needs often pale in comparison.

Once you understand and accept the True Reality of existence, you won't want to go back to Materialism, Darwinism, Naturalism, and Atheism. These things are boring and weak in comparison to the human Spirit or the human Consciousness in pursuit of its NIRVANA.

NIRVANA to BioPsychoSocial to the Ultimate Model of Reality

The word "Spirit", with a capital letter, is defined as a Psyche and Spirit Body combination, and signifies a person or a personality or living individual. The word "spirit" in lower case is often defined as spirit matter and sometimes defined as a spirit body. But, I and others don't always follow this convention in our writing, so every time we encounter the word "spirit", we have to stop and try to determine what type of spirit the author might be talking about.

While I was participating in my Introduction to Psychology course in 2015 and 2016, it was glaringly obvious to me that these people were deliberately excluding PSYCHE from their Science and their Philosophy of Science. I developed this NATURE vs. NURTURE vs. NIRVANA Model to compensate for their oversight. In this model, NIRVANA represents PSYCHE.

Eventually, during my ongoing study of Psychology, I was introduced to the BioPsychoSocial Model of reality and therapy, and quickly realized that the BioPsychoSocial Model is an established version of my NATURE vs. NURTURE vs. NIRVANA Model of Reality. During my studies and research, I was forced to switch away from this NATURE vs. NURTURE vs. NIRVANA Model over to the BioPsychoSocial Model.

In 2016, I even had a Clinical Psychologist tell me that at the University of Utah Medical Center, it has ALL become BioPsychoSocial and they work as a team of Medical Doctors, Psychiatrists, Clinical Psychologists, and Licensed Social Workers based upon the BioPsychoSocial Model treating mental illness from all angles possible. Although, he didn't mention any Priests or Clergy being on their team, but that should probably happen as well if the whole process is to truly become BioPsychoSocial.

This NATURE vs. NURTURE vs. NIRVANA Model, the BioPsychoSocial Model, Near-Death Experiences (NDEs), Shared-Death Experiences (SDEs), Out-of-Body Experiences (OBEs), all Spiritual Experiences, the revelations of Jesus Christ and the revelations from the Biblical God as found in the *Bible*, *Book of Mormon: Another Testament of Jesus Christ*, the *Doctrine and Covenants*, and the *Pearl of Great Price*, Quantum Mechanics, Quantum Nonlocality, Intelligent Design and Creation, Dark Matter (Spirit Matter), Dark Energy (the Zero-Point Field of Light or the Light of Christ), and even Light, Magnetism, and Gravity collectively pointed me

226

to this PSYCHE ONTOLOGY and this Ultimate Model of Reality wherein Psyche is the Ultimate Cause and Psyche is the fundamental unit of Reality and Existence.

Without Psyche or Intelligence, there is NO existence! ONLY Psyche can do Ontology, Reality, Experience, Intentionality, Choice, Intelligence, and Existence in any realm of Reality; and then, remember having done so. It ALL points to Non-Locality and a Psyche Ontology, which becomes the Ultimate Model of Reality.

Part II — MATERIALISM IS A DENIAL OF REALITY

I have had Materialists tell me that Quantum Mechanics has nothing to do with Spirit and Spirituality and that Quantum Mechanics is purely a physical theory.

How can the Materialists possibly know this?

By their own admission, the Materialists know absolutely nothing about Spirit or Spirituality and have actually chosen to believe that there are no such things as a Spirit or a Spirit Realm. So, what makes the Materialists magically believe that they are the preeminent experts on Spirit, Spirituality, and the Spirit Realm so as to believe that they are in the superior position to officially declare for all of us that there is no such thing as Spirit?

Isn't that the very definition of self-centered arrogance, ignorance, and pride?

How can the Materialists possibly know that Quantum Mechanics has nothing to do with Spirit or the Spirit Realm when in fact they know absolutely nothing about Spirit, Spirituality, and the Spirit Realm?

They can't!

The Materialists are all bluster, pomposity, arrogance, and posturing without any substance or science to back up their claims.

It has taken me a while to do so, but I eventually learned that if there is something that an Atheist, Darwinist, Naturalist, or Materialist truly believes to be true, then it is almost certainly FALSE. Therefore, the Materialistic claims that Quantum Mechanics has nothing to do with Spirit or the Spirit Realm, that there is no such thing as Spirit and the Spirit Realm, and that Physical Matter is the only thing that exists in this universe ARE most certainly FALSE.

The Materialists, Atheists, Darwinists, and Naturalists seem to have an in-built barometer for detecting, finding, or manufacturing LIES; and then choosing to believe that those LIES are true. That's what I have observed and learned from Materialism in general. The Materialists are ALWAYS WRONG!

I discuss Quantum Mechanics or Spiritual Mechanisms in much greater detail in my book entitled, "Quantum Mechanics: From a Non-Physical Spiritual Perspective" by Mark My Words; so I won't do so here in this book.

Materialism Is Useless and False and Boring

Materialism IS Creation by Physical Matter or Creation by ROCKS. An important question to ask is, "Can these things design, manufacture, and create Genomes and Life Forms from scratch?" If they can't, then Materialism is FALSE.

Once you get rid of that Materialism garbage, then suddenly everything starts to make sense.

In fact, during my pursuit of The Truth, I developed a foolproof test that seems to work every time. If the Materialists, Darwinists, and Atheists believe in it and are promoting it, then it's most certainly FALSE. If the Materialists, Darwinists, and Atheists are collectively ridiculing it, mocking it, and rejecting it, then that greatly enhances the likelihood that it is Scientifically Accurate, REAL, and TRUE.

The Materialists are extremely shallow, illogical, and ALWAYS have the worst possible explanation for the Scientific Evidence; but, the Materialists seemed to have perfected the ART of picking the Falsehood or the Lie and then choosing to believe in it and choosing to believe that it is true.

In a book about Creation, Materialism cannot be avoided, because the Materialists assure us that their Model of Creation or Origins is the correct model and the most scientifically accurate model and that all of the rest of us are ignoramuses and wrong. The Materialists label themselves as the Real Scientists and call all of the rest of us frauds, hacks, and fakes.

They are right in that my Model for Origins stands in direct opposition to their Materialistic Model for Origins. I lost any respect that I had for Materialism during the past couple of years; and, it ain't coming back!

Materialism is extremely boring! I mean, once Mr. Scientist has declared that there is no such thing as Spirit, Consciousness, Nonlocality, Perception, Intelligence, or Personality, then what more is there to say or do? These people just stop looking and stop doing Science completely satisfied that there is no evidence to be found when it comes to the Spiritual or the Non-Physical or the Non-Local. The Materialists stop searching for evidence, stop looking, and stop doing Science.

Whenever the Materialists are presented with evidence that supports the existence of the Non-Local or the Spiritual or the existence of God, they say, "I can't look at that, because it's not science," meaning of course that they can't look at it because it isn't Materialism and doesn't fit with their personally chosen Materialistic Worldview or Religion. Materialism is a complete and total rejection of the Spiritual Sciences, along with any type of evidence that supports the Spiritual Sciences. Materialism is a denial of REALITY.

SCIENCE is supposed to be the search for The Truth, All Truth; but, the Materialists stop looking for The Truth about Spirit, Consciousness, Nonlocality, Perception, Intelligence, and Personality. These people stop looking and instead waste all of their time and waste all of our time by producing documentaries extolling the virtues and miraculous wonders of Creation by Evolution, Creation by Chance, and

Creation by Physical Processes — none of which can actually design and create. Materialism ends up being THE DECEPTION and THE LIE that they are trying to ram down our throats and make us believe in. Materialism IS NOT SCIENCE. Materialism is Dogma and Religion, and it is a very boring Religion at that. I have learned to despise Materialism.

This extended essay and collection of essays about "Materialism" in part three of this book IS the CORE FOUNDATIONAL MATERIAL in all of my books about Creation and Evolution and THE SCIENTIFIC METHOD. I had to figure out for myself how to handle Materialism scientifically and logically BEFORE I could pursue further light and knowledge about Spirituality, the Spiritual Sciences, the Metaphysical, the Quantum, Nonlocality, Consciousness, Creation, Evolution, Origins, and God.

In my Pursuit of the True Reality of All Things, I had to be willing to abandon any residual belief in Materialism, Darwinism, Naturalism, Behaviorism, and Atheism, BEFORE I was able to find THE TRUTHS that I sought.

Belief in a LIE cannot help us to find THE TRUTH. That's just the True Reality of the situation.

There was a time in my life when I was a Materialist and an Atheist. I realized one day that I didn't like where that road was taking me, so I decided to turn around and go the other way. My life has been getting better and better ever since. My life has gotten infinitely more interesting as a result.

The purpose of THE SCIENTIFIC METHOD is to help us to find THE TRUTH, through a preponderance of the evidence.

THE SCIENTIFIC METHOD has no value to us if we use it to convince ourselves that a LIE is TRUE, as the Materialists and Darwinists always seem to do.

I had to get rid of My Materialism, My Atheism, Darwinism, and The Theory of Evolution BEFORE I could successfully start to study the Spiritual Sciences and achieve noticeable results.

That's what I discovered during my Pursuit of the True Reality of All Things, and during my usage of THE SCIENTIFIC METHOD.

I, personally, had to get rid of My Atheism and My Materialism and any residual belief in Darwinism and Naturalism BEFORE I was finally FREE to pursue the Spiritual Sciences or something like Spirituality, Creationism, Consciousness, Spiritualism, Nonlocality, or God.

That's why I put these essays in this book, because they were essential for getting me past my unbelief and to the point where I was actually willing and able to study a concept like Creationism or Creation.

Materialism IS Creation by Rocks. Materialism doesn't WORK! When was the last time you saw rocks or Physical Matter designing and creating and manufacturing things from scratch? The Materialists have way too much Blind Faith in the creative powers of Physical Matter. I'm not that kind of creationist!

The Theory of Evolution is Creation by Evolution and Creation by Chance. The Darwinists and Materialists have way too much Blind Faith in the creative powers of evolution, random mutations, and natural selection. I'm not that kind of creationist!

Materialism is Creation by Rocks and Creation by Chance. The Materialists have way too much Blind Faith in the creative powers of Chance and Rocks. I'm not that kind of creationist!

Atheism is Creation by NOTHING. The Atheists have way too much Blind Faith in the creative powers of NOTHING. I'm not that kind of creationist!

If you are going to have faith in something, then choose to believe in something that actually WORKS!

That's what I try to do.

Materialism Is a Denial of Reality

Materialists have a hard time seeing the Light.

You see, Materialistic Atheists in principle have conceptual problems understanding and accepting the reality of the Quantum Sea of Light or the Zero-Point Field of Light, because such a thing is sub-Quantum, Immaterial, Non-Physical, holographic, conscious, and Spiritual. Such a concept is hypothetical and experimental and often mathematical; and, the Materialists deliberately cripple themselves a priori by excluding anything Non-Physical or Spiritual from their worldview, their personal religion, and their SCIENTIFIC EXPLORATION.

Instead, the Materialists will be found online mocking the Quantum Sea of Light or the Light of Christ, because they don't understand it and can't understand it. It's beyond their current level of comprehension and acceptance.

If you don't ever look for it, then you will never find it. That's the reality of Science! That's the reality of Life! The Materialists never look for anything sub-Quantum or Spiritual, so they never find anything Spiritual or Immaterial because they prevent themselves from doing so.

Ironically, ALL light is Immaterial, Non-Physical, and Spiritual. Light of any kind has NO mass and NO matter. Light is Immaterial, Non-Physical, and Spiritual. That's why some scientists and theoreticians have chosen to call the "Zero-Point Field" the "Quantum Sea of Light", because they truly believe that the Zero-Point Field IS some type of light. In fact, some of the scientists have taken to calling it the "Zero-Point Field of Light". Spirit is light. Light is spirit. Light is Immaterial. The existence of Light PROVES that Materialism is FALSE. The existence of Time PROVES that Materialism is FALSE, but that's another story.

Obviously, such concepts make Materialists extremely nervous, because the existence of the Zero-Point Field and the existence of Light proves empirically that Materialism is wrong. Light is Spiritual, and spirit is light. And, the existence of Light proves that the Materialists are wrong at a basic fundamental level. The Materialists literally refuse to see the light.

Can you see how Materialists and Atheists completely cripple themselves and their chances for discovering something new simply by the philosophical point of view that they have chosen to believe in? The Materialists literally make the "existence of Light" completely out of the question, when it comes to their personal worldview!! The Materialists DEFINE light or the Spiritual completely out of existence, hobbling themselves in the process. Very unscientific!

It has been hypothesized by some scientists and mathematicians that this Zero-point Field of Light exists throughout ALL of that "empty space" between the nucleus of an atom and that atom's electrons. In fact, many scientists believe that it is this Quantum Sea of Light which actually holds the electrons in their orbits and prevents the electrons from crashing into the nucleus of their atom. The Materialists say that 99.999% of an atom is empty space; and, others say that that space really isn't empty because it is filled with a Quantum Sea of Light or a Zero-Point Field

232

(Forces and Fields of Spirit, Consciousness, and Light) which actually gives the atom structure and substance.

So, who is right — the Materialists or the Spiritualists?

My bet is now on the Spiritualists, because I have slowly discovered that there are millions of Non-Physical or Immaterial things that still have substance or influence — things like forces, fields, waves, strings, thoughts, dreams, dark energy, and LIGHT. The existence of radio waves and television waves PROVES Materialism FALSE!

Searching for and finding evidence for God's existence is the most interesting thing that we can study and learn about. It can be fascinating to identify all of the different kinds of "breadcrumb trails" that God deliberately left behind for us to find and follow, which will lead us directly to Him who is waiting for us at the end of each trail.

It reminds me of the board game "Clue". At the beginning of your search, it's not the least bit clear "Who Dunnit". But, through a process of elimination and through a process of asking questions and getting answers, you finally settle upon convincing proof of "Who Dunnit". At the end of your search, you just KNOW "Who Dunnit"; and, you can even identify who was lying to you along the way if someone did in fact lie to you.

At the end of my Search for Reality, I did indeed discover that it was the Materialists and the Darwinists who have been lying to me all my life. Materialism is demonstrably FALSE, which means that it is a LIE.

—

SPIRIT IS LIGHT, AND LIGHT IS SPIRIT

The Materialists DENY the existence of Light. That's how "in the dark" the Materialists really are!

Don't let them fool you into believing that they make a special exception for Spirit or Light, because I have caught the Materialists online mocking and ridiculing the Quantum Sea of Light, the Zero-Point Field of Light, and the Light of Christ. The Materialists also mock and deny the existence of the Living Lights, or what we typically call Spirits.

Materialism IS by definition the philosophical belief or religion, which teaches and believes that Matter (and the associated energy which matter can be converted into) is the ONLY thing that exists in the universe.

The ONLY light that the Materialists might be willing to make an exception for is the visible light that they can see with their eyes, possibly trying to associate it with energy; but they can't have it BOTH WAYS! That's cheating! The Materialists can't deny the existence of Spirit yet at the same time accept the existence of Light, because Spirit is Light! So technically, a hard-core Materialist will DENY the existence of all light if he is true to his creed, because the very existence of light PROVES that Materialism is FALSE.

Are you starting to see how STUPID and ILLOGICAL Materialism really is?

Materialism only makes sense to an Atheist, because he needs Materialism so that he can deny the existence of Spirit, the Spiritual, the Living Lights, the Holy Spirit, and God. Often, you will actually catch an Atheist looking at the Light and literally denying that it exists. The Materialists are online right now mocking and ridiculing and laughing at the different types of light that they can't see with their physical eyes.

Spirit IS Light. Light IS Spirit. It's just that — what we typically classify as Spirits — exist or live at a frequency of light which isn't visible to the naked human eye. Typically, the entities that we call Spirits are in fact Living Lights. The Living God has a Spirit and a Spirit Body. It's the Spirit or the Light that gives us Life. That's why the Biblical God is often called The Living God. He IS Life, and He gives Life! He is the Father of Lights.

God the Father and Jesus Christ have gained the types of Spirits and Spiritual Power that has the ability to levitate and manipulate Physical Matter at the atomic level. Their thoughts penetrate into all of that empty space between quarks and between the nucleus of an atom and its electrons. Their thoughts reside in the Quantum Sea of Light or the Zero-Point Field of Light, what they call the Light of Christ.

There are different types of "spirit" or light.

There seems to be some type of Spirit Matter, which is not alive but is simply acted upon. The Spirits that are alive are Living Lights. We humans are the spirit children of God the Father and Heavenly Mother. They sired and birthed our Spirit or Light. At our very core, we are Living Lights.

—

It has also been hypothesized by some Scientists that the Zero-Point Field or the Quantum Sea of Light exists in all that "empty space" between us and the Andromeda Galaxy, and between us and the end of this Universe. Those who are of this opinion truly believe that the Zero-Point Field has become something like a Cosmic Internet or a Cosmic Consciousness which pervades the whole of this Universe and ALL of this Physical Reality.

In the Doctrine and Covenants, the Biblical God Jesus Christ calls it the "Light of Christ" that fills all things and fills the immensity of space, confirming that this Zero-Point Field or Quantum Sea of Light or Dark Energy does indeed exist universally within every particle of matter and within all of that "empty space" in this universe. Christians as a whole tend to call this Quantum Sea of Light or the Zero-Point Field the Holy Ghost or the Holy Spirit. Why? It's because light is Immaterial, Non-Physical, and Spiritual; and, even we humans have found different ways to communicate through light. The Holy Ghost is a communicator and a revelator and a recorder, among other things.

The very existence of light of any kind proves Materialism wrong. Spirit is Light; and, Light is Spirit.

Some scientists have hypothesized that this Quantum Sea of Light or Zero-Point Field is in fact what makes up gravity, because as we all know or should know, each atom within our physical body is connected gravitationally with every other atom in this universe, including the atoms on the opposite side of this universe. It's all interconnected gravitationally, instantaneously, and simultaneously from one end of the universe to the other through the Zero-Point Field or Quantum Sea of Light or Gravity.

To the religious believers, particularly the Christians, this Quantum Sea of Light is the Spirit Realm — the Realm of the Living Lights. Out-of-Body Travelers KNOW from direct personal experience that when they are IN SPIRIT, they can travel to anywhere on this earth or to anywhere in this universe instantaneously at the speed of thought. They think of a place, and their spirit is just there instantaneously. There is no speed limit when it comes to Pure Consciousness, Pure Spirit, or Pure Light.

David Bohm has called matter "frozen light" or "condensed light".

If you take Pure Light or Pure Consciousness or Pure Thought and slow it down a tiny fractional bit, then it becomes Spirit Matter — something a bit different than Spiritual Consciousness or Pure Thought which has no speed limit. The associated or linked Spirit Matter lags just a tiny little bit behind the Thought. It's barely perceptible. If you slow light down a lot more, then it becomes packets of photons and can only travel at the "speed of light". There are visible photons and invisible photons, and God has slowed them down to what we call the "speed of light". If you slow light down to zero speed, then it becomes frozen light or Physical Matter. Everything is made up of light.

Are you starting to see the light?

This is COOL STUFF to study and learn about! The Theory of Evolution pales in comparison! PURE LIGHT or TRUE LIGHT is the stuff that God and Spirits are made of!

For me, this is the most interesting chapter in this whole book, because here we are talking about things that truly matter — the things of Eternal importance!

—

Physical Matter Is Enlivened By Light

A huge portion of the "Pistis Sophia" is all about Light. It can be found for free online by looking for it!

Physical Matter is enlivened or animated by Spirits or Living Lights.

In his book, "Temple and Cosmos" page 152, Hugh Nibley emphasizes this REALITY.

> "Matter without light is inert and helpless," says the Pistis Sophia.
>
> The rays from the worlds of light stream down to the earthly world, for awakening mortals.
>
> Sometimes the column of light joins heaven to earth, as in our Facsimile No. 2 (a very important principle), even as the divine plan is communicated to distant worlds by a spark. According to Carl Schmidt, it is the dynamics of light from one world that animates another.
>
> The spark is also called "the drop"; the Egyptians call it the prt ("drop"). It is the divine drop of light that man brought forth with him from above, the spark that reactivates bodies that have become inert by the loss of former light. It's like a tiny bit of God himself.
>
> Christ calls upon the Father to send light to the apostles.

End of quotes from "Temple and Cosmos".

The Living Light or Living Spirit is also at times called The Spark.

The Transformer movies talk about the AllSpark. It's this AllSpark which gives the Transformers a Spirit, or Life, or Living Light.

From the Wikipedia, "The AllSpark is an ancient artifact or object capable of creating new Transformer life by bestowing machinery with sparks."

God the Father (as well as the Biblical God Jesus Christ) is often called THE LIVING GOD, because He and His Spirit is the AllSpark from which all of our Spirits or Living Lights derive.

These are the things which the Materialists actively ridicule, mock, and laugh at online whenever they start trying to preach their religion and ram it down our throats. Yet, in comparison to this, Materialism is extremely boring!

—

Dark Energy

Not all of what makes up this universe is matter!

Astrophysicists and Astronomers have taken to calling the Quantum Sea of Light or the Zero-Point Field by the name "Dark Energy", and they have demonstrated empirically through observational measurements that this Dark Energy comprises 72.1 percent of this universe! This Dark Energy or Spirit Realm or Spiritual Construct comprises 72.1 percent of our Universe's Cosmic Density! All of that empty space isn't empty after all! This Dark Energy or Quantum Sea of Light is stretched throughout the whole of this universe, or stretching and expanding the whole of it!

On page 36 of "Why the Universe Is the Way It Is" by Hugh Ross, he put up a table entitled, "Inventory of All the Stuff that Makes Up the Universe". That table tells us what God put into this universe, giving us the chance to speculate why it is there and what it all means. It's a listing of what comprises this Universe's Cosmic Density.

First of all, we have Dark Energy (the self-stretching property of the cosmic space surface), which many believe to be the Zero-Point Field or the Quantum Sea of Light that stretches an atom thus keeping an atom from imploding. This Dark Energy makes up 72.1 percent of his universe's "cosmic density"; and, Dark Energy completely fills the immensity of space. Dark Energy or PURE LIGHT is the Universal Construct. This is the Light of Christ or the Spirit Realm or the Realm of the Living Lights, where the Living Lights (your spirit and mine) live or reside naturally. Dark Energy or the Zero-Point Field or the Quantum Sea of Light IS the Primal Construct of this physical universe. Dark Energy is the forces and fields that keep an atom and a proton from imploding into a singularity and taking all the rest of us with it.

There is no such thing as empty space, because the whole of space is based-upon and upheld or sustained by Dark Energy and is completely filled with Dark Energy or this Zero-Point Field or Quantum Sea of Light. This Dark Energy or Light of Christ is the Universal Construct that prevents this universe from imploding and keeps this universe stretching and expanding. That's why this Dark Energy comprises 72.1 percent of this universe's cosmic density, because it is literally stretched throughout the whole of this universe's "empty space" keeping this universe expanded like a balloon. If God were to pull out the Dark Energy or Light of Christ from this universe, then this universe would deflate or implode back into the singularity from whence it came.

Imagine what you could do if you were to set up a Network or an Internet onto this thing, as God has already done. God is in the Light, particularly the Quantum Sea of Light. In the Doctrine and Covenants Sections 88 and 93, the Biblical God calls this thing the Light of Christ, and God says that it fills the immensity of space. This Light of Christ or Quantum Sea of Light is what God uses to communicate with us, and this Dark Energy is how God hears all of our prayers and thoughts instantly and simultaneously. The dude has some serious bandwidth!

This Dark Energy or Zero-Point Field is indeed the Cosmic Consciousness or the Cosmic Internet. God can literally beam thoughts into our minds, into all of that "empty space" between the nucleus of an atom and its orbiting electrons. Cool, huh? Of course, that might also explain why God is bothered by and complains about all of our unholy and impure thoughts. He wants us to learn how to control ourselves so that He doesn't have to listen to all of our rot whenever we get our thoughts into a rut.

This Dark Energy goes by many different names! It MUST BE extremely important if it comprises 72.1 percent of this universe; and, it is important and essential because Dark Energy or PURE LIGHT is the thing that keeps this universe expanding and expanded. If you were to pull out this PURE LIGHT, or Zero-Point

Field of Light, or Quantum Sea of Light, then this whole universe would implode into a singularity.

This PURE LIGHT or Dark Energy doesn't seem to have any speed limit, being simultaneously everywhere at all times — in all things, through all things, and filling the immensity of space. This PURE LIGHT or PURE THOUGHT or PURE CONSCIOUSNESS is the Primal Construct. This whole universe is ONE GREAT THOUGHT.

—

Laughing at What They Know Not

In their complete and total ignorance, the Materialists and Atheists mock and laugh at the Quantum Sea of Light or Light of Christ, not knowing what it is or what it does. They mock it and ridicule this Dark Energy or Light of Christ because it is Immaterial or Spiritual, and they can't get their hands on it. Yet, these Materialists and Atheists are mocking and laughing at something that comprises 72.1 percent of this universe's cosmic density, because their deliberate ignorance makes them dense enough to do so.

Ironically, these very same Materialists and Atheists are NEVER caught laughing at and mocking Evolution or Chance, even though Evolution and Chance are infinitely more insubstantial, Immaterial, non-existent, and laughable than Dark Energy or the Zero-Point Field can even begin to be!

Why do they do it? Why mock something like the Zero-Point Field or the Light of Christ for which there is abundant necessity and evidence, yet admire and worship something like Creation by Evolution or Creation by Chance for which there is absolutely NO evidence whatsoever? Why the selective amnesia or hypocrisy?

These Materialists, Atheists, and Darwinists admire and worship Creation by Chance or Creation by Evolution, because these non-existent things are telling the Materialists exactly what they want to hear, that God does not exist. That's what this whole controversy is all about. The Materialists are more than willing to believe in Immaterial and non-existent things like Creation by Evolution or Creation by Chance, so long as those non-existent immaterial things are telling them that God does not exist and that they don't have to repent of their sins.

Interesting is it not, how human psychology works? We are in denial about the things that we don't want to know about or hear about. The Materialists are in denial about 95% of this universe's cosmic density because it is Spiritual, Non-Physical, and Immaterial! How dense can they get? Well, they can get pretty dense, because they want to be that way and they want it that way.

They don't want to know about any of this because it suggests that God might exist. It suggests that they might be facing some kind of Final Judgment someday and have to answer to God for all their sins and misdeeds. Materialism is a denial of Reality, because the Materialists don't want to have to face this Reality.

This Light of Christ or Dark Energy IS what keeps this universe stretching and expanding! If God were to pull this Dark Energy or Light of Christ out of this universe, then this universe would implode into the singularity from whence it came.

The astrophysicists call it Dark Energy because it can't be seen with the naked eye; but, it really is comprised of invisible Light of some kind. I find myself wanting to call this stuff PURE LIGHT, the kind of light that isn't weighed down or slowed down by extra baggage. This PURE LIGHT might come in two flavors or types — a sentient living active type and a non-sentient inert reactive type. I'll mention more of this possibility in a later part of this essay.

This Light of Christ or Dark Energy IS also the thing that keeps protons, neutrons, and electrons from decaying and imploding into singularities. Something is preventing those protons from decaying! This Light of Christ or Dark Energy IS the thing that provides forces and fields to keep electrons from merging with the nucleus of their atom. This Light of Christ or Dark Energy must of necessity exist, or we would not exist as physical mortal beings, because it is this Dark Energy or Light of Christ that prevents EVERYTHING from imploding into a singularity. Something or Someone HAS TO BE doing that, otherwise we all would still be in that singularity from whence we came!

If we were to take a Materialist or Atheist and pull ALL of this Light of Christ out of his physical body, his physical body would implode into a singularity and he would cease to exist — only his spirit or Consciousness would remain.

Yet, day-in and day-out online, you will see these very same Materialists and Atheists laughing at and mocking the Quantum Sea of Light, the Zero-Point Field, Dark Energy, or the Light of Christ because in their deliberate ignorance the Materialists do not know what it is, or what it does, or why it is necessary for it to exist. They laugh at what they know not! But, in comparison to all of this information about Light, the Theory of Evolution is extremely boring!

Materialism is a very shallow, limited, and even laughable worldview or religion, because it is so Immaterial and unsubstantial.

—

Exotic Dark Matter

Next on the list of items that make up our Universe's Cosmic Density, we have what they call Exotic Dark Matter (particles and/or energy that weakly interacts with ordinary particles and light).

For anyone who has studied physics, metaphysics, and the nature of a spirit body, they quickly realize that "spirit" comes in at least two types, Spirit Matter and spirit energy, because physicists have demonstrated that matter of any kind can be converted into energy, and energy can be converted back into matter.

Many of us believe this Exotic Dark Matter to be Spirit Matter — the material and energy that makes up our spirit bodies. Exotic Dark Matter or Spirit Matter comprises 23.3 percent of this universe's cosmic density. Matter of any kind is the kind of stuff that God designed to be acted upon. Matter simply reacts to whatever a living spirit or a living light tells it to do. Matter is reactionary. Thought or Consciousness or Intelligence or Life is what makes that matter react. Consciousness or Life is what does the choosing, desiring, and willing. Matter of any kind was designed to obey the commands of Living Spirit or Living Light. That's why your physical body does what your spirit tells it to do, as long as your physical body is functioning properly.

There are different types of Spirit or Light, many different types.

Apparently, this Spirit Matter is NOT alive, because it is a more refined type of matter, and matter was designed by God to be acted upon, not to act. It's the LIGHT that is active, alive, sentient, aware, and conscious; and, not the matter.

This Exotic Dark Matter or Spirit Matter is said to be concentrated into halos around galaxies preventing the Ordinary Matter in galaxies from flying apart and dispersing evenly or homogenously throughout the whole universe.

After the Big Bang 13.8 billion years ago, all Physical Matter should have distributed itself evenly throughout the universe because every particle was moving away from every other particle; and, this universe should be nothing more than one big homogeneous ball of gas as a result. But, someone forced huge chunks of that Ordinary Matter to gather into galaxies and then later to gather into stars and planets. Someone was driving the bus. The Biblical God says that He is the one who did this for us. Who else was living back then who could have done the job?

It's the SPIRIT or the LIGHT which controls and manipulates matter — both Physical Matter and Spirit Matter. SPIRIT or LIGHT moves this matter and puts this matter where it wants this matter to be. Your spirit or mind might have had a hand in gathering ordinary Physical Matter into galaxies, stars, and planets; and, your spirit or light or Consciousness might have had a hand in organizing and moving all of that Exotic Dark Matter or Spirit Matter into halos around the different galaxies to keep the galaxies from flying apart.

Exotic Dark Matter or Spirit Matter or some kind of "Immaterial Matter" comprises 23.3 percent of this universe's cosmic density, so it must have some purpose, or some use, or some reason for being. There are probably other uses for it that we haven't even thought of yet. But for me, the first thing that comes to mind is that it's massive enough and influential enough to keep Ordinary Matter in check or in line. Something or Someone has to be keeping galaxies from flying apart and becoming homogenous.

When it comes to speed of travel, this Exotic Dark Matter or Spirit Matter seems to lag ever so slightly behind the SPEED OF THOUGHT.

Of course, there is another possibility to consider. I like to think that there are always possibilities. It's possible that this Spirit Matter actually doesn't travel at all. In other words, it's possible that it is PURE THOUGHT or PURE LIGHT or PURE

CONSCIOUSNESS that does the traveling, and then that PURE THOUGHT builds its reality out of the Spirit Matter that exists at its chosen destination.

Out-of-Body Travelers have stated that there are consensus realities that are designed, built, and maintained by spirit beings through a collective Consciousness or a common consent; and then, there are non-consensus realities in the Spirit Realms where the reality actually molds itself to your thoughts and your desires.

So, is it the Exotic Matter or Spirit Matter that travels thus creating that slight lag, or is it the PURE THOUGHT or PURE LIGHT that does the actual instantaneous traveling with the lag being caused by the fact that a SPIRIT or PURE LIGHT has to build or construct its reality out of Spirit Matter when that SPIRIT reaches its chosen destination? Obviously, questions still remain; and, only God knows the answers for sure.

Exotic matter, dark matter, Spirit Matter, consensus matter, and non-consensus matter — that's way too much matter for the Materialists! Imagine it! Matter that the Materialists cannot get their hands on; therefore, it must not exist. Yet, it shows up in the Cosmic Density of this Universe. So, something's the matter with Materialism.

Any way you choose to look at it, this is RADICALLY ADVANCED SCIENCE! Materialism pales in comparison! Materialism is a DENIAL OF REALITY.

—

Ordinary Matter

Next on the chart in Hugh Ross' book is Ordinary Dark Matter (Quantum Objects that interact strongly with light). These little buggers seem to be controlled by sentient living light! It's the Sentient Living Light that's driving the bus; and of course, you make busses out of Ordinary Matter.

Ordinary Matter is the stuff that the Materialists and Atheists worship and admire, because they can get their hands on it. They make idols out of it!

This Ordinary Dark Matter is what our Consciousness and our spirits act upon and control here in the Physical Realm. Ordinary Dark Matter comprises 4.35 percent of our universe's cosmic density. This Ordinary Dark Matter is what our physical bodies are made of. However, the astrophysicists like to break this Ordinary Dark Matter into three types. The type here called Ordinary Dark Matter, which seems to be distributed unorganized throughout space, the type that becomes Ordinary Bright Matter (stars), and the type that goes into making earths and physical bodies like ours. I just call the whole thing "Ordinary Matter".

Then on the chart in Hugh Ross's book comes the Ordinary Bright Matter (stars and star remnants). These make up 0.27 percent of the universe's cosmic density. This "star stuff" that Carl Sagan worshipped only makes up 0.27 percent of this universe. Carl Sagan said that this Ordinary Matter is all that there ever was and all that there will ever be. Impressive ignorance was it not?

And finally, on Hugh Ross's chart, we have planets, humans, and other living organisms — the part that the Materialists really choose to believe in, which is a subset of Ordinary Dark Matter; and, we make up only 0.0001 percent of this universe's cosmic density. In a sense you could clearly say that the Materialists only believe in 0.0001 percent of this universe — the stuff that they can really get their hands on. That's how limited the Materialistic and Atheistic worldview can become! Even if you want to be generous to them, the Materialists still only believe in 4.62 percent of this universe — the Ordinary Matter. The Materialists DENY the existence of over 95 percent of this Universe and its Cosmic Density. The Materialists are so dense, that they deliberately eliminate 95 percent of this universe from their worldview or religion. Impressive ignorance, is it not? That's a really HUGE denial of Reality! No wonders the Atheists and Materialists are always in a State of Denial.

The Materialists deny the reality of all the rest of this universe's cosmic density, because they can't see it with their physical eyes nor get their hands on it with their physical instruments. When it comes to cosmic density, the Materialists are out of their depth and really dense! Materialism is a very shallow and BORING philosophy of life or world religion.

—

A Multitude of Different Kinds of Light or Luminosity

There is now another possibility to consider. I like to think that there are always possibilities. There is the possibility that there might in fact be something in existence that DOES NOT register at all in this Universe's Cosmic Density. Think about it! This would be the kind of stuff that dreams are made of.

If Dark Energy, comprising 72.1 percent of this universe, is made up of PURE LIGHT or PURE SPIRIT, then anything that exists in this universe that doesn't register at all in this Universe's Cosmic Density would be the PUREST OF LIGHT or the PUREST OF CONSCIOUSNESS AND THOUGHT. Think about it! PUREST THOUGHT, or PUREST LIGHT, or PUREST CONSCIOUSNESS, or PUREST INTELLIGENCE — the hypothetical stuff that exists but doesn't weigh in at all in our Universe's Cosmic Density.

So now we come to the stuff that is truly Immaterial, Non-Physical, and completely Spiritual or sub-spirit-matter because it doesn't figure at all into the universe's cosmic density! This stuff is truly the stuff that dreams are made of.

This would be the PUREST and most refined of all existence. This stuff IS thought, Consciousness, the construct of light, awareness, or Intelligence. This stuff IS TRUTH, what we truly are at our very core. Obviously, this PUREST STUFF can travel faster than the "speed of light". In fact, many people have experienced the out-of-body situation, where their spirit or Consciousness can travel instantly to any place in the universe or anywhere on this world simply by thinking about it or willing it to be so. Others while IN SPIRIT have actually experienced being two or more places at once, simultaneously. Think about it! Fascinating, is it not?

242

Consciousness or Thought is like Gravity and exists simultaneously everywhere in this universe — all interconnected in some way and instantly accessible. Some of the scientists call this Cosmic Consciousness or Universal Consciousness. The Doctrine and Covenants and Book of Abraham call this PUREST of stuff "INTELLIGENCE" or "LIGHT AND TRUTH". Others call it The Mind of God.

Light itself is Immaterial and has no mass. Light PROVES that Materialism is false. Thanks to light, all you have to do is open your eyes and see that Materialism is wrong. But, what is light made of? The Doctrine and Covenants and Book of Abraham at times seem to indicate that LIGHT or SPIRIT is made-up-of and/or occupied-by Consciousness, or Intelligence, or Thought.

LIGHT or SPIRIT and SPIRIT MATTER seem to be made-up-of and/or driven by Intelligence, Consciousness, or Thought. Each photon of light is conscious, and aware of its environment, and knows its destination before it launches into space.

Which came first, Intelligence or Light, Consciousness or Spirit? We don't know. They are probably co-eternal and intricately bound to each other. Is there really any difference between them? We don't know, but there might be, especially if one registers in our Universe's Cosmic Density and the other one does not. If Thought, or Consciousness, or INTELLIGENCE does not weigh into our Cosmic Density, then how would we ever know it to be so unless God were to actually reveal that knowledge to us? God hasn't told me anything about it. How about you?

Both Spirit Matter and Consciousness seem to be made up of light — possibly different types of light. Thoughts or Intelligence or Awareness seems to be made up of some kind of Light also. Forces, and fields, and gravity are theoretically made up of different types of light; and, most light is not visible to the naked eye. Spirit Matter and spirit bodies are made up of some type of light. The vibrating strings in String Theory are also made up of light. There are different types of light and different frequencies of light.

David Bohm said that matter is condensed or frozen light. Photons are concentrated light, and during the concentration process that particular type of light slows down to what we call the "speed of light". Light slows down the more "material" or "Materialistic" that it becomes — the closer it comes to manifesting in our Physical Realm. Slow light down to ZERO SPEED, and it becomes Physical Matter. But, there are other types of light which have no speed limit — PURE LIGHT or PURE THOUGHT, the kind of light that makes up Dark Energy and 72.1 percent of this Universe's Cosmic Density, and possibly the kind of light that doesn't weigh-into our Universe's Cosmic Density at all.

Spiritualism is the belief that LIGHT or SPIRIT exists distinct from matter and that SPIRIT or LIGHT is the fundamental component or primal construct of reality.

Which types of LIGHT show-up within our universe's cosmic density, and which types do not? We don't know. The best we can do is to speculate that Thought, Consciousness, Sentience, Self-awareness, or INTELLIGENCE might be the type of light that does NOT weigh into our universe's Cosmic Density. When we climb down that rabbit hole, we speculate and assume that we will find THOUGHT or

INTELLGENCE or THE PUREST OF LIGHT at the very bottom of it, but it too will indeed be some type of LIGHT — SENTIENT LIVING LIGHT.

—

Speculating about THE PUREST LIGHT

I'm going to speculate here that there IS something even more refined than Dark Energy or PURE SPIRIT or PURE LIGHT. This PUREST LIGHT would be what Dark Energy, or the Zero-Point Field is made-of or made-by. This PUREST LIGHT would in fact be PURE SENTIENCE or PURE THOUGHT or PURE INTELLIGENCE or PURE LIFE.

I went down the rabbit hole, and now I'm trying to figure out how far down it goes!

There appears to be a type of light or Consciousness or thought that is even more primal and elemental than Dark Energy or the Quantum Sea of Light — a type of light or intelligence that doesn't even show up in our Universe's Cosmic Density. It's possible that there is NO sentience, or awareness, or life anywhere in our Universe's Cosmic Density. It's possible that PURE LIFE or the PUREST OF LIGHT might not register at all or weigh-in at all in our Universe's Cosmic Density. I like to consider this possibility, because it has certain advantages.

You see, the Biblical God has revealed to us in the Doctrine and Covenants Section 93 that Intelligence or Consciousness is what He calls "LIGHT AND TRUTH". Your thoughts or Consciousness at a very fundamental level is nothing more than light and what you TRULY ARE; and as we have already established, light or Consciousness fills the immensity of space, and light is Immaterial and Non-Physical; and, this sentient or conscious or intelligent type of Light theoretically might not figure into the cosmic density of this universe at all. This THOUGHT LIGHT or INTELLIGENT LIGHT or PUREST LIGHT would be Trans-Dimensional or Extra-Dimensional, and not a part of this physical universe at all. There would be no way to detect it from within this universe — the best we could do is to observe its effects or its influence. It would truly be the Driver of the bus.

According to this theory, Dark Energy or the Zero-Point Field would not be sentient and self-aware. Instead, this Dark Energy or PURE LIGHT or Quantum Sea of Light would be part of the primal construct of this universe comprising 72.1 percent of this universe — kind of like a network or telephone exchange.

This Dark Energy or Light of Christ could in fact have been designed and created by THOUGHT LIGHT or INTELLIGENT LIGHT to be used both as a stabilizing or structural feature of Physical Matter and as a universal communication system. This Dark Energy or Light of Christ could in fact be the Utility System that was built into this universe — the wiring and plumbing so to speak.

The REAL SENTIENCE or the PUREST LIGHT, Consciousness or Intelligence, wouldn't even register at all in this universe's cosmic density but would in fact permeate or occupy the whole of this universe simultaneously. It would be Trans-

Dimensional or Extra-Dimensional. It would be THE LIGHT and THE LIFE of this universe.

Heady stuff, huh? It leaves Materialism behind in the dust! Doesn't it?

—

At your very essence, at your very eternal immaterial core, you ARE conscious sentient light — YOU ARE PURE LIFE or the PUREST OF LIGHT and INTELLIGENCE. You have always existed, and you will always exist. The Gods might be able to disassemble your spirit body, but they can't touch or destroy your inner Consciousness, or light, or intelligence. The Gods can't force your inner light to be good or to be evil, either. You choose that for yourself.

The Biblical God calls this stuff, this Innermost Light, "INTELLIGENCE" in His scriptures, particularly the Doctrine and Covenants and the Book of Abraham. God defines Intelligence as "Light and Truth". Your Immortal Immaterial Intelligence is what you TRULY are at your very core. Your Intelligence IS your light, and your life, and your truth.

As mentioned, Consciousness or Thought is made up of a special kind of light, and that kind of light has no speed limit. That's why God can hear your thoughts and prayers instantaneously no matter where He might happen to be in this universe at any given point in time. Let's see Materialism do that!

Thought or Consciousness is Trans-Dimensional or Extra-Dimensional and thus has NO physical limitations or Spiritual limitations either. Thought or Consciousness existed before the Spiritual, which existed before the physical.

1 John 1: 5: "God is light, and in Him is no darkness at all." If you notice carefully, God has the type of Spirit or Light that can actually levitate, move, and manipulate Physical Matter telekinetically.

God would be the GREATEST LIGHT, the MOST GLORIOUS LIGHT, or THE PUREST OF PURE LIGHTS. God would be THE LIGHT and THE LIFE of this Universe. This universe would be a GREAT THOUGHT in the Mind of God.

Have you heard of any of this before? It all fits together perfectly, doesn't it?

One of my gifts or talents is an ability to notice patterns or trends; and, I love a good mystery. What greater mystery is there than trying to figure out how the Non-Physical Realms or the Spirit Realms might be organized and setup by God? This is something that a Materialist or an Atheist won't touch with a ten-foot pole, which makes it even more appealing once you have finally seen the Light.

I know something they don't. I learned something that they refuse to look at and study. How cool is that? It makes a person feel good, especially after taking a ton of ridicule and abuse from the Atheists, Darwinists, and Materialists online. The Materialists are the "blind and the ignorant" calling all of the rest of us blinder and stupider and "ignoranter" than they are.

I will leave it up to you to decide for yourself which one of us has our head in the sand. But, I got tired of interacting with the Materialists, because they are so shallow and dense most of the time; and, their constant ridicule, name-calling, and contempt grates on you after a while. You find yourself wishing at times that you could pull the Quantum Sea of Light out of them just to see what happens to them. Would it still be considered murder if you were to kill them by removing something from them that they say does not exist?

When it comes to the PUREST OF PURE LIGHTS, I'm definitely not it. I still have a bit of the devil in me; and, that dude seems to be dark light or black light. Still, the boy in me (who liked to fry ants under a magnifying glass) would like to see what would happen to a couple of these Materialists if God were to pull the "unseen", the "untouchable", or the "Immaterial" out of them. It would be fun to watch them implode, wouldn't it? For them, the Immaterial really wouldn't exist anymore, now would it? And neither would their Physical Matter. Their Physical Matter would implode — no longer being sustained or upheld by spirit or light. It would be interesting to see what part (if anything) of them remained, if God were to pull the Spiritual or the light completely out of them.

Is there something more primal than SPIRIT or LIGHT that might in fact remain if the SPIRIT or the LIGHT were to be pulled out of them? I like to think that there are possibilities.

—

Materialism IS the Study of Reactions

Think about it logically and dispassionately.

Materialism of any kind is the Study of Reactions.

Whether we are talking about chemical reactions or reactions from natural selection and random mutations and other natural processes, we are NOT talking about design and creation!

Creation by Reactions IS scientifically and logically IMPOSSIBLE!

Reactions do NOT design and create. They instead REACT to designers, engineers, manufacturers, and creators.

Darwinism or the Theory of Evolution is the Study of Reactions. The Atheistic Darwinists try to remove the Designers and the Creators from the equation, and try to get us to focus exclusively on the REACTIONS, trying to trick us in the process into believing that Reactions can somehow magically design and create.

But the Darwinists and the Materialists are WRONG.

ONLY ACTORS — designers, creators, manipulators, manufacturers, builders, engineers, planners, programmers with hands and spirits and minds — can design and create. Manipulation and manufacture imply HANDS or some kind of Telekinesis. Designing, programming, planning, and organizing imply SPIRITS or

MINDS. It's Actors with Hands and Minds who do the actual designing and creating — NOT Materialistic reactions! Chemical reactions and evolutionary reactions CANNOT design and create anything at all because they are purely physical reactions, and there are NO hands or minds there. Reactions cannot design and create! And, the Materialists LIMIT themselves and try to limit us purely to PHYSICAL REACTIONS, telling us that nothing else exists.

Materialism of any kind is a very limiting and UNSCIENTIFIC philosophical worldview or Religion. The Materialists try to tell us that man-made idols and physical reactions can design and create. Materialism is just a modern-day form of idolatry — a worship of false gods. It's a fairy tale! Everyday online, we see Materialists, Darwinists, and Atheists worshipping chemical reactions, evolutionary reactions, and their own knee-jerk reactions.

God designed and organized two types of things in this Universe — the living things that were designed and organized and meant to ACT, and the inert physical or Materialistic processes and Physical Matter which was designed and created and meant to REACT. It is the living spirit that ACTS or CREATES; and, it is the Physical Matter and the physical processes that REACT or are ACTED UPON. Physical Reactions cannot design and create, although the Materialists and Darwinists will assure us that they do.

The Materialists and Darwinists limit themselves exclusively to the things that were designed and meant to REACT, while completely ignoring the spirit or life or living light which God organized and intended to do all of the ACTION or CREATION in this physical universe.

The Materialists are right about one thing. Physical Processes, Chemical Reactions, and all other Things That React CANNOT give revelations and visions, CANNOT function as Lawgivers, CANNOT function as Holy Spirits, CANNOT perform atonements or reconciliations with God or provide salvation for our immortal spirit, and CANNOT predict or prophecy of future events that they will have a hand in bringing about. But the Materialists don't seem to realize or understand that (for the very same reasons) Physical Reactions and Chemical Reactions CANNOT function in the role of Designers, Creators, Revelators, Spirits, Causal Agents, Planners, Actors, and GOD.

Materialism is a reaction, and thus doesn't require a lot of active thought, which is why God tells us that we must worship Him with all of our heart, might, mind, and strength. God calls us to action, because God organized us to be Actors and Creators!

Lehi was a prophet and a scientist, and the Biblical God Christ Jehovah revealed the following things to him 600 B.C. Lehi wrote these words:

2 Nephi 2: 13-16:

13 And if ye shall say there is no law, ye shall also say there is no sin. If ye shall say there is no sin, ye shall also say there is no righteousness. And if there be no righteousness there be no happiness. And if there be no righteousness nor happiness there be no punishment nor misery. And if these

247

things are not there is no God. And if there is no God we are not, neither the earth; for there could have been no creation of things, neither to act nor to be acted upon; wherefore, all things must have vanished away.

14 And now, my sons, I speak unto you these things for your profit and learning; for there is a God, and he hath created all things, both the heavens and the earth, and all things that in them are, both things to act and things to be acted upon.

15 And to bring about his eternal purposes in the end of man, after he had created our first parents, and the beasts of the field and the fowls of the air, and in fine, all things which are created, it must needs be that there was an opposition; even the forbidden fruit in opposition to the tree of life; the one being sweet and the other bitter.

16 Wherefore, the Lord God gave unto man that he should act for himself. Wherefore, man could not act for himself save it should be that he was enticed by the one or the other.

Source:

https://www.lds.org/scriptures/bofm/2-ne/2?lang=eng

The whole chapter is a fascinating read, especially if you are scientifically inclined and want to know how things really work.

There are two different types of THINGS in this universe, the People or Living Spirits or Living Lights who were organized and meant to be ACTORS and CREATORS, and then the physical things that were designed and created to be ACTED UPON or TO REACT. As their Creator and Organizer, the Biblical God knows about these THINGS, which is why He periodically revealed these things to His chosen prophets, turning them into scientists in the process.

ACTION and REACTION! ACTORS and THINGS TO BE ACTED UPON! That's what this physical universe is all about.

Materialism is the Study of Reactions — chemical reactions, physical reactions, and evolutionary reactions.

Theology is the Study of Designers, Actors, and Creators. Theology and/or Spiritualism is also the Study of Spirits, Living Lights, Angels, and Gods.

Design and Creation by Reactions IS scientifically and logically IMPOSSIBLE. Thus, Creation by Chance Reactions and Creation by Evolutionary Reactions is scientifically and logically IMPOSSIBLE. Chance and Evolution cannot design and create, because they are purely REACTIONS! There's NO Actor there!

How often have I said to you that when you have eliminated the impossible, whatever remains, *however improbable*, must be THE TRUTH? — Sherlock Holmes

My goal in this book is to eliminate the IMPOSSIBLE so that we are left staring at THE TRUTH. Reactions CANNOT design, create, and act; therefore, they should be

ELIMINATED from Science as possible Designers and Creators and Causal Agents. It's only logical. THIS IS SCIENCE! This is SCIENTIFICALLY ACCURATE because it matches with REALITY! Evolution and Chance have NEVER been caught in the act of designing and creating anything! Intelligent Beings have! Therefore, Creation by Chance and Creation by Evolution should be eliminated from Science as possible Designers and Creators, because Evolution and Chance CANNOT design and create anything at all. It's elementary my Dear Reader!

Once you have eliminated the IMPOSSIBLE, the Materialism, then it all points to God as the Designer and Creator of this universe and all the genomes and life in this universe, because God is a Living Designer, Actor, and Creator. This makes logical sense to me; whereas, Creation by Chance or Creation by Evolution or Creation by Reactions does not. God makes sense to me as the Origin and/or the Originator of all the life in this universe. Evolution does not. Materialism is FALSE; therefore, Materialism is UNSCIENTIFIC! Materialism is NOT SCIENCE. It is a philosophy or religion. The scientist in me wants to go with what's actually POSSIBLE! Materialism is NOT it!

You will have to decide for yourself if I met my Burden of Proof, or not. Obviously, my goal here is to go with the preponderance of the evidence. I employ a type of deductive reasoning in which I attempt to eliminate the IMPOSSIBLE PREMISES so that we are left staring at THE TRUTH or the CORRECT CONCLUSION.

My goal here was to put an end to My Materialism and My Atheism making it impossible for me to go back. I have accomplished my goal. I have seen too much, experienced too much, and learned too much to go back to Darwinism and My Atheism. I have decided which side of the fence I want to stand on and defend. I have made my choice; and now, I'm determined to make the best of it.

—

Providing a Capstone to Spiritualism and a Gravestone for Materialism

The Materialists, Darwinists, and Atheists can be extremely stupid, ignorant, uneducated, and dense when it comes to the parts of SCIENCE that they have chosen not to believe in. I know, because I used to be one of them.

If you don't ever go searching for the Spiritual, then you will never find it. The Spiritual hides behind a Materialist's unbelief.

Spiritualism is the view that Spirit or Living Light is the prime element of reality.

Materialism is a denial of that reality! The Materialists try to convince us that there is no such thing as Consciousness or living light. The Materialists deny Spirit, Light, and Life.

Materialism is a stupid and idiotic idea — the dumbest, the most limiting, and the shallowest idea that human beings have ever created out of thin air. There's no substance to it!

But, there is an important distinction that should be made here.

Just because Materialism is the dumbest idea that has ever been devised, it doesn't automatically follow that Materialists and Darwinists and Atheists are the most stupid people that you will ever encounter. You see, I used to be an Atheist and a Materialist; and, although I did indeed harbor some stupid ideas at the time, I'm not exactly stupid or dense as you might be able to sense. Materialists and Atheists can repent or change their minds, especially once they realize how stupid and limiting and boring Materialism and Atheism really are. Even Materialists and Darwinists can learn to be more well-rounded people if they truly want to do so. Once they get over their fear of God, their fear of Spirituality, their fear of Repentance, their fear of God's Commandments, their fear of God's Scriptures, and their fear of Religion, then anything is possible.

Atheism and Materialism are made up of a great deal of suppressed or subconscious fear and prejudice; and, sometimes it isn't all that well-hidden. I have watched the Atheists and Darwinists in action during various debates, and I can sense a great deal of fear coming off them or from them — it typically manifests itself as anger, hatred, vitriol, pride, ridicule, contempt, and rage. Richard Dawkins has often been described as "everyone's favorite madman". He can't seem to control himself at times. Sometimes he acts as if he is a trapped animal.

In contrast, LOVE or CHARITY casteth out all fear. These kinds of Christian or Charitable people are actually kind and fun to be around.

—

I sense that there is still a lot that we don't know about the unseen realms, but it's amazing what we can piece together from the Revelations of God — particularly those found in the Doctrine and Covenants and the Pearl of Great Price.

God has deliberately limited the scope and range of our Consciousness, spirit bodies, and physical bodies; but, He has placed no limits on Himself. As mortal beings, we are in a probationary state or a testing state of existence, to prove to ourselves who and what we really are at our innermost core. God will have a proven people before sharing His powers and abilities with them.

The book "The Holographic Universe" by Michael Talbot speculates a lot about how our Consciousness or Thoughts or Intelligence is holographic in nature, thus allowing our Consciousness or Intelligence to store an infinite amount of information in basically no space at all. Cool, huh?

—

Materialists can mock the Zero-Point Field or the Quantum Sea of Light all they want, but it only serves to demonstrate and reveal to all the rest of us the depth of their ignorance, which they have imposed upon themselves. Materialists become very shallow people indeed; and, they do it deliberately to themselves! There's no depth to Materialism, and Darwinism is purely a Materialistic philosophy.

The Materialists only permit themselves to think about half of science at most and only 5% of reality at most. Materialism is extremely shallow, boring, and limited. Materialism is like putting on blinders before going sight-seeing. The Materialists and Atheists do this to themselves, because they are trying to hide from God. Materialists are like the infant toddler who covers his eyes truly believing that if he can't see it, then the scary monster doesn't exist anymore.

I sometimes have a hard time believing that I used to be an Atheist and a Materialist. Materialism is UNSCIENTIFIC or ANTI-SCIENCE, irrational, illogical, unrealistic, stupid, limited, shallow, and boring; but, you only realize that after you have gotten rid of it.

—

Ironically, we Materialists and Atheists are more than willing to believe in things that are IMPOSSIBLE or things that DON'T EXIST, as long as those things are telling us that God does not exist and that we will never have to repent of our sins or suffer for our sins.

We Materialists and Atheists desperately want something to tell us that God does not exist and that we will never have to answer to God for our sins, vices, and misdeeds; so, we create out of thin air something that does not exist — something like Creation by Chance or Creation by Evolution or Creation by Reactions — to tell us that God does not exist, so that we don't have to think about God and our Final Judgment anymore. Materialism is a state of denial! Materialism is one of the ways that we tell ourselves that we can sin as much as we like, and we will never have to answer for it. KNOW THYSELF!

—

The things I mention in this chapter are really cool stuff — COOL SCIENCE — but you are not going to be able to see it or understand it or accept it until long after you have chosen to open your mind and your eyes enough to actually be able to see it. In my case, I had to get rid of My Materialism and My Atheism, before I was able to see and understand and accept any of these things. I'm a completely different person now than I was five years ago.

Here's one of my favorite quotes from Master Yoda:

> Master Yoda: "Size matters not. Look at me. Judge me by my size, do you? Hmm? Hmm? And well you should not. For my ally is the Force, and a powerful ally it is. Life creates it, makes it grow. Its energy surrounds us and binds us. Luminous beings are we, not this crude matter. You must feel the Force around you; here, between you, me, the tree, the rock, everywhere, yes. Even between the land and the ship."

Science Fiction isn't ALWAYS fiction! In fact, some of the coolest Science Fiction ends up becoming Science Fact. Life or Thought creates this FORCE or FIELD.

There's a force or a field within us connecting every particle in our physical bodies with every other particle in this universe. Some of the scientists call this the Zero-Point Field or the Quantum Sea of Light or Gravity. Astrophysicists call it Dark Energy because it can't be seen with the naked eye but can only be detected indirectly within the Cosmic Density of this Universe and fills the immensity of space. Christians call it the Holy Spirit or the Light of Christ. I have taken to calling it PURE LIGHT. Yoda called it The Force.

Granted, you might not be able to use This Force to lift X-Wing fighters out of a swamp; but, we definitely can use This Force to pray to God and get an instantaneous answer in return. This Force is like a Cosmic Internet or a Cosmic Telephone System. The Holy Ghost can speak thoughts and answers to our prayers directly into our minds using This Force or This Cosmic Phone System.

Luminous beings are we — Spiritual beings or beings of light — not this crude matter which the Materialists limit themselves to and worship. The Materialists and Atheists actually believe that we cease to exist when our Physical Matter dies. As Master Yoda implies, that's a crude understatement of the True Reality of things.

It doesn't matter if he is a fictional character, or not. I'm with Yoda on this one.

As you can tell, I got a little bit excited about all of this, and I studied it and kept studying it until I finally understood it. I wanted to figure out how it all fits together. The more I studied it, the more I concluded that it must be true (God's Truth), because I have never found a better explanation for life, the universe, and everything. It's infinitely better and more informative than "42".

People will MOCK you for finding any of this information useful, informative, interesting, or believable. It's what they do, because they don't have anything better going on in their lives. Just ignore them. The goal here is to acquire information that will make your life better, happier, more interesting, more productive, and more successful.

Information and speculation about the Non-Physical, unseen, Spirit Realms is INFINITELY more interesting than Materialism can even begin to be. So, let them MOCK. Meanwhile, go and learn something new that you have never thought about before. It will make your life a lot more fun and interesting.

—

Light, Truth, and Intelligence

For those who are still interested in knowing more, I tried to save you some time by providing you with some of the revelations from the Biblical God explaining to us what Light or Intelligence really is. Don't let the Materialists or Atheists make you ashamed to learn about and study these kinds of things. They just want to keep you in the dark where they currently are. I know, because I used to be an Atheist and a Materialist. Just ignore them and go out and learn something new. Open your eyes and see the LIGHT! In fact, one of my mottos is to "Learn Something

New Every Day"; and, I just as well start with the Spiritual and Non-Physical things because they are the most interesting.

I'm a completely different person now than I was when I was an Atheist and a Materialist. Some people think I'm a better person. Some people even post and say that they admire my courage or knowledge.

Notice how consistent all of the following is, both with itself and with what I have presented before this:

1 John 1: 5: "God is light, and in Him is no darkness at all." If you notice carefully, God has the type of Spirit or Light that can actually levitate, move, and manipulate Physical Matter telekinetically.

D&C 88: 7-13:
7 Which truth shineth. This is the light of Christ. As also he is in the sun, and the light of the sun, and the power thereof by which it was made.
8 As also he is in the moon, and is the light of the moon, and the power thereof by which it was made;
9 As also the light of the stars, and the power thereof by which they were made;
10 And the earth also, and the power thereof, even the earth upon which you stand.
11 And the light which shineth, which giveth you light, is through him who enlighteneth your eyes, which is the same light that quickeneth your understandings;
12 Which light proceedeth forth from the presence of God to fill the immensity of space —
13 The light which is in all things, which giveth life to all things, which is the law by which all things are governed, even the power of God who sitteth upon his throne, who is in the bosom of eternity, who is in the midst of all things.

Cool, huh?

This Light of Christ is in EVERYTHING giving it substance, structure, presence, and girth! It's the power by which things are made! It's the power by which things are sustained. It's the power that keeps protons from decaying and imploding. This LIGHT is in everything sustaining and upholding everything, and it proceeds forth from the presence of God. If this Light of Christ weren't there, then this whole universe would implode back into the singularity from whence it came. If God were to remove His influence and His presence and His Light from this universe, then it would implode back into the singularity from whence it came.

This Dark Energy or Light of Christ is what gives this universe presence, substance, girth, structure, expansion, LAW, order, contact with God, and life. That's why this Zero-Point Field comprises 72.1 percent of this Universe's Cosmic Density; and, the Biblical God Jesus Christ knew about its existence long before we did, which is why Jesus Christ was able to tell Joseph Smith about it over a 150 years ago!

God is the ULTIMATE SCIENTIST!

This Light of Christ is like the Dark Energy that the Astrophysicists talk about, or the Quantum Sea of Light that the New Agers talk about, or the Zero-point Field or Gravity Field or Strong Forces and Weak Forces or Holographic Universe that physicists talk about. This Light of Christ is also the Life Force. There's also talk of Quantum Consciousness, Morphic Fields, and a Universal Consciousness. ALL of this stuff is Immaterial or Non-Physical! It all exists as Immaterial SPIRIT or LIGHT! God knows what He is talking about, because He's the one who set it all up.

The Materialists don't have the prerequisites for this kind of SCIENCE course! It's RADICALLY MORE ADVANCED than the Materialists can even begin to understand. So, all you will see them doing is mocking it because they don't understand it — what it is, what it does, and why it is necessary for it to exist. Just ignore them, because they don't have a clue what's really going on in this universe. How can the Materialists know what's going on around them and within them, because they refuse to look at it and learn anything new from it! All they can do is mock it and ridicule it. They definitely are NOT going to learn something from it. The Materialists are the worst kind of scientists, if they can be called scientists at all.

—

Abraham 3: 11-12:

11 Thus I, Abraham, talked with the Lord, face to face, as one man talketh with another; and he told me of the works which his hands had made;
12 And he said unto me: My son, my son (and his hand was stretched out), behold I will show you all these. And he put his hand upon mine eyes, and I saw those things which his hands had made, which were many; and they multiplied before mine eyes, and I could not see the end thereof.

Abraham 3: 17-19:

17 Now, if there be two things, one above the other, and the moon be above the earth, then it may be that a planet or a star may exist above it; and there is nothing that the Lord thy God shall take in his heart to do but what he will do it.
18 Howbeit that he made the greater star; as, also, if there be two spirits, and one shall be more intelligent than the other, yet these two spirits, notwithstanding one is more intelligent than the other, have no beginning; they existed before, they shall have no end, they shall exist after, for they are gnolaum, or eternal.
19 And the Lord said unto me: These two facts do exist, that there are two spirits, one being more intelligent than the other; there shall be another more intelligent than they; I am the Lord thy God, I am more intelligent than they all.

Abraham 3: 21-28:

21 I dwell in the midst of them all; I now, therefore, have come down unto thee to declare unto thee the works which my hands have made, wherein my wisdom excelleth them all, for I rule in the heavens above, and in the earth beneath, in all wisdom and prudence, over all the intelligences thine eyes have seen from the beginning; I came down in the beginning in the midst of all the intelligences thou hast seen.

22 Now the Lord had shown unto me, Abraham, the intelligences that were organized before the world was; and among all these there were many of the noble and great ones;

23 And God saw these souls that they were good, and he stood in the midst of them, and he said: These I will make my rulers; for he stood among those that were spirits, and he saw that they were good; and he said unto me: Abraham, thou art one of them; thou wast chosen before thou wast born.

24 And there stood one among them that was like unto God, and he said unto those who were with him: We will go down, for there is space there, and we will take of these materials, and we will make an earth whereon these may dwell;

25 And we will prove them herewith, to see if they will do all things whatsoever the Lord their God shall command them;

26 And they who keep their first estate shall be added upon; and they who keep not their first estate shall not have glory in the same kingdom with those who keep their first estate; and they who keep their second estate shall have glory added upon their heads for ever and ever.

27 And the Lord said: Whom shall I send? And one answered like unto the Son of Man: Here am I, send me. And another answered and said: Here am I, send me. And the Lord said: I will send the first.

28 And the second was angry, and kept not his first estate; and, at that day, many followed after him.

Online Source for the Book of Abraham:

https://www.lds.org/scriptures/pgp/abr?lang=eng

God ORGANIZED these Intelligences in the beginning. God did NOT create them! These Intelligences ARE eternal and indestructible. They are Trans-Dimensional or Extra-Dimensional and preceded Spirit Matter and Physical Matter. It's these Intelligences who designed and created Spirit Matter and then later Physical Matter. Spirit Matter and Physical Matter are hosts for Intelligences — the things that Intelligences act upon. The Apostle Paul called our Physical Bodies the Temples of the Gods, another way of saying that our physical bodies are tabernacles or temples for our Intelligences or Living Lights.

God is God because He is more intelligent than all of the rest of us combined.

—

Doctrine and Covenants 93:29: Man was also in the beginning with God. Intelligence, or the light of truth, was not created or made, neither indeed can be.

These Intelligences cannot be created or made. They can only be organized.

Doctrine and Covenants 93:30: All truth is independent in that sphere in which God has placed it, to act for itself, as all intelligence also; otherwise there is no existence.

There is NO existence or life without these Intelligences. It's these Intelligences that bring Spirit Matter and Physical Matter into existence and enlivens these different types of matter.

Doctrine and Covenants 93:36: The glory of God is intelligence, or, in other words, light and truth.

Doctrine and Covenants 130:18: Whatever principle of intelligence we attain unto in this life, it will rise with us in the resurrection.

Our Intelligence doesn't die when our physical body dies. Our Intelligence doesn't die should our spirit body die.

Doctrine and Covenants 84:45: For the word of the Lord is truth, and whatsoever is truth is light, and whatsoever is **light is Spirit**, even the Spirit of Jesus Christ.

Whatever is Light is Spirit. Spirit is Light; and, Light is Spirit.

Doctrine and Covenants 93:28: He that keepeth God's commandments receiveth truth and light, until he is glorified in truth and knoweth all things.

Doctrine and Covenants 88:7: Which truth shineth. This is the light of Christ. As also he is in the sun, and the light of the sun, and the power thereof by which it was made.

This Light of Christ is what keeps our sun and our earth from imploding into a singularity.

Doctrine and Covenants 93:37: Light and truth forsake that evil one.

Doctrine and Covenants 93:40: But I have commanded you to bring up your children in light and truth.

Psalms 43:3 O send out thy light and thy truth: let them lead me; let them bring me unto thy holy hill, and to thy tabernacles.

Light and Truth, or Intelligence or Thought, is the primal construct.

John 3:21: But he that doeth truth cometh to the light, that his deeds may be made manifest, that they are wrought in God.

Doctrine and Covenants 93:9: The light and the Redeemer of the world; the Spirit of truth, who came into the world, because the world was made by him, and in him was the life of men and the light of men.

This Spirit of Truth is the light and life of men, and also the Redeemer of the world.

Doctrine and Covenants 93:39: And that wicked one cometh and taketh away light and truth, through disobedience, from the children of men, and because of the tradition of their fathers.

Doctrine and Covenants 93:42: You have not taught your children light and truth, according to the commandments; and that wicked one hath power, as yet, over you, and this is the cause of your affliction.

Doctrine and Covenants 88:6: He that ascended up on high, as also he descended below all things, in that he comprehended all things, that he might be in all and through all things, the light of truth.

This Light of Truth or Light of Christ is in all things and through all things. The Light of Christ IS the Dark Energy that expanded our universe and keeps our universe expanding.

Doctrine and Covenants 124:9: And again, I will visit and soften their hearts, many of them for your good, that ye may find grace in their eyes, that they may come to the light of truth, and the Gentiles to the exaltation or lifting up of Zion.

Joseph Smith History 1:25: So it was with me. I had actually seen a light, and in the midst of that light I saw two Personages, and they did in reality speak to me; and though I was hated and persecuted for saying that I had seen a vision, yet it was true; and while they were persecuting me, reviling me, and speaking all manner of evil against me falsely for so saying, I was led to say in my heart: Why persecute me for telling THE TRUTH? I have actually seen a vision; and who am I that I can withstand God, or why does the world think to make me deny what I have actually seen? For I had seen a vision; I knew it, and I knew that God knew it, and I could not deny it, neither dared I do it; at least I knew that by so doing I would offend God, and come under condemnation.

Ether 4:12: And whatsoever thing persuadeth men to do good is of me; for good cometh of none save it be of me. I am the same that leadeth men to all good; he that will not believe my words will not believe me — that I am; and he that will not believe me will not believe the Father who sent me. For behold, I am the Father [of Truth], I am the light, and the life, and THE TRUTH of the world.

Satan is known as the father of lies. Jesus Christ is the Father of Truth.

Alma 38:9: And now, my son, I have told you this that ye may learn wisdom, that ye may learn of me that there is no other way or means whereby man can be saved, only in and through Christ. Behold, he is the life and the light of the world. Behold, he is the word of truth and righteousness.

Jesus Christ, our Anointed Savior, is the light, life, and truth of this world. If you notice carefully, God has the type of Spirit or Light that can actually levitate, move, and manipulate Physical Matter telekinetically.

Doctrine and Covenants 85:7: And it shall come to pass that I, the Lord God, will send one mighty and strong, holding the scepter of power in his hand, clothed with light for a covering, whose mouth shall utter words, eternal words; while his bowels shall be a fountain of truth, to set in order the house of God, and to arrange by lot the inheritances of the saints whose names are found, and the names of their fathers, and of their children, enrolled in the book of the law of God.

Online Source for the Doctrine and Covenants:

https://www.lds.org/scriptures/dc-testament?lang=eng

Online Source for the Book of Mormon:

https://www.lds.org/scriptures/bofm?lang=eng

Online Source for the Pearl of Great Price:

https://www.lds.org/scriptures/pgp?lang=eng

—

Intelligence

Intelligence has several meanings, three of which are:

(1) It is the light of truth which gives life and light to all things in the universe. It has always existed.
(2) The word intelligences may also refer to spirit children of God.
(3) The scriptures also may speak of intelligence as referring to the spirit element that existed before we were begotten as spirit children.

Intelligence cleaveth unto intelligence: D&C 88:40.
Intelligence was not created or made: D&C 93:29.
All intelligence is independent in that sphere in which God has placed it: D&C 93:30.
The glory of God is intelligence: D&C 93:36-37.
Intelligence acquired in this life rises with us in the resurrection: D&C 130:18-19.
The Lord rules over all the intelligences: Abr. 3:21.
The Lord showed Abraham the intelligences that were organized before the world was: Abr. 3:22.

—

The Light of Christ or Dark Energy

Divine energy, power, or influence that proceeds from God through Christ and gives life and light to all things. It is the law by which all things are governed in heaven and on earth (D&C 88:6-13). It also helps people understand gospel truths and helps to put them on that gospel path which leads to salvation (John 3:19-21; 12:46; Alma 26:15; 32:35; D&C 93:28-29, 31-32, 40, 42).

The light of Christ should not be confused with the Holy Ghost. The light of Christ is not a person. It is an influence that comes from God and prepares a person to receive the Holy Ghost. It is an influence for good in the lives of all people (John 1:9; D&C 84:46-47).

One manifestation of the light of Christ is conscience, which helps a person choose between right and wrong (Moro. 7:16). As people learn more about the gospel, their consciences become more sensitive (Moro. 7:12-19). People who hearken to the light of Christ are led to the gospel of Jesus Christ (D&C 84:46-48).

The Lord is my light: Ps. 27:1.
Let us walk in the light of the Lord: Isa. 2:5; (2 Ne. 12:5).
The Lord shall be an everlasting light: Isa. 60:19.
The true Light lighteth every man that cometh into the world: John 1:4-9; (John 3:19; D&C 6:21; D&C 34:1-3).
I am the light of the world: John 8:12; (John 9:5; D&C 11:28).
Whatsoever is light, is good: Alma 32:35.
Christ is the life and the light of the world: Alma 38:9; (3 Ne. 9:18; 3 Ne. 11:11; Ether 4:12).
The Spirit of Christ is given to every man that he may know good from evil: Moroni 7:15-19.
That which is of God is light, and groweth brighter and brighter until the perfect day: D&C 50:24.
The Spirit giveth light to every man: D&C 84:45-48; (D&C 93:1-2).
He that keepeth God's commandments receiveth light and truth: D&C 93:27-28.
Light and truth forsake that evil one: D&C 93:37.

—

Source for these Scriptures and Ideas:

https://www.lds.org/?lang=eng

If you notice carefully, God has the type of Spirit or Light that can actually levitate, move, and manipulate Physical Matter telekinetically.

—

Abductive Reasoning seeks for the best and most parsimonious interpretation or explanation for the available evidence. I also employ deductive reasoning. By eliminating ALL of the FALSE PREMISES, we can deduce the CORRECT CONCLUSION. It ALL points to God.

—

Our Spirits or Our Living Lights Seek NIRVANA, BLISS, and PEACE

In my books, I tended to define NIRVANA as the bliss, happiness, will, and blessedness of our own internal Light or Spirit. Spirit IS Light. Light IS Spirit.

259

It is the Spirit or Living Light or NIRVANA within each one of us that has the inherent ability to design and create — NOT our Physical Matter! Our Physical Matter can't do anything except to react to whatever our Spirits tell it to do.

A rock has Physical Matter, but a rock does NOT have the kind of spirit within it that has the ability to design and create new things. Even though a rock has a lot of Physical Matter, you will NEVER catch a rock in the act of designing and creating new things from scratch. It isn't the Physical Matter that designs and creates! Physical Matter was designed to be molded and acted upon by Spirits or Living Lights. Physical Matter was designed to be occupied and enlivened by Spirits or Living Lights. Inert Physical Matter cannot design and create!

There are different types of Spirit or Light. While in Spirit, people have seen the Consciousness or awareness of rocks and nails. But, there isn't a designing and creating type of Spirit within a rock. There isn't a loving and charitable Spirit within a rock either. A rock is a Spiritual recording device, nothing more.

It's our Human Spirit or Living Light which gives us our desires, intelligence, will, and ability to choose — NOT our Physical Matter. Living Intelligent Spirits like ours are Actors, Designers, and Creators. In contrast, Physical Matter was designed to be acted upon. Spirits or Living Lights act. Physical Matter reacts and is acted upon.

The reason we humans can design and create is that Our Spirits inherited that particular ability from the Parents of Our Spirits — God the Father and Heavenly Mother. The Spirits or Living Lights within human beings is unique among God's creations, because we are the Children of God.

Our Living Light or Spirit or NIRVANA enlivens our Physical Matter, giving that formerly inert Physical Matter purpose, intelligence, desires, goals, and in the case of humans the ability to design, plan, repent, and create.

—

Matching With Reality Makes It Scientifically Accurate

Something is said to be Scientifically Accurate if it matches with Reality.

What's really cool is that the Scriptures from the Biblical God match perfectly with our current understanding of SCIENCE! It's as if the Scriptures from the Biblical God match perfectly with Reality and thus are completely Scientifically Accurate!

What are the odds?

God thought of these things and experienced these things BEFORE we did, which is why the Biblical God was able to tell Joseph Smith about them over 150 years ago before our scientists even began to think about them.

God is the ULTIMATE SCIENTIST! And God is NOT a Materialist, or an Atheist, or a Darwinist!

God KNOWS that He didn't have to rely upon or wait upon evolution, chance, random errors, and random reactions to design and create things for Him. God is like us. If we want something done, we don't wait for evolution, chance, random blind luck, or chemical reactions to do it for us. We just go out and do it — do the Science or do the Manufacturing or do the Construction or do the Work that we want to have done.

We are Actors and Creators and Builders, not blind random processes or Materialistic reactions. We have an inner light or an inner spirit which gives us the ability to design and create. Evolution does not!

If you notice carefully, God has the type of Spirit or Light that can actually levitate, move, and manipulate Physical Matter telekinetically. Intelligent Glorified God-like Spirits can design and create and manipulate Physical Matter with their minds. Evolution cannot!

Creation by Evolution, Creation by Chance, and Creation by Random Reactions IS IMPOSSIBLE! The impossible should be eliminated from Science as being unscientific. That's what the True Science and the ULTIMATE SCIENTIST is telling us.

—

Materialism Is THE LIE That Everyone Wants to Believe In

The Materialists and Atheists won't be able to teach you anything about the Spiritual Sciences because they are unaware that there is such a thing. How can the Materialists and Atheists teach you something that they don't know anything about?

Once you get rid of any residual Materialism, the True Reality of EVERYTHING suddenly becomes obvious and clear. It's like taking the blinders off of your eyes!

Darwinism is a form of Materialism.

Materialism is THE LIE that millions of people want to believe in.

Materialism is THE LIE that Satan whispers into our ears, because Satan wants us to believe that he doesn't exist, that God doesn't exist, and that the Holy Spirit doesn't exist. THE LIE is the opposite of THE TRUTH.

Jesus Christ says that He is The Way, THE TRUTH, and The Life; and that, none of us will get back into Heaven and the Presence of the Father except through Him.

Materialism is the MOST SUCCESSFUL LIE in human history, because it is THE LIE that most people have chosen to believe in and desperately want to be true.

Materialism is THE BANE and THE CURSE against Scientific Exploration and Scientific Discovery, especially when it comes to the Spiritual Sciences.

Isn't it ironic and interesting how the Materialists and Darwinists employ a non-existing, Immaterial, Spiritual, Non-Physical, philosophical, non-entity as their designer, creator, and manufacturer within Materialism and the Theory of Evolution? That's a contradiction in terms, and VIOLATES the Law of Non-Contradiction. It's also a CATEGORY ERROR to turn philosophical and scientific concepts such as evolution and chance and natural reactions into PEOPLE who can design and create and do science at will!

Materialism was designed to keep us in the Dark Ages, especially where Spirituality or righteousness of any kind is concerned. Materialism was designed to drive God out of our lives! Get rid of your Materialism and suddenly the whole world becomes a completely different place, a much better place! Materialism will NEVER lead us to God's Mercy and God's Love, because Materialism is a deliberate and knowing rejection of God and a rejection of ALL the good that God has to offer us.

I slowly got rid of My Atheism and My Materialism, and my life has gotten better and more interesting ever since! Materialism and Atheism are BORING!

I once was blind, but now I see.

Keep the best and get rid of all the rest! That's what I try to do.

In a sense, that's what THE SCIENTIFIC METHOD is all about — keep THE TRUTH, and get rid of all the false premises, false hypotheses, and deliberate lies such as Materialism, Creation by Evolution, and Creation by Natural Reactions.

"Adventures Beyond the Body: How to Experience Out-of-Body Travel" by William Buhlman.

This is one of my many Reviews PROMOTING the book "Adventures Beyond the Body" by William Buhlman. This book single-handedly put an end to Materialism for me!

This book is one of my top-ten most favorite books because of all the different things that it taught me about spirit, Spirituality, and the Non-Physical Realms. This is one cool book! I have purchased multiple paperback copies for me and my friends, and I also own a Kindle copy. My intent here is to point EVERYONE to this book and tell them to read it, especially if they have any lingering feeling that Materialism or Naturalism might be true.

Spirit or the Spiritual has to be experienced first-hand in order to know that it is real and true. There's really no other way, unless you are willing to trust other people's experiences of the Spiritual, which I am willing to do. I have learned to trust this person's Spiritual experiences, because they fit in extremely well with everything else that I have learned about spirit and Spirituality so far.

The Copyright for this book permits "quotations embodied in critical articles and reviews". So, that's what I am going to do here. I'm embodying quotations in this CRITICAL ARTICLE and this REVIEW of his book, "Adventures Beyond the Body" — except it should be noted that I'm NOT being critical of Buhlman and his book, but instead trying to promote it. I USE his books to CRITICIZE the Materialists and the Atheists who mock and ridicule Spiritual things in their "critical reviews" online.

Because Buhlman doesn't seem to be promoting any particular Organized Religion, he comes across as an unbiased authority on spirit and out-of-body travel, which some Atheists and Materialists might actually appreciate, since most Atheistic Materialists are deathly afraid of Organized Religion. Buhlman seems to take a secular approach to Spirituality, which is really cool if you happen to be someone who is on the fence about all of this.

For PROMOTIONAL purposes and preemptive defense, here are a couple of my most favorite quotes from Buhlman's book which are germane to the topic at hand. Whenever I'm taking ridicule and heat from the Materialists and Atheists online, I love to drop these quotes and this book on them:

> Twenty years ago, I firmly believed that the physical world we see and experience was the only reality. I believed what my eyes told me — life possessed no hidden mysteries, only countless forms of matter living and dying. The facts were clear; there was no evidence or proof of nonphysical worlds or our continued existence after death. I questioned the intelligence of anyone softheaded enough to accept the illogical concepts of heaven, God, and immortality. In my mind these were fairy tales created to comfort the weak and manipulate the masses. For me, life was simple to understand: the world consisted of solid matter and form, and the concepts of life after death and heaven were feeble human attempts to create hope where none existed.

I possessed the arrogant knowledge of a man who judges the world with his physical senses alone. I supported my conclusions with the overwhelming observations provided by science and technology. After all, if something mysterious was there, science would certainly be aware of it.

My firm convictions of reality and life continued until June of 1972. During a conversation with a neighbor, our discussion turned to the possibilities of life after death and the existence of heaven. I proceeded to present my agnostic viewpoints with vigor. To my surprise my neighbor didn't contest my conclusions; instead, he related an experience that he had had several weeks before. One evening just after drifting to sleep, he was shocked to discover himself floating above his body. Completely awake and aware, he became frightened and instantly fell back into his physical body. Excited, he told me it wasn't a dream or his imagination, but a fully conscious experience.

Intrigued by his experience, I decided to investigate this strange phenomenon for myself. After several days of research, I discovered numerous references to out-of-body experiences throughout history. With some searching I found a book on the topic that actually described how out-of-body experiences are induced. The entire subject seemed extremely weird, and I considered the book the result of an overly active imagination.

Out of curiosity, I decided to try one of the out-of-body techniques before sleep. After repeated daily attempts, I began to feel a little ridiculous. In three weeks, the only thing I experienced out of the norm was an increase in my dream recall. I became more and more convinced that this entire subject was nothing more than an intense or vivid dream stimulated by the so-called out-of-body techniques.

Then, one night about eleven o'clock I drifted to sleep during my out-of-body technique and began dreaming that I was sitting at a round table with several people. They all seemed to be asking me questions related to my self-development and state of Consciousness. At that moment in the dream I began to feel extremely dizzy, and a strange numbness, like from Novocain, began to spread throughout my body. Unable to keep my head up, I passed out, hitting my head on the table. Instantly I was awake, fully conscious, lying in bed facing the wall. I could hear an unusual buzzing sound and felt somehow different. Extending my arm, I reached for the wall in front of me. I stared in amazement as my hand actually entered the wall; I could feel the vibrational energy of it as if I was touching its very molecular structure. Only then did the overwhelming reality hit me, *My God, I'm not in my body.*

Excited, my only thought was, *It's real. My God, it's real!* Lying in bed, I stared at my hand in disbelief. When I tried clenching my fist, I could feel the pressure of my grip; my hand felt completely solid, but the physical wall in front of me looked and felt like a dense, vaporous material with form.

Determined to stand, I began to move effortlessly to the foot of my bed, my mind racing with the reality of it all. Standing, I quickly touched my arms and legs, checking to see if I was solid, and to my surprise I was completely solid, completely real. But around me, the familiar physical objects in my room no longer appeared completely real or solid; instead, they now looked like three-Dimensional mirages. Glancing down, I noticed a large lump in my bed. Amazed, I could see that it was the sleeping form of my physical body silently facing the wall.

As I focused my vision on the opposite side of the room, the wall seemed to fade slowly from view. In front of me I could see a wide, green field extending far beyond my room. Looking around, I noticed a figure silently watching me from about ten yards away. It was a tall man with dark hair, a beard, and a purple robe. Startled by his presence, I became frightened and instantly "snapped back" into my physical body. With a jolt I was in my body, and a strange feeling of numbness and tingling faded as I opened my eyes. Excited, I sat up, my mind exploding with the realization of what had just occurred. I knew it was absolutely real, not a dream or my imagination. My entire ego awareness had been present.

Suddenly, everything I had ever learned about my existence and the world around me had to be reappraised. I had always seriously doubted that anything beyond the physical world existed. Now my entire viewpoint changed. Now I absolutely knew that other worlds do exist and that people like myself must live there. Most important, I now knew that my physical body was just a temporary vehicle for the real me inside, and that with practice I could separate from it at will.

Excited about my discovery, I grabbed a pen and paper and wrote down exactly what had occurred. A flood of questions filled my mind. Why is the vast majority of the human race unaware of this phenomenon? Why aren't the various sciences and religions investigating it? Is it possible that this unseen world is the "heaven" referred to in religious texts? Why isn't our government exploring this apparent parallel energy world? Is it possible that our overwhelming dependence on physical perceptions has led us to overlook an incredible avenue of exploration and discovery?

As the initial shock of my first experience sank in, I realized that my life could never be the same again. The more I pondered the significance of my experience, the more profound I realized it to be. All my agnostic beliefs had been swept away in a single night. I knew that I had to reappraise everything that I had learned since childhood, everything that I had assumed to be true. My comfortable conclusions about science, psychology, religion, and my existence had obviously been based on incomplete information. I felt excited, but also uneasy — my familiar concepts of reality no longer seemed relevant. Increasingly, I felt in a void. On several occasions when I talked to friends about my experience, they found it too bizarre to take seriously. In 1972 the term *out-of-body experience* had not even been coined; back then, the most common description was astral projection. No one that I knew at the time had even heard of astral projection, and if you told people you had left your body, they immediately thought that you were on drugs or losing your mind. I quickly discovered that I had to keep my experiences to myself or face some degree of disbelief and even ridicule.

After my first out-of-body experience, my mind was overflowing with endless possibilities and questions. Desperate for information and guidance, I spent several weeks in libraries and bookstores searching for whatever knowledge was available on the topic. I quickly found that little was available; only a handful of books had been written on the subject, and some of these were decades old and out of print. By the end of July 1972, I realized that I was on my own.

I decided to focus on the one technique that had worked for me before. This technique involved visualizing a physical location that I knew well as I drifted off to sleep. As before, I pictured my mother's living room with as much detail as possible. At first it seemed difficult, but after a few weeks I could picture the room's details with increasing clarity; the furniture, patterns in fabrics, textures, even small imperfections in wood and paint began to be clear in my mind. I realized that the more I pictured myself within the room interacting with the physical objects, the more detailed my visualizations would become. With practice I learned to physically walk around the room and memorize specific items that it contained. I also learned the importance of "feeling" the environment with my mind: the feel of carpet on my feet; the sensation of sitting in a chair, walking, turning on a lamp, or even opening the door. The more detailed and involved I was within my visualization, the more effective were my results. Although it was challenging at first, after a while it became fun to make my visualizations come alive in my mind. At this point I decided to keep a journal to record my out-of-body experiences.

Buhlman, William L., "Adventures Beyond the Body: How to Experience Out-of-Body Travel" (pp. 3-8). HarperCollins. Kindle Edition.

The Materialists, Darwinists, and Atheists can be extremely stupid, ignorant, inexperienced, uneducated, and dense when it comes to the parts of SCIENCE that they have chosen not to believe in. I know, because I used to be one of them.

—

Journal Entry, October 2, 1982

I hear the buzzing, engine-like sounds and will myself out-of-body. I step to the bedroom door and automatically request "Clarity now!" My vision improves and I step through the door, into the living room. Still feeling a little out of sync, I verbally repeat my request with more emphasis, "Clarity now!" I feel my awareness and vision snap into place.

My thoughts are clear, and I make a verbal demand, "I need to see the form I'm in now!" Instantly I feel an intense sensation of being drawn within myself. I'm suddenly different, weightless as though I'm floating in space. As I look forward I see a sparkling, bluish white form. For some reason, I seem to know that I'm looking at my nonphysical body from a different perspective. I stare in amazement at this form before me that shines and

266

flows with energy and light. It looks like an energy mold created from a million tiny points of light; it radiates a bluish glow but appears to have a defined outer structure. The body of light before me is naked and is identical to my physical form. Even though my body looks firm, there is a noticeable energy motion and radiation present. I can see what appears to be an ocean of blue stars throughout my body. It's difficult to describe because the stars are stable yet moving at the same time; the light and energy of my body appear to change and flow almost like the waves of an ocean.

As I stare at the body of light, it hits me that I must be in another body. Yet I can't perceive any form or substance; I'm like a viewpoint in space without shape or form of any kind. As I reflect upon my new state of being, I feel a sensation of rapid motion and I'm instantly back within my physical body.

Lying still and reviewing my experience, I'm struck by an inescapable conclusion: I must possess multiple energy-bodies. The form I just experienced was noticeably lighter (less dense) than even my second nonphysical body. I realize that the traditional view of our possessing two bodies — a physical body and a Spiritual body — is far too simplistic; we are much more complex than this. Just as there are multiple nonphysical energy dimensions within the universe, each of us must consist of multiple energy-bodies or vehicles of expression.

Now I seriously wonder just how many nonphysical bodies or forms this involves. I suspect that there must be one within each dimension of the universe and that all of these are interrelated and connected, just as the physical body is connected to its first nonphysical (Spiritual) body.

Buhlman, William L., "Adventures Beyond the Body: How to Experience Out-of-Body Travel" (pp. 34-35). HarperCollins. Kindle Edition.

END OF QUOTES.

—

If you want to know more, then go out and buy a copy of William Buhlman's book for yourself.

In my book here, the most I can do is to tell you where to go and look; and, I do make it a point to PROMOTE the books and the people that I found most helpful and useful during my research. Clearly, I have given this whole thing some study and thought. This book from William Buhlman gave me solid irrefutable evidence that Materialism is FALSE. Whenever an Atheist or Materialist is being caustic, mean, and rude online, I love to drop these two quotes on them. They don't know what to do with them except to mock them and dismiss them out-of-hand; or even better, they don't respond at all.

These two quotes from William Buhlman have become an essential part of my defense package whenever I find myself engaging the Borg (Materialists) online.

The Darwinists and Materialists have accused me of carpet bombing the threads that I started online. These two quotes are a couple of bombs that I love to drop on them from time to time!

Take note that William Buhlman has other books for sale, but this one was my favorite one so far.

William Buhlman discovered first-hand that Materialism and Atheism are false.

The only way to KNOW that the Spiritual is real is to experience it first-hand for yourself, or to trust someone who has experienced it first-hand. I have had trust issues for most of my life, so it would have been nice to have had some first-hand Spiritual experiences of my own like these that Buhlman has had. Unfortunately, most Atheistic Materialists like me block ourselves from pursuing and having Spiritual experiences like Buhlman's because we don't believe that such a thing is possible, so we never even try it. Materialism is a curse that ends up being self-confirming and self-reinforcing — believing Materialism to be true makes it true in one's own life.

For most Materialists and Atheists, they need some kind of jump-start or extreme experience to get them into Spirituality and pursuing God. But, once you have had some convincing Spiritual experiences of your own, then typically you don't want to go back to the limitations and darkness and blindness of Materialism.

It can be a very interesting (and sometimes scary) to experience that internal paradigm shift from Materialism to Spirituality or Spiritualism. It's like a light going on, and then suddenly you just KNOW that Materialism is false and why it is false. When you see the Light, you just know that your whole life before as a Materialist and an Atheist was completely false and a self-deceptive lie. It can be a bit of a shock to the system at first; but then, you find yourself getting very excited about all of the new and interesting things that have suddenly opened up for you.

Once you know that Materialism is false, then you KNOW that Darwinism is false, because Darwinism is a Materialistic philosophy. Once you know that Darwinism is false, then you KNOW that goo-to-you Darwinian evolution is FALSE also because the Theory of Evolution is fruit from the poisoned tree.

When it comes to Science of any kind, Materialism IS the ultimate bane of that Science. Materialism IS the poisoned tree from which all falsehoods are produced. Materialism ALWAYS produces the WORST explanation or interpretation of scientific evidence.

I know that some of this makes the Materialists and Atheists and Darwinists uncomfortable, but who cares? What have the Materialists and Atheists done for us? They take without giving anything of value in return. What good have their lies and self-deceptions done for any of us? What thing of value have they actually contributed to our intellectual and Spiritual growth? What thing of value have they done for society? During the 20th century, they exterminated at least 200 million of their opponents in the name of Darwinism (Fascism) and/or some kind of Militant Enforced Atheism (Communism). Materialism has not been good for society — it's too selfish to be of any value to the rest of us. And, Materialism IS BAD SCIENCE!

Creation by Purely Materialistic Processes is scientifically impossible, because Physical Processes require some kind of intelligent being to get them going, coax them or force them in the right direction, and then deliberately weed out the failures while keeping the successes. Scientists are required to DO SCIENCE or to DO MANUFACTURING in order to make Materialistic Processes organize into new and useful things.

Spirit beings or living beings design, program, plan, engineer, create, manufacture, and deploy new things — NOT MATERIALISTIC PROCESSES! There's no brain, no mind, no intelligence, and no hands for manipulation and manufacturing when it comes to purely physical processes or random luck. Spirit or sentience or life designs and creates new things, not random material processes. This IS a Scientific Reality, thus making Materialism scientifically FALSE.

If Materialism is FALSE, then God must of necessity exist in order to do all of the different things and all of the science that needed to be done which "Materialism" or random material processes could never have done. This is logical common sense. This is reasoning, deduction by a process of eliminating False Premises, and even science!

This Scientific Reality can be hard to put into words at times, but random material processes and random chemical reactions cannot design and create anything from scratch. It requires some kind of intelligent mind in order to design and create something new from scratch, and actually make it work. Anytime living breathing Scientists DO SCIENCE in a science lab, then it is NO longer Creation by Evolution or Creation by Natural Processes or Creation by Chance, but is instead Creation by Intelligent Beings. Intelligent Beings can design and create and manufacture new and unique things — Evolution, Chance, Chemical Reactions, and Mindless Directionless Physical Processes cannot.

Darwinian molecules-to-man evolution IS scientifically impossible. Darwinian ape-to-man evolution IS scientifically impossible. Creation by Chance IS scientifically impossible. Abiogenesis IS scientifically impossible. Creation of new functional genomes through step-by-step Mutation/Selection IS scientifically impossible. Macro-evolution of any kind IS scientifically impossible. Creation of new functional genomes by Micro-evolution or Macro-evolution IS scientifically impossible. Creation by Evolution IS scientifically impossible. Technically, Design and Creation by any kind of Mindless Materialism or any kind of Mindless and Directionless Materialistic Process IS scientifically impossible. My goal here is to eliminate the impossible.

How often have I said to you that when you have eliminated the impossible, whatever remains, *however improbable*, must be THE TRUTH? — Sherlock Holmes

When you have eliminated Evolution of every kind from the design and creation process, then what remains? It's elementary my dear Watson.

Follow the evidence!

Keep the best and get rid of all the rest!

My Favorite Near-Death and Out-of-Body Testimonies

P. B. says:

Mark hallucinates: "Any time that you can demonstrate to yourself that spirit exists or that God exists, it reveals Materialism for the lie that it is."

Therefore, Materialism must be true, because nobody has ever objectively proved that spirit exists, or God exists.

—

Says who? You?

P. B., just because YOU haven't had any Spiritual experiences doesn't mean that nobody else has. That's YOUR problem, not ours. When it comes to Spiritual experiences, YOU have to go and get your own in order for YOU to know that they are real and true. YOU have to objectively experience them for yourself in order to prove to yourself that spirit exists, and that God exists.

When it comes to Spiritual experiences, I have had sufficient for my needs; and apparently, you haven't. I have had enough Spiritual experiences to KNOW that Materialism, Naturalism, and Darwinism are FALSE.

Remember, it's no longer blind-faith for the people who have seen and touched our resurrected Lord Jesus Christ. In contrast, you will NEVER get Darwinian Chance to make a physical appearance in the flesh.

P. B., anytime you use extreme modifiers such as NOBODY, it automatically makes your statement FALSE, because there are indeed people out there in this world who have experienced the Spiritual first-hand, and there are also people who have seen and TOUCHED our resurrected Lord Jesus Christ. Just because you haven't doesn't mean that they haven't.

Instead, what you have done here is reveal to the WORLD how narrow and limited your Materialistic Worldview has made you, which is the exact point that I was trying to make in the first place.

My primary purpose here is to demonstrate that Materialism of any kind is a LIE, because there are people in this world who have experienced the Spiritual and experienced God first-hand and KNOW that they are REAL and TRUE.

—

For your benefit, here are a few of my favorite near-death experiences and out-of-body Spiritual experiences from YouTube:

Ian McCormack:
https://www.youtube.com/watch?v=HbTAmN4m2lQ

Dr. Mary Neal:
https://www.youtube.com/watch?v=DX473dF7ChY

John Ramirez:
https://www.youtube.com/watch?v=_M_8lI0-7b0

Dr. Richard Eby:
https://www.youtube.com/watch?v=IInRiIi-zQw

Howard Storm:
https://www.youtube.com/watch?v=UPj4wci_bcI

These people have told their story more than once, so after you have found and experienced my favorite version of their story, then you can explore the other versions of their story from the links that will pop up on the side.

—

I will place other favorite Spiritual experiences at the following link, as time allows and when I come across them:

http://markme.us/forums/forum/favorite-Spiritual-experiences/

You can lead a horse to water, but you can't make him drink. You can lead an Atheist to God, but you can't make him read and think.

—

The purpose of THE SCIENTIFIC METHOD is to help us to find THE TRUTH, through a preponderance of the evidence.

THE SCIENTIFIC METHOD has no value to us if we use it to convince ourselves that a LIE is TRUE, as the Materialists and Darwinists always seem to do.

That's what I discovered during my Pursuit of the True Reality of All Things, and during my usage of THE SCIENTIFIC METHOD.

The Inadequacy of Evolution Proves that God Must Exist

The Theory of Evolution proves that God Exists!

Who would've thought it possible?

Certainly not me back when I was an Atheist and in an Atheistic frame of mind.

However, the more I study the details and the "evidence" associated with the Theory of Evolution, the more Evolution itself has proven to me that God must exist in order to do all of the different things that Evolution could NEVER have done. The Darwinists demand that I study "their science" telling me that I need to study some Real Science. The more that I do as they say and study "their science", the more that "their science" proves to me that Evolution couldn't have done ANY of the creative acts which the Darwinists assure us that Evolution did.

In order for the Theory of Evolution and Darwinian goo-to-you evolution to be true, it would have to be able to design and create anything and everything, without any help from any of us including God. But, Evolution can't do so, because it's not alive and it can't create. Evolution has no hands and no mind. It can't design and create anything! Evolution or Change ONLY works on the things which God has already designed and created in the first place.

The Theory of Evolution is its own worst enemy. It's a counter-intuitive and illogical theory that violates THE SCIENTIFIC METHOD in a wide variety of different ways. The more that I study the Theory of Evolution the clearer it becomes to me that Evolution could never have done any of the miraculous and magical creative acts which the Darwinists say that Evolution did. The more that I study the Darwinian "sciences" or the Darwinian interpretations of the Scientific Evidence, the clearer it becomes to me that Darwinism and the Theory of Evolution are nothing more than pseudo-science masquerading as some kind of real science. Evolution CANNOT do the molecules-to-man evolution which the Darwinists say that it did! That's what studying the Darwinian "sciences" has taught me!

The Darwinists and Evolutionists tell me that Evolution designed and created my eyes and my brain. I don't see how Evolution could have ever done so, because Evolution is NOT a person, personality, living entity, or even a real thing. Evolution has no hands, mind, spirit, or intelligence.

Only a blind and loyal Darwinist will actually believe that Darwinian Chance is real and true, and can truly create everything from scratch.

Who would have thought it possible that the Theory of Evolution could have proven to someone that God must exist? But, that's exactly what happened to me while studying the amazing, miraculous, and impossible claims of Darwinian Evolution.

—

The Battlefield

In this book, I document some of the reasons why I am no longer an Atheist, and why I no longer believe in the Theory of Evolution. THE TRUTH is all that you really want to know, because knowing THE TRUTH sets you free to make an informed decision or choice!

Darwinists and Evolutionists typically use the Theory of Evolution as a proof of God's non-existence.

Their Darwinian Thesis: If the Theory of Evolution is true, then God does NOT exist.

My Counter-Thesis: If the Theory of Evolution is FALSE in any way, then God must exist in order to have done all of the different things that Evolution could NEVER have done.

This book uses Common Sense and their "Darwinian Science" against them and ends up being a Scientific Proof of God's Existence as a result.

Ironically, you don't need to know much science to convince yourself through common sense how and why the Theory of Evolution must be FALSE. Why is this so? It's because one soon discovers that there is actually NO scientific evidence whatsoever supporting the Theory of Evolution, which means that Darwinism is purely a philosophical worldview or a religion. Consequently, Darwinism can be proven false and dispatched philosophically, because ALL of the Scientific Evidence proves that the Theory of Evolution is FALSE.

Once you know WHY evolution of any kind COULD NEVER HAVE DONE any of the miraculous, magical, and creative things that the Darwinists say that evolution did, then you KNOW ALL that you really need to know about the Theory of Evolution. You KNOW that Creation by Evolution is IMPOSSIBLE.

The failures, falsehoods, and inadequacies of Darwinian Evolution have proven to me that God must exist, in order to do all of the SCIENCE and other things which evolution could NEVER have done.

My Observation: Each time Evolution fails to do what the Darwinists say that it does or say that it did, that FAILURE becomes yet another miniature Scientific Proof of God's Existence, because God must of necessity exist in order to do all of the different things that Evolution did not do or could not do. This is logical common sense.

This REALITY or TRUTH is yet another example of God hiding in plain sight where the Atheists and the Darwinists cannot see Him or find Him. God is standing there staring them in the face, but they can't see Him because they don't want to see Him.

Materialism of any kind ends up becoming a self-fulfilling prophecy.

—

Applying THE SCIENTIFIC METHOD

The purpose of THE SCIENTIFIC METHOD is to help us to find THE TRUTH, through a preponderance of the evidence.

THE SCIENTIFIC METHOD has no value to us if we use it to convince ourselves that a LIE is TRUE, as the Materialists and Darwinists always seem to do.

That's what I discovered during my Pursuit of the True Reality of All Things, and during my usage of THE SCIENTIFIC METHOD.

THE SCIENTIFIC METHOD can be used and has been used TO PROVE beyond a shadow of a doubt through a preponderance of the evidence that Creation by Evolution, Creation by Chance, Materialism, and the Theory of Evolution are IMPOSSIBLE and FALSE.

The Theory of Evolution IS Creation by Evolution.

Meanwhile, all aspects of THE SCIENTIFIC METHOD and the Rules of Science have PROVEN definitively and conclusively that Design and Creation BY Intelligent Beings is REAL and TRUE.

What you do with this information is up to you.

ALL of the Evidence IS on the side of "Creation by Intelligent Beings". SCIENCE, THE SCIENTIFIC METHOD, ALL of our OBSERVATIONS, Abductive Reasoning, and Deductive Reasoning make it clear and obvious that it is so.

In contrast, there is NO Evidence and can be NO Evidence to support Creation by Evolution, Creation by Chance, Design and Creation by Natural Processes, and Creation by Materialism, because these things cannot Design, Manufacture, and Create! It's elementary my Dear Reader!

I'm going with the AVAILABLE EVIDENCE, the PREPONDERANCE OF THE EVIDENCE, the SCIENCE, and THE SCIENTIFIC METHOD on this one. By doing so, it has become abundantly obvious to me that GOD MUST EXIST in order to have done ALL of the SCIENCE, design, manufacturing, creation, and organizing which had to be DONE that Evolution and Chance could NEVER have done 3.8 billion years ago, when all of that WORK needed to be DONE. When it comes to Design and Creation, Evolution and Chance DO NOT WORK! I want something to WORK if I'm actually going to believe in it.

I have chosen to FOLLOW THE EVIDENCE wherever it might lead me, even if that Evidence should lead me to God. I'm no longer afraid of The Evidence like I used to be.

In the Bible, the Book of Mormon: Another Testament of Jesus Christ, the Doctrine and Covenants, and the Pearl of Great Price, the Biblical God Jesus Christ has repeatedly REVEALED Himself to us and repeatedly CONFESSED to designing and creating This Universe, This Earth, and ALL of the Genomes and Life Forms on this planet. That's good enough for me!

I have chosen to FOLLOW THE EVIDENCE wherever it might lead me.

You can do whatever you want to do. That's what this Life is all about — making a decision and a choice!

Have a nice day and a wonderful life.

Mark My Words!

Talking About What Chance Can Do for You

Here, I will let the professionals tell you about what Chance can do for you.

BEGINNING OF QUOTE:

> Our next question is crucial. How much influence or effect does chance have on the coin's turning up heads? My answer is categorically, "None whatsoever." I say that emphatically because there is no possibility, real or imagined, that chance can have any influence on the outcome of the coin toss.
>
> Why not? Because chance has no power to do anything. It is cosmically, totally, consummately impotent. Again, I must justify my dogmatism on this point. I say that chance has no power to do anything because it simply is not anything. It has no power because it has no being. I've just ventured into the realm of ontology, into metaphysics, if you please. Chance is not an entity. It is not a thing that has power to affect other things. It is no thing. To be more precise, it is nothing. Nothing cannot do something. Nothing is not. It has no "isness." Chance has no isness. I was technically incorrect even to say that chance is nothing. Better to say that chance is not. What are the chances that chance can do anything? Not a chance. It has no more chance to do something than nothing has to do something. It is precisely at this point that equivocation creeps (or rushes) into the use of the word chance. The shift from a formal probability concept to a real force is usually slipped in by the addition of another seemingly harmless word, by. When we say things happen "by chance," the term by can be heard as a dative of means. Suddenly chance is given instrumental power. It is the means by which things come to pass. This "means" now assumes a certain power to effect change. Something that in reality is nothing now has the ability or power to do something.
>
> Sproul, R. C.; Mathison, Keith (2014-08-12). Not a Chance: God, Science, and the Revolt against Reason (Kindle Locations 195-208). Baker Publishing Group.

END OF QUOTE.

The Darwinists take a non-entity, Chance, and magically transmute it into a being that can act as a causal agent. Something that in reality is nothing suddenly and magically in the hands of the Atheistic Darwinists gains the power and the ability to do everything and create everything. It's MAGIC!

> "When scientists attribute instrumental power to chance, they have left the domain of physics and resorted to magic. Chance is their magic wand to make not only rabbits but entire universes appear out of nothing." (Sproul, R. C.; Mathison, Keith. "Not a Chance". Kindle Locations 229-231.)

In the hands of a Darwinist and Evolutionist, Chance is their magic wand to make entire genomes and entire life forms appear out of nothing. The Theory of Evolution is nothing more than a different version of Creation Ex Nihilo. Creation out of nothing is FALSE no matter where you might encounter it. It's illogical, and not the least bit scientific, because it is magic!

There is NO such thing as creation out of nothing — Creation Ex Nihilo.

The singularity that preceded our big bang and this universe contained within it everything in this universe — ALL time, ALL space, ALL Consciousness or Intelligence, ALL Spirit Matter, ALL light, ALL matter, ALL energy, ALL judgement, ALL mercy, ALL law, ALL potential order, ALL futurity, ALL of YOU, and everything else that would ever exist in this universe. That singularity was SOME THING!

That singularity did NOT originate by Chance. Nothing originates by Chance! Someone put that singularity together, removed all of the entropy from it, and filled it full of potential BEFORE triggering it and creating this universe.

—

The Theory of Evolution Violates the Law of NonContradiction

BEGINNING OF QUOTES:

> To argue that something comes from nothing requires the denial of the law of noncontradiction. The law states simply that A cannot be A and non-A ($\neg$ A) at the same time and in the same relationship. Something can be A and B at the same time but not in the same relationship. I can be a father (A) and a son (B) at the same time, but not in the same relationship. For something to come from nothing it must, in effect, create itself. Self-creation is a logical and rational impossibility. For something to create itself it must be able to transcend Hamlet's dilemma, "To be, or not to be." Hamlet's question assumed sound science. He understood that something (himself) could not both be and not be at the same time and in the same relationship. For something to create itself, it must have the ability to be and not be at the same time and in the same relationship. For something to create itself it must be before it is. This is impossible. It is impossible for solids, liquids, and gasses. It is impossible for atoms and subatomic particles. It is impossible for light and heat. It is impossible for God. Nothing anywhere, anytime, can create itself.
>
> Sproul, R. C.; Mathison, Keith (2014-08-12). Not a Chance: God, Science, and the Revolt against Reason (Kindle Locations 256-264). Baker Publishing Group. Kindle Edition.

—

Chance is not an entity.
Nonentities have no power because they have no being.

To say that something happens or is caused by chance is to suggest attributing instrumental power to nothing.
Something caused by nothing is in effect self-created.
The concept of self-creation is irrational and violates the law of noncontradiction.
To persist in theories of self-creation one must reject logic and rationality.

Sproul, R. C.; Mathison, Keith (2014-08-12). Not a Chance: God, Science, and the Revolt against Reason (Kindle Locations 266-271). Baker Publishing Group. Kindle Edition.

END OF QUOTES.

Over and over again, the Materialists will tell you that God does not exist and that this universe and all the life on this planet has no need for a Creator; and then, they will turn around and tell you that Chance or Evolution designed and created it all.

The Materialistic Darwinists designed the Theory of Evolution to eliminate the need for a Creator; and, the Darwinists openly and boldly say that the Theory of Evolution eliminates any need for a Creator, mocking and ridiculing Creationists in the process. Then these very same Darwinists turn around and invoke Chance and submit Random Mutations as the creative element in the Theory of Evolution. The Atheistic Darwinists tell us that Evolution and Chance created it all, thus making Evolution or Chance the Creator of All things, while at the same time assuring us that there is NO Creator and no such thing as a Creator. This completely violates the Law of Noncontradiction! It IS a blatant contradiction!

The whole Theory of Evolution is based exclusively upon logic fallacies such as these. If the Theory of Evolution is based exclusively upon logic fallacies and falsehoods, then it is by definition in principle FALSE. The Theory of Evolution goes nowhere in the realm of logic and rationality and science, because the whole Theory of Evolution or Creation by Evolution is FALSE to begin with!

Abiogenesis is creation out of nothing, or Creation by Chance. Macro-evolution is Evolution by Abiogenesis or Creation by Chance. Evolution by Abiogenesis and Macro-evolution are Spontaneous Generation. Spontaneous Generation out of nothing, life spontaneously springing from non-life or life creating itself, is Macro-evolution; and, Macro-evolution violates the Law of Noncontradiction.

Something cannot be created out of nothing, because that also violates the Law of Noncontradiction! Nothing cannot make itself into something! Chance cannot design and create anything! It's logical. It's science. Science is the search for suitable, logical, and sufficient causes. Chance is not a suitable, logical, or sufficient cause for anything! Chance intrinsically violates the Law of Noncontradiction by introducing an unlimited number of contradictions into any subject, equation, or concept. Don't believe me? Then try getting Chance to design and create something for you and see how long you have to wait before Chance gets the job done.

Invoking Chance as a Creator and a Causal Agent produces an infinite number of logic fallacies and contradictions into your life and existence, and completely destroys science, rationality, and logic in the process. Yet, the Darwinists invoke Chance as the creative "entity" in the Theory of Evolution. The Darwinists deify Chance and imbue Chance with creative God-like powers in order to eliminate the need for a Creator and a God. That's the very definition of contradiction!

Employing Chance as the creator of this universe and as the creator of all the life on this planet violates the Law of Noncontradiction, and rejects logic and scientific rationality in the process. To persist in Evolution by Abiogenesis, Evolution by Spontaneous Generation, or Macro-evolution one MUST reject logic and rationality!

Without any backing from Logic and Rationality and Science, the Theory of Evolution devolves into nothing more than a philosophical assumption.

—

Evolution relies upon CHANCE to do its MAGIC.

Chance is synonymous with Magic.

Since the Theory of Evolution relies exclusively upon Chance to do ALL of its creation, the Theory of Evolution is logically FALSE, because Chance cannot design and create anything! The Theory of Evolution has no foundation upon which to build. Chance is not a person or an entity. Chance is NOT a causal agent of any kind. Chance cannot do any of the things that the Atheistic Darwinists say that it does, because Chance does not in reality exist. Try getting Chance to build you a house, a computer, a car, a bridge, a genome, or a road. How long will you have to wait until Chance builds one of these things for you? You can wait for all eternity, and it will NEVER happen!

It's only logical. It's science. Science is the search for sufficient and adequate causes. Chance is NOT a sufficient reliable cause for anything! You can test it in real time and see that it is so. Try getting Chance to create something for you, anything, and see how well that goes for you. Try getting Chance to do something for you or to answer your prayers.

Since Chance is NOT a suitable or sufficient replacement for the Biblical God, God must of necessity exist in order to do ALL of the different things that Chance could never have done.

But, can you get an Atheistic Darwinist to understand and believe any of this?

Not a chance!

Each Scientific Discipline IS Proof of God's Existence

EACH Scientific Discipline is an example of God hiding in plain sight where nobody can see Him and nobody can find Him, unless we go looking for Him.

Wrap your mind around that reality, and suddenly an avalanche of truth will come rushing into your life!

I discovered that EACH Scientific Discipline is a witness or a testament of God's existence; but, with EACH Scientific Discipline, you have to become open-minded enough, smart enough, and educated enough to be able to see how God fits into that particular picture or discipline. It can take a while to come up to speed; and, you will NEVER see it or understand it if you don't want to see it or understand it. When I was an Atheist, I couldn't see any of this or understand any of this, because I didn't want to. A desire to see THE TRUTH and to find THE TRUTH, a desire to see God and find God, is the most powerful and useful gift that you can give to yourself, because it will open your eyes to vistas and realities that you have never considered before.

In my case, I was unlucky because I was a skeptic and an Atheist, but I was lucky in the fact that I am multi-disciplinary. I have education and experience in almost every Scientific Discipline. Once I was willing to open my mind and see, I slowly began to realize that EACH Scientific Discipline IS a proof of God's existence. God gave us Science so that we could prove to ourselves that He exists. That realization has changed my life for the better! I once was blind, but now I see!

The question, "Who Created God", is a smoke screen that the Atheists and Darwinists put up in order to hide their ignorance and hide the REAL Scientific Questions!

God does NOT need a creator, because God has always existed.

The Gods ARE the self-existing Ones, or the self-sustaining Ones.

In principle, by definition, the Gods have always existed; and thus, the Gods need NO creator.

God must exist, or you would not exist!

A God created the first physical body, and then inhabited it. There is NO other way for a genome and a physical body to come into existence! That first physical body didn't just pop into existence out of nothing! It had to have a Creator and it had to be designed and created. Physical objects need to be created or organized by intelligent Spiritual beings, in order to take form.

EVERY living organism either has a Creator or Parents! That's what Science teaches us!

You have to have a Creator, or new physical things cannot come into existence. That's logical common-sense Science. Darwinists do a little magic trick by eliminating God as the Creator and putting Random Chance or Darwinian Chance in

His place as the creator of all things. The Darwinists lie and cheat in order to make their case for the Theory of Evolution.

Are you truly willing to believe what Science is telling you?

Without God, nothing would exist. That's what Science is telling us!

Resurrection from the Dead, Immortality, and Eternal Life are ONLY possible if God actually exists! Revelations from God and Revelations of God are ONLY possible if God exists. Your physical existence is ONLY possible if God exists. God must reveal Himself or remain forever concealed. God has revealed Himself to us throughout ALL the various Physical Sciences and Philosophical Sciences!

Those who have seen God know that He exists. Science is ALL about making personal observations. Observation — seeing and experiencing and understanding THE EVIDENCE — is what Science is all about!

I find the scientific proofs of God's existence a lot more convincing than the philosophical proofs of God's existence; and, despite what the Atheists and Skeptics will try to tell you, there are some very convincing proofs of God that come to us from scientific research and scientific evidence. It was a patent-holding scientist and a few other PhD scientists who convinced me that God really does exist — mostly by using Science itself to do so. Think about this for a while!

The very fact that we exist, you and I, is proof of God's existence.

There has been an infinite eternity that has gone on before us, before we ever arrived here on this earth.

According to the Law of Entropy, EVERYTHING should have burned out an eternity ago, and we should be sitting at heat death right now. Yet, here we are.

This reality is logical and empirical PROOF that there is something or someone out there who actually knows how to reverse entropy and has actually reversed entropy in order to produce our universe here, in the first place. We exist here and now, because something or someone out there KNOWS how to reverse entropy!

GOD MUST EXIST, OR YOU WOULD NOT EXIST!

Even the Theory of Evolution IS a proof of God's existence!

The fact that there is NO such thing as Macro-evolution, Evolution by Spontaneous Generation, or Evolution by Abiogenesis IS proof of God's existence! EVERY living organism in this Physical Realm needed a Creator or Parents! God must exist in order to have done all of the Science that evolution could never have done.

The fact that random mutations and natural selection CANNOT produce new genetic programming, new biological information, new biological designs, new genomes, and new life forms IS a proof of God's existence!

Why?

It's because ONLY a God could do these things for real!

The insufficiency and inability of evolution, natural selection, and random mutations to create new life forms or any life form IS a proof of God's existence!

Without God, nothing would exist.

Science properly understood IS a proof of God's existence!

This is a new and recent discovery for me. My Atheism, my unbelief, and my lack of knowledge prevented me from seeing and understanding that EACH Scientific Discipline is a proof of God's existence. It took me decades to start to see it and understand it; and even then, it still took additional years of study and prayer before I actually started to believe it to be true.

For me, this has been a hard-won victory. It didn't come easily; and, there were years when I didn't want it to come.

The fact that we can't get our hands-on Darwinian Chance or Darwinian Evolution and slit its throat IS scientific proof enough that the thing isn't real or true. Think about it! The fact that we can't get a hold of Darwinian Evolution and crucify it to death IS scientific proof that the thing doesn't exist in the first place.

In contrast, they did indeed get their hands on the Designer and Creator of this universe, the Biblical God Jesus Christ; and, they did indeed crucify Him on a cross. It doesn't get more real or true than that!

The Darwinian claim that some kind of Chance designed and created it all BROKE the Theory of Evolution for me. That claim is so illogical and so unscientific that I can't stomach it anymore. I can't go back to my ignorance of this subject. I can't put that genie back into the bottle, nor do I want to.

If you don't want to understand these things, then you NEVER will. That's just the reality of the situation. God is easily and successfully hidden from those of us who don't want to find Him or know Him. God can hide in plain sight, and those of us who don't want to see Him will NEVER see Him.

Self-deception works, and it works every time! God set this whole thing up so that you actually have to work, and work very hard, in order to find Him and know that He exists. Knowledge comes while searching for it; and, there is a lot of hidden knowledge out there waiting for us to find it!

For some of us, it can take a lifetime to see, understand, discover, and accept what Science is really trying to tell us — that God exists.

The purpose of THE SCIENTIFIC METHOD is to help us to find THE TRUTH, through a preponderance of the evidence.

THE SCIENTIFIC METHOD has no value to us if we use it to convince ourselves that a LIE is TRUE, as the Materialists and Darwinists always seem to do.

That's what I discovered during my Pursuit of the True Reality of All Things, and during my usage of THE SCIENTIFIC METHOD.

My Scientific Theory

"A little science estranges a man from God. A lot of science brings him back." — ATTRIBUTED to Francis Bacon, but not written by Francis Bacon.

Science itself became the Main Cure for My Atheism!

I'm a scientist, and the Scientific Evidence eventually started to speak for itself.

My Scientific Theory: Once we arrive at THE TRUTHS hidden within ANY Scientific Discipline, then that Scientific Discipline suddenly becomes a proof of God's existence.

You have to get down into THE TRUTH of a Scientific Discipline, before it will start to testify of God's existence; but, once you do, suddenly you find yourself staring into the face of God.

That has happened to me a few times already with multiple different Scientific Disciplines. My Scientific Theory is that it's possible to achieve this with ANY Scientific Discipline. Now I have some work to do in order to determine if My Theory is true.

The Corollary:

God gave us Science so that we could prove to ourselves that He exists.

My Hypothesis:

Each Scientific Discipline IS a Proof of God's Existence!

EACH Scientific Discipline is an example of God hiding in plain sight where nobody can see Him and nobody can find Him, unless we actually go looking for Him.

My great discovery is that EACH Scientific Discipline is a witness or a testament of God's existence; but, with EACH Scientific Discipline, you have to become open-minded enough, smart enough, and educated enough to be able to see how God fits into that particular picture or discipline. It can take a while to come up to speed; and, you will NEVER see it or understand it if you don't want to see it or understand it.

When I was an Atheist, I couldn't see any of this or understand any of this, because I didn't want to. A desire to see THE TRUTH and to find THE TRUTH, a desire to see God and find God, is the most powerful and useful gift that you can give to yourself, because it will open your eyes to vistas and realities that you have never considered before.

Science, when properly understood, becomes a proof of God's existence!

Conclusion:

Every Scientific Discipline, when you finally get down to THE TRUTHS hidden within it, becomes a proof of God's existence. God gave us Science so that we could prove to ourselves that He exists. When you finally have THE TRUTH in hand, it will ALL fit together perfectly, and it will ALL point to God. This is My Scientific Theory. Now, I get to spend the rest of my life taking each Scientific Discipline and trying to see if I can prove my theory correct. I have already done so to my satisfaction with over a dozen different Scientific Disciplines, but I still have a lot more to go.

Once we understand what it's really all about, we suddenly realize that it all points to God.

God reveals Himself to us in two separate ways — through Revelation or Scripture, and through Nature or Science.

That's the way it is!

God has to find some way to reveal Himself to us, or He will remain forever concealed and forever unknown.

NOT A CHANCE

While writing the book "The Theory of Evolution Proves that God Exists: Why I Am No Longer an Atheist and Why I No Longer Believe in the Theory of Evolution", I came late to "Not a Chance: God, Science, and the Revolt against Reason" by R. C. Sproul and Keith Mathison.

Their book provided the final piece to the puzzle; and, I thank them for it.

Their book ended up putting a capstone on my research about the Theory of Evolution, explaining to me that CHANCE does not exist; therefore, it couldn't have designed and created anything, let alone designing and creating everything as the Materialists claim that it did.

If you read that book, you will see that they explain why CHANCE couldn't have designed and created this universe; but, whatever they say about CHANCE in their book applies equally as well to Darwinian Chance. You find out that the dude doesn't exist, whenever you try to look him up.

I quoted very sparingly from their book in my review per FAIR USE copyright laws, since my primary purpose was to point you to their book and their claim that CHANCE cannot design and create anything, because CHANCE doesn't exist. I don't need to quote the whole book in order to get that point across.

—

My motto and deductive reasoning for this book came from Sherlock Holmes:

How often have I said to you that when you have eliminated the impossible, whatever remains, *however improbable*, must be THE TRUTH?
— Sherlock Holmes

The purpose and theme of this book has been to eliminate the IMPOSSIBLE, so that we are left staring at THE TRUTH.

Creation by Darwinian Chance or Creation by Evolution is IMPOSSIBLE, so we have to find another explanation for the origin of all the life on this planet. It's elementary my dear Watson!

Abductive Reasoning seeks for the best and most parsimonious interpretation or explanation for the available evidence. I also employ deductive reasoning. By eliminating ALL of the FALSE PREMISES, we can deduce the CORRECT CONCLUSION. It ALL points to God.

—

CHANCE is a non-person or a non-entity.

CHANCE DOES NOT EXIST.

CHANCE IS NOTHING.

When was the last time that you got NOTHING to do something for you?

The Darwinists have placed ALL of their faith, hope, and trust into Darwinian Chance or Macro-evolution, which doesn't exist.

The Darwinists have placed ALL of their faith, hope, and trust into NOTHING.

The Darwinists literally believe that NOTHING designed and created everything.

What are the odds that NOTHING could have designed and created everything in this universe? NOT A CHANCE! NOTHING by definition in principle does not exist; and, something that does not exist cannot design and create anything.

Goo-to-you evolution IS NOTHING. It does not exist, never existed, and will never exist. It would take a God to bring such a thing into existence.

The Theory of Evolution is NOTHING, because evolution of any kind cannot design and create anything at all. Evolution doesn't exist as a person or an entity. Mutation and Selection working together cannot design and create anything. Knowing that this is true is EVERYTHING that you really need to know about the Theory of Evolution. The Theory of Evolution is NOTHING, meaning that the Theory of Evolution or Creation by Evolution is FALSE. What are the odds that Evolution could have designed and created it all? NOT A CHANCE!

If you are a Darwinist, or a Materialist, or an Atheist, you have literally given your life away for NOTHING and turned your life over to NOTHING, receiving NOTHING in the return. What will you have accomplished from it all after your life is over? NOTHING! What do I have to show for My Atheism? NOTHING!

When you finally remove Darwinian Chance or Macro-evolution from the picture, you are left staring into the face of SOMEONE real and true, the Biblical God Jesus Christ, who actually says repeatedly to us that He is the One who designed and created everything in this universe.

So, which idea is more rational, logical, scientific, and believable — the idea that NOTHING designed and created it all or the idea that the Biblical God Jesus Christ designed and created it all? Which idea is the most parsimonious and fits best with Occam's razor? Which one is actually SUFFICIENT to the task of being a designer, creator, and causal agent? Which one is alive, and which one doesn't exist? If your life actually depended upon it which one would you choose to help you and save you?

If you are honest with yourself, then you just simply KNOW that NOTHING has nothing to offer you, because NOTHING doesn't exist and NOTHING can't do anything at all.

This little thought experiment is really SOMETHING, isn't it? Only an Atheistic Materialist will get NOTHING out of it.

—

The Darwinists employ Chance Mutations as their Designer and Creator. That's silly and illogical, because Chance cannot design and create anything at all! Employing Chance as a Designer and Creator IS a Category Error, which is a logic fallacy. The whole Theory of Evolution is based exclusively on these kinds of logic fallacies! Garbage in, then garbage out!

It's too late for me now. I can't go back to Atheism, Materialism, and Darwinism because I can't think of any way to defend them and promote them anymore.

—

Once again, I thank these individuals for being my teachers in this journey of exploration.

The purpose of THE SCIENTIFIC METHOD is to help us to find THE TRUTH, through a preponderance of the evidence.

THE SCIENTIFIC METHOD has no value to us if we use it to convince ourselves that a LIE is TRUE, as the Materialists and Darwinists always seem to do.

That's what I discovered during my Pursuit of the True Reality of All Things, and during my usage of THE SCIENTIFIC METHOD.

Evolution is dumb and blind without hands and a mind!

Each failure of Evolution to do what that Darwinists say that it does becomes yet another miniature proof of God's existence, because God MUST EXIST in order to do all of the Science and Creation that evolution or chance could NEVER have done.

Cool, huh?

THE SCIENTIFIC METHOD

It is said that Scientists, at least the good ones, seldom speak of proof — they always leave room for doubt or for further upcoming discoveries.

In contrast, the Darwinists and Materialists often say that they have proven the Theory of Evolution to be true, but all that really means is that they have convinced themselves that the theory is true. However, their alleged "proof" doesn't apply to the rest of us who KNOW that the Theory of Evolution is FALSE and why it is FALSE.

Still, the claim that nothing can be proven to be true is also false. For example, it can be proven to be true that if you keep sticking paper clips into electrical sockets, you are going to get electrocuted unless you take steps to protect yourself. It can also be proven to be true that if you stick your hand on a hot stove, you are going to get burned. There are some scientific claims that can indeed be proven to be true beyond a shadow of a doubt.

However, whenever the Scientists start saying that nothing can be proven to be true, what most of them (being Materialists) are trying to say is that God or Spirit cannot be proven to exist. They are right in one respect — there is NO way for one person to prove to another person that God exists. However, they are wrong in that God can appear to you or speak to you directly and thereby prove to you that He exists. In other words, the existence of God can be proven to you through an appearance of God to you. God's existence can also be proven to you through a preponderance of the evidence.

However, God doesn't seem to be in the habit of appearing to anyone and everyone; so, the question becomes, "Are there other ways to prove that God exists besides direct personal appearances to you?"

And now we come to an extremely important part of THE SCIENTIFIC METHOD and Scientific Methodologies — namely, demonstrable proof through a preponderance of the evidence.

There are many things that cannot be proven directly through empirical observation or scientific experimentation, yet they are still true and still exist. These are things like crime scene investigation, thoughts, dreams, altered states of consciousness, revelations from God, spirituality, out-of-body experiences, near-death experiences, psychic experiences, contact with the dead, and God. These things cannot be proven through replicable direct observation nor through scientific experimentation; but, they can indeed be proven through direct personal experience and a preponderance of the evidence, as long as we are willing to accept that evidence as being true.

For example, even though it is impossible to replay the murder and thereby prove who did it, it is indeed possible through a preponderance of the evidence to prove beyond all reasonable doubt who committed the murder.

Likewise, it is impossible to rerun the tape and observe who designed and created this universe and all the life on this planet; but, it is possible to demonstrate or prove through a preponderance of the evidence that it was the Biblical God who did the job, especially when He starts appearing to lots of different people confessing to having done the job. Confessions are admissible in a court of law.

—

I am a scientist. I study and try to employ Psychology and Philosophical Logic. Psychologists consider themselves to be scientists, and they use THE SCIENTIFIC METHOD in order to pursue and document their science. However, Psychologists and professors of The Science of Philosophy run into a difficult and insurmountable problem whenever they try to train their scientific instruments onto something like "Psyche" or "Spirit" or "Thought". These things by definition in principle cannot be detected, recorded, nor measured with physical instruments.

When dealing with the 'non-physical", the scientists are forced to abandon the Standard Materialistic Scientific Testing Methodology and forced to develop and to use different Testing Methodologies within THE SCIENTIFIC METHOD, in order to explore and do science whenever they are dealing with the non-physical. In other words, because of the existence of "Psyche" or "Thought" and other things non-physical like Light, the Scientists were forced to develop additional Scientific Testing Methodologies before they could then examine these things scientifically using THE SCIENTIFIC METHOD. Intrigued? I know that I was when I was first introduced to this scientific reality.

The best professional discussion regarding this scientific reality, that I have encountered so far, is in the book "Introduction to Psychology 9th Edition" by James W. Kalat. He is an evolutionist and tries his best to cater to the Naturalists, but he can't avoid this scientific reality and was forced to deal with it in his book. Chapter 2, "Scientific Methods in Psychology", is particularly revelatory.

I quote from page 29:

> "The philosopher Karl Popper argued that because no observation proves a theory to be correct, the purpose of research is to find theories which are incorrect. That is, the point of research is to falsify the incorrect theories, and a good theory is one that withstands all attempts to falsify it. In other words, it wins by a process of elimination. A well-formed theory, therefore, is falsifiable — that is, stated in such clear, precise terms that we can see what evidence could count against it — if of course such evidence existed. Falsifiable means that we can imagine something that would count as evidence against the theory. However, when Popper wrote that research is always an attempt to falsify a theory, he went too far."

END QUOTE.

Why did Karl Popper go too far?

It's because there are some things in this world that are not falsifiable. The existence of Mind, Psyche, Consciousness, or Spirit is not falsifiable nor is it detectable in a science lab. Therefore, some adjustments must be made to THE

SCIENTIFIC METHOD or Scientific Methodologies so that we can actually consider Psychology (the Study of the Human Spirit or Human Mind) to be a Science, and then do real Scientific Research into Psychology.

Karl Popper went too far by limiting science to Naturalism or Materialism, and ONLY to the things which can be falsified or proven false using our physical scientific instruments! Too much Materialism and not enough logic and science!

There are things in this world that cannot be falsified nor detected with scientific instruments; yet, we want to employ science in order to discover and identify and explain these things. That means that scientists had to develop new and different and more sophisticated Scientific Methodologies that could then be employed within THE SCIENTIFIC METHOD in order to explore the non-physical realities of our existence.

Psychology considers itself to be a Science; and, psychologists and philosophers like me consider themselves to be scientists. Do you see the problem here that we have to deal with? Probably not if you are a Naturalist or a Darwinist or an Atheist and have deliberately blinded yourself to the non-physical. Psychologists run into difficulties and effectively kill their Science dead if they deliberately limit themselves to Naturalism and limit themselves exclusively to materialistically falsifiable evidence.

Psychologists were forced to develop other Scientific Methodologies that they could then use to study "Psyche" or "Thought" scientifically using THE SCIENTIFIC METHOD.

There's no way determine what a person is thinking, with our scientific instruments. If you want to know what a person is thinking, you have to ask them. Our physical instruments can indeed detect brain activity, but they cannot detect what thoughts or experiences are taking place within that brain activity. We can't read or detect a person's mind or thoughts with our scientific instruments. Mind, or consciousness, or thoughts, or spirit are not detectable with physical instruments; and thus, they are not falsifiable, because they can't be detected in the first place. How can you falsify something if you can't detect it?

The original definition for Psychology was "the study of the spirit" or "the study of the mind". Literally, the study of the psyche. The human psyche is not detectable with our physical scientific instruments, so how can you study such a thing scientifically? How can you study the non-physical or the spiritual scientifically?

Well, first of all, you have to drop any pretense to Naturalism. Naturalism or Materialism is a bane to science — it stunts and kills scientific exploration and scientific discovery across a wide range of different sciences and scientific endeavors.

Second, you have to shift over to a different mode of scientific evidence and scientific discovery. You have to shift over to the "Burden of Proof" method of doing THE SCIENTIFIC METHOD, or the "preponderance of the evidence" method of doing THE SCIENTIFIC METHOD, whenever you want to discuss or pursue scientific discoveries regarding the non-physical or the spiritual realms, because you can no longer rely upon the ability to falsify the evidence using the various physical instruments at our disposal.

My psychology texts tell me that in order to study such a thing as spirit or mind scientifically, the scientists have to get very creative with their scientific experiments. They have to step up their game! In contrast, the Naturalists and Materialists simply run away and hide, by stating that Intelligent Design, or The Study of the Spirit, or the Study of Thoughts and the Mind, or the study of Consciousness, or the study of God are NOT SCIENCE and NOT SCIENTIFIC.

I can't count the number of times that I have heard the "NOT SCIENCE" claim from a Naturalist or a Darwinist or an Atheist. But, they really have nothing better to offer in defense of their beliefs. They have to "in principle" delete and deny anything that they can't detect with their physical instruments, because they truly believe that the physical is all that there is and all that exists. It's a very crippling and limiting worldview that the Naturalists, Materialists, and Atheists adopt, so that they don't have to deal with things like spirit, spirituality, intelligent designers, and God.

In contrast, the message which I get from the Intelligent Design people is that the only type of evolution that actually works is Intelligently Designed Evolution. Think about it, because it is true. The only kind of evolution that actually works is Intelligently Designed Evolution or Genetic Engineering. It's the Genetic Engineering which is the Real Science, and not Creation by Darwinian Chance.

The Naturalists, Materialists, and the Atheists try to take all of the fun out of life, and try to force us not to look at the spiritual side of life. After all, think about it logically, happiness, joy, pleasure, love, friendship, and feelings are to one degree or another triggered by non-physical thoughts. The effects or the results of those feelings can indeed be detected with physical instruments, but the precise thought that triggered them cannot. A person could in fact be exhibiting a great deal of happiness and joy while he is thinking about killing his worst enemy. Physical instruments cannot detect or determine the thought that is causing the happiness or joy. It could be anything! You never know what the person is really thinking about, unless he or she tells you honestly what he or she is thinking about.

So, how do psychologists study such a thing scientifically since it is impossible to get our scientific instruments to detect thoughts, or spirit, or consciousness, or mind, or psyche? How can you study these things if they can't be falsified? Those are questions that psychologists keep asking themselves all the time, because Psychology is a science and Psychologists consider themselves to be scientists.

On page 29 of "Introduction to Psychology 9th Edition" by James Kalat, we have this quote:

> "Instead of insisting that all research is an effort to falsify a theory, another approach is to discuss Burden of Proof, the obligation to present evidence to support one's claim. In science, the Burden of Proof is on anyone who makes a claim that should be demonstrable if it is true."

These claims about Psyche or Thought or Light or the Existence of God should be demonstrable if true. Burden of Proof is the obligation to present evidence to support one's claim! Burden of Proof or meeting one's Burden of Proof is a different type of Scientific Methodology or Scientific Testing within THE SCIENTIFIC METHOD. It is a legitimate Scientific Methodology that is used all the time in criminal cases and courts of law. Since we have no ability to read a prisoner's thoughts with our scientific instruments, we have to develop other ways of getting at the truth — we have to develop other Scientific Methodologies in addition to the "Falsifiable Materialistic Methodology". Burden of Proof is a very powerful and essential Scientific Methodology. The goal is to go with the Preponderance of the Evidence. Juries do so all the time. Whether a person is guilty or innocent is demonstrable, at least in theory.

I eventually learned that we can and we should be applying THE SCIENTIFIC METHOD to Religion, Spirituality, and our Chosen God just as much and just as often as we apply THE SCIENTIFIC METHOD to the physical realities of our lives. THE SCIENTIFIC METHOD loses much of its value unless we learn to apply it to our chosen Religion and our Spiritual Pursuits. We should also be applying Rationality and Logic to our chosen Religion and Spiritual Pursuits.

Intrigued?

Well, I certainly was the first time that all of this started to dawn on me.

I decided for myself that if I'm going to do that religion thing, then I'm going to go with the best and jettison all the rest. I eventually chose the religion that has the most depth, is the most interesting, makes the most demands, makes the most sense, and promises the most rewards. I haven't found anything else that even comes close to it. I have learned to go with the best and get rid of all the rest. I wanted the best for me and mine.

There are LOGICAL REASONS why billions of people believe in the Biblical God Jesus Christ. Billions of people believe in Him, because He has appeared to many of them and spoken to them, answered their prayers in noticeable and predictable and replicable ways, forgave them instantly when they repented of their sins and asked Him for forgiveness, and even came in person and rescued some of them and pulled them out of Hell.

Billions of people believe in the Biblical God Jesus Christ because He has PRODUCED RESULTS for them, come to their aide when they needed Him, helped them to break their addictions and overcome their sins, answered their prayers in timely and noticeable ways, and even come to them in person and pulled them out of Hell when they called upon His Name for help. Should you ever find yourself in Hell, then KNOW that the Biblical God Jesus Christ can get you out of there and into something infinitely better if you call upon Him to do so!

The Biblical God even shows respect for SCIENCE and THE SCIENTIFIC METHOD literally inviting us to EXPERIMENT on His Words, His Teachings, and His Commandments and to see for ourselves if our lives are better as a result. THE

SCIENTIFIC METHOD is all about PRODUCING REPLICABLE RESULTS! With the Biblical God Jesus, WE can literally take Him through ALL of the steps of THE SCIENTIFIC METHOD and get Him to PRODUCE noticeable and replicable and reliable RESULTS.

Billions of people believe in Jesus Christ and have faith in Jesus Christ, because HE PRODUCES RESULTS for them! He's not exactly hiding from us. He's there instantly whenever we call upon Him for help and really need Him for the salvation of our immortal soul. The words "Jesus Christ" mean "Anointed Savior", and HE literally lives-up-to His Name, if in fact it is your immortal spirit or your immortal soul that you are wanting Him to save. WE believe in Him because HE comes through for us! That's what you want — a Religion and a God that PRODUCE RESULTS for you.

A question that I love to ask myself from time to time is, "Has God met His Burden of Proof?" I believe that the Biblical God has.

Why?

Well, let's run this little thought experiment.

If you input the Bible, the Book of Mormon: Another Testament of Jesus Christ, the Doctrine and Covenants, and the Pearl of Great Price as the PREMISES, then it is logical to CONCLUDE that the Biblical God, our Resurrected Lord Jesus Christ, exists and really did rise from the dead and continues to appear to people in the flesh here and now in the physical world. The conclusion follows logically from the evidence or from the PREMISES.

That's deductive reasoning! If the PREMISES are true, then the CONCLUSION must be true also. The Scriptures that the Biblical God had a hand in writing and producing prove to us that He is real and that He truly exists, just as this book here proves to you that I am real and that I truly exist. The conclusion follows logically from the premises. That's a form of SCIENCE! It's a form of meeting one's Burden of Proof!

Cool, huh?

The Biblical God isn't exactly stupid, you know! He wants us to find Him and get to know Him; so, He deliberately left behind traces of His existence for us to find. Obviously, He did so on His own terms and not ours, but that doesn't change the fact that He has spoken to us and given us commandments which He wants us to find and obey, because He knows that keeping His commandments will eventually lead us back to Him and will also make our lives better as a RESULT in the meantime.

I believe that the Biblical God has met His Burden of Proof within the Bible, the Book of Mormon, the Doctrine and Covenants, and the Pearl of Great Price. You might choose to believe differently which is fine if it works for you, but it won't change what I have experienced first-hand and now believe and know to be true. Within those books, the preponderance of the evidence states, demonstrates, and proves that God exists. The existence of the Biblical God is demonstrable through the Scriptures and through the modern-day revelations of God and revelations from God!

Belief in God and a belief in God's existence are not falsifiable; but, they are indeed demonstrable through a preponderance of the evidence! The scientific prediction is that if you take God's Word to His Chosen Prophets seriously and treat His Word as truth, God's Truth, and keep His Commandments, then you will find God in the process.

Jesus Christ through the Book of Mormon even asks you to run that Science Experiment for yourself, a number of different places within that book. Jesus Christ makes the same request of us in the Bible, asking us to make an experiment on His words to see if obedience to His words produces desirable fruits as a RESULT.

Jesus Christ is the ULTIMATE SCIENTIST, and He actually asks us to EXPERIMENT with His teachings and His words. Experimentation is an essential part of THE SCIENTIFIC METHOD! God doesn't want us taking Him on blind faith. God wants us digging in and doing the scientific research, running experiments on His Words, and praying for ourselves so that we find Him for ourselves when we are done!

I believe that the Biblical God deliberately and knowingly meets His Burden of Proof through a preponderance of the evidence in the Bible, Book of Mormon: Another Testament of Jesus Christ, Doctrine and Covenants, and Pearl of Great Price. God employed a legitimate Scientific Methodology to prove to us through a preponderance of the evidence that He exists. He had to do so, or NONE of us would have ever heard of Him and NONE of us would ever know that He exists.

There's actually some demonstrable proof for God's existence. In contrast, there is NONE for the Theory of Evolution or Creation by Evolution.

God has to find some way to reveal Himself to us, or He would remain forever concealed and unknown. However, God's existence is not falsifiable. There's no way for us to trap God in a cage and then examine Him and falsify Him with our scientific instruments. God is not replicable in a science lab either. However, the existence of God is indeed demonstrable through a preponderance of the evidence.

This concept of Burden of Proof and the Preponderance of the Evidence changed the way that I look at THE SCIENTIFIC METHOD and Scientific Methodologies for the better. Whenever Scientists encounter something that isn't falsifiable or directly observable on demand, then the goal becomes to make it demonstrable through a preponderance of the evidence.

A lot of things in Psychology, Forensics, Archeology, Religion, Theology, and Crime Scene Analysis are not falsifiable or reliably replicable. However, they are demonstrable by a preponderance of the evidence. Police and Forensic Scientists examining crimes scenes have to meet their Burden of Proof through a preponderance of the evidence in order to convict a criminal of a crime. Scientists dealing with spirit or mind or "thoughts" have to meet their Burden of Proof also, through a preponderance of the evidence. The Biblical God deliberately went out of His way to meet His Burden of Proof by appearing to His chosen prophets and speaking to them, as recorded in His Scriptures — the Bible, Book of Mormon,

Doctrine and Covenants, and Pearl of Great Price. Even God had to meet His Burden of Proof, if He were to expect any of us to actually believe that He exists.

However, when it comes to some of the other areas of Psychology and Philosophy and Theology, it can be difficult (or impossible) at times for the scientists to meet their Burden of Proof, but that doesn't stop them from trying.

Hugh Ross PhD and his "Reasons to Believe" team have developed Scientific Proofs of God's existence, through a wide variety of different sciences and scientific evidence. I believe that they successfully meet their Burden of Proof from time to time. Their different books using science to prove that God exists can be a fascinating read. I have learned to trust Hugh Ross to give me a scientifically accurate interpretation of Biblical verses.

Gerald L. Schroeder PhD. does the same thing with some of his books, including one of his books entitled "The Science of God".

The Science of God is not falsifiable. However, it is demonstrable. It's based upon the pursuit of the preponderance of the evidence. Drop in sometime and read some of the books from Hugh Ross or Gerald L. Schroeder, and then decide for yourself whether they meet their Burden of Proof, or not.

When it comes to God, each person has to take a leap of faith — either to conclude that He does not exist or to conclude that He does exist. However, a study of the scientific evidence for God's existence, the philosophical evidence for God's existence, and the revelatory evidence for God's existence does indeed make it easier to be sure that your leap of faith is based upon True Evidence, rather than being based upon false evidence or no evidence.

Since everyone has to take a leap of faith when it comes to God, you want to make sure that you are taking the right one. A demonstrable preponderance of the evidence and THE SCIENTIFIC METHOD makes that possible. The Biblical God deliberately meets His Burden of Proof through the philosophical evidence, the scientific evidence, and the revelatory theological evidence.

There is NO evidence that proves that God does not exist. There can be NO evidence that proves that God does not exist — it's philosophically impossible. ALL of the evidence is on God's side, and proves through a demonstrable preponderance of the evidence that God exists. And, to some of us, the Biblical God will choose to make an appearance in the flesh; and, then we will have an even more direct and sure knowledge that God is real and truly exists.

Because the Biblical God deliberately met His Burden of Proof in His Scriptures and because God has deliberately left behind traces of Himself in Science and in Nature, we can in fact use THE SCIENTIFIC METHOD to demonstrate through a preponderance of the evidence that God does indeed exist.

Let the research and experimentation begin!

PART III — THE ULTIMATE MODEL OF REALITY

Introductory Note: This part entitled, "THE ULTIMATE MODEL OF REALITY", is PART I of my books, "Putting Psyche Back into Psychology: Restoring Science to Consciousness" and "The Ultimate Model of Reality: Psyche Is the Ultimate Cause". This part of those books serves as an introduction to those books. These books comprise my magnum opus – my primary contribution to philosophy, quantum realities, science, psychology, and the reality of our lived experiences as a race of human psyches.

My goal is to keep this part tight, as an introduction to psyche and ultimate causality; but, there is still a lot to cover, and there will be times when I will try to go in-depth and other times when I will duplicate some of what was said before and discuss it from a different perspective. In all things, it should point back to Psyche and a Psyche Ontology, when we are done. ALL of the evidence keeps pointing me to Psyche. It got to the point where I couldn't deny it anymore.

My thesis is that Psyche or Non-local Consciousness is the Ultimate Cause of everything that has ever been brought into existence or organized from scratch, including physical matter. In these books, I develop and present the Ultimate Cause Model of Reality for your consideration.

My Ultimate Personality Theory is that Psyche, Personality, Identity, and our Memories survive separation from our physical body, bodily death, and brain death according to the empirical evidence from Near-Death Experiences (NDEs), Out-of-Body Experiences (OBEs), Shared-Death Experiences (SDEs), and other types of Spiritual Experiences (Spiritual Empiricism and Lived Experiences).

In these books, "Putting Psyche Back into Psychology: Restoring Science to Consciousness" and "The Ultimate Model of Reality: Psyche Is the Ultimate Cause", I employ procedural evidence, validated scientific evidence, empirical evidence, experiential evidence, common sense, lived experiences, and knowledge as proof that Psyche is the Ultimate Cause of everything that is real and true, and as proof that Psyche is the Ultimate Cause of everything that has ever been brought into existence from scratch.

I do have a presence online:

The Associated Facebook Page:

https://www.facebook.com/MarkMyScience/

The Associated Twitter Page:

https://twitter.com/Mark_Me_Words

However, I don't participate all that much online nowadays because the Materialists and Atheists have labeled me as spiteful and vindictive and banned me from their websites. You don't last very long online with the message that I have to share with the world, before they find some way to shut you down and lock you out. It goes with the territory, because I'm not telling them what they want to hear. I'm telling them the truth instead; and, they don't like that.

Abstract – Psyche Is the Ultimate Cause

The purpose of this essay is to introduce a fifth cause or an Ultimate Cause into philosophy, metaphysics, science, application, lived experience, and psychology.

Aristotle proposed four main causes responsible for a physical object's existence – namely, **material cause** (the substance from which it is made or built), **efficient cause** (the motions needed to get it manufactured or built), the **formal cause** (the blueprint, plan, or design from which it was built), and the **final cause** (the purpose, goal, or reason for which it was built).

I introduce a fifth cause or an **ultimate cause**, which is the Person, Individual, Builder, Manufacturer, Mover, Agent, Contractor, Consciousness, Designer, Architect, Awareness, Meaning-Maker, Planner, Desirer, Observer, Intelligence, Life, Light, Soul, Spark, and Psyche – the personal, living, individual, immaterial, non-local consciousness **who** makes the ultimate decision or the final choice between competing alternatives. This fifth cause or Ultimate Cause is Psyche, or Living Supernatural Non-Local Immaterial Trans-Dimensional Quantum Consciousness. This Ultimate Cause is the Prime Mover, the first cause behind ALL other causes.

Ultimate Cause or **Psyche** is the originating cause behind every other cause and everything that has ever been ordered and organized, including the ordering or organizing of spirit matter, the calling into existence and organization of physical matter, the design and creation of physical universes, and the design, programming, engineering, and manufacturing of all genomes and life forms in a timely and efficient manner.

My goal in this essay is to present and develop The Ultimate Model of Reality. I want to subsume everything that is useful, efficacious, and true while at the same time eliminating the philosophies and pseudo-sciences that are not demonstrable nor evidential and have to be taken on blind-faith as being true. In order to make a case for Psyche as the Ultimate Cause, one has to understand how and why Materialism and Naturalism are false. The false models of reality have to be eliminated in order to make room for the True One or the Real One. As a scientist and a philosopher, I have been looking for such a construct all my life. I had to make my own model of reality, because as far as I can tell such a thing doesn't exist yet. I introduce this Ultimate Cause Model of Reality for your consideration.

Keywords: ultimate cause, psyche, psychology, psyche-therapy, philosophy, science, fifth cause, formal cause, final cause, NDEs, OBEs, SDEs, lived experiences, spiritual experiences, spiritual empiricism, radical empiricism, phenomenology, hermeneutics.

Aristotle's Four Causes and a Fifth

Aristotle proposed four main causes responsible for a physical object's existence – namely, **material cause** (the substance from which it is made or built), **efficient cause** (the motions needed to get it manufactured or built), the **formal cause** (the blueprint, plan, or design from which it was built), and the **final cause** (the purpose, goal, or reason for which it was built).

Aristotle's four causes are typically defined in terms of what it took to get a physical object made. In other words, they are defined in a materialistic and mechanistic fashion. So, imagine my surprise when someone on Google defined **efficient cause** as the manufacturer or the construction crew who puts the physical matter together into useful items – for example, the carpenter is the **efficient cause** of the table, cabinets, and chair. Well, that definitely messed things up, because suddenly we were talking about **who** made the chair and not just the physical processes involved in building the chair, while we were talking about the **efficient cause** of the chair. But, it certainly caught my attention.

I'm not the first person to realize that some kind of Psyche or Ultimate Cause must exist behind all science, manufacturing, engineering, production, and creation; otherwise, there would be no science experiments, no manufacturing, and no creation of new objects taking place. Philosopher Edward Feser employed this unique definition for **efficient causality** in his books:

> The *material cause* or underlying stuff the ball is made out of is rubber; its *formal cause*, or the form, pattern, or structure it exhibits, comprises such features as its sphericity, solidity, and bounciness. In other words, the material and formal causes of a thing are just its matter and form, considered as two aspects of a complete explanation of it. Next we have *the efficient cause*, that which actualizes a potency and thereby brings something into being. In this case that would be the actions of the workers and/or machines in the factory in which the ball was made, as they molded the rubber into the ball. Lastly we have the *final cause* or the end, goal, or purpose of a thing, which in the case of the ball might be to provide amusement to a child. (Feser, *Aquinas*, p. 16).

After such repeated input from many different sources, I found myself adjusting or tweaking the definitions for Aristotle's four causes as follows:

Aristotle proposed four main causes – namely, **material cause** (physical matter), **efficient cause** (the manufacturer or construction crew who puts the physical matter together into useful items and forms), the **formal cause** (the blueprint, plan, or design), and the **final cause** (the purpose, goal, or reason for choosing to act, or for choosing to manufacture an item and choosing to use an item).

Then on my own, as my research continued, I just kind of automatically found myself extending the "**person**" or the "**who**" or the "**personality**" to formal cause and final cause, because it was so easy to do. Thus, **formal cause** became the Designer or the Architect behind the blueprints, plan, or design. And, **final cause** became The Person or The Agent who wanted an item manufactured, for whatever reason or purpose that this person had in mind. In this manner, my concept of **ultimate cause** was born, because ultimate cause is the person or the psyche **who** is behind the efficient cause, the formal cause, and the final cause of any item, idea, or thought that has ever been manufactured or brought into existence.

It dawned on me one day like an epiphany that there is an **ultimate cause** (or psyche) behind each one of Aristotle's four causes, and suddenly it was clear to me that we needed a fifth cause. Ultimate cause explains **who** is responsible for Aristotle's four causes. Psyche or ultimate cause has always been there behind everything that has ever been designed, manufactured, and brought into existence; but, **ultimate cause** had never been defined nor employed like it should have been. I figured that it was time to rectify the situation, because the empirical evidence for the existence of psyche is literally exploding across this world right now. (Buhlman, 1996; Gibson, 2006; & Rivas, 2016).

The only thing that was lacking was that I didn't know how **material cause** fit in with **ultimate cause**, at first. This required a bit more thought. But, I knew that it must be so.

Eventually, based upon the need for a Conscious Observer in Quantum Mechanics in order to convert spiritual quantum waves into solid physical matter, it all became clear to me; and, I found myself applying **ultimate cause** or psyche to physical matter and **material cause** because of Quantum Mechanics or Spiritual Mechanics; and thus in my Ultimate Model of Reality, **material cause** became **The Person** or the Conscious Observer or the Psyche **who** collapses the quantum wave function thus changing a quantum object from some kind of non-local spiritual wave (or spirit matter) into a physical particle located here in our 3D physical space-time. According to Quantum Mechanics, Psyche or Non-Local Consciousness must of necessity exist in order to bring physical matter into existence in the first place. Without Psyche, there would be NO physical matter. This is what Science is trying to teach us! But, few people are actually listening.

Ultimate cause or a conscious observer is the only thing that can convert spirit matter into physical matter, according to the Science of Quantum Mechanics. Thus, I had finally identified the **ultimate cause** behind every material cause and behind physical matter itself – psyche or the conscious observer. Ultimate Cause, the material cause and efficient cause of physical matter, is the Conscious Observer or Psyche which Quantum Mechanics tells us must exist. (Van Lommel, 2010). According to Quantum Mechanics, if Psyche didn't exist, then there would be no physical matter!

After reading Pim van Lommel's book, *Consciousness Beyond Life: The Science of the Near-Death Experience*, for the first time in my life I finally understood Quantum Mechanics in a way that actually made logical sense to me – from a non-local, non-physical, spiritual perspective. I finally had something useful to build upon. If you are going to read one book about Quantum Mechanics, that's the book to read. Quantum Mechanics never made any sense from a physical or materialistic perspective; but, Quantum Mechanics makes a whole lot of sense from a non-local or a spiritual perspective. (Mark My Words, 2016, *Quantum Mechanics from a Non-Physical Spiritual Perspective*).

In consequence of all of this, within this Ultimate Cause Model of Reality, the **material cause** sort of merged with **efficient cause** into a combined **mechanistic cause** and became The Person or The Psyche **who** observed or called all the physical matter into existence in the first place and **who** organized or manufactured physical matter into computer components, plastic, woodworks, car parts, bricks, cement, lumber, steel, candy, genomes, and the other material things that went into the construction and manufacture of the physical objects which are currently around us. Thus, **material cause** became not only the substance or essence from which something is made but also **The Person** or the psyche behind the organization and the origination of that substance or essence, including planets, stars, galaxies, and physical matter. Ultimate cause is the person or the psyche behind everything that has ever been designed, organized, built, manufactured, and created including physical matter and physical universes. Ultimately, there is an **ultimate cause**, or psyche, or person behind each and every material cause, efficient cause, formal cause, and final cause.

Ultimate cause has always been there behind each and every one of Aristotle's four causes waiting to be found, identified, and used; but, ultimate cause is typically taken for granted and completely ignored as if it doesn't even exist. In fact, the Materialists and Atheists don't want psyche or ultimate cause to exist. Ultimate Cause or Psyche is the antithesis of Materialism and Naturalism. The existence of Psyche or ultimate cause actually proves Materialism and Naturalism false. Psyche by definition in principle is supernatural. Ultimate Cause and Naturalism are mutually exclusive. If one can be demonstrated to be true, then the other one has been demonstrated to be false. Now I finally had something upon which to build.

Eventually, Aristotle's four main causes came to mean – **material cause** (the person who brought the substances, building materials, or physical matter into existence or into some kind of order if they already existed), **efficient cause** (the person who organized the spiritual substances, building materials, and/or physical matter into useful and interesting forms and objects and life forms), the **formal cause** (the person who designed and created the blueprints, plans, software, and genomes), and the **final cause** (the person who chose to act on the plan in order to get things done for whatever purpose he or she desired).

I eventually started to interpret Aristotle's four cause in terms of **ultimate cause** – namely, the person or the psyche who did all of these different things. Our philosophical theories are supposed to be useful and compelling, after all; and, I find this Ultimate Model of Reality useful and compelling. For me, this became the Ultimate Personality Theory and the Ultimate Ontology, a Psyche Ontology wherein Psyche is the fundamental unit of reality.

So, there you have an explanation for the origin of this Ultimate Cause Model of Reality and this Ultimate Cause Personality Theory. I eventually observed that there is an **ultimate cause** or psyche behind each one of Aristotle's four causes; and thereby, I realized that there are excellent reasons for adding a fifth cause onto Aristotle's four causes. Ultimate cause or this fifth cause answers the question of "**who**" did the other four causes. Psyche or ultimate cause can do science, design things, do programming, manufacturing, construction, engineering, thinking, writing, language, philosophy, and creation; whereas, the rocks or raw physical matter cannot. That's what Science is trying to teach us.

For thousands of years, the Philosophers and the Materialists have focused all of their attention on the material thing or the material object which was the cause or the reason for something's existence. But, Materialism is dying and coming to an end; so, it is time for a new focus, or a new paradigm, or a new model of reality. Without even realizing it, this world is slowly switching over to psyche or ultimate cause, within Science itself. It was doing so without me; and now, it can do so with me. There is a paradigm shift taking place right now, but it goes mostly unnoticed because it has been slow in coming. (For an example see, Tart, 2009, *The End of Materialism: How Evidence of the Paranormal Is Bringing Science and Spirit Together*. Tart's book was over 50 years in the making). Many people are ten or twenty years ahead of me in making this paradigm shift within their theorizing and scientific research; but, I finally got there in the end which is all that really matters.

Thanks to these recent insights, I have started to define Aristotle's four causes in terms of **ultimate cause**, as The Person or The Psyche **who** is behind each and every one of Aristotle's four causes. So, I find myself asking, "**Who** was the substance maker or substance organizer, and thus the true **material cause** behind this object or this particle of physical matter?" Or, "**Who** was the manipulator or manufacturer or true **efficient cause** behind this finished product?" Or, "**Who** was the Architect or Designer and thus the real **formal cause** behind this blueprint, plan, computer program, life form, or genome?" Or, "**Who** was the person or the psyche or the **final cause** who desired and requisitioned the final product, or the person **who** chose to bring this concept or idea into existence in the first place?"

In other words, I have redefined Aristotle's four causes in terms of psyche or **ultimate cause**, thereby creating a new, more expansive, more robust, and all-inclusive personality theory and model of reality in the process – the Ultimate Personality Theory and the Ultimate Cause Model of Reality. Ultimate cause attempts to answer the question of **who** performed or **who** is behind the material cause, the efficient cause, the formal cause, and the final cause – a question which the Materialists and Naturalists refuse to ask.

Psyche Is the Ultimate Cause

Psyche is the ultimate cause.

302

Personality is psyche; and, psyche or personality is the part of us that survives bodily death and brain death according to the Empirical Sciences of Near-Death Experiences (NDEs), Shared-Death Experiences (SDEs), Out-of-Body Experiences (OBEs), and other spiritual experiences (Spiritual Empiricism or Lived Experiences). As I see it, this is the Ultimate Personality Theory, an Ultimate Cause Personality Theory. This is a personality theory that will actually survive one's bodily death and continue to be true in the afterlife.

I had cause for developing this Ultimate Model of Reality; and, I ended up doing it for a good cause. I was trying to get at the ultimate truth of all things, as any good scientist should. I was looking for the meaning of Life, the Universe, and Everything; and, it ended up being a whole lot more than just 42. It ended up being Psyche or Ultimate Cause.

Ultimate Cause or Psyche is the reason for our existence; and, our personality or psyche will continue to exist and continue to thrive long after our physical body and physical brain are dead and gone, according to the empirical evidence from NDEs, SDEs, OBEs, and other spiritual experiences.

This Ultimate Cause Model of Reality or Psyche Ontology is a holistic model of reality based upon the Lived Experiences of the human race. In order to make it possible, though, one has to be willing to get rid of ALL the exclusionary philosophies and exclusionary models of reality such as Materialism, Naturalism, Scientism, Nihilism, Darwinism, and Atheism – the models of reality which exclude things like psyche, spirit matter, and non-locality from consideration. Exclude the exclusionary, and suddenly you find yourself staring at the truth!

I observed and discovered that one has to exclude the Exclusionary Philosophies if he or she really wants to get at the truth of our existence. I also discovered during my research that Truth and Knowledge are comprised of our Lived Experiences, including our out-of-body experiences, theophanies, and spiritual experiences.

Ultimately, Truth and Knowledge are based upon the Lived Experiences of the human race and the human psyche, wherein people experience Reality directly for themselves. Experiencing Reality should take precedence over philosophical and scientific interpretations of reality.

Why?

In Philosophy and Metaphysics and Scientific Interpretation, they guess at the Truth. That's not a good foundation upon which to know the truth. It's much better to live the truth and experience the truth directly for yourself, or to choose to trust someone who has done so.

Furthermore, it has been shown that it is impossible to use the scientific methods to prove the Truth, because scientific methods begin and end with philosophical assumptions and philosophical interpretations of the data, and because scientific methods are based upon the *affirming the consequent* logic fallacy whereby it becomes possible to make almost any interpretation of the scientific evidence fit with any piece of scientific data that we produce through experimentation. Scientific methods cannot be used to prove the Truth, which means that scientific methods are no better at getting at the Truth than philosophical speculation is, because scientific methods are based exclusively upon the philosophical interpretation of scientifically derived evidence.

We can't get at the Truth through Philosophy, Science, and the Scientific Methods.

The ONLY way to KNOW the Truth is to experience the truth and live the truth directly for ourselves, or to trust the people who have done so. The ONLY way to KNOW the Truth is through Lived Experience or Psyche Experience, because Psyche is the only thing that can do living, and experiencing, and the knowing of Truth. A Psyche Ontology is the Ultimate Ontology because only the human psyche can do ontology, metaphysics, philosophy, and science. Furthermore, only Psyche can do Lived Experiences, because our psyche is the only part of us that's truly ALIVE, in every sense of the word.

Psyche is truly the Ultimate Cause!

Introduction to Ultimate Cause

In this essay, I develop my own Theory of Personality, which becomes an Ultimate Cause Personality Theory; and, I also develop the Ultimate Model of Reality which I call the Ultimate Cause Model of Reality or a Psyche Ontology. This Ultimate Model of Reality is the antithesis of Scientific Naturalism and Materialism, and therefore a fully complete synthesis with Science, Philosophy, Psychology, Observation, and Lived Experience as a whole. This Ultimate Cause Model of Reality becomes the full embodiment of Science – no limitations and nothing excluded from consideration, as Science should be.

It is my thesis that Psyche is Personality, and that Personality is Psyche; and, Psyche or Consciousness is the Ultimate Cause behind everything that has ever been created, organized, or brought into existence from scratch. Psyche is Intelligence, in other words Light, Knowledge, and Truth. Psyche is life. Psyche is the fundamental unit of our existence and our reality; and therefore, Psyche is the fundamental unit of all Knowledge and Truth, because ONLY Psyche can do knowledge, truth, and lived experiences – and then remember these things in order to be able to use them and share them with others.

Personality or Psyche is the part of us that survives separation from our physical body, separation from our spirit body, death of our physical body, and brain death according to the Empirical Sciences of Near-Death Experiences (NDEs), Shared-Death Experiences (SDEs), Out-of-Body Experiences (OBEs), and Spiritual Experiences (Spiritual Empiricism). (See Alexander, 2012: Buhlman 1996; Durham 1998; Gibson, 2006; Hinze, 1997; Long, 2016; Moody, 1975; Moody, 2010; Neal, 2011; Ring, 1999; Rivas, 2016; Sharkey, 2008; Storm, 2000; and van Lommel, 2010).

Quantum Mechanics tells us that the human psyche or the conscious observer must exist in order for physical matter to be brought into existence. (See van Lommel, 2010). NDEs, SDEs, OBEs, and Spiritual Empiricism tell us that the human psyche or non-local consciousness does exist. (See Buhlman, 1996; and Rivas, 2016). A strong Personality Theory requires a strong empirical base. Follow the evidence. That's what I finally chose to do.

My Ultimate Personality Theory is that psyche or personality is the part of us that survives brain death and bodily death, according to the empirical evidence from NDEs, SDEs, OBEs, and other spiritual experiences. Psyche is the ultimate memory storage device, because the physical body doesn't have lived experiences and doesn't make memories while the Psyche is away from or separated from the physical body, according to the Lived Experiences of the Out-of-Body Travelers.

The major premises of Materialism and Scientific Naturalism have no evidence supporting them and will never have any evidence supporting them; whereas, their antitheses "Ultimate Cause" and "Psyche as the Ultimate Cause" have tons of empirical evidence supporting their truthfulness and usefulness.

Theologians often talk about a Prime Mover, an Uncaused Cause, or a First Cause. These are subsumed by Ultimate Cause. **Psyche is the ultimate cause** behind everything that has ever been brought into existence from scratch, including physical matter and physical universes.

This Ultimate Personality Theory involves psyche or ultimate cause. After studying Personality Theories quite extensively, it is my conclusion that this fifth cause or Ultimate Cause is necessary to explain what's truly going on, in all the realities of our existence. This Ultimate Cause is in fact The Cause which the majority of the naturalistic theoreticians deliberately choose to avoid and reject, a decision and limitation that makes their theories and philosophies incomplete and unsatisfying in the end.

Psyche or Ultimate Cause is the choice-maker, the meaning-maker, the final-cause generator, the deal-breaker, and the tie-breaker affirming or rejecting each and every alternative which comes its way. Psyche, or Ultimate Cause, or Non-Local Consciousness is what brought ALL of the other causes into existence in the first place. Psyche by definition in principle can negate a decision, an opportunity, or a choice by choosing to say "No". That's what makes psyche so unique and powerful. Psyche can choose not to respond to a stimulus. Raw physical matter cannot. Physical matter is obligated to obey and obligated to respond to any stimulus that comes its way. Psyche is NOT!

The Ultimate Cause – immaterial psyche, or consciousness, or intelligence – designed and created the first particle of physical matter and brought order (efficient cause) to pre-existing spirit matter (the 'material cause' of the spirit realm or quantum realm), became the manufacturer behind the physical universe and the ordering (efficient cause) of physical matter into useful items and forms such as planets and stars, produced the various different consensus realities in the spirit realm or non-local realm (efficient causes of a spiritual nature), designed the blueprints, plans, and software behind all genomes and life forms and physical items on this planet (formal cause), and purposefully chose to set this whole plan into motion for its own reasons, purpose, and goals (final cause and teleology).

Ultimate Cause or Non-Local Consciousness is, well, the ultimate cause behind all of the other causes, whether we are talking about this physical 3D space-time realm or the non-local trans-dimensional quantum realm. Everything that has ever been organized, ordered, created, or brought into existence (including physical matter) was brought into existence by some kind of non-local ultimate cause or non-local consciousness, according to the Science of Quantum Mechanics.

According to Quantum Mechanics, psyche or consciousness or ultimate cause is non-local, non-physical, trans-dimensional, immaterial, supernatural, and spiritual in nature and origin. I'm going with the Science on this one, and not the materialistic or naturalistic philosophies. Materialism is based upon a denial of Reality and a refusal to look at contradictory evidence; whereas, this Ultimate Model of Reality is based upon a full embrace of the existential experiences of the whole of mankind. Every experience and truth reported by human beings throughout the whole of recorded history comes under the purview of this Ultimate

Cause Personality Theory and this Ultimate Model of Reality, or at least that's its purpose, goal, and intent. ONLY Psyche can do life, experience, knowledge, and truth.

I eventually decided to take people at their word whenever they report a spiritual experience or a supernatural experience to me, which is something that the Materialists and the Scientific Naturalists refuse to do. I'm no longer an Atheist, and I'm no longer a Materialist. I am now willing to look for evidence of Psyche and Ultimate Cause. My eyes have been opened, and I have seen the light. My goal here is to include or to subsume every truth, every experience, and every reality that I can find into this Ultimate Model of Reality. It's existential and experiential. It's based upon the Lived Experiences of the human race.

This Ultimate Cause Model of Reality attempts to be all-inclusive and attempts to include everything that we humans have ever experienced or thought about, throughout the whole of human history. All my life, I have been looking for the true model of reality, a model which will explain everything and lay a foundation for everything which human beings (human psyches) have ever experienced, created, manufactured, or thought about. I believe that I have finally found what I was looking for, even though I had to create it for myself.

Our Experiences of Reality do and should take precedence over our philosophical speculation and our interpretations of any scientific evidence. In other words, our philosophical guesswork and our interpretations of scientific data should be forced to match with Reality or forced to match with the Lived Experiences or Psyche Experiences of the human race, rather than the other way around as the Materialists and Naturalists and Atheists do with their science and philosophy.

Materialism and Naturalism violate and deny the existence of the Lived Experiences or the Spiritual Experiences of the human psyche. That's Bad Philosophy, Bad Form, and Bad Science – the refusal to let eye-witness testimony into evidence, or the refusal to permit personal experiences into evidence. Materialism, Naturalism, and Atheism of any kind are based upon a refusal to look at the evidence. Materialism and Naturalism are Exclusionary Philosophies and are based upon a Denial of Reality. These exclusionary philosophies should be excluded from Science and Philosophy, especially if we want to experience Truth and Knowledge for ourselves.

Truth and Knowledge are based upon our Lived Experiences wherein we (our psyches) experience Reality directly for ourselves; and therefore, a Lived Experience Epistemology or a Psyche Epistemology is vastly superior to any epistemology that is based exclusively upon the scientific methods or philosophical guesswork. ONLY Psyche can do epistemology, ontology, lived experiences, truth, knowledge, science, philosophy, and reality. The rocks, which the Materialists and Naturalists idolize and worship, need not apply.

The Best Example Which I Currently Have of This Ultimate Reality

Empirical Knowledge, Experiential Knowledge, or Lived Experience is superior to scientific hypotheses, materialistic interpretations of scientific data, and philosophical guesswork.

A single Lived Experience is of more value to me than all of the philosophical speculation in the world combined. The materialistic and naturalistic philosophers tell us that psyche and spirit do not exist; however, the Lived Experiences of the human race tell us that the Materialists and Naturalists are wrong.

What was there in existence BEFORE the first particle of physical matter was designed, created, manufactured, and brought into existence? WE WERE! Our Non-Local Consciousness, Psyche, or Personality was there BEFORE the "Big Bang" or Prime Event was used to bring this physical universe and physical matter into existence. Everything that has a beginning has a Beginner, an Ultimate Cause, who set it in motion. It is illogical to think otherwise. Only something like Psyche or Non-Local Consciousness has always existed and will always exist. ONLY Psyche can act as the ultimate cause of physical matter. ONLY Psyche is infinite in nature and scope. Psyche resides in the realm of the infinite.

According to the Science of Quantum Mechanics, every particle of physical matter required an Observer, or a Non-local Consciousness, or an Ultimate Cause in order to bring it into existence. In other words, according to Quantum Mechanics or Spiritual Mechanics, every particle of physical matter had a beginning; and thus, had a Beginner or an Ultimate Cause behind its construction or its transformation from a quantum wave or a spiritual wave into a physical particle. That's what the Science of Quantum Mechanics is trying to tell us, which explains why the Materialists and Naturalists are unable to understand and accept Quantum Mechanics at face-value. Quantum Mechanics is Spiritual Mechanics, or the way that spirit matter really works.

I use William Buhlman's out-of-body experiences as Empirical Evidence and Living Proof that a Psyche Ontology is in fact the true reality of our existence. There is indeed a Fifth Cause or an **Ultimate Cause** behind ALL things which exist and behind ALL life forms; and, this **Ultimate Cause** is Psyche, or our Non-Local Immaterial Consciousness, or Intelligence.

William Buhlman Wrote

Journal Entry, October 2, 1982

I hear the buzzing, engine-like sounds and will myself out-of-body. [He left his physical body behind.] I step to the bedroom door and automatically request "Clarity now!" My vision improves and I step through the door, into the living room. Still feeling a little out of sync, I verbally repeat my request with more emphasis, "Clarity now!" I feel my awareness and vision snap into place.

My thoughts are clear, and I make a verbal demand, "I need to see the form I'm in now!" Instantly I feel an intense sensation of being drawn within myself. I'm suddenly different, weightless as though I'm floating in space. As I look forward I see a sparkling, bluish white form. For some reason, I seem to know that I'm looking at my nonphysical body from a different perspective. I stare in amazement at this form before me that shines and flows with energy and light. It looks like an energy mold created from a million tiny points of light; it radiates a bluish glow but appears to have a defined outer structure. The body of light before me is naked and is identical to my physical form. Even though my body looks firm, there is a noticeable energy motion and radiation present. I can see what appears to be an ocean of blue stars throughout my body. It's difficult to describe because the stars are stable yet moving at the same time; the light and energy of my [spirit] body appear to change and flow almost like the waves of an ocean.

As I stare at the body of light, it hits me that I must be in another body. Yet I can't perceive any form or substance; I'm like a viewpoint in space without shape or form of any kind. [He, his immaterial viewpoint in space, is pure psyche or intelligence or consciousness.] As I reflect upon my new state of being, I feel a sensation of rapid motion and I'm instantly back within my physical body.

Lying still and reviewing my experience, I'm struck by an inescapable conclusion: I must possess multiple energy-bodies. The form I just experienced was noticeably lighter (less dense) than even my second nonphysical body. I realize that the traditional view of our possessing two bodies — a physical body and a spiritual body — is far too simplistic; we are much more complex than this. Just as there are multiple nonphysical energy dimensions within the universe, each of us must consist of multiple energy-bodies or vehicles of expression.

Now I seriously wonder just how many nonphysical bodies or forms this involves. I suspect that there must be one within each dimension of the universe and that all of these are interrelated and connected, just as the physical body is connected to its first nonphysical (spiritual) body.

Buhlman, William L. (1996). *Adventures Beyond the Body: How to Experience Out-of-Body Travel* (pp. 34-35). New York: HarperCollins. (HarperCollins permits quotes from their books in Critical Reviews and Promotional Reviews. I definitely promote this book any chance I get).

Notice that it's the immaterial viewpoint in space who does all of the thinking, observing, choosing, planning, feeling, and living – not the spirit body which he was looking at, and certainly not his physical body which he left behind in his bed. Only some kind of psyche or immaterial non-local consciousness can do teleology and life. Only some kind of psyche can do purposeful, deliberative, meaningful choosing, planning, decision-making, final causality, design and creation, and agentive moral action. It's that immaterial psyche, spark, or viewpoint in space who is alive and doing all the observing and thinking and choosing, not the spirit body and not the physical body.

William Buhlman's Psyche, Personality, Intelligence, or Non-Local Consciousness is the immaterial viewpoint floating in space looking at his most refined or inner-most Spirit Body or Non-Physical Body. It's this Immaterial Viewpoint, Intelligence, or Non-Local Consciousness which is in fact the **Ultimate Cause** of ALL things that exist, including the first particle of physical matter that was designed and created (material causes), the manufacturers who build or organize new things out of raw materials of both a spiritual and a physical nature (efficient causes), the (formal cause) designs and blueprints and computer programs and genomes, as well as the (final cause) purposes, goals, reasons, choices, intentions, affirmations, and ultimate decisions and destinations.

Personally, I don't believe in the parallel universe theory, wherein there are multiple versions of me doing different things. I personally believe that our single Psyche and single Spirit Body can manifest in all the multiple nonphysical energy dimensions within this universe. In other words, I believe that we have one spirit body and one psyche, but that these things can function in and interact with ALL the dimensions of this universe. They adjust as needed. Our Psyche and Spirit Body can even have a presence within and influence upon this Physical Dimension and our Physical Body.

The Psyche or Non-Local Consciousness is the **Ultimate Cause**, because the Psyche brings ALL of the other Causes into existence. The Psyche or Personality or Consciousness is the part of us that survives separation from the physical body, separation from the spirit body, physical death, and brain death according to the Empirical Science of Out-of-Body Experiences and Near-Death Experiences. I rely upon the Lived Experiences of the human race in order to establish the truth and validity of this Ultimate Model of Reality. I require evidence, lots and lots of evidence. I'm no longer willing to take things on blind-faith, like I used to do when I was a Materialist and an Atheist.

I own literally hundreds of different books about Near-Death Experiences (NDEs), Shared-Death Experiences (SDEs), Out-of-Body Experiences (OBEs), Quantum Consciousness, Lived Experiences, and Spiritual Experiences (Spiritual Empiricism) including most of the books listed by Gibson in *They Saw Beyond Death: New Insights on Near-Death Experiences*; and, William Buhlman's book is the one I chose to quote first in support of my thesis or theory about an **Ultimate Cause**, a "fifth cause", a First Cause, or a Primal Cause. Buhlman was the first person to convincingly teach me that our Psyche is completely separate from our Spirit Body.

Psyche, personality, individuality, intelligence, awareness, intentionality, and non-local consciousness is the **ultimate cause**. I have my own personal library which I have been reading from and drawing upon, possibly as many as 7,000 different books, mostly digital in nature, but hundreds of them are hardback and paperback. Collectively, these books not only testify that psyche exists, but these books also testify that psyche must of necessity exist in order to successfully explain all of the different various phenomena which the human race has experienced and observed throughout the whole of human history.

Psyche is the ultimate causal agent; yet, I refused to look at any evidence for psyche and God while I was a Materialist and an Atheist, and thus I truly believed that there wasn't any evidence supporting the existence of God and psyche, and that there never would be. I was wrong. Self-deception works, and it works every time! However, the books I now have on hand and hundreds of YouTube videos have met their burden of proof and successfully convinced me that psyche and God do indeed exist; and, the evidence just keeps pouring in. I choose to believe that what the evidence is telling me is real and true; and, one of the things which these books and videos are telling me is that Materialism and Scientific Naturalism are false. (See Richards & Bergin; & Goswami). Ultimately, I chose to believe in the Lived Experiences of the human race rather than the philosophical guesswork of the Materialists, Naturalists, and Atheists.

Materialism is the pinnacle of self-deception and the worst form of prejudice. Materialists, Naturalists, and Atheists are anti-psyche, anti-belief, anti-trust, anti-Christ, anti-God, anti-science, anti-reality, anti-rationality, and anti-discovery. The hard-core dedicated Materialists and Naturalists never make any ground-breaking paradigm-shifting scientific discoveries because these people refuse to ask the questions which lead to new scientific discoveries. The non-local sciences, or the spiritual sciences, or the quantum sciences are the final frontier in Science; yet, the Materialists and Naturalists refuse to go there.

My personality theory is based upon **ultimate cause** or psyche. And, based upon the empirical evidence which I now have at hand, it is my theory and belief that personality or psyche is the part of us that survives bodily death and brain death according to the Empirical Sciences, the lived experiences, and the empirical evidence from NDEs, SDEs, OBEs, and Quantum Mechanics.

This Ultimate Cause Model of Reality is based upon the collective Lived Experiences of the human race. The ONLY way to KNOW the Truth is to live it and experience it for yourself, or to choose to trust someone who has.

Again, that immaterial and formless viewpoint in space, or non-local consciousness, or non-local psyche, or immaterial intelligence and awareness which Buhlman mentioned in his personal out-of-body experience is **THE ULTIMATE CAUSE** of everything that has been ordered, organized, or brought into existence including spiritual consensus realities, physical matter, stars, planets, genomes, life forms, this physical universe, and the physical objects which have been engineered and manufactured.

A spirit body localizes a Consciousness or Intelligence or Psyche there in the spirit realm or the quantum realm; and, a spirit body gives us an identity, a form, and a heritage. It is clear from Buhlman's OBE that his spirit body is in the image of the Gods, which means that his spirit body is one of the spirit children of the

Gods. That's quite a heritage, to have all the potential of a God. (Richards and Bergin; Holy Bible; & Bishop, 1998).

As most of us already know, our physical body massively limits us to sub-light speeds. Because our physical body has physical mass, there is no way to accelerate it to the speed of light and beyond. Physical matter by definition in principle exists at sub-light-speed velocities – if you accelerate physical matter faster than the speed of light then it becomes spirit matter or a non-local quantum object, potentially infinite in frequency and scope and potentially instantaneous in velocity. Sub-light-speed limitations don't apply to the immaterial or the spiritual, such as our spirit bodies and our non-local consciousness. Spirit matter or quantum objects in their wave-like format exist at velocities greater than the speed of light, which is why we can't capture them nor detect them with our physical instruments. Quantum objects or spirit matter have to be converted into physical particles by conscious observation before we are then able to detect them with our physical instruments. This is what Quantum Mechanics or Spiritual Mechanics is trying to teach us.

Most of our physical bodies here on earth are mortal, although there are indications that the Gods who have proven themselves trustworthy have access to physical bodies that are immortal, trans-dimensional, and glorified in some way. Of course, a physical body is the best way to experience a physical reality and to go through this physical-schooling process in what it's like to have some real limitations. A physical reality is the ultimate consensus reality, and a mortal physical body is the ultimate experience in what it's like to have some real limitations.

Physical consensus realities do have their advantages, though. Here in this physical consensus reality, I can count on this paper being there on my flash drive and hard drive tomorrow when I turn on my computer and go looking for it. I can count on my land and some version of my house being there when I drive home from work today; and, I'm sure that my dog is going to be there wanting some attention and love. Due to all the limitations of a physical reality, a semi-sense of permanence, stability, control, and predictability comes from a physical reality or a physical consensus reality, which you don't necessarily get from a spirit reality or a quantum reality, even a spiritual consensus reality which was put together by a community of spirits. A physical reality is the ultimate consensus reality. Buhlman in his book was the first person to teach me about consensus realities and non-consensus realities.

I have observed that all of the unsolvable "problems" created by Materialism (such as the mind-brain problem, the nature vs. nurture problem, the binding problem, and the free-will vs. determinism problem) are instantly solved by getting rid of Materialism, Determinism, and Scientific Naturalism and replacing them with some type of teleology, quantum non-locality, final causality, agency, psyche, quantum consciousness, ultimate causality, and a Psyche Ontology. Materialism and Naturalism are archaic, primitive, useless, and worthless in comparison to all of this. You won't get any information about nonlocality or spirituality or non-consensus realities from a Materialist or an Atheist, because they have nothing like

312

this to give you. That has been my scientific observation and my personal experience.

Now you know one of the reasons why I am no longer a Materialist and no longer an Atheist. I now own hundreds of different books about NDEs, OBEs, Quantum Consciousness, and other types of spiritual experiences. The empirical evidence, which I have on hand of Psyche as the Ultimate Cause is so overwhelming, that I no longer have blind-faith enough to be a Materialist or an Atheist. It's too late for me now. I can't go back to my ignorance; and, why should I want to? Quantum Immateriality, Infinite Singularities, Quantum Consciousness, Non-local Consciousness, and Psyche are extremely interesting. Materialism is boring and very limited in comparison.

Astrophysicists have observed that during the Prime Event or "Big Bang", only 4% of this universe got converted into physical matter or "observed into existence" as physical matter. The other 96% of this universe remained spiritual, non-local, and quantum in nature as dark matter (spirit matter) and dark energy (the zero-point field of light or the Light of Christ). The Materialists and Naturalists are wrong about 96% of this universe. (Ross, 2008). So, why should we continue to believe them and their philosophical guesswork?

Choose the best and get rid of all the rest. That's what I decided to do.

Proof of Concept

This Proof of Concept was based upon Joseph Rychlak's books, class notes from Edwin Gantt, and the following article by Williams and Slife.

Slife, B. D. & Williams, R. N. (1995). Science and Human Behavior. In *What's Behind the Research? Discovering Hidden Assumptions in the Behavioral Sciences*, (pp. 167–204). Thousand Oaks, CA: SAGE Publications.

http://mypsyche.us/science/

Materialism is the philosophical belief that ONLY physical matter exists, and that spirit and psyche do not exist.

Naturalism is the philosophical belief that psyche, non-local consciousness, the spiritual, spirit matter, the immaterial, the supernatural, and the non-local realm or spirit realm do not exist.

I was being repeatedly told online that Science and the Scientific Methods cannot be used to prove anything. I was being chastised right and left for claiming that the Scientific Method had proven to me that God exists, or had proven to me that the Theory of Evolution is false. Of course, I wasn't telling these people what they wanted to hear, so they weren't willing to give me a hearing.

However, it used to bug me to no end whenever these people told me that I can't use the Scientific Method to prove things, because I was using the Scientific Method to prove things all the time. I'm a scientist; and, I remember thinking, "What good is it, if it can't be used to prove anything?"

Technically, these people were just as blind as I was whenever they made the claim that the Scientific Method can't be used to prove anything. When it comes to the Scientific Methods, it all comes down to what one is trying to prove! The Scientific Methods cannot be used to prove anything true, at least not directly.

However, in theory and in practice, the Scientific Methods can be used and ARE BEING USED to prove things false. And, I was falsifying things right and left at the time! Furthermore, through repeated falsification of various different scientific theories, we slowly arrive at the truth through a process of elimination. This is what Sherlock Holmes does.

How often have I said to you that when you have eliminated the impossible, whatever remains, *however improbable*, must be the truth? — Sherlock Holmes.

Sherlock Holmes slowly arrives at the truth by falsifying his theories and his imagined evidence. We can do the same with the Scientific Methods. The Scientific Method isn't completely worthless; but, it also isn't as powerful and convincing as I once thought it was. Nevertheless, it has its uses, as long as it is used for what it's good for and for what it was designed for.

Verification and Validation

First, let's discuss how Science can go wrong. We can't use Science and the Scientific Methods to verify and know the Truth. Scientific Methods cannot be used to produce Knowledge.

Due to the *affirming the consequent* logic fallacy, and the *jumping to conclusions* logic fallacy, and the *category error* logic fallacies which are built into the scientific methods, it is impossible to use Science and the Scientific Methods to know the truth and to prove the truth.

If your ultimate goal in life is to KNOW THE TRUTH and prove the truth, then Science and the Scientific Methods are in fact one of the worst ways for accomplishing that task.

This is how scientific verification works in practice.

The following logical argument outlines the basic approach that has been taken by scientists:

If Theory X is true, then we will observe Y.

We observe Y.

Therefore, Theory X is true.

This sort of thinking, however, reflects a logical fallacy called *affirming the consequent*. Here's a comparable example to demonstrate:

We hypothesize: If Sally's pet is a cat, it will have a tail.

We observe: Sally's pet has a tail.

We conclude: Therefore, Sally's pet is a cat.

We can easily see that this logic is fallacious. Just because we observe Y (a pet with a tail) that does not mean that our theory X (the pet is a cat) is true. After all, dogs and lizards have tails too. (Rychlak; Slife; Edwin Gantt; Mark My Words).

Yet, **this IS the scientific method**, and this is exactly how scientists use the scientific methods to demonstrate and prove the truth. The whole enterprise is based upon a logic fallacy or two.

Begging the question consists of making our conclusion one of the axiomatic premises, and effectively results in *jumping to conclusions*. The Materialists and Darwinists do this all the time, without realizing it. Now, let's run a science experiment using the Scientific Method, based upon a *jumping to conclusions* or *begging the question* logic fallacy.

Premise or Hypothesis: All cats have tails.

Observation or Test of Hypothesis: My pet has a tail.

Scientific Conclusion: My pet is a cat.

The premise seems to match with the conclusion; but, this scientist jumped to the wrong conclusion based upon his scientific observations or scientific evidence – unless of course his pet was really a cat, then he simply got lucky. But, his pet theory is in fact a dog, so he got this one wrong.

Let's try another one. This one combines *categorization errors* along with *jumping to conclusions*.

Premise or Hypothesis: Truth matches with reality; and, the fossil record is a part of reality.

Observation or Test of Hypothesis: Darwin's Tree of Life and the Theory of Evolution match with the fossil record.

Scientific Conclusion: Therefore, Darwin's Tree of Life and the Theory of Evolution are the Truth.

Most scientists can't see anything wrong with any of this, because **this IS the scientific method**, which they use each and every day to find and demonstrate the truth of their science. But, this scientific method and logic fallacy demonstrated here are based upon "jumping to conclusions" or "begging the question", along with "false equations" or "categorization errors". These people find some way to input their desired conclusion as an axiomatic indisputable premise; and, this is *begging the question*.   This is what the Darwinists, Materialists, Naturalists, and Atheists do all the time; and, they don't even know it. They can't even see it or understand it, because they have been trained not to see it or understand it. They truly believe that their logic is sound, when it is not.

Jumping to conclusions and *affirming the consequent* is precisely what the Darwinists and Atheists DO each and every time they publicly declare that Science has proven the Theory of Evolution true.

Let's demonstrate their error first by *affirming the consequent*:

If the Theory of Evolution is true, the fossil record will match with Darwin's Tree of Life.

We observe that the fossil record matches with Darwin's Tree of Life.

Therefore, the Theory of Evolution and Darwinism are true.

This is *affirming the consequent*. This is how the Darwinists and Naturalists prove the theory of evolution to be true. They do it every day. You can go online, and you will find them online right now *affirming their conclusions* or *jumping to conclusions*.

Most of the scientists can't see anything wrong with this argument. They use it every day to prove to themselves that the Theory of Evolution is true.

Now let's demonstrate their logic error by *jumping to the conclusion* or *begging the question*, along with a *categorization error* or two:

Premise or Hypothesis: If our theory matches with and is based upon the fossil record, then our theory matches with reality and is true.

Observation or Test of Hypothesis: We observe that Darwin's Tree of Life and the Theory of Evolution match with the fossil record; and, we observe that the fossil record is truly there and really exists.

Scientific Conclusion: Therefore, the fossil record proves that Darwin's Tree of Life and the Theory of Evolution are true.

This is the argument which I see them using everyday online and in their books. They can't see anything wrong with it! I couldn't either, when I was a Materialist and an Atheist.

We can pull this one off multiple different ways, because these people are simply *jumping to the conclusion* that they want to affirm:

Premise or Hypothesis: All truth will match with the fossil record.

Observation or Test of Hypothesis: Darwinism, Darwin's Tree of Life, and the Theory of Evolution match with the fossil record.

Scientific Conclusion: Therefore, Darwinism, Darwin's Tree of Life, and the Theory of Evolution are true.

Can you see anything wrong with these arguments?

I couldn't, when I was a Materialist and an Atheist! I fell for it hook, line, and sinker!

These people are jumping to conclusions throughout the whole process, and can't even see that it is so. Their conclusions require a leap of faith and a confounding of the evidence (premises), and their arguments contain hidden assumptions, which the Materialists and Atheists are unable to see or understand and are more than willing to provide. Confounded! These people mix up something with something else so that the individual elements become something different and something difficult to distinguish from one another. They take that leap of faith and equate everything with everything else, and then equate it with the truth as well.

There are *category errors* taking place here, wherein these people are equating their philosophical theories with the fossil record; but, what gives them the right to consider their philosophical explanations of the fossil record to be superior to other philosophical explanations which contradict their chosen point of view? There's no logical or rational reason why they should do so, except for the fact that they are prejudiced and biased.

Nevertheless, this IS the way that the Darwinists, Materialists, and Naturalists use Science and the Scientific Method to prove that the Theory of Evolution is true. They do it every day of their lives, because they are "scientists". They can't see anything wrong with their logic. They simply pick the conclusion they want, and make the evidence and their interpretation of the evidence fit their conclusion. There's nothing wrong with that! It's Science!

BUT, the whole enterprise is based upon severe logic fallacies, which the Materialists, Naturalists, Darwinists, Atheists, and scientists in general are unwilling and unable to see and understand. In other words, these people suffer from Confirmation Bias.

Using their logic, we can prove ANYTHING to be true.

Premise or Hypothesis: If our theory matches with and is based upon the fossil record, then our theory matches with reality and is true.

Observation or Test of Hypothesis: We observe that Intelligent Design Theory matches perfectly with the fossil record and matches much better than the Theory of Evolution ever did; and, we observe that the fossil record is truly there and really exists.

Scientific Conclusion: Therefore, the fossil record proves beyond a shadow of a doubt that Intelligent Design Theory is real and true.

Using the logic of the scientific method, we can prove anything we want to be true. The Materialists and Naturalists have been doing that for thousands of years!

In other words, using the Scientific Methods, any conclusion we want can be proven to be true; and, any interpretation of the evidence that we want can be made to fit the scientific evidence. ONLY God knows which interpretation is the BEST interpretation or the RIGHT interpretation.

How's this possible?

It's due to all the different logic fallacies which are built into the Scientific Methods! The Scientific Methods are actually a very poor way of finding and knowing the truth; but, most scientists never make this realization throughout the whole of their careers. I never did; and, I considered myself to be a scientist. I went for 55 years BEFORE I learned anything about any of this! The Materialists and Atheists don't teach us about any of this stuff in college, for some strange reason. ONLY the Theists seem to care about the truth; and, most of them don't know where to find it.

Finally, the ultimate weakness of the Scientific Methods comes from the fact that every science experiment <u>begins</u> with a philosophical guess or hypothesis and <u>ends</u> with some kind of philosophical interpretation of the scientific evidence or the scientific data, which was produced by the science experiment. In other words, the Scientific Methods are based exclusively on philosophical speculation and a philosophical interpretation of the scientific data; and, Philosophy is typically considered to be the worst way or the most unreliable way of finding and knowing the truth, because with Philosophy various different logic fallacies, trickery, deception, self-deception, and sophistry abound and are at work. After all, Materialism, Naturalism, and Atheism are NOT science. They are philosophy, sophistry, self-deception, and metaphysics; and, these philosophies have taken over the whole of science.

When it comes to scientific data and scientific evidence, you can have ANY interpretation that you want; and, that's the fundamental weakness of the Scientific Methods.

Ironically, Darwin's Tree of Life does NOT match with the fossil record that we have on hand! The Materialists, Naturalists, and Darwinists simply affirm the consequent, jump to the conclusion, and assume that it does.

Affirming the consequent is a form of logical error which occurs when we uncritically accept as affirmed or confirmed the results of an "if-then" sequence of reasoning or scientific exploration. *Affirming the consequent IS jumping to the conclusion*.

> This is why philosophers are wont to point out that for any given fact pattern which can be demonstrated or "discovered" empirically, an infinite number of theoretical explanations are possible. This is also why we say theories remain theories, even after they have been validated. All validating evidence can establish convincingly is the *negation* of a theoretical proposition. Having postulated an '*If A, then B*' sequence, when our researches *fail* to confirm this sequence, we can logically reject the theoretical relation originally postulated. [*Negating the consequent.*] Now this situation bothers philosophers. A datum accrued is not always knowledge gained, because sometimes genuine understanding is lacking. One can have a great deal of information and still suffer for lack of knowledge. (Rychlak, 1981, *A Philosophy of Science for Personality Theory*, pp. 81-82).

Imagine it! There are an <u>infinite number</u> of possible explanations or interpretations for every piece of validated scientific evidence or scientific data that we generate. Picking the explanation or interpretation you like the most and treating it as the God-given truth IS "affirming the consequent" or "affirming your chosen conclusion". This is what the scientists do all the time! We can have a ton of scientific evidence at hand, yet still suffer from a lack of knowledge and truth.

Once again, using the logic of the scientific method, we can prove anything we want to be true. The Materialists and Naturalists have been doing so for thousands of years!

The idea or "interpretation of the fossil record", which claims that Intelligent Beings or Intelligent Psyches terraformed this earth and seeded this planet in a systematic and progressive manner, matches infinitely better with the fossil record than Darwin's Tree of Life does, and is infinitely more plausible, logical, reasonable, and believable than claiming that physical matter designed and created the genomes and life forms on this earth as the Materialists, Naturalists, and Atheists claim.

But, the Naturalists, Materialists, and Atheists can't see this because they don't want to see it or understand it. Their ultimate goal is to reject it and to deny it. Anything other than Materialism and Naturalism and Atheism, they declare to be "NOT SCIENCE". I KNOW, because I have had them use that "NOT SCIENCE" argument on me – it's Not Materialism, and it's Not Naturalism, and it's Not Atheism; therefore, it's NOT SCIENCE – as a trump card against me every time I have interacted with them. It's petty, and it's unconvincing; but, it's their bread and butter, and their way of doing Science. Nevertheless, their science stinks; and, it's not science. It's philosophy.

I interacted with the Darwinists, Materialists, Skeptics, anti-Christs, and Atheists online for many years. At first, I was curious to know what I was. I think I was open-minded about it all. As my understanding began to grow and I started to understand what these people really believe, I found myself disagreeing with them at times. When one of my favorite friends turned against me and sicked his Flying Spaghetti Monster on me, I was hooked. I found myself diving into a study of the whole thing, studying it as much as I possibly could.

I started with Antony Flew, allegedly the world's most notorious atheist, and his book, *There Is a God: How the World's Most Notorious Atheist Changed His Mind*. I found out that he gave up his atheism because of Intelligent Design Theory. His reason for giving up his atheism was different than mine; and, I started to take Intelligent Design Theory seriously.

Eventually, I debated about the Intelligent Design Theory with a person online, and the ONLY argument he was ever able to produce against it is that "IT'S NOT SCIENCE" – meaning that it's not Atheism, not Naturalism, and not Materialism. It was then when I fully realized how weak these people's case really is. It was then that I realized that these peoples' philosophical arguments are not science. I began to learn exponentially ever since. These people jump directly to the conclusion that they want, and exclude everything else that they don't want to hear about, accept, or believe. That's the way these people do science; but, that procedure is philosophy, not science!

In summary, when it comes to scientific data and scientific evidence, due to the philosophical nature of the whole enterprise, you can have ANY interpretation that you want for the results of a science experiment; and, that's the fundamental weakness of the Scientific Methods. Consequently, Science ends up being an extremely weak way of finding and knowing the truth, and no better than philosophy.

In contrast, God KNOWS how life came to be on this planet, because He was there, and He experienced it first-hand. He lived it, so He knows it. The truth is KNOWN by living it and experiencing it first-hand. Whenever the Biblical God Jesus Christ appears to mankind and claims to be the ONE (the Psyche) who designed and created this physical universe, this earth, and all of the life on this earth, I choose to believe Him because He was there and He knows how it was really done. The Materialists, Naturalists, Darwinists, and Atheists don't know and can't know – they are simply guessing, jumping to conclusions, and affirming the specific consequent or the specific conclusion which they desire most to affirm.

The Scientific Methods cannot be used to prove the truth and know the truth. If we want the truth, then we have to get at the truth through some other way or method. Does such a method exist? I say that it does. I KNOW that it does, because I have lived it and experienced it for myself. Through our Lived Experiences it is possible to find the truth, live the truth, experience the truth, and to KNOW THE TRUTH first-hand, which is something that cannot be accomplished with philosophical speculation, scientific methods, scientific experimentation, jumping to conclusions, affirming the consequent, categorization errors, and philosophical interpretations of the scientific data.

Scientific Methods aren't any good for finding and knowing the truth. Living the truth and experiencing the truth is an infinitely better way of KNOWING THE TRUTH.

Now, let's discuss what the Scientific Methods might be good for.

Falsification

In theory, Scientific Methods can be used to falsify theories or to prove theories false. The falsification approach is much more logically sound than the verification approach to the Scientific Methods. The falsification approach does not involve egregious logic fallacies, although it does involve various different types of philosophical interpretation, hypotheses, and philosophical assumptions!

The falsification approach is called, *negating the consequent*; and, it is logically sound, although it still suffers from philosophical interpretations and the fact that it's typically impossible to control ALL of the variables in a science experiment. For example, the human psyche is a hidden variable and an uncontrollable variable in EVERY science experiment, because ONLY the human psyche can do science experiments! The human psyche is mixed up in every science experiment and in the interpretation of the scientific data as well. That's one of the primary weaknesses of the Scientific Methods – personal interpretation of the scientific evidence.

There is a different argument structure that scientists have begun to use instead of verificationism. It is called the falsificationism approach to science. This approach is much more logically sound than the verificationism approach.

The argument structure for falsificationism, or *negating the consequent*, looks like this:

If Theory X is true, then we will observe Y.

We don't observe Y.

Therefore, Theory X is not true.

Rather than verifying a theory, this approach falsifies a theory. Unlike the verification argument, this argument is logically sound. The idea is that although we can't ever know (based on empirical data alone) when a theory is true, we can know (based on empirical data) when a theory is false.

In other words, we develop a theory, make predictions based on that theory, and if those predictions come true, rather than claim that we have confirmed our theory, we more humbly claim that our theory has not yet been proven false. We prove theories false by *negating the consequent*. Eventually, we arrive at the truth by a process of elimination. (Slife; Gantt; Mark My Words).

In effect, this is what Sherlock Holmes does in order to get at the truth of a matter. He centers in on the truth through a process of elimination, or a process of falsification.

How often have I said to you that when you have eliminated the impossible, whatever remains, *however improbable*, must be the truth? — Sherlock Holmes.

The Scientific Methods can be used to falsify theories and to prove things false. *Negating the consequent* is logically sound.

The following is how falsifying a theory works in practice. This example is a direct application of *negating the consequent*:

Scientific Definitions and Principles: By definition in principle, Materialism, Naturalism, and Darwinism are design and creation by physical matter, because the Materialists and Naturalists as a matter of principle by definition and fiat have chosen to believe that only physical matter exists and that the non-local spirit realm and psyche do not exist. For the Materialists and Naturalists, physical matter is all that there is; and therefore, physical matter is the only thing available that can design and create things for us. Technically, Materialism, Darwinism, and Naturalism ARE Creation by Rocks. It's important to get the fundamental and the correct definition for things!

Premise or Axiom or Law: The Scientific Methods can be used to falsify theories and to prove things false.

Scientific Hypothesis: If Materialism, Naturalism, and Darwinism are true, then we will observe the rocks and physical matter designing, creating, and manufacturing genomes and life forms from scratch.

Scientific Observations: Spontaneous generation and Abiogenesis have been proven false – Louis Pasteur proved Materialism, Darwinism, and Creation by Rocks false in 1859. We don't ever observe the rocks designing and creating anything. They can't.

Scientific Conclusion: Therefore, Materialism, Naturalism, and Darwinism are not true.

Simple truth!

Can you feel the truth of it?

I just falsified Materialism, Naturalism, and Darwinism! And, my argument is logically sound. It's also valid and true. This is precisely what I did the first time that I falsified the Theory of Evolution. It works, because it's true.

By getting the proper and the most fundamental definition for Materialism, Naturalism, and Darwinism, I have just successfully used the scientific method to falsify Materialism, Naturalism, and Darwinism. Said another way, I simply took the Materialists, Naturalists, and Darwinists at their word by assuming along with them that ONLY physical matter exists. Then based upon their assumption or scientific hypothesis, I proceeded to falsify their assumption or their hypothesis using the scientific method and observation, which is easy to do because raw physical matter has never been caught in the act of design and creation, and never will be.

Scientific Definition: Materialism, Naturalism, and Darwinism are design and creation by physical matter.

Scientific Hypothesis: If Materialism, Naturalism, and Darwinism are true, then we will observe the rocks and physical matter designing and creating genomes and life forms.

Scientific Observation: We have NEVER observed the rocks designing and creating anything; and, we NEVER will.

Scientific Conclusion: Therefore, Materialism, Naturalism, and Darwinism are false.

Materialism, Naturalism, and Darwinism have been falsified or proven false thousands of different ways. Consequently, if we want to KNOW the Truth, then we have to look someplace else besides Materialism, Naturalism, and Darwinism. This argument is logically sound, because the falsification of these scientific theories is logically sound.

According to the Materialists and Naturalists, raw physical matter was the ONLY thing in existence before the first genomes and the first life forms were designed and created; therefore, the raw physical matter or the rocks must have designed and created the first genomes and the first life forms on this planet. Such a materialistic and naturalistic claim VIOLATES the lived experiences, observations, and common sense of the human race, because we all KNOW for a fact that the rocks can't design and create anything at all. Our idols can't function in the role of Psyche or God. The rocks can't do formal cause and final cause.

In contrast, intelligent beings, human beings, human psyches, or psyche beings have been caught in the act of design and creation trillions of trillions of different times. Intelligent Design and Creation by Intelligent Psyches has been experienced and observed trillions of different times in trillions of different situations and ways. Follow the evidence! It ALL points to Psyche or Intelligence, because raw physical matter cannot design and create.

This is also an excellent Proof of God's Existence and an excellent proof of psyche's existence. God must of necessity exist in order to have done all of the design, creation, science, engineering, manufacturing, diversification, and systematic deployment that needed to be done, which the rocks and physical matter could NEVER have done, while the genomes and the life forms were being designed, created, and placed upon this earth.

This falsification approach to Science and the Scientific Methods is much more logically sound than the verification approach, because the falsification approach isn't based upon an egregious logic fallacy or two. In other words, my falsification of Materialism, Naturalism, and Darwinism IS logically sound! It also seems to be true, in that it actually matches with reality; whereas, Materialism and Naturalism don't.

Now, let's take this to its logical conclusion. I have observed that Materialism and Naturalism and even Atheism have been falsified trillions of trillions of different times in thousands upon thousands of different ways. There are so many different ways to falsify Darwinism, Naturalism, Atheism, Nihilism, and Materialism that it's impossible to count all the different ways that this has been done and can be done. Furthermore, the falsifications of these philosophical concepts continue to pour into the records of science, and into the records of the human race as well. In fact, every time you have a thought, or see a ray of light, or feel the press of gravity, or witness the effects of magnetism, or have an out-of-body experience, or have a spiritual experience, you have just falsified Materialism and Naturalism and Darwinism. It's easy to do! Thoughts and dreams are spiritual experiences, because there's no way possible to record our thoughts and dreams with our physical instruments. Our very thoughts and dreams – our Psyche Experiences – falsify Materialism and Naturalism and Darwinism.

In contrast, design and creation by Psyche or Intelligence – sometimes called Intelligent Design – has NEVER been falsified and never will be. In fact, Design and Creation by Psyche or Intelligence has been observed and verified and validated trillions upon trillions upon trillions of times, with an infinite number of more times to go. Follow the evidence. Think logically and rationally. The truth is staring us in the face; but, most of us refuse to see it.

This whole falsification process explains why the Scientists and Philosophers, whom I have been reading and studying the past couple of years, refuse to believe in Materialism, Naturalism, and Darwinism on a blind leap-of-faith that these things are axiomatically true, because these people too have used logic and the scientific methods to falsify Materialism, Naturalism, and Darwinism in one way or another. It's easy to do. Where there's a will, there's a way.

In contrast, it's literally impossible to falsify THE TRUTH using the Scientific Methods. The scientists will never be able to falsify Intelligent Design Theory, or design and creation by Intelligence, because it's true. Design and Creation by Intelligence or Psyche can be verified and replicated on demand! The truth never fails and is never falsified! ONLY Psyche – ONLY Intelligence – can design and create things, and do science. Psyche or Intelligence is the ONLY thing we have ever observed doing so. You can verify this right now for yourself, by choosing to design and create something for someone, because the rocks can't do it for you. It's elementary my dear friend.

See also:

Slife, B. D. & Williams, R. N. (1995). Science and Human Behavior. In *What's Behind the Research? Discovering Hidden Assumptions in the Behavioral Sciences*, (pp. 167–204). Thousand Oaks, CA: SAGE Publications.

http://mypsyche.us/science/

The Ultimate Alternative for Knowing the Truth

I think that the ultimate discovery of my Science Career came when I finally observed, realized, and accepted the fact that Lived Experience IS THE BEST WAY of finding and knowing the truth. In fact, it's the ONLY way to find and know the truth, because ONLY Psyche can do Science, Philosophy, Lived Experiences, finding, knowing, reality, existence, and truth. This Reality and Truth is what I call a Psyche Epistemology, which ends up being the BEST METHOD for finding and knowing the truth.

Lived Experience or Psyche Experience is the Ultimate Method for finding and KNOWING the TRUTH! Finding and knowing the truth is what Psyche does. Psyche or Intelligence is Light and Truth and Knowledge.

We KNOW from the Lived Experiences of the human race that Psyche or Non-Local Consciousness does indeed exist. (Buhlman; Durham; Storm; Alexander; Sharkey & McCormack; Neal; Van Lommel; Moody; Gibson; Long; Hinze; Ring; Rivas; Dispenza; Eccles; & Penfield). Psyche IS the fundamental unit of Reality. A Psyche Ontology is the Ultimate Ontology, because ONLY Psyche can do ontology and lived experiences. Consequently, a Psyche Epistemology is the Ultimate Epistemology, because ONLY Psyche can do epistemology.

There is a way to KNOW the TRUTH! We know the truth by living the truth and experiencing the truth for ourselves, or by choosing to trust someone who has.

Unlike the Scientific Methods, Philosophical Speculation, and the Philosophical Interpretation of Scientific Data, which technically can NEVER be used to find the truth, verify the truth, and prove the truth, we can KNOW THE TRUTH by living the truth and experiencing the truth directly for ourselves or by choosing to trust someone who has.

The Lived Experiences of the human race, both our physical experiences and our spiritual experiences, ARE the repository of Truth and Knowledge for the human race. Living the truth and experiencing the truth ARE the best and most efficient method for KNOWING the TRUTH.

Our Philosophical Speculation and our Scientific Interpretations have to match with the Lived Experiences of the human race; otherwise, our philosophies are wrong, and our scientific conclusions are incorrect. Our interpretations of scientific evidence have to match with Reality, or our interpretations are false! A materialistic, naturalistic, and Darwinian interpretation of scientific evidence is always false, because Materialism, Naturalism, and Darwinism don't match with the Lived Experiences of the human race and therefore don't match with Reality.

ONLY Psyche can have Lived Experiences; and therefore, ONLY Psyche can KNOW the TRUTH by directly experiencing, living, and remembering the truth for itself.

God, or God's Psyche, is KNOWN by direct experience and by what God chooses to do for us or chooses to reveal to us. God is KNOWN to us by doing for us the things which we could never have done for ourselves – such as by designing and creating our genomes, the first of our cells, and our physical bodies.

I KNOW from direct observation and experience that the rocks can't design, create, and manufacturing anything. Therefore, I KNOW that God must of necessity exist in order to have designed, created, engineered, and manufactured all of the genomes and life forms in the first place. I KNOW that God and God's Psyche must of necessity exist in order to have done all of the Science that needed to be done in order to bring life to this planet. I KNOW that ONLY Psyche can do Science.

I also KNOW from Quantum Mechanics that Psyche must exist in order to bring physical matter into existence – in order to convert spirit matter into physical matter – in order to convert non-locality into locality. In a sense, Science itself has proven to me that God exists; however, it was in fact my lived experiences and the lived experiences of the human race which proved to me that God does in fact exist. Unlike Philosophy and the Scientific Method which can never be used to prove the truth or know the truth, we can prove the truth and KNOW THE TRUTH through our Lived Experiences, by living the truth and experiencing the truth for ourselves, or by choosing to trust someone who has. Experiential Knowledge is vastly superior to Science, Scientific Methods, and the Philosophical Interpretation of Scientific Data.

An epistemology is a way of knowing the truth. A Psyche Epistemology is better than any Philosophical Epistemology, Naturalistic Epistemology, or Scientific Epistemology, because Lived Experiences are a better way of knowing the truth. In fact, Lived Experiences are the ONLY way to know the truth. Consequently, a Psyche Epistemology is the Ultimate Epistemology, because a Psyche Epistemology subsumes or includes ALL other epistemologies, because ONLY Psyche can do epistemology.

An ontology is a fundamental unit of reality. A Psyche Ontology observes and concludes that Psyche is the fundamental unit of reality. Psyche is axiomatic, foundational, fundamental, essential, necessary, self-evident, lawful, tautological, immanent, infinite, obvious, ultimately causal, and self-sustaining, because ONLY Psyche can do psyche. It's elementary my dear friend!

In other words, we come full circle by observing and accepting the fact that we KNOW from the Lived Experiences, Psychic Experiences, Spiritual Experiences, and Psyche Experiences of the human race that Psyche or Non-Local Consciousness does indeed exist.

ONLY Psyche can have Lived Experiences and remember these experiences in order to share them with others at a later date; and therefore, ONLY Psyche can KNOW the TRUTH by directly experiencing and living the truth for itself. A Psyche Epistemology is the Ultimate Epistemology. A Psyche Ontology is the Ultimate Ontology, because ONLY Psyche can do epistemology and ontology.

Lived Experience or Psyche Experience is the Ultimate Method for finding and KNOWING the TRUTH! In order to find the truth and KNOW the truth, I had to get rid of My Materialism, My Atheism, My Naturalism, and My Nihilism. We have to get rid of all the falsehoods in our lives, in order to find the truth and KNOW the truth. It's elementary my dear friend!

You, your psyche, will determine for yourself whether you believe any of this to be true or not. For you personally, your psyche is the ultimate arbiter of truth. You, your psyche, will decide if any of this is useful to you or not. You are your own punishment, and you are your own reward. I believe that I have met my burden of proof; but, you are free to decide differently if you want to.

My hope and my expectation is that if you have reached this far into my introduction to Psyche as the Ultimate Cause and to Psyche as the Ultimate Ontology, then you are starting to take me seriously. If not, please read it again. After reading it again, if you're still not getting it then it's probably not for you – your biases and prejudices are too strong and are preventing you from being able to see it and understand it. Try again some other time when you have changed your mind.

As you can probably tell, I love this stuff. I'm passionate about all of this, because I feel as if I have finally found the truth. Truth feels good and tastes good. We get goosebumps or we get animated and enlivened, whenever we speak the truth, write the truth, or are in the presence of truth, unless of course we have desensitized ourselves to the truth.

All of this is infinitely more interesting than My Materialism and My Atheism ever were!

This Ultimate Model of Reality is just a great deal of fun to think about, write about, and study!

Definitions Explaining Ultimate Cause

Truth Is Knowledge: Knowledge is truth, or at least it should be. It's kind of impossible to actually know the truth by choosing to believe in something that it false. Furthermore, knowledge of falsehoods doesn't do us any good, UNLESS we actually know that these things are false and know why they are false. Back when I was a Materialist, Nihilist, and Naturalist, I didn't even know what those terms actually meant. At the time, I didn't know what a Materialist, Naturalist, and Nihilist was. I simply was one. All I knew at the time was that I was an Atheist. My Atheism was based upon my lived experiences, wherein I believed at the time that God was completely absent from my life and that as a result of my experiences (or lack of experiences) I was convinced and believed that God does not exist. I was without God in the world. From my perspective during those days of my life, I seemed to know something that nobody else around me seemed to know – I knew that God doesn't exist. I had taken that leap of faith. Of course, I was wrong; but, I didn't know that, while I was right in the middle of it all. It was only after I – with God's help – had successfully gotten rid of My Materialism, My Nihilism, My Atheism, and My Skepticism that I was finally able to see and understand all of the different miraculous interventions which had taken place in my life to keep me alive, to keep me safe, to help me heal, to bring me back to sanity and sobriety, and to help me learn from my mistakes and the falsehoods that I had chosen to believe in. God had been helping me all throughout the whole process, sometimes in amazingly miraculous ways, but I couldn't see it at all when I was a Materialist and an Atheist and living right in the middle of it. Thank God He kept me alive long enough to see and understand the errors of my ways. God is merciful, gracious, forgiving, and kind; but, I wasn't able to see that when I believed that He was not. We blind ourselves to the truth whenever we choose to believe in falsehoods. That's just the way it works. We KNOW the truth, the real TRUTH, by living it and experiencing it for ourselves. We know it, when we OWN it and take full responsibility for it. Nowadays, I even KNOW and understand how and why Materialism, Naturalism, Nihilism, Darwinism, and Atheism ARE false. I KNOW the truth about these things! It's amazing how that works! God was schooling me and teaching me by letting me experience these things for myself. I personally needed to know what it's like to be a Skeptic, Materialist, Nihilist, and Atheist – I needed to know and understand the hell that is typically associated with these worldviews, so that I could more fully appreciate and understand their opposite. The pain makes it real. The lessons are actually learned for real. Pain is our greatest teacher. It motivates us to learn. Through pain's ministrations, we OWN the knowledge that is received through the experience. We experience the knowledge. We experience the truth. We come to KNOW it. It becomes an integral part of us. I now have no desire whatsoever to go back to My Nihilism and My Atheism, because now I KNOW precisely what they are and what they did to me. It was hell. I can see that now; whereas, I couldn't see it before. I had to live it and experience it in order to KNOW it and understand it. Our lived experiences are our greatest and most effective teachers. I can see that now. I KNOW it, because I lived it.

Psyche Ontology or the Ultimate Ontology: With this Ultimate Model of Reality, I'm introducing a Psyche Ontology in which psyche is the fundamental unit of reality and the ultimate maker of meaning and reality. This Ultimate Cause Model of Reality is a Psyche Ontology. ALL the evidence led me to this truth. Brent Slife and friends expended a lot of effort developing a Relational Ontology in which relationships are the fundamental unit of reality; but, it quickly became clear to me that ONLY Psyche can do relationships and friendships, so it became obvious to me that Psyche is in fact the fundamental unit of Reality. Of course, I easily adopted this stance, because I have been developing and promoting Psyche as the Ultimate Cause for the past half year or so, in my theorizing and writing. I find it interesting that nobody else seems to have developed and presented a Psyche Ontology before. Everyone seems to be programmed and conditioned to overlook Psyche automatically; but, ever since I read Buhlman's book at the end of 2015, I have KNOWN that Psyche exists and have basically known what Psyche is. I have been looking at Psyche ever since!

Psyche Epistemology or the Ultimate Epistemology: Eventually, it became clear to me that I was also looking for the Ultimate Epistemology to go along with this Ultimate Model of Reality. George Kelly's Scientism Epistemology or Constructive Alternativism was my original preference, because I considered myself to be a scientist more than anything else. Eventually I found myself looking at Joseph Rychlak's focus upon Teleology and Final Cause, and some kind of Telic Epistemology. Then I developed Spiritual Empiricism as an epistemology, because that seemed to fit better with Psyche as the Ultimate Cause. Then I started reading Brent Slife and friends. In their writings, these people toyed with William James' Radical Empiricism (all of our experiences including our spiritual experiences), Levinas' Phenomenology (phenomena and events and lived experiences), and Hermeneutics (studying the interpretation and the meaning of experiences and events). Eventually it dawned on me that I was looking for a Psyche Epistemology, because ONLY Psyche can do epistemologies. ONLY Psyche can do science, constructs, relationality, spirituality, teleology, lived experiences, phenomenology, and hermeneutics. Therefore, I introduce my reader to a combined Teleological Epistemology, Hermeneutic Epistemology, Radical Empiricism Epistemology, Spiritual Empiricism Epistemology, Phenomenological Epistemology, and Traditional Empiricism Epistemology which is based upon the Lived Experiences of the human psyche and the Lived Experiences of the human race as a whole. This is a Psyche Epistemology – the ultimate way of knowing the truth. ONLY Psyche can do epistemology. ONLY Psyche can find and know the truth. The ultimate purpose of my books is to introduce my reader to a Psyche Ontology in which immortal psyche is the fundamental unit of Reality and therefore the Ultimate Reality. This leads naturally to a Psyche Epistemology, wherein the Lived Experiences of the human psyche become the ultimate best way of finding and knowing the truth. The truth is known by living it and experiencing it first-hand, whether we are talking about a spiritual reality or a physical reality. This is a Psyche Epistemology, the Ultimate Epistemology. In ALL of my theorizing, I kept adjusting this Ultimate Model of Reality until it matched with the Lived Experiences or the Psyche Experiences of the human race. I wanted to get this model as close to the Truth and Reality as I possibly could.

Radical Empiricism: This Ultimate Cause Model of Reality fully embraces what William James called Radical Empiricism, which means that this Ultimate Model of Reality subsumes or includes and encourages ALL epistemologies or all forms of knowing the truth, including philosophy, logic or rationalism, dialectical and demonstrative reasoning, physical empiricism, the scientific methodologies, scientific experimentation, qualitative methodologies, application or practice, quantum mechanics or spiritual mechanisms, lived experiences, hermeneutics, phenomenology, spiritual experiences of all kinds, theophanies or revelations of God, and revelations from God. Nothing is off-limits nor out-of-bounds, which is why I chose to call it the Ultimate Model of Reality. It is a Psyche Ontology – a sort of Holistic Ultimology. Its study is the whole of Reality, and not just a small part of reality. The Ultimate Model of Reality also subsumes a Psyche Epistemology, because ONLY Psyche can do epistemology. (James, W. (1912). *Essays in Radical Empiricism*. https://www.amazon.com/dp/B004TS16TI/).

Lived Experience: Truth is based upon Lived Experiences, and not philosophical abstractions! From the Wikipedia: Pragmatism rejects the idea that the function of thought is to describe, represent, or mirror reality. Instead, pragmatists consider **thought** [psyche] an instrument or tool for prediction, problem solving, and action. Pragmatists contend that most philosophical topics — such as the nature of knowledge, language, concepts, meaning, belief, and science — are all best viewed in terms of their practical uses and successes. The philosophy of pragmatism "emphasizes the practical application of ideas by acting on them to actually test them in **human experiences**". Pragmatism focuses on a "**changing universe** rather than an unchanging one as the Idealists, Realists, and Thomists had claimed". (The preceding was taken from the Wikipedia.) The pragmatic focus of this Ultimate Cause Model of Reality is upon Lived Experiences, human experiences, or psyche experiences – the things which the human psyche actually experiences and acts upon, both while in the physical body and while separated from the physical body. Thought or Psyche is the fundamental unit of reality. Psyche as the Ultimate Cause is based upon Radical Empiricism, or Lived Experiences, or Phenomenology. ONLY Psyche can do life and lived experiences. Thought is a function of psyche, because we continue to think and perceive after our physical body is dead and our physical brain has gone offline, according to the Science of NDEs and OBEs. Every time you have a thought, you have had a spiritual experience. Psyche produces thoughts, and remembers them. Psyche also remembers all of our choices and all of our experiences. Pragmatism is the Philosophy of Thoughtful Practice – practical uses, choices, lived experiences, spiritual empiricism or spiritual experiences, evidentiary pluralism, and successful acts. William James treats Pragmatism or Thought as an epistemology – a way of knowing the truth. For me, all of this started to point to a Psyche Epistemology! William James's Radical Empiricism or Lived Experience, Methodological Pluralism, and Pragmatism are briefly summarized in this article: http://brentslife.com/article/upload/evidence/EBPP%20-%20APA%20Taking%20Sides%20final.pdf. William James was a hundred years ahead of his time. The rest of the scientists are finally starting to catch up.

Phenomenology and Hermeneutics: Phenomenology is the philosophical and scientific study of Lived Experiences, events, and phenomena. ONLY Psyche can do life, consciousness, awareness, and lived experiences. Matter of any kind, which has no psyche residing within it, is incapable of experiencing life and remembering events. Psyche is Life. Psyche is our personality and our memories. Psyche is consciousness and awareness and personality. Psyche is the Ultimate Truth. Hermeneutics is the philosophical and scientific study of interpretations, explanations, and meaning. Teleology is the philosophical and scientific study of purpose, intentions, reasons, desires, wishes, final causality, and goals. Psychology is supposed to be the study of psyche. ONLY Psyche or Non-Local Consciousness can do hermeneutics and teleology and psychology. ONLY Psyche can do interpretation, meaning, purpose, reasoning, thinking, dreams, planning, and final causality. The rocks don't think, reason, nor dream unless they have some kind of living psyche residing within them. ONLY Psyche can do Phenomenology, or Lived Experiences, or Radical Empiricism.

Nonlocality: Although there are dozens of other observed and experienced scientific discoveries which support Psyche as the Ultimate Cause, an understanding and acceptance of Quantum Nonlocality, Quantum Entanglement, and Quantum Consciousness seems to be the most crucial and essential Science for understanding Psyche as the Ultimate Cause. Nonlocality means non-physical, or not located in our physical 3D space-time universe. Nonlocality by definition means that quantum objects (the wave form of quantum objects) and consciousness are trans-dimensional, non-physical and/or immaterial, and not located in our physical 3D space-time reality. Once a non-local trans-dimensional quantum object, or quantum wave, or light wave, or energy wave is consciously observed or receives the Word of Command, then its wave function collapses and it becomes a physical particle or a photon located here in our 3D physical space-time. Again, non-local means non-physical, or trans-dimensional and spiritual, in nature and origin. Quantum Entanglement has in effect proven Quantum Non-Locality to be real and true, because every science experiment regarding Quantum Entanglement has verified Quantum Entanglement to be real and true. Computer Science and the makers of computer chips are completely reliant upon Quantum Mechanics and limited by Quantum Mechanics (or spiritual mechanisms) in what they can do. They can only go so small before the thing ceases to be physical and starts to become spiritual and uncertain instead. The closer we get to the spiritual or the quantum, the more psyche's agency or choice comes into play, and the more unpredictable, unreliable, and uncertain things become. The physical realm is a highly reliable and highly predicable consensus reality, the ultimate consensus reality; but, a consensus reality is made up of non-consensus realities, quantum realities, or spiritual realities where agency, free-will, self-determination, uncertainty, and psyche reign supreme. I have stress-tested computer chips, and had them leak through or break past their physical limitations, and become unpredictable – seeming to develop a mind of their own. Quantum Mechanics or Spiritual Mechanics is real stuff with real-world applications. Quantum Mechanics is verified and validated science, possibly the best-proven and most-used science that we currently have; but, Quantum Mechanics only makes sense from a spiritual or a non-local perspective. Quantum Mechanics when properly understood proves that

Materialism is false (falsifies Materialism) and essentially demonstrates and "proves" that some kind of psyche, conscious observer, and non-local consciousness must of necessity exist. For most people, the Science of Quantum Mechanics and Quantum Non-Locality have proven beyond a shadow of a doubt that some kind of non-local or spiritual conscious observer is the ultimate cause behind each and every physical particle that has ever been brought into existence. Remember, it's possible to falsify theories or to prove them false using Science and the Scientific Methods; therefore, it's theoretically possible to center in on the truth through a process of elimination or falsification. In other words, Quantum Mechanics has always been verified. It has never been falsified. That reality gives us the impression that Quantum Mechanics has been proven true; and, Quantum Mechanics explains how Psyche and Spirit Matter really work. Again, Quantum Mechanics is the best proven and most-used Science that we currently have because it hasn't been falsified; and, Quantum Mechanics when properly understood proves through a process of elimination the existence of the non-local, the non-physical, the immaterial, and the spiritual, as well as "proving" that Psyche is necessary and must exist. In other words, Quantum Mechanics when properly understood falsifies Materialism and Naturalism and proves that they false; and through the process of elimination, indirectly "proves" that psyche or non-local consciousness must exist. That's the primary lesson of Quantum Mechanics or Spiritual Mechanics, because Quantum Mechanics is the science of how spirit matter and the spirit realm really work. Of course, we KNOW from the Lived Experiences of the human race that Psyche, Spirit Matter, the Spirit Realm, Spirit Bodies, and the Biblical God Jesus Christ exist and that Jesus rose from the dead; but, Quantum Mechanics becomes a Scientific Way of demonstrating that Psyche and Spirit Matter and God must exist. ALL TRUTH is mutually sustaining, because all truth will match with Reality and the Lived Experiences of the human psyches or the human race. We also learn from Quantum Mechanics that a wave-like quantum object or spirit matter is a completely different substance or essence than non-local consciousness, psyche, intelligence, awareness, life, or mind. Consciousness or Psyche chooses and acts; whereas, spirit matter or the wave form of quantum objects is acted upon by consciousness. Consciousness and spirit matter are two completely different things in the Non-Local Realm, or Quantum Realm, or Trans-Dimensional Realm, or Spirit Realm. (Van Lommel, 2010; Mark My Words, 2016, *Quantum Mechanics from a Non-Physical Spiritual Perspective*).

Psyche: Psyche is your spirit, soul, mind, consciousness, life, spark, individuality, intentionality, awareness, intelligence, personality, memories, immaterial entity, infinite singularity, living light, existence, long-term memory storage, and non-local life form. Although Psyche or Intelligence is often confounded or equated with spirit matter, spirits, and our spirit body, it has been observed that they are two completely different things. Psyche or non-local consciousness is an immaterial, limitless, infinite singularity or viewpoint in space. Psyche is your core being, your essence, your existence, and your life. Psyche or consciousness is the ultimate cause behind everything else that has ever been organized or brought into existence, including any organization of spirit matter and physical matter, and the existence of physical matter in the first place. Psyche is the eternal and the indestructible part of us. It has always existed and will always exist. ONLY Psyche can do lived experiences, life, memories, epistemology, truth, knowledge, and ontology. Psyche is where the memories of our lived experiences are stored; and, our memories go with us into the next life after our physical body and physical brain are dead and gone. (Buhlman, 1996, *Adventures Beyond the Body: How to Experience Out-of-Body Travel*.)

Psychology: Psychology is the study of the human psyche. Psychology is philosophy with a practical, useful, interesting, compelling, and realistic application. Psychology is also an attempt to study psyche scientifically through observation and empirical evidence. One of the goals of this Ultimate Cause Model of Reality is to put psyche back into Psychology. Psychology has been defined as the study of behavior and experience. The Materialists, Naturalists, and Atheists focus all of their attention on observable physical behavior; and, the rest of us study the Lived Experiences of human beings or human psyches. The one is infinitely more expansive than the other. I have learned to love the Study of Psyche. It's infinitely more interesting than My Materialism and My Atheism ever were!

Darwinism: The philosophical and religious belief taken on blind faith that physical matter designed and created the first functional genome on this planet and the first life form on this planet from scratch. In its purest and most basic form, Darwinism is Creation by Chance, Creation by Rocks, Abiogenesis, or Spontaneous Generation. Darwinism is another form of Materialism, and the foundation of Scientific Naturalism. The Theory of Evolution is Creation by Rocks. Darwinism or Creation by Rocks is false, because we all instinctively know that the rocks or physical matter cannot design and create anything at all. The rocks have never been caught in the act of doing so; and, they never will, because rocks can't design and program genomes and because rocks cannot create and manufacture cells and life forms. The evolutionary functions (or change functions) of natural selection and random mutations did not exist until AFTER some kind of Ultimate Cause or Ultimate Psyche designed and created the first genome and the first life form on this planet in the first place. This is common sense logic, which the philosophers behind Darwinism and Naturalism choose to completely ignore. Science itself has demonstrated that natural selection and random mutations cannot design and create anything at all, but the Materialists and Darwinists refuse to look at, study, and accept this evidence. Darwinism is fascinating to study, but only for purposes of debunking it and falsifying it. It's fun to see how many different ways we can falsify Darwinism and the Theory of Evolution. Remember, Scientific Methods can be used to falsify theories like the Theory of Evolution; and, it's a great deal of fun to do so and to learn how to do so. (Sanford, 2013; Sanford, 2014).

Materialism: To know what something truly is, you have to know what it is not. This Ultimate Cause Model of Reality is NOT Materialism. Ultimate cause or psyche is anti-Materialism, and vice versa. Scientific Naturalism is incompatible with Theism, Psyche, Final Cause, and Ultimate Cause. Materialism and Psyche are mutually exclusive – they both can't be true at the same time. We are forced to choose between them. Materialism let me down. Materialism almost killed me. Materialism is incapable of supporting a robust, believable, and all-inclusive Science. Materialism is the philosophical idea or religious metaphysical idea that spirit, psyche, the non-physical, the non-local, consciousness, and the immaterial do not exist. Materialism is the chosen belief that physical matter is all there is and that nothing else exists or matters. When applied to science, Materialism is often called Scientific Naturalism. Materialism is metaphysics, or a philosophical assumption, not a science. Materialism is a denial of empirical evidence and scientific evidence, and a refusal to look at that evidence. Materialism is the polar opposite of NDEs, SDEs, and OBEs; and, Naturalism is a denial that such Out-of-Body Experiences actually happen for real. There is no scientific evidence and can be no scientific evidence to support the claims or major premise of Materialism. Finding scientific support and evidentiary support for a Denialism, such as Materialism, is philosophically and logically impossible. Materialism is design and creation by rocks or by physical matter. Raw physical matter by definition in principle cannot design and create. ALL of the empirical evidence, observational evidence, and experiential evidence stands firmly against Materialism, because rocks or physical matter cannot design, create, nor manufacture anything at all. It is obvious to all of us, except for the atheists (atheists are only 3% of the population), that Materialism is false, doesn't match with reality, and doesn't hold water. Technically, Materialism is a denial of the existence of psyche, psychology, quantum non-locality, light, gravity, dark energy, dark matter, consciousness, spirit matter, forces, fields, time, magnetism, thoughts, dreams, space, and everything else that is non-physical which exists. Materialism is incompatible with the Science of Psychology, the study of psyche. Materialism is incompatible with Science, because Materialism is unscientific and doesn't match with reality. Materialism is the very definition of Bad Science and Pseudo-Science, because Materialism is in fact metaphysics or philosophy masquerading as science, but not science. (Richards & Bergin, 2005). A strong Personality Theory and a solid model of reality must match with Reality. Materialism and Scientific Naturalism do not match with reality, thus making them incomplete and unscientific models of reality. Materialism is an unsubstantiated model of reality that doesn't hold up under scrutiny. Darwinism is just another form of Materialism. They both reduce to Creation by Rocks or Creation by Chance, which are scientifically impossible and have never been observed in the wild. At their most fundamental, Darwinism and Materialism are Creation by Rocks or Creation by Physical Matter; but, we all know instinctively that the rocks cannot design and create anything at all, let alone fully functional genomes, computers, software, and complex life forms. Spontaneous generation, Creation by Rocks, Materialism, and Darwinism were proven false by Louis Pasteur in 1859, the same year that Charles Darwin published "On the Origin of Species". Furthermore, Quantum Mechanics when properly understood proves beyond a shadow of a doubt that Materialism and Naturalism are false. The Science of Quantum Mechanics trumps the philosophies of Materialism, Naturalism, and Darwinism proving Materialism of any type to be false. Once we have eliminated Materialism or Naturalism and its derivatives from our worldview, then we are obligated to try to put something much more viable and believable in its place. I suggest that psyche or ultimate cause should be the thing we use to replace the falsehoods generated by Materialism and Naturalism. (Van Lommel, 2010; Mark My Words, 2016, *Quantum Mechanics from a Non-Physical Spiritual Perspective*).

Buddhism: The Buddhists have chosen to believe that psyche or the soul does not exist and that our personality and memories cease to exist when we die.

Such an idea is not compatible with Psyche as the Ultimate Cause. Buddhism is "spirituality" for Materialists, Naturalists, and Atheists. Buddhism is Atheism, although there is some debate as to whether Buddhism is Materialism and Naturalism, because Buddhism is internally inconsistent. Parts of Buddhism are deliberate non-sense, such as the koan. Materialism and Scientific Naturalism are incompatible with Theism, and therefore more amenable to something like Atheism or Buddhism. This Ultimate Cause Model of Reality is incompatible with Materialism, Scientific Naturalism, and Atheism; and therefore, incompatible with the fundamental philosophical beliefs of Buddhism. In contrast, the Hindus believe in atman or psyche; therefore, Hinduism would be supportive of and compatible with Psyche as the Ultimate Cause. Islam is theism; and, the Muslims believe in an afterlife and the continuation of the soul or psyche into the afterlife. So once again, Islam would be compatible with and supportive of Psyche as the Ultimate Cause. Christ and the Bible promote an afterlife for the psyche or the soul. Most of the world's religions and most of this world's population would be supportive of Psyche as the Ultimate Cause; but not the Materialists, Naturalists, Atheists, and Buddhists. The law of non-contradiction states that they both can't be right. Either psyche exists, or it does not. "The existence of psyche" and the "non-existence of psyche" can't be true at the same time, because that would be a violation of the law of non-contradiction. We are each forced to choose what it is that we want to believe. We are forced to choose between the two. Belief or faith is a decision or a choice, hopefully based upon lots of evidence. When choosing between psyche and the non-existence of psyche, I have chosen to go with the one that has the most evidentiary, scientific, empirical, observational, logical, and experiential support because there is no evidence and can be no evidence to support "the non-existence of psyche" or any other such Denialism like Materialism, or Naturalism, or Atheism, or Buddhism. ALL of the evidence is on the side of Psyche as the Ultimate Cause. I finally chose to follow the evidence, which any good scientist should do.

The Scientific Method: Although Scientific Methods can never be used to prove the truth, the Scientific Methods can be used to falsify theories or to prove theories false. Science and the traditional Scientific Methods prove Materialism and Naturalism false many different ways. First of all, there is no way possible to design a science experiment which demonstrates and proves that the non-physical or the spiritual does not exist. Instead, ALL of the science experiments which we have done over the centuries keep revealing to us the existence of the immaterial or the non-physical in some form or another – such as gravity, forces, fields, extra-dimensionality, trans-dimensionality, quantum non-locality, quantum entanglement, light, dark energy, dark matter, the zero-point field, magnetism, thoughts, dreams, space, time, non-local consciousness, and a whole host of other things which are immaterial and non-physical in origin and nature. In other words, Science and Common-Sense Logic as a whole prove Materialism, Naturalism, and Darwinism false. Remember, we can use the Scientific Methods to prove theories false or to falsify theories, and we have done so when it comes to Materialism, Naturalism, and Darwinism. Second of all, the various different forms of Materialism are in fact Creation by Rocks or Creation by Chance; and, there is no way to design and create a science experiment wherein we capture rocks or raw physical matter in the process of designing and creating genomes and life forms

335

from scratch. It can't be done. Therefore, once again, we can't use the Scientific Method to demonstrate and prove that Materialism, Darwinism, and Naturalism are true. Instead, in 1859, Louis Pasteur used the Scientific Method to prove that spontaneous generation or Creation by Rocks is false. Therefore, the net effect of all of this is that the Scientific Method itself proves through Common Sense Logic, Experimentation, Procedural Evidence, and the Reality of the Situation that Materialism, Darwinism, and Naturalism are in principle false and patently non-starters. There's no way to use the Scientific Method to prove the philosophies of Materialism, Naturalism, and Darwinism true; consequently, the Scientific Method effectively proves Materialism, Darwinism, and Naturalism false. Furthermore, there is no evidence and can be no evidence to support the major premises of Materialism and Scientific Naturalism stating that psyche or spirit does not exist; whereas, there are thousands of books and videos and millions of personal experiences which provide empirical evidence supporting psyche and ultimate cause, in the form of NDEs, SDEs, OBEs, and other types of spiritual experiences or Spiritual Empiricism. If we choose to follow the evidence, we can easily prove to ourselves that Materialism and Scientific Naturalism are false. That's what happened to me. I used to be a Materialist and an Atheist, but then I finally started looking at all the evidence. Lastly, the Scientific Methods and Common Sense have already been used to prove (verify) Quantum Mechanics, Quantum Nonlocality, Quantum Entanglement, light, gravity, forces, fields, dark energy, and dark matter to be real and true, in that these things have always been verified and never been falsified; and, each one of these things proves Materialism and Scientific Naturalism false while at the same time being supportive of Psyche or Consciousness as the Ultimate Cause. Therefore, the Scientific Method has already been used to prove Materialism, Scientific Naturalism, and Creation by Rocks (the Theory of Evolution) false; but, the Materialists and the Atheists refuse to look at the evidence. Whenever we use the Scientific Method the way that it is supposed to be used, in an unbiased and unemotional and scientific manner, it proves beyond a shadow of a doubt that Materialism of any kind is false. Remember, we can indeed use the Scientific Methods to prove theories false. Such a procedure is logically sound. Once we know that Materialism and Naturalism and Darwinism are false, then we find ourselves looking for a better and more believable Model of Reality which takes into account all of the evidence at our fingertips. I submit this Ultimate Cause Model of Reality or Psyche Ontology as a possible solution. (Mark My Words, 2016, *Using the Scientific Method: To Eliminate the Usual Suspects and to Prove the Truth*; Mark My Words, 2016, *The Scientific Method: Proves That the Theory of Evolution Is False*; Mark My Words, 2016, *The Theory of Evolution Proved to Me that God Exists: Why I Am No Longer an Atheist and Why I No Longer Believe in the Theory of Evolution*).

Material Cause: Material cause is one of Aristotle's four causes. Material cause is, traditionally, the materials or the physical matter from which different items are made here in this physical realm. This is the kind of cause that the Materialists have actually chosen to believe in. Material cause is the essence of a thing – what it is made of. In the spirit world, however, consensus realities and non-consensus realities are made up of spirit matter, which would become their material cause in that trans-dimensional realm, or non-local spirit realm. In the non-local realm, spirit matter and consciousness are the essence of that domain or the material cause of that domain. According to Quantum Mechanics, psyche or non-local consciousness is the material cause and efficient cause behind the conversion of spirit matter into physical matter through conscious observation. In a sense, spirit matter is the material cause of physical matter; whereas, psyche is the efficient cause of converting spirit matter into physical matter. In this physical 3D space-time realm, physical matter is the essence by which things are made; and therefore, their material cause. (See Rychlak. Anything from Joseph Rychlak will give a solid philosophical foundation into Aristotle's four causes, which are in fact the philosophical foundation for Psyche as the Ultimate Cause).

Efficient Cause: Efficient cause is one of Aristotle's four causes. The carpenter is the efficient cause of the table, cabinets, and chair. The engineer and manufacturer are the efficient cause of your computer. In our philosophy books, we also typically see efficient cause presented in physical terms as the motion of past events which combine to make up our current physical reality. Efficient cause is therefore typically presented as the cause-effect model of our physical reality. However, things are a bit different when we start applying efficient cause to the Non-Local Realm, Quantum Realm, or Spirit Realm. According to the Science of Quantum Mechanics, observation from a non-local consciousness is required to bring each particle of physical matter into existence. Observation from a psyche is the thing which transforms a quantum wave into a physical particle, and therefore is its efficient cause. Every physical particle was brought into existence by the observation or the will of a non-local consciousness, according to the Science of Quantum Mechanics. Psyche or non-local consciousness is the efficient cause of physical matter. (Van Lommel, 2010; Mark My Words, 2016, *Quantum Mechanics from a Non-Physical Spiritual Perspective*). In the spirit world or trans-dimensional realm, the psyche is the efficient cause of turning a non-consensus reality into a consensus reality, because it is the psyche who organizes the spirit matter according to its desire and will. That's the power of consciousness or psyche. (Buhlman, 1996). Rocks or physical matter cannot act in the role of carpenter, or manufacturer, or genetic engineer, or psyche, or designer, or creator, or ultimate cause. They have never been observed doing so. Rocks and raw physical matter cannot be the efficient cause of genomes, life-forms, meaningful and useful DNA, and computer programs despite what the Darwinists claim. In contrast, Psyche can act in the role of efficient cause. It has been observed doing so. Physical matter cannot choose to act. Physical matter was designed to react. Deleterious mutations outnumber beneficial mutations thousands to one, and at times even a million to one. Species will go extinct thousands of times over before one of them will get around to evolving something useful, beneficial, and productive. Evolution doesn't work. Mutation/selection cannot design and create genomes and life forms from scratch (Sanford, 2014). This means that only God knows which mutations are beneficial and where those beneficial mutations should be placed into the genome. If we had waited to evolve from some chimp-like ancestor, we and the chimpanzees would all be extinct by now. It has been estimated that it would take trillions of years for 10,000 chimpanzees to evolve into humans "naturally", and that is with God preventing ALL deleterious mutations and with God keeping that population of chimpanzees alive for trillions of years so that they can evolve into humans "naturally". It would take trillions of trillions of years for 10,000 chimpanzees to evolve into humans naturally without God's help, protection, and intervention. Trillions of trillions is synonymous with impossible. Mutation and selection cannot design and create (cannot do formal cause and final cause), nor can they do efficient causation (manufacturing and construction from blueprints). Evolution has NO psyche; therefore, evolution cannot do the conversion of spirit matter into physical matter (material cause). Evolution cannot do formal cause (design), nor final cause (purposeful telic creation). ONLY Psyche can do these kinds of things! Evolution is a process or a reaction within some types of efficient causality, but evolution of any kind can't do the manufacturing and engineering aspects of efficient cause. ONLY Psyche can do that. Yes, evolution, mutation, and natural selection do exist; but, just because these things exist doesn't make Creation by Rocks (the Theory of Evolution) true, as the Darwinists and Materialists claim. I have encountered many Darwinists and Materialists who simply believe, based upon the fact that evolution (change) happens or mutation exists, that the Theory of Evolution, or Darwinism, or Creation by Rocks has been proven true. These people are willing to make that leap of faith and eagerly do so; but, their blind faith in Darwinism is based upon a wide variety of logic fallacies and no empirical evidence whatsoever. In contrast, we ALL simply KNOW from experience that the rocks can't design and create anything at all. Knowledge of these realities gives us a clear understanding why Evolution or Creation by Rocks is insufficient, unscientific, and incapable of designing, creating, and manufacturing genomes, information, and life forms from scratch. Evolution of any kind can't do Psyche or Ultimate Cause. (Sanford, 2014; Sanford, 2013; Mark My Words, 2016, *The Theory of Evolution Proved to Me that God Exists*).

Formal Cause: Formal cause is one of Aristotle's four causes. The formal cause is the form, pattern, design, outline, or blueprint behind each and every physical object. In psychology and philosophy, a personality type, logical consistency, and lifestyle are considered to be a formal cause. (Rychlak, 1981, p. 7). It is empirically obvious that some kind of psyche or mind is behind each and every design, pattern, software program, genome, and blueprint. Only the Materialists deny this reality; and, it is obvious that the Materialists are wrong. The Materialists are always wrong, especially when it comes to the non-local sciences or the spiritual sciences. Rocks or physical matter cannot design and create – they have never been observed doing so. Rocks or raw physical matter cannot act in the role of formal cause. Spirit, psyche, or living consciousness can. Psyche, or consciousness, or intelligence has actually been observed doing so. In recent days, I like to think of formal cause as the Designer, the Genetic Engineer, the Architect, or the Programmer.

Final Cause: Final cause is one of Aristotle's four causes – the reason, end purpose, or goal why a line of behavior is being carried out or a physical object is under creation. Meaning, purpose, intentions, hopes, wishes, desires, expectations, reasons, aims, goals, choices, plans, teleologies, "for the sake of which", and even "just for the fun of it" are all some type of final cause or telic purpose. (Rychlak, 1981, p. 7 & p. 300). Psyche is the efficient cause of formal causes and final causes. In other words, ONLY Psyche can do formal cause and final cause. Psyche is the ultimate cause of formal cause and final cause. Rocks or physical matter cannot act in the role of final cause, because the rocks cannot make plans and goals. They have never been observed doing so. The rocks have no purpose in mind that we know of. The rocks can't choose and act, in any way that we humans would find noticeable. If rocks were to have a purpose, it would in fact be some kind of psyche who would give them that purpose. When was the last time you caught a rock thinking, reasoning, planning, and acting? Rocks or raw physical matter have never exhibited teleology or purpose. Only psyche or self-aware consciousness can function in the role of final cause. This reality in and of itself is enough to prove Materialism false, which is why the Materialists, Determinists, and Radical Behaviorists deny the existence of psyche, free-will, agency, teleology, and final cause. ONLY Psyche can do final cause or teleology.

Ultimate Cause: Psyche or non-local consciousness is the ultimate cause. The ultimate message from material cause, efficient cause, formal cause, and final cause is that behind every material cause or particle of physical matter there is a psyche or an ultimate cause, and that behind every efficient cause or manufacturer there is an ultimate cause, and that behind every formal cause or pattern or blueprint or design there is a designer or an ultimate cause, and that behind every final cause or goal or choice or reason for acting there is an agent, or a psyche, or an ultimate cause. There is an ultimate cause behind each and every one of Aristotle's four causes. Everything reduces to psyche or ultimate cause. Our whole existence and reality is in fact a Psyche Ontology, wherein Psyche is the fundamental unit of reality and existence, and the ultimate cause of everything. In contrast, the Materialists and Naturalists have chosen to believe that only physical matter exists; and, these people deny the existence of psyche, nonlocality, spirit bodies, and the spirit realm. Consequently, the Materialists and Naturalists in principle deliberately exclude formal cause, final cause, and ultimate cause from their models of reality because we all instinctively know that the rocks or physical matter can't do formal cause (design), final cause (purposeful goals or thoughtful reasoning or deliberate choice or telic creation), nor ultimate cause (psyche or living consciousness or lived experiences). In fact, rocks or inanimate physical matter can't do Science nor manufacturing either; consequently, the rocks can't even do the engineering type of efficient causality. Physical matter can't design and create physical matter. Once we fully understand causality and how it really works, then we just know that Materialism and Darwinism are false, because we know why they are impossible and false. Follow the logic and follow the evidence! That's what I finally chose to do. According to Quantum Mechanics, ultimate cause is the Observer, the non-local consciousness who organizes spirit matter into consensus realities and brings physical matter into existence. According to Psychology, ultimate cause is psyche or personality or intentionality. According to Philosophy, ultimate cause is the Agent and is synonymous with free-will, agency, morality, ethics, and choice. The ultimate cause is the person or personality who chooses the outcome of each event. According to metaphysics, ultimate cause or non-local consciousness has always existed and will always exist; there's something out there that has to have always existed, and Psyche is it. Ultimate cause is the primal construct and the Prime Mover. According to the Empirical Science of Near-Death Experiences, Shared-Death Experiences, and Out-of-Body Experiences, ultimate cause or psyche or personality is the part of us that survives separation from the physical body, separation from the spirit body, death of the physical body, and brain death. According to theology, ultimate cause is the Mind of God. Every time we choose to use a personal pronoun (with the exception of a mindless "it"), we witness ultimate cause or a mind of some kind in action. Psyche is the ultimate cause. Obviously, psyche is not rocks, nor physical matter, nor some kind of inanimate non-living object. Psyche is life! Psyche animates matter, both spirit matter and physical matter! Psyche is some kind of non-local trans-dimensional living consciousness, an Observer, according to the Science of Quantum Mechanics. These people call it Quantum Consciousness. Psyche is consciousness, awareness, intelligence, and life. Everything that I encounter testifies to the truthfulness, usefulness, and existence of Psyche and Ultimate Cause. Clearly, I'm going with the Science and the Empirical Evidence on this one, and not the Materialists and their mindless philosophy. The Materialists are simply guessing, and they guessed wrong. The ultimate cause or psyche is the efficient cause of formal causes and final causes, and any other type of spiritual cause or psychical cause as well. Ultimate cause is the efficient cause of physical matter, because Psyche or a Conscious Observer is needed to convert spirit matter into physical matter. Any way we choose to look at it, Psyche is the Ultimate Cause.

Mechanists: Mechanism is another form of Materialism. Mechanists treat human beings (human psyches) as if they are robots or machines. The Mechanists purposefully limit themselves exclusively to material causes and efficient causes, which are often called mechanistic causes; and, these people formally reject formal causes, teleology, humanism, psyche, predication, agency, soul, mind, and final causes. In contrast, this Ultimate Cause Model of Reality is willing to include the whole of human experience into the mix, making for a richer and more robust Science and Model of Reality than what can be produced by Materialism, Determinism, Behaviorism, and the Mechanists.

Entity Dualism: Dualism, as used in this Ultimate Model of Reality, is the philosophical belief and the Observational Knowledge that the spirit body and the physical body are two completely separate entities. Furthermore, psyche and its assigned spirit body are two completely separate and different entities, according to the Lived Experiences of the human race. Psyches control spirit bodies, and spirit bodies can be assigned to or housed within physical bodies. Psyche as the ultimate cause relies upon dualism and is based upon the observation or out-of-body experiences, wherein the person or psyche observes that his spirit body is a completely separate and different entity than his physical body. Furthermore, the SOUL is a holistic unit comprised of the psyche, spirit body, and physical body united as one functional unity or entity. In other words, the Psyche controls the whole Unit, which we call a Soul.

Alas, classical dualists unfortunately tend to believe in a type of Deism, in which God's spirit cannot be functioning at the same time that natural laws are functioning. They can't wrap their minds around the integration of spirit matter and physical matter, nor comprehend the idea that the physical body is the temple for the human spirit to reside within. They refuse to accept the integration – the idea that a spirit or soul (psyche/spirit body) can control a physical body.

Dualistic strategies [of the Deism variety] assign one realm or world to theistic assumptions (usually religion, values, and subjectivity) and another to naturalistic assumptions (science, facts, and objectivity). In a [classical or traditional] dualism the two types of worlds cannot, in principle, interact, so again, no integration here. Theism is the belief that God is active and involved in all aspects of the world. (Brent S. Melling, from class notes).

This Ultimate Cause Model of Reality is compatible with Theism, the idea that not only does God exist but also that God or the Ultimate Cause is active in this world and can intervene in this world at will. Thus, one could say that this Ultimate Model of Reality is based upon Theistic Dualism, because psyche has to be able to be active, intervene, and interact with this physical world in order to be an ultimate cause in this physical world. It's elementary my dear friend. We KNOW from the Lived Experiences of the human race that Psyche exists. Furthermore, the Non-Local Psyche has to exist in order to bring physical matter into existence, according to the Science of Quantum Mechanics. This Ultimate Model of Reality is a Psyche Ontology, wherein psyche is the fundamental unit of reality.

Existentialism: "I AM. I exist. What more needs to be said? Forget all the fighting about monism and dualism. I choose my own being and what I am becoming. Good enough!" The philosophy of existentialism can be made compatible with this Ultimate Cause Model of Reality, because existentialism tries to take into account the whole of human existence and experience. However, in my various models of reality, I now replace the existentialist "threat of non-being" or the existentialist concern about death, nihilism, ceasing-to-exist, and annihilation with the empirical evidence and the experiential hope of the survival of the psyche after bodily death as demonstrated and proven by NDEs, SDEs, and OBEs. Remember, we can KNOW the truth and prove the truth by experiencing and living the truth for ourselves, or by choosing to trust someone who has. Knowing the truth, experiencing the truth, IS proof of the truth! Existentialism is often parodied as nihilism, thus portraying existentialism as morbidly sad and depressing. But, my brand of hopeful existentialism is willing to include psyche and ultimate cause into the mix resulting in the continuation of the human psyche after bodily death and brain death. The existentialists are heavily into free-will, choice, freedom, meaningfulness, teleology, final causality, and moral responsibility. These are in fact some of the logical conclusions of Psyche as the Ultimate Cause. There are theistic existentialists and atheistic existentialists. Obviously, this Ultimate Model of Reality would be most compatible with theism, because God does indeed qualify as an Ultimate Cause. In contrast, this Ultimate Cause Personality Theory would technically be incompatible with the atheistic and nihilistic forms of existentialism – the depressing forms of existentialism which people like to laugh at and joke about. Jokes about death are some of the funniest, but the psyche or non-local consciousness cannot die. Since psyche has no physical or material or corruptible component, psyche is eternal and immortal.

Spiritual Empiricism: Typically, whenever the word "empiricism" is used by the scientists and British Empiricists, they do in fact mean "physical empiricism" and try to limit their empiricism to physical evidence of some kind – British Empiricism is Traditional Empiricism. This Ultimate Cause Model of Reality not only fully embraces physical empiricism, but it also embraces spiritual empiricism, or non-local empiricism, or Radical Empiricism. The Materialists and British Empiricists are determined to limit and cripple their science; but, I am not. Been there and done that; and, I don't want to go back. Spiritual Empiricism is in fact spiritual experiences of any kind, including NDEs, SDEs, OBEs, and modern-day revelations from the Biblical God Jesus Christ. Evidence of non-locality and spirit has exploded exponentially since the year 2000 to the extent that the evidence is so overwhelming now that it is impossible to deny. There are now hundreds of different videos about NDEs and OBEs on YouTube, thousands of respondents to Dr. Long's online surveys about NDEs and seeing God while out-of-body (Long, 2016), and many different studies of hundreds of hospital patients who have had NDEs and OBEs (Gibson, 2006). Recent polling indicates that millions of people on this planet, as much as 10% of the human population has experienced some kind of OBE and/or NDE, and as much as 30% of this world's population has experienced a life-changing spiritual experience (See: http://mypsyche.us/how-common-are-ndes/). There's no question now that the NDE phenomenon is real and has happened to millions of different people on this planet – one would have to have the blind-faith and blind-loyalty of a Materialist or an Atheist in order to be able to deny all of the spiritual evidence or Spiritual Empiricism that we now have access to as a race. I couldn't do it anymore. I no longer have faith enough to be an Atheist or a Materialist. I now own hundreds of different books about NDEs, SDEs, OBEs, and other kinds of spiritual experiences. The evidence became too overwhelming to deny. I had to make a paradigm shift in my own thinking processes. I had to switch to a new and different model of reality. I can no longer straddle the fence like I tried to do for most of my life. Any kind of spiritual experience, psychic experience, spiritual evidence, trans-dimensional evidence, transpersonal evidence, and Quantum Non-locality is fair game within this Ultimate Cause Model of Reality. That's why I chose to call it the Ultimate Model of Reality, because it's not limited to Materialism and Naturalism.

Nihilism: Nihilism is the philosophical belief that "all we are is dust in the wind". Nihilism is another one of the Denialisms, a denial that there is an afterlife and a denial that life has any purpose or meaning. Denialisms, such as Nihilism, have to be taken on blind-faith as being real and true, because the various Denialisms contradict ALL of the empirical evidence that we have on hand as a race. From the Wikipedia, "Nihilism (from the Latin nihil, nothing) is a philosophical doctrine that suggests the lack of belief in one or more reputedly meaningful aspects of life. Most commonly, nihilism is presented in the form of *existential nihilism*, which argues that life is without objective meaning, purpose, or intrinsic value. Moral nihilists assert that morality does not inherently exist, and that any established moral values are abstractly contrived [man-made]. Nihilism can also take epistemological, ontological, or metaphysical forms – meaning that, in some aspect, [truth does not exist], knowledge is not possible, or that reality does not actually exist." (End of wiki quote). Nihilism is a type of relativism which claims that truth does not exist. Truth is knowledge, and knowledge is truth; and, Nihilism claims that truth and knowledge do not exist. Nihilism is a denial of reality. Nihilists deny the existence of an afterlife. The Atheists have chosen to believe in "Creation by Nothing", which takes a great deal of faith to believe to be true. The

Nihilists believe in nothing. Nihilism leads a person to moral relativism, angst, depression, ennui, and Atheism. In contrast, this Ultimate Cause Model of Reality goes in pursuit of the True Reality of our existence, as revealed to us by empirical evidence, spiritual experiences, lived experiences, first-hand observations, scientific experimentation, NDEs, SDEs, OBEs, scriptural evidence or revealed evidence, and the like. This Psyche Ontology proves that Nihilism is false by falsifying Nihilism in many different ways.

Pursuing the Ultimate Psychotherapy: A study of Personality Theories will reveal that most, but not all, of the dozens of different Personality Theories and all of the major Personality Theories were based upon a psychotherapeutic treatment method of one sort or another. Since I'm a philosopher and a scientist, and not a psychotherapist, I'm not going to develop and promote a separate treatment method for this Ultimate Personality Theory. Instead, I prefer to adopt what has already been proven to work and has proven efficacious. Cognitive Therapy was designed by Atheists to treat and cure Atheists; and, it works for the treatment of depression. Cognitive Therapy will work for you whether you are an atheist or not. David Burns' book, "Feeling Good", is the ultimate self-help Cognitive Therapy book for treating depression; and, it worked for me. Other therapy techniques that seemed to be developed by Agnostics for Agnostics, which could be made compatible with the spirit and the feel of this Ultimate Cause Model of Reality, are William Glasser's Choice Therapy and Carl Rogers' Client-Centered Therapy. In my most recent round of psychotherapy, I found Mindfulness helpful and useful – learning to be present in the present, stop ruminating about the past which can't be changed, and stop fearing the future which isn't here yet. Live for the moment in the present moment! Depressed people live in the past and beat themselves up over their past mistakes; and, anxious people are constantly afraid of the future. Stop it! Learn to live in the present, here and now. Mindfulness is an Americanized adaptation from Buddhist Psychotherapy. Existential Psychotherapy as presented by Rollo May and Irvin Yalom in an article by the same name could be made compatible with the spirit of this Ultimate Cause Personality Theory. Existential Psychotherapy is supposed to be a useful supplement for any other type of psychotherapy, which one might choose to employ. Existential Psychotherapy emphasizes taking responsibility for one's own life and one's own future, a process which I found extremely helpful during my last round of psychotherapy. Existentialists believe in free will, choice, and moral responsibility which are all functions of Psyche. I went to a seminar by a clinical psychologist; and, he said that in his clinic the Biopsychosocial Model has become the standard model for therapeutic treatment – a holistic approach treating whatever currently ails the client. This Ultimate Cause Personality Theory is completely compatible with the Biopsychosocial Model of Reality, including the "psycho" or psyche part of the model. This Ultimate Cause Personality Theory would be amenable to an Eclectic Psychotherapy – pursuing whatever works to help and benefit the client. As already mentioned, I see Biopsychosocial Psychotherapy as the pinnacle of an Eclectic Psychotherapy – psychotherapy and treatment tailor-made to the needs of the client. The last time I got sick, dozens of different things went wrong; and in order to get better, dozens of different things had to go right. An eclectic approach and a holistic approach were the right way to go. Though I haven't experienced it

343

personally, Theistic Psychotherapy as presented in the second chapter of *Casebook for a Spiritual Strategy in Counseling and Psychotherapy* and the fifth chapter of *Spiritual Strategy for Counseling and Psychotherapy* (2nd ed.) looks interesting and promising. Theistic Psychotherapy is something which I wish that I could have experienced or been involved with in my life. Even so, I can testify that bringing the healing power and the psychotherapeutic power of the Atonement of Christ (Christ's mercy, forgiveness, love, and grace) into my life has had massive life-changing and extremely positive psychotherapeutic and physical effects. As I see it, the Atonement of Christ is the Ultimate Psychotherapy. Christ can cure anything that ails you. The only question is whether He will do so, or whether He has other plans in mind for you. If I were forced to choose a single psychotherapy for this Ultimate Model of Reality, I would choose the Atonement of Christ, because it works and it works best, from all that I have experienced in my life so far.

The Ultimate Epistemology: While developing this book, I didn't even know what my ultimate goal was. I was just putting out commentary explaining how the various Personality Theorists led me to Psyche as the Ultimate Cause. In a class I took, I was called upon to develop a Personality Theory of my own, and I took it to heart, eventually developing this Ultimate Cause Model of Reality and a Psyche Ontology to go with it. In recent days, it dawned on me that I was in fact looking for the Ultimate Epistemology while developing my books. Epistemologies are various different ways of Knowing the Truth or coming to KNOW the Truth. During my research and commentary, I was in fact looking for the BEST Personality Theorist who had found the best way of uncovering and knowing the truth. Ironically, I didn't find a single one; so, I had to become that one, and develop the Ultimate Epistemology all on my own. It took half a year for me to figure out what I was doing, what I was looking for, and what is in fact the Ultimate Epistemology. The EVIDENCE kept pointing me to what I now call a Psyche Epistemology, wherein the Lived Experiences of the human race or the Lived Experiences of the human psyche becomes the BEST and the most efficient way of KNOWING the Truth. I simply kept looking at the EVIDENCE, which I personally found most convincing; and, ALL of that evidence was comprised of some kind of Psyche Experience or Spiritual Experience or Lived Experience; and then, it became obvious to me that our Lived Experiences are in fact the BEST, the most convincing, and the most effective way we have of finding and knowing the Truth about our Reality. Lived Experience IS a Psyche Epistemology, because ONLY Psyche can do lived experience, truth, and knowledge. A Psyche Epistemology is based strongly upon the Phenomenological, Radically Empirical, and Hermeneutic Epistemologies. I also observed that a Psyche Epistemology subsumes ALL other epistemologies, because ONLY Psyche can do ALL of the other epistemologies. It was obvious and clear. I had finally found what I was searching for. A Psyche Epistemology is the Ultimate Epistemology, because ONLY Psyche can do epistemology and ONLY the human psyche can do epistemology to its fullest extent. It's elementary my dear friend. Scientism, Materialism, Naturalism, Nihilism, Darwinism, and Atheism ARE religions. The practitioners of those religions have chosen to believe that Scientific Methods and Scientific Experimentation ARE the best way of finding and knowing the truth. For the first fifty-five years of my life, I too was of that opinion. I relied heavily and exclusively on Science and Science Methods, because I truly believed that they were indeed the BEST way of finding and knowing the truth. Mine was a Scientism Epistemology. Even when I had successfully abandoned My Materialism and My Atheism, I continued to maintain my belief in Scientism. I knew of no other way. If you are going to tear something down, you are obligated to put something better in its place; and, for most of my life I had NOTHING better to put into the place of My Scientism. But now I do. Psyche is the Ultimate Epistemology and therefore the ultimate way of finding and knowing and living the truth. ONLY Psyche can do Science, Scientism, scientific experimentation, scientific interpretation, and scientific methodologies. A Psyche Epistemology subsumes ALL other epistemologies, because ONLY Psyche can do epistemology!

The Ultimate Personality Theory and Ultimate Ontology: While developing my books, I started by searching for the Ultimate Personality Theory. In one of my classes, I was assigned the task of developing my own Personality Theory. I eventually came to the conclusion that NONE of the Personality Theorists had the BEST personality theory; so, I decided to develop one of my own, which I thought was better than all the rest. Thus, this Ultimate Personality Theory was born. This Ultimate Personality Theory states that Psyche or Personality and our memories survive bodily death, separation from the spirit body, and brain death. Although I didn't know it at first, I was of the opinion that the Ultimate Personality Theory should in fact be based upon the Fundamental Unit of Reality. Over time, it became clear and obvious to me that Psyche is the Ultimate Causal Agent, and thus Psyche is the Fundamental Unit of Reality. As a result, this Psyche Ontology and this Psyche Personality Theory were born. I wanted to find and go with the BEST, and I believe that I have done so. I can't think of anything more fundamental and essential to our lives, our reality, and our existence than Psyche or our own personal Non-Local Consciousness. A Psyche Ontology is the Ultimate Ontology, because Psyche IS the fundamental unit of our own Personal Reality. Consequently, this Psyche Personality Theory is the Ultimate Personality Theory, because ONLY Psyche can do personhood and personality. Furthermore, we KNOW from the lived experiences of the human race that Psyche exists. Psyche or personality is the part of us that survives separation from our spirit body, separation from our physical body, and the death of our physical body. It's elementary my dear friend. It doesn't get more fundamental than Psyche. I now treat Psyche and Ultimate Cause as LAW, because it has become clear to me that Psyche is the Ultimate Ontology, that Psyche is axiomatic, and that Psyche IS the Hidden Assumption or the Invisible Variable in ALL of our theorizing, philosophizing, hypothesizing, experimentation, science, lived experiences, spiritual experiences, empiricism, and scientific interpretations. A Psyche Ontology subsumes ALL other ontologies, because ONLY Psyche can do ontology and metaphysics. This is the Ultimate Model of Reality, after all. In their many free articles, Brent Slife and friends present a compelling case for a Relational Ontology, wherein relationships and friendships are the fundamental unit of reality. That's all fine and well, but I noticed that Psyche is more fundamental than relationships, because ONLY Psyche can do relationships and friendships. This Psyche Ontology nicely subsumes or includes a Relational Ontology, because a Psyche Ontology subsumes or includes every other ontology because ONLY Psyche can do ontology. The goal was to find the fundamental unit of reality, and I believe that I have, because ONLY Psyche can do Reality and Lived Experiences and actually remember having done so. Go with the BEST and subsume all the rest – that's what I chose to do. I find this whole thing amazing. I don't see a Psyche Ontology anywhere in the literature. Everyone seems to have blinded themselves to the possibility. I didn't see it nor understand it either, until after it was revealed to me. It's so clear and obvious to me now, that I sometimes wonder how the whole world has been able to overlook it throughout the whole of human history. I sometimes wonder how I was able to overlook it for the first 55 years of my life; but then again, I wasn't looking for it, now was I? No seeking, then no finding! And, I never received any help or encouragement from my school teachers, either. A Psyche Ontology is definitely something that you will NEVER get from your materialistic, naturalistic, and atheistic college professors. Now is it?

The Ultimate Psychotherapy: All throughout my life, I have submitted to different types of Psychotherapy, because I really needed it at the time. While developing my books, I observed that ALL of the Personality Theorists developed their own unique brand of Psychotherapy, technique, and treatment for their clientele. As I put this Ultimate Personality Theory together, I suddenly found myself looking for the Ultimate Type of Psychotherapy. After examining my personal experiences and observing what worked BEST for me and what had the BEST and the most long-lasting benefits and effects, it became obvious and clear to me that the Atonement of Christ is the Ultimate Psychotherapy. There is nothing better or more effective! ONLY the Atonement of Christ can touch, influence, and change the Psyche directly; thus, it became obvious and clear to me that the Atonement of Jesus Christ IS the best and most effective form of Psyche-Therapy. It's elementary my dear friend. I went looking for the Ultimate Model of Reality, and I believe that I have found it.

The Ultimate Cause Model of Reality: This Ultimate Cause Model of Reality is a holistic model of reality, which means that it is automatically incompatible with the exclusionary models of reality that deny the existence of some aspect of Reality. The exclusionary models of reality that deny the existence of Psyche or Non-Local Consciousness cannot be made compatible with Psyche as the Ultimate Reality or Psyche as the Fundamental Reality of our existence. The exclusionary models of reality are Denialisms, which faithfully claim that some aspect of Reality does not exist. These Denialisms are philosophies such as Naturalism, Materialism, Scientism, Positivism, Physical Empiricism, Nihilism, and Atheism. These Denialisms, or exclusionary models of reality, deny the existence of some aspect of Reality. These exclusionary philosophies can NEVER be made compatible with holistic models of reality, because Denialisms always exclude some aspect of Reality. This is the Ultimate Model of Reality because it attempts to include and embrace the Whole of Reality. Nothing is excluded! In ALL of my theorizing, I kept adjusting this Ultimate Model of Reality until it matched with the Lived Experiences or Psyche Experiences of the human race. I wanted to get this model as close to the Truth and Reality as I possibly could.

A Lesson in Quantum Mechanics or Spiritual Mechanics

Quantum Mechanics explains the Supernatural – what it is and how it works.

As their mathematical equations approach infinity and break down, their equations are in fact approaching Psyche, who lives and works in the realm of the infinite. Whether we are talking Relativity, Calculus, Trigonometry, Geometry, Quantum Mechanics, or any other type of math, the Materialists, Naturalists, and Atheists freak out whenever their equations go infinite and break down on them; but, where else are their equations going to go? These people are thinking materialistically and spatially, rather than transdimensionally and psychically. Their naturalistic thinking greatly limits them.

Nevertheless, as their equations become infinite, their equations become Psyche, who IS infinite in every respect. Whether we are talking about the infinitely large, or the infinitely small, we are talking about Psyche, who is at home in the infinite and resides in the infinite. We are talking about God. (Please see the **y = tan(x)** model in my book, *The Ultimate Model of Reality: Psyche Is the Ultimate Cause*, for a graphic explanation of how all of this works in practice.)

When it comes to infinity, it's all the same to Psyche, because Psyche IS infinite. Psyche is an infinite being. This Reality is so glaringly obvious, that I sometimes wonder how we all managed to overlook it for the whole of human history. Nevertheless, the Materialists and Naturalists have trained us not to see it. Their teachings are ingrained right into us and are hard to get rid of. I KNOW, because I used to be a Materialist and an Atheist.

The Materialists and Naturalists cannot deal with infinity nor handle the infinite.

Why?

It's because these people limit themselves to the physical; and, the physical is finite. The physical is finite! It's limited! As a result, these people and their science are limited and finite. It can't be helped, except by getting rid of their Materialism and Naturalism.

But, their math isn't lying to them. Everything moves towards the infinite. In other words, everything moves towards Psyche, because Psyche IS infinite in every respect.

When the math equations move towards the limit, or move towards an infinite sampling, or move towards infinity in general, they are in fact moving towards Truth, Knowledge, Reality, and Psyche. Anyone who truly understands Calculus understands this Reality. With an infinite sampling, the math moves towards the limit, and Psyche resides at the limit, because Psyche IS Truth, Knowledge, Existence, and Reality.

Naturalism and Materialism mess everything up, making them incomprehensible. This applies to Quantum Mechanics as well. Thanks to

Materialism and Naturalism, there's a ton of confusion associated with Quantum Mechanics. Many mathematicians and scientists believe that Quantum Mechanics has been falsified because their equations break down or go infinite on them. However, they have NOT falsified Quantum Mechanics. Quantum Mechanics and the math are telling them the truth! Reality and the math do indeed move towards Psyche or the Infinite. What these people have indeed done is falsify the Finite! Remember, Materialism and Naturalism are Finite. In other words, these people have successfully falsified Materialism and Naturalism with their mathematical equations that go to infinity and break down.

It's all a matter of one's personally chosen interpretation of the math and the scientific data! There's more than one way to look at these things; but, the Materialists and Naturalists choose ONLY one way, a very finite and limited way, to look at science and reality. It holds them back. They can't deal with it or handle it, because it's Psyche and it's Infinite. As the math moves towards the limit or moves towards the infinite, it is in fact moving towards Psyche who lives and resides at the infinite. Materialism and Psyche are opposites. Materialism or physical matter is finite. Psyche IS infinite!

I'm passionate about Quantum Mechanics. Quantum Mechanics has been verified or validated dozens of different times in dozens of different ways. It has been estimated that a third of our economy in the United States is based upon Quantum Mechanics to one extent or another – its practical utility is undeniable. It has been claimed that Quantum Mechanics has never been falsified, and that no aspect of Quantum Mechanics has ever been falsified. Instead, Quantum Mechanics has been repeatedly verified through scientific experimentation. Quantum Mechanics is Spiritual Mechanics – the way that spirit matter really works. Quantum Mechanics IS the science behind consciousness, psyche, and near-death experiences! (Van Lommel, *Consciousness Beyond Life: The Science of the Near-Death Experience*).

Although Science and the Scientific Methods can't be used to prove the truth, Quantum Mechanics is probably the closest we have ever come to a Proven Science as a race, due to the fact that Quantum Mechanics has never been successfully falsified. Furthermore, Quantum Mechanics is the closest we have ever come to using Science to prove the existence of Psyche, Non-Locality, the Spirit Realm, and Non-Local Consciousness, because Quantum Mechanics teaches us that a Psyche or Conscious Observer is necessary in order to bring physical matter into existence in the first place. Furthermore, Quantum Mechanics is Spiritual Mechanics, or the way that spirit matter works. Quantum Mechanics explains to us how spirit matter gets converted into physical matter through Conscious Observation or the Word of Command. (Van Lommel; and Mark My Words, 2016, *Quantum Mechanics from a Non-Physical Spiritual Perspective*.)

We KNOW from the Lived Experiences of the human race that Psyche exists. Quantum Mechanics lends evidentiary and scientific support to that KNOWLEDGE.

It had to be pointed out to me by my best friend, because I wasn't looking for it at the time, but there are now thousands of different Near-Death Experiences and Out-of-Body Experiences on YouTube just waiting for you to find them and watch them. Most of these verify the existence of our spirit body as a separate object distinct and different from our physical body – a spirit body, human in shape, comprised of some sort of light, photons, stars, or particles of spirit matter. Spirit matter is light, and light is spirit matter. The rest of these NDEs and OBEs on YouTube verify the existence of our Psyche or Intelligence as an immaterial, disembodied, third-person viewpoint in space – a spark of life, an infinite singularity, who is able to look at its assigned spirit body from a third-person vantage point outside of that spirit body and also from the third-eye perspective which is more of a first-person perspective.

One lady on YouTube said that when she passed away, she was floating up near the ceiling watching her physical body. Then she said that she went to look for herself; and, there was nothing there. She said, "I was just a spark on the ceiling". Spark is another word for Psyche, a viewpoint in space. Our spark is our life or our psyche. She had apparently separated from both her spirit body and her physical body. It can be done. (For another example see: Buhlman).

Psyche or non-local consciousness is some kind of matterless infinite singularity, quantum singularity, or unlimited singularity (not limited by physical laws, nor limited by spiritual laws). Psyche, or this immaterial and formless viewpoint in space which Buhlman mentioned, is the polar opposite of a physical 3D space-time universe – making psyche an infinite universe taking up no space whatsoever, or a universe in a viewpoint – infinite, non-physical, immaterial implosion having no size whatsoever.

Imagine it! No size and no dimension! Zero is a different type of infinity. Think of the other times when we might have heard of the concept of an "infinite singularity" being discussed and used in Science. I'm not the first person to come up with the idea, although I might be the first person to apply the idea to the immaterial, disembodied, and non-local Psyche.

Psyche is Intelligence, Light, Truth, and Knowledge. ONLY Psyche is capable of having and remembering lived experiences and knowing the truth. ONLY Psyche can do life, lived experiences, truth, and knowledge.

Psyche is infinite memory and information storage that takes up no space whatsoever. Psyche is our own personal truth – what we truly are at our very core. Psyche is the fundamental unit of reality. Psyche is Light – a Living Light – a Light which exists at an infinite frequency thus making it or transforming it into something that is completely different than light or spirit matter. It becomes consciousness and life.

Just as combining hydrogen gas with an oxygen gas transforms the mixture into a completely different substance, a liquid that we call water; taking light and ramping it up to an infinite frequency, infinite velocity, omnipresent, infinite potential, infinite capacity, and infinite implosion state completely transforms that light into a whole new unity or substance that we call Psyche, Consciousness, and Life. By taking on all those properties, Psyche becomes something completely different than it was before. It transforms into a whole other unique entity. It can see, without needing eyes. It can think without needing a brain. It can speak and hear without needing a mouth and ears. It can sense and feel without needing skin. It can live without needing sustenance. Psyche can move and interact psychically, telepathically, and telekinetically with its surroundings without needing a body to do so.

Psyche or living consciousness is a third type of thing, completely different than spirit matter and completely different than physical matter because non-local consciousness seems to be completely immaterial, both here in this physical realm and there in the spirit realm. That's what makes Quantum Consciousness or non-local consciousness so powerful, so unlimited, and so infinite and omnipresent and instantaneous in scope and range, because it doesn't have any material component at all limiting it and tying it down, neither a spirit matter component nor a physical matter component. Psyche has the potential to be omnipotent, omnipresent, infinite, and eternal or everlasting.

Pure Psyche or Pure Intelligence can function transpersonally or psychically instantaneously at infinite distances at an infinite velocity because it has no matter associated with it – no spirit matter and no physical matter. Psyche can actually be all-knowing and omnipresent.   Psyche or Quantum Consciousness is a third type of thing because it is the only thing that can collapse a quantum wave thereby converting spirit matter into physical matter, according to the Science of Quantum Mechanics.

Matter of any kind, whether spirit matter or physical matter, provides limitations as well as new possibilities and new opportunities. Psyche or Quantum Consciousness is not a quantum object (matter), because consciousness has no spirit matter and no physical matter, although psyche or consciousness can be assigned to and attached to a spirit body and eventually to a physical body as well. Matter of any kind imposes limitations or boundaries or determinants, whether we are talking about spirit matter or physical matter; but, matter also opens up new and different opportunities.

Since psyche or non-local consciousness is completely immaterial, in its pure form and original form it is an infinite singularity with infinite possibilities and no limitations whatsoever. Quantum Consciousness is life – eternal life. Since psyche or consciousness is completely immaterial in nature and origin, it can neither be created nor destroyed. It just is. It's life. Non-local Consciousness is existence and life, intelligence and consciousness, light and truth in its pure and original form. These are fascinating concepts which I was never able to get from my materialistic and atheistic friends and teachers because they had limited themselves exclusively to the study of physical matter.

Classical Physics explains the laws behind physical matter; and, Quantum Mechanics explains the laws associated with spirit matter. They are different, because spirit matter can theoretically exist and function at an infinite velocity and spirit matter has innate velocities which are at least faster than the speed of light; whereas, physical matter is spirit matter which has been reduced to sub-light speeds. Spirit matter is light; and light is spirit matter. Physicist David Bohm describes physical matter as frozen light, or light (spirit matter) which has been slowed down to sub-light speeds and thereby localized into our physical space-time realm. Conscious observation or the Word of Command is the thing that slows down spirit matter or light or an energy wave transforming it into physical matter or a photon thereby freezing it into physical matter or locality, and thereby localizing it into our physical 3D space-time universe.

We know that thoughts and dreams are spiritual experiences or quantum experiences, because our thoughts and dreams cannot be captured, measured, nor recorded by our physical instruments. If you have ever had a thought or have ever had a dream, then you have had a spiritual experience, a conscious experience, and a non-local experience. Even the Materialists and the Atheists have spiritual experiences, but they never recognize them as such because of their cognitive bias and chosen worldview which limits them to thinking of things only in physical terms.

Matter of any kind provides boundaries, determinants, and limitations along with a whole host of new and different opportunities and perspectives. In a similar manner, the philosophy of Materialism limits people to an extremely scaled-down and lobotomized form of science, a pseudo-science or semi-science of sorts. We have to be willing to break free from Materialism in order to make quantum leaps in our scientific understanding of Reality. Materialism holds us back and keeps us in elementary school. However, as part of Reality, we must also realize and acknowledge the fact that an understanding of the Physical Laws has supplied us with much of our technology and infrastructure; so, an understanding of the physical isn't completely worthless. A physical reality is the ultimate consensus reality; and, this is the Ultimate Model of Reality after all.

According to Quantum Mechanics, a **quantum object** has two different complementary states of existence, a spirit-like wave-like non-local infinite-velocity simultaneously-everywhere quantum wave state and a localized sub-light-speed finite physical particle state, after the quantum object has been observed by a living consciousness. Quantum objects can be either in spirit matter format or in physical matter format, but NOT in both formats at the same time according to the quantum law of complementarity. A quantum object manifests as either a spiritual non-local wave or as localized space-time particle, but not as both at the same time. Most scientists get this one wrong thinking that a quantum object is simultaneously wave-like and particle-like, which is not the case. A quantum object is in either one state or the other at any given point in time. (Van Lommel, 2010; & Bishop, 1998).

Interestingly enough, if spirit matter or a quantum wave can be converted into a physical particle through conscious observation or the Word of Command, then the reverse process is theoretically possible as well – converting a physical particle back into spirit matter or a quantum wave. We just have to learn how to do it; and, it ultimately won't be a physical process; it will be some kind of spiritual or conscious process instead – something ultimately caused by a Psyche. In other words, in order for it to happen, God has to allow it to happen or cause it to happen. God's Psyche holds the keys to these kinds of things. That's why He is God, and we are not.

Some people believe that there are quantum objects, such as electrons and subatomic particles, who pop back and forth between the wave-state and the physical-state at will. That would suggest that each electron has a spark of psyche residing within it, because ONLY Psyche can do agency and will. There's a lot more going on there than meets the eye.

Non-local consciousness or living psyche is a completely different animal and only has an immaterial format – it's not spirit matter and it's not physical matter, which means that non-local consciousness is not a quantum object. Non-local consciousness is the thing which has the power and the ability to collapse a quantum wave function and slow down a quantum object, thus converting that quantum object from spirit matter into physical matter. Quantum objects were designed to be acted upon; whereas, Non-Local Consciousness or Intelligence is innately The Actor and The Observer – always has been and always will be.

In the Book of Abraham in the *Pearl of Great Price*, God tells us that He is more intelligent than all the rest of us combined, which is why He is God and the rest of us are not. God apparently has a complete omniscient and omnipotent understanding of non-local consciousness, psyche, infinite singularities, quantum mechanics, and classical physics, which we do not. God knows how to convert one of those infinite singularities or quantum singularities into a physical 3D space-time universe like ours. And when it is time to do so, I imagine that God also knows how to convert our physical universe back into a quantum singularity or an infinite singularity once again.

Each psyche or infinite singularity or point of consciousness has a different degree of light, glory, and intelligence than the one next to it. It's possible for there to be one of these infinite singularities inside each and every electron with plenty of room to spare. Physicists like David Bohm believe that each electron and each sub-atomic particle is conscious, alive, and aware of its surroundings and its existence, and that they can communicate with each other instantaneously across great distances. (Bishop, 1998). Therefore, it's theoretically possible that the consciousness or the infinite singularity associated with an electron could one day be upgraded to become the controlling consciousness of a spirit body or a physical body, or one day become a God. Each point of consciousness is capable of enlargement, or capable of a greater stewardship and responsibility.

Furthermore, if an electron is alive and aware of its surroundings, it's because it has a psyche of some sort within it. Raw physical matter and spirit matter by itself is incapable of having lived experiences and forming memories, as witnessed by Out-of-Body Travelers who have observed that their physical bodies stop having lived experiences and stop forming memories while their spirit body and psyche are separated from their physical body. While outside their physical body looking at their physical body, they can no longer feel their physical body's pain and their physical body doesn't remember that pain. It's only when they go back inside their physical body that the pain returns. Also, while outside their spirit body, their spirit body isn't making and storing memories either! Only the Naturalists, Materialists, and Atheists are incapable of wrapping their minds around concepts such as these!

Psyche or immaterial non-local consciousness appears to be a whole universe within an infinite, timeless, immaterial singularity with no physical limits and no spiritual limits whatsoever. Imagine what such a thing could design and create, having no limits and no limitations whatsoever. Such a thing could design and create universes, like our physical 3D space-time universe and comprehend the whole thing all at once. And, the Materialists and Darwinists want us to believe that we descended from pond scum, fish, monkeys, and apes. We humans are the children of the Gods – co-equal and co-eternal. That's what Quantum Mechanics and Buhlman's OBEs are trying to teach us, or at least that's the message I got from Quantum Mechanics and Buhlman's OBEs. (Buhlman; Van Lommel; Mark My Words, 2016, *Quantum Mechanics from a Non-Physical Spiritual Perspective*).

Again, Matter is a Quantum Object. A full understanding of both aspects of matter opens up a whole new world of possibilities and application. A Quantum Object, or matter, can either exist as spirit matter or as physical matter; but, it cannot exist in both states simultaneously, according to the Quantum Law of Complementarity. A Quantum Object is either in its spirit matter phase or in its physical matter phase; but, it can't be in both phases simultaneously.

We KNOW from Quantum Mechanics (or Spiritual Mechanics) that what the Bible calls the WORD – the Word of Command – and what the Quantum Physicists call a Conscious Observer or Non-Local Consciousness is necessary to convert spirit matter into physical matter. That's powerful and useful science there!

What else does this reality tell us?

It tells us that, if the Word of Command or God's Psyche can convert spirit matter into physical matter at will, then the same Word of Command should be able to convert physical matter back into spirit matter at will. Think about it! That's powerful science! It's much better than anything you will ever get from the Materialists and the Atheists.

It had to be pointed out to me by my best friend, because I wasn't looking for it at the time, there are now thousands of different Near-Death Experiences (NDEs) and Out-of-Body Experiences (OBEs) on YouTube just waiting for you to find them and watch them.

Many different people on YouTube have observed and stated that their spirit body can teleport from one location to another instantaneously. Their spirit body can levitate, and fly, and pass through ceilings and walls. They can simply walk, while in the spirit realm, if they want to. Their spirit body can be in two places at once. In non-consensus realms, their spirit body can take any form they like, and their psyche can conjure up any type of reality and surroundings that it wants. Their spirit body and/or psyche can hear the thoughts and prayers of others. Spirit beings communicate telepathically. The Angels have NO wings. Their spirit body can transform into trees and animals and other things; and, these people can spend time experiencing what it is like to be a bear, or an alien grey, a snake, or one of the blues or Nordics, or some such. One man on YouTube said that as he walked up a flight of stairs, he could see from the third-eye view that his spirit body began to dissolve beneath him. He said that by the time he got to the top of the stairs, his spirit body had transformed into an orb of light; and, he was apparently looking at this orb of light from a third-person perspective or psyche perspective. Others have said that their spirit bodies are comprised of millions of points of light – like an ocean of stars.

Physical matter is condensed light or frozen light – light which has been slowed down to below the speed of light. Physical matter is the ultimate consensus reality – highly reliable and stable and sure. Spirit matter is light; and, light is spirit matter. Spirit matter exists at velocities and frequencies faster than the speed of light. Spirit bodies can take upon themselves any shape they want. Spirit bodies are malleable or can manifest in many different shapes and forms.

Most of the truly unidentified UFO's and aliens that people have seen are in fact spirits, who have transformed themselves into a different shape besides a human shape. UFO's function like spirit bodies, or spirit orbs, or quantum objects, or light. The alien spaceships which some people have seen and gone into are spirit realms or communities which have been transformed into alien spaceships by demons – Satan's followers – evil spirits – a group of human spirits who can take on any shape they desire and can create any kind of reality they desire, in an effort to deceive us and distract us from the truth. Demonic "spaceships" made out of spirit matter can travel much faster than the speed of light and teleport – or seem to. The spirit realm in the hands of evil spirits becomes a full holodeck experience, the matrix, and a complete star trek experience; and, it is infinitely more real than any of our science fiction movies can ever be. (Ross, Samples, & Clark).

Should we ever fall under or submit to Satan's power and influence, he and his demonic followers can run us through any holodeck experience of their choosing, just for the fun of it. While out-of-body, the human spirit can therefore have the full alien-autopsy experience under the ministration of demons who have made themselves look like aliens; and, the resulting spiritual experience will be as real as anything these people have ever experienced in their mortal lives or physical lives. Many demons or evil spirits have learned how to interact with and influence the physical realm telekinetically – with laser beams, pyrotechnics, UFO's, explosions, burns, levitation, cutting, slaps, punches, sound effects, dragon's breath, and the whole works. (Ross, Samples, & Clark).

People who use drugs, LSD, peyote, ayahuasca, alcohol, oxygen deprivation, witchcraft, sorcery, whirling, crystal balls, and hypnosis to force an NDE or OBE to happen often experience some really strange things, because they have unknowingly opened themselves up to Satan's influence and demonic influence. Satan is the master deceiver.

For example, under hypnosis we can form false memories and manufacture whole new realities, such as past lives that we allegedly lived as an ape, or a tree, or an ant, or Caesar, or Cleopatra, or an alien grey. We can also go and live in consensus realities in the spirit world, such as Atlantis or Lemuria and experience complete lifetimes of memories. We can contact aliens living on other worlds or supposedly flying in spaceships through space, and receive revelations from them. Or we can talk with snakes, bears, and other animals. They seem to be real aliens or animal spirits, but they are in fact Satan and his demonic followers giving us the holodeck experience of our choosing.

False memories, false lives, and manufactured lives and realities feel every bit as real and memorable as our own lives, if not more so. Most people who have OBEs and NDEs state that being Out-of-Body feels infinitely more real and clear and solid than being in a physical body. Anything becomes possible while out-of-body or in an altered state of consciousness; and, it will seem more real than our physical life when we are done. There's some really funny stuff that can happen as well.

This is one of the funniest NDE and OBE stories that I have ever watched:

https://www.youtube.com/watch?v=yjiSbpOJhZo

We learn from NDEs and OBEs – the lived experiences of the human race – how spirit matter and spirit bodies work in the non-local realm, quantum realm, or spirit realm. There are thousands upon thousands of these OBEs and NDEs online now. Most of them have gone online during the past five years. The evidence is impossible to deny, unless of course you are determined to do so. It's now April 2017 as I write this. God has started to reveal these things to us exponentially during the past five years or so.

We know from the Scriptures and from the Out-of-Body Travelers that a spirit body can travel much faster than the speed of light to remote destinations in other parts of this universe, like passing through a tunnel or a wormhole; yet, that same spirit body can come back to its physical body instantaneously – snap! We also know from the scriptures that Resurrected Beings can travel across this universe at the speed of thought, can levitate or stand in the air, and can walk through walls; yet, that same being can materialize and become fully physical, is able to eat and drink and be touched by mortal beings like us, when that Resurrected Being reaches its destination. The science fiction terms for these kinds of capabilities are called "teleportation," "levitation," and "phase-shifting".

Said another way, at the Word of Command, Resurrected Beings can phase-shift or transmute their physical bodies back into spirit matter, and then teleport at the speed of thought to anywhere in this universe. When this Resurrected Being reaches its destination, then at the Word of Command, this person can phase-shift back into being a physical body once again. If it's possible to convert spirit matter into physical matter through Conscious Observation or the Word of Command, then it should be equally possible to convert physical matter back into spirit matter by the same Word of Command. That's how the Resurrected Beings in the scriptures can do all of the different things that they can do, yet be fully and completely physical when they reach their final destination. Everything becomes possible once a person or psyche is granted full control over matter, or Quantum Objects.

If God were to give us humans the ability (or the permission) to phase-shift and teleport at will, we could travel to Alpha Centauri or Vega in a matter of seconds, and then rematerialize at our destination. Have you ever seen the movie *Contact*? I have watched and read NDEs and OBEs, wherein people (psyches) have connected with other humans in other parts of this galaxy and universe, while out-of-body and/or near-death. Through the Holy Spirit, the righteous spirits and mortal Saints of God can be in contact with each other, even though they reside on different sides of this universe.

I know people and I know of people who have phase-shifted and teleported – mortal beings like you and me. God intervened, decided to protect them, and decided to save them. My best friend has phase-shifted and teleported and slingshotted out of harm's way, many different times. He didn't actually consciously do it, though – God did it for him. These kinds of things can and do happen to us – the serendipitous and the miraculous – but they can't be replicated at will, because they are God's doing, not ours. God is KNOWN by what He does for us – things that we can't do for ourselves.

I have phase-shifted and teleported as well, because there is no other way to describe and explain what happened to me; but, it was nothing that I did. It was ALL God's doing and God's intervention. There is no other explanation for it.

There was no place to go when that car pulled out in front of me. I was going too fast and I was right on top of it. I hit the brakes. Time stopped. I was suddenly able to sense everything around me 360 degrees. It was like the whole of time had imploded into me, and I had an eternity of time to look into the mirrors, to look around, and to assess the situation. Time had simply stopped. Then, I saw a way out or I saw where I wanted to be. Snap! Time resumed and I was instantly in the middle of the lawn to the right and front of me, my car was completely stopped, and my car was dead or turned off. I had no memory of how I got there. I remember hitting the brakes and having time stop; and then snap, I just teleported or woke up at my destination twenty or thirty feet away with no memory of the intervening travel that took place and no knowledge of how the car had gotten stopped and turned off.

The guy who came running up to see if I was all right said that that was the coolest thing he had ever seen in his life, and that there was no way that I could have missed that other car, but I did.

There had been no contact, and there was no scratch or anything like that on my car. I was simply on that other car's side and bumper one instant and the next instant I was in the middle of the lawn completely stopped with the engine shut down, as if I had teleported or phased right through that other car. And it wasn't just me; it was my car as well that had phase-shifted and teleported. The only logical way that I can explain it is that God or the Spirit phase-shifted me or teleported me, in order to prevent me from merging with and becoming one with that other car.

The Materialists and Atheists will tell you that I imagined it all, or that it was some kind of hallucination, or that there must be some other logical and rational explanation. I KNOW, because I used to be a Materialist and an Atheist. I've talked myself out of it a few times before, so I know how it goes. I've also forgotten about the experience for long periods of time. People can become so separated from God and Spirit that they can no longer remember the things that they used to know. I KNOW this is true, because I have experienced this as well when I was addicted, a Nihilist, a Materialist, and an Atheist. These people can think what they want; but, they weren't there. I started the car, backed out into the street, and went off to work as if nothing had happened. There hadn't even been enough time for my heart to start racing or for the adrenaline to start flowing. The whole experience had been calm, peaceful, and serene, as if I had been in a bubble of protection or in the dream realm the whole time.

Ever since then, I have known (but haven't always wanted to admit or remember) that time can stop and that God can intervene in our lives and can phase-shift us, or teleport us, or miraculously save us if He wants to do so. I know others who have had similar experiences. These people have phased or teleported or slingshotted through no conscious will of their own. A Divine hand simply intervened in their lives. God is KNOWN by doing for us the kinds of things that we can't do for ourselves. Simply put, God is KNOWN by what He does for us. Jesus Christ has come and gotten many different people out of hell – just for the asking. God is KNOWN by living Him and experiencing Him. It all comes down to the psyche experiences, the spiritual experiences, the theophanies, and the lived experiences of the human race.

In conclusion, this Ultimate Model of Reality is the antithesis (the polar opposite) of Materialism and Scientific Naturalism. This Ultimate Cause Model of Reality is everything that Materialism and Naturalism are not, yet subsumes an understanding of physical matter into the mix as well. This Psyche Ontology is a holistic model that's all-inclusive. That's why I decided to call it the Ultimate Model of Reality, because it attempts to include everything that human beings have ever experienced, studied, learned, thought about, researched, or observed throughout the whole of human history. I'm no longer limiting myself to physical matter. I have completely abandoned My Materialism, My Nihilism, and My Atheism. I'm now a Quantum Scientist to the fullest extent of the word! I'm a student of the Quantum Realm or the Spirit Realm.

What could be more exciting or interesting?

The Philosophical, Scientific, and Empirical Foundation Supporting Ultimate Cause

A strong Personality Theory and solid Model of Reality requires a strong philosophical base. It also requires strong empirical evidence.

Obviously, I have a lot of empirical evidence or KNOWLEDGE and many convincing reasons for choosing to believe that this Ultimate Cause Model of Reality is correct, useful, and true. William Buhlman's book is just a start. In ALL of my theorizing, I kept adjusting this Ultimate Model of Reality until it matched with the Lived Experiences or the Psyche Experiences of the human race, including Near-Death Experiences (NDEs), Out-of-Body Experiences (OBEs), Shared-Death Experiences (SDEs), and all the other types of Spiritual Experiences. I wanted to get this model as close to the Truth and Reality as I possibly could; so, I kept choosing to follow the evidence wherever it might lead me.

Because of his reliance on Aristotle's four causes, Joseph Rychlak's books and Logical Learning Theory provide a strong philosophical base for this Ultimate Cause Personality Theory and Ultimate Model of Reality. Logical Learning Theory (LLT) deliberately rejects Materialism and mechanistic exclusivity by including formal causes and final causes into all its theories, experiments, and models of reality. LLT is supportive of and compatible with this Ultimate Cause Model of Reality and is subsumed by this Ultimate Personality Theory.

I took instruction and inspiration from others as well, especially Brent D. Slife. Everything ever written and experienced by mankind supports Psyche as the Ultimate Cause, except for Naturalism and its derivatives such as Materialism which by definition in principle deny the existence of psyche, spirit, mind, consciousness, soul, and reality itself.

My personality theory is that psyche is personality and that psyche is the ultimate cause of everything that has ever been created, organized, and brought into existence from scratch. Furthermore, our psyche or personality or memories survive bodily death and brain death according to the KNOWLEDGE gained from NDEs, OBEs, SDEs, Radical Empiricism, Theophanies, and the Lived Experiences of the human race.

I submit the following books into evidence as proof of the truthfulness and usefulness of this **Ultimate Personality Theory** and this **Ultimate Cause Model of Reality**. As you can tell, I'm no longer an Atheist and no longer a Materialist, because I no longer refuse to look at evidence. I'm no longer afraid of the evidence for psyche and spirit and God, like I used to be. The preceding definitions and this **Ultimate Cause Theory of Personality** were informed strongly by the books in the following informal annotated bibliography:

Buhlman, W. L. (1996). *Adventures Beyond the Body: How to Experience Out-of-Body Travel*. New York: HarperCollins Publishing. (Out-of-Body Experiences). This book was my first exposure to someone who can replicate out-of-body experiences on demand. Buhlman has turned the whole thing into a

Science, and takes a non-denominational and secular approach to it all. He is non-religious and seems to be mildly against organized religion of any kind. This is a Science book, full of observations and empirical evidence and how-to suggestions, and not a religious book. I have encountered many other people who have this spiritual gift and can leave their physical body at will; but, this book remains my most favorite comprehensive introduction to the subject of OBEs. Best of all, Buhlman doesn't use drugs or hypnosis to go Astral – he does it naturally. The Materialists, Atheists, Naturalists, and Mechanists refuse to look at this kind of evidence, because it proves that their philosophies are wrong. This book proves that psyche exists, and that psyche or personality survives separation from the physical body and bodily death. This book also proves that we humans have a spirit body that's in the same shape or form as our physical body. A careful reading of this book also proves that Psyche is something completely different than our spirit body and our physical body. We can KNOW the truth and prove the truth simply by living the truth and experiencing the truth first-hand for ourselves, or by choosing to trust someone who has. This book provides KNOWLEDGE and PROOF of the truth.

Gibson, A. S. (2006). *They Saw Beyond Death: New Insights on Near-Death Experiences*. Springville, UT: Horizon Publishers. (Near-Death Experiences). This book is an attempt to list and briefly discuss ALL of the books about Near-Death Experiences that were written by 2004. It's a good and semi-comprehensive introduction to the empirical or experiential Science of Near-Death Experiences. The Materialists, Atheists, Naturalists, and Mechanists refuse to look at this kind of evidence, because it proves that their philosophies are wrong. NDEs provide evidence in support of this Ultimate Cause Personality Theory, because it is the contention of this Ultimate Personality Theory that psyche or personality is the part of us that survives bodily death and brain death according to the empirical evidence from NDEs, OBEs, and SDEs.

Mark My Words. (2016). *Quantum Mechanics from a Non-Physical Spiritual Perspective*. Kindle. (Reference: https://www.amazon.com/dp/B01J023TGU). When it comes to Science, it's all about perspective or one's chosen interpretation of the available scientific data. The Materialists, Atheists, Naturalists, and Mechanists choose to interpret ALL scientific data from a physical perspective or a local perspective; and thus, they come up with completely different results and explanations of the scientific evidence than the people who choose to interpret all of the same scientific evidence and scientific data from a non-local, trans-dimensional, and spiritual perspective. Materialism provides a very limited view of Science and Reality, and an extremely scaled-down interpretation of scientific data. Quantum Mechanics becomes a completely different, and infinitely more comprehensive, Science when one looks at it from a spiritual or non-local perspective. Quantum Mechanics is better and makes more sense from a spiritual perspective than from a materialistic perspective. The Mechanists and Materialists are at a loss as to how to explain Quantum Mechanics from a physical perspective; but, the Spiritualists and Theists take to Quantum Mechanics like a fish to water, because Quantum Mechanics explains how spirit matter really works. Quantum Mechanics is in fact

Spiritual Mechanics. Spirit is light; and, light is spirit, after all. The spirit realm is the quantum realm, the realm of light.

McTaggart, L. (2002). *The Field: The Quest for the Secret Force of the Universe*. New York: HarperCollins. (Quantum Mechanics, and the Immaterial or Non-Physical Zero-Point Field). This is an extremely popular book, and for good reason! In preparation for this book, Lynne McTaggart interviewed seventy-five different scientists, studying their scientific research which the Materialists and Naturalists ridicule, mock, censor, censure, and formally reject. In *The Field* McTaggart wrote, "For a number of years, while researching *The Field* and subsequent work carried out in this area, I was patiently tutored in quantum physics by some seventy-five frontier scientists. I badgered, cajoled, demanded, and wheedled from each one of them countless hours, up to twenty interviews apiece, teasing out explanations, eventually wresting some crude translation for concepts that often exist for the physicist as pure mathematics. *What exactly is quantum coherence? Why does the Zero Point Field exist?* I would take their frequently incomprehensible answers and play them back via a metaphor until we could both agree on a lay approximation." (From the Preface to the 2008 Paperback Edition). McTaggart demanded to be taught the things which the Materialists and Naturalists refuse to study, accept, or acknowledge. There are many other books from Michael Talbot, Bernard Haisch, Gregg Braden, Rupert Sheldrake, and Amit Goswami (see the References for a few recommendations) who consider and study the hidden, immaterial, and unseen aspects of Quantum Mechanics which the Materialists and Atheists refuse to study and think about; but, *The Field* by Lynne McTaggart is considered to be the best of the bunch. Anything which deals with the Zero-Point Field of Light, the Quantum Sea of Light, the Light of Christ, Biophotonics, Dark Energy, Dark Matter, Spirit Matter, Trans-dimensionality, Parapsychology, Transpersonal Psychology, and Quantum Non-Locality will lend evidentiary support to psyche, non-local consciousness, and this Ultimate Cause Model of Reality. All one really needs to know is that these things exist and that Materialism is false, in order to have all the support one needs for this Ultimate Model of Reality. Anything that explains why Materialism is wrong will generally be supportive of Psyche as the Ultimate Cause. Psyche and ultimate cause have always existed, just waiting for one of us to identify and label them as a causal agent in the overall scheme of reality. Alas, most of the pundits and theoreticians approach Quantum Mechanics and Consciousness from a physical perspective, because that's a no-brainer; but, that process ends up being completely worthless when we are trying to establish evidence for psyche, spirit, consciousness, or ultimate cause. The truly ingenious and creative people try to get a handle on Quantum Mechanics from a non-local or a spiritual perspective. These are the people McTaggart interviewed and reported on in her book. The non-local or the spiritual is by definition in principle non-physical, which means that it has to be inferred to exist or experienced in person first-hand because we will NEVER be able to measure it directly with our physical instruments. Spirit matter is light; and, light is spirit matter. We can't measure the wave-form of light. We can only get a handle on it when it has slowed down, and become a physical photon, and left a mark. And, there are different types of light. According to Quantum Mechanics, only non-local immaterial consciousness has the ability to convert spirit matter into

360

physical matter by collapsing the quantum wave function or the spiritual wave function thereby localizing that particular quantum object or spiritual object into our local 3D physical space-time universe. Spiritual objects or quantum objects only become detectable by our physical instruments after their quantum wave function has collapsed, they have slowed down, and they have become physical particles or photons instead of spiritual waves. That's what the Science of Quantum Mechanics is trying to teach us. If trying to establish evidence for psyche or ultimate cause, Michael Talbot's science books come highly recommended, because he was psychic and was surrounded by the supernatural while growing up. Michael Talbot, therefore, approached the Science of Quantum Mechanics from a psychical or a spiritual perspective which was rather unique at the time while he was alive and doing so. In the same vein, Dean Radin and Charles Tart approach Quantum Mechanics from the perspective of Parapsychology, Transpersonal Psychology, and psychical research. Rupert Sheldrake approaches the unseen and the non-local from a biological perspective, which gets the Atheists, Materialists, and Darwinists stirred up to no end directing a lot of persecution his way. Amit Goswami believes that consciousness, not matter, is the foundation of all existence, holds that the universe is self-aware, teaches that God is the source of "downward causation" (which is similar to first cause or ultimate cause), and claims that consciousness creates the physical world. Goswami uses Quantum Mechanics to establish and prove the existence of God. You know that's not going to sit well with the Atheists, Naturalists, and Materialists. But, the people who find evidence for the non-local, the paranormal, and the supernatural from Quantum Mechanics are geniuses, which is something the Materialists and Naturalists are not. The Materialists come across as lazy in comparison. The only thing the Materialists are interested in is limiting what we are permitted to study and research as scientists, psychologists, and theoreticians. The Materialists typically do their "science" by mocking and ridiculing the people who are doing the Real Science on the cutting-edge – the people whom Lynne Taggart interviews for this book. It doesn't take much ingenuity or creativity to ridicule and mock the people who are doing what you can't do. It's petty, actually. But, it's their modus operandi. Regarding his opponents and religious people, Atheistic Darwinist Richard Dawkins said at the Reason Rally, "Mock them! Ridicule them! In public! They need to be ridiculed with contempt!" Online, Richard Dawkins posted, "I lately started to think that we need to go further: go beyond humorous ridicule, sharpen our barbs to a point where they really hurt. I think we should probably abandon the irremediably religious precisely because that is what they are – irremediable. I am more interested in the fence-sitters who haven't really considered the question very long or very carefully. And I think that they are likely to be swayed by a display of naked contempt. Nobody likes to be laughed at. Nobody wants to be the butt of contempt." That's how Richard Dawkins does science, logic, and reason; and, he is not alone. Most of the Materialists and Atheists are this way – they do their science through mocking and ridicule rather than theorizing and research. While writing *The Field*, McTaggart went to the scientists (who were being mocked and ridiculed by the Atheists and Materialists) for her information and scientific understanding. An amazing, expansive, and popular science book was the result of her efforts. The Materialist's and Atheist's response to this book? "It's not science." That's their default response to everything that isn't Materialism, Naturalism, and Atheism.

Moody, R., & Perry, P. (2010). *Glimpses of Eternity: Sharing a Loved One's Passage from This Life to the Next*. New York: Guideposts. (Shared-Death Experiences). Dr. Raymond Moody introduced the world to NDEs in his ground-breaking book *Life After Life*. Decades later, after having a Shared-Death Experience of his own, Raymond Moody introduces us to SDEs, which are solid and convincing proof that psyche, personality, perception, and memories survive separation from the physical body, because in an SDE a healthy living person accompanies the dead person on the first part of the dead person's afterlife journey and life review. The living person's brain isn't starved for oxygen, and thus can't be hallucinating as a result. SDEs are solid and convincing proof that psyche or personality survives brain death and bodily death, because the dead person is also there having the same out-of-body experience that the live people are having. And with some SDEs, there are two or more live people there with the dead person when the dead person passes on into the next life. In other words, there's independent confirmation of the same event from multiple different sources and perspectives. SDEs on YouTube are interesting to study and observe also.

Richards, P. S., & Bergin, A. E. (2005). *A Spiritual Strategy for Counseling and Psychotherapy* (2nd ed.). Washington DC: American Psychological Association. This book introduces a Theistic Spiritual Perspective, Spiritual Personality Theory, Theistic Personality Theory, and Theistic Psychotherapy, which means that this book is fully compatible with this Ultimate Cause Personality Theory. God fully qualifies as an Ultimate Cause, after all. Consequently, this book from Richards and Bergin provides a psychotherapeutic model, scientific evidence, clinical application, applied science, empirical evidence, and philosophical support for this Ultimate Cause Model of Reality. Any book that lends evidentiary support for theism, theophanies, revelations from God, and spiritual experiences will also lend evidence to this Ultimate Cause Model of Reality. The scriptures which the Biblical God Jesus Christ had a hand in writing and producing would be such a thing. For case studies supporting Theistic Psychotherapy, see the companion book, *Casebook for a Spiritual Strategy in Counseling and Psychotherapy* by Richards and Bergin. In contrast, any psychotherapy based exclusively on Materialism and Scientific Naturalism would technically be incompatible with Psyche and this Ultimate Cause Model of Reality. In these books, Richards and Bergin do an excellent job explaining what's wrong with Scientific Naturalism.

Ring, K., & Cooper S. (1999). *Mindsight: Near-Death and Out-of-Body Experiences in the Blind* (2nd ed.). Kearney, NB: Morris Publishing. (Near-Death Experiences and Out-of-Body Experiences). The people who were born blind can see while out-of-body during their NDEs and OBEs. This is solid evidence that psyche, memories, and personality survive separation from the physical body, bodily death, and brain death.

Rivas, T., Dirven, A., & Smit, R. H. (2016). *The Self Does Not Die: Verified Paranormal Phenomena from Near-Death Experiences*. Durham, NC: IANDS Publications. (Near-Death Experiences). This book turns NDEs into a verified and validated Science, by providing external confirmation that the NDEs and OBEs really took place. The truthfulness and usefulness of this Ultimate Cause Model of Reality hinges upon the fact that the human psyche survives bodily death and brain death according to the empirical sciences of NDEs, SDEs, and OBEs. If there were no empirical evidence for psyche surviving bodily death and separation from the physical body, then this Ultimate Cause Model of Reality would be no better than Materialism and Naturalism, which also have no evidence to support their primary premises. But, since there is plenty of evidence to support Psyche as the Ultimate Cause and no evidence to support Materialism's claim that psyche or spirit does not exist, this Ultimate Cause Model of Reality ends up being infinitely superior and a whole lot more complete, compelling, believable, and useful than Naturalism or Materialism will ever be, as a model of reality. Materialism and Naturalism don't match with reality, so they make for a very poor model of reality. Something is said to be scientifically accurate if it matches with Reality. Since Materialism and Naturalism don't match with reality, they are by definition in principle unscientific – nothing but pure philosophy, sophistry, self-deception, and metaphysics.

Rychlak, J. F. (1970). The Human Person in Modern Psychological Science. *British Journal of Medical Psychology*, 43(3), 233–240. (Retrieve from: http://mypsyche.us/rychlak/). This article provides an excellent introduction to Joseph Rychlak and his use of Aristotle's four causes in psychological science. The rocks or physical matter cannot do science, teleology, predication, nor final-cause; but, the human psyche or human person certainly can. Personality is psyche.

Rychlak, J. F. (1981a). *A Philosophy of Science for Personality Theory* (2nd ed.). Malabar, FL: Robert E. Krieger Publishing Company. Since one of my purposes in this essay is to introduce a new and unique Personality Theory, the Ultimate Personality Theory, this book from Rychlak serves as the perfect philosophical foundation for this Ultimate Cause Personality Theory, which I introduce in this essay. My Theory of Personality is that psyche is personality and that psyche is the part of us that survives bodily death and brain death, according to the Empirical Sciences of NDEs, SDEs, and OBEs. Under this Ultimate Personality Theory, psyche becomes the ultimate cause of everything that has ever been created, organized, and brought into existence from scratch, including genomes and life forms and physical matter. Any of Rychlak's books can be used to supply a philosophical introduction to Aristotle's four causes, which are necessary to understand, because ultimate cause is a fifth-cause meant to explain the "who" behind each of Aristotle's four causes.

Rychlak, J. F. (1981b). *Introduction to Personality and Psychotherapy: A Theory-Construction Approach* (2nd ed.). Boston, MA: Houghton Mifflin Company. Since one of my purposes in this essay is to introduce a new and unique Personality Theory, the Ultimate Personality Theory, this book from Rychlak serves as the perfect introduction to Personality Theory and is a good foundation for this Ultimate Cause Personality Theory, which I introduce in this essay. I will take up this book in much greater detail in other books that I'm working on.

Rychlak, J. F. (1988). *The Psychology of Rigorous Humanism* (2nd ed.). New York: New York University Press. In this book, on pages 8 to 31, Joseph Rychlak goes through 104 philosophers and scientists who had a hand in laying the foundation for the Science of Psychology and Science in general; and, Rychlak identifies the philosophers and theoreticians who employed formal cause and final cause in their theories; and, those who did not. Fourteen of the 104 basically limited themselves to mechanistic causes, material cause and efficient cause. The other 90 slipped over into some kind of formal cause (design theory). This is a clear case where the minority seized control of Science and made the arbitrary rule that only material causes and efficient causes should be considered and allowed into evidence while doing Science. The Mechanists and Materialists are the people who lobotomized Science in an attempt to enforce their philosophical worldview or personal religion onto the rest of us. It's the atheistic, mechanistic, and materialistic philosophers and scientists, who reject teleology and final-cause, who have forced their way into controlling our public schools, scientific research labs, the tenure and peer-review process, and our college textbooks, so that we the paying public are prevented from ever learning about final-cause, teleology, psyche, mind, and soul. It worked. Most of us don't have a clue. The Materialists had to hijack Science and censor Science, because once a person understands and accepts teleology and final cause, then he or she just automatically knows why Materialism and Naturalism are false. Raw physical matter can't do teleology, design, formal cause, final cause, nor ultimate cause. Physical matter is an effect, a Quantum Effect; and therefore, physical matter can't be the ultimate cause for anything nor the first cause of anything. Of the 104 philosophers and scientists that Rychlak researched and documented, 58 of them employed some type of teleology or final cause in their theorizing. That's quite a coup for the Mechanists, Materialists, and Atheists; wherein, the minority clearly dominated and tried to exterminate the majority and pretty much succeeded in doing so where Science and Scientific Evidence are concerned. Rychlak's goal, as a Rigorous Humanist or Experimental Humanist, was to put teleology, predication, purpose, meaning, and final cause back into the Science of Psychology. The ultimate goal of this Ultimate Cause Model of Reality is to put psyche back into science, philosophy, clinical application, and psychology. It seems like the humane and right thing to do. These goals are in sync with each other. The goal is to get at the truths which the Materialists, Naturalists, and Atheists are trying to hide from us.

Rychlak, J. F. (1994). *Logical Learning Theory: A Human Teleology and Its Empirical Support*. Lincoln, NB: Nebraska University Press. This book from Rychlak contains empirical support or scientific evidence for a human teleology, or the human psyche's innate ability to function, predicate, and cognate in formal-cause and final-cause roles. The rocks and physical matter cannot do formal-cause design nor final-cause choice and creation; but, the human psyche can. I see this book as providing scientific evidence and experimental evidence for Psyche as the Ultimate Cause of everything that has ever been created, organized, and brought into existence. I submit this book into evidence as scientific support and empirical support for this Ultimate Cause Model of Reality, which I develop in this essay. For case studies supporting Logical Learning Theory, see, *Personality and Life-Style of Young Male Managers: A Logical Learning Theory Analysis*, by Joseph Rychlak. For a "Reader's Digest" version of Logical Learning Theory (LLT) and a simplified introduction to all of this see, *The Human Image in Postmodern America*, by Joseph

Rychlak. Rychlak's book *Discovering Free Will and Personal Responsibility* is also a good introduction to his Logical Learning Theory, because ONLY Psyche can do discovery, free will, personality, and responsibility. I feel very lucky to have discovered Joseph Rychlak and his books, because Rychlak is one of the people that the Materialists, Mechanists, and Naturalists are trying desperately to keep hidden from us, and have succeeded in doing so.

Rychlak, J. F. (1997). *In Defense of Human Consciousness*. Washington DC: American Psychological Association. I debated about whether to include this book in this list, because it is Rychlak's Logical Learning Theory yet again, but from the perspective of consciousness. Alas, as of 1997 Rychlak is actually very weak and lukewarm when it comes to psyche and non-local consciousness, leaving open what he apparently believes might be the possibility that the psyche or mind or consciousness is an epiphenomenon (a side-effect or a derivative) of the physical brain, an idea which is incompatible with the empirical evidence provided by NDEs, SDEs, and OBEs. Of course, the exponential explosion in books about NDEs and OBEs didn't start until about 2006; and, I didn't get onboard with the whole OBE thing until the end of 2015 after I read William Buhlman's book *Adventures Beyond the Body: How to Experience Out-of-Body Travel*; so, I can understand why Rychlak didn't mention NDEs in his books, because we were all in the same boat back then in the mid-1990's with limited NDE and OBE evidence at hand. In this book, *In Defense of Human Consciousness*, Rychlak employs *Logos* as the grounds for "psychic consciousness" making the two basically synonymous; yet also stating that Logos is not physical and that "psychic consciousness is not mysterious or spiritual". In other words, Rychlak implies that psyche (logos) is not physical and that it's not spiritual either. So what is it? Well, Rychlak employs "psychic consciousness" as a philosophical concept, or a psychological construct, or a "mental" construct; and, he seems to leave it at that. In his books, Rychlak states that he is not going to take an official stand on the mind-brain problem, only stating that his Logical Learning Theory (LLT) can be made compatible with the convictions of those who have chosen to believe that the mind is a completely separate entity from the physical brain. Consequently, I turn to others for procedural evidence, validating evidence, and empirical evidence of psyche or mind being a non-local quantum phenomenon completely separate from the physical brain. I use Joseph Rychlak primarily for the philosophical, logical, and procedural foundation of this Ultimate Cause Model of Reality; but, his books do indeed provide experimental evidence supporting Humanism, Logos, Teleology, Predication, Final Causality, and this Ultimate Cause Personality Theory; and, only psyche or non-local consciousness can do final cause, choice, and predication. I turn to NDEs, SDEs, OBEs, revelations from the Biblical God Jesus Christ, and other spiritual experiences for empirical, experiential, and observational evidence supporting this Ultimate Cause Model of Reality. I turn to Pim van Lommel and his cutting-edge (2010) interpretation of Quantum Mechanics, from a spiritual perspective or an NDE perspective, for the primary scientific foundation of this Ultimate Cause Model of Reality. Gravity, dark matter, dark energy, forces, magnetism, and the zero-point field of light, as non-physical immaterial trans-dimensional spiritual phenomena, also lend evidentiary support to this Ultimate Model of Reality. The physical brain is a transceiver for the non-local living consciousness, who is in fact the broadcaster to the physical brain, the recorder of information coming from the physical brain, and the director of the physical brain. Psyche is the driver, and the physical brain is the machine that gets driven. The psyche or non-local consciousness and our memories survive bodily death and brain death, according to the empirical evidence from NDEs, SDEs, and OBEs. I finally chose to follow the evidence. Rychlak, for whatever reason, chose to straddle the fence and chose to leave the mind-brain problem unresolved in his theorizing; whereas with a couple of extra decades of empirical evidence to draw upon, I quickly noticed that the mind-brain problem is instantly and immediately solved once Materialism or Scientific Naturalism is eliminated from the equation and taken out of the picture with extreme prejudice. Along the same lines of inquiry, you might be interested in Rychlak's book, *Artificial Intelligence and Human Reason: A Teleological Critique*, which makes a solid case demonstrating that computers and artificial intelligence cannot do teleology nor final cause; but, the human psyche or "psychic consciousness" certainly can.

Sanford, J. (2014). *Genetic Entropy* (4th ed.). Cornell University: FMS Foundation. (Scientific Proof and Common-Sense Proof that All Types of Materialism or Naturalism Are False). This book provides convincing procedural evidence, mathematical modeling evidence, scientific evidence, and logical common-sense evidence demonstrating beyond a shadow of a doubt that Natural Selection and Random Mutations cannot design and create genomes and life forms from scratch as the Darwinists and Materialists claim. In fact, Natural Selection and Random Mutations didn't even exist at all, until after God or Psyche designed and created the first fully functional genome and life form on this planet, in the first place. A fully functional genome and life form had to be produced by some kind of Psyche or Consciousness or Intelligence, before Natural Selection, Random Mutations, and other Physical Processes could come into play; therefore, Mutation/Selection (Evolution) cannot be the origin of life on this planet. When fully understood, this book from Sanford provides proof of Psyche as the Ultimate Cause of everything that has ever been created, organized, and brought into existence from scratch, including physical matter, genomes, and life forms.

Van Lommel, P. (2010). *Consciousness Beyond Life: The Science of the Near-Death Experience*. New York: HarperCollins. (Quantum Mechanics from a Non-Local or Non-Physical Perspective). This just may be the most popular, most read, and most useful book about Quantum Mechanics and Near-Death Experiences that has ever been written. This book is Science, and it demonstrates clearly and conclusively that psyche or mind survives separation from the physical body and brain death. Quantum Mechanics is Spiritual Mechanics, the way that spirit matter really works. Classical Physics tells us how physical matter works; and, Quantum Mechanics tells us how spirit matter works. Quantum Non-Locality, Quantum Mechanics, and Spiritual Mechanics are the Science behind Psyche, Non-Local Consciousness, and Near-Death Experiences. I refer to this book almost constantly in all of my writings and theorizing.

Goswami, A. (2008). *God Is Not Dead: What Quantum Physics Tells Us about Our Origins and How We Should Live*. Charlottesville, VA: Hampton Roads. I completely designed and wrote this Ultimate Cause Model of Reality BEFORE reading any of Amit Goswami's books. After writing most of this essay, I started reading from *God is Not Dead*, and it quickly became apparent to me that Goswami's theories and ideas are compatible with Psyche as the Ultimate Cause. Of course, that shouldn't be too surprising, because anyone who blasts Materialism as heavily as Goswami does will end up by definition in principle being compatible with this Ultimate Cause Model of Reality. By the time Goswami gets done with Materialism, there's nothing left but a smoking crater. Ultimate Cause is the exact opposite of Materialism. Goswami mentions Near-Death Experiences (NDEs) in his book. Goswami makes his ideas compatible with Intelligent Design Theory. Goswami also tries to salvage the theory of evolution by describing evolution as a consciously directed process, which is something that I no longer feel the need to do. In my theorizing, I have observed that the various versions of evolution or Darwinism are synonymous with Materialism, which is synonymous with Creation by Rocks. Psyche or ultimate cause is the antithesis of Materialism and Darwinism; but technically, the Theory of Evolution can be salvaged if psyche or consciousness is employed to direct and do the evolution. Goswami defines God as "quantum consciousness", which is an idea that is completely compatible with Psyche and Ultimate Cause. Goswami explains evolution and creation as "downward causation" or top-down causation or psyche causation, which is completely compatible with **ultimate cause**, which I purposefully designed to be compatible with and explanatory of Aristotle's four causes in an attempt to give this Ultimate Cause Model of Reality a solid philosophical base. Goswami also fully embraces Rupert Sheldrake's *morphic resonance* and morphogenetic blueprints, which are spiritual blueprints or non-local quantum designs (formal causes) upon which life forms are built while developing in the womb, because there is not enough information in DNA to build a living organism from scratch, but instead just enough information in the DNA to keep a living organism alive and functioning after it has been built from its spiritual blueprint or morphogenetic blueprint. Cell differentiation, the arbitrary turning on and off of certain genes during the development of the fetus (just the right genes being turned on or off in each and every cell with one cell initially being completely different than the cell next to it or the cell from which it came) is not done by DNA but is instead done by this spiritual blueprint or quantum blueprint or the human spirit. Your spirit provided the design or blueprint from which your physical body was made, not your DNA. That's the lesson of epigenetics and morphogenetic blueprints. Goswami also repeatedly emphasizes that physical matter and physical machines like computers cannot do meaning and cannot process meaning, a reality which signifies that physical matter can't do final causality. Goswami covers all the bases necessary to make his unnamed model of reality compatible with this Ultimate Cause Model of Reality. Not being a Christian, Goswami has some strange ideas about Jesus Christ, though. Personally, I would adjust and tweak some of Goswami's ideas to make them match more fully with the modern-day revelations which we have received from the Biblical God Jesus Christ, because I prefer to get information about Jesus Christ straight from Jesus Christ himself rather than from Goswami's speculations about Jesus Christ; but, it's doable! With occasional adjustments, Goswami's ideas can be made compatible with the modern-day revelations from Jesus Christ as found in the *Book of Mormon: Another Testament of Jesus Christ*, the *Doctrine and Covenants*, and *Pearl of Great Price*. Where else is God going to reveal Himself to us besides the books He had a hand in writing and producing? In these books, the Biblical God Jesus Christ tells us that He created everything spiritually (quantumly), before it was organized physically. In other words, our physical bodies develop in the womb and our cells differentiate according to the spiritual blueprint or the quantum blueprint which the Biblical God designed and put together BEFORE this physical universe was called into existence by quantum consciousness, or psyche, or God. Amit Goswami has many other books about Quantum Consciousness, which I own, but haven't read yet. Based upon what I have read so far in *God Is Not Dead*, it looks promising. Goswami puts another nail into Materialism's coffin, which automatically lends support to this Ultimate Model of Reality and to Psyche as the Ultimate Cause of everything that has ever been brought into existence from scratch. Every truth will be compatible with every other truth; and, every truth will be incompatible with Materialism and Naturalism. That is my observation.

Bishop, B. G. (1998). *The LDS Gospel of Light*. USA: Ponce de Leon. (Consciousness, Energy, Waves, Light, Spirit, NDEs, Light of Christ, God is Light, and Quantum Mechanics). This is another book that I started reading after writing this essay; and, the whole book seems to be compatible with Psyche as the Ultimate Cause of our Reality. Of course, I already realized that this Ultimate Model of Reality would be compatible with the *Bible*; with the Gospel of Christ as presented to us in the *Bible*, the *Book of Mormon: Another Testament of Jesus Christ*, the *Doctrine and Covenants*, and the *Pearl of Great Price*; with the revelations of the Biblical God and the revelations from the Biblical God; and with Christ's Gospel of Light or the Light of Christ. Every truth lends evidentiary support to every other truth; and, every truth explains to us why Materialism and Naturalism are false. That has been my experience and my observation.

These were the main books which helped me to create this **Ultimate Cause Model of Reality** and that caused me to believe that this **Ultimate Personality Theory** is real and true.

This Ultimate Cause Model of Reality is an all-inclusive model of reality; and, its goal is to subsume all truth and every human experience into its ranks. It was actually difficult to pick the very best books to represent and support this Ultimate Cause Model of Reality, because everything in science, philosophy, and experience supports psyche as the ultimate cause, except for the books and people who deliberately limit themselves to Materialism, Mechanism, and Naturalism. Materialism and Naturalism are incompatible with the rest of Science, Philosophy, Religion, Theology, and Human Experience including this Ultimate Cause Model of Reality. Materialism and Naturalism are extremely limited and exclusive, so they can't be made compatible with the rest of Science, Philosophy, Applied Psychology, Empirical Evidence, and Lived Experiences because Naturalism and Materialism are based upon a refusal to look at evidence and a refusal to accept evidence. I'm not interested in excluding evidence from this Ultimate Model of Reality – been there and done that already, when I was an atheist and a materialist. I don't want to go back to My Materialism and My Atheism, because I have found something infinitely better to take their place.

Psyche is the ultimate cause. Joseph F. Rychlak provided the core philosophical foundation and the personality theory foundation for this Ultimate Cause Model of Reality. Richards and Bergin provided the psychological and psychotherapeutic foundation for this Ultimate Model of Reality. Sanford reveals the biggest error that has been made so far in the physical sciences, an error or falsehood that keeps on giving and keeps holding people back. One has to eliminate the falsehoods, deceptions, and lies before he or she can pursue the true reality or the ultimate reality of our existence.

Pim van Lommel, Lynne McTaggart, Amit Goswami, and Quantum Mechanics provide the scientific foundation for Psyche or Non-Local Consciousness as the Ultimate Cause. Hugh Ross and Astrophysics explains to us that only 4% of this universe got converted to physical matter during the Prime Event or "Big Bang", and that the other 96% of this universe is still in its original spiritual state of existence. Near-death Experiences (NDEs), Shared-Death Experiences (SDEs), Out-of-Body Experiences (OBEs), and Spiritual Experiences in general provide the empirical evidence, experiential evidence, and observational evidence needed to prove that this Ultimate Model of Reality is correct and true. The truth can be KNOWN and PROVEN by experiencing it and living it first-hand, or by choosing to trust someone who has. After having given it years of study, research, and thought, I build this Ultimate Cause Model of Reality on a solid foundation.

Every Spiritual Experience Supports Psyche as the Ultimate Cause

Empirical Knowledge, or Experiential Knowledge, or Lived Experience is superior to scientific hypotheses and philosophical guesswork.

Whenever I encounter people who are sharing with me some of the spiritual experiences, visions, psychic experiences, out-of-body experiences, and near-death experiences which they have had, I choose to believe them and take them at their word because I have had spiritual experiences of my own and know that the phenomenon is real and truly happens from time to time. Furthermore, I have friends who have seen and talked with the spirits of their dead relatives. I know of many people who have seen and talked with Jesus Christ while out-of-body. Quantum Mechanics has also taught me that the Non-Local or the Spiritual has to exist in order to explain all of the scientific evidence and empirical evidence associated with Quantum Mechanics. Psyche must exist, or physical matter would not exist.

The Materialists and Naturalists can't explain Quantum Mechanics in a way that makes logical sense, just like these people can't explain spiritual experiences in a way that is parsimonious and makes sense. You will never learn anything useful about the spiritual or the non-local from a Materialist or an Atheist. I have observed that once people have had spiritual experiences of their own, then they are no longer Materialists or Naturalists; and, once these people have seen God, then they are no longer Atheists.

The following YouTube Videos about OBEs and NDEs had an influence on my philosophical worldview and this Ultimate Cause Model of Reality; therefore, I submit them into evidence. These spiritual experiences and out-of-body experiences are sensational. Mine have been quite mellow in comparison. Nevertheless, my best friend is a prophet, seer, and revelator; and, he has had many different types of spiritual experiences, visions, out-of-body experiences, miraculous experiences, and near-death experiences and he has testified to me while watching these videos with me that these things are real and true and match perfectly well with his own Lived Experiences while out-of-body in the spirit realm. My friend told me that these OBEs and NDEs and Theophanies have all the signatures of authenticity. The ONLY way to KNOW the truth is to live it and experience for yourself, or to choose to trust someone who has. I trust my friend; and, I trust these people as well.

Let's start with Howard Storm. Howard Storm was an atheist, died, and went to hell; and, after finally calling upon Jesus Christ to save him, Jesus came and rescued Howard Storm from hell. Should you ever find yourself in hell, remember that Jesus Christ can get you out of there for the asking.

https://www.youtube.com/watch?v=UPj4wci_bcI

https://www.youtube.com/watch?v=Vm647n1360A

Ian McCormack was an atheist, went to hell, and was saved from hell by Jesus Christ. This is currently my most favorite version of Ian McCormack's NDE. There are many others:

https://www.youtube.com/watch?v=HbTAmN4m2lQ

Dr. Mary Neal had a very vigorous near-death experience, which kind of makes it hard to believe that she was hallucinating the whole time. I'm interested in the NDEs wherein the individual gets to see God.

https://www.youtube.com/watch?v=DX473dF7ChY

https://www.youtube.com/watch?v=ULsl92H-Noc

Each one of these people has written and/or published a book or two about their Near-Death Experience, which I own and have been reading from. Check the References for a list.

The Hell Experience of an Ex-Satanist, such as John Ramirez, can be informative and also provide empirical evidence for Psyche as the Ultimate Cause:

https://www.youtube.com/watch?v=I11L71PD3Lw

There are hundreds, if not thousands, more on YouTube right now. Anything from Raymond Moody or Eben Alexander proves fascinating. NDEs, SDEs, OBEs, and Spiritual Experiences provide empirical evidence that this Ultimate Cause Model of Reality is correct and true. Psyche is the Ultimate Cause. I could never find any supporting evidence for My Materialism and My Atheism; but, I have found reams of evidence for Psyche as the Ultimate Cause, and each day I come across more and more evidence which proves to me that Psyche exists. I finally chose to follow the evidence.

Remember, Empirical Knowledge, or Experiential Knowledge, or Lived Experience is superior to scientific hypotheses and philosophical guesswork. There's no substitute for KNOWING the truth, having lived it and experienced it for yourself.

The Ramifications of Psyche and Ultimate Cause

Human beings are composite entities having both an immaterial spiritual component and a physical body component. While here in mortality, a person is a spiritual being having a physical experience. In the afterlife, before the resurrection from the dead which Christ says is going to happen to all of us, a person is a spiritual being.

Personality is psyche; and, psyche is personality. Psyche is typically defined as one's spirit, soul, mind, consciousness, individuality, intelligence, awareness, spark, and life. Personality or psyche is the part of us that survives separation from our physical body, separation from our spirit body, death of our physical body, and brain death according to the empirical sciences of Near-Death Experiences (NDEs), Out-of-Body Experiences (OBEs), Shared-Death Experiences (SDEs), and other types of Spiritual Empiricism (spiritual experiences). Psyche is the Ultimate Cause of every contingent thing that has ever been brought into existence from scratch.

I define the Self as the Psyche, in Cartesian Dualism fashion, with the psyche or consciousness being an immaterial substance completely different than and separate from the physical brain. I rely upon Quantum Mechanics, Astrophysics, and Brain Stimulation in Neuroscience for the scientific evidence necessary to confirm the truthfulness and usefulness of this Ultimate Cause Model of Reality and the existence of Psyche. I rely upon NDEs, OBEs, and SDEs for the necessary empirical evidence. I also rely upon scriptural evidence or revelatory evidence for this Ultimate Cause Personality Theory.

Doctrine and Covenants 88: 15: "And the spirit and the body are the soul of man."

Doctrine and Covenants 93: 33-34: "For man is spirit. The elements are eternal, and spirit and element, inseparably connected, receive a fullness of joy; and when separated, man cannot receive a fullness of joy."

This is the Ultimate Model of Reality because it attempts to conform to all known truths and to the whole of human experience and knowledge.

What is truth, and how do we know it?

The truth is knowledge of things as they really are. If something is true, then it has always been true, and it will always be true. Even better, truth is Lived Experience! The best way to KNOW the truth is to live the truth and experience the truth for yourself, or to choose to trust someone who has.

There are at least three ways of knowing the truth.

1. The least effective and least reliable way of knowing the truth is philosophically or though logical common sense – sometimes called procedural

evidence. Philosophy is guesswork or the making of hypotheses – a comparing of ideas and points of view. Philosophy is used to explore the ideas and concepts that are unseen, intuitive, immaterial, and/or unknowable by our physical senses and physical instruments. Ideally, Philosophy is supposed to be the pursuit of the truth; but, many people (sophists and materialists) use Philosophy in an attempt to make their lies seem true, which is easy to do since Philosophy is an abstract mental activity. Philosophy is rife with deliberate deception, including self-deception. Philosophy is worthless if it is used to prove a lie true, as happens in the case of Materialism and Naturalism. Materialism, Naturalism, and Atheism are philosophy, because they have no evidence to support them and have to be taken on blind-faith as being true. Materialism or Naturalism is metaphysics and religion, not science.

2. A more effective way of knowing the truth (at least where physical reality is concerned) is the Scientific Method or Scientific Experimentation. However, the Scientific Method has an inbuilt logic fallacy called "affirming the consequent". What this means is that the final step of the Scientific Method calls for an interpretation of the scientific data or an explanation of the scientific evidence. This is where the flaws come into play with the Scientific Method, because the scientific data calls for a philosophical best explanation of the scientific evidence or a philosophical best interpretation of the scientific data. The Materialists and Naturalists simply guess wrong, and interpret the scientific data incorrectly from an exclusively physical perspective. It requires faith to believe that one's chosen interpretation is correct, right, and true. It's very easy to provide the wrong interpretation or the worst possible explanation to scientific data and scientific evidence. The Materialists, Naturalists, and Darwinists do so all the time. The Materialists interpret everything in terms of Creation by Physical Matter or Creation by Rocks; yet, we all know that the rocks cannot design and create anything at all.

3. The best way of knowing the truth is to experience it first-hand or first-person for yourself, or to choose to trust someone who has. I have taken to calling it The Art of Knowing, or Knowledge. Upon further research, I learned that most philosophers call it Lived Experience or Phenomenology. Knowing or certain knowledge trumps philosophical guesswork and scientific interpretation every time. Physical Empiricism and Spiritual Empiricism, experiencing the physical world and the spirit world directly for yourself, is the best way of knowing how these things really work. The people who have been to the spirit world, seen God, and talked with the angels of heaven during their Near-Death Experiences (NDEs), Out-of-Body Experiences (OBEs), Shared-Death Experiences (SDEs), and other types of Spiritual Empiricism (spiritual experiences) simply KNOW that God exists and that the human psyche survives separation from the physical body, bodily death, and brain death. My best friend has experienced many of these things directly, so he KNOWS that they are real and true. First-hand experience and first-person observation is the best way of knowing the truth. The scientists call this method of knowing the truth, Observation. Philosophers tend to call this method of knowing the truth, Experience or Empiricism, a branch of Epistemology; and, many people call this method of knowing the truth, Direct Revelation, Radical Empiricism, or Lived Experience. Knowledge of the truth, experiencing the truth and living the truth, is the best way of knowing the truth. Truth is knowledge.

Doctrine and Covenants 93: 24: "And truth is knowledge of things as they are, and as they were, and as they are to come." Truth is knowledge. John 8: 32: "And ye shall know the truth, and the truth shall make you free." Truth is synonymous with knowledge; and even though it's tautological, simply KNOWING THE TRUTH or experiencing the truth first-hand is in fact the BEST way of knowing the truth. It sets you free! It's elementary my dear reader.

Every time I pick up a book or article about Science, Quantum Mechanics, Quantum Objects, Quantum Entanglement, Quantum Nonlocality, Trans-Dimensionality, Action at a Distance, Forces, Fields, Magnetism, the Zero-Point Field, Conscious Observers, Particle Physics, Gravity, Dark Matter, Dark Energy, Light, Faster than Light Travel, Cosmological Constants, Intelligence, Thought, Dreams, Psychology, Philosophy, Psyche, Life, Consciousness, Mind-Over-Matter, the Placebo Effect, Time, Space-Time, Universal Constants, Physical Laws, Causality, Spirituality, Heaven, Hell, Revelations, Visions, NDEs, OBEs, SDEs, Spiritual Experiences, or God, I quickly realize that NONE of these things are possible if Materialism is true. Materialism or Naturalism precludes and excludes the existence of these kinds of things. If Materialism were really true, it would prevent these things from happening and make them impossible. You and I would not exist if Materialism or Naturalism were 100% true. Since these things have been experienced and thereby proven to exist, we KNOW that Materialism and Naturalism are false. Truth can be KNOWN by living it and experiencing it for ourselves. Truth falsifies lies such as Materialism, Naturalism, and Darwinism.

Materialism or Scientific Naturalism is a worthless and useless philosophy masquerading as Science; but, it's not Science. Materialism is nothing more than philosophy, pseudo-science, sophistry, and metaphysics. Philosophy, speculation, hypotheses, and the spreading of known falsehoods are the weakest way of knowing the truth, because they can actually prevent us from knowing the truth. All you really want is the truth, unless of course you are a Materialist, Naturalist, or Atheist – then any old lie will do. You'll even be eager to share that lie with someone else if you are a Materialist or an Atheist because you can make some good money if you do.

Ramifications of this Ultimate Cause Personality Theory

This Ultimate Model of Reality and Psyche Ontology is based heavily upon our Lived Experiences as a race, including our theophanies, spiritual experiences, visions, revelations from God, near-death experiences, and out-of-body experiences. ONLY Psyche can do Lived Experiences both in the spirit realm and here in this physical realm and actually remember those experiences when it's done. This chapter answers some of the questions that I was asked about this Ultimate Personality Theory or Psyche Personality Theory.

Psyche, or one's immaterial non-local trans-dimensional consciousness, is the basic structure of personality and self, because according to NDEs, OBEs, SDEs, and other types of Spiritual Empiricism our self, or psyche, or consciousness, or memories survive separation from the physical body, separation from our spirit body, death of the physical body, brain death, and resurrection from the dead.

Under this Ultimate Cause Personality Theory, going to hell is the ultimate cause of an abnormal personality, psychological illness, spiritual illness, or mental illness. It's possible to go to hell while here in mortality. I did, so I know of what I speak. There are many applicable definitions for hell. Hell is being completely alone, with no friends and no social support and no reason to live. Hell is being immobilized and trapped in anxiety, depression, hopelessness and fear year after year after year. Hell is having no purpose in life. Hell is being addicted to substances and thereby having no control over one's life and destination. Hell is constant never-ending anxiety and fear. Hell is being trapped in sin or addicted to sin, unable to get out. Sin is anything which prevents us from reaching our full potential. Sin is damnation and hell – being stopped in our progress, fulfillment, and actualization. Consequently, the Atonement of Jesus Christ and getting out of hell is the ultimate therapy for one's spirit, body, mind, and soul.

Psychological illnesses should be addressed or treated with a combination of the Atonement of Christ, repentance or change for the better, spiritual comfort and support through prayer, hope, judicious careful drug therapy if one's mental illness can be demonstrated to have a physical component, and Friendship Therapy or Social Interest. We are not afraid of our friends, and we look forward to seeing our friends. Making friends and supporting each other through the hard times is our primary reason for existence here in mortality. Try to make a new friend every day. The purpose of life is to make friends, including making friends with God.

What is the process of normal development under this Ultimate Personality Theory? The human psyche has multiple stages of development similar to a butterfly; but, the psyche is a bit more complex than an insect because psyche or consciousness has an immaterial, non-local, trans-dimensional, non-physical basis according to Quantum Mechanics and Lived Experiences.

In the beginning, psyche is pure immaterial consciousness. This stage is pre-egg, meaning that at this stage psyche is pure intelligence or pure thought, an unformed actuality. Eventually, psyche can be assigned to occupy spirit matter (a type of egg stage), and subsequently assigned to occupy physical matter (a type of larval or caterpillar stage capable of interacting directly with the physical world). In time, that larva or caterpillar will "die" and go into an underworld, or a cocoon, or a state of physical dormancy; yet, its psyche or mind or consciousness or individuality will go on. According to the scriptures which the Biblical God Jesus Christ had a hand in writing and producing, at some point in the future our psyche will be resurrected from the dead and emerge as a glorious immortal being (similar to the butterfly rising from the dead, the cocoon or chrysalis, after its heavenly transformation).

According to Jesus Christ, one of the main purposes of mortal life is to follow Him and to become like Him. Families are known to sup together. Christ invites us to become a part of His family.

Revelation 3: 20-22: "Behold, I stand at the door, and knock: if any man hear my voice, and open the door, I will come in to him, and will sup with him, and he with me. To him that overcometh will I grant to sit with me in my throne, even as I also overcame, and am set down with my Father in his throne. He that hath an ear, let him hear what the Spirit saith unto the churches."

During the normal process of development under this Ultimate Cause Model of Reality and this Ultimate Personality Theory, if successful in our development, we will eventually sup with Christ and sit down with Christ on His Father's throne.

Consequently, under this Ultimate Cause Model of Reality, the good life is ultimately defined as becoming like Christ and then sitting down with Him on His Father's throne. Meanwhile, here in mortality the ultimate goal is peace of mind.

Doctrine and Covenants 59: 23: "Learn that he who doeth the works of righteousness shall receive his reward, even peace in this world, and eternal life in the world to come."

The good life consists of peace in this world, and eternal life in the world to come. In other words, the good life consists of getting out of hell and staying out of hell. Likewise, psychological health implies having a sound mind, which is impossible if one is in hell.

2 Timothy 1: 7: "For God hath not given us the spirit of fear; but of power, and of love, and of a sound mind."

Within this Ultimate Cause Model of Reality and this Ultimate Personality Theory, psychological health consists of getting God's Spirit within us, so that His Spirit can fill us with hope, love, and a sound mind while at the same time vanquishing all anxiety, depression, addiction, sin, and fear. Psychotherapy consists of getting out of hell and into God's presence. Psychotherapy consists of finding the peace that surpasses all understanding. Psychotherapy is learning how to bring the full effects of the Atonement of Christ into our lives.

Remember! If it's motivated by guilt, force, anger, hatred, pride, profit, selfishness, competition, jealousy, or fear, then it's Satan's work which you are doing even if you are doing a good thing or trying to do a good thing. In order for the full blessings, happiness, joy, and peace to accrue, it must be motivated by friendship, charity, compassion, and love. We must learn to do things for the right reasons in order to achieve the best results both for ourselves and for others.

There is an element of force, fear, and intimidation in classical conditioning and operant conditioning. Conditioning works on the animals, but humans tend to rebel whenever someone tries to condition them or force them to comply. The human psyche is a choosing organism, an agent in the fullest sense of the word; and, the human psyche doesn't respond favorably to force, fear, guilt, hatred, and intimidation.

Each psyche is unique, a type of infinite singularity, originally without form or matter of any kind. This also means that every psyche or intelligence has its own unique set of traits, temperaments, likes, dislikes, and dispositions. Whenever presented with an opportunity or some kind of dialectical opposition, the psyche is the thing that chooses between the alternatives; and, the psyche tends to choose in conformity to its desires, likes, goals, and dispositions; yet, it can also choose to experiment and try-out alternative courses in an attempt to discern if its likes and dispositions and goals have changed.

Zion is meant to be a community affair. From an LDS Perspective, Paradise is the Celestial Kingdom or the City of Zion in the spirit world. The other part of the spirit world has been called by various names including purgatory, hell, Gehenna, the spirit prison, hades, Sheol, and outer darkness; and, it is probably composed of different degrees or different levels and types of existence – some people will be completely and utterly alone or self-absorbed, and others will have gathered into communities. Furthermore, the Latter-day Saints believe that our spirit body is literally the offspring of God the Father and Heavenly Mother, which means that all of us are related to each other as brothers and sisters.

Doctrine and Covenants 76: 22-24: "And now, after the many testimonies which have been given of him, this is the testimony, last of all, which we give of him: That he lives! For we saw him, even on the right hand of God; and we heard the voice bearing record that he is the Only Begotten of the Father — that by him, and through him, and of him, the worlds are and were created, and the inhabitants thereof are begotten sons and daughters unto God."

This scripture from the Biblical God Jesus Christ indicates that our spirit bodies are literally the sons and daughters of God the Father and Heavenly Mother, which means that in the spiritual plane of existence, we are in fact brothers and sisters. We are related to each other.

This Ultimate Cause Model of Reality is also contained within Abraham 3: 17-28:

17 There is nothing that the Lord thy God shall take in his heart to do but what he will do it.

18 He [God] made the greater star; as, also, if there be two spirits [psyches], and one shall be more intelligent than the other, yet these two spirits [psyches], notwithstanding one is more intelligent than the other, have no beginning; they existed before, they shall have no end, they shall exist after, for they are gnolaum, or eternal.

19 And the Lord said unto me: These two facts do exist, that there are two spirits [psyches], one being more intelligent than the other; there shall be another more intelligent than they; I am the Lord thy God, I am more intelligent than they all.

21 I dwell in the midst of them all [the intelligences or psyches, and their assigned spirit bodies]; I now, therefore, have come down unto thee [Abraham] to declare unto thee the works which my hands have made, wherein my wisdom excelleth them all, for I rule in the heavens above, and in the earth beneath, in all wisdom and prudence, over all the intelligences

[psyches] thine eyes have seen from the beginning; I came down in the beginning in the midst of all the intelligences [psyches] thou hast seen.

22 Now the Lord had shown unto me, Abraham, the intelligences [psyches] that were organized [into spirit bodies] before the world was; and among all these there were many of the noble and great ones;

23 And God saw these souls [His spirit children] that they were good, and he stood in the midst of them, and he said: These I will make my rulers; for he stood among those that were spirits, and he saw that they were good; and he said unto me: Abraham, thou art one of them; thou wast chosen before thou wast born.

24 And there stood one among them that was like unto God [Jehovah or Jesus Christ], and he said unto those who were with him: We will go down, for there is space there, and we will take of these materials, and we will make an earth whereon these may dwell;

25 And we will prove them herewith, to see if they will do all things whatsoever the Lord their God shall command them;

26 And they who keep their first estate [their spiritual pre-mortal life] shall be added upon; and they who keep not their first estate [Satan and the demons] shall not have glory in the same kingdom with those who keep their first estate; and they who keep their second estate [mortal life or physical life] shall have glory added upon their heads for ever and ever.

27 And the Lord said: Whom shall I send? And one answered like unto the Son of Man: Here am I, send me. And another answered and said: Here am I, send me. And the Lord said: I will send the first.

28 And the second [Satan] was angry, and kept not his first estate; and, at that day, many followed after him.

These things only make sense in the light of psyche and ultimate cause. All of these things are silliness and foolishness from the perspective of Materialism and Naturalism. The perspective, worldview, philosophy of life, paradigm, or model of reality which we each choose to embrace makes all the difference in the world to the outcome and the results that we will be able to achieve.

Remember! If it's motivated by guilt, force, anger, hatred, pride, profit, selfishness, competition, materialism, naturalism, jealousy, or fear, then it's Satan's work that you are doing even if you are doing a good thing or trying to do a good thing. In order for the full blessings, happiness, joy, and peace to accrue, it must be motivated by friendship, charity, compassion, and love. We must learn to do things for the right reasons in order to achieve the best results both for ourselves and for others.

Remember, there is an element of force, fear, and intimidation in classical conditioning and operant conditioning. Conditioning works on the animals, but humans tend to rebel whenever someone tries to condition them or force them to comply. The human psyche is a choosing organism, an agent in the fullest sense of the word; and, the human psyche doesn't respond favorably to force, fear, guilt, hatred, and intimidation.

Conclusions Regarding this Ultimate Model of Reality

Consciousness or psyches "have no beginning; they existed before, they shall have no end, they shall exist after, for they are gnolaum, or eternal".

Our spirit bodies had a beginning, when our Heavenly Mother gave birth to our spirit bodies and our psyche or consciousness was assigned to our spirit body. It was at this point in time, during the birth of our spirit body, that our psyche was assigned a form, a gender, a heritage, a potential to become a God, and some spirit matter to occupy and control; and, we became the sons and daughters of God the Father.

God's Psyche is the Ultimate Cause of all other contingent realities that have ever been organized and brought into existence from scratch. The good life consists of keeping one's first estate, and then keeping one's second estate. The good life consists of getting out of hell and staying out of hell, and getting into Paradise and the Kingdom of God instead. The good life consists of peace of mind in this world, and eternal life in God's presence as the ultimate end of our journey in the world to come. As I see it, this is the Ultimate Model of Reality because it conforms to the whole of human experience and knowledge.

Psyche Is the Ultimate Cause

The existence of **ultimate cause** is so obvious and so necessary that **we** tend to take it for granted and completely overlook it; however, every time that **you** use a personal pronoun in one of **your** sentences, **you** are in fact invoking and applying some type of ultimate cause. Whether expressly written or simply implied, ultimate cause is the subject of every sentence that **you** write or speak. Where **you** are concerned, **you** or **your** psyche is the ultimate cause of everything having to deal with **you**, including every sentence that **you** think about, read, or write. **You** wouldn't exist without it! Wherever **you** go, there it is. What do **you** think? Do **I** have a point or not? Only the personal pronoun "it" can be used by a psyche to refer to something that is not-psyche; but, it is always a psyche **who** writes sentences using the personal pronoun "it". The rocks or physical matter can't write books for **us** to study and learn from.

If you read, study, understand, and accept everything that has been written in the books and articles which I have mentioned so far in this essay, then you will have a good and solid understanding of ultimate cause, why it's necessary, why it must exist, and why it must be true.

Material Cause answers the question of **what** something is made of, its essence. Things can be made of consciousness, spirit matter, and/or physical matter. The lower ones can reside within the higher ones on the list – consciousness can be housed in a spirit body, and a spirit body can be housed in a physical body. Efficient Cause and Formal Cause are an attempt to answer the question of **how** something was made or **how** something came into being or existence. Formal cause also explains **what** something is. Final Cause is an attempt to answer the question of **why** something was made or **why** something was brought into existence.

Ultimate Cause is always an attempt to answer the question of **who** wrote this sentence, or **who** brought order and structure to this physical object or that particular physical genome, or **who** did that particular material cause or efficient cause or formal cause or final cause, or **who** wrote that computer program or genome, or **who** designed and created and produced the first functional genome and the first physical life form, or **who** brought physical matter and this physical universe into existence.

This **Ultimate Cause Model of Reality** is an umbrella model that subsumes everything that stands in opposition to Materialism and Naturalism. **Psyche as the Ultimate Cause** is so obvious that after its existence dawned on me like a lightning bolt, I have found myself wondering ever since why nobody has ever thought of it before. I found myself wondering how I could have overlooked it for the first fifty-five years of my life. Psyche as the Ultimate Cause has been staring us in the face for the duration of human history, yet nobody seems to have noticed it. It's the core foundation of our Reality, but everyone seems to ignore it or take it for granted.

This Ultimate Model of Reality is the **Model of Lived Experience**. It is a Psyche Ontology. It subsumes or includes the whole of Reality and the whole of human experience, including the revelations of God and the revelations from God. I don't think it can get more expansive or all-inclusive than that. If it can, I'll have to let someone else figure out how.

I hope that what I have written to this point in this essay serves as a solid, useful, and constructive introduction to this **Psyche as Ultimate Cause Model of Reality**.

This essay, which I prepared for a class on Personality Theory that I was taking, serves as the introduction to a book I have written entitled, "The Ultimate Model of Reality: Psyche Is the Ultimate Cause". It will also be used as the introduction to my book, "Putting Psyche Back into Psychology: Restoring Science to Consciousness".

In the subsequent parts of those books, I discuss the various applications and ramifications of the **Ultimate Cause Model of Reality**, thereby hopefully bringing the items in the preceding paragraphs into sharper relief. It's all about finding evidence to support one's thesis, because models of reality such as Materialism and Naturalism are completely worthless because it's impossible to find any evidence to support their major premises which state that psyche, the non-physical, the immaterial, and quantum non-locality do not exist. Instead, ALL of the evidence we have on hand proves beyond a shadow of a doubt that Materialism and Scientific Naturalism are false, and that Psyche does indeed exist.

Within "Putting Psyche Back into Psychology: Restoring Science to Consciousness" and "The Ultimate Model of Reality: Psyche Is the Ultimate Cause", I explore many of the practical applications of this Ultimate Cause Model of Reality, as well as bring to light more of its evidentiary support. Ultimate cause is huge, with lots of evidence and logical common sense to support it, once a person is willing to start looking for that evidence. In contrast, I could never find any compelling or convincing evidence to support My Materialism and My Atheism. I eventually had to let them go. Materialism and its derivatives are the ultimate fiction; and, I was looking for something real and true.

Concluding Note: This part entitled, "THE ULTIMATE MODEL OF REALITY", is PART I of my books, "Putting Psyche Back into Psychology: Restoring Science to Consciousness" and "The Ultimate Model of Reality: Psyche Is the Ultimate Cause". Within those books, I eventually bring Doctorate of Theoretical and Philosophical Psychology, Brent D. Slife, into the mix by mentioning and/or discussing all of his free articles about Psychology and Philosophy for a more full and robust treatment of this Ultimate Model of Reality. Hold on to your hats! These books comprise my magnum opus – my primary contribution to philosophy, quantum realities, science, psychology, and the reality of our lived experiences as a race. If this topic interests you, then please look at those books for a continuation of this theme – Psyche is the Ultimate Cause. This first part is just a drop in the ocean compared to what I was able to accomplish with this theme in those books.

I hope you found this interesting and useful. I certainly did when it was first revealed to me. It came to me in a flash of insight; and, it has taken me months to put it into words. I discovered and realized that everything, all the evidence, points to Psyche as the Ultimate Cause. Only the things such as Materialism, Naturalism, and Atheism – things which have no evidence supporting them – fail to point to Psyche. These falsehoods fail to point to a lot of other truths and realities as well. It's in their nature to do so, because they are exclusionary philosophies and are based upon a refusal to look at evidence and based upon a denial of reality.

I eventually realized that if I wanted to get at the truth and KNOW the truth, then I had to get rid of My Atheism, My Materialism, My Nihilism, and any type of Naturalism including Darwinism. The truth cannot be built upon falsehoods such as these. That's the TRUTH, and I KNOW it!

Core Set of References

Alexander, E. (2012). *Proof of Heaven: A Neurosurgeon's Journey into the Afterlife*. New York: Simon & Schuster.

Alexander, E., & Moody, R. (2013). *Conversations Beyond Proof of Heaven*. Reference: https://www.amazon.com/dp/B00LYRYFCC/.

Bannister, D., & Fransella, F. (1986). *Inquiring Man: The Psychology of Personal Constructs* (3rd ed.). Dover, NH: Croom Helm.

Bishop, B. G. (1998). *The LDS Gospel of Light*. USA: Ponce de Leon.

Boeree, C. G. (2006). *Personality Theories*. Psychology Department: Shippensburg University.
(Retrieved from http://webspace.ship.edu/cgboer/perschapterspdf.html).

Braden, G. (2007). *The Divine Matrix: Bridging Time, Space, Miracles, and Belief*. Carlsbad, CA: Hay House.

Buhlman, W. L. (1996). *Adventures Beyond the Body: How to Experience Out-of-Body Travel*. New York: HarperCollins Publishing.

Denton, M. (1986). *Evolution: A Theory in Crisis*. Chevy Chase, MD: Adler & Adler.

Dispenza, J. (2014). *You Are the Placebo: Making Your Mind Matter*. USA: Hay House Inc.

Durham, E. (1998). *I Stand All Amazed: Love and Healing from Higher Realms*. Orem, UT: Granite Publishing and Distribution.

Eccles, J. C., & Popper, K. R. (1977). *The Self and Its Brain: An Argument for Interactionism*. New York: Routledge.

Eccles, J., & Robinson, D. N. (1984). *The Wonder of Being Human: Our Brain and Our Mind*. New York: The Free Press.

Eccles, J. (1985). *Mind and Brain: The Many-Faceted Problems*. New York: Paragon House Publishers.

Engler, B. (2009). *Personality Theories: An Introduction* (8th ed.). Boston, MA: Houghton Mifflin Harcourt Publishing Company.

Feser, E. (2009). *Aquinas (A Beginner's Guide)*. Oxford, England: Oneworld Publications.

Gibson, A. S. (2006). *They Saw Beyond Death: New Insights on Near-Death Experiences*. Springville, UT: Horizon Publishers.

Goswami, A. (2008). *God Is Not Dead: What Quantum Physics Tells Us about Our Origins and How We Should Live*. Charlottesville, VA: Hampton Roads.

Haisch, B. (2006). *The God Theory: Universes, Zero-Point Fields, and What's Behind It All*. San Francisco, CA: WeiserBooks.

Hinze, S. (1994, 1997). *Coming from the Light*. New York: Pocket Books.

Kalat, J. W. (2008). *Introduction to Psychology* (9th ed.). Belmont, CA: Wadsworth, Cengage Learning.

Kelly, E. F., Kelly, E. W., Crabtree, A., Grosso, M., & Gauld, A. (2007). *Irreducible Mind: Toward a Psychology for the 21st Century*. Plymouth, United Kingdom: Rowman and Littlefield.

Long, J., & Perry, P. (2016). *God and the Afterlife: The Groundbreaking New Evidence for God and Near-Death Experience*. New York: HarperOne.

Mark My Words. (2016). *Quantum Mechanics from a Non-Physical Spiritual Perspective*. Kindle. (Retrieve from: https://www.amazon.com/dp/B01J023TGU).

Mark My Words. (2016). *The Scientific Method: Proves That the Theory of Evolution Is False*. Kindle. (Retrieve from: https://www.amazon.com/dp/B01IAAIRT2).

Mark My Words. (2016). *The Theory of Evolution Proved to Me that God Exists: Why I Am No Longer an Atheist and Why I No Longer Believe in the Theory of Evolution*. Kindle. (Retrieve from: https://www.amazon.com/dp/B01HZYBZ7K).

Mark My Words. (2016). *Using the Scientific Method: To Eliminate the Usual Suspects and to Prove the Truth*. Kindle. (Retrieve from: https://www.amazon.com/dp/B01J6STHP0).

Marshall, P. D., Kelly, E. F., & Crabtree A. (Eds.). (2015). *Beyond Physicalism: Toward Reconciliation of Science and Spirituality*. London, United Kingdom: Rowman and Littlefield.

McTaggart, L. (2002). *The Field: The Quest for the Secret Force of the Universe*. New York: HarperCollins.

Meyer, S. C. (2010). *Signature in the Cell: DNA and the Evidence for Intelligent Design*. New York: HarperCollins.

Meyer, S. C. (2013). *Darwin's Doubt: The Explosive Origin of Animal Life and the Case for Intelligent Design*. New York: HarperCollins.

Moody, R. A. (1975, 2015). *Life After Life: The Bestselling Original Investigation That Revealed "Near-Death Experiences"*. New York: HarperCollins.

Moody, R., & Perry, P. (2010). *Glimpses of Eternity: Sharing a Loved One's Passage from This Life to the Next*. New York: Guideposts.

Neal, M. C. (2011). *To Heaven and Back: A Doctor's Extraordinary Account of Her Death, Heaven, Angels, and Life Again: A True Story*. Colorado Springs, CO: WaterBrook Press.

Penfield, W. (1978). *The Mystery of the Mind: A Critical Study of Consciousness and the Human Brain*. Princeton, NJ: Princeton University Press.

Radin, D. (2006). *Entangled Minds: Extrasensory Experiences in a Quantum Reality*. New York: Paraview Pocket Books.

Richards, P. S., & Bergin, A. E. (2004). *Casebook for a Spiritual Strategy in Counseling and Psychotherapy*. Washington DC: American Psychological Association.

Richards, P. S., & Bergin, A. E. (2005). *A Spiritual Strategy for Counseling and Psychotherapy* (2nd ed.). Washington DC: American Psychological Association.

Ring, K., & Cooper S. (1999). *Mindsight: Near-Death and Out-of-Body Experiences in the Blind* (2nd ed.). Kearney, NB: Morris Publishing.

Rivas, T., Dirven, A., & Smit, R. H. (2016). *The Self Does Not Die: Verified Paranormal Phenomena from Near-Death Experiences*. Durham, NC: IANDS Publications.

Ross, H. (1991). *The Fingerprint of God: Recent Scientific Discoveries Reveal the Unmistakable Identity of the Creator* (2nd ed.). Orange, CA: Promise Publishing Co.

Ross, H. (1996). *Beyond the Cosmos: The Extra-Dimensionality of God: What Recent Discoveries in Astronomy and Physics Reveal about the Nature of God*. Colorado Springs, CO: NavPress.

Ross, H. (2008). *Why the Universe Is the Way It Is*. Grand Rapids, MI: Baker Books.

Ross, H., Samples, K. R., & Clark, M. (2002). *Lights in the Sky & Little Green Men: A Rational Christian Look at UFOs and Extraterrestrials*. Colorado Springs, CO: NavPress.

Rychlak, J. F. (1979). *Discovering Free Will and Personal Responsibility*. New York: Oxford University Press.

Rychlak, J. F. (1981a). *A Philosophy of Science for Personality Theory* (2nd ed.). Malabar, FL: Robert E. Krieger Publishing Company.

Rychlak, J. F. (1981b). *Introduction to Personality and Psychotherapy: A Theory-Construction Approach* (2nd ed.). Boston, MA: Houghton Mifflin Company.

Rychlak, J. F. (1982). *Personality and Life-Style of Young Male Managers: A Logical Learning Theory Analysis*. New York: Academic Press.

Rychlak, J. F. (1988). *The Psychology of Rigorous Humanism* (2nd ed.). New York: New York University Press.

Rychlak, J. F. (1991). *Artificial Intelligence and Human Reason: A Teleological Critique*. New York: Colombia University Press.

Rychlak, J. F. (1994). *Logical Learning Theory: A Human Teleology and Its Empirical Support*. Lincoln, NE: Nebraska University Press.

Rychlak, J. F. (1997). *In Defense of Human Consciousness*. Washington DC: American Psychological Association.

Rychlak, J. F. (2003). *The Human Image in Postmodern America*. Washington DC: American Psychological Association.

Sanford, J. (2014). *Genetic Entropy* (4th ed.). Cornell University: FMS Foundation.

Sanford, J. C., Marks, R. J., Behe, M. J., Dembski, W. A., & Gordon, B. L. (Eds.). (2013). *Biological Information: New Perspectives*. Hackensack, NJ: World Scientific.

Sharkey, J. (2008). *A GLIMPSE OF ETERNITY: One man's story of life beyond death. Ian McCormack's Story*. Orewa, New Zealand: Arun Books.

Slife, B. D. & Williams, R. N. (1995). Science and Human Behavior. In *What's Behind the Research? Discovering Hidden Assumptions in the Behavioral Sciences*, (pp. 167–204). Thousand Oaks, CA: SAGE Publications.

Storm, H. (2000, 2005). *My Descent into Death: A Second Chance at Life*. USA: Random House.

Talbot, M. (1991, 2011). *The Holographic Universe: The Revolutionary Theory of Reality*. New York: HarperCollins.

Talbot, M. (1993). *Mysticism and the New Physics*. New York: Penguin Books.

Tart, C. T. (2009). *The End of Materialism: How Evidence of the Paranormal Is Bringing Science and Spirit Together*. Oakland, CA: New Harbinger Publications.

Van Lommel, P. (2010). *Consciousness Beyond Life: The Science of the Near-Death Experience*. New York: HarperCollins.

Wells, J. (2000. *Icons of Evolution: Science or Myth? Why Much of What We Teach About Evolution Is Wrong*. Washington, DC. Regnary.

PART IV — DEFINING THE LIGHT

A Study in Light!

Within all the different Models of Reality that I have developed during the past few years, I have repeatedly observed that photons and psyches are massless and therefore non-physical. In fact, they seem to be the same thing, from what I can tell. Whenever out-of-body travelers have observed a psyche at the quantum level while in the spirit world or quantum realm, a psyche is always observed as being a pinpoint of light or a photon.

A photon is defined as a massless particle, which means that at the speed-of-light, a photon is obviously non-physical, spiritual, supernatural, or quantum in nature. A photon has to slow down to zero in order to manifest or show up in our physical realm; otherwise, it remains as a non-physical quantum object.

Clearly, photons and massless particles exist; therefore, it is obvious that the non-physical exists. The proven and verified existence of anything non-physical FALSIFIES Materialism, Naturalism, Darwinism, Nihilism, Atheism and their derivatives.

Mark My Words

The End from the Beginning

This is a study in light.

ALL LIGHT is conscious, sentient, intelligent, aware, and alive. This means that ALL LIGHT is capable of obeying God's Laws and God's Commands. LIGHT IS LIFE. LIGHT IS INTELLIGENCE. LIGHT IS CONSCIOUS AND AWARE. SPIRIT IS LIGHT. MATTER IS MADE OF LIGHT. It's ALL made of light! The Natural Laws are in fact God's Laws; and, it's all made of light. God's Psyche or God's Light is the organizing factor in this universe. God is in the Light!

LIGHT is alive, to one degree or another, which means that Light is intelligent to one degree or another. When it comes to Light or Intelligence, God's Psyche or Intelligence is more intelligent than all of the other intelligences combined, which is why He is God, and we are not.

On this model, a photon KNOWS where it's going to land before it launches; otherwise, it doesn't launch. From the photon's perspective, there is no distance and there is no time. Even if its destination is on the other side of this universe, it KNOWS where it's going to land before it launches, or it never goes. This means that LIGHT is non-local or spiritual in nature and origin.

In its native state, or pure primal state, LIGHT doesn't have any limitations, although light can take on forms or states which do have limitations as well as new possibilities. LIGHT can take upon itself God's Laws and God's Commands. When one door closes, another one opens. Whenever LIGHT takes upon itself certain restrictions or limitations, NEW POSSIBILITIES open up to it. God's Psyche or God's Light brings order and organization to ALL other types of Light. God's Psyche or God's Light gives all other types of light purpose and meaning – a reason for being or a reason for existing.

Seeing the Light

EVERYTHING IS MADE OF LIGHT! EVERYTHING, including spirit matter and physical matter. It's ALL light.

LIGHT or ENERGY seems to have two main states of existence.

First, LIGHT or ENERGY can manifest as Matter, both spirit matter and physical matter. Matter is simply an organized form of LIGHT, or condensed light. Matter REACTS. In order to be able to REACT to Natural Laws and God's Commands, Matter or this type of Light needs to be minimally conscious, sentient, intelligent, aware, and alive. Matter is reactionary and is acted upon by a different type of Light, which I call Psyche or Intelligence or Non-Local Consciousness. But, Matter has to be minimally conscious and aware; otherwise, it wouldn't KNOW how and when to REACT to Psyche's command. Matter is made of light, a type of living light. ALL Light is alive, conscious, and aware of its surroundings.

In my books, I describe Matter as a Quantum Object. Matter or Quantum Objects have two known aspects or states of existence – spirit matter or physical matter. Due to the Quantum Law of Complementarity, a single Quantum Object can either be spirit matter or physical matter, but it can't exist in both states simultaneously. A Quantum Object has to be in one state or the other. According to the Science of Quantum Mechanics, ONLY Psyche or Non-Local Consciousness can convert spirit matter into physical matter by conscious observation or by the Word of Command. This means that Psyche is a completely different type of light than Matter.

While dealing with different physical illnesses and psychological issues, I have found it helpful to tell myself that I am NOT my physical body. Psyche is something completely different than our spirit body and our physical body.

Often in my books, I portray Matter as if it were dead, inert, and ineffective. It's easy to do, but this portrayal isn't completely accurate. Since ALL Matter is made of light, including both spirit matter and physical matter, there is a minimal degree of intelligence and consciousness within Matter. There has to be, or Matter wouldn't be able to REACT to God's Commands and Matter wouldn't be able to FOLLOW or OBEY God's Laws – what we humans typically call Natural Laws.

However, Matter is not alive in the same way that Psyche is alive. Matter is ONLY capable of reacting. Whereas, only Psyche can ACT. There's a clear difference between the level of intelligence found in Matter and the level of intelligence or light found in Psyche. Psyche can command and control. Matter, whether we are talking about spirit matter or physical matter, simply REACTS.

I make this emphasis on the inert or the reactionary nature of Matter, in order to EMPHASIZE that Matter of any kind cannot design and create and do science. This is VERY important to understand. Matter, including both spirit matter and physical matter, cannot program software such as genomes and cannot manufacture hardware such as life forms or what we call living organisms. According to Scientific Observations or Lived Experiences, Matter lacks the innate inherent ability to design, program, engineer, create, manufacture, and deploy genomes and life forms. Matter has NEVER been caught in the act of doing these kinds of things, because Matter cannot act. Matter can ONLY REACT. That's the LAW!

An understanding and acceptance of this Reality becomes Scientific Proof of God's Necessity or Scientific Proof of God's Existence. Your genome IS God's Signature. Your physical body IS God's handiwork. We KNOW this to be true because we KNOW from Scientific Observations or Lived Experiences that Matter of any kind cannot design and create. Lived Experience IS the BEST kind of Scientific Evidence! With Lived Experience or Direct Observation, we human beings or human psyches can go straight to KNOWING the TRUTH by living it and experiencing it for ourselves.

In a similar vein, Quantum Non-Locality is a verified science whereas Materialism and Naturalism are falsified "sciences". Non-locality is the Spirit Realm or the Transdimensional Realm. In contrast, Materialism and Naturalism make the philosophical claim that the supernatural or the non-local does not exist, and that psyche does not exist. In a very real sense, Materialism and Naturalism make the absurd claim that Light does not exist, because Light is supernatural, immaterial, and non-local in nature and origin. Spirit IS light, whether we are talking about the Psyche, Spirit Matter, or Spirit Bodies. It's all Light; and, Materialism makes the claim that it does not exist.

Either the spiritual exists, or it doesn't. If the spiritual or the non-local can be demonstrated to exist – if light can be demonstrated to exist – then Materialism and Naturalism have been falsified. If you can read this, then you KNOW that light exists; and, you have just falsified Materialism and Naturalism, or proven to yourself that Materialism and Naturalism are false. Materialism and Naturalism have been falsified trillions of trillions of times in thousands of different ways. In contrast, through our Lived Experiences or Direct Observations – what I call Scientific Observations – Psyche, Intelligence, Consciousness, and Light have been verified trillions of trillions of times with an infinite number of more verifications to go.

Technically, if done correctly, the Scientific Methods can't be used to falsify the TRUTH. The Scientific Methods or Scientific Observations will ALWAYS verify the truth and point us to the TRUTH. The Scientific Methods always point us to Psyche, or Light, or Intelligence. Intelligence or Psyche is needed to do science in the first place! The Non-Local Realm or the Spirit Realm is always needed to make sense of Quantum Mechanics, Uncertainty, Action at a Distance, and the Conscious Observer.

Second, LIGHT can be Psyche or Intelligence, which means that it ACTS and therefore has Agentic Capabilities, including the Word of Command. Psyche or the Conscious Observer has the innate ability to ACT and to choose by the WORD OF COMMAND to convert spirit matter into physical matter, and back again into spirit matter. ONLY Psyche or Intelligence can design and create! ONLY Psyche can do science! ONLY Psyche or Non-Local Consciousness can convert spirit matter into physical matter! This is what Science and Quantum Mechanics are trying to teach us, but few people are actually listening.

Psyche or Intelligence is the type of LIGHT that we typically associate with LIFE, entities, and living organisms. Psyche IS an independent Living Agent. Psyche ACTS and COMMANDS. In contrast, Matter REACTS.

Psyche IS the Ultimate Force! Psyche IS the Ultimate Causal Agent! It doesn't get more fundamental than Psyche or Non-Local Consciousness. Psyche is a special kind of LIGHT. This IS a Psyche Ontology wherein Psyche is the fundamental unit of reality. This IS the Ultimate Model of Reality, the Ultimate Paradigm. Everything is based upon Psyche or Intelligence.

Matter is the type of LIGHT which we typically associate with inanimate objects like rocks and metal. Inanimate Matter, all matter really, is reactionary. Matter REACTS to commands from Psyche or Intelligence. Even what we term "living organisms" or "physical organisms" were designed to react to a controlling Psyche or Intelligence. Psyche is a completely different state of existence or type of LIGHT than Matter. ONLY Psyche can ACT and DO the Word of Command.

Matter is the type of LIGHT which lacks the innate inherent ability to design, create, program, engineer, build, and manufacture. Matter simply REACTS. Therefore, there is no way possible for Matter to have designed and created the first genomes and life forms on this planet, as the Materialists and Darwinists claim. Matter can only react, not act! This reality becomes Scientific Proof of God's Necessity or Scientific Proof of God's Existence. The Ultimate Scientist must exist, or there would be NO Science! The Ultimate Scientist must exist, or there would be NO genomes and NO life forms on this planet. The Ultimate Scientist must exist, or there would be NO physical matter. This is what Science is trying to teach us; but, few people are actually listening.

According to Quantum Mechanics, some kind of Ultimate Psyche or Ultimate Scientist must of necessity exist, in order to convert spirit matter into physical matter. If God didn't exist, there would be NO localized physical matter. It would ALL still be non-local spirit matter. Spirit matter and the light within spirit matter has ALWAYS existed. In contrast, physical matter has to be brought into existence by Psyche and the Word of Command.

The light within physical matter has always existed; but, physical matter hasn't always existed. What was there before the first particle of physical matter was designed and created? WE WERE! Our psyche, intelligence, or light was there in existence before the first particle of physical matter was commanded into existence. Psyche functions as if it were a god. This reality becomes Scientific Proof of God's Necessity or Scientific Proof of God's Existence. The Ultimate Conscious Observer or the Ultimate Conscious Commander must exist, or there would be NO physical matter!

Furthermore, according to Science as we KNOW it, matter or rocks cannot design, create, program, manufacture, engineer, nor deploy anything because matter cannot ACT. Matter can ONLY REACT. Therefore, according to Scientific Observations and the Lived Experiences of the human race, we KNOW that some kind of Ultimate Psyche or Ultimate Scientist must of necessity exist, or there would be NO genomes and NO life forms on this planet. ONLY Psyche or Intelligence can design, create, program, engineer, and manufacture things out of physical matter. ONLY Psyche can do science!

Spirit matter, or the light within spirit matter, has always existed; but, spirit matter is the type of light that REACTS to Psyche or Conscious Command. There must of necessity be some kind of Conscious Commander or God's Psyche; otherwise, spirit matter and reality would be completely unorganized and totally chaotic. Once again, this reality and TRUTH becomes Scientific Proof of God's Necessity or Scientific Proof of God's Existence, because there would be no order and no organization and no organisms without God's Psyche and God's Word of Command. It would all be chaos. This is what Science is trying to teach us, but few people are actually listening.

This Whole Reality is elementary and obvious my dear friend; but, most people refuse to think about and accept these kinds of things. They prefer instead to remain in darkness. Only a few people are willing to see the light.

Everything Is Made of Light

ALL LIGHT is conscious, sentient, aware, intelligent, and alive. Even the LIGHT within spirit matter and physical matter is conscious and aware of God's Commands and God's LAWS, or what we call Natural Laws.

A close observation of Quantum Mechanics seems to indicate that physical particles are conscious and aware of their surroundings. Matter has to be conscious and aware in order to be able to REACT to God's Commands and God's LAWS; otherwise, Matter would do nothing at all, no matter what we tried to do to it. If Matter wasn't conscious and aware of its surroundings or environment, then it would NEVER react to anything and it would therefore be unable to REACT to natural laws, God's Laws, our commands, and our interventions.

It would do no good to try to mold matter into a table or chair if Matter didn't know that it's supposed to respond to stimuli, and didn't know that it's supposed to obey natural laws or physical laws, and didn't know what those laws were and meant. Furthermore, without God's Laws or what we call Natural Laws or Physical Laws, Matter wouldn't know what it's supposed to do or how it's supposed to react. It would be chaos! Once again, thanks to the order and the organization that exists within this universe, this reality and TRUTH becomes Scientific Proof of God's Necessity or Scientific Proof of God's Existence. If God didn't exist, then Natural

Laws or Physical Laws would not exist; and, the LIGHT in the Matter wouldn't know what to do with itself or wouldn't know how to react. If God didn't exist, this whole universe would be Chaos.

It ALL points to the existence of the Ultimate Scientist or the Ultimate Psyche, without which NONE of this would exist. That's what Science is trying to teach us, but not everyone is listening. You can lead a horse to water, but you can't make him drink, because he is an independent Psyche or an independent Agent. He has his own spark or light within him. But for most of us, that light is nothing but darkness.

There Are Other Types of Light

There are other types of light besides Matter and Psyche.

This is important to understand, because everything is made of LIGHT.

Forces, Fields, Gravity, Magnetism, Strong Nuclear Forces, and Weak Nuclear Forces, Electrons, Protons, Neutrons, and Quarks are ALL some type of LIGHT, which means that each one of them has a certain degree of intelligence or consciousness within it, which means that ALL of these things are capable of REACTING intelligently and responsibly to God's Commands. These things couldn't properly REACT, though, unless they were aware of their surroundings and their environment, aware of external stimuli, and aware of God's Commands or God's Laws.

Remember, ALL LIGHT is conscious, sentient, intelligent, aware, and alive to one degree or another. This means that ALL LIGHT is capable of obeying God's Laws and God's Commands. ALL LIGHT in this universe KNOWS what the natural laws are or what God's Laws are; and therefore, ALL LIGHT KNOWS how it should REACT or how it should "behave".

Gravity is a type of light that pulls matter together. Gravity is universal in range and scope.

Strong nuclear forces and weak nuclear forces are a type of light which keeps the electrons from imploding into the protons. These are the types of light which create the Space between electrons and the nucleus of an atom. Nuclear forces are local in range and scope.

Magnetism is a type of light that both attracts and repels depending upon its polarity.

Electricity is a type of energy or light, which we use to power things.

ALL LIGHT is intelligent and aware to one degree or another, because it KNOWS what its assigned task is and how it's supposed to REACT. It's ALL LIGHT, but with different assignments or tasks.

Dark Energy is a type of light which functions as the stretching factor in this universe. Dark Energy is also universal in range and scope. In a lot of our Science and science fiction, Dark Energy is often called the Zero-Point Field of Light. New Agers call Dark Energy the Quantum Sea of Light. In the scriptures which the

Biblical God had a hand in writing and producing, God calls this Dark Energy or Zero-Point Field the "Light of Christ which fills the immensity of space". This type of Light is power; and, it also seems to function as the universal internet. Like Gravity, the influence of Dark Energy or the Light of Christ seems to permeate the whole of space. Dark Energy is the counteracting force to Gravity. Gravity pulls together Matter; whereas, Dark Energy creates Space.

On this model of reality, Dark Matter is Spirit Matter or Non-Local Matter. Spirit Matter or Dark Matter is just a different type of light, which was designed to REACT a certain way to God's Commands.

Scientists often call these other forms of Light, "Natural Laws". Natural Laws or God's Laws provide order and structure to this universe.

BUT, it's ALL made of light! Even Matter is made of Light.

Light itself has always existed. It cannot be created nor destroyed. There has to be some part of this Universe and this Reality which has always existed; otherwise, this Universe and this Reality would not exist. Light IS the part of this universe and this Reality which has always existed and will always exist.

Psyche IS the Ultimate Type of Light. Without Psyche or Intelligence, there IS NO existence! Without Psyche, there is no command and control. Without God's Psyche, this whole thing would be nothing but Chaos; and, we human beings or human psyches would not exist as physical human beings. Psyche provides order, organization, and LAW to the types of Light which REACT. This once again becomes Scientific Proof of God's Necessity, or Scientific Proof of God's Existence.

Lived Experience or the REAL Scientific Evidence tells us how things really are; but, few people are actually listening.

Psyche Is the Most Interesting Type of Light
PSYCHE or Non-Local Consciousness is the most interesting type of LIGHT to study, because it has the innate or inherent ability to COMMAND and to ACT.

The Biblical God Jesus Christ and His prophets tend to call this type of Light, "Intelligence" or "Spirit" or "Ghost" or "Life". It's all the same thing. Psyche is a living Agent. Psyche is a conscious Commander. Psyche is an Actor. Without some kind of Ultimate Psyche or Ultimate Commander, there would be NO physical matter, NO order, NO organization, NO software, NO genomes, NO hardware, NO science, and NO physical life forms.

The whole of Science points to this TRUTH and this REALITY, but few people are actually listening, because this is a message that they don't want to hear. They are afraid of God or the Ultimate Psyche. I KNOW, because I used to be one of them. I used to be a Materialist, Nihilist, and Atheist; and, I was afraid of God at the time, even though I didn't want to admit it. God is in the Light. God is Light. But, I was afraid of that type of light. I preferred to remain in darkness instead. I got my wish, until I changed my mind.

The Materialists and Naturalists and Atheists are the most ignorant people on this planet – they are ever-learning but never able to come to a knowledge of the truth. These people are typically highly educated ignoramuses. Their learning and their education actually blinds them to the TRUTH. These people have perfected the art of teaching and preaching and believing falsehoods and lies. Self-deception works, and it works every time. The Materialists and Naturalists are deceiving themselves. I KNOW, because I used to be one of them. I was a highly educated and highly indoctrinated idiot.

For me personally, Scientism was my chosen religion. I gave it precedence over everything else. I had a great deal of faith – blind faith – in Science and the Scientific Methods. It can be extremely fascinating to study the Psychology of the highly educated ignoramuses – the Materialists, Naturalists, Darwinists, Nihilists, and Atheists. We can learn from these people how to do faulty logic, self-deceptive philosophy, and BAD SCIENCE. These people can teach us what we don't want to be and what we don't want to become. These people are slaves to Satan and his lies. These people have gone where we don't want to go. I KNOW, because I used to be one of them. It was hell. Ignorance is hell.

Psyche IS a different type of Light.

Psyche or Non-Local Consciousness seems to be completely immaterial. Psyche is PURE LIGHT. Psyche is NOT spirit matter, and it's definitely not physical matter. Psyche can't be Matter, because Matter can ONLY REACT. Matter lacks the ability to ACT. ONLY Psyche can ACT. ONLY Psyche can function as a Living Agent. Even though Matter is technically alive and aware and sentient, Matter cannot function as a Living Agent. Matter cannot do Psyche! Matter cannot ACT. Matter can ONLY REACT. That reality is what gives this universe structure and order. Matter REACTS to God's Commands. According to Science and Quantum Mechanics, Psyche or Non-Local Consciousness has to be something completely different than Matter, because ONLY Psyche can convert spirit matter into physical matter through the Word of Command or Conscious Observation.

Some Out-of-Body Travelers have experienced their Psyche separate from their spirit body. While their Psyche is separate from their spirit body looking at their spirit body, their Psyche seems to be completely immaterial – simply a "viewpoint in space". Their Psyche or Non-Local Consciousness is alive and aware and remembers the experience of being separate from its spirit body and its physical body; but, this Psyche or Personality or Individual seems to be completely immaterial and invisible. There seems to be nothing there whenever Psyche goes looking for itself. Obviously, there is something there because Psyche is aware, but there doesn't seem to be anything there. Psyche is an infinite singularity. Psyche is completely immaterial, although it completely matters. Without Psyche or Intelligence there would be NO existence. ONLY Psyche can ACT and construct Reality. In contrast, matter of any type can ONLY REACT.

The Other Types of Light also seem to respond to and defer to Psyche. Psyche is the Commander and the Controller – the Order and the Organizer – the Structure and the Law. The other types of light RESPOND to Psyche. Psyche is the

Master. Psyche is the Ultimate Type of Light. The other types of light are servants. The Whole of Reality and Existence points to Psyche, particularly God's Psyche. If God's Psyche did not exist, then there would be NO physical matter. Everything in this universe would still be spirit matter, and a very chaotic spirit matter at that! Psyche IS Order, Organization, and Life. If God or the Ultimate Scientist didn't exist, then this whole universe would be nothing but Chaos.

Experiencing the Light

After a couple of years of research, I finally realized that Lived Experience is a much better way of finding and KNOWING the TRUTH than the Scientific Method.

The following Lived Experience changed the way that I look at Reality. I'm no longer a Materialist, Nihilist, and Atheist. In the following Out-of-Body Experience, William Buhlman exited his physical body and went to the non-local realm or spirit realm. There in the spirit realm or transdimensional realm, he exited his spirit body, looked at his spirit body, and described his spirit body.

This Lived Experience is Scientific Evidence or Observational Evidence that our physical body, our spirit body, and our Psyche or Intelligence or Non-Local Consciousness ARE three completely different entities; and, this Lived Experience provides Scientific Evidence and Observational Proof that it's the immaterial Psyche who does all of the observing, thinking, and remembering. Through Lived Experiences, we human beings or human psyches can go directly to KNOWING the TRUTH and directly to PROOF of the TRUTH. Our Lived Experiences ARE the best and most efficient type of Scientific Evidence! Just because you and I haven't had some of these types of experiences doesn't mean that nobody else has.

Journal Entry, October 2, 1982

I hear the buzzing, engine-like sounds and will myself out-of-body. I step to the bedroom door and automatically request "Clarity now!" My vision improves and I step through the door, into the living room. Still feeling a little out of sync, I verbally repeat my request with more emphasis, "Clarity now!" I feel my awareness and vision snap into place.

My thoughts are clear, and I make a verbal demand, "I need to see the form I'm in now!" Instantly I feel an intense sensation of being drawn within myself. I'm suddenly different, weightless as though I'm floating in space. As I look forward I see a sparkling, bluish white form. For some reason, I seem to know that I'm looking at my nonphysical body from a different perspective. I stare in amazement at this form before me that shines and flows with energy and light. It looks like an energy mold created from a million tiny points of light; it radiates a bluish glow but appears to have a defined outer structure. The body of light before me is naked and is identical to my physical form. Even though my body looks firm, there is a noticeable energy motion and radiation present. I can see what appears to be an ocean of blue stars throughout my body. It's difficult to describe because the stars are stable yet moving at the same time; the light and energy of my body appear to change and flow almost like the waves of an ocean.

As I stare at the body of light, it hits me that I must be in another body. Yet I can't perceive any form or substance; I'm like a viewpoint in space without shape or form of any kind. As I reflect upon my new state of being, I feel a sensation of rapid motion and I'm instantly back within my physical body.

Lying still and reviewing my experience, I'm struck by an inescapable conclusion: I must possess multiple energy-bodies. The form I just experienced was noticeably lighter (less dense) than even my second nonphysical body. I realize that the traditional view of our possessing two bodies — a physical body and a spiritual body — is far too simplistic; we are much more complex than this. Just as there are multiple nonphysical energy dimensions within the universe, each of us must consist of multiple energy-bodies or vehicles of expression.

Now I seriously wonder just how many nonphysical bodies or forms this involves. I suspect that there must be one within each dimension of the universe and that all of these are interrelated and connected, just as the physical body is connected to its first nonphysical (spiritual) body.

Buhlman, William L., "Adventures Beyond the Body: How to Experience Out-of-Body Travel" (pp. 34-35). HarperCollins. Kindle Edition.

You won't get anything like this from the Materialists, Atheists, Naturalists, and Darwinists because they don't have anything like this to give you. These people have chosen to believe that the Scientific Method is the only way to find and know the truth; but, they are wrong. Materialism and Atheism of any kind are based upon a refusal to look at evidence. Ironically Lived Experience, including our spiritual experiences, is an infinitely better source of Scientific Evidence than the Scientific Methods will ever be. The Scientific Methods are based upon logic fallacies and are fundamentally flawed. Not so with Lived Experience! Through Lived Experiences, we human beings or human psyches can go straight to KNOWING the TRUTH. Powerful, huh?

The real William Buhlman, his immaterial Psyche, is "like a viewpoint in space without shape or form of any kind". His Psyche doesn't have any "form or substance". It's completely immaterial. His Psyche is a force or a field, and NOT MATTER. His Psyche is PURE LIGHT or PURE CONSCIOUSNESS. His Psyche is the Ultimate Force. His Psyche is the Ultimate Causal Agent.

His spirit body is "energy and light", "identical to his physical form", and "like an ocean of stars". His physical body is back home sleeping in his bed. For me, this is Scientific Evidence and Scientific Proof that Psyche is something completely different than our spirit body, and that our spirit body is something completely different than our physical body. They are ALL made up of light, but different types of Light with different assignments or tasks or aspects!

After reading this Lived Experience, I got the feeling that I had finally found the TRUTH about our existence and our reality as human beings and human psyches. Psyche or Intelligence or Non-Local Consciousness is something completely different than spirit matter and spirit bodies. Furthermore, our spirit body is something completely different than our physical body. They are ALL made of light, but light that is in different states of being or different states and phases of existence.

Nowadays, I treat Lived Experience as Scientific Evidence. Lived Experiences do have some of the same weaknesses and limitations that the Scientific Methods have, though. Lived Experiences, whether we are talking about physical experiences or spiritual experiences, aren't always replicable on demand. Furthermore, Lived Experiences or Direct Scientific Observations are subject to biases and faulty interpretations just as much as the Scientific Methods are – some people come up with some very strange interpretations and explanations for their spiritual experiences. But, in every other respect, Lived Experiences or Psyche Experiences are vastly superior to the Scientific Methods.

I really liked William Buhlman's book, because he turned the whole Out-of-Body or Astral Projection thing into a Science and experimented with it while he was out of body.

I got really excited about this information and this KNOWLEDGE, and I started looking for independent confirmation of this Ultimate Reality or Psyche Ontology. Seek, and ye shall find. Knock, and it shall be opened unto you.

Evidence of Psyche and Signs of Life

While researching the Lived Experiences of the human race, I now keep my eyes open for Evidence of Psyche. I now treat Lived Experience as Scientific Evidence, as any Real Scientist should. Lived Experience is the BEST type of Scientific Evidence, because with Lived Experience or Direct Observation we can go directly to KNOWING the TRUTH by experiencing and living the truth for ourselves. The TRUTH is KNOWN by living it and experiencing it for yourself, or by choosing to trust someone who has.

Lived Experience or Direct Observation IS the BEST and most efficient way for doing Science! I'm talking about a Paradigm Shift here, a Psyche Ontology and Psyche Epistemology, wherein Lived Experience or Psyche Experience becomes our primary source of Scientific Evidence and our primary way of doing Science. Lived Experience subsumes or includes Science Experiments and Scientific Observations of Science Experiments, because these are Lived Experiences as well. The major advantage is that Lived Experience also includes our non-local experiences or our spiritual experiences as Scientific Evidence. This all-inclusive Psyche Ontology, based upon Lived Experience, is the Ultimate Paradigm. This Ultimate Paradigm becomes the Ultimate Model of Reality.

Close your eyes. Where are you? The REAL YOU. You can sense where you are, where your living spark or your Psyche resides. It's not in the back of your head. It's not in your ears. It's not in your nose. It's not in your hands or your feet. It's not in your stomach. Your Psyche or Consciousness resides behind the bone in your forehead, in the Third Eye position. Yes, you have a spirit body. Yes, you have a physical body. But the REAL YOU, your personality, consciousness, intelligence, awareness, and life is a tiny immaterial Spark residing and emanating from the Third Eye Position in your forehead.

Every now and then, Out-of-Body Travelers will exit their spirit body and will look at their spirit body (or look at their physical body) from an external third-person perspective or vantage-point. This immaterial viewpoint in space, or spark on the ceiling, or pinpoint of light is Psyche or Non-Local Consciousness. It's your person and your personality, complete with all of your intelligence and memories. It's an infinite singularity. Psyche is infinite memory storage capacity taking up no space whatsoever. Whenever Psyche goes looking for itself, there's nothing there to be seen. Psyche is completely immaterial. Psyche IS the Ultimate Force. Psyche is consciousness and life. Psyche is Intelligence. Psyche IS Light.

Astral Travelers have observed that their spirit body and their physical body don't have experiences and don't remember anything while their Psyche is away from their spirit body and physical body. It's the Psyche who does the experiencing and the remembering, not the spirit body and certainly not the physical body. Psyche is the light of life or the spark of life. Your Psyche is your Ultimate Reality. This is the Ultimate Model of Reality or a Psyche Ontology, wherein Psyche is the fundamental unit of reality. As far as I know, I'm the only person on the planet to have developed and presented a Psyche Ontology. I sometimes wonder how everyone else could have missed it. But, they miss it because they don't want to see it, don't want to understand it, and don't want to accept it as being true. I KNOW, because I used to be one of them. I used to be a Materialist, Nihilist, and Atheist until I learned better.

Obviously, I use William Buhlman's out-of-body experiences as Scientific Evidence or Observational Evidence that Psyche is something completely different than one's spirit body.

On YouTube, you can find thousands of Near-Death Experiences (NDEs) and Out-of-Body Experiences (OBEs). Every now and then, you will find someone who has left their spirit body and has experienced their independent immaterial Psyche directly. We KNOW from the Lived Experiences or Scientific Observations of the human race that Psyche exists and that Psyche is something completely different than our spirit body.

One woman described leaving her body and floating up near the ceiling. She said that she went to look for herself and couldn't see anything. She said that she was just a spark on the ceiling. Most people who go looking for themselves while out-of-body explain seeing parts of their spirit body. A few, however, see nothing at all. They are pure psyche, or pure light, or pure consciousness. They have no material form. They are just a spark on the ceiling, or a viewpoint in space, or a pinpoint of light. The accounts of those who experience their Psyche or Non-Local Consciousness separate from their spirit body are rare, but they do exist.

During my research, I have discovered that Lived Experience is the BEST type of Scientific Evidence, because when it comes to Lived Experience or Direct Observation, we KNOW the TRUTH by living the truth and experiencing the truth directly for ourselves. In fact, in most instances, Lived Experiences or Out-of-Body Experiences are the only way to gather Scientific Evidence from the Non-Local Realm or the Spirit Realm.

When Dr. Anthony Cicoria was struck by lightning, he died. He, his spirit body, was outside his physical body. As he walked up the stairs, his spirit body dissolved beneath him and transformed into an orb of light. He, his Psyche, was looking at his spirit body and later that orb of light from an immaterial third-person perspective or that Third Eye Perspective. This NDE is Scientific Evidence or Empirical Proof that Psyche or Intelligence is a different entity than our spirit body and our physical body.

https://www.youtube.com/watch?v=WUXzj0Tczz4

When people like Dr. Anthony Cicoria describe looking at their spirit body from within their spirit body, they describe the process from the Third-Eye Perspective. The interesting thing to note is that the Psyche can not only see, but that the Psyche can also see events from different perspectives, both from within the spirit body and from outside the spirit body. Psyche can project, which means that Psyche can exit its assigned spirit body. It's always the Psyche who does the experiencing and the remembering, and not the spirit body and not the physical body.

Often the Psyche is called the Spark. It's the Spark of Life. This indicates that the Psyche is made of energy or light. The following videos describe the Spark of Life that takes place at Conception:

https://www.youtube.com/watch?v=UUZfWB95I1A

https://www.youtube.com/watch?v=1fAfQlpPKK0

https://www.youtube.com/watch?v=DfH4dP26WsE

ONLY Psyche can do telepathy. Psyche, or mind-over-matter, is involved in physical healing. Psyche is a supernatural phenomenon; and, there are often many different supernatural healings and supernatural events which take place during Near-Death Experiences, that lend evidence to Psyche and the supernatural. These are Empirical Experiences which take place under the direction of God's Psyche; but, are impossible for the physical body and physical matter to perform or to do on their own. Psyche IS the Ultimate Causal Agent. Psyche is the Prime Mover or the Ultimate Cause.

The following Near-Death Experience (NDE) talks about the Bubble of Protection and a Physical Teleportation to safety. This is NOT something that we consciously do. It's something that God does for us.

https://www.youtube.com/watch?v=DmBnCTuQUOc

I have experienced this Bubble of Protection and Physical Teleportation myself, so I KNOW that this is real and can truly happen to us, under God's direction and protection.

I was going too fast and couldn't miss the car that had pulled out in front of me. Time stopped, and I seemed to have 360-degree awareness of my surroundings, and I saw or felt where I wanted to be. It felt as if the whole of eternity had gathered into me; and then, snap! I and my car were someplace else in the middle of a lawn twenty or thirty feet away with the car completely stopped and the engine shut off. I have NO memory of traveling the distance, stopping the car, and shutting the car off. I was on that other car's bumper one instant and someplace else entirely the next instant. Amazing! Teleportation is not anything that we do for ourselves or can do for ourselves as mortal fallen beings. Teleportation to safety is something that God has to do for us.

In the next NDE, Ana Christina was murdered, saw Christ, and was returned completely healed. God can heal our physical body miraculously, if we have been killed and He wants us to return to life.

https://www.youtube.com/watch?v=GDmtj6KHJNk

A favorite Ted talk. Anita Moorjani died of cancer. After we die, God can miraculously heal our physical body if He wants us to return to life, or God can teach our Psyche how to do so. Psyche is the part of us that learns and remembers, not our spirit body. Our spirit body is just another shell. Our spirit body and our physical body don't have experiences and don't make memories while our Psyche is away from our spirit body and our physical body.

https://www.youtube.com/watch?v=rhcJNJbRJ6U

Kim Rives died of cancer. Jesus absorbed all of her pain. ONLY Psyche can do telepathy.

https://www.youtube.com/watch?v=EnRZGLkp2M4

The following from Ian Wiltshire is the funniest NDE that I have come across so far:

https://www.youtube.com/watch?v=yjiSbpOJhZo

In this NDE, Ian describes himself [psyche] watching himself [spirit body] watch himself [physical body]. It's kind of confusing, unless you understand the fact that Psyche is something completely different than our spirit body and that our spirit body is something completely different than our physical body.

In the next NDE, Barbara Wilcox explains not having a body, or spirit, or a soul but being Total Intelligence.

https://www.youtube.com/watch?v=JQzap_jFXT8

In the following NDE, Jessica describes being a Point of Light, a Small Spec, and a Pinpoint of Light during her near-death experience. She, too, is describing Psyche or Non-Local Consciousness. Jessica also described the fact that we don't see everything and don't understand everything which is there to be seen and understood while out-of-body. Instead, there is ever-growing awareness and understanding, with still more to go that is never reached. Jessica could sense the presence of others, but didn't engage with them during her NDE as much as others seem to do. Hers is the most "self-centered" NDE that I have encountered so far. Jessica's Psyche is very much the center of her universe.

https://www.youtube.com/watch?v=Ve6RG9K3qrA

God gives each person a piece of the puzzle, and we have to find a way to put all of those pieces together if we want to see the Whole Picture, or the True Reality of our existence.

There's tons of evidence for the existence of the spirit body. But, you have to pay close attention and do a bit of research to find evidence of Psyche existing and functioning separately from the spirit body. It's there to be found, but it's not as common.

I have faith that there are many more people who have experienced their Psyche, Intelligence, or Spark separate from their spirit body. That's one of my talents or gifts – the ability to notice patterns and trends. For me personally, this is one of my greatest and most useful Scientific Discoveries or Scientific Observations, namely that Psyche is something completely different than our spirit body and spirit matter, and something completely different than the typical forces, fields, and LAWS. Psyche is a god. This KNOWLEDGE and TRUTH is one of the greatest things that God has revealed to me through the Lived Experiences or the Scientific Evidence of the human race as a whole.

Yes, it's ALL made of light – our psyche, our spirit body, and our physical body – and every type of light has a bit of consciousness or awareness and can respond to stimuli. However, spirit matter and physical matter are the types of light that REACT, while the Psyche is the type of light that ACTS, CHOOSES, THINKS, EXISTS, REMEMBERS, and is CONSCIOUS, AWARE, SENTIENT, and fully ALIVE. Psyche or Non-Local Consciousness is our true essence.

$E = mc^2$. Matter contains tons of energy or light – physical matter more so than spirit matter. Remove the mass or the energy from physical matter, and it becomes spirit matter again. Whenever God chooses to pump some of His Light or Life or Consciousness or Energy into spirit matter, it takes on mass and transforms into physical matter. All Matter is made of light. Forces and Fields are also a type of energy or light, and complete this picture.

This whole thing is a Psyche Ontology wherein Psyche is the fundamental unit of Reality. Psyche is immaterial light, or Pure Light. Psyche is an infinite singularity. We could also call this a Light Ontology, because Psyche IS Light. Either way, this is the Ultimate Model of Reality and the Ultimate Paradigm.

God gives each person a piece of the puzzle. When we combine all of our pieces together, we find ourselves looking at the Whole Truth or the True Reality of Our Existence. That's why I call it the Ultimate Model of Reality, because it is based upon and built upon the Lived Experiences or the Direct Observations of the human race as a whole, including our Non-Local Experiences or our Scientific Observations while in the spirit realm. Within this Ultimate Model of Reality or Psyche Ontology, I treat Lived Experiences or Psyche Experiences as Scientific Evidence. This is Real Science, and not the brain-dead junk which the Materialists and Atheists try to force upon us.

Go with the best and get rid of all the rest. That's what I try to do.

The Ultimate Model of Reality: Y = TAN(X)

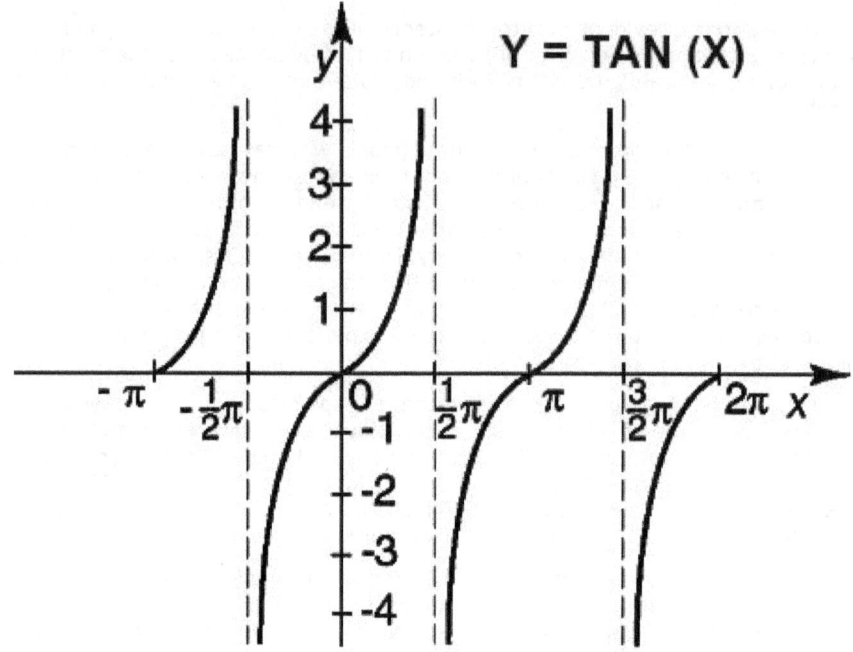

I use **y = tan(x)** to mathematically model this Ultimate Cause Model of Reality, quantum objects or particles of matter, infinite singularities, spirit bodies, universes, psyche, and non-local consciousness. This model is a lot more powerful than its simplicity implies. The nuances multiply, the more that one thinks about it.

The domain between -π/2 and π/2 is the measure of a single quantum object or a single particle of matter. A quantum object is a particle of matter.

It is matter all the way down, with one important exception.

Below (0, 0), the quantum object or particle of matter is in a spiritual state of existence; in other words, it is spirit matter. Astral Travelers have observed that spirit matter exists at different frequencies, wave-lengths, phases, or levels. There are theoretically an infinite number of different frequencies at which spirit matter can exist or reside.

Above (0, 0), the quantum object or particle of matter is in a physical state of existence; in other words, it is physical matter. Sticking with the phase or level or frequency analogy, this implies that physical matter can exist at different phases or can phase-shift. In other words, there could theoretically be an infinite number of physical worlds like ours existing in the exact same space where you are right now, and we wouldn't even know it because these different physical worlds would be at a different phase or at a different level or frequency. This is science fiction at its best.

In fact, the spirit world exists in the exact same space in which you currently exist as a physical being, but it's simply in a different phase of existence. That's how your psyche and your spirit body can reside within your physical body in the exact same space! Your spirit body exists at a different phase than your physical body. Your Psyche exists at a different phase than your spirit body and your physical body. They can all exist simultaneously in the same space.

In the domain between $-\pi/2$ and $\pi/2$, this single quantum object or particle of matter can theoretically exist at any frequency or phase or level. The domain between $-\pi/2$ and $\pi/2$ thus becomes a measure of this particle's physicality or materiality or density. It has been theorized and observed that as more energy is poured into a particle or an atom, the levels or the orbits of its electrons rise or increase. Likewise, if one were to find a way to pull energy or light out of a hydrogen atom, the orbit of its electron would decrease and that hydrogen atom would become a hydrino, which is dark matter or spirit matter.

A Quantum Object is a particle of matter. MATTER IS LIGHT that exists at different frequencies or different wavelengths. Matter IS light that exists in different phases. Some have suggested that matter is vibrating strings of light. Physicist David Bohm said that physical matter is frozen light or condensed light. That means that spirit matter is high frequency and high velocity unfrozen light. The frequency and the velocity of the matter decreases from nearly infinite at the $-\pi/2$ asymptote to nearly zero and nearly frozen at the $\pi/2$ asymptote. When measuring the velocity of this quantum object or particle of matter, the speed of light is at $(0, 0)$.

That means that physical matter exists at velocities slower than the speed of light; and, it means that spirit matter exists at velocities faster than the speed of light. This also means that spirit matter exists and travels faster than the speed of light, which means that spirit matter can theoretically travel over nearly infinite distances instantaneously or travel at nearly infinite velocities. This also means that physical matter cannot travel faster than the speed of light, unless it is first transmuted back into spirit matter. Physical matter has to have the mass and inertia and extra energy pulled out of it, before it can become spirit matter and travel faster than the speed of light. According to Quantum Mechanics, spirit matter becomes physical matter through the Word of Command or conscious observation. God pours a bit of Himself, His energy or light, into that particle of spirit matter transforming it into a physical atom.

Physical matter is frozen or condensed light. Spirit matter is light that exists at velocities and frequencies faster than the speed of light. Spirit matter exists in a more refined phase or state than physical matter. Physical matter has massive limitations placed upon it and is placed under God's control so that it can be used to create the Ultimate Consensus Reality – a reality which can be relied upon to function according to LAW. The beauty of a physical consensus reality is that I can rely upon the fact that this book will be there on my flash drive and hard drive tomorrow morning when I turn on my computer and go looking for it. A physical reality is reliable, predictable, and controllable.

Warp drive is impossible, because it is impossible to push physical matter faster than the speed of light. However, a Quantum Drive, Blink Drive, Jump Drive, Phase-Shifting, or Teleportation is not only possible but has actually been experienced. I have blinked, or jumped, or teleported, or tunneled, or some such. Actually, it was nothing that I consciously did. God did it for me, in order to save me. God did it to prevent me from becoming one with the car that had pulled out in front of me. I teleported. Time stopped; and when time resumed, I and my car were in the middle of a lawn twenty or thirty feet away with my car completely stopped and the engine completely off. I have NO memory of the intervening travel. I just blinked or teleported. The whole thing was calm and peaceful. I didn't even have time to get my adrenaline to start pumping. Teleportation is something that I have experienced and therefore I KNOW that it is real and can truly happen for real.

I had another friend caught in a fire, and it didn't burn him. It burned the clothes off of him, but didn't burn him and his underwear. Again, it was nothing that he did. God did it for him. I have a couple of friends who raised their children from the dead using the priesthood of God. Again, it was nothing that they did. God did it for them. The child had drowned and was dead, but there was no water in her lungs when God brought her back to life.

My best friend has phase shifted. Again, it was nothing that he did consciously. God did it for him, in order to save him from becoming one with the elk that he ran into at 70 mph. The elk passed right through him and his truck unscathed. He and his brother actually saw the elk pass through them. That's phase shifting.

LIVED EXPERIENCE is the best kind of scientific evidence. Through Lived Experience, we can go directly to knowing the truth. The truth is KNOWN by living it and experiencing it for yourself, or by choosing to trust someone who has.

Under God's control or through God's intervention, quantum objects or matter can phase-shift and teleport. That's how resurrected beings travel from one end of this universe to the other in the blink of an eye. It's infinitely more efficient than a warp drive or a wormhole will ever be.

Now for the special exception. Should the LIGHT reach the $-n/2$ asymptote, it becomes infinitely immaterial and infinitely non-physical, which means that it transmutes, transforms, and becomes something completely different. It becomes a unity and a singularity and takes on NEW PROPERTIES. It comes alive. Existing at an infinite frequency and an infinite velocity, it comes alive. It's completely immaterial and totally conscious and alive. It has moved completely to the Limit and has become Integral. It becomes living light, a spark of life. It becomes

Psyche, Intelligence, or Non-Local Consciousness. Psyche has zero mass, zero size, zero materiality, and takes up zero space. It's an infinite singularity. At the asymptote, the LIGHT or ENERGY or PSYCHE is no longer matter. It's completely immaterial. It is something else. It transmutes and takes on new and different properties and abilities.

The same thing happens to it should the LIGHT reach the п/2 asymptote. It becomes infinitely solid or infinitely condensed and infinitely real. It's a paradox which helps to bring the thing alive. In this state, Psyche has infinite scope, infinite range, infinite influence, infinite contact, and can permeate the whole of space. It becomes an infinite singularity, with infinite potential. Once again, it comes alive. It becomes Psyche, Intelligence, or Non-Local Consciousness. It gains the Word of Command and the ability to ACT on a universal scale. In other words, if you understand the math, the -п/2 and the п/2 asymptotes are one and the same.

ALL of our math moves towards infinity or moves towards zero (which is another type of infinity or unity). $Y = TAN(X)$ is one eternal round. The infinite and the asymptote points to Psyche. This means that ALL of our math points to the existence of Psyche, Intelligence, or Non-Local Consciousness. The math isn't lying! The Materialists freak out whenever their mathematical equations start moving towards infinity or towards zero; but, where else are they going to go? The mathematical equations will go to Psyche and the Truth every time!

Most scientists are trapped by Materialism. When I told my scientist friend that Psyche is completely immaterial, he complained that that's impossible – "it has to be something," he responded. Oh, it's SOME THING! It's infinite LIGHT. It's infinite TRUTH. It's infinitely INTELLIGENT. It's infinitely ALIVE, CONSCIOUS, and AWARE. Its infinite memory storage capacity taking up NO SPACE whatsoever. It's the Ultimate Reality! It's an infinite singularity. It's the ether. It's existence! It's big enough to fill the universe, but small enough to dwell in your heart. It's a paradox, which is what makes it so powerful and what makes it come alive. It's LIVING LIGHT. It's the glory of God. It's the Light of Christ. It's everything rolled into one. It's the Spark of Life! ALL of our math keeps pointing to it!

It has been suggested and observed that there is life, consciousness, and awareness in ALL types of light – in all phases of matter. Light is alive, so technically matter is alive to one extent or another.

However, THE LIGHT existing on the asymptotes IS a whole other level of life, intelligence, knowledge, awareness, unity, infinity, agency, potential, and truth. It has reached the limit. It has integrated. It has arrived! That kind of infinite LIGHT is the master and the living agent. That kind of LIGHT is Psyche and it has the innate ability to Command and to Act, rather than just reacting as matter does.

Psyche is an Infinite Singularity. At the -п/2 asymptote, the emphasis is on Singularity or Unity. At the п/2 asymptote, the emphasis is on the Infinite. At the asymptotes, the LIGHT is no longer a quantum object. It's no longer matter. It transforms and transmutes. It's a living entity! It's a living Agent. It's Psyche, Intelligence, or Non-Local Consciousness. Psyche acts. Matter reacts. The ability to COMMAND and the ability to ACT is what happens to LIGHT when it reaches the asymptote or becomes the asymptote. It goes down the rabbit hole and climbs the stairway to heaven all at once! Cool, huh?

Psyche, or Intelligence, or Agency, or Non-Local Consciousness, or Quantum Consciousness is the most interesting Science to study through the scientific evidence which is provided to us by Lived Experiences or Direct Observations.

Lived Experience or Direct Conscious Observation IS the BEST and most effective type of Scientific Evidence. Through Lived Experience, or Psyche Experience, or Lived Science, we human beings go directly to KNOWING the TRUTH by living and experiencing the truth for ourselves, or by choosing to trust someone who has. Human beings or human psyches are unique, in that we can write down our Lived Experiences or Scientific Observations and then share them with other human psyches. This is the best way for doing science! It's much better and more reliable than the Scientific Method.

The Biblical God Tried to Teach Us about the Different Types of Light
A 150 years BEFORE our scientists and out-of-body travelers started to figure out some of these things, the Biblical God Jesus Christ told us about all these different things, what they are, and how they work.

The Biblical God Jesus Christ provided independent confirmation that these things are REAL and TRUE. Of course, few people are willing to look and listen, because they don't want to learn anything from God nor anything about God. But, it has been there in the record for nearly two hundred years now.

It's powerful and really cool, but ONLY if you are looking for such a thing and are willing to accept it when you find it. However, the whole thing is simply annoying, if you desperately want Materialism and Naturalism to be true. I KNOW, because I used to be a Materialist, Nihilist, and Atheist.

Lessons from the Biblical God are precisely the types of things which the Materialists, Naturalists, Nihilists, Darwinists, and Atheists ridicule and mock. I KNOW, because I have been ridiculed and mocked by these people whenever I have presented this kind of EVIDENCE to them. But, ridiculing, mocking, name-calling, intimidating, censoring, and banning is the way that these people do science. Their Scientism and their religion are based upon a refusal to look at EVIDENCE. Ridiculing and mocking Lived Experiences, Direct Observations, or Real Scientific Evidence, and refusing to look at that Evidence, is the way that these people do science. Even though Lived Experience is the BEST type of Scientific Evidence, the Materialists and Atheists categorically reject this kind of evidence, especially if it is spiritual in nature and origin. Instead, these people mock and ridicule this type of Empirical Evidence or Experiential Evidence. It's what they do. These people are afraid of this kind of EVIDENCE. I KNOW, because I used to be one of them.

But, I'm no longer afraid of spiritual experiences or psyche experiences. I have seen the light. One day I finally realized that Revelations from God are also Lived Experiences.

Here are some modern-day revelations from the Biblical God Jesus Christ about this subject.

Doctrine and Covenants 131: 7-8:

7 There is no such thing as immaterial matter. All spirit is matter, but it is more fine or pure, and can only be discerned by purer eyes;

8 We cannot see it; but when our bodies are purified we shall see that it is all matter.

Ironically, most people interpret this scripture from a materialistic perspective and choose to believe that God is telling us that everything is Matter and that Matter is the only thing that exists. One of my friends used this scripture to claim that Psyche or Intelligence is spirit matter. That's the way these verses are typically interpreted – from a materialistic perspective. But, I have learned to force all my interpretations of Observational Evidence, Revelatory Evidence, or Scientific Evidence to match with the Lived Experiences of the human race. In this case, I depart from the typical materialistic interpretation of this scripture, and realize that God is trying to tell us that all Matter is matter, both spirit matter and physical matter are Matter. Matter is matter all the way down. Even psyche is matter, being the ultimate point-particle.

I have observed that whenever the Biblical God talks about the Spirit or the Ghost, He is in fact talking about our spirit body which is made of spirit matter. The Spirit or spirit body is made of Matter, a more refined type of matter which we call spirit matter. In contrast, whenever God is talking about Psyche exclusively, God calls it Intelligence, Light, and Truth. Whenever God is talking about the Psyche and Spirit Body combination, God simply calls it Spirit or Ghost and treats it as a united whole. Finally, whenever God talks about "element" or "body", He is talking about our physical body. This whole thing can easily get confusing, unless we force our definitions for these things to match with the Lived Experiences of the human race.

Notice how I make my own personal interpretation of these various terms match with the Lived Experiences and Out-of-Body Experiences of the human race. I treat Lived Experience as Scientific Evidence, and use Lived Experience to define and explain everything else.

NONE of this makes any sense to someone who has made Materialism his religion and has chosen to believe that Psyche and the Non-Local do not exist.

Psyche IS an Agentic and Agentive Being. The human psyche – the descendants of the Gods – is the ultimate Moral Agent. ONLY Psyche can do Agency. ONLY Psyche can do ethics and morality. ONLY Psyche can ACT. ONLY Psyche is free to choose among alternatives. In contrast, Matter must always obey.

2 Nephi 2: 11-16, 25-30:

11 For it must needs be, that there is an **opposition** in all things. If not so, my firstborn in the wilderness, righteousness could not be brought to

pass, neither wickedness, neither holiness nor misery, neither good nor bad. Wherefore, all things must needs be a compound in one; wherefore, if it should be one body it must needs remain as dead, having no life neither death, nor corruption nor incorruption, happiness nor misery, neither sense nor insensibility.

12 Wherefore, it must needs have been created for a thing of naught; wherefore there would have been no **purpose** in the end of its creation. Wherefore, this thing must needs destroy the wisdom of God and his eternal purposes, and also the power, and the mercy, and the justice of God.

13 And if ye shall say there is no **law**, ye shall also say there is no sin. If ye shall say there is no sin, ye shall also say there is no righteousness. And if there be no righteousness there be no happiness. And if there be no righteousness nor happiness there be no punishment nor misery. And if these things are not there is no God. And **if there is no God we are not**, neither the earth; for there could have been no creation of things, neither **to act** nor **to be acted upon**; wherefore, all things must have **vanished away**.

14 And now, my sons, I speak unto you these things for your profit and learning; for there is a God, and he hath created all things, both the heavens and the earth, and all things that in them are, both **things to act** and **things to be acted upon**.

15 And to bring about his eternal purposes in the end of man, after he had created our first parents, and the beasts of the field and the fowls of the air, and in fine, all things which are created, it must needs be that there was an opposition; even the forbidden fruit in opposition to the tree of life; the one being sweet and the other bitter.

16 Wherefore, the Lord God gave unto man that he should **act for himself**. Wherefore, man could not **act for himself** save it should be that he was enticed by the one or the other.

25 Adam fell that men might be; and men are, that they might have joy.

26 And the Messiah cometh in the fulness of time, that he may redeem the children of men from the fall. And because that they are redeemed from the fall they have become **free forever**, knowing good from evil; **to act for themselves and not to be acted upon**, save it be by the punishment of the law at the great and last day, according to the commandments which God hath given.

27 Wherefore, **men are free** according to the flesh; and all things are given them which are expedient unto man. And they are **free to choose** liberty and eternal life, through the great Mediator of all men, or to choose captivity and death, according to the captivity and power of the devil; for he seeketh that all men might be miserable like unto himself.

28 And now, my sons, I would that ye should look to the great Mediator, and hearken unto his great commandments; and be faithful unto his words, and **choose** eternal life, according to the will of his Holy Spirit;

29 And not choose eternal death, according to the will of the flesh and the evil which is therein, which giveth the spirit of the devil power to captivate, to bring you down to hell, that he may reign over you in his own kingdom.

30 I have spoken these few words unto you all, my sons, in the last days of my probation; and **I have chosen** the good part, according to the words of the prophet. And I have none other object save it be the everlasting welfare of your souls. Amen.

Psyche ACTS. Matter REACTS and is ACTED UPON. Any living physical organism which has a Psyche within it is a thing that God designed and created to ACT. Matter was designed and created by God to be ACTED UPON. Psyche is free to choose. Matter is not free to choose, but must obey the Word of Command or conscious observation. Materialists and Naturalists deny the existence of Agency, because these people deny the existence of Psyche or Non-Local Consciousness. Isn't it interesting how these people – these psyches – choose to use their Agency to deny the existence of Agency?

Satan or the devil is a Spirit, which means that he has a psyche and a spirit body. In contrast, a Soul is a spirit body, psyche, and physical body combination. All LAW comes from God, including what our scientists call Natural Laws.

Without God's Psyche, there would be NO LAW! Without God's Psyche, nothing would be organized, nothing would exist, and there would be NO physical matter. Without God's Psyche, we wouldn't exist, and this earth wouldn't exist. It would all be a "compound in one" or Chaos! Without God's Psyche there would be no purpose and no reason for being. Without God's Psyche, this whole universe would implode and vanish away. There's some HEAVY-DUTY SCIENCE in these verses of scripture! The Whole of Reality is explained in these verses.

Everything, the whole of existence, points to God's Psyche. In English, the word Psyche is a mid-17th century invention, coming from the Greek through Latin. Instead, of calling it God's Psyche as I'm doing, they simply called it God. It's the same thing. Without Psyche or Intelligence, there is NO existence.

The Biblical God's words are consistent, make sense, and fit together perfectly no matter what time period He chooses to say them. The Biblical God IS the Ultimate Scientist, which means that He is the Ultimate Psyche.

I find it fascinating and amazing that God knew these truths and was teaching these truths to His prophets thousands of years BEFORE our scientists started to discover them. In fact, most of our scientists still haven't discovered these truths for themselves. The Materialists, Naturalists, Darwinists, Nihilists, and Atheists don't have a clue. These people claim to be scientists, but they still haven't discovered these things for themselves. These people have NO understanding of the kind of light that ACTS and the kind of light that REACTS.

409

These people have NO clue that Psyche or Non-Local Consciousness even exists. It's interesting to realize that the Prophets of God were more scientifically advanced thousands of years ago than our PhD Materialists and PhD Atheists currently are today.

The Materialists, Naturalists, Darwinists, and Atheists claim to be scientists, but they really aren't. They are philosophers, and bad ones at that.

I find it fascinating that, hundreds if not thousands of years later, some of our scientists are finally starting to discover the very same things which the Biblical God was telling His prophets about thousands of years ago.

I realized one day that when we finally discover the TRUTH, it's going to be consistent with itself across the board, or we are interpreting it incorrectly. The TRUTH is KNOWN by living it and experiencing it for yourself, or by choosing to trust someone who has. All TRUTH is going to be consistent with itself – it's going to fit together seamlessly, as long as we understand it correctly.

I choose to make my interpretations of Scientific Data match with Reality or the Lived Experiences of the human race. Therefore, whenever the Scriptures tell us that God is Spirit, I take that to mean that God has a spirit body. Whenever the Scriptures tell us that God is Light, I take that to mean that God is Psyche or that God has a Psyche or a Mind. Whenever the Scriptures tell us that God is a personage of flesh and bones, I take that to mean that God has a physical body. In other words, God is a lot like us. Of course, that would make sense, because God is the Father of our spirit bodies and also the Father or the progenitor of our physical bodies.

1 John 1: 5: "This then is the message which we have heard of him, and declare unto you, that God is light, and in him is no darkness at all." **God is Psyche or Light.**

Doctrine and Covenants 129: 1-3: "There are two kinds of beings in heaven, namely: Angels, who are resurrected personages, having bodies of flesh and bones — for instance, Jesus said: 'Handle me and see, for a spirit hath not flesh and bones, as ye see me have.' Secondly: the spirits of just men made perfect, they who are not resurrected, but inherit the same glory." **God has a spirit body and God has a physical body. A personage is a Psyche or a personality. "Just Men" are Psyches. Spirits are spirit bodies that have a psyche residing within them. Resurrected beings and bodies of flesh and bone are physical bodies.**

Doctrine and Covenants 130: 22: "The Father has a body of flesh and bones as tangible as man's; the Son also; but the Holy Ghost has not a body of flesh and bones, but is a personage of Spirit. Were it not so, the Holy Ghost could not dwell in us." **Every intelligent being has a Psyche, including the Father, the Son, and the Holy Ghost. The Father and the Son have both a spirit body and a physical body. The Holy Ghost only has a spirit body. A Spirit or a Ghost is a spirit body that has been occupied by an intelligent Psyche.**

According to the Bible, Adam (his physical body) was the son of God.

Luke 3: 38: "Which was the son of Enos, which was the son of Seth, which was the son of Adam, which was the son of God."

I take this literally, and you should too. The scriptures come alive if you take them literally, rather than interpreting them materialistically and atheistically as most people seem to do. It ALL points back to God and to God's Psyche. Psyche IS the fundamental unit of reality. Without Psyche or Intelligence, there would be NO physical matter, NO genomes, NO physical life forms, and there would be NO existence whatsoever. Without God's Psyche, the whole universe would still be spirit matter and nothing but Chaos.

The whole thing centers on Psyche or Intelligence, because ONLY Psyche can have Lived Experiences in ALL realms of existence and remember having done so. ONLY Psyche can do the Word of Command, which means that ONLY Psyche can ACT. ONLY Psyche can do science!

Doctrine and Covenants 93: 29-30, 36:

29 Man [psyche] was also in the beginning with God. Intelligence [psyche], or the light of truth [psyche], was not created or made, neither indeed can be.

30 All truth [psyche] is independent in that sphere in which God has placed it, **to act for itself**, as all intelligence [psyche] also; otherwise there is no existence.

36 The glory of God is intelligence [psyche], or, in other words, light and truth.

PSYCHE is light and truth. Psyche is intelligence. Psyche is glory. Psyche ACTS for itself! Psyche is eternal and immortal. Psyche or Non-Local Consciousness has always existed. It can't be made, and it can't be destroyed. Psyche is some type of light. Psyche IS independent and can act for itself. Without Psyche or Non-Local Consciousness, there IS NO existence. God is Psyche. God is Light. God is Truth. God is Intelligent. God's Psyche is an Agent and an Actor. Psyche IS the true reality of our existence – what we truly are at our very core. We are Psyche. We are alive. Psyche IS the fundamental unit of reality. Again, without Psyche, there is no existence. This is a Psyche Ontology, and the Ultimate Model of Reality. This is the Ultimate Paradigm.

The glory of God is intelligence, or light and truth. God IS Light; and, God IS in the Light! The Light radiates from God. God, Jesus Christ, IS the being of light whom people encounter during their near-death experiences. Jesus Christ is the Savior of their Psyche, as well as their spirit body. Psyche IS Intelligence.

Abraham 3: 11, 21-23:

11 Thus I, Abraham, talked with the Lord, face to face, as one man talketh with another; and he told me of the works which his hands had made;

411

21 I [God] dwell in the midst of them all; I now, therefore, have come down unto thee to declare unto thee the works which my hands have made, wherein my wisdom excelleth them all, for I rule in the heavens above, and in the earth beneath, in all wisdom and prudence, over all the intelligences [psyches] thine eyes have seen from the beginning; I came down in the beginning in the midst of all the intelligences [psyches] thou hast seen.

22 Now the Lord had shown unto me, Abraham, the intelligences [psyches] that were organized [placed into spirit bodies] before the world was; and among all these there were many of the noble and great ones;

23 And God saw these souls [psyche and spirit body combinations] that they were good, and He stood in the midst of them, and He said: These I will make my rulers; for He stood among those that were spirits [psyche and spirit body combinations], and He saw that they were good; and He said unto me: Abraham, thou art one of them; thou wast chosen before thou wast born.

Over and over again, the Atheists, Materialists, Nihilists, Darwinists, and Naturalists ridicule and mock the people like Abraham who "talk with the Lord, face to face, as one man talketh with another".

BUT, the Lord is the BEST source of Scientific Evidence, Scientific Truth, and Scientific Knowledge. God KNEW about these things billions if not trillions of years BEFORE our scientists began to discover them!

God organized the Psyches. As a part of that organization, God sired spirit bodies for some of those Psyches and placed those Intelligences into those spirit bodies; and, they became Living Spirits – the literal sons and daughters of God and a Heavenly Mother. God IS the Father of the Lights.

We humans understand gravity to a certain degree, because we have EXPERIENCED its effects. Gravity is some type of universal light or force which is infinite and instantaneous in range and scope, which means that it travels infinitely faster than the speed of light. Every particle of matter in this universe is connected instantaneously and simultaneously with every other particle of matter in this universe through gravity. The existence of gravity FALSIFIES Materialism and Naturalism, because gravity is immaterial and gravity is supernatural.

What we don't fully understand, and what is a relatively new discovery for human beings is Dark Energy. However, Dark Energy is some type of universal light or force – the complement of gravity – and God was talking about this thing 150 years before our scientists began to discover it.

Dark Energy is the stretching force in this universe. Dark Energy creates space! Dark Energy fills the immensity of space, just like gravity does. On the quantum scale, this Dark Energy functions in part as the nuclear forces preventing atoms from imploding and keeping electrons in orbit thus creating space within the atom. The scientists tell us that the atom is 99.999% empty space. Well, it isn't exactly empty space. It's filled with this Dark Energy, which keeps the electrons in orbit and keeps the atom from imploding back into the singularity from whence it came.

Astrophysicists call it Dark Energy. Quantum Mechanists sometimes call it the Zero-Point Field of Light. Classical Physicists tend to call it the Strong and the Weak Nuclear Forces. New Agers call it the Quantum Sea of Light. The Biblical God calls it the Light of Christ. It's all the same thing!

The estimate changes each time our scientists send up another probe; but, it is estimated that 72.1% of this universe is comprised of Dark Energy or the Light of Christ.

The **Light of Christ** is a special type of light. It proceeds forth from God Himself, and is a part of God's Psyche or a part of God's Glory and Intelligence. The existence of Order and Physical Life points us directly to God's Psyche. Without God's Psyche, physical matter would not exist and this whole universe would be nothing but Chaos.

Like I said, Christ was talking about this thing 150 years before our scientists started to discover it.

Doctrine and Covenants 88: 4-17, 41-42:

4 This Comforter [Psyche] is the promise which I give unto you of eternal life, even the glory of the celestial kingdom;

5 Which glory is that of the church of the Firstborn, even of God, the holiest of all, through Jesus Christ his Son —

6 He that ascended up on high, as also he descended below all things, in that he comprehended all things, that he might be in all and through all things, **the light of truth**;

7 Which truth shineth. **This is the light of Christ.** As also he is in the sun, and the light of the sun, and the power thereof by which it was made.

8 As also he is in the moon, and is the light of the moon, and the power thereof by which it was made;

9 As also the light of the stars, and the power thereof by which they were made;

10 And the earth also, and the power thereof, even the earth upon which you stand. [Physical matter has this Light of Christ within it. Dark Energy or the Light of Christ creates Space.]

11 And the light which shineth, which giveth you light, is through him who enlighteneth your eyes, which is the same light that quickeneth your understandings;

12 **Which light proceedeth forth from the presence of God to fill the immensity of space —**

413

13 **The light which is in all things, which giveth life to all things, which is the law by which all things are governed, even the power of God who sitteth upon his throne, who is in the bosom of eternity, who is in the midst of all things.**

14 Now, verily I say unto you, that through the redemption which is made for you is brought to pass the resurrection from the dead.

15 And the spirit and the body are the soul of man. [The psyche/spirit body combination is what God calls the Spirit or the Ghost. The physical body God calls element or the body. The soul of man is the Ghost and the Physical Body combined.]

16 And the resurrection from the dead is the redemption of the soul. [The redemption of the Spirit and the Physical Body].

17 And the redemption of the soul is through him that quickeneth all things, in whose bosom it is decreed that the poor and the meek of the earth shall inherit it.

41 He comprehendeth all things, and all things are before him, and all things are round about him; and he is above all things, and **in all things**, and is through all things, and is round about all things; and all things are by him, and **of him**, even God, forever and ever.

42 And again, verily I say unto you, he hath given a law unto all things, by which they move in their times and their seasons.

This scripture tells us precisely what Dark Energy is and how it works. Dark Energy is the Light of Christ, and proceeds forth from the presence of God to fill the immensity of space.

I find it fascinating every time that God tells us that the sun, moon, stars, and atoms are MADE OF HIM. God pours a small part of His Psyche, His Glory, His Light, His Presence, His Consciousness, or His Life into each and every physical atom that He brings into existence. This Dark Energy or Light of Christ is what stretches an atom and makes it physical, giving it size and presence and making it take up space. That's powerful science there, which the Materialists, Naturalists, and Atheists completely reject and ignore.

23.3% of this universe is made up of Exotic Dark Matter, or Spirit Matter, or Non-Local Matter, or Trans-Dimensional Matter. Remember, 72.1% of this universe is comprised of Dark Energy or the Light of Christ. Combining Dark Matter and Dark Energy together, we discover that 95.4% of this universe is STILL spiritual and non-local in nature. ONLY 4.6% of this universe was transformed or transmuted into physical matter during the Prime Event or the Big Bang. The Materialists, Naturalists, Nihilists, and Atheists are WRONG about 95.4% of this universe! Doesn't it make you wonder what else they got wrong?

Of course, Dark Matter or Spirit Matter is going to be rejected by the Materialists, Nihilists, Naturalists, and Atheists. It's what these people do. Rejecting evidence is how these people do science.

However, the Biblical God Christ Jehovah has been telling us about the pre-mortal existence of Psyche and Spirits for thousands of years now. Jesus Christ has also renewed or refreshed that information in modern times, because the Greek Philosophers who took over Christ's Primitive Church were Materialists and Naturalists and rejected the pre-mortal existence of Psyche and Spirit Bodies, so that information was lost for 1,500 years. They call it the Dark Ages for a reason. During the Dark Ages, God's Light or Truth was rejected by this world. It had to be restored.

Doctrine and Covenants 76: 22-24:

22 And now, after the many testimonies which have been given of him, this is the testimony, last of all, which we give of Him: That He lives!

23 For we saw Him, even on the right hand of God; and we heard the voice bearing record that He is the Only Begotten of the Father —

24 That by him, and through him, and **of him**, the worlds are and were created, and the inhabitants thereof are begotten sons and daughters unto God.

The inhabitants of these worlds, you and I, are the sons and daughters of God the Father and a Heavenly Mother.

Our spirit bodies are literally the sons and daughters of God.

Our physical bodies are also descended from the Gods, because Adam was a Son of God, and Eve was his wife.

The human psyche is unique, because it is housed in both a spirit body and a physical body which are descended from the Gods. Our surname IS God. We are gods. ONLY the human psyche can do Science and the full range of Moral Agency.

God puts a little piece of Himself into everything that He organizes or creates!

Your spirit body is a child of God. Your genome IS God's signature. Your physical body IS God's handiwork. Each physical atom is made of God. Each atom in your physical body has a small part of God's Psyche, or God's Consciousness, or God's Power, or God's Life within it keeping it from imploding.

God KNOWS how all these things work, because God is in it all sustaining it all. Should God ever choose to remove His presence and His sustaining influence from this universe, the whole thing would implode back into the singularity from whence it came. It would vanish away. Dark Energy, or the Light of Christ, is what's stretching this universe and creating space and physical matter. Without God, there would be NO space and NO physical matter!

Furthermore, we KNOW from Quantum Mechanics that ONLY Psyche or Non-Local Consciousness can convert spirit matter into physical matter through the Word of Command. Once again, without God's Psyche or God's conscious command, there would be NO physical matter. The existence of order and life points us directly to God's Psyche!

ALL of this becomes Scientific Proof of God's Necessity and Scientific Proof of God's Existence, but ONLY if you are willing to accept these Lived Experiences or Scientific Observations into Evidence, which is something that the Materialists, Naturalists, Nihilists, Darwinists, and Atheists aren't willing to do. It's their loss. I KNOW, because I used to be one of them.

What Is Light?

Studying the Metaphysics of Light is a fascinating subject to think about.

From the Wikipedia:

> Light is electromagnetic radiation within a certain portion of the electromagnetic spectrum. The word usually refers to visible light, which is visible to the human eye and is responsible for the sense of sight.

—

Light is electromagnetic radiation.

What do we know about radiation?

We know that there are certain types of light or radiation that can pass right through us — gamma rays and x-rays.

Gamma rays and x-rays are a different type of light.

There are also micro-waves and radio waves, which are also part of the electromagnetic spectrum; and, these as well are a type of light.

The existence of gamma rays, x-rays, micro-waves, radio waves, and light PROVE conclusively that Materialism is FALSE, because these things exist in reality and we can actually use them to send information, but they are NOT comprised of physical matter. In fact, there are certain types of light that can pass right through physical matter as if the physical matter wasn't even there.

Light can travel along fiber optic cables. The fiber optic cables are indeed physical in nature, but the light is NOT! Light can pass through a prism. Once again, the prism is physical in nature, but the light is NOT!

The existence of light of any kind PROVES scientifically, logically, and conclusively that Materialism is FALSE.

Spirit is Light. Light is Spirit.

God has a spirit body. God is a Living Light.

It has also been observed by out-of-body travelers and those who have had Near-Death experiences that our thoughts are a type of light which has no speed limits associated with it. Thought, Consciousness, Spirit, Intelligence, Sentience, or Awareness is a type of Living Light which has NO spiritual limitations, NO physical limitations, NO sense of time or time restrictions, and NO speed limitations in its native form. We KNOW this to be so from the accounts of those who have had out-of-body experiences and Near-Death experiences. The prophets of the Biblical God have also experienced this reality and have stated that it is so.

While IN SPIRIT, out-of-body travelers have noticed that they can think of any location on this earth, and they are instantly there. They travel at the speed of thought.

Photons are packets of light, and they are limited to what physicists call "the speed of light". Their wave or frequency, or their particle nature, seems to limit them to the speed of light.

But, the spiritualists, clairvoyants, mystics, and out-of-body travelers have noticed that there is a PURER type of Light such as Thought, Intelligence, Spirit, or Consciousness which is extra-dimensional or trans-dimensional in nature and thus has no speed limits associated with it and thus has no speed limits while manifesting in the physical realm either. This is a type of Light that we can't get our physical instruments onto. Spirit or Living Light is too refined and can't be detected with our physical instruments and instead passes through the physical as if it weren't there, or in case of a spirit body actually enlivens and embodies a physical body giving it purpose and life.

There is a type of Living Light or Consciousness that cannot be detected by our physical instruments — only its effects on the physical can be detected. Scientists can detect that our physical brains are in the process of thinking or dreaming, but the scientists can't use their physical instruments to detect what we are in fact thinking about or dreaming about, because our thoughts and reams are spiritual in nature and thus the result of LIVING LIGHT.

Spirits are LIVING LIGHTS.

God has a Spirit. God is also a LIVING LIGHT or a THINKING LIGHT.

There is also a type of light that is used to make spirit matter. Condense light or slow it down the tiniest bit, and it becomes spirit matter.

There is a different type of light that is used to make up the Zero-Point Field Light, or the Quantum Sea of Light, or Dark Energy, or the Light of Christ. This thing seems to be a Quantum Consciousness, or Universal Consciousness, or a Cosmic Internet, or a Cosmic Communication System; and, it's all made up of some type of light. God is literally in the Light, the Quantum Sea of Light or the Light of Christ. It is how He communicates with us, hears our thoughts, and hears our prayers. There is NO speed limit associated with this type of light. It is universally everywhere simultaneously. Gravity seems to be comprised of this type of light also.

Forces and Fields are made up of different types of light. The existence of Forces and Fields also PROVES Materialism FALSE. These Forces and Fields are also a different type of spirit or light.

Physicist David Bohm has stated that physical matter is condensed light or frozen light.

—

So, what God does is take the Universal Light or the Light of Christ which has NO speed limit; and, God slows it down a bit to make spirit matter out of it. It the Spiritual Creation which the Biblical God often talks about.

God also takes some kind of light and slows it down a lot more to the speed of light and creates visible photons out of it. This is the ONLY type of light that the Atheists and Materialists make any kinds of allowances for, because they really can't deny its existence; and, Materialism is in fact at its core a state of denial and a denial of reality.

God also takes some type of light and slows it down to zero speed or freezes it so that it becomes physical matter.

—

Thought, or Consciousness, or Pure Spirit is a type of light that has NO speed limit. Thought or Pure Spirit is transcendent and extradimensional, which means that it has NO physical limitations and NO speed limit whatsoever, and has a presence simultaneously in all of the different dimensions.

It's all Light or Thought to begin with and has NO speed limit; but, by slowing that light down or limiting it to different speeds, it becomes different types of light — it becomes spirit matter at a slightly slower speed than the speed of thought; and it becomes photons of visible light at the "speed of light"; and if the light is slowed down to zero speed or is frozen in place, then it becomes physical matter.

Light or Thought or Pure Spirit is the Primal Construct or the Primal Element upon which everything else is based — spirit matter, photons of visible light, vibrating strings, gravity, or physical matter — it's all based upon Thought or Light or comprised of Thought and Light. This whole universe is One Great Thought in the Mind of God.

Some types of light can pass through physical matter as if the physical matter isn't even there. Other types of light can occupy physical matter and live within the physical matter. Some types of light can be detected with physical instruments and manipulated by physical instruments. And some types of light IS physical matter or frozen light.

The existence of light and electromagnetic radiation PROVES that Materialism is FALSE, and PROVES that some type of spiritualism or spirituality is TRUE. Spirit is

418

Light. Light is Spirit. God is a Living Light. God is in the Light and works through the Light. We each are Living Lights at our very core, and as human beings, our Living Lights or our Spirit Bodies are literally the offspring of God the Father and Heavenly Mother. We have a little bit of their light within each one of us! As human beings, and as the spirit children of God, we are NOT descended from apes. We are in fact, descended from the Gods! God the Father is the Father of Lights. We are those Lights. That's why we have the inherent ability to design, manufacture, write, and create. We are just like the Gods, but on a much more limited physical, mortal scale. We are mortal lights having a physical experience. The Gods are Immortal Living Lights existing in transcendent extra-dimensional realms, with NONE of our physical limitations. Their Light hasn't been slowed down to the "zero speed" of our physical realm. Their Light also hasn't been slowed down to the "speed of light," which we typically associate with photons or packets of visible light. When it comes to the Gods, their Light or their Thoughts or their Presence has NO speed limit. The Gods and their Thoughts or Spirits or Lights are like gravity — instantaneously and simultaneously everywhere all at once. The Gods can read our thoughts and hear our prayers and be everywhere simultaneously, because the Gods haven't been slowed down to the speed of "frozen light" or the speed of "Physical matter" as we have.

We are physical, mortal beings, and the Gods are NOT. God IS Light, and God is in the Light; and, God's Light is Pure Spirit or Pure Light or Pure Thought which has NO limitations whatsoever. God's Light can and is everywhere simultaneously, within everything, and through everything, filling the immensity of space, and filling up that 99.999% of so-called "empty space" between the nucleus of an atom and its electrons.

In fact, in the book "The Holographic Universe" by Michael Talbot, he came up with evidence that our Thoughts are in fact holographic in nature; and, holographic light can store so much information is so little space that it's theoretically possible for ALL of the knowledge and information in this universe to be stored holographically within that 99.999% of "empty space" within an atom that the Materialists are always talking about. Since light is immaterial and doesn't take up space at all, it's theoretically possible to store an infinite amount of information and knowledge in literally NO SPACE at all, using light and holography to do so.

Some types of light appear to be trans-dimensional or extradimensional, which means that they take up NO SPACE whatsoever. Can you imagine the amounts of information that can be stored holographically if that information is stored in light, which takes up NO SPACE whatsoever?

People laugh at the Prophets of God and their "peep stones"; but, it's theoretically possible for God to put an infinite amount of information into a single stone if that information is stored in that stone holographically using light. According to the Materialists, there is lots of "empty space" in a stone where God can store anything that He wants, especially if he is storing that information as light which takes up NO SPACE at all.

God is in the Light, which means that He has NO limitations at all.

419

This is by far one of the most interesting subjects to study. It's infinitely more interesting than Materialism, or Atheism, or Darwinism. Those things are like frozen light in comparison to the kinds of Light that the Gods have access to.

Once we see the light, we realize that light is in everything and that light IS everything. When it comes to the Light, God has NO restrictions whatsoever. God can do anything He wants to with the Light, because God is in the Light.

PART V – FIXING SCIENCE WITH QUANTUM MECHANICS

Quantum Mechanics, or Syntropy, or Psyche is the answer to life, the universe, and everything.

One of the most useful things I ever did was to use Quantum Mechanics to upgrade science from Materialism and Naturalism TO Observation and Experience. I upgraded my science to Science 2.0. Rather than defining science as Materialism, Naturalism, Darwinism, Nihilism, and Atheism as most college professors do, I now define science as Observation and Experience.

Unless it has been experienced and observed by Someone Psyche somewhere sometime, it really isn't science. The major premises or hidden assumptions of Materialism, Naturalism, Darwinism, Nihilism, and Atheism – which claim that the quantum or the supernatural does not exist – have not been and cannot be experienced and observed. Therefore, Materialism and Naturalism are NOT science, because their major premises or hidden assumptions cannot be experienced nor observed but have to be taken on blind faith as being real and true.

The observed, experienced, and verified existence of Quantum Mechanics, Psyche, the Biblical God Jesus Christ, and Action at a Distance FALSIFIES Materialism, Naturalism, Darwinism, Nihilism, Atheism, and their derivatives.

The truth falsifies the false; and, the truth has been repeatedly verified, experienced, and observed.

Mark My Words

—

Reference

Science 2.0: I Upgraded My Science.

https://www.amazon.com/dp/B0771K6WTX

Evolution Is Entropy

Web Page:

Now pay attention!

Despite the FACT that I have and will have debunked and falsified the Theory of Evolution thousands of times in hundreds of different ways, don't ever once let me convince you that there is NO such thing as evolution or random mutations.

Evolution is REAL, very REAL. Evolution or random mutation is entropy or the second law of thermodynamics. It's REAL. Entropy is very REAL at the physical level; and, it works as advertised at the physical level. Entropy produces death and extinction.

One of my all-time most significant scientific discoveries came to me when I first realized that the Theory of Evolution is based upon entropy.

Obviously, I'm not the first person to discover entropy; but, I seem to be the first person on the planet to realize that Materialism, Naturalism, Darwinism, Nihilism, Atheism, Determinism, Physicalism, Behaviorism, Scientism, Classical Physics, Chemical Evolution, Macro-Evolution, Natural Selection, Random Mutations, and the Theory of Evolution are based exclusively on entropy; and, I seem to be the first person to realize what that truly means for science in general.

You see, these people have chosen to believe that physical matter and entropy are the ONLY thing that exists. They based ALL of their science exclusively on entropy.

Are you starting to see the significance of this reality? Can you see the problem?

Well, ask yourself, "What is entropy?"

Entropy is corruption, disease, disorder, chaos, death, and extinction; and, the Theory of Evolution is based exclusively on entropy. Are you starting to see the problem? It's a BIG ONE! This is HUGE!

Evolution of any type is entropy; and, evolution of any type is based upon entropy. Entropy is death and extinction. When was the last time that you caught **death** in the act of designing, creating, and producing new and unique life forms from scratch? It has never happened, and it will never happen. Entropy or evolution prevents it from happening.

Entropy literally prevents the Theory of Evolution from becoming true. Evolution produces entropy. The different types of evolution produce death and extinction. Entropy prevents the different types of evolution from designing, creating, and producing something new. Entropy cannot design and create, which means that evolution of any kind cannot design and create. Entropy can only destroy, which means that the different types of evolution can only destroy. This reality is so obvious, that I sometimes wonder why nobody has ever thought of it before.

Life, design, creation, organization, order, information, intelligence, proteins, genes, genomes, computer programs, hardware, cars, buildings, computers, phones, televisions, physical matter, and physical life forms REQUIRE an infusion of Syntropy in order to come into existence. We KNOW that Syntropy must exist, or there wouldn't be any entropy. We KNOW that Syntropy must exist, or there wouldn't be any physical matter. We KNOW that Syntropy must exist, or there wouldn't be any life – the whole thing would be death, or chaos, or entropy. We KNOW that Syntropy must exist, or we wouldn't be here to think about it.

The Theory of Evolution can't work as advertised. It's prevented from doing so by entropy, or death and extinction. Evolution is entropy; and, evolution prevents the Theory of Evolution from becoming true.

The Fruits of Evolution or Entropy

Whenever the Materialists, Naturalists, Darwinists, Nihilists, Behaviorists, and Atheists start talking about evolution, they get most everything wrong because evolution or entropy cannot design and create. However, these people do indeed get one thing perfectly right. Evolution, or random mutation, or entropy is indeed the CAUSE of ALL of our heritable diseases, developmental diseases, and heritable mental illnesses.

Remember, the Theory of Evolution is FALSE because random mutations or entropy cannot design and create genes, proteins, and life forms. However, evolution or random mutation or entropy is very REAL; and, it can indeed destroy genes, proteins, and life forms. Do you see how that works? It's important to understand.

The Theory of Evolution is a fictional story that they made up out of thin air. There is NO empirical evidence supporting any version of it. In fact, ALL of the empirical evidence and experimental evidence that we have on hand as a race FALSIFIES the different versions of the Theory of Evolution and VERIFIES Quantum Mechanics instead. Quantum Mechanics is Supernatural. Quantum Mechanics is the Priesthood Power of God. Quantum Mechanics and Psyche are Pure Syntropy. There is NO entropy in the spirit realm, the non-local realm, the quantum realm, or the transdimensional realm. Psyche, Spirit Matter, and Quantum Mechanics ARE Syntropy.

The very existence of something like the Orthodox Interpretation of Quantum Mechanics from Henry P. Stapp FALSIFIES Materialism, Naturalism, and the various versions of the Theory of Evolution. Quantum Mechanics FALSIFIES Materialism, Naturalism, and their derivatives. Fictional stories like the Theory of Evolution cannot stand in the light of truth.

The Theory of Evolution is the very pinnacle of fictional ad hoc just-so story telling; and, *ad hoc just-so stories* are logic fallacies. The Theory of Evolution is fictional because it never happened – none of it happened! The chemical evolution of proteins and genes from atoms is physically impossible. Abiogenesis, spontaneous generation, and the various different forms of macro-evolution are physically impossible. They are prevented from happening by entropy.

The Theory of Evolution is science fiction. The fictional nature of the story becomes most egregious whenever they try to guesstimate how many millions of years it took for evolution to do something for us, because evolution or entropy can't do anything for us – not ever. They are making the Theory of Evolution up as they go along. It's a fictional story, and nothing more.

Since the whole Theory of Evolution is nothing but a fictional story, you can successfully and rightfully make up fictional stories of your own to debunk it. That's the way fiction works!

Scientific Inference

Comparative Psychology and Evolutionary Psychology are based upon Darwinism and the Theory of Evolution, which means that they too are nothing more than *fictional ad hoc just-so stories* that these people have made up out of thin air.

In fact, in his book *Biopsychology*, John Pinel tells us as much when he tells us that his "evolutionary perspective" is based upon *scientific inferences*. *Scientific inferences* are fictional ad hoc just-so stories. *Scientific inferences* are logic fallacies. The whole of their evolutionary perspective, evolutionary psychology, and comparative psychology is based upon *scientific inferences* or stories that they have manufactured out of thin air. Making up stories is what makes being a scientist fun, according to John Pinel.

From *Biopsychology* page 13, John Pinel writes:

> **Scientific inference is the fundamental method of biopsychology and of most other sciences – it is what makes being a scientist fun. This section provides further insight into the nature of biopsychology by defining, illustrating, and discussing scientific inference.**
>
> **The scientific method is a system for finding things out by careful observation, but many of the processes studied by scientists cannot be observed. For example, scientists use empirical (observational) methods to study ice ages, gravity, evaporation, electricity, and nuclear fission – none of which can be directly observed; their effects can be observed, but the processes themselves cannot. Biopsychology is no different from the other sciences in this respect. One of its main goals is to characterize,**

424

through empirical methods, the unobservable processes by which the nervous system controls behavior.

The empirical method that biopsychologists and other scientists use to study the unobservable is called <u>scientific inference</u>. Scientists carefully measure key events they can observe and then use these measures as a basis for logically inferring the nature of events that they cannot observe.

Like a detective carefully gathering clues from which to recreate an unwitnessed crime, a biopsychologist carefully gathers relevant measures of behavior and neural activity from which to infer the nature of the neural processes that regulate behavior.

The fact that the neural mechanisms of behavior cannot be directly observed and must be studied through scientific inference is what makes biopsychological research such a challenge – and as I said before, so much fun. (*Biopsychology*, p. 13.)

Pinel, J. (2014). *Biopsychology* (9th ed.). New York: Pearson.

John Pinel is trying to lay a scientific foundation for his belief in Evolution, Materialism, Naturalism, and Darwinism; and, he is using *scientific inferences* to do so. He wrote this section of his book as empirical proof that the Theory of Evolution is true. *Scientific inferences* are how the Materialists, Naturalists, and Darwinists prove that the Theory of Evolution is true. So, what are *scientific inferences*?

He defines *scientific inference* as an empirical method or a scientific method. Then he uses *scientific inference* to prove that the Theory of Evolution is true. Because he has never studied the Philosophy of Science, he has NO idea how faulty, fallacious, and weak *scientific inferences* really are. He has NO idea that *scientific inferences* are logic fallacies. Instead, he literally treats *scientific inferences* as empirical evidence, so that he can prove that the Theory of Evolution is true.

Notice carefully how he defines Evolution, Materialism, Naturalism, and Darwinism as events that we cannot observe. These are events that have NEVER been observed. So, we are going to use *scientific inferences* to manufacture "empirical evidence" for these events that cannot be observed and that have never been observed. *Scientific inference* is a logic fallacy; yet, he erroneously calls it an empirical method. They ALL do that in one way or another, without even realizing that they are using logic fallacies to prove that Evolution, Materialism, Naturalism, and Darwinism are true. These people tack on the word "scientific" to make it seem like science, but it is nothing more than "inference" or personal interpretation masquerading as empirical fact. These people are *begging the question* and *jumping to conclusions*.

Inference is the *affirming the consequent* logic fallacy in action. If you ever study the Philosophy of Science and get a professor like Joseph Rychlak who knows what he is talking about, he will teach you that for every single piece of scientific evidence, there are literally an infinite number of interpretations or inferences that

425

can be given to that single piece of scientific evidence or that single event; and, many of them will seem plausible or believable, but only one of them can be true.

Scientific inference is the affirming the consequent logic fallacy in action.

Scientific inference is the jumping to conclusions logic fallacy in action.

Scientific inference involves begging the question or circular reasoning, which are also logic fallacies. Scientific inference is a logic fallacy; but, John Pinel is choosing to use scientific inferences as empirical evidence and as an empirical method or a scientific method. It's a logic fallacy to do so; but, he doesn't know that. He's never been taught that scientific inference is a logic fallacy. Scientists typically don't study the Philosophy of Science in college and grad school; and if they do, they are taught the philosophy of science by Materialists, Naturalists, Darwinists, and Atheists who don't know what they are talking about.

When it comes to the origin of life, there are an infinite number of hypotheses or inferences that are being used to explain the origin of life on this planet. In fact, the Intelligent Design Theory is infinitely more plausible and believable than the Theory of Evolution because intelligent design has actually been OBSERVED trillions of trillions of times, whereas the evolution of genes and proteins from atoms has NEVER been observed and NEVER will be because it's physically impossible. Evolution of any kind has NEVER been observed doing spontaneous generation, abiogenesis, macro-evolution, or the creation of new life forms from scratch because it is physically impossible for them to do so.

Intelligent Design Theory is an infinitely more plausible, believable, credible, and parsimonious scientific inference than the Theory of Evolution is because the evolution of atoms into genes, proteins, eyes, brains, and life forms is physically impossible. If you are going to use logic fallacies and are determined to use scientific inferences as empirical evidence, then you owe it to yourself to go with the BEST explanation or the BEST scientific inference – one that has actually been OBSERVED and caught in the ACT – Intelligent Design Theory. Get rid of the ones that have NEVER been observed such as Materialism, Naturalism, Darwinism, Nihilism, Determinism, and Atheism.

The whole Theory of Evolution is based upon scientific inferences or fictional ad hoc stories that these people have manufactured out of thin air. There's nothing empirical about it! There is NO empirical evidence demonstrating the creative powers of entropy, evolution, or random mutations. Chemical evolution of proteins and genes from atoms is physically impossible thanks to entropy or the second law of thermodynamics. It can't happen, which means that it never happened. Evolution is entropy, or the second law of thermodynamics. It can't design and create. It's physically impossible.

Furthermore, evolution (genetic change), random mutations, and natural selection didn't even exist until AFTER God designed and created the proteins, genes, genomes, brains, eyes, and life forms in the first place. It's physically impossible for something that doesn't even exist yet to design, program, engineer, manufacture, and create proteins and genes out of thin air. Spontaneous generation or abiogenesis is physically impossible. Entropy or the second law of thermodynamics prevents it from happening.

Evolution is entropy, which means that it can't design and create anything. Evolution or entropy can only deteriorate and destroy things. It can't design and create. It's physically impossible for evolution or entropy to design and create proteins, genes, genomes, brains, eyes, and life forms. That's just the way it is, because evolution of any kind is entropy.

It took me years, even decades, to discover that evolution or random mutation is entropy or the second law of thermodynamics. That discovery also came with a powerful gift. I discovered Syntropy! Since evolution or entropy cannot design and create, some type of Syntropy must have done the job.

I finally realized that Quantum Mechanics is the exact opposite of Materialism, Naturalism, Darwinism, Classical Physics, and Entropy. Quantum Mechanics and Psyche are Pure Syntropy. Quantum Mechanics is the Power of God or the Priesthood Power of God. Psyche or Quantum Non-Local Consciousness is the only thing that can control Quantum Mechanics at the quantum level or the psyche level. Physical matter and entropy can't touch nor control the quantum level. Only the smaller can dwell within and control the larger. Psyche is Syntropy. Quantum Mechanics is Syntropy.

There is NO aging or entropy in the Quantum Realm, Psyche Realm, Spirit Realm, or Transdimensional Realm. Transdimensional means non-physical and non-local – not located in our physical 3D space-time realm. Everything in the Quantum Realm is Pure Syntropy. It is endless, timeless, eternal, and everlasting because there is NO entropy which means that nothing ages, gets old, or dies in the Non-Local Realm or the Syntropy Realm.

The Gods create physical matter by infusing a particle of spirit matter with space-time and the ability to acquire entropy. The Gods create a particle of physical matter by taking a particle of spirit matter, filling it full of space, slowing it down to sub-light speeds, and making it subject to entropy or the passage of time. According to the theory of relativity, the particles of spirit matter existing at velocities faster than the speed of light experience NO passage of time, meaning that they do not age and are not subject to entropy. Entropy is a function of time or an aging process. Spirit matter and physical matter are the same thing – they are quantum objects. However, spirit matter is pure syntropy; whereas, physical matter has been slowed down by being infused with space-time and made subject to entropy or the passage of time.

What is the difference between a hypothesis and a theory?

A hypothesis is an idea. A theory is a hypothesis that has observational evidence supporting it. It really should be called the "hypothesis of evolution" because there is NO observational evidence supporting macro-evolution, chemical evolution, abiogenesis, and spontaneous generation; and, there NEVER will be. Thanks to entropy, random diffusion, or the second law of thermodynamics, it's physically impossible for atoms to spontaneously generate into functional genes, proteins, genomes, eyes, brains, and life forms. It can't be done, which means that it wasn't done.

A hypothesis is a prediction. A hypothesis is a testable proposition.

In their quest for insight, social psychologists propose *theories* that organize their observations and imply testable *hypotheses* and practical predictions.

A theory is an integrated set of principles that explain and predict observed events.

To a scientist, facts and theories are [the same difference]. **Facts are agreed upon statements about what we observe. Theories are ideas that summarize and explain facts.** (*Social Psychology*, p. 17.)

What are the Theory of Evolution's hypotheses or predictions?

According to the science fiction that I have watched on television – *Star Trek*, *Babylon 5*, and *Earth: Final Conflict* – a million years from now human beings are going to evolve into energy beings. There's a serious problem with that prediction. There has been life on this earth for billions of years, and NONE of it has ever evolved into an energy being. It's not going to happen because it's physically impossible. It won't be done because it can't be done.

The theory of evolution makes NO realistic or useful predictions. It just catalogues what happened in the past according to the fossil record, and then it forces a personal interpretation or a scientific inference onto the fossil record after-the-fact. *Scientific inference* is a logic fallacy. *Personal interpretations* introduce a wide variety of logic fallacies into science. The theory of evolution is based upon a wide variety of scientific inferences, logic fallacies, personal interpretations, and wishful thinking.

Macro-evolution, chemical evolution, abiogenesis, and spontaneous generation have NEVER been observed because they are physically impossible thanks to entropy, random diffusion, and the second law of thermodynamics.

Technically, the theory of evolution makes NO testable predictions because ALL of the observational evidence, empirical evidence, experimental evidence, and experiential evidence FALSIFIES the theory of evolution.

The FACT is that the spontaneous generation of atoms into functional genes, proteins, genomes, eyes, brains, and life forms is physically impossible. Since macro-evolution, spontaneous generation, chemical evolution, and abiogenesis are physically impossible, that means they never happened because they couldn't happen.

The FACT is that the major premises or primary hidden assumptions of Materialism, Naturalism, Darwinism, Nihilism, Behaviorism, Determinism, and

428

Atheism – claiming that the quantum or the supernatural does not exist – are FALSIFIED due to a complete lack of observational evidence or a complete lack of supporting evidence. Furthermore, they are FALSIFIED by observational evidence and experimental evidence. The verified and proven existence of Quantum Mechanics and Action at a Distance FALSIFIES Materialism, Naturalism, and their derivatives.

The primary assumptions or hidden premises associated with Naturalism and Darwinism are NOT testable.

There are NO observed events and NO observed facts associated with chemical evolution, design and creation by random mutations, abiogenesis, spontaneous generation, macro-evolution, or design and creation by genes, proteins, RNA, and amino acids. Stand-alone atoms cannot spontaneously generate into functional genes, proteins, genomes, eyes, brains, and physical bodies because spontaneous generation is physically impossible thanks to entropy or random diffusion.

There are NO observed facts associated with creation by evolution, which means that the "theory of evolution" is in fact a falsified hypothesis and NOT a verified theory.

The theory of evolution is based exclusively upon correlation, and NOT observation. There is a huge difference between the two. Correlation does NOT prove causation. The Darwinists have carefully correlated their hypotheses with the fossil record. The Darwinists have deliberately correlated the fossil record with Darwin's Tree of Life. However, NOBODY except for the Biblical God Jesus Christ actually observed the production of the fossil record. Our fossil record could have been produced in many different ways; and with hindsight, all of these different ways or hypotheses can be made to correlate with the fossil record. Only God knows which one of those ways is the actual way by which He produced the fossil record.

Observation trumps correlation, or at least it should. However, when it comes to the theory of evolution, the Materialists, Naturalists, Darwinists, and Atheists make their correlations trump ALL the observational evidence that falsifies their pre-chosen beliefs. They cheat in order to make their case.

ORIGIN: Probability of a Single Protein Forming by Chance

https://www.youtube.com/watch?v=W1_KEVaCyaA

http://www.originthefilm.com/mathematics.php

http://science-2-0.com/wp-content/uploads/2018/03/THE-MATHEMATICS-OF-ORIGIN.pdf

https://www.youtube.com/watch?v=cQoQgTqj3pU

https://www.youtube.com/watch?v=_zQXgJ-dXM4

http://bio-complexity.org/ojs/index.php/main/article/view/BIO-C.2010.1/BIO-C.2010.1

http://science-2-0.com/wp-content/uploads/2018/03/Case-Against-Darwinian-Origin.pdf

http://science-2-0.com/wp-content/uploads/2018/03/Chemical-Evolution-Is-Impossible.zip

http://science-2-0.com/wp-content/uploads/2018/04/Origin-Of-Life.zip

Once you have eliminated everything that is FALSE and everything that is IMPOSSIBLE, then ONLY the Truth remains. It's elementary.

The theory of evolution has been tested and falsified. There's NO practical application for creation by rocks, chemical evolution, or the theory of evolution because spontaneous generation is physically impossible. There are NO observations when it comes to the theory of evolution. The theory of evolution makes NO testable and verifiable predictions because ALL of observed evidence falsifies Materialism, Naturalism, Nihilism, Atheism, and the Theory of Evolution.

Materialism, Naturalism, Darwinism, Nihilism, Behaviorism, Determinism, and Atheism are FALSIFIED due to a complete lack of observational evidence or a complete lack of supporting evidence for their hidden assumptions or major premises which claim that the quantum or the supernatural does not exist. In contrast, it is said that Quantum Mechanics or Action at a Distance is the most-verified, best-proven, and most-used science that we currently have.

Quantum Mechanics and Quantum Neuroscience

On page 25 of *Biopsychology*, John Pinel quotes another worshipper of the theory of evolution:

> **Evolution is both a beautiful concept and an important one, more crucial nowadays to human welfare, to medical science, and to our understanding of the world than ever before. It's also deeply persuasive – a theory you can take to the bank. The supporting evidence is abundant, various, ever increasing, and easily available in museums, popular books, textbooks, and a mountainous accumulation of scientific studies. No one needs to, and no one should, accept evolution merely as a matter of faith (Quammen, 2004, p. 8).**

Ironically, the mountains of evidence supporting the theory of evolution are correlational – designed and manufactured after-the-fact to fit the fossil record. NONE of that evidence is observational. Correlation is not causation. Instead, ALL of the observational evidence and experiential evidence that we have on hand as a race FALSIFIES Materialism, Naturalism, Darwinism, Nihilism, Behaviorism, Scientism, Determinism, and Atheism. Correlation cannot be used to prove causation. All of these falsified philosophies have to be taken on blind-faith as being true because there is NO observational evidence supporting them. The theory of evolution is bankrupt. There's no logical reason to believe in the theory of evolution, except for the fact that everybody else has been deceived by it and has chosen to believe in it.

I talk about all of this in great detail in my book, *Quantum Neuroscience: The Answer to Life, the Universe, and Everything*. If a comparison between Evolution and Quantum Mechanics interests you, I recommend you take a look at that book:

https://www.amazon.com/dp/B079Z6QQQB

The book, *Quantum Neuroscience: The Answer to Life, the Universe, and Everything*, makes a detailed comparison between Neuroscience and Quantum Neuroscience, which means that it makes a detailed comparison between Classical Physics and Quantum Mechanics.

When properly understood, Quantum Mechanics FALSIFIES Classical Physics, Materialism, Naturalism, Darwinism, Nihilism, Scientism, Behaviorism, Determinism, and even Atheism. Quantum Mechanics is Supernatural. Quantum Mechanics or Syntropy is the exact opposite of Classical Physics, Entropy, Random Mutations, Materialism, Naturalism, Darwinism, and the Theory of Evolution.

Quantum Mechanics is a proven and verified science. Quantum Mechanics is the best-proven and most-used science that we have. In contrast, the Theory of Evolution has NO empirical evidence supporting it. In fact, ALL of the empirical evidence that we have on hand as a race, including Quantum Mechanics, FALSIFIES Materialism, Naturalism, Darwinism, and the Theory of Evolution. Do you see how that works? It's important to understand.

Quantum Mechanics, Spirit Matter, and Quantum Non-Local Consciousness (Psyche or Intelligence) are PURE SYNTROPY. The Syntropy has to exist somewhere someplace somehow because, according to the Law of Entropy and the Second Law of Thermodynamics, the physical Multiverse should have burned out and suffered heat death an eternity or two ago; and, there should be NO more physical universes anywhere, but here we are nonetheless. The very existence of this physical universe – it's beginning full of syntropy and its ongoing existence billions of years later – is positive proof that Someone Psyche knows how to do syntropy or Someone Psyche is syntropy. Quantum Mechanics is syntropy or the Priesthood Power of God. God's Psyche knows how to do syntropy or Quantum Mechanics; otherwise, this physical universe would not exist.

This is what Quantum Mechanics or Transdimensional Physics is trying to teach us. Quantum Mechanics, Spirit Matter, and Psyche are pure syntropy. Evolution, Random Mutations, Physical Matter, and Classical Physics are entropy.

Quantum Mechanics is an infinitely better theory than Materialism, Naturalism, Darwinism, Nihilism, Atheism, Classical Physics, and the Theory of Evolution. Old theories are supposed to fall by the wayside whenever a better theory is proposed to account for the findings. Intelligently designed quantum machinery is an infinitely better theory for the origin of life than the theory of evolution because Quantum Mechanics and Intelligence have been experienced, observed, replicated, verified, and proven to be real and true through a preponderance of the evidence. Design and creation by entropy or evolution has NOT. Instead, design and creation by entropy or evolution has been falsified. It's physically impossible.

Evolution Is Entropy

Random mutations are entropy.

It's physically impossible to produce a functional protein by throwing dice. It can't be done, which means that it wasn't done.

Furthermore, designing, engineering, and making a matching gene to go along with a functional protein requires deliberation, planning, engineering, and intelligence. It can't happen through random chance or luck. It's physically impossible. The production of a protein, a matching gene, as well as a functional genome requires a programmer, an engineer, a designer, a fine-tuner, and Syntropy of the highest order. Someone Psyche has to infuse syntropy or order into the equation in order to make it happen because entropy naturally prevents it from happening. Your genome is God's Signature.

Random chance produces disorder, disorganization, chaos, cancer, death, extinction, and entropy.

Random mutations are the mechanism of change when it comes to evolution, genetic change, genetic drift, or genetic entropy – not natural selection. Natural selection doesn't touch our genes and proteins. Natural selection is worthless when it comes time to change our genes and proteins.

Natural selection is also a product of chance or a function of luck. The ultimate product of natural selection is death and extinction. Death and extinction are entropy. The physical mechanism that we call natural selection causes entropy, death, and extinction. Natural selection cannot design and create because natural selection doesn't touch our genes and proteins.

Since the very beginning of the theory of evolution, Darwinists and Evolutionists have erroneously stated that natural selection is the causative agent behind genetic change – the origin of species by means of natural selection. They are wrong. They don't even understand their own theory. Selection or sexual activity determines whether our genes get passed on to the next generation, or not; but, natural selection or selection of any kind does NOT change our genes. Selection or sexual activity does NOT touch our genes!

Natural selection or survival of the fittest doesn't do anything except result in death and extinction. Death and extinction are entropy, not syntropy, design, and creation. Natural selection is NOT a mechanism of change. Natural selection doesn't do anything. Natural selection is death, extinction, and entropy. Natural selection doesn't touch our genes. It can't. It's physically impossible. There's no invisible person called "Natural Selection" who is reaching into our genes and making them come alive.

The majority of our heritable mutations come from the genetic recombination process that takes place during the production of our gametes (egg and sperm). Genetic recombination is the process that shuffles our genes and introduces errors or mutations into the system; and, genetic recombination or the production of gametes takes place BEFORE any type of selection or sexual activity. Remember, natural selection doesn't touch our genes. God is the person who designed and created our proteins and the matching genes to go along with them, which means that God is the one who designed the physical process that we call genetic recombination. Genetic recombination is the primary mechanism of change driving evolution, random mutations, and genetic drift. Random mutations are entropy, an integral part of physical matter.

Combine random mutations with natural selection and we end up with genetic entropy – the devolution of our genome to the eventual point where our species goes extinct. It all ends in death or entropy. Technically, the "theory of evolution" is the origin of species by means of genetic entropy, random mutations, and extinction, which is physically impossible. Natural selection has nothing to do with the origin of species. Death and extinction are the natural result of evolution or genetic change.

[See: Sanford, J. (2014). *Genetic Entropy* (4th ed.). Cornell University: FMS Foundation.]

Evolution or Entropy Puts on the Brakes

Evolution of any kind cannot design and create. It's physically impossible. It's prevented from happening by entropy.

Chemical evolution, abiogenesis, spontaneous generation, or macro-evolution is physically impossible. Entropy prevents these things from happening, thereby making them physically impossible. That's the way it really is in the natural world. The chemical evolution of functional genes and proteins is physically impossible. It can't happen, which means that it didn't happen.

ORIGIN: Probability of a Single Protein Forming by Chance.

https://www.youtube.com/watch?v=W1_KEVaCyaA

I find all of this fascinating. I used to be a Materialist, Naturalist, Nihilist, and Atheist; so, I find it fascinating to observe how wrong I really was in my chosen conclusions at the time. But, I'm not alone. Millions have fallen for the deceptions and the lies. Nevertheless, we have had the truth all the way along, if we were willing to look for it, find it, and accept it.

Spontaneous generation, abiogenesis, chemical evolution, or macro-evolution was FALSIFIED in 1859 by Louis Pasteur – the very same year that Charles Darwin published "On the Origin of Species". We've known that the Theory of Evolution is false from the very beginning; but as a race, we chose to ignore the evidence.

In fact, evolution (genetic change), random mutations, and natural selection didn't even exist until after God designed and created the proteins, the matching genes, the genomes, the eyes, the brains, and the physical bodies in the first place. It's so obvious that I sometimes wonder how I managed to overlook it for the first fifty years of my life; but, I wasn't looking for it, so I wasn't able to see it. Evolution is entropy. Each species has been de-evolving or degenerating ever since God created its genome, thanks to evolution or entropy.

Entropy prevents a genome from spontaneously generating out of thin air. Spontaneous Generation or Chemical Evolution was FALSIFIED in 1859 by Louis Pasteur, which means that Materialism, Naturalism, Darwinism, Abiogenesis, Macro-Evolution, and the Theory of Evolution were FALSIFIED at the same time in 1859. Isn't it fascinating to observe that we KNEW that the theory of evolution is false from the very beginning; but, our scientists chose to go along with it anyway? The truth is that genomes don't spontaneously generate out of thin air. They NEVER have, and they NEVER will. The same can be said of proteins.

Your genome is programming code. Your genome is God's Signature. I was a computer programmer for a decade. I know for a fact that programming code doesn't spontaneously generate out of thin air. That's physically impossible. Entropy or evolution cannot do computer programming. You cannot do computer programming or genome programming by throwing dice, shuffling the deck, random mutations, or random chance. I also know what happens whenever there is a bug in the code. The results are often fatal; and, if the program manages to survive, the results from a bug in the code are less than optimal. Disease, deformity, cancer, pain, suffering, death, and extinction are the natural result of entropy, or random mutations, or evolution.

Finally, take note that entropy or random mutations prevent a genetically compatible Mr. and Mrs. Mutant from coming into existence at the same time in the same place. Thanks to entropy or the randomness of mutations, it's physically impossible for chimp-like ancestors to evolve into chimpanzees and humans, because it's physically impossible to produce the requisite Mr. and Mrs. Mutant each and every time that one is needed, so as to make a sexually reproducing species evolve naturally into some other species. Macro-evolution of this type is prevented from happening by entropy when it comes to a sexually reproducing species.

Remember, the theory of evolution was made to fit the fossil record after-the-fact in hindsight, and it's constantly adjusted every time another fossil is dug up. The theory of evolution is reactive and not predictive. The theory of evolution is correlational, and not observational. Correlation cannot be used to prove causation. The missing links really are missing. There are NO observations supporting any aspect or any version of the theory of evolution. Design and creation by evolution has NEVER been caught in the act. There are infinitely better and more plausible explanations for the origin of life than random chance or blind luck. The only way to make the theory of evolution work as advertised is to get God to intervene and force it to work; and, God has infinitely better and infinitely faster ways of designing and creating life than random mutations and natural selection.

The theory of evolution is wishful thinking in action. The theory of evolution is the product of Confirmation Bias, which is the psychological tendency to search only for information that confirms one's preconceptions or one's pre-chosen conclusions. These people automatically reject and dismiss anything that falsifies the Theory of Evolution, Materialism, and Naturalism. They cheat, in order to make their case. They employ a wide variety of logic fallacies to make their case for the theory of evolution.

The Summary

So, what have we learned?

Evolution is entropy. Entropy cannot design and create. These people fail to meet their burden of proof, because the preponderance of the evidence falsifies Materialism, Naturalism, Darwinism, Nihilism, Atheism, and the Theory of Evolution.

A hypothesis is an idea. A theory is a hypothesis that has observational evidence supporting it. It really should be called the "hypothesis of evolution" because there is NO observational evidence supporting macro-evolution, chemical evolution, abiogenesis, and spontaneous generation; and, there NEVER will be. Thanks to entropy, random diffusion, or the second law of thermodynamics, it's physically impossible for atoms to spontaneously generate into functional genes, proteins, genomes, eyes, brains, and life forms. It can't be done, which means it wasn't done.

Isn't it fascinating how we can find and know the truth simply by thinking about things critically, logically, impartially, and rationally?

I don't make my living by preaching and teaching evolution, so I'm free to see through all of its deceptions and lies. I'm not motivated by confirmation bias where the theory of evolution is concerned, so I'm free to allow ALL of the evidence into evidence and free to pursue a preponderance of the evidence.

I have observed that the preponderance of the evidence falsifies Materialism, Naturalism, Nihilism, Atheism, and the Theory of Evolution. I have also observed that there are NO observations wherein any aspect of the theory of evolution has been caught in the act. Evolution of new life-forms from pre-existing life-forms has never been observed, which means that the theory of evolution has never been verified. In contrast, design and creation by Intelligent Beings or Intelligent Psyches has been observed, verified, and experienced trillions of trillions of times.

435

So, which explanation for the origin of life is true – the one that has been falsified or the one that has been verified?

Repeated falsification proves that a theory is false. Constant verification implies that a theory is true.

Entropy FALSIFIES the claims of Materialism, Naturalism, Darwinism, and the Theory of Evolution, because evolution is entropy, and entropy cannot design and create anything. Entropy can only degradate and destroy. Entropy cannot design, program, fine-tune, field-test, manufacture, engineer, and create, which means that evolution of any kind cannot design, program, fine-tune, and create. Evolution is entropy.

So, where did life come from since it obviously didn't come from evolution or entropy?

Life came from Syntropy.

Life is syntropy. Psyche is syntropy. Intelligence is syntropy. Physical matter originated from syntropy or spirit matter. Quantum Mechanics is syntropy. The Priesthood Power of God is Syntropy. Quantum Mechanisms are the Priesthood Power of God. Programming code or genomes are the result of syntropy. The arrow of progression in the fossil record, or the ever-increasing complexity of life in the fossil record, is the result of syntropy. WE KNOW this is so because entropy or evolution would prevent it from happening. God is Syntropy.

Life came from Syntropy, not entropy or evolution. Evolution or entropy cannot do life. Evolution or entropy can only do disorder, disease, chaos, cancer, death, and extinction. It takes an infinite amount of blind-faith to believe that entropy or evolution can design and create, when ALL of the physical evidence we have on hand is telling us that entropy or evolution can only do disorder, disease, deformity, death, and extinction. The Darwinists and Evolutionists say that evolution is a beautiful concept, but there's nothing beautiful about cancer, disease, death, and extinction. Cancer, disease, deformity, death, and extinction are the result of evolution or random mutations.

Ironically, the very existence of entropy or evolution is Scientific Proof of God's Necessity; and therefore, Scientific Proof of God's Existence. Evolution is entropy, which means that evolution of any kind cannot design and create anything. The chemical evolution of proteins, and the matching genes to go along with them, NEVER happened because entropy prevents it from happening. The development of new species through random mutations and natural selection never happened because entropy prevents it from happening. Consequently, we have to look someplace else besides entropy or evolution for the origin of life on this planet. What's the opposite of entropy or evolution? Syntropy is the opposite of entropy. Syntropy is order and organization. Syntropy is intelligence. God is Syntropy.

It's elementary.

God has to exist, because Someone Psyche had to be there in the first place to wind up the clock or wind up this physical universe with Syntropy in the first place. We couldn't have all of that subsequent entropy without in initial infusion of syntropy. The initial syntropy within this physical universe had to come from someplace, and it came from God and the quantum realm. There's no other logical explanation for its origin. The Syntropy had to come from the syntropy realm, or quantum realm, or psyche realm, or spirit realm because it clearly doesn't exist here in this physical realm. The order, organization, or syntropy couldn't have come from evolution because evolution is chaos and entropy. The order, organization, programming, and structure within 3D proteins, genes, eyes, brains, and genomes had to come from Someone Psyche or Someone Syntropy.

This physical realm and classical physics are based upon entropy. Entropy cannot do design and creation. The psyche realm, quantum realm, and Quantum Mechanics are based upon syntropy, order, and intelligence. Syntropy, order, organization, psyche, and intelligence have the innate ability or the inherent capability of doing design and creation.

Remember, evolution of any kind is entropy; and, evolution of any kind is prevented from happening by entropy. The production of functional genes and proteins from atoms is prevented from happening by entropy. Evolution or entropy cannot produce order or syntropy. It's physically impossible. Therefore, we are looking for a non-physical explanation for the origin of life or the origin of syntropy.

Order and Organization are Syntropy. Life is Syntropy. Psyche is Syntropy. Intelligence is Syntropy. Quantum Mechanisms are Syntropy. Functional genomes and functional proteins are the result of Syntropy. Your genome is God's Signature. God is Syntropy.

Quantum Mechanics is Syntropy, which means that quantum mechanisms are supernatural in nature and origin. Quantum Mechanics is the Priesthood Power of God. Quantum Mechanics is observed, proven, and verified science. It's time that we start using Quantum Mechanics to explain how things really work. The very existence of Quantum Mechanics, Syntropy, Transdimensional Mechanisms, or Supernatural Mechanisms FALSIFIES Materialism, Naturalism, Darwinism, Nihilism, the Theory of Evolution, and even Atheism.

Materialism, Naturalism, Darwinism, Entropy, and Classical Physics completely lack explanatory power when it comes time to explain what the Human Psyche and Nature's Psyche are doing in the quantum realm and how they work in the quantum realm or the syntropy realm. For that explanation, we need quantum mechanisms, supernatural mechanisms, or psyche mechanisms.

Quantum Mechanics is our best-proven, most-verified, and most-used science that we currently have. It's time that we start using Quantum Mechanics or Transdimensional Physics to explain what's happening in the quantum realm, syntropy realm, or psyche realm.

If your interpretation of Quantum Mechanics cannot explain what the Human Psyche and Nature's Psyche are doing at the quantum level in order to get things done for us at the physical level, then your interpretation of Quantum Mechanics is worthless because it's based upon Materialism, Naturalism, Nihilism, and Classical

Physics. Naturalism and Classical Physics lack explanatory power when it comes to the psyche realm or the quantum realm because Materialism and Naturalism deny the existence of psyche or syntropy.

You can't use something that is incomplete or false to demonstrate and verify the truth. Materialism, Naturalism, Darwinism, Nihilism, Behaviorism, Determinism, Atheism, and Classical Physics cannot be used to verify the truthfulness and usefulness of Quantum Mechanics, Syntropy, and Psyche. Quantum Mechanics is supernatural in nature and origin, which means that Naturalism and Quantum Mechanics are mutually exclusive. If the one is true, then the other is automatically false.

You are going to have to decide for yourself which one is true, and which one is false; but, I KNOW for myself which one the preponderance of the evidence verifies and which one the preponderance of the evidence falsifies; and, that's good enough for my needs. All I ever really wanted to know is the truth, and now I do.

Mark My Words

—

Source Material

1. *Scientific Proof of God's Existence: Finding God Where the Atheists Refuse to Look for Him*.

   https://www.amazon.com/dp/B07B26CRHX

2. Myers, D. G. (2010). *Social Psychology* (10th ed.). New York: McGraw-Hill.

3. Pinel, J. (2014). *Biopsychology* (9th ed.). New York: Pearson.

4. Sanford, J. (2014). *Genetic Entropy* (4th ed.). Cornell University: FMS Foundation.

References

1. *Quantum Neuroscience: The Answer to Life, the Universe, and Everything*.

   https://www.amazon.com/dp/B079Z6QQQB

2. *NATURE vs. NURTURE vs. NIRVANA: An Introduction to Reality*

   https://www.amazon.com/dp/B01JWRCSVA

   https://www.amazon.com/dp/1521132615

3. ***BioPsychoSocial: Including Psyche or Light into our Theoretical Models***

https://www.amazon.com/dp/B0713NDHVW

4. ***Science 2.0: I Upgraded My Science***.

https://www.amazon.com/dp/B0771K6WTX

Syntropy vs. Entropy

Syntropy is Mind-Over-Matter or the Placebo Effect. If we are paying attention, we will notice that it shows up everywhere in our physical universe. Syntropy or Mind-Over-Matter shows up a lot in Social Psychology. Obviously, it should because Social Psychology is the study of the interaction between the Human Psyche and all those other psyches, on every level of existence.

Syntropy vs. Entropy

In the cyclical nature of the Universe there are two opposing principals, Entropy and Syntropy. To understand syntropy, it is useful to understand its opposing force, entropy.

The Principle of Entropy has effects that are easy to understand and observe. According to this Principle, all organized forms of matter require more energy than those that are less organized. These organized forms will lose their order and initial energy unless they are constantly nourished by energy. Plants need water and sunlight to grow (energy); when that is deprived they begin to breakdown. A new car will fall apart unless energy is applied to maintain it, and a business that doesn't put energy into maintaining itself will close its doors. As life forms age their systems lose energy becoming less efficient allowing for disease and illness to set in. The tendency of nature, systems, and organisms to lose energy and become disorganized over time, essentially the law of death is entropy.

Entropy's opposing force, however, is Syntropy. This Principle is much harder to notice, the best example of which, is life itself. It is the law of order and organization, finality and differentiation, the ability to attract, evolve and bring together ever-increasing complex forms creating something new. A new galaxy forming from the ruins of an older galaxy, building a business, and the bringing together of individual cells to create an organism are all examples of the syntropic nature of the Universe. Consciousness focusing energy to create and maintain a system; the law of life is Syntropy.

In an organism the degree of internal disorganization is entropy and the level of internal organization is syntropy. The sum of these two quantities represents the level of health of an organism in the present and the transformation potential for the evolution of that system or organism.

As far as health is concerned, the more energy we have available to maintain and balance our various systems, the greater our health benefits will be. But if we are overly stressed, in chronic pain, mentally and emotionally drained, most of our energy is being used to sustain these negative states. This means much less energy for our healing and growth processes. Over time an organism's systems begin to break down, dis-ease sets in and entropic forces overtake the syntropic forces until death occurs.

The holistic healing model integrates the whole person in the healing process: body, emotion, mind, and spirit, all of which are different forms of energy. A holistic approach utilizes non-invasive treatments and natural healing techniques to ultimately change the energy associated with disease and dysfunction, bringing balance and a higher degree of homeostasis to the patient.

Aside from practitioner's treatments, a major aspect of holistic healing is the patient's active participation in the process. The outcome is subject to the patient's personal desire, intent, and follow-through in their healing process. Self-healing by means of a balanced diet, exercise, adequate rest, and stress release techniques to name a few are imperative for anyone wanting to heal themselves and become more whole.

Syntropy Energetics is an integrative approach to health and wellness, which utilizes holistic therapeutic modalities and life practices to bring about change in a client. These healing techniques work with different energies of the body at the different layers and levels in which they manifest. *Primal Reflex Release Technique* (PRRT), works with the physical body, utilizing reflexes to quickly release pain, tension, and stress. *Subtle energy healing* accesses our subtle energy fields to create instant change at these primary levels of organization. *Biodynamic Cranial Touch* is a means to access the different levels of consciousness associated with energy, allowing *it* to return the client to a state of wholeness.

http://syntropyenergetics.com/Syntropy_vs._Entropy.html

I'm not promoting syntropy energetics per se; but, I am indeed promoting Syntropy or Mind-Over-Matter. The Placebo Effect is REAL, documented, proven, verified, observed, and powerful. The Placebo Effect or Mind-Over-Matter is an integral part of Science 2.0, Spirit Matter, and Syntropy. Likewise, Action at a Distance or Transpersonal Psychology is an integral part Quantum Mechanics, Transdimensional Physics, Psyche, and Syntropy. It has been experienced and observed, which means that it's REAL and truly exists.

The physicists have observed that it requires a huge infusion of energy to elevate electrons from a lower level to a higher level. That's how spirit matter or dark matter is converted into physical matter – by a huge infusion of energy, or syntropy, or God's glory, power, and light. If that energy is released from the physical matter, then it goes back to being spirit matter. Where does this huge infusion of energy come from? It comes from the Quantum Realm, the Syntropy Realm, or the Zero-Point Field which is the Light of Christ. The explanatory power of Syntropy or Quantum Mechanics is through the roof!

Begging the question and applying *circular reasoning* – which are logic fallacies – the Materialists, Naturalists, Darwinists, Nihilists, and Atheists define "science" as Materialism and Naturalism. Materialism, Naturalism, Darwinism, and their derivatives are based upon a wide variety of different logic fallacies. Materialism and Naturalism are based exclusively on entropy. There's NO Syntropy where Materialism, Naturalism, Darwinism, Nihilism, and Atheism are concerned.

441

I have upgraded my science to Science 2.0. Science 2.0 defines "science" as observation and experience, both at the quantum level and at the physical level. Unlike Materialism and Naturalism, Science 2.0 works at both levels of existence – both the psyche level and the physical level. Science 2.0 defines "science" as Syntropy. Quantum Mechanics is Syntropy. It works at both the quantum level and the physical level to explain what's happening at both levels of existence. Quantum Mechanics, Transdimensional Physics, or Syntropy has infinitely greater explanatory power than Materialism, Naturalism, Darwinism, Nihilism, Atheism, and Classical Physics. Materialism, Darwinism, and Naturalism are powerless to explain the Placebo Effect or Mind-Over-Matter.

Please let me demonstrate.

To consider the answer, consider how we interpret and label our bodily states.

The principle seemed to be: *A given state of bodily arousal feeds one emotion or another, depending on how the person interprets and labels the arousal.*

A frustrating, hot, or insulting situation heightens arousal. When it does, this arousal, combined with hostile thoughts and feelings, may form a recipe for aggressive behavior. (*Social Psychology*, p. 368).

We = the Human Psyche.

The person = the Human Psyche.

It's the very same physical input from the environment, the very same physical body, and the very same physical genes making up the physical cells in the physical body; but, we are observing two completely different chosen responses or two different chosen behaviors depending upon how the same Human Psyche decides to interpret and label the incoming physical stimulus.

If the physical harm is interpreted and labeled as deliberate, mean, intentional, and hostile, anger or retaliation tends to ensue.

If the physical harm is interpreted and labeled as accidental, unintentional, non-hostile, and coming from a friend who typically means us well, then compassion and forgiveness tend to be the result.

Everything in these experiments remains the same, except for the way that the Human Psyche chooses to label and interpret the physical harm or the physical stimulus.

Even a dog can tell the difference between a deliberate kick and an accidental trip, even though the physical behavior is precisely the same and has the same physical effect of doing harm to the dog. It's the Psyche that interprets and labels the physical action, and that makes ALL the difference in the world.

This is the smoking gun, where Psychology, Psyche, and Science are concerned. It's the same physical input, but a completely different chosen response or a completely different chosen interpretation, depending upon the way that the Human Psyche or the dog's psyche chooses to interpret and label the physical assault.

Our chosen interpretation of the evidence helps to determine the emotional feelings and the behavioral outcome of every physical event.

Guns prime hostile thoughts and punitive judgments. What's within sight is within mind. This is especially so when a weapon is perceived as an instrument of violence rather than a recreational item. For hunters, seeing a hunting rifle does not prime aggressive thoughts, though it does for non-hunters. (*Social Psychology*, 369.)

This is the smoking gun, where Psychology, Social Psychology, Behaviorism, Biology, and Genetics are concerned. We have the very same physical stimulus, but two completely different chosen interpretations given to that physical stimulus.

How's that possible?

It's NOT according to Determinism, Behaviorism, Darwinism, Genetics, Materialism, Naturalism, Classical Physics, and the Theory of Evolution.

Here we are observing the same people, with the same genetics as before the experiment or observation, with the same physical stimulus, but two completely different chosen interpretations or two completely different chosen responses. It's the Human Psyche who is making the choices and doing the interpretation and the labeling – NOT their genes. Their genes have NO way of telling the difference between a gun that is used for recreational purposes and a gun that is used to kill people; but, the Human Psyche does and can.

The society or environment provides the gun; but, it's the Human Psyche who decides what that gun means to that particular Human Psyche. Some Human Psyches decide that the gun means recreation and fun. Other Human Psyches decide that the gun is a good way to kill other people and commit suicide. It's the very same gun, but a completely different meaning given to that gun by a Human Psyche.

Do you see how that works?

It's called Psyche Determinism or a Psyche Ontology, wherein Psyche is the fundamental unit of reality. Psyche is the Ultimate Cause. Psyche is the ultimate causal agent. It is the Psyche's chosen interpretation or chosen label for the physical stimulus that determines how that gun is used and perceived by that specific Psyche who is doing the using, choosing, and the perceiving. Their chosen interpretation of the evidence helps to determine the behavioral outcome of their chosen actions and chosen beliefs.

The Human Psyche can literally transform a physical object typically used for recreation and fun into a murder instrument if it chooses to do so. The Human

443

Psyche can transform this computer into a murder instrument with enough planning and forethought or anger and rage.

Anger, rage, and hate are chosen into existence by the Human Psyche. So are justice, mercy, kindness, friendship, and love. The very same physical object transforms and morphs into something completely different depending upon whether the Human Psyche chooses to love it or chooses to hate it. This reality and truth FALSIFIES Behaviorism, Materialism, Naturalism, Determinism, and even Classical Physics.

The Human Psyche can trump its nature and nurture at will.

The choice or decision as to how to interpret and label the meaning of physical objects is done by the Human Psyche, and NOT the genes. The Human Psyche is being impacted and influenced by the physical input, and NOT the genes. Decisions and choices change the Human Psyche, NOT the genes.

Arousal tends to spill over: One type of arousal energizes other behaviors.

Other research shows that viewing violence *disinhibits*. In Bandura's experiment, the adult's punching of the Bobo doll seems to make outbursts legitimate and to lower the children's inhibitions. Viewing violence primes the viewer for aggressive behavior by activating violence-related thoughts. Listening to music with sexually violent lyrics seems to have a similar effect.

Media portrayals also evoke *imitation*. (*Social Psychology*, p. 377.)

Thoughts and memories are a product and a function of the Human Psyche.

How do we know?

We KNOW because our thoughts, memories, individuality, and personality survive the death of our physical brain, according to the empirical evidence from Near-Death Experiences (NDEs), Out-of-Body Experiences (OBEs), Shared-Death Experiences (SDEs), and our after-death Life Reviews.

The viewer in these experiments is the Human Psyche, NOT our genes.

Being a cognitive behaviorist, an atheist, a materialist, and a determinist, Bandura denies the existence of the Human Psyche or the human spirit, but the evidence tells us that he is wrong to do so.

We can't explain what's really happening, both here in the physical realm and there in the spirit realm or quantum realm when we remove the Human Psyche from the equation.

Genes and physical matter don't make choices and decisions for us; and, a careful study of the evidence makes this abundantly clear.

Television's Effects on Thinking

We have focused on television's effect on behavior, but researchers have also examined the cognitive effects of viewing violence: Does prolonged viewing *desensitize* us to cruelty? Does it give us mental *scripts* for how to act? Does it distort our *perception* of reality? Does it *prime* aggressive behavior? (*Social Psychology*, p. 378.)

Cognitions or thoughts are a function of the Human Psyche. Mental scripts and aggressive thoughts are a function of the Human Psyche, NOT our genes. Our genes can't make us do anything we don't want to do. Our genes can't make us believe anything we don't want to believe. Our reinforcement history cannot make us do anything we don't want to do. Our beliefs are chosen into existence by the Human Psyche.

Chosen actions or chosen behaviors are a function and product of the Human Psyche. It's the Human Psyche who is getting desensitized, NOT our genes. Mental scripts and anything within the mind have to do with the Human Psyche and NOT our genes. Physical input or physical stimulus comes in from the physical environment and registers on the physical brain, where it is then transferred or transmitted by the physical brain to the Human Psyche. Then the Human Psyche decides how to interpret and label that physical input, and the Human Psyche chooses how to respond to that physical input or chooses to ignore it. Physical stimuli take place at the physical level, and spiritual stimuli take place at the quantum level; but, our **perception of reality** is done by the Human Psyche at every level or dimension of existence.

Do you see how that works?

Our chosen behaviors or chosen actions are chosen into existence by the Human Psyche. The meaning, or label, or interpretation that is given to a physical object or a physical stimulus is chosen into existence by the Human Psyche. Our beliefs are chosen into existence by the Human Psyche. Love or friendship does not exist until it is chosen into existence by the Human Psyche.

You can grind down the whole physical universe into its constituent parts, and you will never find a single molecule or gene of justice, mercy, kindness, friendship, or love. There's no such thing as justice, mercy, kindness, friendship, or love until it is chosen into existence by Someone Psyche. Furthermore, once these things are chosen into existence by the Human Psyche, these things continue to exist long after your physical genes, physical body, physical society, physical world, and physical brain are dead and gone.

That's just the way things work at every level of existence.

Syntropy is Endless and Eternal without a beginning of days or an end of years. Quantum Mechanics and Psyche are Syntropy. Compared with the Quantum Realities or Eternal Realities associated with the Human Psyche, evolution and evolutionary psychology are completely worthless and inadequate, because evolution and evolutionary psychology are restricted and limited exclusively to the physical realm and classical physics. They don't touch the quantum realm.

445

We can't use the theory of evolution, evolutionary psychology, behaviorism, and determinism to explain what's happening at the psyche level and what's being done at the quantum level by Syntropy or Psyche to get things accomplished for us here at the physical level. However, we can indeed use Syntropy, Quantum Mechanics, and Psyche to explain these things at every level of existence or every dimension of existence, including both the psyche level and the physical level.

When it comes to science in general, Syntropy, Psyche, and Quantum Mechanics or Transdimensional Physics has infinitely more explanatory power than Materialism, Naturalism, Darwinism, Nihilism, Scientism, Behaviorism, Determinism, Evolutionary Psychology, and Atheism. Syntropy, Psyche, and Transdimensional Physics can be used to explain and understand every type of science and every level of existence that comes our way.

Mark My Words

—

Source

Quantum Mechanics from a Non-Physical Spiritual Perspective

https://www.amazon.com/dp/B01J023TGU

https://www.amazon.com/dp/1521132380

References

Quantum Mechanics from a Non-Physical Spiritual Perspective

https://www.amazon.com/dp/B01J023TGU

https://www.amazon.com/dp/1521132380

Interpretations of Quantum Mechanics

There are literally thousands of different interpretations of Quantum Mechanics, and that is in fact the weakness of Quantum Mechanics. Quantum Mechanics or Transdimensional Physics is a proven and verified science. What hasn't been proven is "what it all means". You cannot prove an interpretation. Interpretation is a function of the Human Psyche, and you cannot use the physical sciences to prove the truthfulness or the falseness of things at the quantum level or the psyche level. The existence of things at the quantum level can only be inferred at the physical level by the effects that these things have on physical matter. That's just the way these things work.

The main problem with Quantum Mechanics is that it has spent decades under the command and control of the Materialists, Naturalists, and Atheists. These people don't have a clue as to how things really work. I KNOW, because I used to be a Materialist, Naturalist, Nihilist, and Atheist. These people neutralize, defoliate, dumb-down, limit, and castrate every theory and idea that comes their way – especially when it comes to science.

The psyche level or the quantum level has REAL effects on physical matter, but Psyche and Quantum Mechanisms cannot be observed directly by our physical instruments, because they are sub-atomic. Remember, when it comes to physics, the smaller can dwell within and control the larger. Transdimensional means non-physical. The sub-atomic or the quantum waves can dwell within and control the quarks, strings, electrons, bosons, gluons, and atoms because the sub-atomic or psyche is smaller than these things. Psyche transmits, receives, and stores quantum waves. Psyche is the fundamental unit of reality – the ultimate elementary particle – because it is the smallest thing that exists.

Materialism, Naturalism, Darwinism, Nihilism, Atheism, Scientism, Behaviorism, Determinism, Atheism, and Classical Physics create unsolvable problems in science.

I was specially prepared to resolve these unsolvable problems in science.

First of all, I successfully FALSIFIED Materialism, Naturalism, Darwinism, and Atheism by using the Scientific Method and *negating the consequent* in order to do so. I was in fact the first person to do so, as far as I know. I learned how to do this by studying the Philosophy of Science, *affirming the consequent*, and *negating the consequent*. Then I used the Scientific Method and negating the consequent in all of my books to FALSIFY Materialism, Naturalism, Darwinism, and Atheism. I observed that ALL of the unsolvable problems in Psychology are immediately resolved by getting rid of Materialism, Naturalism, Darwinism, Nihilism, and Atheism. Materialism and Naturalism caused the binding problem, the mind-brain problem, the nature versus nurture problem, and the free-will versus determinism problem. These problems are automatically solved by eliminating Materialism, Naturalism, Darwinism, and their derivatives. Simple. Fundamental. Logical. Parsimonious.

Second, I was the first person to develop a Psyche Ontology, wherein I demonstrate that Psyche is the fundamental unit of reality. I called it the Ultimate Model of Reality. Look for it in some of my other books. If there is such a thing as a point particle, Psyche or Non-Local Consciousness is it. A point particle would be the smallest unit of reality – the fundamental unit of reality – an infinite singularity. I have hypothesized many times that Psyche, or Quantum Non-Local Consciousness, is an infinite singularity. As such, it is theoretically capable of storing an infinite amount of information in something that has NO physical size and takes up NO physical space whatsoever. The smaller can dwell within and control the larger. Psyche transmits, receives, and stores quantum waves. It's Someone Psyche who chooses the frequency at which each string will vibrate. Thoughts and memories are quantum waves or the vibrations within strings. Simple. Fundamental. Parsimonious. Logical. All of the complex math points to the existence of these types of things.

Third, I was the first person to develop Science 2.0. Science 2.0 is the way that science should have been done but wasn't. Science 2.0 allows ALL of the evidence into evidence and then pursues a preponderance of that evidence. Under Science 2.0, the BEST and FASTEST way to find and know the truth is to observe it, live it, and experience it, or to choose to trust someone who has. Under Science 2.0, the second-best way to find and know the truth is to use the Scientific Methods and *negating the consequent* to eliminate everything that is false. If you successfully eliminate everything that is false, then ONLY the truth will remain. This is logical, parsimonious, simple, and true. It's fascinating to observe and study what remains, after you have successfully falsified and eliminated Materialism, Naturalism, Darwinism, Nihilism, Behaviorism, Scientism, Determinism, and Atheism. I observed when you successfully eliminate everything that is false, then only the VERIFIED or the OBSERVED or the EXPERIENCED remains. Syntropy, Quantum Mechanics, and Quantum Non-Local Consciousness (Psyche) remain after you have successfully eliminated everything that is false, unfruitful, and unproductive. The verified and the observed remains. What you have actually experienced for yourself remains.

Fourth, based upon Science 2.0, I developed a new science discipline called Quantum Neuroscience. When it comes to Quantum Neuroscience, I chose to allow ALL of the evidence into evidence, and then I pursued a preponderance of that evidence. The discoveries were stunning, and quite unbelievable at first. There's nothing like it on the planet that I know of. Quantum Neuroscience taught me how thoroughly brainwashed into Scientism, Naturalism, Nihilism, and Atheism I really was.

Fifth, I observed that if your interpretation of Quantum Mechanics cannot explain what the Human Psyche and Nature's Psyche are doing at the quantum level, then your interpretation of Quantum Mechanics is absolutely worthless. Materialism, Naturalism, Darwinism, Nihilism, Behaviorism, Scientism, Determinism, Atheism, and Classical Physics CANNOT explain what the Human Psyche and Nature's Psyche are doing at the quantum level, and how they work at the quantum level to get things done for us at the physical level. Interpretations of Quantum Mechanics based upon Classical Physics, the Indeterminacy Principle,

Random Diffusion, Materialism, Naturalism, and Entropy CANNOT explain what the Human Psyche and Nature's Psyche are doing at the quantum level or the psyche level. Materialism and Naturalism have NO explanatory power at the quantum level or the psyche level, with means that they are worthless when it comes time to explain these things.

I found two interpretations of Quantum Mechanics that actually explain what Nature's Psyche and the Human Psyche are doing at the quantum level and how they work and function at the quantum level in order to get things done for us at the physical level. I discovered the interface or the bridge between the quantum level and the physical level. The BEST interpretation of Quantum Mechanics is the Orthodox Interpretation of Quantum Mechanics by Henry P. Stapp. It's the interpretation that has the most explanatory power at the quantum level or the psyche level.

Henry P. Stapp explained to me that the mathematics for what the Human Psyche is doing at the quantum level are completely different from the mathematics for what Nature's Psyche is doing at the quantum level. The Human Psyche makes choices and requests of Nature's Psyche. It is Nature's Psyche who collapses the wave function thereby producing a single physical reality as a result.

Heisenberg's Uncertainty Principle is worthless and wrong because it's based upon Classical Physics, Random Diffusion, and Entropy. I, Henry Stapp, and some of his followers modified the Uncertainty Principle by redefining it as "Infinite Possibilities". It's the same difference, but it's a huge difference. Out of ALL the infinite possibilities, the Human Psyche makes a choice. Then, depending upon whether it's physically possible and whether God allows it or not, Nature's Psyche reaches out and collapses all of those infinite possibilities into ONE single physical reality. At that point, it's a done deal. At that point, it's physically real.

Rinse and repeat.

Once again, we start with infinite possibilities. The Human Psyche makes a choice deciding what it wants to do with its physical body. Then Nature's Psyche reaches out into those infinite possibilities and collapses them into one single physical reality. By redefining the Uncertainty Principle as "Infinite Possibilities", we convert Quantum Mechanics from classical physics and entropy over to Syntropy. Infinite Possibilities is Syntropy. In contrast, the Uncertainty Principle is classical physics, random diffusion, and entropy. You have to switch over to Syntropy and Quantum Mechanics if you want to explain what the Human Psyche and Nature's Psyche are doing at the quantum level in order to get things done for us at the physical level. The Uncertainty Principle is a dead-end. Entropy is a dead-end. In contrast, Infinite Possibilities are infinite, endless, and eternal – the very definition of Syntropy. Suddenly, you can explain everything, simply by switching over to Syntropy, Quantum Mechanics, and Psyche.

Sixth, my ultimate science discovery is that there are NO BYTES of programming code or memory within a physical brain. I didn't believe it at first. I refused to believe it at first. My science colleagues rejected it at first, because we have been trained, conditioned, and brainwashed to do so. However, there are NO wires in our brain connecting neurons together into functional BYTES of programming code and memory. A neuron is a switch. It is either ON or OFF. ALL 7,000 synapses on a single neuron get compressed into a single Post-Synaptic Potential (PSP), a single BIT of information, which then determines whether a neuron fires or not. NO messages are transmitted through a synaptic cleft. A synaptic-cleft scrambles everything that comes its way in terms of information or memories. Information transmitted through synapses gets reduced to a single BIT of information, which means that the postsynaptic neuron either fires or it doesn't.

I kept waiting for the Neuroscientists to tell me where the programming code is being stored and how it is being executed. They NEVER did. I kept waiting for the Neuroscientists to tell me where the data or the memories are being stored and how this information is being accessed. They NEVER did. I kept waiting for the Neuroscientists to tell me where all the CPUs and RAM are being stored within the physical brain. They NEVER did. The ONLY thing these people would ever say is that we don't know how these things work, but they do. Clearly, they work. Clearly, they exist. But, we don't know how they work or where they exist. These Materialists and Naturalists couldn't answer these questions and explain this science, because there are NO functional BYTES anywhere within a physical brain at the physical level. You can't answer questions when there are no answers – when your chosen paradigm or chosen model prevents you from providing the answers. By design, Materialism, Naturalism, and Classical Physics have NO explanatory power at the quantum level or the psyche level.

If there really are BYTES of programming code and memory within our brain, then it's ALL being stored and processed at the quantum level or the psyche level, because it doesn't exist at the physical level. It can't exist at the physical level. Why? It's physically impossible to construct functional BYTES out of stand-alone switches with NO wires to connect them. Neurons are stand-alone switches. This means that it's physically impossible to construct functional BYTES of programming code and memories out of neurons at the physical level, because there are NO visible physical wires connecting the neurons or switches together into a functional whole. There are no BYTES observable within a physical brain, which means that those BYTES have to exist and must exist at the quantum level or the psyche level, because clearly they do exist judging by the effect that they have on our physical brain and physical body.

You can only explain what's happening at the quantum level by choosing to allow Syntropy, Quantum Mechanics, and Psyche into the mix. Instantly, everything can be explained, once you have allowed Syntropy, Quantum Mechanics, and Psyche to come in to play.

Seventh, I fell in love with Quantum Maps of Physical Functionality due to the massive explanatory power that it provides. Science is only valuable when it has explanatory power. If your science can't explain what Nature's Psyche and the Human Psyche are doing at the quantum level and how they work at the quantum level, then your science is worthless. By choosing to allow Syntropy, Quantum Mechanics, Transdimensional Physics, Quantum Processing, and Psyche in to play, I discovered that I can explain everything in science in a logical and realistic manner.

The output from an infinite number of quantum computers existing at the quantum level can be MAPPED to a single physical neuron or concept cell.

Even with physical supercomputers running the jobs and the image recognition software, it still takes a ton of time for a computer to "recognize" a face or "identify" a person. The Human Psyche does so instantly. On the input side, our neurons are registers of physical information. That information is transferred directly to Nature's Psyche where it is processed at the speed of thought, and the output from those infinite number of quantum computers is sent to a single concept cell or neuron which Nature's Psyche has MAPPED for that physical functionality at the quantum level. That single neuron or concept cell then fires unilaterally "out-of-the-blue" thereby giving us a physical feeling of recognition.

ALL of the processing and memory storage are done by Nature's Psyche completely outside the conscious awareness of the Human Psyche. The only thing the Human Psyche experiences is the physical feeling of recognition. "Ah, I know that person." The Human Psyche-Quantum Processing-Nature's Psyche Symbiosis is vastly superior to anything that our physical computers are able to do. Nature's Psyche MAPS our physical brain at the quantum level, and then uses those MAPS to get things done for us at the physical level. Instantly, the explanatory power of our science goes through the roof! Suddenly, we KNOW how everything is done and how everything works. Suddenly, we have the answer to life, the universe, and everything.

Materialism, Naturalism, Darwinism, Nihilism, Atheism, Scientism, Behaviorism, Determinism, Atheism, and Classical Physics create unsolvable problems in science.

By allowing Quantum Mechanics and Psyche in to play, ALL of the unsolvable problems in science are automatically solved and go away. Through Quantum Mechanics and Psyche, you can answer every scientific question and scientific hypothesis that comes your way. The explanatory power of Syntropy, Quantum Mechanics, Quantum Processing, Quantum Memory Storage, and Psyche are through the roof. There is NO limit to what you can explain. Everything can be solved. Everything is solved.

Mark My Words

—

Source

Quantum Mechanics from a Non-Physical Spiritual Perspective

https://www.amazon.com/dp/B01J023TGU

https://www.amazon.com/dp/1521132380

References

Quantum Mechanics from a Non-Physical Spiritual Perspective

https://www.amazon.com/dp/B01J023TGU

https://www.amazon.com/dp/1521132380

Transdimensional Physics

Transdimensional means Non-Local, Non-Physical, or Not-Located in our physical 3D space-time realm. Transdimensional means quantum, or syntropic, or spiritual.

Quantum Mechanics is Transdimensional Physics. Thoughts and memories are quantum waves. Psyche is intelligence or personality. They survive the death of our physical brain, according to the empirical evidence from Near-Death Experiences (NDEs), Out-of-Body Experiences (OBEs), Shared-Death Experiences (SDEs), and our after-death Life Reviews.

According to Quantum Mechanics, you (your psyche) are everywhere in the universe simultaneously, and your spirit body is never far behind. Quantum Tunneling is faster than light. Quantum Teleportation is instantaneous.

Quantum Tunneling and Quantum Omnipresence also apply to physical matter, but not as much as they apply to spirit matter and psyche because Nature, the Physical Laws, Entropy, and/or God deliberately and greatly limit the "omnipresence" of physical matter.

Is Quantum Tunneling Faster than Light?

https://www.youtube.com/watch?v=-IfmgyXs7z8

https://syntropy.site/wp-content/uploads/2018/04/Is-Quantum-Tunneling-Faster-than-Light.zip

In this video, while talking about Quantum Tunneling and your physical body, he states:

Wouldn't it be nice to be everywhere at once? According to Quantum Mechanics, you are. At least a little bit.

Talking about your physical body he says:

Any material object is really a matter wave. You're everywhere in the universe, but not very much.

Remember this!

According to Quantum Mechanics, you are everywhere at once. According to Quantum Mechanics, you are omnipresent. So, ask yourself why you aren't omnipresent right now. What has been done to you, particularly what has been done to your physical body, so that you are NOT omnipresent right now? This IS the answer to life, the universe, and everything. Something has been done to physical matter and your physical body so that you are mostly right where you are right now. Done by whom? And, what precisely has been done?

In this video, he actually answers the question and explains why your physical body isn't everywhere all at once. He also indirectly, though not explicitly, explains why your spirit body and particularly your psyche WERE everywhere at once or omnipresent but aren't anymore.

It has to do with the De Broglie Wavelength of each of these different objects – psyche, spirit body, and physical body. They each have a different De Broglie Wavelength.

This De Broglie Wavelength defines how well determined an object's position is. A large wavelength means a highly uncertain position [omnipresence]. **A small wavelength means a well-defined position** [the locality of physical matter]. **That's true of subatomic particles, and it's sort of true of anything. Right now, I'm mostly right here. But there's also a small chance that I'm here, here, or here. There's an infinitesimal chance that I'm on the moon. Observe me, and you'll collapse my wave function and probably find me pretty much exactly where you expect to.**

Of course, he is only talking about physical matter because these people don't believe in the existence of spirit matter and psyche; but, this De Broglie Wavelength applies equally as well to spirit matter and psyche as it does to physical matter. Quantum Mechanics works at ALL levels or dimensions of existence. The explanatory power of Quantum Mechanics, Psyche, and Spirit Matter are infinitely greater than anything we can get from being limited exclusively to classical physics, physical matter, Materialism, and Naturalism.

Subatomic particles ARE spirit matter and/or psyche. Think about it! It's true! It's only when we get to the atoms that we are actually talking about "pure physical matter" if there is such a thing.

According to the Orthodox Interpretation of Quantum Mechanics by Henry P. Stapp, the Human Psyche makes ALL of the choices as to what it wants to do with its spirit body and its physical body, while it is Nature's Psyche or God's Psyche who actually collapses the wave function thereby making the Human Psyche's choices actual and real both in the quantum realm and here in the physical realm.

During the making of this physical universe, God took a small portion of dark matter or spirit matter, slowed it down to sub-light speeds, gave it a short De Broglie Wavelength, and made it subject to entropy or space-time. God imposed limitations or physical laws upon it. God imposed structure and order upon it. Remove those restraints, and it would go back to being dark matter or spirit matter.

It's all defined and explained by the De Broglie Wavelength.

A physical body has a short De Broglie Wavelength, which means that it has been localized or made physical and present in this 3D space-time realm. It's mostly here, rather than being everywhere. Physical matter has been made subject to entropy or the passage of time. Entropy is a function of space and time. Entropy doesn't exist in the Quantum Realm or Syntropic Realm. Physical matter has been localized into both space and time. Spirit matter and psyche have not.

In contrast to physical matter, a spirit body has an innately long De Broglie Wavelength, which means that it tends toward being omnipresent all at once. A psyche or intelligence or consciousness seems to be capable of an infinite De Broglie Wavelength, or TRUE omnipresence.

Physical beings such as us are comprised of three different entities, and each one of them has a different De Broglie Wavelength. The Law of Quantum Superposition allows for all three to exist in the same space at the same time in an additive fashion, because they each are out of phase with the other or exist in a different dimension than the other. These realities explain everything.

This is cool science!

In its native original state, your Psyche, Quantum Non-Local Consciousness, or Intelligence experiences Quantum Omnipresence. It's capable of being simultaneously everywhere in the universe. According to Quantum Mechanics, your Psyche IS everywhere in the universe all at once, in its native original state. A Psyche is a pinprick of light; but, its thought waves or its quantum waves are omnipresent, or its De Broglie Wavelength is infinite. The spirit matter in your spirit body has many of the same capabilities. A large De Broglie Wavelength means a highly uncertain position or Quantum Omnipresence.

A Psyche or Intelligence is restrained by and limited by being assigned to a spirit body; but, out-of-body explorers have observed and experienced the fact that the Psyche or Consciousness can separate from the spirit body and observe the spirit body from an immaterial viewpoint in space. It's the Psyche who has the experiences and forms new memories, and NOT the spirit body, and definitely NOT the physical body.

Being attached to a physical body with a silver cord greatly limits and restrains a spirit body and a psyche even more by making the spirit or the soul subject to a short De Broglie Wavelength or a physical body. God – the Maker of physical matter, physical universes, and physical bodies – deliberately forces physical matter to be entropic, limited to sub-light speeds, and made to have a very short De Broglie Wavelength. We will never be able to propel physical matter

454

faster than the speed of light because God deliberately prevents it from happening by the physical limits that He has placed upon physical matter.

However, as the Maker and Enforcer of these physical laws or physical limitations, God doesn't place and hasn't placed these physical limitations on His own physical body. His physical body is fully capable of Quantum Tunneling, Teleportation, and even Omnipresence. With no physical limitations being imposed upon it, even a physical body is capable of teleportation, quantum tunneling, and omnipresence just like a spirit body and psyche. It's amazing to think about, isn't it?

Most scientists spend little or no time thinking about the spiritual side of the equation, because they have convinced themselves that it does not exist. However, their chosen beliefs don't change the fact that the Psyche and the Spirit Body have been experienced and observed by millions of different people.

Psyche leads, and your spirit body follows; but, both seem capable of experiencing Quantum Omnipresence or a very long De Broglie Wavelength, as well as Quantum Tunneling or Teleportation at will, while they are separated from the physical body. Psyche and a spirit body are capable of an infinite velocity or teleportation because they have little or no mass. The capabilities of Psyche and a Spirit Body seem to be limitless.

However, while contained within or married to a physical body, a spirit body and a psyche are also limited to a short De Broglie Wavelength. They have to temporarily separate from the physical body in order to experience a longer De Broglie Wavelength and omnipresence.

God deliberately and knowingly designed physical matter to significantly limit, inhibit, restrain, and block the effects of Quantum Tunneling, Quantum Teleportation, Quantum Omniscience, and Quantum Omnipresence. Physical matter has been given a short De Broglie Wavelength, which means that it has locality or presence, which means that it has physical limitations and a well-defined position or location in space-time. Those physical limitations or physical laws show up in Classical Physics, the Laws of Thermodynamics, and the Theory of Relativity, which state that it is physically impossible to accelerate physical matter faster than the speed of light; but, those physical limitations don't apply to the quantum realm or the spirit world.

Violation of relativity, teleportation, quantum tunneling, quantum omnipresence, omniscience, and faster-than-light travel seem to be an inherent and natural part of the quantum realm or psyche realm. In contrast, physical matter and physical laws greatly reduce quantum uncertainty thereby significantly increasing predictability, dependability, stability, and control. The result of all this is the Ultimate Consensus Reality, a physical reality.

Furthermore, as the designer and creator of these physical laws and physical limitations, God granted himself an exemption allowing Him to continue to teleport or quantum tunnel His physical body instantaneously anywhere in the physical universe that He wants to go. God holds the KEYS to Quantum Tunneling or the teleportation of physical matter; and for obvious reasons, God only gives those abilities or KEYS to those whom He trusts. God also continues to experience Quantum Omnipresence and Quantum Omniscience. God's Psyche and Spirit is capable of being everywhere at once and capable of multitasking. The same reality apparently applies to His resurrected physical body as well.

If our Psyche is natively omnipresent and omniscient, then why in the universe would we ever allow ourselves to be limited by a spirit body or a physical body? We've answered some of this already, but it's still a great deal of fun to think about.

As I see it, the answer lies in Order, Organization, Standardization, Structure, and Consensus. Limitations, rules, and laws open up new possibilities. Spirit body limitations and physical body limitations allow us to experience gender and sexual activity, which really isn't possible if you were to remain as a pinprick of light or a Psyche. A physical reality is the Ultimate Consensus Reality. I can depend on my dog, car, wife, and house being there after I get home from work. In a physical reality, I can depend upon this essay being there on my hard drive and in the cloud when I go looking for it tomorrow. None of that would be possible if I were to have chosen to remain as a Psyche or a pinprick of light. It's hard to interact with physical objects if you have no hands. Limitations, structure, order, commandments, or laws provide new opportunities. A physical body does have its advantages.

God deliberately designed physical matter to greatly limit your quantum mechanical nature and properties. A physical reality is the Ultimate Consensus Reality, meaning that it is highly stable, predictable, controllable, reliable, and dependable. You can actually do science in a physical reality. God deliberately designed physical matter and this physical reality to greatly limit, inhibit, control, restrain, and contain your innate quantum capabilities. It's necessary to provide the order, stability, and predictability of this physical realm. In a physical realm, our actions and choices actually have serious consequences. It's REAL because it hurts. It's all theory, until after it has been experienced and observed. The physical makes it real.

God created this physical universe and your physical body to keep your spirit body and your psyche mostly where you currently are – localized here in this physical 3D space-time realm; but, a physical realm is NOT our original innate state of existence. Transdimensional Physics or Quantum Mechanics is the NORM; and, a physical reality is a highly limited and restricted sub-set of that NORM.

Again, according to Quantum Mechanics, you (your psyche or spirit) exist everywhere in the universe simultaneously. God designed this physical universe and your physical body as a school-ground to keep your spirit or soul caged or penned mostly within your physical body so that you can learn from the experience what it is like to have limitations.

Why is this important?

It's because limitations are the ONLY way to provide us with order, structure, predictability, control, law, stability, organization, creativity, and some type of physical life or physical existence. The Ultimate Consensus Reality, a physical reality, is ONLY possible through limitations, structure, order, organization, laws, and agreed upon restrictions. A consensus reality is the opposite of random chaos and omnipresence. A single physical reality is the opposite of infinite possibilities. We learn best and learn fastest while in a physical body. The goal is to learn self-control and responsibility – to learn how to limit, restrain, direct, respect, and choose our actions and behaviors.

Imagine what your physical life would be like without the restraints being imposed by God and His physical laws. Without restrictions such as entropy, space-time, locality, and the theory of relativity, the atoms within your physical body could quantum tunnel away on you at will, before long leaving you without a physical body. How would you feel if your physical body were to quantum tunnel away on you one atom at a time as you slowly dissolve into thin air? Your physical body has the innate capability to quantum tunnel away on you; but, God prevents it from doing so by the physical laws or short De Broglie Wavelength that He imposes on your physical body.

How would you feel if your hard drive, or your car, or your house, or your spouse were to quantum tunnel away from you one atom at a time? If such a thing were to happen as a regular part of our lives, then this physical reality would no longer be the Ultimate Consensus Reality. God uses the physical laws or the physical limitations of this physical universe to prevent this from happening to you.

Do you see why this is important? Do you see now why the commandments of God, rules, laws, limitations, orders, and restraints are important to you and your physical body?

I find it fascinating to try to figure out how things really work at every level of existence or reality, and why they are made to work that way.

Transdimensional Physics Has NO Physical Limitations

Transdimensional means non-physical, non-local, and not located in our Physical Consensus Reality. Transdimensional means not located in our local physical 3D space-time realm. Transdimensional Physics is Quantum Mechanics – the way that matter ordinarily "acts" in the complete absence of physical laws to restrain it and constrain it.

Remember, physical matter was designed to limit and restrain the effects of Quantum Mechanics. Physical matter was designed by God to localize Action at a Distance or to shorten the De Broglie Wavelength. Physical matter was designed and created to form the Ultimate Consensus Reality – reliable, predictable, dependable, and controllable.

In its original native state, spirit matter or dark matter or quantum matter had full access to all aspects of Transdimensional Physics or Quantum Mechanics. Physical matter retains those same innate capabilities. So, who or what prevents the atoms in your physical body from quantum tunneling away on you? Who or what prevents you from dissolving into thin air?

It's God's physical laws, God's physical restraints, God's physical limitations, or God's control over physical matter that keeps your physical body from quantum tunneling away on you one atom at a time to the nether reaches of the universe. It's that short De Broglie Wavelength which God has placed upon your physical body that keeps it from quantum tunneling away from you at will.

By creating the Ultimate Consensus Reality, a physical reality, God uses the physical laws to dampen, restrain, and limit the Quantum Mechanics or the Transdimensional Physics that are an innate part of all matter. By converting spirit matter into physical matter, God infuses space-time, locality, entropy, physical restraints, physical limitations, a short De Broglie Wavelength, and physical laws into the newly organized physical matter thereby greatly increasing the improbabilities or decreasing the probabilities that the physical matter will teleport away on you one atom at a time.

Quantum Mechanics or Transdimensional Physics is an innate capability of ALL matter. The Physical Laws or Classical Physics greatly dampen, restrain, limit, and minimize the quantum mechanisms that are an innate part of all matter.

So, who or what is preventing your car, your home, your wife, your computer, and your dog from quantum tunneling away on you one atom at a time? It's God, the Physical Laws, and the short De Broglie Wavelength that are dampening, limiting, and restricting the quantum laws or transdimensional physics so as to provide us with the Ultimate Consensus Reality, a physical reality.

A physical reality was designed by God to be reliable, predictable, and controllable. Once again, imagine what would happen to you and your life if the atoms in your hard drive, your physical body, your car, and your home retained the ability to quantum tunnel or teleport at will anywhere in the universe that they wanted to go. The Physical Laws that God put into place prevents them from doing so.

Nevertheless, the physical atoms and physical molecules do indeed retain their innate ability to quantum tunnel or teleport; and, these physical particles can and do quantum tunnel whenever and wherever God permits them to do so.

Fascinating, is it not?

What would it be like if God were to remove the physical limitations or physical restraints from your physical body, yet leave you here on this earth? What would it be like if God were to give you complete psychic control over your own physical body? What would your life be like? What would you be able to do?

Obviously, you would be able to teleport your physical body anywhere you wanted it to go.

What else would you be able to do?

Your life would be the ultimate in mind-over-matter. You would be able to heal your physical body at will.

You would be able to phase-shift or walk through walls.

You would be able to levitate or walk on water.

If you had complete psychic control over every atom in your physical body, you would be indestructible. They wouldn't be able to kill you. If they were to surprise you and blow you up, you could just reassemble yourself. Theoretically, you should even be able to force the atoms in your physical body to stay in place or stay coherent even while you are standing in the middle of a nuclear explosion or in the middle of the sun.

You should be able to turn off all pain signals within your physical body.

You could even let people kill you, and then reassemble yourself and teleport out of there later on if you wanted to do so.

Are there human beings who have demonstrated or experienced any of these quantum capabilities or transdimensional capabilities?

Yes, indeed there are.

My best friend has experienced phase-shifting. An elk passed through his truck unphased. Either God wanted my friend to live, or God wanted that elk to live. His friend also experienced phase-shifting – a whole car and family passed through their car and his family during their vacation. Obviously, God wanted those families to live for whatever reason.

I have experienced quantum tunneling or teleportation. I and my mother's car were teleported to safety. That was an interesting experience, let me tell you. It was nothing that I did. It was something that God did for me in order to save me. Obviously, in the process of saving me from physical harm, God also taught me that it's possible for Him to quantum tunnel physical bodies and physical cars at will.

Both my best friend and I have experienced that "bubble of protection" wherein God makes you an indestructible super-hero for a few seconds.

The Bubble of Protection, Quantum Tunneling, and Teleportation:

https://www.youtube.com/watch?v=DmBnCTuQUOc

This stuff sounds like science fiction; but, thanks to Quantum Mechanics or Transdimensional Physics, it's very real indeed. Quantum Mechanics is the best-proven and most-used science that we have. It's the Materialism, Naturalism, Nihilism, and Atheism that are the science fiction in all of this – whether we realize it or not.

Quantum Weirdness

So, ask yourself, "Do you want the truth, or do you want a convenient fiction? Do you want infinite explanatory power where your science is concerned, or do you prefer to be kept in ignorance and darkness where your science is concerned?" This is the million-dollar question.

When I was a Nihilist and Atheist, I wanted the convenient fiction. I wanted to die and cease to exist. I wanted annihilation. I was a sinner, and I was in a very bad situation. You can't in good conscience end your life when you believe that God exists, and you believe that there is going to be some kind of afterlife or judgment. I desperately needed the convenient fiction at the time; and, that's precisely what I received.

We Materialists, Naturalists, Darwinists, Nihilists, and Atheists call it "quantum weirdness" or "spooky action at a distance". But, there's nothing weird or strange about it once a person chooses to interpret Quantum Mechanics from a spiritual perspective or a non-physical perspective.

Quantum Mechanics from a Non-Physical Spiritual Perspective

https://www.amazon.com/dp/B01J023TGU

https://www.amazon.com/dp/1521132380

Nevertheless, I used to be a Materialist, Naturalist, Nihilist, and Atheist back in 2012. God let me experience that and be that so that I would know what it is and what it is like. At the time, when it came to Transdimensional Physics or Quantum Mechanics, I wasn't seeing it, and I wasn't having it. I didn't believe it. I didn't want it. More than anything else, I didn't want God to exist, and I didn't want there to be an afterlife. I truly wanted to die and cease to exist. I wanted annihilation. That is what I wanted most. And, in a very real sense, I got my wish. That person or personality eventually died and ceased to exist.

Boy, am I seeing things differently now! I can see now that the explanatory power of Transdimensional Physics is through the roof. The explanatory power of Quantum Mechanics is limitless and infinite. It explains everything. There's nothing that you can't explain when you finally choose to bring Quantum Mechanics, Psyche, and Syntropy into the mix.

Eventually, you realize that Quantum Mechanisms or Supernatural Mechanisms have been experienced and observed in real life by real people. They have been caught in the act. They are real; and, they are true. I've even experienced them; but, I didn't know what they were at the time, so I tried to explain them away as a figment of my imagination or thought that maybe I had passed out or some such.

I experienced Quantum Teleportation of my physical body and the car that I was driving. One instant I was traveling 45 mph and about to broadside a car that had pulled out in front of me; and, the next instant I was 30 feet away in the middle of a lawn, with the car completely stopped and the engine turned off. I felt time stop; and, I have NO memory of the intervening travel. It's as if I had passed out and woke up in the middle of the lawn with the car stopped and the engine turned off. I had experienced Quantum Tunneling or Teleportation; but, I didn't know what it was at the time. I explained it away as having passed out, because I didn't have a scientific explanation for it. For all I know, I might have experienced phase-shifting as well and actually passed through that other car unphased. I remember hitting the brakes; but, I have no memory of how I got into the middle of the lawn.

To physical mortal beings like us, Quantum Mechanisms or Supernatural Mechanisms seem like magic or miracles, but only because God uses physical laws or our physical consensus reality to deliberately limit and restrict the normal and natural effects of Quantum Mechanics or Transdimensional Physics while we are here in this physical realm. Physical matter and classical physics put a deliberate damper on Quantum Mechanisms or Supernatural Mechanisms or Spiritual Mechanisms. The physical limitations are real, and they truly have an impact because they were designed to be that way. The physical makes everything seem real, and the physical makes everything dependable and reliable and predictable. When it comes to the physical, cause and effect truly come into play and are obviously real.

However, God doesn't apply the same physical limitations to Himself.

Jesus Christ has been observed walking on water, levitating into the air, teleporting, and walking through walls with His physical body. These are all quantum mechanical phenomena or transdimensional phenomena. Remove the physical limitations, and even physical matter can be made to do these quantum mechanical functions. Quantum Mechanics and Quantum Tunneling remain an innate part of physical matter, although God typically prevents them from happening as a regular every-day part of our lives.

I'm not going to apologize for finding Quantum Mechanics or Transdimensional Physics interesting. It is science after all, or it should be. It has been experienced and observed. Quantum Mechanics is found all throughout the Bible. Transdimensional Physics has been caught in the act.

Luke 4: 28-30 New International Version (NIV):

28 All the people in the synagogue were furious when they heard this.

29 They got up, drove him out of the town, and took him to the brow of the hill on which the town was built, in order to throw him off the cliff.

30 But he walked right through the crowd and went on his way.

Jesus phase-shifted and/or teleported to safety. This really happened. It was experienced and observed.

461

Luke 4: 28-30 King James Version (KJV):

28 And all they in the synagogue, when they heard these things, were filled with wrath,

29 And rose up, and thrust him out of the city, and led him unto the brow of the hill whereon their city was built, that they might cast him down headlong.

30 But he passing through the midst of them went his way.

While here in mortality, Jesus Christ had the ability to phase-shift and teleport. Luke was a physician or a medical doctor, which is one of the reasons why he went out of his way to document the eye-witness accounts of all this quantum mechanical stuff that Jesus could do while here in mortality and after He rose from the dead.

Luke 24 New International Version (NIV)

Jesus Has Risen

24 On the first day of the week, very early in the morning, the women took the spices they had prepared and went to the tomb. 2 They found the stone rolled away from the tomb, 3 but when they entered, they did not find the body of the Lord Jesus. 4 While they were wondering about this, suddenly two men in clothes that gleamed like lightning stood beside them. 5 In their fright the women bowed down with their faces to the ground, but the men said to them, "Why do you look for the living among the dead? 6 He is not here; he has risen! Remember how he told you, while he was still with you in Galilee: 7 'The Son of Man must be delivered over to the hands of sinners, be crucified and on the third day be raised again.'" 8 Then they remembered his words.

9 When they came back from the tomb, they told all these things to the Eleven and to all the others. 10 It was Mary Magdalene, Joanna, Mary the mother of James, and the others with them who told this to the apostles. 11 But they did not believe the women, because their words seemed to them like nonsense. 12 Peter, however, got up and ran to the tomb. Bending over, he saw the strips of linen lying by themselves, and he went away, wondering to himself what had happened.

On the Road to Emmaus

13 Now that same day two of them were going to a village called Emmaus, about seven miles from Jerusalem. 14 They were talking with each other about everything that had happened. 15 As they talked and discussed these things with each other, Jesus himself came up and walked along with them; 16 but they were kept from recognizing him.

17 He asked them, "What are you discussing together as you walk along?"

They stood still, their faces downcast. 18 One of them, named Cleopas, asked him, "Are you the only one visiting Jerusalem who does not know the things that have happened there in these days?"

19 "What things?" he asked.

"About Jesus of Nazareth," they replied. "He was a prophet, powerful in word and deed before God and all the people. 20 The chief priests and our rulers handed him over to be sentenced to death, and they crucified him; 21 but we had hoped that he was the one who was going to redeem Israel. And what is more, it is the third day since all this took place. 22 In addition, some of our women amazed us. They went to the tomb early this morning 23 but didn't find his body. They came and told us that they had seen a vision of angels, who said he was alive. 24 Then some of our companions went to the tomb and found it just as the women had said, but they did not see Jesus."

25 He said to them, "How foolish you are, and how slow to believe all that the prophets have spoken! 26 Did not the Messiah have to suffer these things and then enter his glory?" 27 And beginning with Moses and all the Prophets, he explained to them what was said in all the Scriptures concerning himself.

28 As they approached the village to which they were going, Jesus continued on as if he were going farther. 29 But they urged him strongly, "Stay with us, for it is nearly evening; the day is almost over." So he went in to stay with them.

30 When he was at the table with them, he took bread, gave thanks, broke it, and began to give it to them. 31 Then their eyes were opened and they recognized him, and he disappeared from their sight. 32 They asked each other, "Were not our hearts burning within us while he talked with us on the road and opened the Scriptures to us?"

33 They got up and returned at once to Jerusalem. There they found the Eleven and those with them, assembled together 34 and saying, "It is true! The Lord has risen and has appeared to Simon." 35 Then the two told what had happened on the way, and how Jesus was recognized by them when he broke the bread.

Jesus Appears to the Disciples

36 While they were still talking about this, Jesus himself stood among them and said to them, "Peace be with you."

37 They were startled and frightened, thinking they saw a ghost. 38 He said to them, "Why are you troubled, and why do doubts rise in your minds? 39 Look at my hands and my feet. It is I myself! Touch me and see; a ghost does not have flesh and bones, as you see I have."

40 When he had said this, he showed them his hands and feet. 41 And while they still did not believe it because of joy and amazement, he asked them, "Do you have anything here to eat?" 42 They gave him a piece of broiled fish, 43 and he took it and ate it in their presence.

44 He said to them, "This is what I told you while I was still with you: Everything must be fulfilled that is written about me in the Law of Moses, the Prophets and the Psalms."

45 Then he opened their minds so they could understand the Scriptures. 46 He told them, "This is what is written: The Messiah will suffer and rise from the dead on the third day, 47 and repentance for the forgiveness of sins will be preached in his name to all nations, beginning at Jerusalem. 48 You are witnesses of these things. 49 I am going to send you what my Father has promised; but stay in the city until you have been clothed with power from on high."

The Ascension of Jesus

50 When he had led them out to the vicinity of Bethany, he lifted up his hands and blessed them. 51 While he was blessing them, he left them and was taken up into heaven. 52 Then they worshiped him and returned to Jerusalem with great joy. 53 And they stayed continually at the temple, praising God.

If you had complete control over every atom in your physical body, you could raise yourself from the dead at will.

John 20:19-20 New Living Translation (NLT)

Jesus Appears to His Disciples

19 That Sunday evening the disciples were meeting behind locked doors because they were afraid of the Jewish leaders. Suddenly, Jesus was standing there among them! "Peace be with you," he said.

20 As he spoke, he showed them the wounds in his hands and his side. They were filled with joy when they saw the Lord!

This required some kind of phase-shifting and teleportation.

Transdimensional Physics in Action

John the Beloved was boiled in oil and didn't die.

> **This event that I will now recount is not found in the Bible, but a church writer named Tertullian makes mention of it in the 36th chapter of a book that he authored ... But, it appears that Jesus had plans for the John that was placed into a vat of boiling oil, and when God decides that you will not die, then you simply will not die!**

> https://www.deedsofgod.com/index.php/120-95-ad-the-apostle-john-is-forced-to-bathe-in-boiling-oil-in-rome-mainmenu-405

If God decides that you are not going to die, then you are not going to die.

> **John was allegedly banished by the Roman authorities to the Greek island of Patmos, where, according to tradition, he wrote the Book of Revelation. According to Tertullian (in The Prescription of Heretics) John was banished (presumably to Patmos) after being plunged into boiling oil in Rome and suffering nothing from it.**

> https://en.wikipedia.org/wiki/John_the_Apostle

That's physically impossible!

But, through Quantum Mechanics or Transdimensional Physics, the Human Psyche and God's Psyche can do the physically impossible.

John the Revelator and the Three Nephites were given complete psychic control over the physical atoms within their physical bodies.

3 Nephi 28: 1-40:

> **1 And it came to pass when Jesus had said these words, he spake unto his disciples, one by one, saying unto them: What is it that ye desire of me, after that I am gone to the Father?**

> **2 And they all spake, save it were three, saying: We desire that after we have lived unto the age of man, that our ministry, wherein thou hast called us, may have an end, that we may speedily come unto thee in thy kingdom.**

> **3 And he said unto them: Blessed are ye because ye desired this thing of me; therefore, after that ye are seventy and two years old ye shall come unto me in my kingdom; and with me ye shall find rest.**

> **4 And when he had spoken unto them, he turned himself unto the three, and said unto them: What will ye that I should do unto you, when I am gone unto the Father?**

5 And they sorrowed in their hearts, for they durst not speak unto him the thing which they desired.

6 And he said unto them: Behold, I know your thoughts, and ye have desired the thing which John, my beloved, who was with me in my ministry, before that I was lifted up by the Jews, desired of me.

7 Therefore, more blessed are ye, for ye shall never taste of death; but ye shall live to behold all the doings of the Father unto the children of men, even until all things shall be fulfilled according to the will of the Father, when I shall come in my glory with the powers of heaven.

8 And ye shall never endure the pains of death; but when I shall come in my glory ye shall be changed in the twinkling of an eye from mortality to immortality; and then shall ye be blessed in the kingdom of my Father.

9 And again, ye shall not have pain while ye shall dwell in the flesh, neither sorrow save it be for the sins of the world; and all this will I do because of the thing which ye have desired of me, for ye have desired that ye might bring the souls of men unto me, while the world shall stand.

10 And for this cause ye shall have fulness of joy; and ye shall sit down in the kingdom of my Father; yea, your joy shall be full, even as the Father hath given me fulness of joy; and ye shall be even as I am, and I am even as the Father; and the Father and I are one;

11 And the Holy Ghost beareth record of the Father and me; and the Father giveth the Holy Ghost unto the children of men, because of me.

12 And it came to pass that when Jesus had spoken these words, he touched every one of them with his finger save it were the three who were to tarry, and then he departed.

13 And behold, the heavens were opened, and they were caught up into heaven, and saw and heard unspeakable things.

14 And it was forbidden them that they should utter; neither was it given unto them power that they could utter the things which they saw and heard;

15 And whether they were in the body or out of the body, they could not tell; for it did seem unto them like a transfiguration of them, that they were changed from this body of flesh into an immortal state, that they could behold the things of God.

16 But it came to pass that they did again minister upon the face of the earth; nevertheless, they did not minister of the things

which they had heard and seen, because of the commandment which was given them in heaven.

17 And now, whether they were mortal or immortal, from the day of their transfiguration, I know not;

18 But this much I know, according to the record which hath been given — they did go forth upon the face of the land, and did minister unto all the people, uniting as many to the church as would believe in their preaching; baptizing them, and as many as were baptized did receive the Holy Ghost.

19 And they were cast into prison by them who did not belong to the church. And the prisons could not hold them, for they were rent in twain.

20 And they were cast down into the earth; but they did smite the earth with the word of God, insomuch that by his power they were delivered out of the depths of the earth; and therefore, they could not dig pits sufficient to hold them.

21 And thrice they were cast into a furnace and received no harm.

22 And twice were they cast into a den of wild beasts; and behold they did play with the beasts as a child with a suckling lamb, and received no harm.

23 And it came to pass that thus they did go forth among all the people of Nephi, and did preach the gospel of Christ unto all people upon the face of the land; and they were converted unto the Lord, and were united unto the church of Christ, and thus the people of that generation were blessed, according to the word of Jesus.

24 And now I, Mormon, make an end of speaking concerning these things for a time.

25 Behold, I was about to write the names of those who were never to taste of death, but the Lord forbade; therefore, I write them not, for they are hid from the world.

26 But behold, I have seen them, and they have ministered unto me.

27 And behold they will be among the Gentiles, and the Gentiles shall know them not.

28 They will also be among the Jews, and the Jews shall know them not.

29 And it shall come to pass, when the Lord seeth fit in his wisdom that they shall minister unto all the scattered tribes of Israel, and unto all nations, kindreds, tongues and people, and shall

bring out of them unto Jesus many souls, that their desire may be fulfilled, and also because of the convincing power of God which is in them.

30 And they are as the angels of God, and if they shall pray unto the Father in the name of Jesus they can show themselves unto whatsoever man it seemeth them good.

31 Therefore, great and marvelous works shall be wrought by them, before the great and coming day when all people must surely stand before the judgment-seat of Christ;

32 Yea even among the Gentiles shall there be a great and marvelous work wrought by them, before that judgment day.

33 And if ye had all the scriptures which give an account of all the marvelous works of Christ, ye would, according to the words of Christ, know that these things must surely come.

34 And wo be unto him that will not hearken unto the words of Jesus, and also to them whom he hath chosen and sent among them; for whoso receiveth not the words of Jesus and the words of those whom he hath sent receiveth not him; and therefore he will not receive them at the last day;

35 And it would be better for them if they had not been born. For do ye suppose that ye can get rid of the justice of an offended God, who hath been trampled under feet of men, that thereby salvation might come?

36 And now behold, as I spake concerning those whom the Lord hath chosen, yea, even three who were caught up into the heavens, that I knew not whether they were cleansed from mortality to immortality —

37 But behold, since I wrote, I have inquired of the Lord, and he hath made it manifest unto me that there must needs be a change wrought upon their bodies, or else it needs be that they must taste of death;

38 Therefore, that they might not taste of death there was a change wrought upon their bodies, that they might not suffer pain nor sorrow save it were for the sins of the world.

39 Now this change was not equal to that which shall take place at the last day; but there was a change wrought upon them, insomuch that Satan could have no power over them, that he could not tempt them; and they were sanctified in the flesh, that they were holy, and that the powers of the earth could not hold them.

40 And in this state they were to remain until the judgment day of Christ; and at that day they were to receive a greater change, and

to be received into the kingdom of the Father to go no more out, but to dwell with God eternally in the heavens.

God is Syntropy. If God decides that you are not going to die, then you are not going to die.

These three were transfigured, or translated, or turned into seraphim. They can't be killed. Many of their capabilities and powers are discussed. Their capabilities are quantum mechanical, supernatural, and transdimensional in nature and origin. They seem to violate the physical laws or trump the physical laws, because God permits them to do so.

As a race, we are never going to master teleportation or quantum tunneling unless God permits us to do so. We are never going to be able to travel faster than the speed of light unless God permits us to do so. The physical laws that God put into place prevents us from doing so against His will.

Nevertheless, it has been observed that ALL of these quantum mechanical capabilities return to us when our spirits leave their physical body behind. A spirit or a ghost is a holistic combination of both psyche and spirit body. Psyches and spirit bodies function and exist according to the rules of transdimensional physics or quantum mechanics. In other words, spirit bodies and psyches have NO physical limitations whatsoever.

Out-of-Body Travelers and Near-Death Experiencers have observed that their psyche and their spirit body can teleport or quantum tunnel anywhere at will. At the quantum level or the psyche level, we are dealing with Transdimensional Physics or Quantum Mechanics, and not Physical Limitations or Physical Laws.

Powerful, is it not?

God is Syntropy. If God decides that you are not going to die, then you are not going to die. If God wants to teleport you to safety, He can, and He will. If God wants to translate you or transfigure you, He can, and He will. God can do, and He will do, whatever He needs to do and wants to do in order to accomplish His purposes and His goals. God has NO physical limitations. In their native original state, transdimensional physics, quantum mechanisms, psyche, and spirit matter have NO physical limitations because they are pure syntropy.

God gives certain particles of spirit matter physical limitations thereby converting those particles of spirit matter into physical matter. The physical matter is then used to create or organize the Ultimate Consensus Reality, a physical reality. A physical reality is predictable, controllable, reliable, and dependable. A physical reality is the ultimate in order, law, and organization. A physical reality is powerful because it's limited, ordered, organized, contained, controlled, predictable, dependable, and lawful.

A physical reality, being the Ultimate Consensus Reality, ends up being the opposite of chaos. That's its advantage, and that's also its limitation. Remember, when it comes to a physical reality, its limitations are its advantage. The atoms in your physical body aren't going to quantum tunnel away on you willy-nilly while they are subject to the physical limitations of God's physical laws. The physical limitations or God's Physical Laws make our physical bodies possible, reliable, dependable, controllable, and predictable.

Mark My Words

—

Source

Quantum Mechanics from a Non-Physical Spiritual Perspective

https://www.amazon.com/dp/B01J023TGU

https://www.amazon.com/dp/1521132380

Putting Psyche Back into Psychology: Restoring Science to Consciousness

https://www.amazon.com/dp/B071NC987S

NATURE vs. NURTURE vs. NIRVANA: An Introduction to Reality

https://www.amazon.com/dp/B01JWRCSVA

https://www.amazon.com/dp/1521132615

References

God Is in the Light: God is light, and in Him is no darkness at all.

https://www.amazon.com/dp/B07168S37N

Quantum Mechanics from a Non-Physical Spiritual Perspective.

https://www.amazon.com/dp/B01J023TGU

https://www.amazon.com/dp/1521132380

Quantum Neuroscience: The Answer to Life, the Universe, and Everything.

https://www.amazon.com/dp/B079Z6QQQB

Quantum Mechanics Is Spiritual Mechanics

The goal in all of this is to hook people up and give them a Model of Reality that actually matches with reality. I want to help people find the truth. I put a lot of effort into this so that I would have a Model of Reality that actually matches with the Scientific Evidence and the Observational Evidence that we have on hand as a race.

In contrast, the goal of the Materialists, Naturalists, Darwinists, Nihilists, Behaviorists, Determinists, Physical Reductionists, Atheists, and Classical Physicists is to convince people that certain things DO NOT EXIST. That goal is totally worthless. That's NOT science. That's religion and dogma. It takes an infinite amount of blind-faith to believe that something DOES NOT EXIST. Furthermore, it's philosophically and scientifically impossible to prove that something does not exist. It can't be done, which means that they will NEVER be able to prove that Materialism, Naturalism, Darwinism, and Atheism are true.

These people are tilting windmills, and they don't even know it. These people are fighting imaginary enemies which are nothing more than a figment of their imagination. They are also promoting pseudo-sciences that are also nothing more than wishful thinking which they have manufactured within their minds out of thin air. There's nothing tangible within Materialism and Naturalism that we can get our hands on. The whole thing is purely a Grand Illusion. It doesn't match with Reality, which makes it unscientific to begin with.

When it comes to Scientific Naturalism, Darwinism, and Physicalism, these pseudo-sciences are based exclusively on entropy. Entropy is death, which means that these pseudo-sciences are literally a Dead-End to begin with. There's NO life there! Entropy cannot produce life. Entropy cannot produce order. Entropy can only produce death. Entropy IS death. Evolution is entropy. Evolution is death. The theory of evolution is Creation by Death. Please explain to me how that's supposed to work.

One of my greatest and most useful scientific discoveries is that Quantum Mechanics is Spiritual Mechanics or Supernatural Mechanics. The proven and verified existence of Action at a Distance falsifies Materialism, Naturalism, Darwinism, Nihilism, Behaviorism, Atheism, Classical Physics, and their derivatives. All you really want is the truth, unless you are trying to deceive yourself.

Interpretations of Quantum Mechanics and Science

I used to be a Physicalist, Naturalist, Nihilist, and Atheist. I wasted a lot of time and effort learning concepts that are demonstrably false.

My Atheism was a very dark time in my life. If I can help to prevent even one person from going through what I went through, then it will have all been worth it!

What was the net effect of My Atheism?

It dumbed me down. It made me ignorant. It made me dumb and blind even though I still had a soul or a mind.

I knew that God does not exist.

Atheism makes you believe that you know more than you really do. Atheism makes you ignorant of all the different things that you need to learn and know about. Atheism made me ignorant of Quantum Mechanics or Supernatural Mechanics. Atheism made me ignorant of Syntropy. I'd never heard of Syntropy. I didn't know what it was. Atheism of any kind is based upon a refusal to look at evidence. Atheism makes you ignorant. Atheism makes you think that you are smart when you are not. You become arrogant and condescending towards others. That's the fruits of Atheism, Materialism, Naturalism, Nihilism, and Darwinism. You become dumber than what you used to be. Some people have called it the "dumbing down of America".

That's why it took me over 55 years to figure out that evolution is entropy. The theory of evolution is based upon entropy. That's why it doesn't work as advertised. It can't! Entropy can't design, create, and manufacture anything! Entropy is death. Entropy can only destroy; and think about it, the theory of evolution is based exclusively on entropy. The theory of evolution is Creation by Death. The theory of evolution is Creation by Entropy. The theory of evolution was false to begin with; but, I couldn't see that nor understand that because I had been educated and trained in Entropy, Materialism, Naturalism, Nihilism, Darwinism, Atheism, and Classical Physics. It made me dumb and blind.

The Materialists and Naturalists deliberately remove the quantum non-local sciences from consideration. These people are actively trying to hide Syntropy, Psyche, Quantum Mechanics, and Supernatural Mechanisms from us. The various different types of macro-evolution, or spontaneous generation, or chemical evolution have NEVER been observed in the wild. They have NEVER been experienced nor observed because they are physically impossible. They are prevented from happening by random diffusion or entropy. The Naturalists and Darwinists are trying to hide these scientific truths from us.

The Physicalists, Naturalists, and Atheists ONLY permit correlational research and experimental research into evidence. They ONLY permit physical evidence into evidence. They only permit the correlational method and the experimental method. They deliberately BAN all other sources of information and knowledge. However, there is also a Burden of Proof Methodology that is used in archeology and the forensic sciences. It is based upon the preponderance of the evidence. We have to use a Burden of Proof Methodology when going after spiritual, supernatural, and quantum realities and forces because we can't use our physical instruments to detect and record these things.

Correlations indicate a relationship, but that relationship is not necessarily one of cause and effect. Correlational research allows us

to *predict*, but it cannot tell us whether changing one variable (such as social status) will *cause* changes in another (such as health).

The correlation-causation confusion is behind much muddled thinking in popular psychology.

The great strength of correlational research is that it tends to occur in real-world settings where we can examine factors such as race, gender, and social status (factors that we cannot manipulate in the laboratory). Its great disadvantage lies in the ambiguity of the results. This point is so important that even if it fails to impress people the first 25 times they hear it, it is worth repeating a twenty-sixth time: *Knowing that two variables change together (correlate) enables us to predict one when we know the other, but correlation does not specify cause and effect.* (*Social Psychology*, pp. 20-21.)

One of my greatest and most useful scientific discoveries came when I first realized that correlation-causation confusion is behind much of the muddled thinking that's associated with Darwinism and the theory of evolution. Materialism, Naturalism, Darwinism, Nihilism, Atheism, Behaviorism, Determinism, Physical Reductionism, and the Theory of Evolution are purely correlational! There's NO evidence backing their claims that that Quantum or the Supernatural does not exist; and, there never will be. These falsified philosophies or falsified religions are purely correlational. They have NO evidence supporting them. They have to be taken on blind faith as being real and true. In fact, the verified and proven existence of Action at a Distance or Quantum Mechanics falsifies Materialism, Naturalism, Darwinism, and their derivatives. ALL the EVIDENCE falsifies Physicalism, Naturalism, and Darwinism. The theory of evolution is based exclusively on correlation-causation confusion. Fascinating, is it not?

Scientific experimentation has its weaknesses too. Sometimes laboratory research and science experiments fail to generalize to real-world settings. Manufacturing amino acids in a beaker in a simulated laboratory setting has absolutely NOTHING to do with getting entropy and physical matter to spontaneously generate into amino acids in the wild.

Experimental Research: Searching for Cause and Effect

The difficulty of discerning cause and effect among naturally correlated events prompts most social psychologists to create laboratory simulations of everyday processes whenever this is feasible and ethical.

We need to be cautious, however, in generalizing from laboratory to life.

Generalizing from Laboratory to Life

As the research on children, television, and violence illustrates, social psychology mixes everyday experience and laboratory analysis. Throughout this book we will do the same by drawing our data mostly from the laboratory and our illustrations mostly from life. Social psychology displays a healthy interplay between laboratory research and everyday life. Hunches gained from everyday experience often inspire laboratory research, which deepens our understanding of our experience.

This interplay appears in the children's television experiment. What people saw in everyday life suggested correlational research, which led to experimental research. Although the laboratory uncovers basic dynamics of human existence, it is still a simplified, controlled reality. It tells us what effect to expect of variable *X*, all other things being equal — which in real life they never are! Moreover, as you will see, the participants in many experiments are college students. Although that may help you identify with them, college students are hardly a random sample of all humanity. Would we get similar results with people of different ages, educational levels, and cultures? That is always an open question.

Nevertheless, we can distinguish between the *content* of people's thinking and acting (their attitudes, for example) and the *process* by which they think and act (for example, *how* attitudes affect actions and vice versa). The content varies more from culture to culture than does the process. People from various cultures may hold different opinions yet form them in similar ways. (*Social Psychology*, pp. 24, 28, 29.)

The ONLY way to distinguish between Psyche's thoughts and the observed physical actions is to accept the fact that both exist and that both are REAL. The Psyche's choices do indeed correlate with a person's physical behavior or physical actions; but if we want a fullness of the truth, we must accept the independent evidence or experiential evidence that verifies and proves the existence of Psyche because we are not going to get any of that from physical matter and our physical instruments. We must treat the observations and experiences associated with Near-Death Experiences and Out-of-Body Experiences as Scientific Evidence and allow them into evidence, if we want to find and know the truth about Psyche Mechanics or Quantum Mechanics; otherwise, we choose to remain in ignorance. That's just the way it is.

The laboratory is NOT a real-world setting most of the time. The variables in our lives aren't typically controlled, constrained, and manipulated in an artificial manner. Furthermore, a physical laboratory is slow to reveal the existence of Psyche or Quantum Mechanics. It can be done if the scientist is smart enough and creative enough; but, direct observation and experience is a better way for finding and knowing the truth about Psyche, Quantum Mechanics, and Syntropy.

Thought is a function of the Human Psyche. Thoughts and memories are quantum waves.

How do we know?

We know because our thoughts and memories survive the death of our physical brain. We know because our thoughts and memories can't be detected nor recorded by our physical instruments.

If a thought or memory survives the death of your physical brain, then it's truly a memory in every sense of the word, is it not?

This is just a small portion of what I learned while studying Social Psychology. Social Psychology is the scientific study of the Human Psyche and its interaction with all those other psyches. Social Psychology is valid on both sides of the veil – both here in this physical reality and afterwards in the spirit world. This is radically advanced science when it is no longer being restrained and limited by Naturalism, Darwinism, and Atheism. Social Psychology has whole chapters on the Self or the Psyche. When it comes time to define and explain the Self or the Mind, Psyche or Quantum Non-Local Consciousness ends up being the BEST explanation. That is what I have experienced and observed. Science is observation and experience after all.

In contrast, Atheism is the making of something by "Nothing". Atheism is patently absurd. Creation ex nihilo or spontaneous generation is the making of something from "Nothing". Creation ex nihilo or spontaneous generation is a type of Atheism, and it's just as absurd as Atheism. Creation ex nihilo assumes facts not in evidence. In fact, it's contradicted by logic and by evidence. The same can be said of Atheism, Physicalism, Darwinism, Nihilism, and Classical Physics. They are contradicted or falsified by logic and by evidence. They are falsified by Quantum Mechanics or Supernatural Mechanisms.

Do you want the truth, or do you want the science fiction? You can't have both. The one is true, and the other has been falsified by the truth. The one is true, and the other has been falsified by observation and experience. The false is falsified by the truth; and, the truth is repeatedly experienced and observed. Science is observation and experience – NOT Physicalism and Naturalism. Technically, the "science" behind Materialism, Naturalism, Darwinism, Nihilism, Atheism, and their derivatives has NEVER been experienced nor observed.

There are as many different interpretations of Quantum Mechanics as there are scientists, from what I can tell.

https://en.wikipedia.org/wiki/Interpretations_of_quantum_mechanics

https://en.wikipedia.org/wiki/Minority_interpretations_of_quantum_mechanics

Quantum Mechanics or Supernatural Mechanics is our most-used, most-verified, and best-proven Science that we have; but, nobody seems to know what it means.

I have experienced and observed the fact that interpretations of Quantum Mechanics that are based upon Physical Matter, Materialism, Naturalism, Darwinism, Nihilism, Atheism, and Classical Physics completely lack explanatory power and are therefore absolutely worthless.

Quantum Mechanics or Transdimensional Physics ONLY makes sense from a non-local, non-physical, transdimensional, spiritual perspective.

Mark My Words

Source Material

Quantum Mechanics from a Non-Physical Spiritual Perspective

https://www.amazon.com/dp/B01J023TGU

https://www.amazon.com/dp/1521132380

Myers, D. G. (2010). *Social Psychology* (10th ed.). New York: McGraw-Hill.

—

Entropy or the Passage of Time

According to the theory of relativity, anything traveling faster than the speed of light experiences NO passage of time. Time STOPS at the speed of light or faster.

Do you fully understand what this means?

It means that everything existing or traveling at frequencies faster than the speed of light is Syntropy. It doesn't age because there is NO passage of time. In other words, there is NO entropy. Entropy is a function of time. ONLY physical matter experiences entropy or is subject to entropy. Everything else is Syntropy.

Light is Syntropy. There is NO age limit on the Light that has been traveling for 13.2 billion years to get here from the other galaxies in this universe. There's NO entropy in Light. It doesn't age, get old, and die. From its perspective, it arrives the very moment it launches, even though from our perspective it took 13.2 billion years to get here. In fact, it has been theorized that Light already knows where it is going to land 13.2 billion light years away, or it doesn't launch in the first place.

Likewise, Psyche or Intelligence is Syntropy. It is ageless and timeless. There is NO entropy in Psyche, Intelligence, or Quantum Non-Local Consciousness. It is eternal and everlasting, without a beginning of days or an end of years. It is Syntropy. It doesn't bleed; so, there's no way to kill it or end it.

We will NEVER be able to accelerate or push physical matter faster than the speed of light because then it would no longer be physical matter, and it would no longer be subject to entropy or the passage of time. It would be spirit matter or dark matter instead. It would be Syntropy. It would be spiritual matter.

Physical matter has been caught in the act of Quantum Tunneling; so, it still retains that inherent capability. If you stop the passage of time or temporarily remove the entropy from physical matter, then you can teleport it or quantum tunnel it anywhere you want in the universe instantly. During the quantum tunneling process, it experiences NO passage of time. It doesn't age.

Those of us who have experienced Quantum Tunneling or Teleportation KNOW that time STOPS when you Quantum Tunnel. I felt time stop. It was like I had passed out. And then, I was just instantly somewhere else instead, with NO knowledge or memory of how I got there. I and the car I was driving had just jumped, blinked, or teleported. It was nothing that I did, though. It was something that God did for me in order to save me.

The science is REAL, and it's powerful; and, it's totally consistent, coherent, and harmonious with itself. It ALL makes sense. The Truth matches perfectly with all other truths. The Truth has been experienced and observed.

Quantum Tunneling is REAL. It has been experienced and observed. The only drawback or disappointment when it comes to Quantum Tunneling is that ONLY God has access to it, or ONLY God can do it and trigger it at will. ONLY God holds all the KEYS to Quantum Mechanics or Supernatural Mechanics. Mortal fallen beings like us aren't given direct access to these things. God resides within Syntropy. God is Syntropy. Whereas, we mortal fallen beings are stuck within or placed within entropy and physical matter. We age and die. When we die, our psyche or soul is released back into Syntropy.

God imposes entropy, the aging process, or the passage of time onto physical matter in order to keep the physical matter within your physical body from teleporting or quantum tunneling away from you at will. God keeps the physical matter within your body, house, computer, and car from dissolving into thin air by quantum tunneling to the other side of the universe at will. Physical matter still retains that capability; but, God prevents physical matter from quantum tunneling by making it subject to entropy or the passage of time. God deliberately slows the physical matter down to sub-light speeds so that we can live it, experience it, observe it, and remember having done so.

The explanatory power of this Science is through the roof. By choosing to allow Psyche, Syntropy, and Quantum Mechanics in to play, we can literally explain everything that comes our way.

Mark My Words

—

Quantum Mechanics Is Mind Over Matter

I used to be a Physicalist, Naturalist, Nihilist, and Atheist.

One of my greatest scientific discoveries came when I first realized that Quantum Mechanics is in fact Spiritual Mechanics, or the way that spirit matter really works. When I first realized that Psyche or a Non-Local Conscious Observer and the Word of Command are needed to convert spirit matter into physical matter, then suddenly the whole of Quantum Mechanics became clear to me.

God must of necessity exist in order to have provided the necessary Word of Command or Conscious Observation, which Quantum Mechanics tells us must take place in order to convert spirit matter into physical matter. The Bible actually tells us that Jesus Christ is the WORD or the Word of Command who brought physical matter into existence in the first place. If God's Psyche didn't exist, then there would be NO physical matter.

For me personally, Quantum Mechanics became my first positive Scientific Verification of God's Existence. The scientists keep verifying Quantum Mechanics or Spiritual Mechanics over and over again; so, it must be the truth, because it's impossible to use the scientific methods to falsify the truth. Quantum Mechanics or Spiritual Mechanics has never been falsified, although the materialistic and naturalistic interpretations of Quantum Mechanics have been falsified trillions of times.

After realizing that God's Psyche must of necessity exist in order to convert spirit matter into physical matter, I simply KNEW that God exists.

Here's a list of some of the books and articles that confirmed or verified these scientific observations for me.

See: Van Lommel, P. (2010). *Consciousness Beyond Life: The Science of the Near-Death Experience*. New York: HarperCollins.

http://avalonlibrary.net/ebooks/Pim%20van%20Lommel%20-%20Consciousness%20Beyond%20Life.pdf

https://science-2-0.com/wp-content/uploads/2018/05/Consciousness-Beyond-Life.pdf

https://quantum-neuroscience.com/wp-content/uploads/2018/05/Consciousness-Beyond-Life.pdf

See: Goswami, A. (2008). *God Is Not Dead: What Quantum Physics Tells Us about Our Origins and How We Should Live*. Charlottesville, VA: Hampton Roads.

See also: Mark My Words. (2016). *Quantum Mechanics from a Non-Physical Spiritual Perspective*. Kindle. Retrieve from:

https://www.amazon.com/dp/B01J023TGU

The existence of Light falsifies Materialism and Naturalism, and it ends up pointing us directly to God and God's Psyche.

See: Mark My Words. (2017). *God Is in the Light: God is light, and in Him is no darkness at all*. Retrieve from:

https://www.amazon.com/dp/B07168S37N

Quantum Mechanics is Supernatural Mechanics or Supernatural Mechanisms. Quantum Mechanics or Transdimensional Physics IS Action at a Distance; and, Action at a Distance falsifies the claims of Classical Physics or Naturalism which claim that Action at a Distance is impossible and does not exist.

I find it fascinating to observe that the Real Science – the experienced and verified observations of the human race – repeatedly and steadily FALSIFY the claims of Materialism, Naturalism, Darwinism, Nihilism, and Atheism. EVERYTHING that we have experienced and observed as a race FALSIFIES Materialism, Naturalism, and their derivatives; whereas, there is NO evidence and can be NO evidence supporting the hidden assumptions or major premises of Materialism and Naturalism which claim that the quantum, or the supernatural, or action at a distance DOES NOT EXIST.

Every time we observe, verify, and catch Quantum Mechanisms or Action at a Distance in the act, those scientific observations and experiences FALSIFY Materialism, Naturalism, Nihilism, and their derivatives such as Darwinism and Atheism.

If you successfully eliminate everything that is false, then ONLY the truth will remain. If you successfully eliminate Materialism, Naturalism, Darwinism, Nihilism, Atheism, and their derivatives, then ONLY the truth remains.

Physical matter or entropy cannot explain everything that comes our way. Classical Physics, Physicalism, or Naturalism cannot explain everything that we encounter. Materialism, Naturalism, Darwinism, Nihilism, Behaviorism, Determinism, Physical Reductionism, Atheism, and Classical Physics LACK explanatory power when it comes to Action at a Distance, Psyche, Syntropy, and Quantum Mechanics.

Here are some of the articles and books that convinced me that the Materialists and Naturalists are wrong in their assumptions.

Is Your Brain Really Necessary?

https://quantum-neuroscience.com/wp-content/uploads/2018/05/Is-Your-Brain-Really-Necessary.pdf

A brain really isn't needed for thoughts and memories to take place. Thoughts and memories are quantum waves. We KNOW this is so because it is physically impossible to record our thoughts and memories with our physical devices or physical machines. Thoughts, memories, or quantum waves are NOT a product or epiphenom of physical matter.

Mind Time

https://zodml.org/sites/default/files/%5BBenjamin_Libet%2C_Profess or_Stephen_M._Kosslyn%5D_Min.pdf

https://quantum-neuroscience.com/wp-content/uploads/2018/05/Mind-Time.pdf

https://science-2-0.com/wp-content/uploads/2018/05/Mind-Time.pdf

We continue to think, learn, and remember long after our physical brain is dead and gone according to the scientific evidence that has been obtained from Near-Death Experiences (NDEs), Shared-Death Experiences (SDEs), Out-of-Body Experiences (OBEs), and our after-death Life Reviews. Thoughts and memories are a product of the Human Psyche.

The Mind and the Brain: Neuroplasticity and the Power of Mental Force - Jeffrey M. Schwartz, Sharon Begley (2003):

http://publicism.info/psychology/mind/

https://science-2-0.com/wp-content/uploads/2018/05/The-Mind-and-the-Brain.zip

https://quantum-neuroscience.com/wp-content/uploads/2018/05/The-Mind-and-the-Brain.zip

Quantum Mechanics NEVER made any sense to me while I was a Materialist, Naturalist, Nihilist, and Atheist. Eventually I learned that Quantum Mechanics or Action at a Distance is a non-local, non-physical, immaterial, spiritual, transdimensional phenomenon.

When I finally realized that Quantum Mechanics or Supernatural Mechanics is in fact how spirit matter and psyche work at the quantum level or psyche level, then suddenly it became clear to me what Spiritual Mechanics or Quantum Mechanics is and how it works. Quantum Mechanics only makes sense from a Spiritual Perspective. Classical Physics, Materialism, Naturalism, Darwinism, Nihilism, Behaviorism, Determinism, Physical Reductionism, and Atheism CANNOT explain how Quantum Mechanics or Syntropy works, nor can they explain what Quantum Mechanics or Supernatural Mechanics is.

The Spiritual Brain

https://epdf.tips/the-spiritual-brain-a-neuroscientists-case-for-the-existence-of-the-soul4a910f6e71b0354edde057767fea620d82703.html

https://science-2-0.com/wp-content/uploads/2018/05/The-Spiritual-Brain.pdf

https://quantum-neuroscience.com/wp-content/uploads/2018/05/The-Spiritual-Brain.pdf

These are a few books that you must read and understand if you have a desire to move your Science into the Quantum Age.

These are some of the Science Books that the Materialists, Naturalists, Nihilists, and Atheists are trying to convince you do not exist. They don't want you to find this material because it FALSIFIES Materialism, Naturalism, Nihilism, Atheism, and even Darwinism or the Theory of Evolution.

Statistics seem to suggest that the majority of scientists and philosophers have chosen to believe in and promote Scientific Naturalism, Physicalism, Darwinism, Nihilism, Atheism, and Classical Physics. NO amount of evidence will ever convince them that they are wrong until they are actually willing to look at the evidence and accept it as being real and true.

But, we don't need to convince the scientists that they are wrong. We – the public at-large – can skip right past them and figure out for ourselves why they are wrong and how they are wrong. That's what I eventually chose to do. So long as you have found the truth, it doesn't really matter what the scientists choose to believe; now does it? All they really need is classical physics and the physical sciences to do their work of saving world – or destroying the world. God holds ALL the KEYS to Quantum Mechanics in any case; and, God doesn't give us mere mortals direct access nor direct control over Quantum Mechanics anyway. There's nothing that the scientists can do to get at Quantum Mechanics or Supernatural Mechanics and use that Science against us or against God. Most of them don't need Quantum Mechanics to do their work; so, just let them remain in the dark if they want to.

The beauty of Psyche, Syntropy, and Quantum Mechanics is that you can literally explain everything that happens to you and everything that comes your way once you choose to allow them into play.

Mark My Words

—

The Self Does Not Die

The Self or the Psyche does not die. It can't die.

Why?

This reality and truth just might be one of the most important scientific discoveries that a person can make during his or her scientific career.

The theory of relativity teaches us that at the speed of light or faster than the speed of light TIME STOPS. There is NO passage of time at the speed of light or faster than the speed of light.

Do you know what that means?

Well, being a Materialist, Naturalist, and Classical Physicist, Albert Einstein erroneously concluded that nothing exists faster than the speed of light. He was wrong!

Tachyons or spirit matter exist at frequencies faster than the speed of light.

In a book entitled *Physics of the Impossible* by Michio Kaku, we find the following description for Tachyons.

> **Tachyons live in a strange world where everything travels faster than light. As tachyons lose energy [mass], they travel faster, which violates common sense. In fact, if they lose all energy [or mass], they travel at infinite velocity. As tachyons gain energy [or mass], however, they slow down until they reach the speed of light. (p. 280).**

That's the exact SAME definition for Spirit Matter that I have been giving in all of my books for the past couple of years.

Tachyons ARE Spirit Matter!

Tachyons make perfect sense to me, because I'm no longer a Materialist, Naturalist, Nihilist, and Atheist. At velocities slower than the speed of light, objects take on mass or energy, which slows them down. All of that energy is needed to create the space between the nucleus of the atom and its electrons. Some type of force or field is creating that space-time between the nucleus and the electrons. It has been observed that one has to infuse a lot of energy into an atom in order to raise the electrons to a higher state or a higher orbit. The net result of this slow-down process is that these objects also take on entropy and are forced to experience the passage of time or the aging process. They become subject to space-time. In contrast, if objects lose ALL of their mass, then they are capable of infinite velocities and they experience NO passage of time or NO entropy! Entropy is a function of time.

There's the truth of the matter!

There's the answer to life, the universe, and everything. It's been there all along just waiting for someone to see it, discover it, and declare it.

A photon traveling at the speed of light experiences NO passage of time. From its perspective, it arrives at its destination instantly the very moment it launches. From our perspective, existing at velocities much slower than the speed of light, it took that photon 13.2 billion years to reach us. From its perspective, it took NO time at all.

At the speed of light, TIME STOPS. There is NO passage of time. There is NO entropy. At the speed of light or faster, everything is SYNTROPY, meaning that everything is eternal and everlasting.

The Psyche, Self, Mind, Soul, Intelligence, or Quantum Non-Local Consciousness exists at frequencies faster than the speed of light, which means that it is SYNTROPY. It is eternal and everlasting, without a beginning of days or an end of years.

This just might be one of my most significant and most useful scientific discoveries of all time. It fits and makes sense to me. It's predicted by the theory of relativity.

Entropy is death.

Why is entropy death?

It's because entropy involves the passage of time. Things age and die at velocities slower than the speed of light.

In contrast, SYNTROPY is eternal life. The Psyche is eternal and everlasting because it experiences NO passage of time. It doesn't age. It has NO entropy. That pretty much explains life, the universe, and everything – now doesn't it?

What's the answer to life, the universe, and everything?

Well, it's not 42.

The answer to life the universe, and everything is SYNTROPY or PSYCHE or INTELLIGENCE. The question is, "What is eternal and everlasting, without a beginning of days or an end of years?" The answer is Psyche or Syntropy. SYNTROPY is the fundamental basis for life, the universe, and everything!

Quantum Mechanics is a type of Syntropy. It's eternal and everlasting.

These realities and truths have been experienced and observed. Science IS observation and experience. Science or observation is an infinitely better way for getting at the truth than just to say that psyche or syntropy does not exist, as the Materialists and Naturalists do. Saying that psyche or syntropy DOES NOT EXIST is a cop out. It's the way that these people avoid doing the science and the observations that need to be done in order to establish the truth of the matter. It's an evasion or a dodge, NOT science. These people are avoiding the truth or hiding from the truth rather than seeking for the truth. Scientists are supposed to search for and find the truth; but, these people refuse to do so.

The Self Does Not Die: Verified Paranormal Phenomena from Near-Death Experiences

https://www.amazon.com/dp/0997560800

https://www.amazon.com/dp/B01LYMHXCO

Physicalism, Naturalism, Darwinism, Nihilism, and Atheism are based upon a refusal to look at evidence. I KNOW because I used to be a Materialist, Naturalist, Nihilist, and Atheist. We Naturalists and Atheists have to castrate and lobotomize science in order to prevent people from falsifying our theories and our ideas because ALL of the scientific evidence or observational evidence that we have on hand as a race falsifies the claims of Materialism, Naturalism, Darwinism, Nihilism, and Atheism which state that such falsifying evidence does not exist.

Science itself convinced me that Materialism, Naturalism, Darwinism, Nihilism, and Atheism are FALSE. I believe the Science – the observations and the experiences of real people – and not the philosophical wishful thinking of the Materialists, Naturalists, Darwinists, Nihilists, and Atheists.

Mark My Words

—

Orthodox Interpretation of Quantum Mechanics

The Orthodox Interpretation of Quantum Mechanics by Henry P. Stapp was the FIND of a lifetime. Henry P. Stapp's work is important to know about and understand – everything else pales in comparison. This is the information and the scientific evidence that the Materialists, Naturalists, Darwinists, Nihilists, and Atheists are trying to hide from you; and, they are very good at what they do. I KNOW because I used to be a Materialist, Naturalist, Nihilist, and Atheist. These people are actively banning, blocking, and destroying this information about Syntropy and Quantum Mechanics because it falsifies Physicalism, Naturalism, Darwinism, and Nihilism.

Remember, Materialism, Naturalism, Darwinism, Nihilism, Behaviorism, Determinism, Physical Reductionism, and Atheism are based exclusively on Classical Physics and Entropy. Scientific Naturalism or Physicalism IS Classical Physics. Quantum Mechanics falsifies Classical Physics, or Materialism and Naturalism.

Stapp seems to be giving everything away for free on his website.

https://sites.google.com/a/lbl.gov/stappfiles/

http://www-physics.lbl.gov/~stapp/

He seems to be much more interested in getting his message out to the world than he is in maintaining the copyright for his books.

The Orthodox Interpretation explains what the Human Psyche and Nature's Psyche are doing at the quantum level in order to get things done for us at the physical level.

If your interpretation of Quantum Mechanics can't explain what the Human Psyche and Nature's Psyche are doing at the quantum level in order to get things done for you at the physical level, then your interpretation of Quantum Mechanics is absolutely worthless.

Quantum Mechanics makes absolutely NO sense whatsoever whenever it is reduced to Classical Physics, Materialism, Naturalism, Physicalism, and Nihilism. Most of the interpretations of Quantum Mechanics found online and within our

484

college textbooks have indeed been reduced to Materialism, Naturalism, and Classical Physics; and thereby, they have completely lost their explanatory power and their scientific value. Materialism and Naturalism neuter and destroy everything that comes their way, including Quantum Mechanics.

To demonstrate some of these truths, first I will quote from:

NEUROSCIENCE, ATOMIC PHYSICS, AND THE HUMAN PERSON

http://www-physics.lbl.gov/~stapp/2nd.doc

https://quantum-neuroscience.com/wp-content/uploads/2018/05/Neuroscience-and-the-Human-Person.pdf

http://www-physics.lbl.gov/~stapp/phys.txt

The Quantum Approach.

Classical physics is *an approximation* to a more accurate theory - called quantum mechanics - and quantum mechanics makes mind efficacious. Quantum mechanics *explains* the causal effects of mental intentions upon physical systems: it *explains* how your mental effort can produce the brain events that cause your bodily actions. Thus, quantum theory converts science's picture of you from that of a mechanical automaton to that of a mindful human person. Quantum theory also shows, explicitly, how the approximation that reduces quantum theory to classical physics completely eliminates all effects of your conscious thoughts upon your brain and body. Hence, from a physics point of view, trying to understand the mind-brain connection by going to the classical approximation is absurd: it amounts to trying to understand something in an approximation that eliminates the effect you are trying to study.

Quantum mechanics arose during the twentieth century. Scientists discovered, empirically, that the principles of classical physics were not correct. Moreover, they were wrong in ways that no minor tinkering could ever fix. The *basic principles* of classical physics were thus replaced by *new basic principles* that account uniformly both for all the successes of the older classical theory and also for all the newer data that is incompatible with the classical principles.

Physical theory was turned inside out.

The most profound alteration of the fundamental principles was to bring the consciousness of human beings into the basic structure of the physical theory. In fact, the whole *conception of what science is* was turned inside out. The core idea of classical physics was to describe the "world out there," with no reference to "our thoughts in here." But the core idea of quantum mechanics is to describe *our activities as knowledge-seeking and knowledge-using agents*. Thus, quantum theory involves, basically, not just what is "out there," but also what is "in here," namely "our knowledge." Consciousness is thus introduced into contemporary orthodox physical theory, not as something *whose existence needs to be explained,* but as rather as something whose detailed structure and detailed connection to brain activities needs to be *further* explicated.

Science must bridge the psycho-physical divide.

The basic philosophical shift in quantum theory is the *explicit* recognition that science is about *what we can know*. It is fine to have a beautiful and elegant mathematical theory about an imagined *"really existing physical world out there"* that meets a lot of intellectually satisfying criteria. But the essential demand of science is that the theoretical constructs be tied to the experiences of the human scientists who devise ways of testing the theory, and of the human engineers and technicians who both participate in these tests, and eventually put the theory to work. So, the structure of a proper physical theory must involve not only the part describing the behavior of the not-directly-experienced theoretically postulated entities, expressed in some appropriate symbolic language, but also a part describing the human experiences that are involved in these tests and applications, expressed in the language that we actually use to describe such experiences to ourselves and each other. Finally, we need some "bridge laws" that specify the connection between the concepts described in these two different languages.

Classical physics met these requirements in a rather trivial kind of way, with the relevant experiences of the human participants being taken to be direct apprehensions of various gross behaviors of large-scale properties of big objects composed of huge numbers of the tiny atomic-scale parts. And these apprehensions were taken to be *passive*: they had no effect on the behaviors of the systems being studied. But the physicists who were examining the behaviors of systems that depend sensitively upon the behaviors of their tiny atomic-scale components found themselves forced to go to a less trivial theoretical arrangement, in which the human agents were no longer passive observers, but were *active participants* in ways that contradicted, and were impossible to comprehend within, the general framework of classical physics, *even when the only features of the physically described world that the human beings observed were large-scale properties of measuring devices.*

The two-way quantum psycho-physical bridge.

The sensitivity of the behavior of the devices to the behavior of some tiny atomic-scale particles propagates in such a way that the acts of observation by the human observers of *large-scale properties of the devices* could no longer be regarded as passive: these acts were assigned a crucial *selective* action. Thus, the core structure of the basic general physical theory became transformed in a profound way: the connection between physical behavior and human knowledge was changed from a one-way bridge to a mathematically specified two-way interaction that involves *selections* performed by conscious minds.

This profound change in the principles is encapsulated in Niels Bohr dictum that "in the great drama of existence we ourselves are both actors and spectators." (Bohr, 1963: 15 & 1958: 81) The emphasis here is on "actors": in classical physics we, and in particular our minds, were mere spectators.

This revision must be expected to have important ramifications in neuroscience, because the issue of the connection between mind (the psychologically described aspects of a human being) and brain/body (the physically described aspects of that person) has recently become a matter of central concern in neuroscience.

The Copenhagen formulation.

The original formulation of quantum theory was created mainly at an Institute in Copenhagen directed by Niels Bohr and is called "The Copenhagen Interpretation." Due to the profound strangeness of the conception of nature entailed by the new mathematics, the Copenhagen strategy was to refrain from making ordinary ontological claims, but to take, instead, a fundamentally pragmatic stance. Thus, the theory was formulated *basically* as a set of practical rules for how scientists should go about their tasks of acquiring knowledge, and then using this knowledge in practical ways. Speculations about "what the world out there – apart from our knowledge of it - is really like" were regarded as "metaphysics," and hence outside *real* science.

Copenhagen quantum theory is about the relationships between human agents (called *participants* by John Wheeler) and the systems that they act upon. In order to achieve this conceptualization, the Copenhagen formulation separates the physical universe into two parts, which are described in two different languages. One part is the observing human agent and his measuring devices. That part is described in mental terms - in terms of our instructions to colleagues about how to set up the devices, and our reports of what we then learn. The other part of nature is *the system that the agent is acting upon*. That part is described in physical terms - in terms of mathematical properties assigned to tiny space-time regions.

Von Neumann's Process II.

The great mathematician and logician John von Neumann formulated Copenhagen quantum theory in a rigorous way.

Von Neumann identified two very different processes that enter into the quantum theoretical description of the evolution of a physical system. He called them Process I and Process II (Von Neumann, 1955: 418). Process II is the analog in quantum theory of the process in classical physics that takes the state of a system at one time to its state at a later time. This Process II, like its classical analog, is *local* and *deterministic*. However, Process II by

487

itself is not the whole story: it generates physical worlds that do not agree with human experiences. For example, if Process II were the *only* process in nature then the quantum state of the moon would represent a structure smeared out over large part of the sky.

Process I: A dynamical psycho-physical bridge.

To tie the quantum mathematics to human experience in a rationally coherent and mathematically specified way quantum theory introduces *another process*, which Von Neumann calls Process I. It is a *selection* process that is tied to conscious experience, and it is not determined by the micro-local deterministic Process II. It is a selection made by an agent about how he or she will act or attend.

Any physical theory must, in order to be complete, specify how the elements of the theory are connected to human experience. In classical physics this connection is part of a *metaphysical* superstructure: it is not part of the core dynamical description. But in quantum theory this connection of the mathematically described physical state to conscious experiences is part of the essential dynamical structure. And this connecting process is not passive: it does not represent a mere *witnessing* of a physical feature of nature by a passive mind. Rather, the process is active: it injects into the physical state of the system being acted upon properties that depend upon the intention of the observing agent.

Quantum theory is built upon the practical concept of intentional actions by agents. Each such action is expected or intended to produce an experiential response or feedback. For example, a scientist might act to place a Geiger counter near a radioactive source and expect to see the counter either "fire" during a certain time interval or not "fire" during that interval. The experienced response, "Yes" or "No", to the question "Does the counter fire during the specified interval?" specifies one bit of information. Quantum theory is thus an information-based theory built upon the knowledge-acquiring actions of agents, and the knowledge that these agents thereby acquire.

Probing actions of this kind are performed not only by scientists. Every healthy and alert infant is engaged in making willful efforts that produce experiential feedbacks, and he or she soon begins to form expectations about what sorts of feedbacks are likely to follow from some particular kind of effort. Thus, both empirical science and normal human life are based on paired realities of this action-response kind, and our physical and psychological theories are both basically attempts to understand these linked realities within a rational conceptual framework.

The basic building blocks of quantum theory are, then, a set of intentional actions by agents, and for each such action an associated collection of possible "Yes" feedbacks, which are the possible responses that the agent can judge to be in conformity to the criteria associated with that intentional act. For example, the agent is assumed to be able to make the

488

judgment "Yes" the Geiger counter clicked or "No" the Geiger counter did not click. And he must be able to report. "Yes" the counter is in the specified place, or "No" it is not there. Science would be difficult to pursue if scientists could make no such judgments about what they were experiencing.

All known physical theories involve idealizations of one kind or another. In quantum theory the main idealization is not that every object is made up of miniature planet-like objects. It is rather that there are agents that perform intentional acts each of which can result in a feedback that may conform to a certain criterion associated with that act. One bit of information is introduced into the world in which that agent lives, according to whether the feedback conforms or does not conform to that criterion. Thus, knowing whether the counter clicked or not places the agent on one or the other of two alternative possible separate branches of the course of world history.

These remarks reveal the enormous difference between classical physics and quantum physics. In classical physics the elemental ingredients are tiny invisible bits of matter that are idealized miniaturized versions of the planets that we see in the heavens, and that move in ways unaffected by our consciousness, whereas in quantum physics the elemental ingredients are intentional actions by agents, the feedbacks arising from these actions, and the effects of our actions on the physical systems that our actions act upon.

Consideration of the character of these differences makes it plausible that quantum theory may be able to provide the foundation of a scientific theory of the human person that is better able than classical physics to integrate the physical and psychological aspects of his nature. For quantum theory describes the effects of a person's intentional actions upon the physical world, whereas classical physics systematically leaves these effects out.

An intentional action by a human agent is partly an intention, described in psychological terms, and partly a physical action, described in physical terms. The feedback also is partly psychological and partly physical. In quantum theory these diverse aspects are all represented by logically connected elements in the mathematical structure that emerged from the seminal discovery of Heisenberg. That discovery was that in order to get the quantum generalization of a classical theory one must formulate the theory in terms of *actions.* A key difference between *numbers* and *actions* is that if A and B are two actions then AB represents the action obtained by performing the action A upon the action B. If A and B are actions then, generally, AB is different from BA: the order in which actions are performed matters.

The intentional actions of agents are represented mathematically in Heisenberg's space of actions. Here is how it works.

Each intentional action depends, of course, on the intention of the agent, and upon the state of the system upon which this action acts. Each of these two aspects of nature is represented within Heisenberg's space of actions by an action.

The idea that a "state" should be represented by an "action" may sound odd, but Heisenberg's key idea was to replace what classical physics took to be a "being" by a "doing." I shall denote the action that represents the state being acted upon by the symbol S.

An intentional act is an action that is intended to produce a feedback of a certain conceived or imagined kind. Of course, no intentional act is sure-fire: one's intentions may not be fulfilled. Hence the intentional action puts in play a process that will lead either to a confirmatory feedback "Yes," the intention is realized, or to the result "No", the "Yes" response failed to occur.

The effect of this intentional mental act is represented mathematically by an equation that is one of the key equations of quantum theory. This equation represents, within the quantum mathematics, the effect of the Process I mental action upon the quantum state S of the system being acted upon. The equation is:

$$S \rightarrow S' = PSP + (1-P)S(1-P).$$

This formula exhibits the important fact that this Process I action changes the state S of the system being acted upon into a new state S', which is a *sum* of two parts.

The first part, PSP, represents the possibility in which the experiential feedback called "Yes" appears, and the second part, $(1-P)S(1-P)$, represents the alternative possibility "No", this feedback does not appear. Thus, the intention of the action and the associated experiential feedback are tied into the mathematics that describes the dynamics of the physical system being acted upon.

The action P is important. It represents an action upon the system that is being acted upon by the agent, and it depends on *the intention of the agent*. The action represented by the symbol P, acting both on the right and on the left of S, is the action of eliminating from the state S all parts of S except the "Yes" part. That particular retained part is determined by the intentional choice of the agent. The action of $(1-P)$, acting both on the right and on the left of S, is, analogously, to eliminate from S all parts of S except the "No" parts.

The projection operator P is required to satisfy $P = PP$. This implies $P(1-P) = (1-P)P = 0$, which says that the sequence of these two actions, P and $(1-P)$, in either order, leave nothing.

Thus, the action P is an action in the space in which the physical system is represented, and it reduces to zero all components that correspond to the "No" response, but it leaves intact the components corresponding to the "Yes" response to the intentional action. The action of $(1-P)$ is the analogous action with "Yes" and "No" interchanged. The action of P is the representation of an intentional mental action upon a physically described system.

Notice that Process I produces the *sum* of the two alternative possible feedbacks, not just one or the other. Since the feedback must either be "Yes" or "No = Not-Yes," one might think that Process I, which *keeps* both the "Yes" and the "No" parts, would do nothing. But that is not correct! This is a key point. It can be verified by noticing that S can be written as a sum of four parts, only two of which survive the Process I action:

$$S = PSP + (1-P)S(1-P) + PS(1-P) + (1-P)SP.$$

This formula is a strict identity. The dedicated reader can easily confirm it by collecting the contributions of the four occurring terms PSP, PS, SP, and S, and verifying that all terms but S cancel out. This identity shows that the state S can be expressed as a sum of four parts, *two of which are eliminated by Process I.*

But this means that Process I has a *nontrivial effect* upon the state being acted upon: it eliminates the two terms that correspond neither to the appearance of a "Yes" feedback nor to the failure of the "Yes" feedback to appear.

That is the *first key point*: quantum theory has a specific dynamical process, Process I, which specifies the effect upon a physically described system of an *intentional act* by a conscious agent.

Free Choices.

The second key point is this: the agent's choices are "free choices," *in the specific sense specified below.*

Orthodox quantum theory is formulated in a realistic and practical way. It is structured around the activities of human agents, who are considered able to freely elect to probe nature in any one of many possible ways. Bohr emphasized the freedom of the experimenters in passages such as:

"The freedom of experimentation, presupposed in classical physics, is of course retained and corresponds to the free choice of experimental arrangement for which the mathematical structure of the quantum mechanical formalism offers the appropriate latitude." (Bohr, 1958: 73)

This freedom of action stems from the fact that in the original Copenhagen formulation of quantum theory the human experimenter is considered to stand outside the system to which the quantum laws are applied. Those quantum laws are the only precise laws of nature recognized by that theory. Thus, according to the Copenhagen philosophy, *there are no presently known laws that govern the choices* made by the agent/experimenter/observer/participant about how the observed system is to be probed. This choice is, *in this very specific sense*, a "free choice." It is not ruled out that some deeper theory will eventually provide a causal explanation of this "choice."

Probabilities.

The predictions of quantum theory are generally statistical: only the *probabilities* that the agent will experience each of the alternative possible feedbacks are specified. Which of these alternative possible feedbacks will actually occur in response to a Process I action is not determined by quantum theory.

The formula for the probability that the agent will experience the feedback 'Yes' is Tr PSP/Tr S, where the symbol Tr represents the trace operation. This trace operation means that the actions act in a cyclic fashion, so that the rightmost action acts back around upon the leftmost action. Thus, for example, Tr ABC=Tr CAB =Tr BCA. The product ABC represents the result of letting A act upon B, and then letting that product AB act upon C. But what does C act upon? Taking the trace of ABC means specifying that C acts back around on A.

An important property of a trace is that the trace of any of the sequences of actions that we consider must always give a positive number or zero. Thus, this trace operation is what ties the actions, as represented in the mathematics, to measurable numbers.

[The trace operation, and in fact the operation of multiplying together any two operators, is the quantum analog of the classical process of integrating over all of "phase space," giving equal a prior weighting to

equal volumes of phase space. Thus, the trace operation is in effect a statistical sum over all of the "loose ends" that are not fixed in the expression upon which the trace operation acts.]

Von Neumann's psycho-physical theory of the conscious brain.

The Copenhagen approach separates the world into two parts: "The Observer" which includes the mind, brain, and body of the personal observer together with his measuring devices; and "The System" that this observer is acting upon. "The Observer" is described in psychological terms, whereas "The System" is described in physical/mathematical space-time terms.

This procedure works very well in practice. However, it seems apparent that the body and brain of the human agent, and his devices, are parts of the physical universe. Hence a complete theory ought to be able to include our bodies and brains in the physically described part of the theory. On the other hand, the structure of the theory depends critically also upon the features that are represented in Process I, and that are described in mentalistic language as intentional actions and experiential feedbacks.

Von Neumann showed that it was possible, without significantly disturbing the predictions of the theory, to shift the bodies and brains of the agents, along with their measuring devices, into the physical world, *while retaining. and ascribing to the mind of the agent, those mentalistically described properties of the agents that are essential to the structure of the theory.* The system acted upon by the mind is the brain. Thus, in this von Neumann re-formulation the Process I action is an action of mind upon brain. Hence von Neumann's re-formulation provides us with the core of a science-based dynamical theory of the conscious brain.

It is worthwhile to reflect for a moment on the ontological aspects of Von Neumann quantum theory. Von Neumann himself, being a clear-thinking mathematician, said very little about ontology. But he called the mentalistically described aspect of the agent "his abstract 'ego'." (von Neumann, 1955: 421). This phrasing tends to conjure up the idea of a disembodied entity, standing somehow apart from the body/brain. But another possibility is that consciousness is an *emergent property* of the body-brain. Notice that some of the problems that occur in trying to defend the idea of emergence within the framework of classical physical theory disappear when one accepts the validity of quantum theory. For one thing, one no longer has to defend against the charge that the emergent property, consciousness, has no "genuine" causal efficacy, because anything it does is done already by the physically described process, independently of whether the psychologically described aspect emerges of not. In quantum theory the causal efficacy of our thoughts is no illusion: it's the real thing!

Another difficulty with "emergence" in a classical physics context is in understanding how the motion of a set of miniature planet-like objects, careening through space, can *be a painful experience.* But within the quantum framework the basic physical structure, namely the quantum state, is essentially knowledge or information imbedded in space-time. Hence there is no intrinsic problem with the idea that a sudden increment in a person's knowledge should be represented by a sudden jump in the quantum state of his brain. The identification of conscious actions with physical actions is no longer problematic. This is because the old idea of "matter" has been eradicated and replaced by a mathematical representation of an information-based psycho-physical reality.

In this connection, Heisenberg remarked:

"The conception of the objective reality of the elementary particles has thus evaporated not into the cloud of some obscure new reality concept, but into the transparent clarity of a mathematics that represents no longer the behavior of the particle but rather our knowledge of this behavior." (Heisenberg, 1958).

Conservation of Causality.

The question arises: How can the effect of a psychologically described action be injected into the dynamics of a physically described system without upsetting the causal structure of the latter.

The answer is this: Physicists have discovered an important and unexpected property of nature. It pertains to observable phenomena that depend upon microscopic properties that are *in principle inaccessible to*

493

observation. In such a situation we are *in principle* unable, due to the lack of crucial micro-data, to give a complete causal description of the observable phenomena. However, our principled inability to give a complete causal account of the psychologically described phenomena, due to this inherent gap in the micro-data, can be partially offset by introducing into the theory, *instead of the inaccessible micro-data,* the *psychologically described selection of an action* made upon the system by an agent.

Thus, the loss of causal determination at the microlevel, due to the limitations imposed by Heisenberg's uncertainty principle, allows an alternative (statistical) causal account to be achieved by replacing the inaccessible micro-data by empirically available and controllable data about human selections of actions!

This feature discovered in atomic science should be equally importance in neuroscience. That is because the basic problem in neuroscience is essentially the same as the one in atomic physics. In both cases the problem is to provide a causal account of connections between experiences that depend sensitively upon micro-properties that are in principle inaccessible. But quantum theory shows how the principled loss of information at the microlevel can be partially offset by using, instead, the controllable and reportable variables of the intentional actions of human beings. Nature left open a causal gap for us to occupy.

The Quantum Brain.

The quantum state of a human brain is, of course, a very complex thing. But its main features can be understood by considering first a classical conception of the brain, and then folding in some key features that arise already in the case of the quantum state of a single particle, or object, or degree of freedom.

Source

http://www-physics.lbl.gov/~stapp/2nd.doc

https://quantum-neuroscience.com/wp-content/uploads/2018/05/Neuroscience-and-the-Human-Person.pdf

http://www-physics.lbl.gov/~stapp/phys.txt

https://docslide.com.br/download/link/stapp-mind-matter-and-quantum-mechanics

https://quantum-neuroscience.com/wp-content/uploads/2018/05/Mind-Matter-and-Quantum-Mechanics.pdf

https://science-2-0.com/wp-content/uploads/2018/05/Mind-Matter-and-Quantum-Mechanics.pdf

Stapp, H. P. (2009). *MIND, MATTER, AND QUANTUM MECHANICS* (3rd ed.). Berlin: Springer.

Process 1 consists of the choices that are made by the Human Psyche.

Process 2 happens when Nature's Psyche collapses the necessary wave functions and/or fires the specific neurons needed to make the Human Psyche's choices a physical reality.

Orthodox Quantum Mechanics explains what the Human Psyche and Nature's Psyche are doing at the quantum level in order to get things done for us at the physical level. This is good stuff! It's thousands of years ahead of anything that we have gotten from the Materialists, Naturalists, Darwinists, Nihilists, and Atheists.

If your interpretation of Quantum Mechanics can't explain what Nature's Psyche and the Human Psyche are doing at the quantum level or the psyche level, then your interpretation of Quantum Mechanics is absolutely worthless when it comes time to figure out how Quantum Mechanics or Action at a Distance really works.

Quantum Mechanics explains how your Psyche, Intelligence, Spirit, Soul, Spark, Mind, or Quantum Non-Local Consciousness controls your physical body and your physical brain. That's the purpose of Quantum Mechanics. In other words, explaining Psyche or Non-Local Consciousness scientifically is what Quantum Mechanics is good for. Quantum Mechanics or Supernatural Mechanics completes science, finishes science, and perfects science. Science is incomplete without Psyche or Syntropy.

For me personally, this is one of the most important, most significant, and most powerful scientific discoveries of my science career. It's essential. It's foundational. It changes everything!

If your interpretation of Quantum Mechanics can't explain how the Human Psyche is controlling its physical brain and physical body, then your interpretation of Quantum Mechanics is worthless. That is what I have experienced and observed.

The Human Psyche makes the decisions and the choices. Nature's Psyche collapses the wave functions and/or turns on the specific neurons needed to make the Human Psyche's choices a physical reality. We KNOW that it must be Nature's Psyche (or God's Psyche) who is collapsing the wave functions and firing the neurons because the Human Psyche isn't consciously aware of any of these quantum mechanical processes.

By choosing to allow Psyche, Syntropy, and Quantum Mechanics in to play, we can literally explain everything that comes our way. That is what I have experienced and observed.

I used to be a Materialist, Naturalist, Nihilist, and Atheist. Quantum Mechanics never made any sense to me until after I got rid of My Materialism, My Naturalism, My Nihilism, and My Atheism. Materialism, Naturalism, Darwinism,

Nihilism, and Atheism claim that Psyche, Syntropy, and Supernatural Mechanisms do not exist; but, Quantum Mechanics IS Action at a Distance or Supernatural Mechanisms. Materialism, Naturalism, and their derivatives are FALSIFIED by the verified and proven existence of Action at a Distance or Quantum Mechanics.

Materialism, Naturalism, Darwinism, Nihilism, Behaviorism, Determinism, Physical Reductionism, Atheism, Entropy, and Classical Physics CANNOT EXPLAIN how the Human Psyche and Nature's Psyche are interacting with each other at the quantum level in order to get things done for us at the physical level. Therefore, these "sciences" are worthless, incomplete, and completely lack explanatory power when it comes to the Human Psyche, Nature's Psyche, Quantum Mechanics, Action at a Distance, Invisible Non-Physical Forces and Fields, and Supernatural Mechanisms.

Materialism and Naturalism have NO EXPLANATION for invisible non-physical forces and fields such as Psyche, Gravity, Magnetism, Nuclear Forces, Radio Waves, Microwaves, X-Rays, Dark Energy, Quantum Waves, Thoughts, and our after-death Memories which have been experienced, observed, and caught in the act.

It's time for the world at-large to upgrade science to Quantum Mechanics or Supernatural Mechanics. Once you choose to allow Psyche, Syntropy, and Quantum Mechanics in to play, instantly you can explain everything that comes your way.

Mark My Words

—

The Mindful Universe

The Orthodox Interpretation of Quantum Mechanics by Henry P. Stapp was the FIND of a lifetime. Henry P. Stapp's work is important to know about and understand – everything else pales in comparison. This is the information and the scientific evidence that the Materialists, Naturalists, Darwinists, Nihilists, and Atheists are trying to hide from you; and, they are very good at what they do. These people are actively banning, blocking, and destroying this information because it falsifies Physicalism, Naturalism, Darwinism, and Nihilism.

Stapp seems to be giving everything away for free on his website.

https://sites.google.com/a/lbl.gov/stappfiles/

http://www-physics.lbl.gov/~stapp/

He seems to be much more interested in getting his message out to the world than he is in maintaining the copyright for his books.

Stapp, H. P. (2011). *Mindful Universe: Quantum Mechanics and the Participating Observer*. Heidelberg: Springer.

https://epdf.tips/mindful-universe-quantum-mechanics-and-the-participating-observer-second-edition.html

https://science-2-0.com/wp-content/uploads/2018/05/Quantum-Mechanics-and-the-Participaing-Observer.pdf

https://quantum-neuroscience.com/wp-content/uploads/2018/05/Mindful-Universe.pdf

—

Snippets from the *Mindful Universe*:

http://www-physics.lbl.gov/~stapp/Key.doc

https://quantum-neuroscience.com/wp-content/uploads/2018/05/Mindful-Universe-Key.doc

http://www-physics.lbl.gov/~stapp/Book.doc

https://quantum-neuroscience.com/wp-content/uploads/2018/05/Mindful-Universe-Booklet.doc

http://www-physics.lbl.gov/~stapp/Chapt3.txt

http://www-physics.lbl.gov/~stapp/Chap4-6.txt

—

Quote from the *Mindful Universe*:

The World of Actions

Werner Heisenberg was, from a technical point of view, the principal founder of quantum theory. He discovered in 1925 the completely amazing and wholly unprecedented solution to the puzzle: the quantities that classical physical theory was based upon, and which were thought to be numbers, must be treated not as numbers but as actions! Ordinary numbers, such as 2 and 3, have the property that the product of any two of them does not depend on the order of the factors: 2 times 3 is the same as 3 times 2. But Heisenberg discovered that one could get the correct answers out of the old classical laws if one decreed that certain numbers that occur in classical physics as the magnitudes of certain physical properties of a material system are not ordinary numbers. Rather, they must be treated as *actions* having the property that the order in which they act matters!

This 'solution' may sound absurd or insane. But mathematicians had already discovered that logically consistent generalizations of ordinary mathematics exist in which numbers are replaced by 'actions' having the property that the order in which they are applied matters. The ordinary numbers that we use for everyday purposes like buying a loaf of bread or paying taxes are just a very special case from among a broad set of rationally coherent mathematical possibilities. In this simplest case, *A* times

B happens to be the same as *B* times *A*. But there is no logical reason why Nature should not exploit one of the more general cases: there is no compelling reason why our physical theories must be based exclusively on ordinary numbers rather than on actions. The theory based on Heisenberg's discovery exploits the more general logical possibility. It is called quantum mechanics, or quantum theory.

The difference between quantum mechanics and classical mechanics is specified by Planck's constant, which is a tiny number on the scale of human actions. Thus, this tweaking of laws of physics might seem to be a bit of mathematical minutia that could scarcely have any great bearing on the fundamental nature of the universe, or of our role within it. But replacing *numbers* by *actions* upsets the whole apple cart. It produced a seismic shift in our ideas about both the nature of reality, and the nature of our relationship to the reality that envelops and sustains us. The aspects of nature represented by the theory are converted from elements of *being* to elements of *doing*. The effect of this change is profound: it replaces the world of *material substances* by a world populated by *actions*, and by *potentialities* for the occurrence of the various possible observed feedbacks from these actions. Thus, this switch from 'being' to 'action' allows – and according to orthodox quantum theory demands – a draconian shift in the very subject matter of physical theory, from an imagined universe consisting of causally self-sufficient mindless matter, to a universe populated by allowed possible physical actions and possible experienced feedbacks from such actions. A purported theory of matter alone is converted into a theory of the relationship between matter and mind.

What is this momentous change introduced by Heisenberg?

In classical physics the center point of each physical object has, at each instant of time, a well-defined location, which can be specified by giving its three coordinates (*x, y, z*) relative to some coordinate system. For example, the location of (the center point of) a spider dangling in a room can be specified by letting *z* be its distance from the floor, and letting *x* and *y* be its distances from two intersecting walls. Similarly, the velocity of that dangling spider, as she drops to the floor, blown by a gust of wind, can be specified by giving the rates of change of these three coordinates (*x, y, z*). If each of these three *rates of change*, which together specify the velocity, are multiplied by the *weight* (= mass) of the spider, then one gets three numbers, say (*p, q, r*), that define the *momentum* of the spider. In classical physics one uses the set of three numbers denoted by (*x, y, z*) to represent the position of the center point of an object, and the set of three numbers labeled by (*p, q, r*) to represent the momentum of that object. These six numbers are just ordinary numbers that obey the commutative property of multiplication that we all, hopefully, learned in third grade: $x * p$ equals $p * x$, where $*$ means multiply.

The six-dimensional space of all possible values $(x, y, z; p, q, r)$ is called *phase space*: it is the space of all possible instantaneous 'states' of the particle.

Heisenberg's analysis showed that in order to make the formulas of classical physics work in general, $x * p$ must be different from $p * x$. He found that the difference between these two products must be Planck's constant. (Actually, the difference is Planck's constant divided by $2n$ and multiplied by the imaginary unit I, which is a number such that i times i is minus one.) Thus, modern quantum theory was born by recognizing, or declaring, that the symbols used in classical physical theory to represent ordinary numbers actually represent actions such that their ordering in a sequence of actions matters. The procedure of creating the mathematical structure of quantum mechanics from that of classical physics, by replacing numbers by corresponding actions, is called 'quantization'.

The idea of replacing the numbers that specify where a particle is, and how fast it is moving, by mathematical quantities that violate the simple laws of arithmetic may strike you – if this is the first you've heard about it – as a giant step in the wrong direction. You might mutter that scientists should try to make things simpler, rather than abandoning one of the things we really know for sure, namely that the order in which one multiplies factors does not matter. But against that intuition one must recognize that this change works beautifully in practice: all of the tested predictions of quantum mechanics are borne out, and these include predictions that are correct to the incredible accuracy of one part in a hundred million. There must be something very, very right about this replacement of numbers by actions.

In classical physical theory each elementary particle is asserted to have at each instant of time a definite location, defined by a set of three numbers (x, y, z), and definite momentum, defined by a set of three numbers (p, q, r). In quantum theory one generally considers systems of many particles, but insofar as one can consider one particle alone the state of that particle at any instant of time would be represented by a *cloud of pairs of numbers*, with one pair of numbers (called a complex number) assigned to each point in three-dimensional (position) space. Someone might choose to perform a phenomenologically (i.e., experimentally/experientially) described probing action on this 'particle'. In quantum mechanics each such possible probing action turns out to have an associated set of distinct *experientially distinguishable* possible outcomes. The cloud of numbers *taken as a whole* determines the probability for the appearance of each of the alternative possible outcomes of that chosen probing action. The theory thus gives specified rules for computing the probabilities for each of the distinct alternative possible empirically described feedbacks from each of the alternative possible experimental probing actions that the human experimenter might chose to perform, but no rules that specify which probing action he or she will choose.

In classical physical theory when one descends from the macroscopic world of visible objects to the microscopic world of their elementary constituents one arrives at a world containing the 'solid, massy, hard, impenetrable moveable particles' that Newton spoke of. But in quantum theory one arrives instead at clouds, or quantum smears, of numbers that *taken as a whole* have empirical meaning in terms of probabilities of alternative possible experiences.

Briefly stated, the orthodox formulation of quantum theory (see Appendix D) asserts that, in order to connect adequately the mathematically described state of a physical system to human experience, there must be an abrupt *intervention* in the otherwise smoothly evolving mathematically described state of that system.

According to the orthodox formulation, these interventions are probing actions *instigated by human agents who are able to 'freely' choose which one, from among various alternative possible probing actions, they will perform*. The physically describable effect of the chosen probing action is to separate (partition) the prior physical state of the system being probed in some particular way into a set of component parts. Each physically described part corresponds to one *perceivable* outcome from the set of distinct alternative possible perceivable outcomes of that particular probing action.

If such a probing action is performed, then *one* of its allowed perceivable feedbacks will appear in the stream of consciousness of the observer, and the mathematically described state of the probed system will then jump abruptly from the form it had prior to the intervention to the partitioned portion of that state that corresponds to the observed feedback. *This means that, according to orthodox contemporary physical theory, the 'free' choices of probing actions made by agents enter importantly into the course of the ensuing psychologically and physically described events. Here the word 'free' means, however, merely that the choice is not determined by the (currently) known laws of physics; not that the choice has no cause at all in the full psychophysical structure of reality. Presumably the choice has some cause or reason – it is unreasonable that it should simply pop out of nothing at all – but the existing theory gives no reason to believe that this cause must be determined exclusively by the physically described aspects of the psychophysically described nature alone.*

If one sets Planck's constant equal to zero in the quantum mechanical equations, then one recovers (the fundamentally incorrect) classical mechanics. Thus, classical physics is *an approximation* to quantum physics. It is the approximation in which Planck's constant, wherever it appears, is replaced by zero. In this approximation the quantum smearing does not occur – each cloud is reduced to a point – and one recovers classical physics, along with the physical determinism (the causal closure of the physical) entailed by classical physics.

In the classical approximation there is no need for, *and indeed no room for*, any effect of any probing action. The *uncertainty* – arising from the non-zero size of the quantum cloud – that in the unapproximated theory needs to be resolved by the intervention of some particular probing action is already reduced to zero by the replacement of Planck's constant by zero. Thus, all effects upon the physically/mathematically described aspects of nature's process that are instigated by the actions 'freely' chosen by agents are eliminated by the classical approximation. *Consequently*, any attempt to understand or explain within the framework of classical physics the physical effects of consciousness is irrational, *because the classical approximation eliminates the effect one is trying to study.*

Intentional Actions and Experienced Feedbacks

The concept of intentional actions by agents is of central importance. Each such action is intended to produce an experiential feedback. For example, a scientist might act to place a Geiger counter near a radioactive source, with the intention to see the counter either 'fire', or 'not fire', during a certain time interval. The experienced response, 'Yes' or 'No', to the query 'Does the counter fire?' specifies one bit of information. The basic move in quantum theory is to shift, *fundamentally*, from the airy plane of high-level abstractions, such as the unseen precise trajectories of invisible elementary material particles, to the nitty-gritty realities of consciously chosen intentional actions and their experienced feedbacks, and to the theoretical specification of the mathematical procedures that allow us successfully to predict relationships among these empirical realities.

Probing actions of this kind are performed not only by scientists. Every healthy and alert infant is engaged in making willful efforts that produce experiential feedbacks, and he or she soon begins to form expectations about what sorts of feedbacks are likely to follow from some particular kind of felt effort. Thus, both empirical science and normal human life are based on paired realities of this action–response kind, and our physical and psychological theories are both basically attempts to understand these linked realities within a rational conceptual framework.

A purposeful action by a human agent has two aspects. One aspect is his conscious intention, which is described in psychological terms. The other aspect is the linked physical action, which is described in physical terms; i.e., in terms of mathematical entities assigned to space-time points. For successful living the physically described action should be a *functional counterpart* of the conscious intention: after sufficient empirical honing by effective learning processes the physically described aspect of the felt intentional act should have a tendency to produce the intended experiential feedback.

John von Neumann, in his seminal book, *Mathematical Foundations of Quantum Mechanics*, calls by the name 'process 1' the basic probing action that partitions a potential continuum of physically described possibilities into a (countable) set of empirically recognizable alternative possibilities. I shall retain that terminology. Von Neumann calls the orderly mechanically controlled evolution that occurs between interventions by name 'process 2'. This process is the one controlled by the Schrödinger equation. The numbering, 1 and 2, emphasizes the important fact that the conceptual framework of orthodox quantum theory requires *first* an acquisition of knowledge, and *second*, a mathematically described propagation of a representation of this acquired knowledge to some later time at which a further inquiry is made.

There are two other associated processes that need to be recognized. The first of these is the process that *selects the outcome*, 'Yes' or 'No', of the probing action. Dirac calls this intervention a "choice on the part of nature", and it is subject, according to quantum theory, to statistical rules specified by the theory. I call by the name 'process 3' this statistically specified choice of the *outcome* of the action selected by the prior process 1 probing action.

Finally, in connection with each process 1 action, there is, presumably, some process that is not described by contemporary quantum theory, but that determines what the so-called 'free choice' of the experimenter will actually be. This choice *seems to us* to arise, at least in part, from conscious reasons and valuations, and it is certainly strongly influenced by the state of the brain of the experimenter. I have previously called this selection process by the name 'process 4'; but will use here the more apt name 'process zero', because this process must precede von Neumann's process 1. It is the absence from orthodox quantum theory of any description on the workings of process zero that constitutes the causal gap in contemporary orthodox physical theory. It is this 'latitude' offered by the quantum formalism, in connection with the "freedom of experimentation" (Bohr 1958, p. 73), that blocks the causal closure of the physical, and thereby releases human actions from the immediate bondage of the physically described aspects of reality.

Cloudlike Forms

The quantum state of a single elementary particle can be visualized, roughly, as a continuous cloud of (complex) numbers, one assigned to every point in three-dimensional space. This cloud of numbers evolves in time and, taken as a whole, it determines, at each instant, for each allowed process 1 action, an associated set of alternative possible experiential outcomes or feedbacks, and the 'probability of finding (i.e., experiencing)' that particular outcome.

Heisenberg's uncertainty principle specifies that if one squeezes this spatial cloud – the spatial region in which the numbers are nonzero – into a sufficiently small region, it will violently explode outward when the constricting force is removed.

https://epdf.tips/mindful-universe-quantum-mechanics-and-the-participating-observer.html

https://books.google.com/books?id=yVpyM3dJOqYC&pg=PA24#v=onepage&q&f=false

Stapp, H. P. (2011). *Mindful Universe: Quantum Mechanics and the Participating Observer* (2nd ed.). Berlin: Springer-Verlag.

If any of this is true, and I believe that it is, then it completely falsifies Radical Behaviorism and Hard Determinism. It also falsifies Materialism, Physicalism, Naturalism, and Nihilism. The false is falsified by the truth; and, the truth is repeatedly experienced and observed.

The Materialists, Naturalists, and Classical Physicists are right in that there is no such thing as free will or choice at the physical level. Physical matter doesn't make choices. Physical processes such as random mutations, natural selection, evolution, and entropy don't make choices. Your physical brain doesn't make choices because it can't make choices! Physical matter can't do choice. Physical matter has no choice in the matter.

But, these people don't realize why they are right because they don't realize that choice is a function of Psyche and that every choice takes place at the quantum level or the psyche level instead of the physical level. Instead, these people deny the existence of choice and the existence of Psyche because their naturalistic model for reality can't explain psyche nor choice; consequently, these people erroneously choose to state that psyche or choice does not exist, and that's as far as their science can take them.

Materialism and Naturalism completely lack explanatory power when it comes to Psyche, Syntropy, Non-Locality, Action at a Distance, Quantum Mechanics, and Choice.

Mark My Words

—

On the Nature of Things

"On the Nature of Things: Human Presence in the World of Atoms" by Henry P. Stapp.

http://www-physics.lbl.gov/~stapp/NOT72C.pdf

https://quantum-neuroscience.com/wp-content/uploads/2018/05/On-Nature-of-Things.pdf

Here's a free book from Henry P. Stapp that he is handing out to the public.

The important thing is to get this information out to the public so that they can take advantage of it. Most of the scientists have officially rejected this information, so you won't get any of this information from them. You will have to look someplace else besides our college classrooms and public schools if you want to find this information because it's being banned from our public schools by the Materialists, Naturalists, Darwinists, and Atheists.

—

Quantum Physics in Neuroscience and Psychology

This article was my first introduction to Henry P. Stapp and the Orthodox Interpretation of Quantum Mechanics. I mention and discuss it elsewhere, so I won't do so here.

Schwartz, J. M., Stapp, H. P., & Beauregard, M. (2004). *Quantum Physics in Neuroscience and Psychology: A Neurophysical Model of Mind-Brain Interaction*. Published Online: Phil. Trans. R. Soc. B.

http://www-physics.lbl.gov/~stapp/PTRS.doc

https://quantum-neuroscience.com/wp-content/uploads/2018/05/PTRS.zip

http://www-physics.lbl.gov/~stapp/PTRS.pdf

http://mypsyche.us/wp-content/uploads/2017/10/PTRS.pdf

https://www.researchgate.net/publication/7613549_Quantum_physics_in_ne uroscience_and_psychology_A_neurophysical_model_of_mind-brain_interaction

http://escholarship.org/uc/item/4w8665vk

This paper summarizes everything I have been looking for during the past fifty-five years of my life. That's a long time to wander in darkness looking for the truth; but luckily, our generation finally has the truth, if we know where to find it and recognize it as true when we do find it.

This paper and Henry P. Stapp's papers and books taught me what Stapp calls the **Orthodox Interpretation of Quantum Mechanics**, which explains the interplay and the scientific interface between mind and matter.

Examine these two websites, from Henry P. Stapp:

https://sites.google.com/a/lbl.gov/stappfiles/

http://www-physics.lbl.gov/~stapp/

They explain the Orthodox Interpretation of Quantum Mechanics in great detail. You could spend a lifetime studying them, and still have things to learn.

The Grand Designer

In the book, *The Grand Designer: Discovering the Quantum Mind Matrix of the Universe*, Graham Smetham gathers together some of the very best from Henry P. Stapp into a very useful and interesting section entitled, "Henry Stapp's Mindful Universe and the Psycho-Physical Bridge".

This section is available on Google Books – just page down and up to populate all the relevant pages.

https://books.google.com/books?id=Z0FOAgAAQBAJ&pg=PA158

I also archived it at this link:

https://quantum-neuroscience.com/wp-content/uploads/2018/05/The-Grand-Designer.zip

I quote from Graham Smetham:

Henry Stapp outlines his version of von Neumann's analysis as the following three stages of the quantum experimental process:

Process 1: The 'free choice' of the experimental setup. Heisenberg called this phase "a choice on the part of the 'observer' constructing the measuring instruments and reading their recording". This choice is "not controlled by any known physical process, statistical or otherwise, but appears to be influenced by understandings and conscious intentions." Whilst this process was originally delineated as a phase within the experimental setting, Stapp indicates that such 'free choice' of 'probing actions' is a part of the general human condition:

Probing actions of this kind are performed not only by scientists. Every healthy and alert infant is engaged in making willful efforts that produce experiential feedbacks, and he/she soon begins to form expectations about what sorts of feedbacks are likely to follow from some particular kind of effort. Thus, both empirical science and normal human life are based on paired realities of this action-response kind.

The hugely significant point in this 'process 1' 'free choice' is that it poses a question to which 'reality' [or Nature's Psyche] can feedback a 'yes' or a 'no', and the fact that the choice of the question is free means that the 'free choice' actually determines the nature of the possible feedbacks. Thus the 'free choices' determine the nature of the experienced reality:

The process is active: it injects into the physical state of the system being acted upon the properties that depend upon the intentional chosen action of the observing agent.

Stapp calls this process 1 'a dynamical psychophysical bridge.'

Process 2: The deterministic quantum evolution of the potentialities within the Schrödinger wave function, this is the mathematical description of the development of the probabilities associated the potentialities within the quantum realm.

Process 3: This is what Paul Dirac called a 'choice on the part of nature.' It is the yes or no feedback from the experimental setup – yes, reality is this way or no, reality is not this way; Stapp indicates that complex questions can be reduced to yes-no choices. This delineation of the experimental process, which starts in the classical realm, then evolves within the quantum realm, and finally gives its answer within the classical level again, is forced upon us because we are limited to perceptions and language within the classical realm. This, of course, is a fundamental aspect of our embodied existential situation which was often commented upon by the early quantum physicists. In other words, we can only ask questions and receive answers in the experiential classical domain whilst the processes which determine the nature of the answer take place in the 'ultimate" quantum domain.

The Grand Designer: Discovering the Quantum Mind Matrix of the Universe. Graham Smetham.

This three-step process is the BRIDGE between Psyche and physical reality. It's called a psychophysical bridge. The Orthodox Interpretation of Quantum Mechanics explains how the Human Psyche and Nature's Psyche work together at the quantum level in order to get things done for us at the physical level.

This is the answer to life, the universe, and everything.

Process 1 – Choices are made by the Human Psyche. The Human Psyche chooses what it wants to do with is physical body and physical brain.

Process 2 – Nature's Psyche collapses the wave functions and/or turns on the necessary neurons in order to make the Human Psyche's choices a physical reality. We know that it is Nature's Psyche or "Reality" who is controlling all of this at the quantum level because the Human Psyche isn't consciously aware of any of this quantum mechanical stuff. The firing of neurons and the collapsing of wave functions happen outside our conscious awareness in the quantum realm under the command and control of Nature's Psyche.

Process 3 – Nature's Psyche responds to the Human Psyche's requests by determining what is possible and what is impossible, based upon the physical laws and the quantum mechanical rules which God has

set into place. God's Psyche controls and restricts our access to quantum mechanics, and God's Psyche made the physical laws. Nature's Psyche enforces God's will both at the physical level and the quantum level. God and Nature can read our mind.

God holds ALL the KEYS to both the physical laws and the quantum mechanisms. God is Syntropy.

Try as he might, a fallen mortal physical being can't do anything that God, quantum mechanics, and the physical laws won't let him do. God created and enforces the physical laws in order to prevent us from using quantum mechanisms to destroy ourselves and our physical existence. God created the physical laws to prevent the physical matter in our bodies from quantum tunneling away from us at will. The physical laws make physical life possible.

The physical laws force order, structure, restrictions, and limitations into spirit matter or dark matter thereby slowing it down and converting it into physical matter. Entropy or physical law slows down physical matter to sub-light speeds, makes physical matter subject to time or an aging process, localizes physical matter in space, and prevents the physical matter from quantum tunneling away from us at will. The result is the Ultimate Consensus Reality, a physical reality. It is highly reliable, dependable, controllable, predictable, and stable. In a physical reality, I can actually count on this paper being on my hard drive and cloud drive tomorrow when I go looking for it.

Scientific Observation: Anything that began obviously has Someone Psyche or Someone Intelligent who caused it to begin. There's no such thing as spontaneous generation or creation ex nihilo.

Scientific Observation: Physical matter, physical laws, and entropy began.

Scientific Conclusion: Therefore, physical matter, physical laws, and entropy have Someone Psyche or Someone Intelligent who caused them to begin or made them begin.

It is also logical to conclude that physical matter has Someone Psyche or Someone Intelligent who is currently forcing it or making it obey the physical laws, including the second law of thermodynamics.

The physical laws are NOT an inherent part of spirit matter or dark matter – they ONLY apply to physical matter. Someone Psyche or Someone Intelligent is making the physical laws and entropy apply to physical matter. God's Psyche forced the physical laws and entropy (an aging process) into a small portion of the spirit matter or dark matter thereby converting that spirit matter into physical matter. God deliberately slowed down the spirit matter converting it into physical matter and subjecting it to the passage of time (entropy) so that we can live it, experience it, depend on it, control it, learn from it, and remember having done so.

Remember, at velocities or speeds faster than the Speed of Light, TIME STOPS – there is NO entropy or passage of time. Light doesn't age because it doesn't experience the passage of time, meaning that it has NO entropy. Light is Syntropy. Furthermore, Someone Psyche is forcing the wave functions to collapse into physical realities. This is what Quantum Mechanics is trying to tell us and teach us.

This IS Science at the cutting-edge.

If you consider yourself to be a scientist, then it's extremely important for you to find, read, and understand this material, so that you can bring yourself into the modern age where science is concerned.

My ultimate desire and goal is to set people free so that they can find the truth and know the truth so that they don't have to go through a lifetime of darkness like I did.

Quantum Mechanics or Syntropy is thousands of years ahead of what the Materialists, Naturalists, Darwinists, Atheists, and Classical Physicists have been able to produce.

Mark My Words

—

Synaptogenesis Is Applied Quantum Mechanics

Synaptogenesis, synaptic mapping, synapse pruning, and synapse elimination are dynamic and alive.

Who or what is overseeing and controlling this process? There's nowhere near enough information in our genome to code for synaptogenesis and synaptic pruning. Our genes code for proteins, NOT synapses. Our synaptic mapping is taking place someplace else besides our genes.

Who or what is mapping the synapses and deciding what each synapse means or what each synapse does? A synapse has no meaning at the physical level. The synapses can't communicate with each other at the physical level. Technically, neurons can't communicate with each other at the physical level.

Who or what is using this synaptic map or synaptic network at the quantum level to get things done for us at the physical level? In other words, who is doing this synaptic mapping, and then using these quantum maps of physical functionality at the quantum level or the psyche level in order to get things done for us at the physical level?

It can't be the Human Psyche because the Human Psyche isn't consciously aware of any of these quantum processes. It can't be our genes because our genomes have nowhere near enough information storage capacity within them to code for and control synaptogenesis, synaptic mapping, and synaptic pruning.

We are in fact looking at Nature's Psyche and/or God's Psyche for an answer to the source and the cause of Synaptogenesis, Synaptic Mapping, Synaptic Meaning, Synaptic Purpose, and Synaptic Pruning.

I'm a scientist. I'm also a generalist. I'm good at everything, master of nothing. Among many other things, I have been a computer scientist all of my life. I understand physical storage capacity.

Our genome is capable of storing 725 megabytes to 750 megabytes of information. In order to successfully MAP or NETWORK one quadrillion synapses – the estimated number of synapses in a newborn child – assign a 3D address to each, and then assign a meaning or purpose to each synapse, as well as regulate its synaptogenesis and any synaptic pruning, do REMAPPING or UPDATING or RE-NETWORKING of what remains, while keeping it all straight in one's mind would require petabytes of information storage capacity to accomplish. It's physically impossible to shove petabytes of information into our puny 750-megabyte genome. It can't be done, which means that it isn't being done.

Furthermore, genes ONLY code for proteins. There is enough memory storage capacity within our genome to code for all the different proteins that are used to make up our physical body; but, where is the information that's needed to organize all those different proteins into 3D machines and 3D cells being stored? Where are all the 3D blueprints being stored? These millions of different 3D blueprints telling the proteins how to assemble themselves into working machines can't be stored within our genome because there's simply not enough memory storage capacity within our physical genome to store all of that information. That, too, would require petabytes of information storage that simply isn't possible within our 750-megabyte genome.

Who, within a living cell, decides that that particular cell needs a new protein? Who sends the transcription enzymes to the right location in the genome for the specific gene needed to code for that protein? How is that information transferred throughout the cell? Once the mRNA molecule has been made and is on its way to a ribosome, who tells the activation enzymes the correct amino acid to join to the correct codon during the production of transfer RNA (tRNA), so that each tRNA molecule arrives at the ribosome in the correct sequence and at the right time for protein synthesis and the translation process? How is all of this information being transmitted from atom to atom, and molecule to molecule? How would you get two physical atoms to communicate with each other and coordinate with each other? How would you get two or more proteins to communicate with each other and coordinate with each other so as to assemble into much larger nanomachines at the right time in the correct 3D location?

https://en.wikipedia.org/wiki/Protein_biosynthesis

These are atoms and molecules that we are talking about here. Who is telling these different atoms and molecules where to go during the construction of amino acids, during the construction of mRNA and tRNA, and during the construction of proteins? Who tells these atoms and molecules where to go? Who tells the different enzymes which molecules to make? Who tells the transcription enzyme which protein is needed and which gene to read? Who is telling the different proteins where to go and what type of nanomachine to become when they get there?

If given the assignment, how would you make the atoms and the molecules communicate with each other and share information with each other at a distance? Remember, there's NO wires connecting the different atoms together at a distance. Communication between the atoms and molecules takes place wirelessly at a distance.

There's only one logical answer to all of this.

Have you seen it yet?

Telepathy or quantum waves are WiFi at the quantum level or the psyche level. Quantum waves are the answer to life, the universe, and everything. Psyche or Intelligence produces, transmits, receives, and stores quantum waves.

The Materialists, Naturalists, Darwinists, Nihilists, and Atheists assure us that physical atoms are the ONLY thing that exists. They are either right, or they are wrong. Which is it?

Proteins are like tinker toys. The genes determine which shape of tinker toy to produce; but, who or what decides how the different tinker toys should be assembled so as to construct different structures or machines? There's NOT enough memory storage capacity within a genome to store all the different 3D blueprints for all the different gadgets, structures, and nanomachines that can be made with those tinker toys. The genome only has enough memory storage capacity to code for the construction of the individual tinker toys. So, where is all of that other information being stored for all those different 3D blueprints? How is all that information being transmitted from atom to atom, molecule to molecule, and protein to protein?

Who is transmitting all of this information from atom to atom? What makes these atoms and molecules move from one location to their target location? Who is telling these atoms and molecules where their target is located? How is all of this information being transmitted wirelessly from atom to atom, and molecule to molecule?

Where is all of this COMMAND and CONTROL information being stored, and how is it being transmitted throughout the cell to the different atoms and molecules within the cell?

It can't be our genes that are storing and transmitting petabytes of this invisible command and control information. The genes are one of the targets of that information but can't be the source of that information. Our physical genes can't store and transmit petabytes of information at the physical level through physical mechanisms. There's no mechanism in place for doing so at the physical level.

So, what are we left with since the source of this information can't be our genes?

The MAPPING and NETWORKING information for Synaptogenesis and synaptic organization, as well as the meaning or purpose of each synapse cannot be stored within our genome. Furthermore, all the different blueprints for all the different 3D nano-machines and 3D cells can't be stored within our genome. Command and control information, target location addressing, and assignments for trillions of different cells and quadrillions of different molecules can't be stored within our genome. So, where is all this information being stored because it clearly must exist and must be getting stored someplace? How is this invisible information being transmitted wirelessly from atom to atom and molecule to molecule since it can't be transmitted by the atoms and molecules themselves at the physical level? It can't be our physical genes doing all of this information storage and information exchange because our genes are just another molecule.

So, once again, what are we left with?

Well, think about it logically. Physical information storage or physical memory storage is limited by the size of an atom. Physical atoms take up space. It has been estimated that a genome is about as densely packed as information can be stored in a physical format. It doesn't get much more optimal than that at the physical level. But, we NEED millions or billions of times more memory storage capacity in the same amount of space that atoms occupy in order to achieve petabytes to exabytes of memory storage capacity within something the size of a physical brain.

The ONLY solution to this conundrum is to go sub-atomic or quantum. You have to go sub-atomic or invisible in order to achieve information storage densities and information storage capacities greater than what's possible with atoms at the physical level. Physical matter greatly limits information storage density and therefore information storage capacity because physical matter takes up space. Those limitations disappear if you go sub-atomic or quantum.

If you were to choose to store information as quantum waves, you could theoretically store exabytes of information in all of that "empty space" between the nucleus of an atom and its orbiting electron shells. I mean, ask yourself what type of "invisible information" is holding those electrons in orbit in the first place. Positive and negative attract, so there should be NO physical space within an atom; yet, there is. Why? Who is forcing the electrons to orbit the nucleus rather than merging with the protons in the nucleus? There has to be some kind of invisible force and intelligent force doing so; otherwise, physical matter wouldn't exist.

Entropy is death. Syntropy is eternal life. Syntropy is eternal and everlasting. There has to be some type of Syntropy or Psyche in existence, or all of that subsequent entropy and physical matter wouldn't have been possible in the first place.

Telepathy or quantum waves are WiFi at the quantum level or the psyche level. Quantum waves or telepathy could be used to transmit information from atom to atom. Quantum waves could also be used to store information invisibly within atoms. Quantum waves or telekinesis could move an atom from one location to a different target location. There is NO answer at the physical level; but, there are an endless number of different answers to be found at the quantum level or the psyche level.

I have hypothesized that Psyche or Quantum Non-Local Consciousness is an infinite singularity. The math points to the existence of such a thing. If I'm right, then that means that Psyche is capable of an infinite amount of memory storage capacity within something that has NO size and takes up NO space whatsoever. Now, that's infinitely dense information storage capacity that we are talking about there within an infinite singularity or Psyche. That's infinitely more information storage capacity than what's possible at the physical level.

Mark My Words

—

Source

Quantum Mechanics from a Non-Physical Spiritual Perspective

https://www.amazon.com/dp/B01J023TGU

https://www.amazon.com/dp/1521132380

NATURE vs. NURTURE vs. NIRVANA: An Introduction to Reality

https://www.amazon.com/dp/B01JWRCSVA

https://www.amazon.com/dp/1521132615

Reference Material

Quantum Neuroscience: The Answer to Life, the Universe, and Everything

https://www.amazon.com/dp/B079Z6QQQB

The Ultimate Model of Reality: Psyche Is the Ultimate Cause

https://www.amazon.com/dp/B071NC9JK6

Psyche or Syntropy Is the Best Explanation

Ask yourself how we detect Psyche or Quantum Non-Local Consciousness. How do we catch Psyche in the act? By definition, in principle or practice, Psyche or Intelligence is non-physical which means that it can't be detected nor recorded by our physical instruments. So, how do we know that it exists?

When it comes to the various different invisible and non-physical forces and fields like Psyche, we KNOW that they exist by the effects that they have on physical matter. We KNOW that gravity, wind, magnetism, dark matter or spirit matter, quantum waves, the strong and weak nuclear forces, light, the zero-point field, and dark energy exist by the effects that they have on physical matter. That's just one of the ways that we KNOW that Psyche exists.

As I studied and thought about this subject, I discovered that there are many different ways through which we come to KNOW that the Psyche, Spirit, Self, or Soul does in fact exist.

One of the most obvious Signs of Psyche is CHOICE! It all comes down to choice. Whenever you encounter some kind of choice being made, you have in fact caught some kind of Psyche in the act. Physical matter doesn't have any choice. Physical matter does whatever the different forces and fields make it do. The physical matter does whatever the Human Psyche makes it do.

I have dozens of different things that I could eat for breakfast tomorrow; and, there is absolutely no way for anyone to predict in advance which one I will choose. I don't even know in advance which one I will choose. If you were to make a prediction based upon my reinforcement history that I'm going to eat oatmeal tomorrow – knowing about your prediction I could and probably would go out of my way to make sure that I eat something else just so that I can falsify your theories about Determinism, Behaviorism, and Reinforcement Histories.

Materialism, Naturalism, Darwinism, Nihilism, Classical Physics, Behaviorism, Determinism, Physical Reductionism, and Atheism don't have any explanatory power when it comes to Psyche, Syntropy, Non-Locality, Non-Physicality, Action at a Distance, Quantum Waves, Choice, and Quantum Mechanics. Materialism, Naturalism, and Classical Physics are based upon physical matter, nature, disorder, chaos, entropy, randomness, chance, and luck. Quantum Mechanics is supernatural in nature and origin. Quantum Mechanics is based upon Syntropy, Psyche, Intelligence, Forces, Fields, Order, and Organization. There's a HUGE difference between the two – an obvious difference for anyone who is willing to look, think, and see.

Quantum waves are supernatural and sub-atomic. Thoughts and memories are quantum waves. We KNOW this is true because our thoughts, memories, and personality survive the death of our physical brain according to the empirical evidence or scientific evidence from Near-Death Experiences (NDEs), Out-of-Body Experiences (OBEs), Shared-Death Experiences (SDEs), and our after-death Life

Reviews. If a thought or memory survives the death of your physical brain, then it's truly a memory in every sense of the word, is it not?

Psyche is Syntropy. Quantum Mechanics is Syntropy. The different invisible and non-physical Forces and Fields are Syntropy. Spirit Matter is Syntropy. It's just a different type of Quantum Wave. Syntropy is without a beginning of days or an end of years. Syntropy is eternal and everlasting. Syntropy has always existed and it will always exit.

While studying psychology, I spent a lot of my effort looking for and documenting Signs of Psyche – psychology is supposed to be the study of the Human Psyche after all. That's what the word originally meant. I'm looking for Signs of Psyche. These are situations, phenomena, and scientific evidence where Psyche, Syntropy, Quantum Mechanics, Action at a Distance, and Quantum Non-Local Consciousness provide the best explanation or have the most explanatory power. Science is all about explanatory power, or it should be.

The BioPsychoSocial Model was the first official model that I found that is based upon all three of the different aspects of Science which have been experienced and observed – namely: biology and genetics and physical development (BIO), society and culture and environment (SOCIAL), and Psyche, Syntropy, Action at a Distance, Quantum Waves, Choice, and Quantum Mechanics (PSYCHO).

The BioPsychoSocial Model has massive explanatory power and is powerfully efficacious, as long as we don't let the Materialists and Naturalists destroy it by restricting it exclusively to physical matter, determinism, and the physical sciences.

While studying Neuroscience and Biopsychology, Signs of Psyche were hard to find because my college teachers and the assigned textbooks deny the existence of Psyche and ignore the existence of Quantum Mechanics. For the most part, the focus is exclusively on physical matter; therefore, the other parts of the BioPsychoSocial Model and Quantum Mechanics are completely ignored by these people. I had to work at it, in order to include Psyche into this material.

I did see Signs of Psyche within my Abnormal Psychology course because the authors of that textbook based their theories and ideas upon what they called an Integrative Approach, which was in fact the BioPsychoSocial Model. As a result of their choice, their college textbook had a ton of scientific explanatory power; and, I did see Signs of Psyche from time to time.

Obviously, there were Signs of Psyche in my Personality Theory or Personal Psychology course because some of the theoreticians actually define personality or the self as Psyche. The existentialists are one such group.

I was surprised to find the most compelling and convincing evidence for Psyche within my Social Psychology course. The author of that textbook wasn't afraid of religion. In fact, he had observed many times during his research and science experiments that the religious folk as a whole are a lot happier and more psychologically sound than the Atheists or the non-believers. Believing in God is good for one's psychological health and physical health. It's also good for one's social health or social relationships. It's been scientifically proven to be so.

At the beginning of his college textbook, the author of my Social Psychology textbook even stated up front that Social Psychology is based upon the BioPsychoSocial Model; and, he meant it. All throughout that book, I was seeing Signs of Psyche – instances where Psyche is the best explanation for the scientific evidence at hand.

It's fun to try to identify the true cause behind everything that we encounter; but, that can only be done successfully if we choose to include Psyche into the mix. There are times when Psyche is the BEST explanation; and, in those particular instances, when Psyche is deliberately excluded as a possible explanation or a possible cause, then suddenly we have NO logical explanation for what we are seeing, observing, or experiencing.

Whenever Psyche is the Ultimate Cause of a phenomenon, Materialism, Naturalism, Darwinism, Nihilism, Classical Physics, Determinism, Behaviorism, Entropy, and Atheism have NO explanation for what we are witnessing and experiencing. However, Quantum Mechanics, Action at a Distance, Choice, and Syntropy certainly do.

While studying Psychology or the Human Psyche, there are many instances and phenomena where Psyche is in fact the best explanation and/or the best source for finding a cure to one's mental illness. Whenever you encounter a mental illness that is caused by one's Biology and/or one's Society, then the ONLY treatment and cure available has to come from the Human Psyche or God's Psyche. Addictions and substance-induced psychosis are precisely the types of mental illness that are caused by our Biology and our Society; and, the ONLY treatment and cure that is realistically possible comes from the Human Psyche and the Atonement of Christ, when the individual person or psyche finally chooses and decides to get sober and stay sober, and then asks God for help in doing so.

On my mother's side, we are overly sensitive to drugs and medications; and, there are many alcoholics on that side of the family. My biology produces a diathesis, tendency, or vulnerability towards addiction of any kind. My medical doctors and psychiatrists got me addicted to half a dozen different brain-altering drugs. I experienced psychosis whenever they tried to get me off the drugs. I experienced a substance-induced psychosis during the six months of withdrawal. Substance-induced psychosis is a mental illness. It's listed in the psychology books as such. My Biology and my Society gave me this mental illness. I felt like hell and wanted to die.

The only cure – the only thing that worked – was when I (my psyche) chose to stop taking the drugs and chose to suffer the consequences of that choice. I went cold-turkey, and I went insane; but, I stuck with it. I made a choice! My biology didn't make that choice for me. My biology wanted me to keep taking the drugs. My society and psychiatrists and wife didn't make that choice for me. The psychiatrists in particular wanted me to keep taking the drugs. That's how they make their money. I made the choice – my psyche, spirit, or soul; and, I suffered

515

six months of hellish withdrawal, weirdness, insanity, pain, insomnia, hallucinations, delusions, paranoia, anxiety, fear, racing thoughts, detachment from reality, no sense of time, no sense of right and wrong, no desire to eat, apathy, magical thinking, and so forth. It was a magical place and it was hell all at the same time. That's your brain on drugs.

Psyche is identified by the choices that it makes, especially when those choices run contrary to the choices which its biology and its society want it to make. My biology and my society wanted me to stay on the drugs. My Psyche chose to get off the drugs and stuck with that choice to the bitter end.

Like gravity and magnetism, Psyche is some type of invisible and non-physical force or field that can make physical matter do things which the physical matter can't do or wouldn't normally do by itself when left to itself.

My butt just sits here in the chair, until I (my psyche) decides to make it get up and go do something else. Right now, there's NO pressure from society or my biology making me get out of this chair; so, if I get out of the chair, it's because I have chosen to do so. Psyche is identified by the choices that it makes. One psyche is distinguished from another psyche by the different choices that they make. I'm different than you because I have made different choices than you have made. My biology didn't make those choices for me; and, neither did my society. Yes, my society has forced me to do some things against my will at times and has forced some of their choices onto me; but, if I would have been left alone, I would have chosen differently, or I would have postponed making some of those choices.

Who you really are is determined by the choices that you make when you believe that nobody else is looking and believe that nobody is going to be checking up on you.

Mark My Words

Documenting Signs of Psyche or Syntropy

Love is Syntropy

Myers, D. G. (2010). *Social Psychology* (10th ed.). New York: McGraw-Hill.

Love cures people – both the ones who give it and the ones who receive it. Altruism is a motive to increase another's welfare without conscious regard for one's own self-interests. (*Social Psychology*, pp. 441-443.)

Love is Syntropy. The more you give away, the more you have. That's Syntropy. Syntropy is the exact opposite of entropy. Altruism is a type of Syntropy.

So, who is giving this love away? It's the Human Psyche and God's Psyche! Both in the quantum realm and the physical realm, it's the Psyche who is giving the love away.

So, who is receiving this love? Who is being cured by this love? It's the Human Psyche and all of those other psyches. Sometimes the effects of this love spill over onto and into the physical body; but, it's the transpersonal interaction or psyche-to-psyche interaction that's producing this Syntropy that we call Love.

Even my dog and my cat love to be loved. There's a person or a psyche in there, and it thrives on love.

As far as we know, the physical matter doesn't care about love. It just does what it's told to do by the Human Psyche, Nature's Psyche, those Other Psyches, and God's Psyche. Physical matter and spirit matter simply react, meaning that they do whatever some Psyche tells them to do. The physical matter follows or obeys the physical laws that God gave it.

The explanatory power of this science is through the roof. It explains everything if we let it.

This is Social Psychology at its very core.

The idea that Love is Syntropy, and that the more you give away the more you have, just might be the most important scientific discovery that we human beings can make. Can you think of anything more important to discover and learn?

I can't.

Mark My Words

—

Redemption Is Syntropy

Myers, D. G. (2010). *Social Psychology* (10th ed.). New York: McGraw-Hill.

Atonement or sympathy is a function of Syntropy and a product of the Human Psyche. It has nothing to do with our genes.

> **In one of these experiments, Batson and his associates had University of Kansas women observe a young woman suffering while she supposedly received electric shocks. During a pause in the experiment, the obviously upset victim explained to the experimenter that a childhood fall against an electric fence left her acutely sensitive to shocks. The experimenter suggested that perhaps the observer (the actual participant in this experiment) might trade places and take the remaining shocks for her.**

Previously, half of these actual participants had been led to believe the suffering person was a kindred spirit on matters of values and interests (thus arousing their empathy). Some also were led to believe that their part in the experiment was completed, so that in any case they were done observing the woman's suffering. Nevertheless, their empathy aroused, virtually all willingly offered to substitute for the victim. (*Social Psychology*, pp. 456-457.)

Jesus Christ is the being of light and love whom most people encounter during their Near-Death Experiences (NDEs.)

How was He able to garner or gather such a monumental amount of love?

He did it by giving away love and forgiveness. We love the things the most for whom we sacrifice the most. During the Atonement, Christ went to hell on behalf of each one of us. He took our beatings for us. He suffered the demands of Justice for us. That's how He is able to have an infinite amount of love for us – He demonstrated and gave an infinite amount of forgiveness for each of us. He suffered the most for us. Forgiveness is the pinnacle of love; and, forgiveness involves some kind of sacrifice. That's how Jesus Christ was able to become the being of infinite light and love whom we encounter during our Near-Death Experiences.

During different NDEs as discussed on YouTube, Jesus Christ actually came to hell and got various different sinners and atheists out of hell – just for the asking. That's all it took. They simply had to ask Jesus to come and save them from hell, and He did. Remember, should you ever find yourself in hell, Jesus Christ can get you out of there just for the asking. He is able to do so because His forgiveness and mercy are infinite. This is Syntropy that we are talking about here.

Mosiah 26: 21-26:

21 And he that will hear my voice shall be my sheep; and him shall ye receive into the church, and him will I also receive.

22 For behold, this is my church; whosoever is baptized shall be baptized unto repentance. And whomsoever ye receive shall believe in my name; and him will I freely forgive.

23 For it is I that taketh upon me the sins of the world; for it is I that hath created them; and it is I that granteth unto him that believeth unto the end a place at my right hand.

24 For behold, in my name are they called; and if they know me they shall come forth, and shall have a place eternally at my right hand.

25 And it shall come to pass that when the second trump shall sound then shall they that never knew me come forth and shall stand before me.

518

26 And then shall they know that I am the Lord their God, that I am their Redeemer; but they would not be redeemed.

27 And then I will confess unto them that I never knew them [and they never knew me]**; and they shall depart into everlasting fire prepared for the devil and his angels.**

Who are the people who go to hell, and why do they go to hell?

These are the people who refuse to be redeemed!

They refuse to ask Christ for His forgiveness and help. They refuse to ask Christ for His mercy and love. That's it. They refuse to be redeemed. They don't want to have anything to do with Christ, God, and their mercy, forgiveness, and love. These people have their pride that must be maintained. It's a pride thing. They have yet to learn humility. They have yet to accept a helping hand.

Fascinating, is it not?

The practical applications of these truths are infinite because we are dealing with Syntropy here, and not entropy. Entropy is the enemy. Entropy is death and hell. In contrast, Syntropy is Love, Forgiveness, and Life. Jesus Christ has an infinite supply of Syntropy, Forgiveness, and Love for those who are willing to be redeemed; but, He can't force us to accept His forgiveness and love. That's something that only we, our Psyche, can do for ourselves or choose for ourselves.

Forgiveness and Love are chosen into existence by the Human Psyche; and, so are a willingness to receive Forgiveness and Love.

Mark My Words

—

Reward Theory

Myers, D. G. (2010). *Social Psychology* (10th ed.). New York: McGraw-Hill.

In Social Psychology, the Reward Theory is based upon physical matter and entropy. The idea is that the only reason we do anything good is because we are hoping for some type of reward. Reward Theory or Egoism is the opposite of Altruism, Love, and Syntropy.

REWARDS

We attribute their behavior to their inner dispositions only when we lack external explanations. When the external causes are obvious, we credit the causes, not the person.

There is, however, a weakness in reward theory. It easily degenerates into explaining-by-naming. If someone volunteers for the Big Sister tutor program, it is tempting to "explain" her compassionate action by the satisfaction it brings her. But such after-the-fact naming of rewards creates a circular explanation: "Why did she volunteer?" "Because of the inner rewards." "How do you know there are inner rewards?" "Why else would she have volunteered?" Because of this circular reasoning, *egoism* — the idea that self-interest motivates all behavior — has fallen into disrepute.

To escape the circularity, we must define the rewards and the costs independently of the helping behavior. If social approval motivates helping, then in experiments we should find that when approval follows helping, helping increases. And it does.

INTERNAL REWARDS

So far, we have mostly considered the external rewards of helping. We also need to consider internal factors, such as the helper's emotional state or personal traits. The benefits of helping include internal self-rewards.

(*Social Psychology*, p. 445.)

Notice that Reward Theory or Behaviorism is based upon *circular reasoning*, *hindsight bias*, *labeling*, and *after-the-fact naming* which are logic fallacies.

Materialism, Naturalism, Behaviorism, and Atheism are based upon a wide variety of logic fallacies.

Reward Theory or Behaviorism is also based upon entropy. Entropy is a dead-end. Entropy results in this *circular reasoning* and *confirmation bias* which this author is talking about. Entropy is used to describe, promote, and explain entropy.

In contrast, these internal rewards, internal emotional states, internal factors, and internal personality traits are based upon the Human Psyche; and, the Human Psyche is capable of Syntropy. Syntropy is infinite and eternal. There's no entropic dead-end within Syntropy. Psyche or Syntropy is capable of choice. Internal dispositions are a product of the Human Psyche, and not our genes. The genes or physical matter don't have inner dispositions. Genes aren't seeking for internal self-rewards. The genes, entropy, and physical matter don't care one way or the other what we choose to do.

Once again, an understanding of Psyche and Syntropy provides us with the best explanation for the scientific evidence that we are observing, witnessing, and experiencing.

Mark My Words

—

Genuine Altruism Does Exist

Myers, D. G. (2010). *Social Psychology* (10th ed.). New York: McGraw-Hill.

Genuine altruism does exist.

Everyone agrees that some helpful acts are either obviously egoistic (done to gain external rewards or avoid punishment) or subtly egoistic (done to gain internal rewards or relieve inner distress). Is there a third type of helpfulness — a genuine altruism that aims simply to increase another's welfare (producing happiness for oneself merely as a by-product)? Is empathy-based helping a source of such altruism? Cialdini and his colleagues Mark Schaller and Jim Fultz have doubted it. They note that no experiment rules out all possible egoistic explanations for helpfulness.

But other findings suggest that genuine altruism does exist: With their empathy aroused, people will help even when they believe no one will know about their helping. Their concern continues until someone has been helped. If their efforts to help are unsuccessful, they feel bad even if the failure is not their fault. (Social Psychology, p. 457.)

https://psyche-ontology.com/wp-content/uploads/2018/05/Genuine-Altruism.pdf

Remember, the Human Psyche is capable of Syntropy; and, Syntropy can override and overcome the effects of entropy. The "behavior" of raw physical matter is determined by the physical laws. It takes some type of Syntropy or Psyche to override the physical laws, meaning that it takes some type of Syntropy or Psyche or Love to overcome and override the effects of entropy.

This is science at its very most basic and fundamental foundation that we are talking about here. We are talking about a Psyche Ontology, wherein Psyche or Syntropy is the fundamental unit of reality and existence.

Genuine Altruism or Syntropy is only possible under the command and control of Someone Psyche. Genes and physical matter can't do altruism because altruism requires some type of Choice or Syntropy. Once again, we observe that Psyche or Syntropy ends up providing us with the best scientific explanation for what's really going on in the world and the universe around us. In order to have a complete science – in order to understand what's really going on in this universe – we must understand Psyche or Syntropy and how it affects everything else.

When doing science, we have got to take the Human Psyche into consideration or learn to tolerate not having an explanation for most of the things that we encounter.

Mark My Words

—

Helping Is a Function of the Human Psyche

Myers, D. G. (2010). *Social Psychology* (10th ed.). New York: McGraw-Hill.

What circumstances prompt people to help, or not to help?

Ultimately "helping" is a function of the Human Psyche. We either choose to help or we choose not to help. It's a choice; and, choice of any kind is the ultimate Sign of Psyche.

Why do some people help, and others don't?

It's because "helping" is a function of the Human Psyche which means that it is variable and unpredictable because it is based upon choice! Some days I choose to help and some days I don't. Some days I choose to do my chores, but most days I don't. Some days I choose to weed, and some days I don't. Some days my wife chooses to go grocery shopping, but most days she doesn't.

There's no way on this earth to predict which day that my wife will choose to go grocery shopping. It's different every week; and, there are some weeks when it never happens. The best I can do is to prepare a grocery list and have it ready and sitting there for the day when she finally chooses to go grocery shopping. My wife doesn't want me shopping for groceries because I buy things that she doesn't want me to have. She wants me totally at her mercy when it comes to the groceries, so that's just the way it is. I've learned to accept it. Some things we can change, and some things we can't.

When my wife has made up her mind, there's basically nothing that I can do to change it. When my wife makes up her mind, the discussion is over. Sometimes I choose to comply and/or choose to assist; and sometimes, I choose to rebel and choose to do my own thing my own way. Either way, it's a choice. It all comes down to choice! I rebel and do my own thing if I believe that my way is better, my way is faster, or my way will produce better results in the end. I have my own funds that I use when I choose to rebel and choose to do things my way; and, I'm also more satisfied with the results when I have chosen them into existence. That's just the way it is. Choice makes it real and makes it valuable to us. When we aren't given any choice, we sometimes choose to rebel. In Social Psychology it is called Reactance.

If sad people are sometimes extra helpful, how can it be that happy people are also helpful? Experiments reveal that several factors are at work. Helping softens a bad mood and sustains a good mood. (Perhaps you can recall feeling good after giving someone

522

directions.) A positive mood is, in turn, conducive to positive thoughts and positive self-esteem, which predispose us to positive behavior. In a good mood — after being given a gift or while feeling the warm glow of success — people are more likely to have positive thoughts and associations with being helpful. Positive thinkers are likely to be positive actors. (*Social Psychology*, p. 448.)

Who are these "people" that these scientists are talking about?

They are Human Psyches.

This is psyche-over-matter or mind-over-matter that they are talking about here!

We continue to feel good or feel bad long after our physical brain is dead and gone according to the abundant empirical evidence or scientific evidence from Near-Death Experiences. Your genes and the physical matter within your physical body aren't feeling the warm glow of success; but, your Psyche certainly is.

Moods are a function or a product of the Human Psyche. Your genes and hormones don't have moods. Your genes and hormones can't change your moods either. Genes don't change their minds. The Human Psyche does. The genes don't change; but, the Human Psyche does.

Choosing to help someone is a choice that is made by the Human Psyche. Your genes don't make that choice for you. We KNOW because some days you (your Psyche) choose to help, and other days you choose not to help. There's no way in the universe for anyone to predict which choice you will make because you can choose to help or choose not to help any day that you want to. You don't even know which choice you are going to make until after you have made the choice.

Our actions or our chosen behaviors are chosen into existence by our Human Psyche or Human Intelligence on both sides of the veil.

Observe that Materialism, Naturalism, Darwinism, Nihilism, Behaviorism, Physicalism, Classical Physics, Determinism, and Atheism have NO explanation for choice, chosen behavior, moods, forgiveness, altruism, and changing moods. These philosophies have NO explanatory power when it comes to Psyche, Syntropy, Feelings, Choices, and Moods. These people teach that everything is determined by physical matter and entropy. Psyche, Syntropy, and Choice are not a part of their vocabulary. The materialistic, naturalistic, and atheistic scientists have no idea that Syntropy and Choice even exists. These people have chosen to believe that Psyche, Syntropy, Choice, and Action at a Distance do not exist.

Once again, we see that Psyche and Syntropy provide the best explanation for the scientific evidence that we are observing and experiencing, especially when it comes to our chosen actions, chosen behaviors, and chosen moods. Psyche, Syntropy, and Quantum Mechanics can explain CHOICE. Materialism, Naturalism, Darwinism, Nihilism, Classical Physics, Determinism, Behaviorism, and Atheism cannot.

Once again, we see that Psyche or Syntropy provides the best explanation for most of our scientific observations and most of the scientific evidence. The explanatory power of Quantum Mechanics or Syntropy is through the roof!

Fascinating, is it not?

Well, I think it is because I used to be a Materialist, Naturalist, Nihilist, and Atheist.

Mark My Words

—

Interpretation Is a Function of Psyche

Interpretation is a function of the Human Psyche. Interpretation isn't done for us by our genes, and our society doesn't do it for us either.

This is another one of those smoking guns where Psyche is concerned.

Your environment (Nurture) isn't deciding what the different events mean to you or are doing to you. Your Psyche is making that choice.

Your genes (Nature) definitely are not deciding what that pain in your side means. Your Psyche is making that decision. Your genetic inheritance doesn't change; but, your interpretations of your surroundings and the science experiments certainly do.

People's interpretations also affect their reactions. Interpretations matter. (*Social Psychology*, p. 463.)

The fact that we can actually choose to change our interpretation of a specific event is Proof of Psyche. The fact that we can actually choose to forgive is Proof of Psyche. Interpretation isn't done by our genes, and our society doesn't do it for us either.

This is very simple but very powerful science. Once again, when it comes to our interpretations, Psyche provides the BEST explanation for what is really going on.

Our genes, our environment, our society, our hormones, and our physical matter can't explain our choices and our interpretations of evidence; but, our Psyche certainly can. The explanatory power of Psyche is through the roof. Psyche takes over whenever the other explanations have failed. By including Psyche, Syntropy, and Quantum Mechanics into the mix, we can literally explain everything that comes our way. There's no part of science that remains unexplained.

Mark My Words

Whence Empathy?

Myers, D. G. (2010). *Social Psychology* (10th ed.). New York: McGraw-Hill.

Where does empathy come from? Is empathy produced by our genes and hormones as the Darwinists and Evolutionists claim?

Empathy: The vicarious experience of another's feelings; putting oneself in another's shoes.

When we feel empathy, we focus not so much on our own distress as on the sufferer. Genuine sympathy and compassion motivate us to help others for their own sakes. When we value another's welfare, perceive the person as in need, and take the person's perspective, we feel empathic concern.

In humans, empathy comes naturally. Even day-old infants cry more when they hear another infant cry. In hospital nurseries, one baby's crying sometimes evokes a chorus of crying. Most 18-month-old infants, after observing an unfamiliar adult accidentally drop a marker or clothespin and have trouble reaching it, will readily help. To some, this suggests that humans are hardwired for empathy. Primates and even mice also display empathy, indicating that the building blocks of altruism predate humanity.

Nevertheless, their empathy aroused, virtually all willingly offered to substitute for the victim. (Social Psychology, pp. 255-257.)

Can our genes sense another's feelings? Can our genes put themselves in another's shoes vicariously within their mind's eye?

Most people, I believe, are in agreement that they cannot.

So, where is all of this empathy coming from since it can't be coming from our genes? There are no molecules for empathy and mercy anywhere within this physical universe.

The evidence suggests that human beings are born with empathy. So, where did they get that empathy from since they couldn't have gotten it from their genes?

The only logical answer is the Human Psyche. Psyche is Syntropy, which means that Psyche is without a beginning of day or an end of years. Psyche or Syntropy is eternal and everlasting. Empathy is Proof of Psyche. Empathy goes all the way back to the time before physical matter was made.

Empathy is a function of the Human Psyche. It's not imposed upon us by our genes or biology. In fact, since evolution or entropy results in death, the influence of entropy, evolution, and our genes tend to make us selfish and not empathetic and altruistic. We can't use our genes, physical matter, and entropy to explain empathy and altruism.

Empathy is chosen into existence by the Human Psyche, or it would never exist in the first place where this physical universe is concerned. According to the theory of evolution and survival of the fittest, empathy and altruism shouldn't exist. Evolution, physical matter, and entropy feel no empathy and have no empathy. This truth is one of the hallmarks of the theory of evolution. So, the only place that empathy can be coming from is the Psyche or the Soul; and, the very existence of empathy falsifies Materialism, Naturalism, and the Theory of Evolution.

Once again, we observe that Psyche or Syntropy is the BEST explanation where scientific observations or scientific evidence is concerned.

Mark My Words

—

Social Norms

Myers, D. G. (2010). *Social Psychology* (10th ed.). New York: McGraw-Hill.

Social Psychologists explain helping and altruism by the Social-Responsibility Norm and the Reciprocity Norm.

The Materialists, Naturalists, and Darwinists tell us that these Social Norms are programmed into us by our genes. Our genes made us do it. Our selfish genes made us do it. Our genes are selfish.

Are our genes really selfish? Do our genes really make us do things?

Reciprocity Norm: An expectation that people will help, not hurt, those who have helped them.

Social-Responsibility Norm: An expectation that people will help those needing help.

"If you love those who love you [the reciprocity norm], what right have you to claim any credit? . . . I say to you, love your enemies" (Matthew 5:46, 44).

With people who clearly are dependent and unable to reciprocate, such as children, the severely impoverished, and those with disabilities, another social norm motivates our helping. The social-responsibility norm is the belief that people should help those who need help, without regard to future exchanges. The norm motivates people to retrieve a dropped book for a person on crutches, for example.

In India, a relatively collectivist culture, people support the social-responsibility norm more strongly than in the individualist

West. They voice an obligation to help even when the need is not life-threatening or the needy person — perhaps a stranger needing a bone marrow transplant — is outside their family circle.

Even when helpers in Western countries remain anonymous and have no expectation of any reward, they often help needy people. However, they usually apply the social-responsibility norm selectively to those whose need appears not to be due to their own negligence. (*Social Psychology*, pp. 449-450.)

Do you see signs of Psyche within any of this?

Who is keeping track of whether they have been helped or hurt by someone else? Is it our genes? Of course not!

Are your genes going to love your enemies? Of course not! Your genes don't care one way or the other.

Are our genes feeling that societal expectation that we should help others who need help? Obviously not! Our genes don't have feelings. Our genes could care less what our society expects from us.

Your genes aren't going to choose to donate a kidney or bone marrow to someone outside their family circle. Your genes aren't being conditioned or trained by your collectivist culture or individualist culture to care or to be selfish. The Human Psyche is.

Your genes aren't selective about whom they want to help or ignore. Your genes don't decide if the person's suffering is due to their own neglect. The Human Psyche, your psyche, makes these observations and decisions.

Do you see how that works?

Isn't it cool how the truth always floats to the top if you let it? Your genes remain the same; but, the Human Psyche changes its mind all the time. The explanatory power of Psyche or Syntropy is through the roof. It can explain everything!

Don't you love it? You, the real you, your human psyche, don't you just love it? You are the only one who knows! Your genes could care less what you have chosen to love and chosen to believe.

Notice how they always dance around psyche, but they never go all the way because they are Materialists, Naturalists, and Darwinists and have chosen to believe that Psyche does not exist. But, our genes can't explain altruism, mercy, compassion, friendship, forgiveness, empathy, and love. Consequently, we have to look someplace else for a scientific explanation for these things. The only logical explanation is that Psyche or Syntropy really does exist.

Once again, Psyche or Syntropy ends up providing us with the best explanation where the social sciences are concerned. None of this can be explained by our genes; but, it definitely can be explained by the Human Psyche and all those other psyches.

Mark My Words

—

Comparing Theories of Altruism

Myers, D. G. (2010). *Social Psychology* (10th ed.). New York: McGraw-Hill.

How is altruism explained?

The Materialists, Darwinists, and Naturalists use our genes to explain it.

Are they right, or is there more going on here than meets the eye?

If individual self-interest inevitably wins in genetic competition, then why will we help strangers? Why will we help those whose limited resources or abilities preclude their reciprocating? And what causes soldiers to throw themselves on grenades?

One answer, initially favored by Darwin (then discounted by selfish-gene theorists, but now back again) is *group selection*: **When groups are in competition, groups of mutually supportive altruists outlast groups of non-altruists.** [The militaristic atheists kill and destroy their own people, the Christians less so.] **This is most dramatically evident with the social insects, who function like cells in a body. Bees and ants will labor sacrificially for their colony's survival. To a much lesser extent, humans exhibit in-group loyalty by sacrificing to support "us," sometimes against "them."**

Natural selection is therefore "multilevel," say some researchers. It operates at both individual and group levels.

Donald Campbell offered another basis for unreciprocated altruism:

Human societies evolved ethical and religious rules that serve as brakes on the biological bias toward self-interest. Commandments such as "love your neighbor as yourself" admonish us to balance self-concern with concern for the group, and so contribute to the survival of the group.

Richard Dawkins (1976) offered a similar conclusion: "Let us try to teach generosity and altruism, because we are born selfish. Let us understand what our selfish genes are up to, because we may

528

then at least have the chance to upset their designs, something no other species has ever aspired to".

Comparing and Evaluating Theories of Helping

By now you perhaps have noticed similarities among the social-exchange, social norm, and evolutionary views of altruism. Each proposes two types of prosocial behavior: a tit-for-tat reciprocal exchange and a more unconditional helpfulness. They do so at three complementary levels of explanation. If the evolutionary view is correct, then our genetic predispositions *should* manifest themselves in psychological and sociological phenomena.

Each theory appeals to logic. Yet each is vulnerable to charges of being speculative and after the fact. When we start with a known effect (the give-and-take of everyday life) and explain it by conjecturing a social-exchange process, a "reciprocity norm," or an evolutionary origin, we might merely be explaining-by-naming.

The argument that a behavior occurs because of its survival function is hard to disprove. With hindsight it's easy to think it had to be that way. If we can explain *any* conceivable behavior after the fact as the result of a social exchange, a norm, or natural selection, then we cannot disprove the theories. Each theory's task is therefore to generate predictions that enable us to test it.

An effective theory also provides a coherent scheme for summarizing a variety of observations. With this criterion, our three altruism theories get higher marks. Each offers us a broad perspective that illuminates both enduring commitments and spontaneous help.

Genuine Altruism

After 25 such experiments testing egoism versus altruistic empathy, Batson and others believe that sometimes people do focus on others' welfare, not on their own. Batson, a former philosophy and theology student, had begun his research feeling "excited to think that if we could ascertain whether people's concerned reactions were genuine, and not simply a subtle form of selfishness, then we could shed new light on a basic issue regarding human nature". Two decades later he believes he has his answer. Genuine "empathy-induced altruism is part of human nature".

(*Social Psychology*, pp. 453, 454, 457.)

Altruism and empathy are a part of human nature, meaning that they are a part of the Human Psyche. They don't have anything to do with our genes. Our genes could care less whether they live or die. Our genes could care less whether we live or die. Altruism and empathy are chosen into existence by the Human

Psyche. Where it once didn't exist, suddenly it now exists thanks to the Human Psyche who chose it into existence. That's Syntropy in action!

Notice how the Naturalists and Darwinists go back and forth, back and forth. It's because the theory of evolution is internally inconsistent. It lacks explanatory power. Consequently, they have to make up stories out of thin air; and, they have to keep adjusting those stories after-the-fact in order to make them fit with additional ideas and information that come along over the years.

The commandments and physical laws have nothing to do with natural selection and evolution! They have nothing to do with our genes. It's the people, the psyches, who are making these different commandments and choices. Notice that even Richard Dawkins suggests that "we may" find ways to override our genes and our genetic inheritance. Who is this "we" that he keeps talking about? It's not our genes! Our genes can't make choices. "We" are Human Psyches who are choosing to override our biology and DNA, thereby choosing to teach generosity and altruism to our children.

The social psychologists suggest modeling altruism and empathy to our children. They want us socializing altruism. Why? Modeling altruism and empathy will do absolutely nothing to change our genes! Our genes don't change, so what does? It's the Human Psyche who changes its mind, not our genes. This is so logically obvious that everyone seems to overlook it. How are they able to overlook it? It's because they don't want to see it nor understand it. I KNOW because I used to be a Materialist, Naturalist, Nihilist, and Atheist. I didn't want God and the Human Psyche to exist at the time; so, for me personally they did not exist. They weren't a part of my psyche or worldview.

Do you see how that works?

It works because our beliefs are chosen into existence by the Human Psyche.

"Human societies evolved ethical and religious rules that serve as brakes on the biological bias toward self-interest. Commandments such as 'love your neighbor as yourself' admonish us to balance self-concern with concern for the group, and so contribute to the survival of the group."

According to these people, our genes or evolution gave us these commandments and put brakes on themselves in order to contribute to the survival of the group or the survival of our genes.

Do you buy it?

I don't. When was the last time that your genes commanded you to do something, and you chose either to rebel or to obey? What? It's never happened. Well of course it hasn't happened! Our genes don't give us commandments. Our genes could care less what we choose to do. It is God and our society – those other psyches – who give us commandments, which our Human Psyche then chooses to accept or reject.

When it comes to the Darwinists and Naturalists, it's obvious that these people are making up stories after-the-fact. The whole theory of evolution is nothing but a fictional story. They observe what their own personal psyche is choosing to do and choosing to believe, and then they assign those choices to their genes. It's a *category error* logic fallacy that they are employing here, in order to make their case and prove their point.

The theory of evolution is tailor-made after-the-fact to fit the observational evidence. It doesn't make any predictions. It simply observes what has already happened, and then ascribes it all to our genes. The theory of evolution is *explaining-by-naming*, another logic fallacy. The theory of evolution is correlational and conjectural. The theory of evolution is a fictional *ad hoc just-so story*. The theory of evolution is based upon *hindsight bias* and *confirmation bias*. There will NEVER be any way to test it or verify it because evolution or entropy cannot design and create. They've never been observed doing so. In fact, entropy or the second law of thermodynamics prevents the theory of evolution from becoming true.

When it comes to empathy and altruism, Psyche is the BEST explanation. It really has nothing to do with our genes or our evolutionary history.

Mark My Words

Conclusions Regarding Helping and Altruism

All of this was the result of just ONE chapter within the book *Social Psychology* that I chose to analyze in detail. There are fifteen other chapters that I could analyze and explore looking for Signs of Psyche. I might do so in some other book at some future date. I did this so that I could demonstrate how it is done. I'm looking for scientific observations where Psyche, Syntropy, or Quantum Mechanics end up being the Best Explanation for the scientific evidence.

In the margins of my book, I made a note whenever I encountered Signs of Psyche or Signs of Syntropy; and, my book is peppered with notes. Signs of Psyche are impossible to ignore. They reveal themselves to us whenever we encounter some phenomenon, feeling, choice, or event that our genes could never have done and that genetics cannot explain. Psyche, Syntropy, and Quantum Mechanics explain the physically impossible.

Do you understand the full ramifications of these scientific observations and the scientific research into altruism? Most people don't, because they have chosen not to understand. They don't want to understand it, so they can't understand it.

Observe what becomes possible by choosing to allow ALL of the evidence into evidence, and then choosing to pursue a preponderance of that evidence!

Helping, Altruism, and Choice have nothing to do with our genes. Sometimes people (psyches) choose to stop and help; and, sometimes they don't. It's impossible to predict which course of action they will choose. It has been observed that if a person has a specific appointment and knows that somebody is waiting for him, he will be <u>less likely</u> to stop and help. If a person has nothing better to do, that reality <u>increases the likelihood</u> that he or she will stop and help a stranger in need. Only the Psyche, the person, knows whether he is late for an appointment or not. Our genes are completely clueless. Decisions to help are made on a case-by-case basis by the Human Psyche or the human individual. It has nothing to do with our genes.

Our genes don't change; but, our decisions to stop and help sometimes do. Our genes don't have minds of their own; therefore, they are incapable of making our decisions for us. Likewise, our genes don't change their minds; but, Human Psyches do.

The Behaviorists, Naturalists, Darwinists, Nihilists, Determinists, and Atheists seem to be united in their belief that genes and physical matter cannot do choice. The Behaviorists and Determinists teach that there is no such thing as free will or choice because our physical matter or our genes cannot make choices. They are right. The genes and physical matter cannot make choices. Since these people have chosen to believe that ONLY physical matter exists, then it logically follows that we don't have free will because our physical matter or our genes are incapable of making choices.

So, where are all of these different choices coming from since they obviously cannot be produced by physical matter or our genes? There's only one other logical explanation. It has to be something non-physical like Psyche or Syntropy. Through a process of elimination, we inevitably arrive at the truth.

Isn't that fascinating?

The Materialists, Naturalists, Darwinists, Nihilists, Behaviorists, Determinists, Physical Reductionists, and Atheists are right whenever they make the claim that physical matter or genes cannot make choices. But, these people don't understand the full implications of this knowledge and what it truly means to science. The very existence of CHOICE falsifies Materialism, Naturalism, Darwinism, Nihilism, Behaviorism, and Determinism.

Choice has been observed, lived, and experienced. The observed and proven existence of choice falsifies Materialism, Naturalism, Darwinism, Nihilism, Behaviorism, Determinism, Physical Reductionism, and even Atheism. Since genes and physical matter obviously can't make choices nor do free will, something non-physical must be doing and making all of these choices for us – something like Psyche or Quantum Non-Local Consciousness. Remember, the observed and proven existence of choice falsifies Materialism, Naturalism, and their derivatives. That's what the science is trying to tell us.

Whenever we are dealing with scientific explanations, then observe that Psyche, Syntropy, and Quantum Mechanics take over whenever and wherever our genes begin to fail as a logical explanation for the scientific evidence. By choosing to bring Psyche, Syntropy, and Supernatural Mechanisms into science, we can literally explain everything that comes our way. Nothing remains unexplained.

Materialism, Darwinism, and Naturalism break science leaving most of the universe unexplained. Psyche, Syntropy, and Quantum Mechanics answer every question that comes our way on both sides of the veil.

Mark My Words

—

Source

BioPsychoSocial: Including Psyche or Light into our Theoretical Models

https://www.amazon.com/dp/B0713NDHVW

NATURE vs. NURTURE vs. NIRVANA: An Introduction to Reality

https://www.amazon.com/dp/B01JWRCSVA

https://www.amazon.com/dp/1521132615

PART VI – BOHM'S PHILOSOPHY OF SCIENCE

This part is a late and last addition to this book. I put it up front because I found it interesting. When it comes to science and philosophy, my most recent discoveries always end up being the most interesting. It's an additive process. Once I learn and understand one thing, then it opens the doors to learning and understanding something new and interesting. I write because I discover new and interesting ideas during the process while trying to explain what I have learned to someone else. It's a growth process or a learning process. It works; and, I find it interesting.

David Bohm is shown presenting a new Philosophy of Science within the book, *The Essential David Bohm* – a Philosophy of Science based upon Quantum Mechanics and Consciousness. I had a great deal of fun with Bohm's book.

https://philosophy-of-science.com/The-Essential-David-Bohm

https://epdf.tips/the-essential-david-bohm.html

This part of this book ended up being a 200-page book report about *The Essential David Bohm*.

I had already written a couple of different books about Quantum Mechanics before I found and started reading *The Essential David Bohm*. While reading that book, I found it interesting to observe that he had come to some of the very same conclusions that I had discovered on my own. He too had started to discover all the different stuff that the Materialists, Naturalists, Mechanists, and Atheists ridicule and dismiss from science because they can't get it to pay. These people are looking for the pay-off, or they aren't interested in it.

The problem with Bohm's Super-Implicate Order, Consciousness, Psyche, Action at a Distance, Quantum Field Theory, Quantum Complementarity, the Quantum Zeno Effect, and so forth is that as you travel down the rabbit hole, eventually you reach a point where you can no longer control it and can no longer predict it. When it comes to science, the Materialists, Naturalists, Behaviorists, Determinists, Mechanists, and Atheists are only interested in prediction and control. They want to control it. They want to condition it. They want to own it. They want to enslave it. They want to be paid. They want to make money from it. If they can't get it to pay, then they aren't interested in it.

Their mantra: Show me the money!

For someone like Richard Dawkins, it's all about the money. Atheists, Materialists, and Mechanists will pay a lot of money to be told what they want to hear.

https://www.spectator.co.uk/2014/08/the-bizarre-and-costly-cult-of-richard-dawkins/

https://origin-science.org/The-Bizarre-and-Costly-Cult-of-Richard-Dawkins

When it comes to science, these people want an empirical, physical pay-day; or, they aren't interested in it. However, if you get deep enough into Quantum Mechanics, you eventually encounter a level of existence or reality that can't be predicted and can't be controlled. It's autonomous because it's alive and making choices. Psyche or Intelligence of any kind makes choices between opposing opinions and opposing options. That's the way things really work. It has been experienced and observed. A Psyche or Consciousness or Intelligence either chooses to make a Skinner Box and stuff rats and pigeons into it; or, it doesn't. It's a choice; and, Psyche makes choices. You can't always predict it or control it. I might choose to eat oatmeal every day for breakfast for years on end; but tomorrow, I can always choose to eat something different for breakfast if I want to. There's no way to predict or control that. That's one of the reasons why the Materialists, Naturalists, Nihilists, Darwinists, Behaviorists, Determinists, Physical Reductionists, and Atheists HATE Psyche or Choice so much. They can't predict it and control it. They can't always mold it and shape it and condition it. It has a will of its own. The human psyche can break its conditioning or habits or addictions at will. That's not good news for a drug dealer, or a pharmacist, or a medical doctor, or a behaviorist.

Imagine an ocean. Physical matter is at the surface of the ocean where it can be easily accessed and seen. David Bohm and his Philosophy of Science based upon Quantum Mechanics, with its implicate order and super-implicate order, were all about depth – what's under the surface or supporting the surface. Physicist David Bohm was all about multiple levels of reality.

As a scientist who has gotten deeply involved in the non-physical or the pre-physical aspects Quantum Field Theory and Psyche, parts of Joseph Rychlak's Philosophy of Science and David Bohm's Philosophy of Science have become scripture to me – they have become valuable or sacred, because I can sense the truth and the depth coming from their ideas and their words.

A false and falsified religion or philosophy makes for a false science. If science is interpreted through a false or falsified religion such as Materialism, Naturalism, Nihilism, Mechanism, Behaviorism, Determinism, Physical Reductionism, Atomism, or Atheism, then a false and falsified science will be the result. Scientific interpretation has to be built upon a foundation of truth and reality, or deceptive falsehoods such as the Theory of Evolution will be the result. One can't squeeze the truth out of a string of falsehoods. It doesn't work that way.

I left Joseph Rychlak's Philosophy of Science for other essays that I have written. This essay will primarily focus on David Bohm's Philosophy of Science, as found in part within *The Essential David Bohm*:

https://origin-science.org/The-Essential-David-Bohm

During my research into Quantum Mechanics, I always avoided David Bohm for a multiplicity of different reasons. First, David Bohm was a proponent of Pilot-Wave Theory, and Pilot-Wave Theory has been falsified or has proven to be inadequate because it contradicts the Theory of Relativity.

Pilot-Wave Theory:

https://www.youtube.com/watch?v=RlXdsyctD50

Second, David Bohm wasn't comfortable talking about God and didn't believe in God, per se; and, David Bohm never talked about psyche or soul, but rather hid these concepts behind Latin terms that nobody understood without buying and reading his books. However, David Bohm talked a lot about light and he clearly believed in the existence of energy or light, which is what made him interesting and pertinent.

Bohm started his journey with the falsification of the mechanistic level by pointing out what's wrong with or inadequate about the Philosophy of Mechanism. Later, Bohm called the mechanistic level the Explicate Order. In my writings, I called the mechanistic level or the Explicate Order "the physical level", and I called Mechanism by the name Physicalism, or Materialism, or Naturalism. My "physical level" is self-explanatory, and I think everyone knows automatically what I'm talking about whenever I talk about the physical level of existence and reality.

The Origin of Matter and Time:

https://www.youtube.com/playlist?list=PLsPUh22kYmNCLrXqf8e6nC_xEzxdx 4nmY

Later on, within his research and theorizing, some twenty years later, Bohm created what he called the Implicate Order and the Explicate Order. As stated, the Explicate Order is the physical level. The Implicate Order is what I call the "quantum level" – the level of Quantum Mechanics, Quantum Fields, and Quantum Field Theory.

Quantum Mechanics:

https://www.youtube.com/playlist?list=PLsPUh22kYmNCGaVGuGfKfJl- 6RdHiCjo1

A few years later, David Bohm developed what he called the Super-Implicate Order, which I tend to call the "psyche level". The Super-Implicate Order is a supernatural level that is dominated by Psyche, Intelligence, Information, Consciousness, Energy, Life Force, or Light. Bohm called this level an "organizing principle". As I see it, Someone Psyche or Someone Intelligent has to organize all of that chaotic energy into quantum fields; otherwise, the physical level remains forever impossible, because physical matter needs quantum fields in order to exist, and Someone Psyche has to make or organize the quantum fields out of raw energy in the first place before physical matter can exist. We have to have mind-over-energy BEFORE we can successfully have mind-over-matter, because physical matter is made from energy or light by Someone Psyche. This is Logic 101.

Quantum Field Theory:

https://www.youtube.com/playlist?list=PLsPUh22kYmNBpDZPejCHGzxyfgitj2 6w9

Who designed and made the Quantum Fields? It had to be Someone Psyche, because physical matter didn't exist yet. Physical matter was the last thing that the Gods made, which means that physical matter cannot be the fundamental unit of reality, which means that the Philosophies of Materialism and Naturalism are falsified by common sense physics, quantum fields, and light. It's obvious that the quantum fields were designed and made to fulfill a specific purpose. They didn't just magically spring into existence from nothing as the Materialists and Naturalists claim. It's obvious that quantum fields were designed and made. Psychic Force or Life Force is the only thing that we know of that's capable of manipulating energy at the quantum level.

If you are going to deal with Physics, you are going to have to cover all of this and deal with all of this. Raw primal energy is chaotic. Chaos always remains chaotic unless Someone Psyche or Someone Intelligent comes onto the scene and organizes it. It NEEDS some kind of Organizing Principle – some kind of Super-Implicate Ordering. Some kind of Law Giver had to design, create, and impose the Quantum Laws and the Physical Laws in the first place; otherwise, they wouldn't exist. This reality is obviously true; and, this reality becomes obviously necessary whenever we start dealing with Bohm's Super-Implicate Order or what I like to call "the psyche level" of existence.

Scientific Observation: Anything that is obviously organized obviously has an Organizer who organized it.

Scientific Observation: Quantum Fields were obviously organized.

Scientific Conclusion: Therefore, Quantum Fields obviously have an Organizer or Someone Psyche who organized them or brought them into existence.

All I want is the truth; and, I want it all to make sense when I'm done. Bohm's theorizing and Philosophy of Science progresses through all of these levels or stages and basically arrives at the same destination that I arrived at, except I give these different levels obvious self-explanatory names, and David Bohm gave them Latin names that don't make any sense until after you have read his books.

Whenever I'm talking about Psyche or Soul, I like it to be obvious that I'm talking about Psyche, Mind, Life Force, Consciousness, Intelligence, Information, Awareness, Organizer, Law-Giver, or Soul. Someone Intelligent or Someone Psyche had to play these roles, or quantum fields would not exist, and physical matter would not exist. If I gave you the task to design and create quantum fields from raw chaotic energy, your task would in fact be one of mind-over-energy or psyche-over-energy because physical matter can't exist yet. This is Logic 101. First things first!

Psyche has been experienced and observed. It took me fifty-five years to find evidence sufficient enough and convincing enough to prove to me beyond a shadow of a doubt that this is true. Psyche is experienced first-hand as an immaterial viewpoint in space; and, whenever people are at the quantum level or the non-physical spiritual level, Psyche is seen or observed as a pin-prick of light.

https://psyche-ontology.com/psyche-observed/

https://psyche-ontology.com/psyche-experienced-and-observed/

Buhlman, William L. (1996). *Adventures Beyond the Body: How to Experience Out-of-Body Travel*. New York: HarperCollins.

https://psyche-ontology.com/buhlman/

I'm no longer afraid of these concepts like I used to be when I was a materialist, naturalist, nihilist, and atheist. Observation and experience (science) convinced me that I was wrong. I'm dealing with a comprehensive Philosophy of Science here, and not the neutered or limited stuff that the Materialists, Naturalists, Atomists, and Atheists try to force-feed us through ridicule, intimidation, anger, censorship, groupthink, brainwashing, hatred, deception, and force.

Materialism and Naturalism make the philosophical claim that the non-physical does not exist. Such a claim is easily falsified with Quantum Mechanics, Action at a Distance, and Quantum Field Theory. The proven existence of space, time, dark energy, dark matter (spirit matter), radio waves, microwaves, x-rays, action at a distance, quantum tunneling, telepathy (the quantum Zeno effect), quantum mechanics, and quantum fields FALSIFIES Materialism, Naturalism, and their derivatives such as Darwinism, Nihilism, and even Atheism. The false is always falsified by the truth; and, the truth is repeatedly experienced and observed.

https://www.youtube.com/results?search_query=NDE+Jesus

Eventually, I chose to go with what has been experienced and observed rather than the philosophical speculation and wishful thinking of the Materialists, Naturalists, Nihilists, and Atheists. That's what any scientist is supposed to do.

Mark My Words

Editor's Introduction

The following is an introduction to David Bohm from someone who was NOT trying to do him harm and was NOT trying to ridicule and dismiss him – Lee Nichol:

There are few scientists of the twentieth century whose life's work has generated more excitement and controversy than that of physicist David Bohm. Marginalized by his colleagues in the physics élite for challenging the standard interpretation of quantum theory, forced into *de facto* expatriation for resisting McCarthyism, admired and derided alike for his interest in the phenomenology of consciousness, Bohm's penchant for questioning scientific and social orthodoxy was the natural expression of a rare and maverick intelligence.

Bohm, who in the 1940s refined the theory of plasma as the fourth state of matter, and in the 1950s recast the Einstein–Podolsky–Rosen paradox and shared in the discovery of the Aharonov–Bohm effect, was widely considered one of the most talented and promising physicists of his generation. But his primary work from the 1950s to the 1990s – the ongoing development of his "causal interpretation" of quantum mechanics as an alternative to the standard Copenhagen interpretation – was met with surprising hostility by the majority of the world physics community.

Due largely to a 1994 *Scientific American* cover story and F. David Peat's *Infinite Potential – The Life and Times of David Bohm* (1997), the means by which Bohm's alternative quantum theory had been effectively suppressed came to light, and the general outlines of this alternative were finally presented to a substantial reading public. This theory, developed in collaboration with Prof. Basil Hiley and known in its mature form as the "ontological interpretation" of quantum mechanics, is now widely viewed as a serious critique of the Copenhagen interpretation, and proffers a revisioning of quantum theory in which objective reality is restored and undivided wholeness is fundamental.

Concurrent with the development of the ontological interpretation – and ultimately as an integral part of it – was the evolution of Bohm's philosophy of consciousness. At once ancient and modern, this philosophy employs concepts of contemporary psychology and cognitive science; yet it is deeply resonant with perspectives as diverse as those of Native America, Buddhism, and Vedanta. And while there exists no school of Bohmian philosophy of consciousness, Bohm's insistence that the subtle unity of the "observer and the observed" be brought out of the laboratory and into the domain of immediate experience has contributed significantly to the mainstreaming of cognitive philosophy and consciousness studies.

For many observers there has been an easy temptation to think of these lines of inquiry as two separate and unrelated threads. To many physicists, Bohm squandered precious time and resources in exploring the nature of mind. For those non-scientists who were intuitively drawn to Bohm's philosophy of consciousness, particularly as it manifested in his work with social dialogue, Bohm's physics was the stuff of academia – respectable arcana perhaps, but forever out of reach, and without much concrete relevance.

For Bohm himself, however, these seemingly disparate lines of inquiry were manifestations of an integral primary vision he eventually came to call the *implicate order*. This "order" was a propositional template for plotting the emergence and dynamics of both matter and consciousness. Hinted at by topical analogies such as the hologram, Bohm's vision of the implicate order posited a dynamic "structure-process" from which our coordinate-based experience of three-dimensional space and time is but a derivative, temporary projection. Within that coordinate-based "explicate" order, we may have any number and variety of experiences, yet never come to consider that the very coordinate frame for all those experiences is itself a projection, another experience arising from a deeper "implicate" order.

Consequently, grasping what Bohm was alluding to with the implicate order can be an elusive undertaking, and as it likely was for him, ever a work in progress. Yet without a working sensibility of what an implicate order might be, our conceptions of the various aspects of Bohm's work will necessarily remain limited and incomplete. With such a sensibility in hand, however, not only does the wholeness of Bohm's work become readily apparent, but each facet – be it scientific, sociological, or contemplative – is now illuminated as if from within, infused with new and richer meaning. – Lee Nichol (pp. 1-2.)

Source:

https://epdf.tips/the-essential-david-bohm.html

https://philosophy-of-science.com/The-Essential-David-Bohm

Particles Are Born and Particles Can Die

The primary axiom of Quantum Field Theory states: Particles are born, and particles can die.

Who needs quantum field theory?

Quantum field theory arose out of our need to describe the ephemeral nature of life.

No, seriously, quantum field theory is needed when we confront simultaneously the two great physics innovations of the last century of the previous millennium: special relativity and quantum mechanics. Consider a fast-moving rocket ship close to light speed. You need special relativity but not quantum mechanics to study its motion. On the other hand, to study a slow-moving electron scattering on a proton, you must invoke quantum mechanics, but you don't have to know a thing about special relativity.

It is in the peculiar confluence of special relativity and quantum mechanics that a new set of phenomena arises: Particles can be born, and particles can die. It is this matter of birth, life, and death that requires the development of a new subject in physics, that of quantum field theory.

Let me give a heuristic discussion. In quantum mechanics the uncertainty principle tells us that the energy can fluctuate wildly over a small interval of time. According to special relativity, energy can be converted into mass and vice versa. With quantum mechanics and special relativity, the wildly fluctuating energy can metamorphose into mass, that is, into new particles not previously present.

Special relativity tells us that energy can be converted to matter.

Zee, A. (2010). *Quantum Field Theory in a Nutshell* (2nd ed.). Princeton, NJ: Princeton University Press.

https://epdf.tips/queue/quantum-field-theory-in-a-nutshellf65ca91cbc5fdf00a51fd98ad1007dd18493.html

https://syntropy.site/Quantum-Field-Theory-in-a-Nutshell

This is possibly the most fascinating idea that I have ever encountered in a science book. **Particles are born, and particles can die.** This implies that particles are alive. This implies that particles have within them some type of Life Force, Intelligence, or Psyche.

Quantum Field Theory unites psyche or life force, classical physics, special relativity, and quantum mechanics. The Gods or God's Psyche had to design and make the quantum fields BEFORE the Gods could make and sustain physical matter and other quantum particles. Quantum fields are made from energy; and, Psyche or Life Force is the only thing we know of that can manipulate energy's form at the quantum level. Psyche is made from energy, and every Psyche has a certain amount of energy that's under its control. The Controlling Psyche or Controlling Life Force decides what form the energy under its control will assume. It can assume any form it wants; and, Psyche or Life Force is the only thing we know of that is capable of making that choice. There's the whole of reality in a nutshell.

Mark My Words

The Qualitative Infinity of Nature (1957)

David Bohm is known for producing "unorthodox" interpretations of Quantum Mechanics. They are unorthodox because they are true, because they have actually been verified, experienced, observed, and proven to be true. The "Orthodox" Interpretations of Quantum Mechanics that were produced by the Materialists, Naturalists, Darwinists, and Atheists have all been FALSIFIED. Back in the 1950's, Bohm caught a ton of flak from the Materialists and Naturalists for presenting them with the truth. Since then, Bohm has been completely vindicated; whereas, Materialism, Naturalism, and their derivatives have been FALSIFIED by Science (observation and experience).

Physical matter is the result of energy interactions between different types of quantum fields. The primary axiom of Quantum Field Theory is that particles are born, and particles die. Particles are made by Something Psyche or Something Intelligent; and, then when they have served their purpose, particles are disassembled and reabsorbed back into the Matrix of Quantum Fields from whence they came. It's all made from energy or light. Even the Atheists and Materialists are able to sense this truth on some level.

Ruđer Bošković explained the qualities of matter as a result of an interplay of forces.

Friedrich Nietzsche in his *Will to Power* furthered this idea with some of the following ideas. Matter is a result of an interplay between different forces. Matter is a synthesis of forces. Movement is governed by the power relations between bodies and forces. Forces are a primitive form of the Will. Nietzsche also periodically wrote about a metaphysical will to power rather than a material will to power, even though he formally rejected psyche or spirit in print. Many philosophers find the truth before finally choosing to reject it.

It's all about quantum forces or quantum fields. It's obvious that the Gods or God's Psyche had to design and make the quantum fields BEFORE physical matter could be made and sustained. Psyche, Intelligence, or Life Force is the only thing we know of that would be capable of taking unorganized energy and forming that energy into functional quantum fields with the intention that those quantum fields would then be used to produce and sustain particles and physical matter. There has to be some kind of Organizing Force or Syntropy Force within our universe; otherwise, everything would still be chaos. Chaos or disorder is the primal state of energy; and, it takes Someone Psyche to bring order and organization to disorder and chaos at the quantum level or the energy level. Psyche is the only thing we know of that can influence energy both at the quantum level and at the physical level. Energy is infinitely malleable or infinitely transformable; and, energy or psyche is always conserved. Psyche, too, is made from energy. Everything is made from energy or light. Every Psyche has a certain amount of energy that's under its control; and, the Controlling Psyche decides what form the energy under its control will assume or take. Psyche has to be intelligent, alive, perceptive, and aware because it has a choice to make. Psyche determines the form that the energy under its control will take.

Many scientists and philosophers play with the idea that energy, psyche, power, or light is the fundamental stuff from which everything is made.

Energy is movement or life. Energy is constantly in motion. Energy is the fundamental stuff from which everything is made, including psyche, spirit matter, quantum fields, and physical matter. Energy also contains some type of Psyche or Will. Will or choice is a function and product of Psyche. Force is used as a theoretical concept to explain interactions between physical bodies. Force or energy is non-physical and pre-physical. The Gods had to design and make quantum fields from energy BEFORE physical matter could be made.

In 1957, David Bohm announced his desire and intention to upgrade the Philosophy of Science from the Philosophy of Mechanism to something that he called The Qualitative Infinity of Nature.

I did the same thing in 2017 with my book, *The Ultimate Model of Reality*, sixty years later. We scientists keep re-inventing the same wheel (the same interpretation of Quantum Mechanics) over and over again, because the Materialists, Naturalists, Physicalists, and Atheists who control our public schools won't allow us to standardize our terms and teach them in our public schools because something like the Qualitative Infinity of Nature FALSIFIES Materialism, Naturalism, Nihilism, and even Atheism. Nevertheless, we scientists have to propose and teach something like the Qualitative Infinity of Nature, non-physical Quantum Fields, and some sort of organizing principle like Psyche or Consciousness or Thought if we want to know how things really work.

From the very beginning of time, physicists and scientists have been search for the Ultimate Particle or the most elementary particle. If there is such a thing as a point-particle or "the ultimate particle", it will be Psyche or Intelligence. When it comes to physics, the smaller always dwells within and controls the larger. Psyche has been experienced and observed.

https://psyche-ontology.com/psyche-observed/

https://psyche-ontology.com/psyche-experienced-and-observed/

Buhlman, William L. (1996). Adventures Beyond the Body: How to Experience Out-of-Body Travel. New York: HarperCollins.

https://psyche-ontology.com/buhlman/

Psyche is experienced first-hand as an immaterial viewpoint in space. Psyche is seen or observed as a pin-point of light. We know that Psyche exists because it has been experienced and observed. It's time for the scientists to start to define it, start to identify it, and start to explain how it works.

That's what science is supposed to be – observation and experience – knowledge. Observation and experience teach us that the Ultimate Particle is Psyche or Intelligence. If there is such a thing as a point-particle or an infinite singularity, then Psyche is it. We know that it exists because it has been experienced and observed; and therefore, as scientists, we are obligated to try to figure out what it is and how it works – if we are going to do science for real.

The Ultimate Model of Reality: Psyche Is the Ultimate Cause

https://www.amazon.com/gp/product/B071NC9JK6/

Ever since the discovery of Quantum Mechanics, science has been in desperate need of a whole new Model of Reality that includes the whole of reality and existence. As we approach 2020, a hundred years later, we're still not there yet. Newtonian Mechanics, Materialism, Physicalism, Mechanism, and Naturalism are still holding us back preventing us from making this quantum leap. Materialism, Naturalism, and their derivatives were designed to prevent us from discovering and accepting Quantum Mechanics, Quantum Field Theory, Psyche, Intelligence, and Consciousness. Materialism, Naturalism, and their derivatives work as advertised, keep us ignorant, and prevent us from finding and knowing the truth.

Back in 1957, David Bohm started to go beyond what the Materialists, Naturalists, and Atheists are capable of understanding and accepting. David Bohm started to pursue The Qualitative Infinity of Nature.

In order to find and know the truth, we have to identify and eliminate everything that is false. The false is falsified by the truth; and, the truth is repeatedly experienced and observed. That's the way it works.

Editor Lee Nichol starts the book, *The Essential David Bohm*, by falsifying the Philosophy of Mechanism, or falsifying the Philosophy of Materialism and Naturalism. That's where we scientists always have to start if we are going to make any real progress in science. The Philosophies of Materialism, Naturalism, and Mechanism were designed to falsify Quantum Mechanics, Action at a Distance, and Quantum Field Theory; so, we have to use Quantum Field Theory, Light, Space, Time, Action at a Distance, Quantum Tunneling, the Quantum Zeno Effect, and Quantum Mechanics to falsify the philosophies of Materialism, Naturalism, Nihilism, and Mechanism, if we are to make any progress whatsoever in science.

https://origin-science.org/The-Essential-David-Bohm/

In this essay Bohm argues for a philosophical perspective that transcends classical mechanism, while still utilizing a mechanistic framework when applicable. With such an approach we can sidestep unnecessary "either–or" philosophical choices, and thus maintain maximum flexibility of view with regard to a philosophy of science. Innocuous on its face, this perspective ultimately casts doubt on the widely-held view that the whole of nature can be reduced to a set of laws that are in principle finite and knowable.

To illustrate the basis for such a non-mechanical view, Bohm points to research into the structure of the atom. In spite of perennial claims that the ultimate structure of the atom is about to

be found, the simple fact is that no such structure has yet been discovered. On the contrary, the depth and scale of atomic substructure give no indication whatsoever of being exhaustively determined, giving way to ever greater subtlety and theoretical complexity. An ultimate structure will not be found, says Bohm, because it does not exist. There is no ultimate or final structure to anything at all, including atoms.

Intersecting with the inexhaustible nature of form and structure is what Bohm refers to as the "background." This background is the totality of causes and conditions that contribute to the creation and dissolution of each temporary "thing." Each thing not only has its own relatively stable substructure, but has a potentially unlimited context as well, much like a particle's derivative relation to its field. In Bohm's view, the complex relation of each thing to its background, together with the richness of its substructure, suggest that mechanistic accounts will always be superseded by new levels of complexity and order. If indeed elementary material "things" are thus potentially infinite in their nature, far more so will be the case for the whole of nature. Consequently, we must reformulate our conceptions of the laws of nature so that they account for context and contingency, rather than generalizing them in absolute fashion.

The pairing of substructure and background is an early formulation of the relation between implicate and explicate orders. While the investigation of material substructures is oriented toward increasingly subtle aspects of an explicate three-dimensional order, the notion of an infinite background lays the foundation for conceiving a multi-dimensional implicate order, from which all material structure – of any degree of subtlety – derives its relative existence and sustenance. – (Lee Nichol, *The Essential David Bohm*, pp. 9-10.)

Energy or light is infinitely malleable. Energy, psyche, or light can be formed into anything, including spirit matter and physical matter. God is in the light. There really is NO ultimate or final structure because energy or light can be formed or changed into anything that it can possibly imagine. This truth falsifies Materialism, Naturalism, Physicalism, Atomism, Mechanism, and Nihilism. David Bohm was starting to pursue this truth back in 1957 before I was born. Somebody has got to pursue the truth because you won't get it from the Materialists, Naturalists, Darwinists, Nihilists, Behaviorists, and Atheists who are in control of our public schools and public institutions.

David Bohm taught about different levels of reality or existence. David Bohm had the Manifest Order or the Explicate Order which is what I tend to call the physical level within my different books. Then Bohm had the Implicate Order, upon which the physical level is built. The Implicate Order consists of non-physical quantum fields that permeate the whole of space and form a sort of Matrix. This is

546

the transition point between the quantum level and the physical level; and, it is based upon Quantum Field Theory and Quantum Mechanics. In my books, I tend to call this level the quantum level.

Then later on David Bohm introduces what he calls the Super-Implicate Order. This is the raw energy or light from which the quantum fields are organized and made. Everything is made from light. Energy or light is infinitely malleable. It can be formed into anything – including psyche, quantum waves, quantum fields, spirit matter or dark matter, dark energy, visible light, and physical matter. Everything is made of light; and, this energy is always conserved. It can neither be created nor destroyed. This is the light level or the energy level. I often call it the quantum level within my books, because Quantum Mechanics is the science of how energy or light really works at the quantum level.

Then David Bohm goes one step further and mentions a "significance level", or a "meaning level", or an "active information level", or a "consciousness level" which is even more fundamental and essential than the energy level or the light level. When it comes to physics, the smaller always dwells within and controls the larger. Every type of energy or light has within it some type of intelligence, psyche, or consciousness. Light is sentient and alive. It can think. This final level is a sort of Super-Super Implicate Order. I tend to call it the psyche level within my books. Psyche has been experienced and observed. If there is such a thing as a point particle or an infinite singularity, then Psyche is it.

Most of *The Essential David Bohm* falsifies the Philosophy of Mechanism, or what we tend to call Materialism or Physicalism.

> **In conceiving the implicate order, Bohm was originally concerned with the nature of order per se. Both scientifically and socially, he felt that we are largely unaware of the degree to which inherited orders, or paradigms, dominate our perception and thought. In our current time, it is the order of mechanism that prevails. The essence of a mechanistic view is that all natural phenomena exist in a strictly external relationship to one another and exhibit precise and discernible chains of cause and effect. In such a world of externalized entities, the theory goes, it is only a matter of time before the "ultimate particle" – the basic stuff of which the universe is made – will be discovered and explained.**

> **However, in both theory and experiment this mechanistic order has been undermined by a series of developments in twentieth-century physics. Bohm points to four features of the "new" physics that, taken together, require the formulation of a larger, more encompassing order. The first significant challenge to mechanism came from Einstein, who claimed there were deep contradictions in the very notion of an independently existing particle. He proposed that what we normally think of as a particle is actually a temporary localized pulse emerging from a larger field, very much as a vortex temporarily forms from the dynamic flowing of a stream. In Bohm's**

words, "the field structure associated with two pulses will merge and flow together in one unbroken whole." In this way, the mechanistic notion of the inherent separability of material structures was initially questioned. Yet Einstein ultimately took recourse in a mechanistic view by asserting that the fields themselves were separate and external from one another.

It was to be three subsequent discoveries in quantum mechanics that would more profoundly challenge a mechanistic world view. The first of these is that the movement of quantum entities (photons, electrons, etc.) does not possess the classical (and common sense) attribute of steady, continuous development from one state or position to another. Rather, the movement of quantum entities is *discontinuous*. Action or movement is constituted of indivisible *quanta* – discrete pulses or packets of energy – which "jump" from one energy state to the next, without passing through intermediary energy states. To Bohm, this quantum discontinuity implies a universe woven together in a dynamic, tapestry-like configuration: "If all actions are in the form of discrete quanta, the interactions between different entities (e.g., electrons) constitute a single structure of indivisible links, so that the entire universe has to be thought of as an unbroken whole."

The second quantum discovery is that matter may behave like a solid particle in one context, like a wave in another context, and like both together in yet a third context. This context-dependent aspect of matter violates the mechanistic axiom that *an entity is what it is,* until it irreversibly transforms through disintegration. That an entity (e.g., an electron) can manifest as a wave in one context and a particle in another suggests, according to Bohm, a kind of information exchange that is more akin to transformations in organisms than to the interacting parts of machines. From this perspective, the "particle" would seem to be gathering information about its environment and responding according to the *meaning* of the information. This capacity for transformation – the famous "wave–particle duality" – is vividly illustrated in the double slit experiment.

The third quantum discovery, and perhaps the strangest, is the well-demonstrated non-local connection between particles (e.g., electrons) – referred to as "entanglement" in current terminology. When under certain conditions electrons combine to form a coupled pair, and are then again separated by significant distances, instantaneous influence between the particles is discovered, for which there is no "local" or mechanical explanation. One of the possibilities that arises from this observation is that the speed of light has been violated; at the very least it becomes clear that local causality cannot fully account for the phenomena of quantum mechanics. Thus, either the theory of quantum mechanics is wrong –

very unlikely, based on a century of unqualified experimental validation – or some form of non-locality permeates reality.

For Bohm, a reality structure founded on discontinuous quantized movement, context-dependent form, and non-local action requires the envisioning of a new order that both incorporates and goes beyond the order of mechanism. However, it was Bohm's position that this requirement is understood by the majority of the physics community only via highly abstract mathematical algorithms, with little or no concern for what these rather shocking assaults on common sense might mean. Indeed, the prevailing quantum orthodoxy maintains that any search for meaning or structure behind experimental phenomena is itself pointless and without meaning. Thus, the three quantum discoveries – with their potentially vast implications – are shoe-horned into a tacit metaphysical commitment to mechanism, in spite of the fact that such a commitment is philosophically inconsistent with experimental fact. And as go the assumptions of the scientific community, felt Bohm, so go the assumptions of the culture at large. In this way the philosophy of mechanism continues to trickle into and structure the perception, imagination, and creativity of contemporary society. – Lee Nichol (pp. 3-5.)

It is obvious to most of us that Quantum Mechanics and Quantum Field Theory FALSIFY Materialism, Physicalism, Mechanism, Naturalism, and Nihilism. Quantum Fields are obviously non-physical in nature and origin. The Gods had to design and create the quantum fields first before the existence of physical matter became possible.

The discontinuous nature of quantum tunneling FALSIFIES the efficient causality of the Mechanistic Philosophy. The little buggers often teleport rather than travel. They are also omnipresent in scope and range.

Energy or psyche can change form at will. This is what has been experienced and observed in the quantum realm or the spirit world. This observational fact completely FALSIFIES Materialism and Naturalism, which are highly dependent upon the form remaining the same.

Energy or psyche is telepathic and omniscient in its native format. The Quantum Zeno Effect proves that individual atoms can read your mind, and they know when you are looking at them or paying attention to them. Atoms are intelligent and alive, which means that they have some kind of psyche or life force within them that's also made out of energy.

Energy or psyche is non-physical in nature and origin; and, quantum fields are made out of energy. The proven existence of anything non-physical such as space or time or radio waves FALSIFIES Materialism, Naturalism, Mechanism, Physicalism, Atomism, and their derivatives. All you need is one.

Furthermore, when it comes to human behavior, it's very much a statistical phenomenon. It can be demonstrated through scientific experimentation that under certain given circumstances a majority of people, on average, will choose to act in a certain way; but, a minority on average will not. It has also been demonstrated and observed that when people know that they are being tested or examined in a psychological or behavioral experiment, a large number of people will deliberately act contrary to expectations just to buck the trend. When dealing with free agents, nothing is statistically certain. That's just the way that Quantum Mechanics is. Every type of energy has an innate Psyche or Intelligence and therefore the ability to choose whatever form it wants to assume. Energy is infinitely malleable. It can form itself into anything, including spirit matter and physical matter. We know this is true because it has been experienced and observed. Energy is constantly changing form. What's miraculous is that a small portion of psyche or energy or light actually chooses to stay in a stable physical form.

In order to make progress in science, we each must develop a Philosophy of Science or a view of science that falsifies the philosophies of Materialism, Naturalism, Nihilism, Darwinism, and Mechanism. The false always has to be falsified by the truth. David Bohm started his research by developing a new Philosophy of Science that transcends classical mechanics, thereby allowing for the inclusion of Quantum Mechanics or Transcendental Physics into science. We have to create a non-mechanical view or a non-physical view if we want to successfully include Quantum Field Theory into science, because quantum fields are obviously non-physical and pre-physical.

Back in 1957, David Bohm had to falsify the Philosophy of Mechanism before he could do real science.

For several centuries there has existed a very strong tendency for one form or another of the philosophy of mechanism to be generally adopted among physicists. In the present chapter we shall criticize this philosophy, demonstrating the weaknesses in its basic assumptions, and then we shall go on to propose a different and broader point of view which we believe to correspond more nearly than does mechanism to the implications of scientific research in a wide range of fields. In addition to presenting this broader point of view in some detail, we shall also show how it permits a more satisfactory resolution of several important problems, scientific as well as philosophical, than is possible within the framework of a mechanistic philosophy.

The essence of the mechanistic position lies in its assumption of fixed basic qualities, which means that the laws themselves will finally reduce to purely quantitative relationships.

We shall now review some of the most important criticisms that can be made against the philosophy of mechanism. First of all, the historical development of physics has not confirmed the basic

assumptions of this philosophy, but rather, has continually contradicted them. Secondly, the mechanistic assumption of the absolute and final character of any feature of our theories is never necessary. For the possibility is always open that such a feature has only a relative and limited validity, and that the limits of its validity may be discovered in the future.

Thirdly, the assumption of the absolute and final character of any feature of our theories contradicts the basic spirit of the scientific method itself, which requires that every feature be subjected to continual probing and testing, which may show up contradictions at any point where we come into a new domain or to a more accurate study of previously known domains than has hitherto been carried out. Indeed, the normal pattern that has developed without exception in every field of science studied thus far has been just the appearance of an endless series of such contradictions, each of which has led to a new theory permitting an improved and deeper understanding of the material under investigation. Thus, the full and consistent application of the scientific method makes sense only in a context in which we refrain from assuming the absolute and final character of any feature of any theory and in which we therefore do not accept a mechanistic philosophy.

Lumping all of the above diverse possibilities into the single category of "things," we see that a systematic and consistent analysis of what we can actually conclude from experimental and observational data leads us to the notion that nature may have in it an infinity of different kinds of things. – David Bohm (pp. 10-13.)

First, the Philosophy of Mechanism and the Philosophy of Materialistic Naturalism have continually been FALSIFIED by science itself. Secondly, Materialism and Naturalism are severely reduced and limited. It has been demonstrated that over 95% of our physical universe is in fact non-physical in nature and origin being comprised mostly of dark matter (spirit matter) and dark energy. Energy itself is non-physical and pre-physical in nature and origin. Thirdly, the enforcement of the Philosophy of Materialism is contrary to the spirit or the intent of the scientific method which demands that we follow the evidence wherever it leads us, even into the transdimensional, or the non-physical, or even the quantum and the spiritual. Nature has in it an infinity of different kinds of non-physical things, instead of just physical matter. Materialism, Naturalism, and Mechanism have been falsified by observation, experience, and experimentation. Science itself falsifies them.

Any limited number of qualities, properties, and laws will prove to be inadequate. (p. 16.)

Materialism, Naturalism, or Mechanism is worthless for science because it is inadequate and deliberately limited. That's its main flaw. By definition, in principle or practice, Scientific Naturalism is deliberately reduced, limited, and restricted specifically to the physical domain. It deliberately excludes the non-physical, the non-local, the quantum, the spiritual, and the pre-physical.

The Philosophies of Materialism, Naturalism, Mechanism, Nihilism, and Atheism have been deliberately reduced and limited. Their whole purpose is to falsify and eliminate Psyche and Quantum Mechanics, because the proven and verified existence of Psyche and Quantum Mechanics falsifies them. The Materialists, Naturalists, Nihilists, and Atheists are sophists – they are deliberately trying to trick you and deceive you because they don't want you to find and know the truth.

In 1957, David Bohm suggested that we replace the limited and inadequate Philosophy of Mechanism with something that he calls The Qualitative Infinity of Nature. Of course, people ain't gonna do it, if it falsifies their most cherished beliefs.

Not only can nothing of real value for scientific work be lost if we adopt the notion of the qualitative infinity of nature in the specific form that has been described here, but on the contrary, much can be gained by doing this. For, first of all, we can thereby free scientific research from irrelevant restrictions which tend to result from (and which have in fact so often actually resulted from) the supposition that a particular set of general properties, qualities, and laws must be the correct ones to use in all possible contexts and conditions and to all possible degrees of approximation. Secondly, we are led to a concept of the nature of things which is in complete accord with the most basic and essential characteristic of the scientific method; i.e. the requirement of continual probing, criticizing and testing of every feature of every theory, no matter how fundamental that theory may seem to be. For this view explains the necessity for doing scientific research in just this way and in no other way, since, if there is no end to the qualities in nature, there can be no end to our need to probe and test all features of all of its laws. Finally, as we shall show throughout the rest of this chapter, the assumption of the qualitative infinity of nature leads to a much more satisfactory solution of a number of important problems, both scientific and philosophical, than is possible within the framework of a mechanist philosophy; and this in turn gives further evidence that it is a better point of view for the guidance of scientific research.

In conclusion, then, the notion of the qualitative infinity of nature permits us to retain all the positive achievements that were made possible by the development of mechanism. In addition, it enables us to go beyond mechanism by showing the limitations of the latter philosophy and by pointing towards new directions in which our concepts and theories may undergo further development. Naturally, we do not wish to propose here that the qualitative infinity of nature is a final doctrine, beyond which no further steps can ever be made. Indeed, as science progresses, it seems very likely that the

552

qualitative infinity of nature will eventually be found to fit into some still more general point of view, which in turn retains its positive achievements, and which goes beyond them, much as the notion of the qualitative infinity of nature goes beyond mechanism. But, in this chapter, our purpose is merely to call attention to the many factors that suggest the need for this important step carrying us outside the limits of a mechanistic philosophy, and to show the numerous advantages that come from taking this step. (pp. 16-17.)

As scientists, the only way we can find and know the truth is to eliminate everything that has been falsified or proven false, starting with the philosophies of Mechanism, Materialism, Naturalism, Darwinism, Nihilism, Behaviorism, Determinism, Physical Reductionism, Atomism, and Atheism. If you successfully eliminate everything that is false – then only the truth will remain. The "non-existence of God" cannot be experienced and observed – it cannot be verified; and, that philosophy (Atheism) doesn't make any sense in light of the thousands of times that the Biblical God Jesus Christ has been experienced and observed after He rose from the dead.

https://www.youtube.com/results?search_query=NDE+Jesus

Observation and experience trump philosophical speculation and wishful thinking every time. The Materialists, Naturalists, Nihilists, Mechanists, and Atheists don't want God to exist, so they deny all evidence of His existence. Atheism of any kind is based upon a refusal to look at evidence. That's how these people prove their point – by refusing to look at the evidence that falsifies their chosen point of view. That's not science. That's dogmatism and extremism.

The truth of an event or phenomenon is always established by witnesses. That's how we know that physicists' many-world theory or multi-me theory is wrong – because it has never been experienced nor observed outside of science fiction.

Many-Worlds Theory

https://www.youtube.com/watch?v=dzKWfw68M5U

We human beings as a collective whole repeatedly experience what is real and what is true. In contrast, there is no way to experience or observe the non-existence of God or the non-existence of the non-local non-physical quantum realm. There is no way to experience the non-existence of Psyche, Quantum Mechanics, or Quantum Fields. Your every waking moment is an experience of consciousness or psyche. There's no way to experience the non-existence of Psyche, because it's your Psyche that has and remembers experiences.

Remember, if you eliminate everything that is false, then only the truth will remain. You start by eliminating Materialism, Naturalism, Darwinism, Nihilism, and Atheism. You have no other choice if you want to find and know the truth. I belabor the point in my essays and books because it's extremely important to understand. Science is supposed to identify and eliminate everything that is false; and, that can't be done if one's science is based exclusively on falsehoods such as

Materialism, Naturalism, Darwinism, Nihilism, Determinism, Atomism, and Atheism. The truth cannot be built on a foundation of falsehood.

The false is repeatedly falsified by the truth; and, the truth is repeatedly experienced and observed. The truth is constantly being verified by witnesses. That's the way it works. There are NO witnesses to the non-existence of God. There are NO witnesses to the non-existence of psyche, quantum fields, non-locality, or the non-physical. There are NO witnesses to the non-existence of an after-life. There are ONLY witnesses to these kinds of things.

https://www.youtube.com/results?search_query=NDE

https://www.youtube.com/results?search_query=NDE+Spark

https://www.youtube.com/results?search_query=NDE+God

The Science of Near-Death Experiences and the Science of Out-of-Body Experiences is hitting critical mass on YouTube. I had no idea until I saw it with my own eyes.

When I was near to death, and when my old personality died and my new personality was born, I remember floating in a bright obsidian black empty void and I was completely and totally at peace for the first time in my remembered existence. Based upon my personal experience, the atheists are right. There's nothing there when we die. However, I also realized that the atheists are wrong, because I was there when I died, and I remember it.

We have to get past Materialism, Naturalism, Darwinism, Nihilism, Mechanism, and Atheism if we want to make any progress in science.

Quantum Mechanics never made any sense to me while I was a materialist, naturalist, nihilist, and atheist. Now I know why. The Philosophies of Materialism, Naturalism, Darwinism, Nihilism, Behaviorism, Determinism, Physical Reductionism, Atomism, and Atheism were DESIGNED to falsify and eliminate Quantum Mechanics and Psyche from science. They work by turning Quantum Mechanics and Psyche into non-sense. That's why materialistic interpretations of Quantum Mechanics and Quantum Field Theory NEVER make any sense, because quantum fields are non-physical in nature and origin to begin with. Quantum fields are pre-physical. Quantum fields are needed before physical matter can be made and sustained.

Mark My Words

Thoughts Are Quantum Waves

Thoughts are quantum waves. The Human Psyche produces, transmits, receives, and stores quantum waves or thoughts. That's what Psyches do. That would make "thought" even smaller and more fundamental than Psyche, would it not?

By 1957, David Bohm isn't quite there yet; but, he is well on his way. He is laying the foundation for Psyche, consciousness, memories, mind, and thought.

Each kind of thing is maintained in existence by a balance of opposing processes in its infinite background and substructure, which are tending to change it in different ways.

We conclude, then, that we must finally reach a stage in every theory where we introduce the notion of something with unvarying and exhaustively specifiable modes of being, if only because we cannot possibly take into account all the inexhaustibly rich properties, qualities, and relationships that exist in the process of becoming. At this point, then, we are making an abstraction from the real process of becoming. Whether the abstraction is adequate or not depends on whether or not the specific phenomena that we are studying depend significantly on what we have left out. With the further progress of science, we are then led through a series of such abstractions, which furnish ever better representations of more and more aspects of matter in the concrete and real process of becoming.

Now, when we refer to the process of becoming by the word "concrete," we mean by this to call attention to the quality of being special, peculiar, and unique that one always finds to be characteristic of real things when one studies them in sufficient detail. For example, if we consider any concept (e.g. apples), then this concept contains nothing in it that would permit us to distinguish one apple from another. We may then indicate other qualities which make such a distinction possible (red apples, hard apples, sweet apples, etc.). Evidently, no finite number of such qualities can ever give a complete representation of any specific example of a real apple. Of course, by going deeper (e.g., by giving the physical and chemical state of each part of the apple) we could come closer to our goal. But this process could never end. For even the modes of being of the individual atoms, electrons, protons, etc., inside the apple are in turn determined by an infinity of complex processes in their substructures and backgrounds. Thus, we see that because every kind of thing is defined only through an inexhaustible set of qualities each having a certain degree of relative autonomy, such a thing can and indeed must be unique; i.e., not completely identical with any other thing in the universe, however similar the two things may be.

Carrying the analysis further, we now note that because all of the infinity of factors determining what any given thing is are always changing with time, no such a thing can even remain identical with itself as time passes. In certain respects, this brings us to a deeper notion of the process of becoming than we had before. For at each instant of time, each thing has, when viewed from one side, an enormous (in fact infinite) number of aspects which are in common with those that it had a short time ago. Indeed, if this were not so, it would not be a thing; i.e., it would not preserve any kind of identity at all. On the other hand, when viewed from another side, it has an equally enormous (in fact infinite) number of aspects that are not those that it had a short time ago. For typical sorts of things with which we commonly deal, however, these latter aspects are not essential in the normal contexts and conditions with which we work. In new contexts (e.g., a sub-atomic or a super-galactic time scale) or under new conditions (e.g., very high temperatures), these aspects may, however, take on a crucial importance.

We are in this way led to the conclusion that the process of becoming will necessarily have, at each moment, certain aspects that are concrete and unique. In other words, each thing in each moment of its existence must have certain qualities which, in some respects, belong uniquely to that thing and to that moment. The notion of unvarying and exhaustively specifiable modes of being is then an abstraction obtained, in general, by considering what is common to the same thing at different moments, or to many similar things at the same moment. In doing this, we evidently ignore the differences between these things, which are just as essential a side of them as are their similarities. By abstracting in more detail from these differences, we are then led to see newer but subtler aspects in which these differences contain common or similar relationships that apply to all of these things. Thus, the uniqueness of each thing at each instant of time is reflected in our abstract concepts by the limitless richness and complexity of the concepts that one needs to obtain a better and better abstract representation of matter in the process of becoming, or, in other words, by the inexhaustibility of the qualities that are to be found in nature. (*The Essential David Bohm*, pp. 35-36.)

The Human Psyche is "concrete" in that it is special, peculiar, and unique; and, one always finds the Human Psyche to be characteristic of real things when one studies it in sufficient detail. Each Psyche is unique; i.e., not completely identical with any other thing in the universe, however similar any two Psyches may be. Each Psyche contains a unique set of memories, thoughts, or quantum waves. Evidently, no finite number of such qualities or memories can ever give a complete representation of any specific example of a real Psyche. The fundamental and fatal flaw of Mechanism, Materialism, Naturalism, Nihilism, and Atheism is that it's finite, limited, and restricted. Scientific Naturalism is not capable of describing an infinite

556

singularity or a non-physical entity such as the Human Psyche in any logical or useful fashion.

Ask yourself, "What are the infinite qualities or the infinite number of thoughts and memories that make up each individual unique Psyche?" Inside the Psyche, we experience and observe an infinity of complex processes within the substructures and backgrounds of that Psyche, which we call quantum waves, memories, and thoughts. Thus, we see that because every kind of Psyche is defined only through an inexhaustible set of qualities each having a certain degree of relative autonomy, such a Psyche can and indeed must be unique; i.e., not completely identical with any other Psyche in the universe, however similar those two Psyches may be.

Carrying the analysis further, we now note that because all of the infinity of factors determining what any given Psyche is are always changing with time, no such Psyche can even remain identical with itself as time passes. In certain respects, this brings us to a deeper notion of the process of becoming than we had before. A Psyche is a real thing. In fact, it's the most real thing within us, because it is eternal and everlasting. Your Psyche has always existed, and it will always exist. That's the very definition of truth! Your Psyche is your absolute truth. It's what you always were, and it's what you will always be; and, your Psyche is always in the process of becoming something more than it was before. The inexhaustibility of the qualities that are to be found in nature are in fact found within your Human Psyche. Your Psyche was a part of nature or a part of reality long before the first particle of physical matter was ever conceived or made by Someone Psyche.

So, even though David Bohm never mentioned the words "psyche" or "soul" within his essays, he did indeed develop the science and the theories necessary to explain what a Psyche is and how it works at the quantum level. Even though Bohm avoids mentioning psyche or soul within his work, he was already laying the foundation for a Science of Psyche back in 1957 amidst the ridicule and persecution of the Materialists and Naturalists who surrounded him at work. The way the Atheists and Nihilists do science is through ridicule, name-calling, censorship, and intimidation as they try to enforce their point of view onto the rest of us. It works. We let them get away with it, which is why the majority of scientists are Materialists, Naturalists, Darwinists, Nihilists, and Atheists. Anyway, Materialism and Naturalism are dying because they have been falsified by Science itself; so, it's finally time to try to develop something better to take their place.

While developing Quantum Neuroscience, my great discovery came when I first realized that our after-death Life Review is NOT stored within our physical brain. Huge parts of it seem to be stored within Nature's Psyche rather than the Human Psyche. During a Life Review, the human being who is having the Life Review can at times be seen from an external third-person perspective that was never experienced by the human brain. The Life Review also contains 360-degree surround-vision that's NEVER produced nor experienced by the physical body and physical brain. The evidence is overwhelming and proves that most of our memories are NOT stored within our physical brain.

Scientists have also proven that there are NO memory engrams within a physical brain. There are NO wires within a physical brain. There is NO physical RAM within a physical brain. The physical brain has NO memory storage system at the physical level, which means that our memories are being produced and transmitted at the quantum level and then stored at the psyche level.

Our memories are NOT stored within our synapses (gaps in space) nor within our neurotransmitters, as the Materialists, Naturalists, Nihilists, and Atheists claim. The memories that show up in our after-death Life Review are obviously stored someplace else besides our physical brain. Furthermore, if our memories were in fact stored within our neurotransmitters as the Naturalists and Atheists claim, then our memories would be randomized and scrambled trillions of times per second, because neurotransmitters are dumped randomly into the synaptic cleft. There is no order or organization there whereby memories could ever be stored. Our memories are NOT transmitted by neurotransmitters, NOR are our memories stored within neurotransmitters at the physical level.

Many of the Materialists and Naturalists claim that our memories are stored within our synapses. Synapses are gaps in space. They are NOT physical objects. The synapses are filled with cerebral spinal fluid which facilitates the transfer of neurotransmitters between the synapse, but if our memories are stored within the cerebral spinal fluid or within the water, then all of that is taking place at the quantum level within Nature's Psyche, and NOT at the physical level. There's NO quantum field matrix or RAM at the physical level within the cerebral spinal fluid capable of producing and storing memories. If memories and ideas are in fact being transmitted from one neuron to the next, it's actually taking place at the quantum level, and NOT the physical level.

The physical level and physical matter are the ultimate worst way to transmit and store information. Remember, at the physical level, by definition a single physical bit can only store ON or OFF – YES or NO – OPEN or CLOSED. A neurotransmitter is a physical bit. It is a KEY that OPENS an ion gate. We actually have to move over to something quantum, in order to be able to store more than ON or OFF within a single physical bit. In order to get more than a single bit of information within a single hardware bit, we actually have to switch over to Quantum Mechanics and Qubits. Not only does it take intelligence to make and load a Qubit, but it also takes intelligence or psyche to read and use a Qubit to do computations. Intelligence or psyche is required to properly process information, both at the physical level and at the quantum level; otherwise, random chaos ensues, and nothing will ever get done. Remember, only Psyche can read, understand, process, transmit, receive, and store quantum information at the quantum level.

https://en.wikipedia.org/wiki/Qubit

Our scientists are trying to develop "physical implementations" of Qubits and Quantum Computers. Again, a physical implementation of information transfer and information storage is the ultimate worst way to do memory storage. However, the

Psyche can and does function at the quantum level where there are NO physical limitations. Psyche has direct access to non-physical implementations of Qubits.

When dealing with physical implementations of Qubits, we are limited to "two-state" or "two-level" quantum mechanical objects such as electron spin – up or down. However, Psyche theoretically has direct access to infinite-state quantum mechanical objects or infinite-level quantum mechanical objects; and thereby, the memory storage capacity that a Psyche can access and process at the quantum level goes towards infinity. Remember, there are NO physical limitations at the quantum level, and Psyche is NOT subject to physical limitations at the quantum level. Psyche can make its own quantum computer out of light at the quantum level, and the whole thing could easily fit within an electron and store and process all the information in the universe within that electron faster than the speed of light.

Remember, physical implementations of Quantum Computers will be powerful yet limited; but, a true non-physical or quantum implementation of a Quantum Computer would have NO limitations whatsoever. Physical matter was designed by the Gods to place limits on us; otherwise, there are no limits at the quantum level or the spiritual level, and everything is highly unreliable and highly unpredictable as a result. Physical matter gives us predictable stability. I can depend on this paper being on my computer and cloud drive when I go looking for it tomorrow. In contrast, if my Psyche and Spirit Body were to enter a non-consensus reality in the spirit world or quantum world, and make a nice world for myself from the unformatted spirit matter found there, I could not depend on that world being there the next time I go looking for it; and, I'll have to make it all over again if I want it again.

Physical matter is used to make the ultimate consensus reality. It works because it has been greatly limited and restricted. At the physical level, the Gods slow things down for us so that we can experience them, learn from them, and remember having done so.

In contrast, a non-consensus reality or an unorganized reality is the default state in the quantum realm or the spirit world. Whenever the Human Psyche enters into a non-consensus reality, it is recognized by the other psyches as an Organizer or a God, and the spirit matter organizes itself automatically according to the demands and expectations of the Human Psyche. Nature's Psyche can read your mind and knows what you want it to do and what you want it to become. There are things that act, and there are things that are acted upon. The Human Psyche has advanced to the level where it can ACT; whereas, Nature's Psyche REACTS or responds to the demands of the Human Psyche.

One of my greatest discoveries of Quantum Neuroscience is my discovery that memories are NOT stored at the physical level within the physical brain. Memories are quantum waves, which means that memories are being produced by Something Psyche, transmitted by Something Psyche, received by Something Psyche, and then stored by Something Psyche.

Quantum Neuroscience: The Answer to Life, the Universe, and Everything

https://www.amazon.com/gp/product/B079Z6QQQB/

When it comes to physics, the smaller dwells within and controls the larger. That's the way it works. Remember, quantum fields, physical matter, computers, quantum computers, genes, proteins, genomes, and consensus realities are MADE by Someone Psyche or Someone Intelligent. They don't just spring into existence whole from nothing as the Materialists, Naturalists, Darwinists, Nihilists, and Atheists claim. There's no such thing as Creation Ex Nihilo or Atheism. Creation ex nihilo, chemical evolution, macro-evolution, and spontaneous generation are physically impossible. They can't be done, which means that they weren't done.

Due to the random nature of the water within the cerebral spinal fluid, it's obvious that our memories have to be stored within something that is infinitely more stable and eternal than physical water or physical cerebral spinal fluid. If our memories are indeed stored within our synapses (gaps in space), then they have to be stored at the quantum level and the psyche level because it's physically impossible for them to be stored at the physical level. Physical matter and a physical brain are temporary. Memories stored within a physical brain at the physical level would never be able to survive the death of that physical brain and then show up in an after-death Life Review. This is Logic 101.

https://www.youtube.com/results?search_query=nde+life+review

There are only two parts of us that are in fact eternal and everlasting – the energy that has been assigned to us and is under our control, and our Human Psyche, which is a repository for our memories, dreams, and experiences. In order for our memories to survive the death of our physical brain and show up in an after-death Life Review, those memories have to be stored at the quantum level and the psyche level as quantum waves (organized energy) within Something Psyche or within something capable of storing quantum waves or thoughts at the psyche level.

The false is falsified by the truth; and, the truth is repeatedly experienced and observed. I know what I'm talking about; but, you are going to have to decide for yourself whether you believe me or not.

Mark My Words

Energy Is Infinitely Malleable

Energy is infinitely malleable or infinitely transformable. Psyche can transform the energy under its control into anything that psyche wants it to be! That's the true nature of energy – it's capable of constantly changing its form. Energy, under the control of psyche, can take on an infinite number of different forms. It can become anything that it can possibly imagine; and, imagination is a function of psyche. In physics, the smaller dwells within and controls the larger. Every type of energy is psychic and has some kind of psyche or intelligence within it.

If we are going to destroy something like the Philosophy of Mechanism, then as scientists we are obligated to replace it with something better. David Bohm replaced it with The Qualitative Infinity of Nature.

Back in 1957, David Bohm realized as I have finally realized that energy or psyche is infinitely malleable. Energy or light can change form at will. Energy is intelligent and alive. Energy is omnipresent, omnipotent, and omniscient. Energy, psyche, or light can transform into anything, including quantum fields, spirit matter, and physical matter. This is what the whole of physics is trying to teach us! God is in the light. Energy, psyche, or light is the fundamental stuff of existence and reality. Quantum Mechanics is the scientific study of how energy, psyche, and light really work.

Then came the discovery of a structure for the electrons and protons involving in some as yet poorly understood way the motions of unstable particles such as mesons and hyperons. Still later came the realization that because these latter particles can be "created," "destroyed," and transformed into each other, they too are very likely to have a further structure that is related to the motions of some still deeper-lying kinds of entities the nature of which is not yet known.

An essential characteristic of the rich and highly interconnected substructure of moving matter described above is that not only do the quantitative properties in it continually change but that the basic qualities that define its mode of being can also undergo fundamental transformations when conditions alter sufficiently.

Moreover, in nuclear processes, neutrons can be transformed into protons, either by the emission of neutrinos or of mesons; and of course, as we have seen, mesons are unstable, so that their very mode of existence implies the necessity for their transforming into basically different kinds of particles. Thus, further research into the structure of matter has not only shown what is, as far as we have been able to tell, an unlimited variety of qualities, processes, and relationships, but it has also demonstrated that all of these things

are subject to fundamental transformations that depend on conditions.

Thus far, we have tended to emphasize the inexhaustible depth in the properties and qualities of matter. In other words, we have considered how experiments have shown the existence of level within level of smaller and smaller kinds of entities, each of which helps to constitute the substructure of the entities above it in size, and each of which helps to explain, at least approximately, by means of its motions how and why the qualities of the entities above it are what they are under certain conditions, as well as how and why they can change in fundamental ways when conditions change.

But now we must take into account the fact that the basic qualities and properties of each kind of entity depend not only on their substructures but also on what is happening in their general background [quantum fields]. In physics, research thus far has not tended to stress this feature of the laws of nature as much as it has emphasized the substructure. Nevertheless, the various fields (e.g., electromagnetic, gravitational, mesonic, etc.) that have been introduced into the conceptual structure of physics represent to some extent an explicit recognition of the importance of the background. For, as we have seen, these fields (whose mode of existence requires that they be defined over broad regions of space) enter into the definition of the basic characteristics of all the fundamental particles of current physics.

Moreover, when such fields are highly excited, they too can give rise to qualitative transformations in the particles, while, vice versa, the particles have an important influence on the character of the fields.

Indeed, previous discussion of the quantum theory shows that fields and particles are closely linked in an even deeper way, in the sense that both are probably opposite sides of some still more general type of entity, the detailed character of which remains to be discovered. Thus, the next step in physics may well show the inadequacy of the simple procedure of just going through level after level of smaller and smaller particles, connected perhaps by fields which interact with these particles.

Instead, we may find that the background enters in a very fundamental way even into the definition of the conditions for the existence of the new kinds of basic entities to which we will eventually come, whatever they may turn out to be. Thus, we may be led to a theory in which appears a much closer integration of substructure and background into a well-knit whole than is characteristic of current theories.

The explicit inclusion of the effects of a background [quantum fields] **is essential for the very existence of the** [physical and non-physical] **entities in terms of which our theories are to be formulated** [and the various different particles or entities are to be made]. **– David Bohm** (*The Essential David Bohm*, p. 17-19.)

According to quantum physics, the smaller dwells within and controls the larger. Physical particles have to be made out of something because they are not fundamental stand-alone entities. It's all made out of energy or light. Energy or light is infinitely malleable or infinitely transformable. Entropic physical particles are made out of energy or light, which means that they are infinitely transformable or infinitely malleable as well – they are NOT conserved.

Psyche or Intelligence can form energy or light into anything that that psyche wants it to be. Psyche, or Consciousness, or Life Force is also a type of energy or light. Energy or light of any kind seems to be conscious and alive at the fundamental foundational level, or the psyche level. Whenever energy or the energy within particles transforms into something else, it has to KNOW the nature, the structure, and the laws associated with what it is transforming into; otherwise, the process isn't going to work right. There is Intelligence or Psyche within every type of energy that gives it the innate ability to understand, follow, and obey physical laws. This explains the whole of everything that we humans have ever experienced and observed.

In this essay, Bohm talks a lot about the importance of the background. What's in the background? What permeates the whole of "empty space"? Quantum fields! What do we know about quantum fields? Well, they are non-physical and pre-physical. The Gods had to make the quantum fields BEFORE the existence of physical matter became possible; and, quantum fields are made out of energy or light. Physical matter is also made out of energy or light. It's all made out of the same stuff. Go figure!

Over and over again within his different essays, David Bohm mentions the "background" which is a Matrix of different Quantum Fields. The Materialists and Naturalists claim that the non-physical doesn't exist; so, they call this background "empty space" or "the void" or "non-existence". They've been doing so for thousands of years. Because these people can't see this non-physical Matrix of Quantum Fields with their physical eyes, they simply assume that it doesn't exist. These are the same people who used to assure us that viruses, bacteria, radio waves, quarks, and electrons do not exist. They are wrong; but, these people take great pride in being wrong, and many of them try to force their point of view onto the rest of us through ridicule and intimidation.

Neither causal laws nor laws of chance can ever be perfectly correct, because each inevitably leaves out some aspect of what is happening in broader contexts. Under certain conditions, one of these kinds of laws or the other may be a better representation of the effects of the factors that are dominant and may therefore be the better approximation for these particular conditions. Nevertheless,

with sufficient changes of conditions, either type of law may eventually cease to represent even what is essential in a given context and may have to be replaced by the other.

Thus, we are led to regard these two kinds of laws as effectively furnishing different views of any given natural process, such that at times we may need one view or the other to catch what is essential, while at still other times, we may have to combine both views in an appropriate way. But we do not assume, as is generally done in a mechanistic philosophy, that the whole of nature can eventually be treated completely perfectly and unconditionally in terms of just one of these sides, so that the other will be seen to be inessential, a mere shadow, that makes no fundamental contribution to our representation of nature as a whole. Thus, the notion of the qualitative infinity of nature leads us to the necessity of considering the laws of nature both from the side of causality and from that of chance, as well as more generally from new directions that may go beyond these two limits. (*The Essential David Bohm*, pp. 22-23.)

Physical matter and the physical level are deliberately based upon material causality and efficient causality – the Gods purposefully made them that way so that they would be dependable, predictable, and controllable. In a physical reality, I can depend on my house, my car, my wife, my computer, and my dog being there when I go looking for them after work. In contrast, the non-consensus realities in the spirit realm or quantum realm literally mold or form themselves to the demands and expectations of the individual Human Psyche who enters into them. In the non-consensus realities of the spirit world, the Human Psyche is God; and, the raw unorganized energy can literally form itself into anything that we want it to be.

Quantum Mechanics is based more upon statistical probabilities – chance – than causality, or at least that's the way that the physicists in general tend to treat it. Quantum Mechanics is the science of how energy, or psyche, or light really works.

However, back in 1957, David Bohm left out the third part of reality and existence. Do you know what it is? David Bohm left out meaning, purpose, desire, will, or CHOICE. It wasn't until 1985 that David Bohm started talking about psyche, mind, meaning, significance, thought, and choice. They never talk about CHOICE, until the lights go on and they finally see it and accept the fact that it is real and potent.

Psyche or choice is more fundamental and essential than quantum fields, quantum mechanics, physical matter, causality, and chance. Psyche is the Ultimate Cause. Psyche is the ultimate causal agent. Psyche is what made the quantum fields, quantum laws, physical laws, and physical matter in the first place. Psyche can override causality and chance at will. All types of energy or light are psychic, or intelligent, or conscious, or alive, or perceptive, or aware. Life force is energy or light. Psyche and energy seem to be integrated together into light. Energy and psyche seem to be two sides of the same coin. All energy seems to be intelligent

564

and alive. Energy or light can change form at will. Back in 1957, David Bohm was searching for a common denominator that would tie quantum fields together with physical particles. Psyche, energy, or light is that common denominator. Everything is made out of energy or light, including Psyche or Intelligence. Light or energy is infinitely malleable or transformable. Psyche can literally turn energy or light into anything that it wants it to be.

Aristotle's four "causes" were deliberately limited to Materialism or Mechanism – the physical or observable results of causality. We actually NEED a type of causality that is non-physical and pre-physical in nature and origin; and, Psyche is it. Psyche is the Ultimate Cause. Psyche or Intelligence was the ultimate causal agent. Seeing the need, I resolved this philosophical issue in a number of different books that I wrote.

The Ultimate Model of Reality: Psyche Is the Ultimate Cause

https://www.amazon.com/gp/product/B071NC9JK6/

We have to deliberately replace the falsehoods with truths, or we will always have something false as the philosophical basis of our science.

Back in 1957, David Bohm went beyond what the Materialists, Naturalists, Nihilists, and Atheists are capable of understanding and accepting. I started to do the same thing in 2015, as evidenced by all the different books that I started to write and release in 2016.

https://www.amazon.com/author/science

For me personally, reading William L. Buhlman's book, "Adventures Beyond the Body: How to Experience Out-of-Body Travel" in December 2015 was the turning point. I saw the truth; and, I couldn't go back to the deceptions and the lies. The genie or the psyche got out of the bottle, and I couldn't put it back, nor did I want to. I found the truth way more interesting than all the deceptions and lies, because the truth explains how things really work and what they are.

https://psyche-ontology.com/buhlman/

Energy is infinitely malleable or infinitely transformable. Psyche can transform the energy under its control into anything that psyche wants it to be! That's the true nature of energy – it's capable of constantly changing its form. Energy under the control of psyche can take on an infinite number of different forms.

This is where science and philosophy start to get interesting. It all starts with the First Law of Thermodynamics, or the Conservation of Energy.

The first law of thermodynamics is a version of the law of conservation of energy, adapted for thermodynamic systems. The law of conservation of energy states that the total energy of an isolated system is constant; energy can be transformed from one form to another but can be neither created nor destroyed.

Energy is a conserved quantity; the law of conservation of energy states that energy can be converted in form, but not created or destroyed.

There's the whole of physics in a nutshell. Energy, psyche, or light is eternal or immortal. Energy is pre-physical and non-physical in nature and origin. Energy is the ultimate perpetual motion machine. Energy can be transformed by psyche, choice, intelligence, or will from one form into another; but, energy or light can neither be created nor destroyed. According to astrophysics, 95% of our physical universe is non-physical and still exists in is pre-physical energetic state. Energy is infinitely malleable – it can be formed or transformed into anything that Someone Psyche wants it to be.

Psyche or the Gods transformed some of that energy into entropic physical matter; and, the Gods can transform that entropic physical matter into anything else they want it to be anytime they choose to do so, including spiritual matter or resurrected matter which is physical matter that has no entropy or expiration date built into it. In fact, some of those planets, stars, and galaxies that we see in our telescopes might already be eternal matter or resurrected matter instead of entropic physical matter. We wouldn't be able to tell the difference between the two. They would all look the same to us.

We know from observation and experience that resurrected beings have a tangible physical body that is capable of functioning quantum mechanically or spiritually. The Apostle Paul called a resurrected body a "spiritual body" because it was capable of functioning quantum mechanically or spiritually. A resurrected physical body can eat and drink just like a regular or natural physical body; but, it can also teleport (quantum tunnel), phase-shift (walk through walls), levitate (anti-gravity), enter into a bubble of protection (become indestructible), communicate telepathically psyche-to-psyche, multi-task (be two places at once), as well as die and resurrect at will.

1 Corinthians 15: 44:

42 So also is the resurrection of the dead. It is sown in corruption; it is raised in incorruption:

43 It is sown in dishonor; it is raised in glory: it is sown in weakness; it is raised in power:

44 It is sown a natural body; it is raised a spiritual body. There is a natural body, and there is a spiritual body.

Paul called our ordinary physical body a "natural body"; and because they didn't have quantum mechanics back then, Paul correctly called a resurrected physical body a "spiritual body" because it was capable of functioning quantum mechanically or spiritually. Quantum mechanics is the scientific study of how energy, spirit matter, physical matter, and psyche really work at the quantum level and the psyche level. It has been observed that a resurrected body, or a spiritual body, or a quantum mechanical body has NO physical limitations. It functions according to the laws of Quantum Mechanics.

Energy or light is non-physical in nature and origin; but, the Gods or Psyches have formed some of it into physical matter, including spirit matter (dark matter), physical matter, and resurrected matter.

I haven't figured out the proper relationship between psyche and light yet. They seem to be interchangeable terms – two sides of the same coin. Psyche has been experienced and observed. Psyche is experienced as an immaterial view-point in space. Whenever out-of-body explorers are separated from their spirit body and their physical body, and they go looking for themselves, there is nothing there to be found because psyche is a spark of light or a pin-point of light. It's obvious that psyche is made out of energy. It is also obvious from the Quantum Zeno Effect that all types of energy are psychic or have some kind of psyche or intelligence within them. Energy can be formed into anything that psyche wants it to be – including a pin-point of light or psyche, quantum waves, quantum fields, dark matter or spirit matter, dark energy, gravity, space-time, clocks to measure entropy or the passage of time, and physical matter.

https://syntropy.site/Quantum-Zeno-Effect-Verified/

Henry P. Stapp has been able to distinguish the difference between the Human Psyche (process 1) and Nature's Psyche (process 3) within the mathematics of quantum mechanics; and, quantum mechanics is the science and the math that explains how energy, psyche, and light really work at the quantum level and the psyche level. In contrast, classical mechanics or manifest mechanics explains how energy, psyche, and light work at the physical level – each thing in its own proper place.

In order for the math to have any value to us, every variable within a mathematical equation is supposed to represent something that has actually been experienced and observed and is therefore real and is known to truly exist.

https://sites.google.com/a/lbl.gov/stappfiles/

http://people.bss.phy.cam.ac.uk/~mjd1014/stappr.html

http://www-physics.lbl.gov/~stapp/mdonald.txt

http://www-physics.lbl.gov/~stapp/PTRS.doc

According to the hard-core Classical Physicists, Materialists, Naturalists, and Atheists, consciousness, intelligence, and choice happen "out of the blue" and come from nothing. It's magic! It's creation ex nihilo. For them, conscious choice is an emergent property of physical matter – it has to be, because for them ONLY physical matter exists. These people don't understand Quantum Mechanics, Psyche, Energy, Consciousness, and how these things really work. They refuse to understand. They don't want to understand.

Classical physics or Newtonian physics attempts to enumerate all the different physical limitations that entropic physical matter has been subjected to. Ordinary physical matter or entropic physical matter has been deliberately slowed down and the passage of time or entropy has been deliberately introduced into its constituent parts. The Gods deliberately slowed down entropic physical matter so that we can experience it, observe it, manipulate it, learn from it, and actually remember having done so. At the speed-of-light, there is NO passage of time. There's nothing to experience and observe at the speed-of-light because time stops at the speed-of-light according to the Theory of Relativity. The passage of time literally had to be FORCED into the energy that makes up the constituent parts of entropic physical matter. Someone Psyche or Someone Intelligent had to deliberately slow the stuff down to sub-light speeds in order for entropy or the passage of time to have any effect on the energy that has been frozen into physical matter.

Scientific Axiom: Anything that was obviously and deliberately made obviously has an intelligent Designer and Maker who made it. This is Logic 101. It is obvious. This truth has been experienced and observed.

Scientific Observation: It is obvious that entropic physical matter or ordinary physical matter was obviously designed and made to fulfill that specific function. Entropic physical matter functions too well and has too many limitations or laws associated with it for it to have accidentally come about by chance on a universal scale. The stuff was deliberately made from energy in huge quantities to fulfill a specific purpose. It's obviously organized, rather than random and chaotic.

Scientific Observations: Entropic physical matter is still being made and destroyed on a daily basis. This is what has been experienced and observed. This is science. We humans create new physical matter from energy within our particle accelerators; and, we humans have converted physical matter back into energy within nuclear explosions. In every case, the quantity of energy is always conserved; but, the physical matter obviously is NOT conserved, especially when it is destroyed and converted back into raw energy within a nuclear explosion. These truths are obviously true because they have been repeatedly experienced and observed.

Scientific Conclusion: Therefore, it is logical to conclude that entropic physical matter or ordinary physical matter was designed and made by Someone Psyche or Someone Intelligent to fulfill a specific purpose.

This syllogism is valid. It's properly constructed. You will have to decide for yourself if it is sound or true. Each premise is self-evident and obviously true, which makes the conclusion both valid and sound.

The fact that energy has been observed to be infinitely malleable by Someone Psyche or Someone Intelligent at the quantum level is icing on the cake.

These truths explain everything that we humans have ever experienced or observed.

The Law of the Conservation of Energy: Energy or light is always conserved. It can neither be made nor destroyed. Energy or light is eternal and everlasting. However, energy is infinitely malleable. Energy or light can be formed and transformed into anything, including spirit matter (dark matter), physical matter, dark energy, gravity, radio waves, magnetism, quantum fields, quantum waves, and psyche. Energy or light seems to be the fundamental stuff of existence and reality because it can be formed into anything by Someone Intelligent or Someone Psyche. Quantum Mechanics is the scientific study of how energy or light works at the quantum level, and what makes it work.

Scientific Axiom: Anything that has been obviously organized and made obviously has an intelligent Organizer who organized it and made it. This is Logic 101.

Scientific Observation: At least 24 different types of quantum fields have obviously been designed, made, and deployed. This is what has been experienced and observed. These truths have been verified through scientific observation and experimentation; and, they are an integral part of Quantum Field Theory.

Scientific Observation: Quantum fields are made from energy. Quantum fields are obviously organized energy. This premise is self-evident. Everything seems to be made from energy or light, including physical matter and quantum fields.

Scientific Observation: Quantum fields are essential for the production and maintenance of physical matter. This is the essence of Quantum Field Theory, which some claim is our best-proven and most-used science on this planet. Physical matter cannot exist without the prior existence of quantum fields. Quantum fields are pre-physical. In other words, someone had to make the quantum fields before physical matter became possible and could be made. These truths are self-evident and obviously true.

Scientific Conclusion: Therefore, it is logical to conclude that quantum fields have an intelligent Organizer who designed them, made them, organized them, and deployed them.

These scientific premises are obviously true and mutually supportive; therefore, the conclusion is philosophically valid and obviously sound.

In order to figure out how everything works, we have to figure out how energy or psyche or light works at the quantum level and the psyche level, because entropic physical matter has been built from energy, psyche, and light at the quantum level by someone intelligent existing originally and eternally at the psyche level.

What I have learned to love about all of this is its explanatory power. In contrast, all we get from the Materialists, Naturalists, Mechanists, and Atheists is the claim that the non-physical, the transdimensional, the non-local, quantum fields, quantum waves, thoughts, consciousness, and the psyche DO NOT EXIST. How does that explain anything? It doesn't. It's worthless! I want to know how things really work – not be told that they don't exist. It's obvious that these things exist; so, I want to know what they are and how they work. That's what scientists are supposed to do – figure out what things are and how they work – not deny the evidence, suppress the evidence, ridicule the evidence, and delete the evidence.

I want to know what everything is and how it works. I can't get that from Materialism, Naturalism, Darwinism, Behaviorism, Determinism, Physical Reductionism, and Atheism. I am forced to turn to quantum mechanisms, or spiritual mechanisms, or psychic mechanisms if I want to know what everything is and how it really works.

I'm a Scientist, which means that I'm no longer a materialist, naturalist, nihilist, and atheist because these things are not science. Materialism, Naturalism, Nihilism, and Atheism are dogmatic philosophies or dogmatic religions – not science. Scientists pursue the evidence and the truth wherever it leads them – Materialists, Naturalists, Darwinists, Nihilists, and Atheists don't. Materialists and Atheists deliberately rebel against the truth and try to suppress and destroy the truth. Atheism of any kind is based upon a refusal to look at and accept eye-witness evidence or first-hand science. These atheistic and materialistic dogmatists do their "science" through ridicule, suppression, censorship, groupthink, and the destruction or the deletion of evidence. That is what I have experienced and observed.

Mark My Words

Energy Is the Basis of Everything Else

Energy is movement or motion. Energy or light is alive. Energy or light is the stuff from which everything else was made, including the non-physical quantum fields, non-localized particles, localized particles, and physical matter.

Quantum fields were obviously designed and made from energy to fulfill a specific purpose, which makes it obvious that quantum fields have Someone Psyche or Someone Intelligent who designed them and made them. Quantum Field Theory is the unification of quantum mechanics and the theory of relativity. Quantum Mechanics tells us how energy and psyche function at the quantum level – or the non-physical, non-local, transdimensional, spiritual level. Quantum Mechanics and Quantum Field Theory are repeatedly verified and have effectively been proven to be real and true. Quantum Mechanics is energy mechanics or spiritual mechanics. Quantum Mechanics explains how the non-physical works. Quantum Mechanics explains how energy works, how quantum fields work, how psyche works, and how spirit matter works. It's all made from energy.

Within the documentary, *Everything and Nothing: The Amazing Science of Empty Space*, we have these ideas about energy that I place in **bold**.

If matter and anti-matter meet, they annihilate each other turning back into energy. Within this movie, he repeatedly states that when matter and anti-matter annihilate each other, they disappear completely into nothing; and, that's not true because energy is always conserved. They simply turn back into the invisible and intangible energy from whence they originally came. When it comes to the truth, he is half in and half out, because he is trying to salvage Materialism, Naturalism, Darwinism, and Creation Ex Nihilo (Atheism) in the process.

Energy is the substrate for everything else. Whenever particles annihilate each other or disappear, they don't go into nothing. Instead, they always turn back into the raw energy and quantum fields from which they came. Quantum fields are non-physical in nature and origin. The proven existence of quantum fields falsifies Materialism, Naturalism, Nihilism, Mechanism, and even Atheism.

> **Here, finally, was the answer to the riddle of empty space. Heisenberg's uncertainty principle had suggested that matter could pop into existence for incredibly short periods of time. Now Dirac had provided the mechanism by which matter could be created out of the vacuum** [quantum fields] **and just as quickly disappear again.** (*Everything and Nothing.*)

The Materialists, Naturalists, Nihilists, and Atheists try to make it seem as if these particles come from nothing and then disappear back into nothing like magic, and that simply isn't true. These particles come from the energy within the quantum fields, and when these particles die, their energy is simply returned to the quantum fields from whence they came. The fundamental principle of Quantum Field Theory is that particles are born, and particles die; however, the energy is always conserved. There's no such thing as creation ex nihilo or creation from

571

nothing. Creation ex nihilo is Atheism, and there is no such thing as creation ex nihilo or creation from nothing. The energy has always existed, and it will always exist; and, it is obvious that Someone Psyche or Someone Intelligent organized some of that energy into the universal Matrix of Quantum Fields that permeates the whole of space-time within our physical universe today. The Gods or Overlord Psyches had to design and make the quantum fields BEFORE physical matter could be designed and made. That's the way it works.

Whenever a particle pops out of empty space so simultaneously does its anti-particle. So, whenever you try to remove everything you can from empty space, it's still always a wash with all these quantum fluctuations. Within nothingness, there's a kind of fizzing, a dynamic dance as pairs of particles and anti-particles borrow energy from the vacuum [quantum fields] **for brief moments before annihilating and paying it back again.** (*Everything and Nothing.*)

There you have the truth amidst all the different lies that the Naturalists and Atheists are trying to feed us. Particles and physical matter come from the energy within the quantum fields. Particles and physical matter are born from the energy within the quantum fields, and then when particles or physical matter die, their energy is returned to the quantum fields from whence they came. There's no such thing as creation ex nihilo or Atheism. Energy or psyche has always existed, and it will always exist.

Dogmatic extremisms or dogmatic philosophies such as Materialism, Naturalism, Darwinism, Nihilism, Behaviorism, Determinism, Physical Reductionism, Atomism, and Atheism were deliberately designed to hide the truth from us. However, if you keep pursuing the truth long enough, eventually the truth becomes obvious.

For millennia, the scientists and philosophers have gone back and forth – there's nothing in empty space, to there's a luminiferous ether in space, to there's nothing in empty space, to there's a Matrix of Quantum Fields permeating the whole of space. Once we finally realize that creation ex nihilo or Atheism is false and has been falsified by Psyche and Quantum Field Theory, then suddenly everything becomes clear. We realize that the vacuum isn't empty. Even in a complete and total vacuum, light waves pass through it. The vacuum or space is comprised of a Matrix of Quantum Fields, and quantum fields are made or constructed from energy by Someone Psyche. The quantum fields had to be designed and made BEFORE the first particle of physical matter could be made and sustained. Quantum fields are non-physical and pre-physical. The truth prevails.

Waves are a disturbance traveling through a medium. Quantum waves are a disturbance traveling through quantum fields. There is NO luminiferous ether made out of physical matter. However, the luminiferous ether does in fact exist. It is made from a couple dozen different non-physical quantum fields. Light waves are a disturbance traveling through non-physical quantum fields. This is what we learn from quantum field theory. Particles are made, and particles die. Particles are waves, that travel through quantum fields. Quantum fields are the medium. It's a purely quantum phenomenon and has nothing to do with classical physics. The only time it becomes physical is when a light wave reaches a barrier or its destination; and then, the light wave gathers all of its energy into a bundle and presents itself as a single point at a position of its own choosing. In other words, it's nature's psyche (or the psyche or intelligence or observer within the photon) who collapses the wave function. Nature's psyche or some kind of universal intelligence was making particles and collapsing wave functions long before the first human observer was made. Furthermore, the quantum fields had to be designed and made before the first particle of physical matter could be made and sustained.

Quantum Field Theory is the unification of quantum mechanics and the theory of relativity.

When you bring relativity and quantum theory together, then you have for certain this notion of electron and anti-electron pairs just appearing out of the vacuum [quantum fields]. **So, the vacuum goes from being nothing to being a place absolutely teeming with matter, anti-matter creation.** (*Everything and Nothing.*)

What do we know about creation?

Creation of any kind absolutely requires some kind of Creator; otherwise, all we would have is random chaos. Creation requires Someone Intelligent to design, oversee, and control the whole process. Order and organization don't just spring into existence whole from nothing. Order and organization are deliberately designed and made. This reality is obviously true. It has been experienced and observed.

Who or what would be capable of designing and creating particles out of raw energy and quantum fields? Who or what would be capable of creating quantum fields? Who or what would be capable of creating and sustaining physical matter and physical atoms? Who or what would be capable of choosing what form the energy under its control will take? Who or what can control energy with its mind?

The only logical answer is Someone Psyche, because when it comes to physics, the smaller always dwells within and controls the larger. There has to be something or someone within the energy itself who is capable of creating particles and physical matter out of that energy. Energy is infinitely malleable or infinitely transformable. Energy can be organized into anything. The Controlling Psyche can form the energy under its control into anything that it wants that energy to be. This process requires intelligence, information, decisions, and choice.

Dirac's ideas about empty space were refined and developed into what is known today as quantum field theory; and, these strange fleeting things within nothing became known as virtual particles. So, it seems nothingness is in fact a seething mass of virtual particles appearing and disappearing trillions of times in the blink of an eye. There is activity within apparent nothingness. In order to glimpse it, you have to peer deep within a single atom. Everywhere in the

universe, space is filled with this vacuum that has a deep, mysterious energy. (*Everything and Nothing.*)

It's as if the vacuum is a whole field of energy that is alive with motion, activity, purpose, and intention. These virtual particles that quantum physicists make such a big deal about are simply the transfer of energy and information through the quantum field from one particle to another. They are like radio waves or microwaves. Microwaves transfer energy; and, radio waves tend to be used to transfer information. Virtual particles are communication between organized particles – quantum waves if you will. Psyche, Intelligence, or Consciousness has the innate ability to create, transmit, receive, interpret, and store quantum waves, or these virtual particles that show up in the math of Quantum Field Theory. The scientists have laid the foundation for all of this, but very few of them are willing to take all of this to its logical and rational completion which ends in Absolute Truth, or something that is eternal and everlasting – the Conservation of Psyche or the Conservation of Energy. These people refuse to go there because they have stated dogmatically and axiomatically that Psyche, or Conscious Energy, or Quantum Consciousness does not exist. They can't explain what it is and how it works after they have chosen to believe that it does not exist. These people are blocked from further progress and learning and discovery as a result.

The Materialists, Naturalists, Nihilists, and Atheists assure us that virtual particles don't exist, for the very same reasons that these people tell us that Psyche does not exist. Yet, these same people use virtual particles in Quantum Field Theory because of their powerful explanatory power and their usefulness. It's amazing how something that doesn't exist, like virtual particles, can be so powerful and have such superior explanatory power within something like Quantum Mechanics and Quantum Field Theory.

Whenever these people try to tell us that the non-physical does not exist, these people are always wrong. Virtual particles do exist. They are something real; and as scientists, our job is to figure out what they are and how they work.

Are Virtual Particles A New Layer of Reality?

https://www.youtube.com/watch?v=ztFovwCaOik

Within this video, Matt suggests the following:

Virtual particles can travel-faster than light and outside of time because they aren't physical. Virtual particles don't travel because they are omnipresent. At the speed-of-light, time stops and the distance between particles collapses to zero; therefore, quantum waves, thoughts, or virtual particles are omnipresent or everywhere at once. Virtual particles have a completely undefined position, which is the same thing as saying that they are everywhere at the same time. In other words, information or virtual particles quantum tunnel or teleport from one particle to the next. According to Quantum Field Theory, virtual particles are quantum interactions between particles.

It's all made from energy. Virtual particles are energy interactions between particles through the quantum fields. Virtual particles represent the transfer of information, or quantum waves, or thoughts between particles. Virtual particles are quantum waves of energy – or Wi-Fi at the quantum level. Each particle has an Intelligence or a Psyche controlling it; and, every Psyche has a certain amount of energy that's under its control. These particles are alive and communicating with each other and interacting with each other through quantum waves or virtual particles. Every type or form of energy has a Psyche or an Intelligence that is controlling it. That's how energy is able to understand, follow, and obey physical laws. There's intelligence or psyche or consciousness within the energy. Energy is sentient, perceptive, omniscient, omnipresent, and alive.

Matt always gets close to the truth; but, he never goes all the way. His Materialism, Naturalism, and Atheism prevent him from going all the way to the absolute truth. His associates won't let him go all the way to the truth. Instead of trying to identify what virtual particles are, he simply states that they don't really exist. That's a cop-out that every materialistic and atheistic scientist is trained in school to make. In order to be real, an object has to be a person, place, or thing. Virtual particles are real things. They show up in the math. So, what are they? Virtual particles represent thoughts, or information exchange between particles, at the quantum level. Virtual particles are quantum waves.

In order for the math to have any value to us, every variable within a mathematical equation is supposed to represent something that has actually been experienced and observed and is therefore real and is known to truly exist.

If we arbitrarily define "particles" as being physical or solid, then things like quantum fields, quantum waves, and virtual particles must be defined as being non-physical or spiritual in nature and origin. If we define "particles" as being local objects or local entities, then quantum fields, quantum waves, raw unorganized energy, thoughts, intelligence, information, and virtual particles must be categorized as being non-local in nature and origin. The verified existence of something like light, energy, life force, psyche, quantum waves, thoughts, intelligence, information, or quantum fields FALSIFIES Materialism, Naturalism, Mechanism, Physicalism, and their derivatives such as Darwinism (spontaneous generation) and Atheism (creation ex nihilo).

All of the information and knowledge in the universe is stored within the quantum fields, and that information is transmitted and received through quantum waves or thoughts or virtual particles Psyche-to-Psyche as needed through the quantum fields that have been made and organized by God's Psyche. Nature's Psyche has direct access to all of this information; whereas, the Human Psyche has been placed behind a veil of flesh which keeps us ignorant and prevents us from gaining direct access to this universal internet filled with knowledge and information. Nature's Psyche (or God's Psyche) is infinitely more intelligent and knowledgeable than any Human Psyche.

One of the reasons why we are here in this school within a physical body that has been assigned to us is to learn "mind-over-matter" control. In contrast, if you study the animals, it's obvious that they are functioning mostly in terms of matter-over-mind – they are enslaved to their biology. Human beings don't have to be enslaved to their biology, if they don't want to be. In fact, a human being who submits to matter-over-mind (addictions for example) finds himself enslaved and in prison. He's no longer free. He becomes like an animal. Only when we master the flesh or overcome the flesh – mind-over-matter – do we in fact become truly free to pursue and be what we truly are. The Atonement of Christ was designed by God to help us to overcome "matter-over-mind" and therefore to become truly free. Ye shall know the truth, and the truth shall set you free. Human Psyches can actually learn to plan, think, choose, and act, rather than simply being acted upon or reacting.

The verified and proven existence of something like the placebo effect – mind-over-matter – falsifies Materialism, Naturalism, Atheism, and their derivatives. Choice, intention, thoughts, memories, order, organization, attention, and intelligence are a function of Psyche or a product of Psyche. It has been observed that heart-transplant recipients sometimes receive the memories of their donors. That tells us that your memories are not being stored within your physical brain per se, but instead your memories are being stored holographically or collectively within the atoms that have been assigned to your physical body. According to Quantum Mechanics, the atoms and therefore the cells within your physical body become entangled and start sharing information at the quantum level. Memories are part of the Quantum Zeno Effect, which demonstrates that individual atoms can read your mind and therefore record your thoughts. What within an atom would be capable of reading your mind and recording your memories? The only logical answer is Something Psyche – another mind.

Psyche is the ultimate point-particle, even at the quantum level. Psyche is seen as being a pin-point of light at the quantum level. This spark of light is in fact the spark of life. Psyche is also made from energy. Everything is made from energy or light. Thoughts and memories are quantum waves; and, Psyche is the thing that creates, broadcasts, receives, and stores quantum waves or thoughts or virtual particles at the quantum level. Thoughts, memories, and quantum waves are the product of Psyche.

Out-of-body explorers have observed that whenever they are outside of their spirit body looking at their spirit body, it is in fact their psyche or spark or pinprick of light who is having and remembering that experience, and NOT their spirit body. Psyche is the thing that has experiences and remembers those experiences. It's not our physical brain. We continue to have thoughts and continue to make memories long after our physical brain is dead and gone. It's our Psyche who dreams, after it is temporarily disconnected from its physical body. Have you ever noticed that while you are asleep and dreaming, you can no longer feel any physical pain; but, once you wake up and your Psyche reconnects with your physical body, all of the pain comes rushing back into you? When you first wake up, there is no pain; and, then from head to foot the pain seems to flow into you as your Psyche reconnects with your physical body through your physical brain. If we were nothing but physical matter, such a phenomenon would not be possible. The pain would be there all the time, even while we were "unconscious" or "asleep". In fact, there

576

would be no way to sleep because there would be no way to change the energy waves that are coursing through our body and brain at the quantum level.

Neurons don't communicate with each other through neurotransmitters. They communicate with each other through the electrical waves that show up on an EEG, and through other quantum waves that we can't detect with our physical instruments. A neurotransmitter only transmits one bit of information – OPEN. A neurotransmitter is a single physical hardware bit that tells an ion-gate to open. It's physically impossible to transmit a complex message or a stream of thoughts through a single hardware bit at the physical level. In contrast, it's theoretically possible to transmit an infinite amount of information Psyche-to-Psyche through radio waves and electrical waves at the physical level, and through quantum waves or thoughts at the quantum level. All of this information is being transmitted Psyche-to-Psyche through the various different quantum fields. That's why the quantum fields were designed and made – to facilitate Psyche-to-Psyche communication at the quantum level, as well as the physical level. The Quantum Field Matrix is the universal internet; and, this Quantum Field Internet is the thing that allows Psyches or Intelligences to communicate with each other at the quantum level, and even ultimately at the physical level. It's obvious once a person chooses to look and see and understand.

The truth is hiding in plain sight where the Materialists, Naturalists, and Atheists cannot see it nor find it because they formally and religiously deny its existence. They don't want to know the truth because they don't want it to be true. But, their rejection of evidence and their refusal to look and learn doesn't automatically make them right and bright, as they claim to be.

It's all made from energy. Virtual particles are energy interactions between particles through the quantum fields. Virtual particles represent the transfer of information, or quantum waves, or thoughts between particles. Virtual particles are quantum waves of energy – or Wi-Fi at the quantum level. Each particle has an Intelligence or a Psyche controlling it; and, every Psyche has a certain amount of energy that's under its control. These particles are alive and communicating with each other and interacting with each other through quantum waves or virtual particles. So, where did the quantum fields come from? How were they organized? Who made them?

We need Someone who is smart enough and influential enough to design, create, and sustain quantum fields at a universal scale. That Someone has to be someone or something like God's Psyche or God's Intelligence. It is obvious that something like Psyche or Intelligence must of necessity exist to design, make, and control the quantum fields, quantum waves, and quantum information that surrounds us and runs through us. Every Psyche has a certain amount of energy that's under its control. That Psyche decides what form or shape that the energy under its control will take. The Psyche can decide that the quantum fields or energy under its control will form into a particle or a pair of virtual particles; and then later, that same Psyche can decide to turn the particles under its control back into energy or quantum fields. The whole thing is intelligent and alive.

Scientific Axiom: Anything that is obviously organized and made obviously has an Organizer and a Maker who designed it, organized it, and made it. There is no such thing as creation from nothing. Nothing can't design and create anything.

Scientific Observation: Quantum fields have obviously been designed, organized, and made. Quantum fields obviously exist. Quantum fields are organized energy, and they were obviously designed, made, and organized to fulfil a specific purpose.

Scientific Conclusion: Therefore, it is logical and rational to conclude that quantum fields have Someone Intelligent or Someone Psyche who designed them, made them, and then organized them into the Matrix of Quantum Fields that we have today. Such complex order, organization, fine-tuning, and teleology doesn't just spring into existence from nothing as the Materialists, Naturalists, Nihilists, and Atheists claim. It is purposefully designed and made. The same thing can be said about particles of any kind, including physical matter. There is a universal blueprint telling the psyches within energy how to make physical matter, if that's what they decide to become.

Even though the Materialists, Naturalists, Darwinists, Mechanists, and Atheists don't want us to find and know the truth because the truth falsifies their philosophical beliefs; nevertheless, the truth eventually prevails as long as we are willing to keep searching for the truth. Seek, and ye shall find. Knock, and it shall be opened unto you. It is obvious to us now that the vacuum or the void of space is filled with a Matrix of Quantum Fields. They have to be there, or physical matter couldn't exist. It's obvious.

Our physical universe is derived from the vacuum (from the quantum fields), which means that the quantum fields had to be designed and made before the physical matter could be made.

The theory of quantum mechanics is the most accurate and powerful description of the natural world that we have. We can see these quantum fluctuations, because they are written into the stars. It is quantum reality that has shaped the structure of the universe that we see today. Our universe is just the quantum world inflated many, many times. (*Everything and Nothing.*)

So, who designed and made this quantum reality, the different types of quantum waves or quantum fluctuations (particles) that are allowed to exist, the quantum rules, and the quantum fields? It can't be physical matter or a physical brain, because physical matter didn't exist yet at the beginning of our universe. Someone Intelligent or Someone Psyche had to design and make the quantum fields and the rules of behavior for these quantum fields BEFORE particles of any kind became possible to make. There is no other logical explanation for how the quantum fields were designed, organized, and made. I certainly don't believe that the quantum fields just sprang into existence from nothing fully functional as the Materialists, Naturalists, Darwinists, Mechanists, and Atheists claim.

It now appears as if the quantum world, the place we once thought of as empty nothingness, has actually shaped everything we see around us. The teeming, seething activity of the vacuum, of "nothing", and the quantum fluctuations within it were the seeds. Seeds which grew into the universe we see today. At the beginning of time, the universe sprang from the vacuum creating not only vast amounts of matter but also the strange stuff that was predicted by Paul Dirac, antimatter. But the universe we see today is made of matter. Nearly all of the antimatter seems to have vanished. (*Everything and Nothing.*)

According to these people, the seeds or quantum fluctuations became galaxies; and, the antimatter vanished into thin air like magic. The non-existence of antimatter is a prime example of fine-tuning. Our physical universe was obviously fine-tuned. Remember, anything that is obviously fine-tuned obviously has an intelligent Fine-Tuner who fine-tuned it. Quantum fields and physical matter are also obvious examples of fine-tuning.

What's Wrong with the Big Bang Theory?

https://www.youtube.com/watch?v=JDmKLXVFJzk

There are at least a dozen different things that are wrong with the Big Bang Theory, any one of which is sufficient enough to falsify the Big Bang Theory and eliminate it from science. So, why do we still have the Big Bang Theory? It's because the Big Bang Theory is creation ex nihilo or Atheism. That's the only reason that it survives. It's granted a special pleading or a special exemption because it is Atheism or creation from nothing by nothing. That's the reason that spontaneous generation, chemical evolution, or macro-evolution continues to survive in science, because it's Atheism or Magic. The theory of evolution or spontaneous generation was proven false by Louis Pasteur in 1859; but, it continues to exist in science because it is a form of creation ex nihilo – the design and creation of something by nothing. Creation Ex Nihilo is Atheism – the creation of something from nothing by nothing. Creation Ex Nihilo is falsified by logical common sense. Spontaneous generation, chemical evolution, macro-evolution, and creation ex nihilo are physically impossible.

The thing these people don't realize is that if you believe in the Big Bang Theory, then you have to believe in God or Psyche. Someone Intelligent had to make the Big Bang Singularity from a lot of energy. More importantly, Someone Psyche had to trigger the explosion and then guide and fine-tune the explosion so that everything that we see around us today would become possible. The Big Bang would have required a lot more psychic fine-tuning than simply taking raw chaotic energy and organizing that into a Matrix of Quantum Fields. A Quantum Field Model for Origins (making the quantum fields from pre-existing unorganized energy) is a lot more parsimonious and plausible than making a Big Bang Singularity and then carefully fine-tuning its explosion and expansion.

You have probably already seen the Atheist's Nothing Camera online. The Atheists are sitting around waiting for nothing to make something from nothing.

The problem with the Atheist's Nothing Camera is the fact that it actually requires intelligence or psyche in order to make a camera in the first place. It also takes intelligence, consciousness, or psyche in order to choose to wait for something to happen. Intelligence or psyche obviously exists, or we wouldn't be talking about any of this to begin with. The Materialists, Naturalists, Darwinists, Nihilists, and Atheists are adamant that intelligence or psyche does not exist. That claim is absurd, because it requires intelligence or psyche in order to make the claim that psyche does not exist. Atheism is patently absurd and self-defeating.

The Atheists want the Big Bang to be creation ex nihilo; but, Someone Psyche had to design and make the Big Bang Singularity in the first place, and then carefully fine-tune the explosion so that it would produce quantum fields, physical matter, planets, stars, and galaxies at the right time in the right place. One of the problems with the Big Bang Theory is that it breaks parsimony. It's not the simplest explanation possible; and, it couldn't have been creation ex nihilo because it actually worked.

Due to all the evidence that falsifies the Big Bang Theory, I personally don't believe that the Big Bang happened as they claim. Instead, I know that the quantum fields had to be designed, organized, and made first from raw chaotic energy BEFORE it became possible to make physical matter. I can visualize the construction of the quantum fields starting from a central chosen point within the Chaos Realm, and then the order and organization of the quantum fields being spread outward or stretched outward through the already pre-existing chaotic universe from that central starting point. The order and organization had to start somewhere, sometime, or it wouldn't exist. It would be nothing but random chaos instead. Only Psyche or Intelligence can do order and organization, both at the quantum level and at the physical level.

Remember, Quantum Field Theory, Quantum Fields, Organized Energy, and Quantum Mechanics FALSIFY Materialism, Naturalism, Mechanism, and their derivatives such as Behaviorism, Determinism, Nihilism, and Atheism.

The Psyches who oversaw the construction and the spread of the quantum fields could then use those quantum fields to make quantum particles or physical matter anywhere and anytime they chose to do so, once the quantum fields had been established. The initial order and organization – the creation of the quantum fields in our part of the Chaos Realm – could have taken place 13.7 billion years ago, and as that organization or construction of the quantum fields spread outward in all directions at the speed-of-light, it would appear to us as if the universe was expanding from that central point.

From what we can see, galaxies are born all at once, and then they evolve or change thereafter through natural physical processes. It's physically impossible for expanding clouds of hydrogen and helium gas to coalesce into planets, stars, and galaxies; so, that reality and truth tells us right there that the Big Bang Theory isn't true and that we are still looking for a better theory to explain the origin of our physical universe, along with its planets, stars, and galaxies.

It's obvious that quantum fields are non-physical and pre-physical. It's obvious that the quantum fields were made from energy or light – not physical matter. It's obvious that the Gods or God's Psyche had to make the quantum fields BEFORE they could make physical matter. It's obvious to me that the quantum fields had to be designed and made by Someone Psyche. It's seems logical to me that the construction of these quantum fields would have spread outward from a central point at the speed-of-light, since these quantum fields were made out of energy or light. Once the quantum fields are in place, then the Psyches or the Gods can phase-in physical galaxies into the Matrix of Quantum Fields wherever they see fit to do so. This creation process or organization process is ongoing. It wasn't a one-time event. The primary axiom of Quantum Field Theory is the observation that particles are born, and particles die. Physical atoms are born, and someday they can die by being reabsorbed into the Matrix of Quantum Fields from whence they came. The physical atoms that are in the planets, stars, and galaxies were BORN. They were made. They had a beginning. Anything that is obviously made obviously has a Maker who made it.

Whether the Big Bang happened or not, it doesn't really matter, because it's obvious that the Gods had to design and make the Quantum Fields BEFORE they could make the physical matter. It's also obvious that all the different physical atoms are compatible with each other, which means that they are made from the same set of universal laws or the same universal blueprint. Someone Intelligent or Someone Psyche had to design and make the physical laws and the blueprints for physical matter; otherwise, physical matter wouldn't be reliably reproduced, and physical atoms wouldn't be compatible with each other and wouldn't be able to form into molecules. It's obvious that physical atoms were intelligently designed and made from energy by Someone Intelligent or Someone Psyche. It's also obvious that the physical atoms themselves have some sort of intelligence or psyche within them that gives them the ability to understand, follow, and obey physical laws and universal blueprints. It's likely that the psyches within atoms have access to a lot more information and knowledge than we Human Psyches do. In other words, Nature's Psyche is currently smarter and more educated than the best Human Psyches. Our geniuses are really quite stupid and ignorant when you get right down to it, especially the Materialists, Naturalists, Darwinists, Nihilists, and Atheists. They are deliberately ignorant or stupid because they want to be. They actually take pride in their ignorance and see it as a badge of honor and superiority. They are always making stupid statements that reveal their ignorance.

These people believe in magic, creation ex nihilo, chemical evolution, macro-evolution, creation by luck, design and creation by natural selection, and spontaneous generation which are physically impossible. The theory of evolution or spontaneous generation was falsified in 1859 by Louis Pasteur, the very same year that Charles Darwin published "On the Origin of Species"; but, the Darwinists grant themselves a special pleading or a special exemption so as to keep the theory of evolution alive and well within their minds, hopes, and dreams.

Within Quantum Mechanics, Quantum Field Theory, and Psyche or Intelligence, we find the answer to life, the universe, and everything. This Quantum Field Model for origins seems a lot more believable and plausible to me than the materialistic and atheistic claim that our physical universe sprang into existence from nothing, which is the primary claim made by the Big Bang Theory and its proponents. The non-existence of anti-matter also suggests that the origin of this physical universe didn't happen in the way that these people believe it happened. They have to find some way to explain away the anti-matter because the Big Bang Theory says that it should be there; and, it isn't.

However, if the quantum fields were simply organized from raw chaotic energy from a central starting point, then the Gods or Controlling Psyches could simply pop-in, phase-shift in, or quantum tunnel in the physical galaxies where and when they wanted them, without having to create any anti-matter whatsoever. Furthermore, if there was anti-matter and it simply vanished, then it would have vanished back into the quantum fields from whence it came. Either way, the quantum fields are the KEY. Physical matter can't be made without them.

I obviously know what I'm talking about; but obviously, you are going to have to decide for yourself whether you believe it or not.

According to the materialistic and atheistic physicists, matter simply pops into existence from the vacuum or the void ex nihilo, which means that whole galaxies could pop into existence from the vacuum or the void under the command and control of Someone Psyche even based upon their physical model or naturalistic model. The Materialists, Naturalists, Darwinists, and Atheists propose creation ex nihilo – creation from nothing by nothing – as their model for origins. Creation from something by someone is a whole lot more parsimonious and logical. Everything is made from energy by Someone Psyche. Simple. Logical. Parsimonious. It's infinitely superior to Creation Ex Nihilo or Atheism.

This Quantum Field Model for origins seems a lot more parsimonious and logical to me than the Big Bang Theory, which depends upon magically getting expanding clouds of hydrogen and helium gas to coalesce into the planets, stars, and galaxies that we see around us today. Due to Quantum Mechanics and Quantum Field Theory, the idea of galaxies popping into existence whole from the Matrix of Quantum Fields is a lot more logical and rational than trying to force expanding clouds of hydrogen and helium gas to coalesce into planets, stars, and galaxies as the Big Bangers claim was done.

The Big Bang Theory was designed by the Materialists, Mechanists, Darwinists, Naturalists, and Atheists to eliminate the need for God during the creation process. These people actually teach that quantum fluctuations somehow magically provided the seeds for the construction of whole galaxies from scratch. Does that make sense to you? It doesn't to me. It's like having a random burp or fart magically create a car, computer, airplane, television, or house each and every time that one is needed. It takes a great deal of blind faith to believe that random quantum fluctuations can magically produce complex galaxies out of nothing; but, that's exactly what these people teach and believe.

As is their way, the Naturalists and Atheists who preach the Big Bang Theory deliberately reject and ignore everything that falsifies it. Instead, they launch out in blind faith and choose to believe that nothing can design and create everything, and magically have it come out working perfectly fine in the end. These people truly believe that nothing can design and create a complex physical atom from nothing, and then do so over and over again perfectly trillions of trillions of times. Their motto is, "Anything but God." These people are literally willing to believe in anything but Psyche or God. They believe in the most unlikely and impossible things, just so that they don't have to believe in Psyche or God. It's illogical and irrational, but that's what they do.

The Big Bang Theory was designed to eliminate God or God's Psyche from the equation. The problem is that once you eliminate God from the process, then the Big Bang Theory is no longer plausible, and the Big Bang Singularity is no longer realistically possible. The same can be said for the Theory of Evolution or Darwinism. The Theory of Evolution was designed to eliminate God or God's Psyche from the equation. However, once you have eliminated God from the process, the design, creation, engineering, manufacturing, and deployment of functional genomes is no longer possible. Random processes or quantum fluctuations do NOT magically create order and organization out of nothing. It's physically impossible for information-rich Genomes or Big Bang Singularities to spontaneously generate out of nothing as the Materialists, Naturalists, and Atheists claim was done. Order, organization, fine-tuning, spiritual blueprints, quantum fields, quantum information, intelligence, and even the right quantum fluctuations or quantum waves require the intention and intervention of Someone Psyche or Someone Intelligent; otherwise, it is impossible for them to exist. It's impossible to eliminate God or Intelligence from the equation and then get the results that we need in a timely fashion.

Contrary to all the evidence that falsifies the Big Bang Theory, the Materialists, Naturalists, Darwinists, Nihilists, Determinists, Physical Reductionists, and Atheists simply assume that it is true. These people are relativists, which means that they believe that the truth is determined by vote or by the majority. It's a logic fallacy to assume that the majority are always right, because the majority of the time it can be demonstrated that the majority has actually been wrong throughout much of history. Remember, the assumption that the majority determines what is right and what is true is a logic fallacy, because the majority are often proven wrong in the end. Polling seems to suggest that the majority of scientists believe that the Big Bang Theory is true; but, are they right? Is there in fact a better, more logical, and more parsimonious explanation for the origin of our physical universe other than the Big Bang Theory? The majority of scientists assume that the Big Bang Theory is true, but are they right?

Assumption of truth, wishful thinking, jumping to conclusions, or affirming the consequent is the Logic Fallacy upon which Materialism, Naturalism, Darwinism, Mechanism, and Atheism are based. This Logic Fallacy automatically makes their assumptions false, and these people don't even know it. Still, if we start with their assumption of truth where the Big Bang Theory is concerned, we can still use that assumption of truth to get closer to the truth if we allow ourselves to do so. This in itself can be an interesting thought experiment or philosophical exercise.

Let's assume that the Big Bang really happened. Then we automatically know that Someone Psyche, Someone Powerful, and Someone Highly Influential had to gather all of that energy into that single Big Bang Singularity; and, then that Same Intelligence had to trigger an explosion and fine-tune that explosion so that everything came out perfectly so that this physical universe would be the automatic result. That level of fine-tuning, power, and perfection that went into the design and creation of the Big Bang Singularity, as well as the level of precision and control that went into the subsequent explosion, is obvious proof of God's existence. If we assume that the Big Bang is true, then we KNOW that God exists, because God is the only person that we know of who could have designed, programmed, created, managed, and controlled such a thing as a Big Bang Singularity and the subsequent Big Bang explosion.

Imagine all the force it would take to gather all of that energy into that Big Bang Singularity and imagine all of the micromanagement and fine-tuning it would take to control that explosion so that it actually produced the physical universe that we see around us today. Explosions don't normally produce order and organization, now do they? Explosions just produce additional chaos and destruction. If the Big Bang really happened – and the majority of our scientists believe that it did – then the Big Bang is scientific proof of God's existence. It's unavoidable. Someone Psyche or Someone Intelligent had to make the Big Bang Singularity from energy, and that same Someone Psyche had to control that subsequent explosion so as to force that subsequent explosion to make this physical universe. If the Big Bang actually happened, then it is in fact Scientific Proof of God's Existence.

However, there are more logical, rational, and parsimonious explanations for the origin of our physical universe besides the Big Bang Theory. My Quantum Field Model for origins is one such example. Within the Quantum Field Model, all of the energy and matter already exists in a chaotic state, and all the Gods or Overlord Psyches have to do is to organize it into the physical universe that we see and experience today, as their work of organization and creation moved outwards from a central spot of their own choosing.

In this Quantum Field Model for origins, physical matter and particles of any kind end up being a deliberate conscious act of creation by Someone Psyche who is in control of the energy from which the physical matter and particles are made. The primary axiom of Quantum Field Theory is the idea that particles are born, and particles die. The only way that particles and physical matter can be born is if there are already quantum fields in place to sustain them; and, it's obvious that quantum fields were designed and made to fulfill that specific purpose. Random fluctuations are not going to design and build anything reliably, over and over again for billions of years. The only way that compatible particles and atoms can be reproduced reliably and consistently zillions of zillions of times over billions of years of time is if the Psyches who are building these particles out of energy are using the same exact blueprints every time. Intelligent organization is the only logical way to overcome random chaos. Intelligence or Psyche can override entropy at will; whereas, nothing is not capable of producing anything. Wherever we encounter order and organization, there always has to be an Intelligent Organizer, because order never springs into existence from nothing as the Materialists, Naturalists, Darwinists, Mechanists, and Atheists claim.

Remember, Quantum Field Theory is the unification of quantum mechanics and the theory of relativity.

What we once thought of as the void now seems to hold within it the deepest mysteries of the entire universe. (*Everything and Nothing.*)

The void or the vacuum of "empty space" now holds within it an organized Matrix of Quantum Fields, from which particles and physical matter can be produced and sustained by Someone Psyche. Quantum fields are the obvious sign of deliberate order and organization. Anything that is obviously made obviously has a Maker who made it. It's obvious that quantum fields were designed and made to fulfill a specific purpose. Therefore, it's also obvious that Someone Psyche designed and created and deployed these quantum fields which fill and permeate the void or vacuum of our universe including the innards of every atom of physical matter that has ever been made.

The false is falsified by the truth; and, one thing that I have observed is that the truth is always self-evident and obviously true, because the truth is repeatedly experienced and observed. These quantum fields and the effects of these quantum fields have been experienced and observed ever since the time that they were made by Someone Psyche. Quantum fields obviously exist which means that they are obviously true. Quantum fields had to be designed and made by Someone Intelligent or Someone Psyche, because physical matter and physical brains didn't exist yet. This reality is also obvious and true. The Gods or Controlling Psyches had to make the quantum fields first before physical matter could be made. That's just the way it is. It's obviously true.

My handiwork – the things that I have designed and created – is the surest sign of my existence. The same can be said for the Person who designed and created the quantum fields and the particles that flow through them.

Mark My Words

Physics and Perception (1967)

Bohm ventures a lot into the Science of Psychology. Originally, Psychology was the scientific study of the human psyche or the scientific study of human consciousness, until the Behaviorists got their hands on it and neutered it in the 1920's.

Outside of Quantum Mechanics, David Bohm is heavily steeped in Materialism and Mechanism. David Bohm doesn't successfully carry what he has learned about the non-physical nature of Quantum Mechanics into Psychology, Biology, and the other Origin Sciences. David Bohm still preaches Darwinism, Mechanism, and Physicalism where Neuroscience, Biology, Psychology, and Theology are concerned. He never explicitly brings his Implicate Order or Philosophy of Consciousness into play where these other sciences are concerned.

Bohm is also attached to Relativism, which is the bread and butter of the Materialists, Naturalists, and Atheists. Relativism is the philosophical belief that there is no such thing as absolute truth. Bohm is constantly stating that there is no such thing as absolute truth in the realm of Physics. Of course, as a scientist, I can see and understand how Bohm might get the Philosophy of Relativism from Quantum Mechanics. Quanta do seem to have a mind of their own, and they are indeed unpredictable and uncontrollable. There really does seem to be no absolute truth where quanta are concerned, because they are capable of choosing between competing options. Surprisingly, quanta are psychic, perceptive, sentient, and intelligent. They are alive.

When I was addicted to prescription drugs, my physical body and physical brain seemed to have a mind of their own; and, that's impossible at the physical level because the Materialists and Naturalists assure us and teach us that there is no such thing as psyche or mind. If the Materialists and Naturalists are right, then I experienced the impossible, because my brain and body developed a mind of their own and seemed to take off on their own while I was going through withdrawal, which kind of suggests to me that the Materialists and Naturalists are not right after all.

We need a quantum mechanical explanation for how psyche or mind works because it's definitely NOT a physical phenomenon and it's definitely NOT being controlled by our physical brain. I didn't have much control over my life, my physical body, and my physical brain while I was high on drugs and going through withdrawal. Substance-induced psychosis is not a good way to live. It's confusing as hell when the physical brain starts feeding false information (hallucinations and delusions) or static to the Human Psyche. A psychic explanation or a quantum mechanical explanation ended up being infinitely more explanatory than what my fellow Materialists, Naturalists, Nihilists, and Atheists were able to give me.

My bachelor's degree is in Psychology, and I'm as up-to-date as I can possibly be where Psychology is concerned. I'm at the cutting edge. I'm the creator of Quantum Neuroscience, so I've taken it a bit further than Bohm took it when it comes to Neuroscience and Psychology. Even though I take a shotgun approach to these scientific studies trying to include everything I possibly can as way of introduction to an Ultimate Model of Reality, and therefore I'm not as conservative and selective as I possibly could be, I really am fifty years ahead of David Bohm where Psychology, Perception, Darwinism, and Quantum Neuroscience are concerned.

Back in the beginning of his research into a Philosophy of Consciousness (1960's), like everyone else, David Bohm confounded or confused sensation with perception. Perception is a function of the human psyche. Near-Death Experiencers and Out-of-Body Explorers have observed first-hand that the Human Psyche continues to perceive its environment even though their physical body and physical brain are either dead or offline. Perception is a product of the human psyche. We continue to perceive even when our physical brain is dead and gone. Perception is a part of Bohm's Implicate Order and Super-Implicate Order.

In contrast, sensation is a product and a function of our five physical senses. Sensation is very much a physical, mechanical, materialistic function. Our brain is a machine that was designed by the Gods to transfer these physical sensations to our Human Psyche where they are perceived. There's a difference between these two functions. There's also a difference between the Human Psyche and its physical brain. The one is mortal, and the other is eternal and everlasting. Psyche or energy is always conserved; whereas, entropic physical matter is a temporal phenomenon. There are different types of expiration dates deliberately built into entropic physical matter and our fallen mortal physical bodies. Physical sensations end up being a part of Bohm's Explicate Order.

Within this essay, Bohm doesn't seem to separate these functions but merges them into one thereby giving the impression that physical sensations and psychic perceptions are one and the same thing. You have to dig and think, in order to separate them out for him. Bohm hadn't developed his Implicate Order, or Quantum Order, or Consciousness Order, or Psychic Order yet, when he wrote this particular essay about perception in 1967. Back then, Bohm was still laying the foundation for what was yet to come; and, for a cutting-edge Psychologist or Quantum Neuroscientist like me, this chapter of the book and many of his other pop psychology chapters within *The Essential David Bohm* were painful reading because Bohm isn't there yet. He literally confounds or confuses perception with sensation. He can't yet seem to see the difference between the two.

Bohm never really applied Quantum Mechanics and his Implicate Order to anything else besides Quantum Mechanics while he was alive. Bohm wasn't multi-disciplinary like I am. I'm a generalist. I'm good at everything and master of nothing. Bohm was very much a specialist – a maverick in Quantum Theory, and pretty much like everyone else when it came to everything else, at least back in the 1960's.

I found very little worth quoting, and a lot worth correcting within Bohm's Psychology chapters and books.

I've come a long way in a very short time, because I'm no longer confounding and confusing physical sensations with psychic perceptions. I'm no longer a materialist, naturalist, nihilist, and atheist like I used to be. I finally saw through all the naturalistic and materialistic deceptions and lies to the truths that were hidden down-below at the Quantum Level and the Psyche Level within Bohm's Implicate Order and Super-Implicate Order; and instantly, everything changed for me.

Quantum Mechanics: From a Non-Physical Spiritual Perspective

https://www.amazon.com/gp/product/B01J023TGU/

The Ultimate Model of Reality: Psyche Is the Ultimate Cause

https://www.amazon.com/gp/product/B071NC9JK6/

Science 2.0: I Upgraded My Science

https://www.amazon.com/gp/product/B0771K6WTX/

Quantum Neuroscience: The Answer to Life, the Universe, and Everything

https://www.amazon.com/gp/product/B079Z6QQQB/

Syntropy in Defense of Quantum Mechanics: The Answer to Life, the Universe, and Everything

https://www.amazon.com/gp/product/B07BPT3W8R/

I made the transition from falsehood into truth as quickly as I possibly could, once I successfully falsified My Materialism, My Naturalism, My Nihilism, and My Atheism. I had to grow into it, like everyone else; but, I got there in the end.

> **First of all, we point out that if there are an unlimited number of kinds of things in nature, no system of purely determinate law can ever attain a perfect validity. For every such system works only with a finite number of kinds of things, and thus necessarily leaves out of account an infinity of factors, both in the substructure of the basic entities entering into the system of law in question and in the general environment in which these entities exist. – David Bohm** (*The Essential David Bohm*, p. 21.)

Periodically within this and other essays, David Bohm erroneously claims that NO absolute knowledge is ever to be encountered or perceived. According to Bohm, there is no such thing as perfect validity or perfect truth. David Bohm had adopted a relativistic perspective during his lifetime, which is the philosophical belief that there is no such thing as absolute truth. David Bohm is not looking for absolute knowledge or absolute truth; and, that's a flawed approach to Science that we got from the Materialists, Naturalists, Nihilists, Mechanists, and Atheists. What good is Science if we can never find and know the truth?

Like most everyone else that I have encountered, David Bohm had successfully convinced himself that Science and the Scientific Method cannot be used to find and know the truth. Technically, they are right in that the Scientific Method cannot be used to prove the truth. However, they are wrong because the Scientific Method is perfect for identifying falsehoods and for falsifying and eliminating falsehoods such as Materialism, Naturalism, Darwinism, Nihilism, and Atheism. If you use the Scientific Method (or Observation and Experimentation) to successfully eliminate everything that is false, then ONLY the truth will remain. This is Logic 101. The false is falsified by the truth; and, the truth is repeatedly experienced and observed.

There is such a thing as truth. Anything that is eternal and everlasting is absolutely true. It has always been true, and it will always be true. There are absolute truths. We just have to know where to look for them in order to find them. Ask yourself what part of you is eternal and everlasting, and there you will find the existence of absolute truth.

Doctrine and Covenants 93: 23-24:

23 Ye were also in the beginning with the Father; that which is Spirit, even the Spirit of truth.

24 And truth is knowledge of things as they are, and as they were, and as they are to come.

Think about it! Your psyche, or your absolute truth, was there in the beginning with the Father. Your psyche or intelligence is co-eternal with God's psyche or intelligence. Psyche or energy is always conserved, which means that it's always true or always real. It always exists; therefore, it is absolutely true. Your psyche or energy has always existed, and it will always exist. It cannot be created, and it cannot be destroyed. This is what has been experienced and observed. It is absolutely true.

It's only the Materialists, Naturalists, Darwinists, Nihilists, Mechanists, Relativists, and Atheists who have an absolute need to convince themselves that there's no such thing as absolute truth. The rest of us are free to roll with it and deal with it wherever we might encounter it. The truth really does set us free.

Mark My Words

The Enfolding–Unfolding Universe and Consciousness (1980)

Within his book, *Tales of the Quantum*, Art Hobson writes:

> **Everybody has heard that we live in a world made of atoms. But far more fundamentally, we live in a universe made of quanta. Many things are not made of atoms: light, radio waves, electric current, magnetic fields, Earth's gravitational field, not to mention exotica such a neutron stars, black holes, dark energy, and dark matter. But everything, including atoms, is made of highly unified or "coherent" bundles of energy called "quanta" that (like everything else) obey certain rules. In the case of the quantum, these rules are called "quantum physics." This is a book about quanta and their unexpected, some would say peculiar, behavior.**

What are quanta made from?

Quanta are packets of energy or light. Quanta are made from energy, psyche, life force, or light by Someone Psyche or Someone Intelligent. Things like quanta, quantum fields, the strings of string theory, and the cosmological constants were obviously set at a specific frequency or a specific setting for a specific purpose. They were obviously finely-tuned. Anything that was obviously fine-tuned obviously has a Fine-Tuner who fine-tuned it. In order for quanta to be able to obey certain rules, there obviously has to be a Universal Rule-Maker or Universal Law-Giver. Furthermore, in order for quanta to be able to obey certain rules, quanta must obviously have within them some kind of psyche or intelligence that is capable of understanding, following, and obeying specific rules. In order for quanta to "behave" or act in a certain way, they must have Someone Psyche or Someone Intelligent in the driver's seat. These are truths which the Materialists, Naturalists, and Atheists formally reject, which tells us that we are on the right track, because these people have a knack for finding the truth and then formally rejecting it.

Within this particular chapter of his books, David Bohm defines his Implicate Order. The Implicate Order ends up being the non-physical realm, the quantum level, the energy realm, or the spirit world. Quantum Mechanics or Spiritual Mechanics explains how psyche, spirit matter, and energy work at the quantum level or within Bohm's Implicate Order.

Bohm seldom talks about psyche, spirit, soul, spirit matter, non-locality, non-physicality, and the spirit world. Instead, he includes all of this within his Implicate Order, and then lets you dig in like a detective and try to translate what he is talking about into English.

I assume that Bohm did it this way in order to maintain plausible deniability, because he took a ton of flak from the Materialists, Naturalists, and Mechanists before he learned how to cloak what was saying and hide it where they would never see it nor understand it. The problem with that approach to science and his struggle for legitimacy is that the people, who are sincerely looking for what Bohm had to offer us, can't find what Bohm taught because they can't understand what he taught. The word "implicate order" is NOT self-explanatory; and, I had no idea what it meant nor its significance until after reading *The Essential David Bohm*.

The Materialists, Naturalists, and Atheists created this mess for us by refusing to allow competing theories into academia. There is no standard definition or interpretation for Quantum Mechanics because it is non-physical in nature and origin; and thanks to the Materialists and Naturalists, each scientist and physicist has to create his own. There are as many interpretations of Quantum Mechanics as there are physicists on this planet because the Materialists and Atheists have successfully prevented us from standardizing around a true and correct interpretation that's non-physical in nature and origin.

Bohm was forced to create his own interpretation of Quantum Mechanics because the Materialists and Naturalists are forever unable to provide us with a useful one that actually makes sense and has been experienced and observed.

Reading Bohm can be painful at times because he always uses words that nobody understands. It's clear once a person understands what he is saying that his implicate order is the non-physical realm or the spirit world, and his super-implicate order or "higher level of meaning and significance" consists of Psyche, Soul, or Intelligence. But because of all the crap that the Materialists, Mechanists, and Naturalists dumped on him back in the 1950's, we are forced to dig in and read carefully if we want to truly understand what Bohm is saying. We have to find a way to define what Bohm is talking about before we can understand it. It's brilliant or genius, but Bohm and the Naturalists have successfully hidden it from us where it takes some work to get at it.

Editor Lee Nichol explains:

> **In this chapter Bohm gives an extended account of what is meant by an "implicate order," as well as analogies indicating how implicate orders manifest in the explicate, sensual world. The processes of matter, organic life, and consciousness are all seen as flowing from the reciprocal ordering principles of enfoldment and unfoldment. It is through these ordering principles that the holomovement – the "ground of all that is" – expresses itself in particular forms and experiences.**
>
> **The basic notion of enfoldment is quite straightforward – as when baking a cake eggs are "folded into" the batter. What was once overt and explicit – the eggs – becomes implicit and enfolded in the batter.**
>
> **In similar fashion, claims Bohm, our three-dimensional world – including entangled particles in a laboratory – manifests as a projection from a yet more fundamental multi-dimensional reality. The concept of a multi-dimensional reality – from which explicate "things" unfold and into which they enfold – is closely linked to Bohm's considerations of the nature of space. In this view, space is**

neither inert nor empty. The deep nature of space – of which our "regular" space is but a projection – is understood as a plenum, a highly varied structure-process which includes potentially infinite dimensions. Thus, space itself is understood as a multi-dimensional ordering medium – a higher-order correlate to the mechanical analogy of the glycerin mixing device. In outlining the relationship of consciousness to enfoldment and unfoldment, Bohm first considers consciousness as a "substantial" process.

Bohm suggests that the totality of these material phenomena may be enfolded and unfolded throughout the brain in a process not unlike that of a hologram, creating memory structures that bear a likeness to "original" perceptions. Beyond this substantial aspect of consciousness, Bohm outlines his views of the "essence of conscious experience" – factors such as awareness, attention, and understanding. At the root of any such capacities, says Bohm, is a preconscious "undivided state of flowing movement" – the actual and immediate activity of the holomovement [and the Implicate Order].

Bohm suggests that the "essence" of consciousness is just this flowing movement, arising from the depths of the holomovement itself. The substantial aspects of consciousness – memory, reification, and subject–object polarities – are understood as explicated forms of this deeper unitary movement, much as the two images of the fish unfold from a deeper, single actuality. Thus, the material structures of the objective world, the substantial aspects of consciousness, and the "essence" of consciousness can now be seen as a continuous flow [from the Implicate Order], eliminating any absolute distinction between mind and matter. (*The Essential David Bohm*, pp. 78-80.)

Within this essay, Bohm talks about "ordering principles". Ordering principles are a function of psyche, intelligence, or consciousness whether we are talking about the quantum level of existence or the physical level of existence. Whenever we are dealing with raw chaotic energy in a pre-physical state of existence, it requires Someone Psyche to organize it and remove the chaos from it. This is obviously true. Order and organization are a product of psyche and intelligence. This reality has always been true, and it will always be true. Order and organization don't just spring into existence from nothing as the Materialists, Naturalists, Darwinists, Nihilists, Mechanists, and Atheists claim. Whenever it comes to anything important or essential, the Atheists and Naturalists are always wrong. Reality, science, and intelligence itself falsifies their chosen beliefs.

Bohm makes it clear in this essay that the mechanistic model or materialist model is inadequate when it comes time to explain the non-local phenomena or what we typically call the non-physical phenomena.

The quantum theory presents, however, a much more serious challenge to this mechanistic order, going far beyond that provided by the theory of relativity. The key features of the quantum theory that challenge mechanism are:

1 Movement is in general discontinuous, in the sense that action is constituted of indivisible quanta (implying also that an electron, for example, can go from one state to another, without passing through any states in between).

2 Entities, such as electrons, can show different properties (e.g., particle-like, wavelike, or something in between), depending on the environmental context within which they exist and are subject to observation.

3 Two entities, such as electrons, which initially combine to form a molecule and then separate, show a peculiar nonlocal relationship, which can best be described as a non-causal connection of elements that are far apart (as demonstrated in the experiment of Einstein, Podolsky, and Rosen).

The three key features of the quantum theory given do, however, clearly show the inadequacy of mechanistic notions. Thus, if all actions are in the form of discrete quanta, the interactions between different entities (e.g., electrons) constitute a single structure of indivisible links, so that the entire universe has to be thought of as an unbroken whole. In this whole, each element that we can abstract in thought shows basic properties (wave or particle, etc.) that depend on its overall environment, in a way that is much more reminiscent of how the organs constituting living beings are related, than it is of how parts of a machine interact. Further, the non-local, non-causal nature of the relationships of elements distant from each other evidently violates the requirements of separateness and independence of fundamental constituents that is basic to any mechanistic approach.

It is instructive at this point to contrast the key features of relativistic and quantum theories. Relativity theory requires continuity, strict causality (or determinism) and locality. On the other hand, quantum theory requires non-continuity, non-causality and non-locality. So, the basic concepts of relativity and quantum theory directly contradict each other. It is therefore hardly surprising that these two theories have never been unified in a consistent way. Rather, it seems most likely that such a unification is not actually possible. What is very probably needed instead is a qualitatively new theory, from which both relativity and quantum theory are to be derived as abstractions, approximations and limiting cases.

The basic notions of this new theory evidently cannot be found by beginning with those features in which relativity and quantum theory stand in direct contradiction. The best place to begin is with what they have basically in common. This is undivided wholeness.

Though each comes to such wholeness in a different way, it is clear that it is this to which they are both fundamentally pointing.

To begin with undivided wholeness means, however, that we must drop the mechanistic order. (David Bohm, *The Essential David Bohm*, pp. 82-84.)

All throughout this book, Bohm challenges or falsifies the mechanistic or materialistic order. He talks about an "undivided wholeness". What is the wholeness comprised of? Bohm calls it an implicate order, which doesn't tell us anything about it. Subsequent research and experimentation have proven that this undivided wholeness is comprised of a wide variety of different quantum fields that fill the immensity of space and permeate every physical atom to its very core. Discrete quanta are connected together or meshed together holistically by quantum fields. These quantum fields have proven to be non-physical, non-local, omnipresent, and made of energy and not physical matter. The verified existence of quantum fields and quantum mechanics FALSIFIES the philosophies of Mechanism, Materialism, Nihilism, Darwinism, Naturalism, and even Atheism. The false is falsified by the truth; and, the truth is repeatedly experienced, observed, and verified.

The point you need to grasp and the point I'm trying to make is that Bohm's ideas have constantly been verified for the past sixty or seventy years; whereas at the same time, Materialism, Naturalism, Darwinism, Nihilism, Atheism, and their derivatives have been constantly falsified. Bohm, and his crazy ideas, have for the most part been vindicated; whereas, the Materialists and Naturalists have repeatedly gone down in flames. Quantum fields are made from energy, and not physical matter. Quantum fields are non-physical and pre-physical. The Gods had to design and make the quantum fields BEFORE physical matter became possible to make and sustain. No quantum fields, then no physical matter.

It's obvious to us now that physical matter is built from the non-physical, the non-local, the quantum, or the spiritual. Quantum mechanics is spiritual mechanics or energy mechanics – it's the way psyche, or light, or life force functions at the quantum level. The false is falsified by the truth; and, the truth is repeatedly experienced, observed, and verified. The verified existence of Quantum Mechanics and Quantum Fields and Non-Local Intelligence (Psyche) falsifies Materialism, Naturalism, Mechanism, and their derivatives such as Darwinism, Nihilism, and Atheism. I had to get used to it, because that's the way it is. We each have to decide what it is that we truly want – do we want the truth, or do we want the lies? It's inconsistent to try to have it both ways.

Non-Local Intelligence, Psyche, Consciousness, or Life Force is obviously non-physical in nature and origin – it's made from energy. It is energy. Psyche or energy is always conserved. It has always existed, and it will always exist. Psyche or energy is also pre-physical. It had to exist before it could design and make the first particle of physical matter. Even at the quantum level, Psyche is seen as a pinpoint of light, a spark, or a point-particle. Psyche is the fundamental object or fundamental unit of reality. Every psyche has a certain amount of energy that's under its control. Psyche can form the energy under its control into anything that that psyche wants that energy to be, including quantum fields, spirit matter or dark matter, dark energy, gravity, radio waves, photons, or physical matter. Everything is made from energy or light by the psyches that control that energy or light. The energy or light is alive, because it is conscious, perceptive, intelligent, and aware.

What has been said thus far indicates that the implicate order gives generally a much more coherent account of the quantum properties of matter than does the traditional mechanistic order. What we are proposing here is that the implicate order therefore be taken as fundamental.

Our proposal to start with the implicate order as basic, then, means that what is primary, independently existent, and universal has to be expressed in terms of the implicate order. So, we are suggesting that it is the implicate order that is autonomously active while, as indicated earlier, the explicate order flows out of a law of the implicate order, so that it is secondary, derivative, and appropriate only in certain limited contexts.

Or, to put it another way, the relationships constituting the fundamental law are between the enfolded structures that interweave and interpenetrate each other, throughout the whole of space, rather than between the abstracted and separated forms that are manifest to the senses (and to our instruments).

What, then, is the meaning of the appearance of the apparently independent and self-existent "manifest world" in the explicate order? The answer to this question is indicated by the root of the word "manifest," which comes from the Latin "manus," meaning "hand." Essentially, what is manifest is what can be held with the hand – something solid, tangible, and visibly stable. The implicate order has its ground in the holomovement which is vast, rich, and in a state of unending flux of enfoldment and unfoldment, with laws most of which are only vaguely known, and which may even be ultimately unknowable in their totality. Thus, it cannot be grasped as something solid, tangible and stable to the senses (or to our instruments). Nevertheless, as has been indicated earlier, the overall law (holonomy) may be assumed to be such that in a certain sub-order, within the whole set of implicate order, there is a totality of forms that have an approximate kind of recurrence, stability and separability. Evidently, these forms are capable of appearing as the relatively solid, tangible, and stable elements that make up our "manifest world." The special distinguished sub-order indicated above, which is the basis of the possibility of this manifest world, is then, in effect, what is meant by the explicate order.

We can, for convenience, always picture the explicate order, or imagine it, or represent it to ourselves, as the order present to the

595

senses. The fact that this order is actually more or less the one appearing to our senses must, however, be explained. This can be done only when we bring consciousness into our "universe of discourse" and show that matter in general and consciousness in particular may, at least in a certain sense, have this explicate (manifest) order in common. (pp. 93-94.)

What we are suggesting here is that the implicate order, the non-physical order, the energy order, the psyche order, the spiritual order, or the quantum order be taken as fundamental. What he is suggesting here is that we start with the truth; and, the truth is that the physical is made from the non-physical by Consciousness or by Someone Psyche. In contrast, all the Materialists, Naturalists, Nihilists, Darwinists, Behaviorists, and Atheists can tell us is that the non-physical does not exist. That's worthless! Claiming that the non-physical doesn't exist doesn't tell us what it is and how it works.

According to Bohm, the essence of consciousness resides within the holomovement or the Implicate Order, which is the quantum level or the spirit world. It has to reside someplace, because consciousness has been experienced and observed. It's real. It truly exists. Furthermore, the Explicate Order or our physical reality unfolds or explicates from this much deeper holomovement or Implicate Order. The physical is built from the quantum or the spiritual or the non-physical, just like Quantum Field Theory states and just like the Biblical God Jesus Christ tries to teach us. It would take quite a bit of Intelligence, Information, or Psyche to design and create a quantum field or a holomovement and then convince the energy within it to act that way.

Bohm's holomovement is synonymous with his Implicate Order, which are synonymous with quantum fields and quantum mechanisms. Quantum Mechanics explains how energy, or psyche, or light acts at the quantum level or the implicate level.

Scientific Axiom: Energy or light can take on any form that it desires to assume, including spirit matter and physical matter. Energy or psyche is always conserved, but it can change form at will. This is what has been experienced and observed.

Scientific Observation: Anything that is obviously organized obviously has an Organizer who organized it.

Scientific Observation: Quantum fields were obviously organized from raw chaotic energy.

Scientific Conclusion: Therefore, quantum fields obviously have an Organizer who organized them and taught the energy within them to act that way or to form that way.

Bohm was teaching these very same principles back in 1980. He wrote:

Indeed, if one applies the rules of quantum theory to the currently accepted general theory of relativity, one finds that the

gravitational field is also constituted of such "wave-particle" modes, each having a minimum "zero-point" energy. As a result, the gravitational field, and therefore the definition of what is to be meant by distance, cease to be completely defined. As we keep on adding excitations corresponding to shorter and shorter wavelengths to the gravitational field, we come to a certain length at which the measurement of space and time becomes totally undefinable. Beyond this, the whole notion of space and time as we know it would fade out, into something that is at present unspecifiable. So, it would be reasonable to suppose, at least provisionally, that this is the shortest wavelength that should be considered as contributing to the "zero-point" energy of space.

When this length is estimated it turns out to be about 10^{-33} cm. This is much shorter than anything thus far probed in physical experiments (which have got down to about 10^{-17} cm or so). If one computes the amount of energy that would be in one cubic centimeter of space, with this shortest possible wavelength, it turns out to be very far beyond the total energy of all the matter in the known universe.

What is implied by this proposal is that what we call empty space contains an immense background of energy, and that matter as we know it is a small, "quantized" wavelike excitation on top of this background, rather like a tiny ripple on a vast sea. (p. 98.)

We have the truth on the one hand and the lie on the other hand. Most scientists have chosen to believe in the lie. However, the truth is that it's all made from energy or light – what the new agers call the Quantum Sea of Light. We'll never run out energy that can be converted into physical matter. And, whenever physical matter reaches thermal equilibrium or maximum entropy, the psyches controlling it can turn it back into raw energy anytime they choose to do so. In other words, entropy is a non-starter. Entropy is an illusion. There is more energy in one cubic centimeter of space than in all of the physical matter in the whole universe combined. All it takes is the energy of one cubic centimeter of space to make the whole physical universe that we see around us today. Are you starting to see it yet?

It is being suggested here, then, that what we perceive through the senses as empty space is actually the plenum, which is the ground for the existence of everything, including ourselves. The things that appear to our senses are derivative forms and their true meaning can be seen only when we consider the plenum, in which they are generated and sustained, and into which they must ultimately vanish.

This plenum is, however, no longer to be conceived through the idea of a simple material medium, such as an ether, which would be regarded as existing and moving only in a three-dimensional space.

Rather, one is to begin with the holomovement, in which there is the immense "sea" of energy described earlier. This sea is to be understood in terms of a multidimensional implicate order, along the lines sketched in Section 4, while the entire universe of matter as we generally observe it is to be treated as a comparatively small pattern of excitation. (p. 99.)

Energy is intelligent or psychic. Energy is life force. Energy is non-physical, pre-physical, and primal. The energy is vast. The energy is eternal. The energy is infinite. In contrast, all the physical matter in our universe is just a blip on energy's radar. It's the energy or the psyche that's real, and not the physical matter. Physical matter is simply one of the many emergent properties or emergent forms of energy, or psyche, or life force.

David Bohm is writing this back in 1980. Back then, the scientists hadn't successfully falsified Darwinism or the Theory of Evolution yet. That process started in 1986 with the publication of Michael Denton's book, *Evolution: A Theory in Crisis*. Consequently, Bohm got things wrong whenever he switched over to biology or to other forms of science besides Quantum Mechanics.

Let us begin by considering the growth of a living plant. This growth starts from a seed, but the seed contributes little or nothing to the actual material substance of the plant or to the energy needed to make it grow. This latter comes almost entirely from the soil, the water, the air and the sunlight. According to modern theories the seed contains *information*, in the form of DNA, and this information somehow "directs" the environment to form a corresponding plant. (p. 101.)

We now KNOW that DNA codes for proteins. The actual blueprint or the information for building a physical body and a physical brain would require petabytes of information and possibly an exabyte of information. The human genome can only store around 750 megabytes of information in digital format. All of those other exabytes and blueprints are stored at the quantum level within the energy or the psyche that designed and made these non-physical blueprints in the first place. The seed contains the information for making physical proteins. All of the blueprints describing how those proteins are to be organized cannot be stored within the genome because the genome doesn't have an exabyte of storage capacity. When it comes to physical brain, the blueprints for making the brain and quantum maps of brain functionality have to be stored at the quantum level by Someone Psyche or Someone Intelligent, because there's not enough space at the physical level to store all of that information. Just to map functionality onto the quadrillion synapses found in a child's brain would require a petabyte of data storage at the minimum. Data storage is highly inefficient and takes a lot of space at the physical level.

https://www.backblaze.com/blog/petabytes-on-a-budget-how-to-build-cheap-cloud-storage/

They hadn't realized yet back in 1980 that 99.999999999% of the universe's information is stored at the quantum level by Someone Psyche; and, the only biological information stored at the physical level is the 750 megabytes needed to code for the physical proteins that are based upon the physical genes. Yes, DNA contains exabytes of information; but, the vast majority of that information is being stored as quantum waves of energy at the quantum level, with just a small blip of that information being stored at the physical level within the physical genome. Remember, the information or the blueprint as well as the quantum maps of physical functionality are stored within the spirit or the psyche at the quantum level – within the energy waves or the quantum waves. DNA primarily codes for proteins, and it requires Someone Psyche to know which protein is needed and which gene needs to be read in order to produce that protein. It's all being controlled by energy or psyche. It's all being controlled by information or intelligence. The energy is real. It has actually been experienced and observed, both at the quantum level and at the physical level.

Psyche or life force is the KEY to understanding the whole of reality and existence. David Bohm started to catch onto this back in 1980.

> **It may indeed be said that life is enfolded in the totality and that, even when it is not manifest, it is somehow "implicit" in what we generally call a situation in which there is no life.**

> **The above does not mean, however, that life can be reduced completely to nothing more than that which comes out of the activity of a basis governed by the laws of inanimate matter alone (though we do not deny that certain features of life may be understood in this way). Rather, we are proposing that as the notion of the holomovement was enriched by going from three-dimensional to multidimensional implicate order and then to the vast "sea" of energy in "empty" space, so we may now enrich this notion further by saying that in its totality the holomovement includes the principle of life as well. Inanimate matter is then to be regarded as a relatively autonomous sub-totality in which, at least as far as we now know, life does not significantly manifest. That is to say, inanimate matter is a secondary, derivative, and particular abstraction from the holomovement (as would also be the notion of a "life force" entirely independent of matter). Indeed, the holomovement which is "life implicit" is the ground both of "life explicit" and of "inanimate" matter, and this ground is what is primary, self-existent and universal. Thus, we do not fragment life and inanimate matter, nor do we try to reduce the former completely to nothing but an outcome of the latter.** (p. 102.)

The physical matter or inanimate matter is just a small fraction of the Quantum Sea of Light, the Life Force, or the Matrix of Quantum Fields that fill the immensity of space. God or Psyche is in the light. The intelligence is within the light. Life force is within the light. The energy within just one cubic centimeter of "empty space" is enough energy to make all the physical matter found in our physical universe today. How's that possible? It's because physical matter was designed to take up space. In contrast, pure energy or raw energy at the quantum level basically takes up no space at all. You can store ALL of the physical information within our whole physical universe within a single cubic centimeter of "empty space". In fact, it's theoretically possible to store all of the physical information found within our physical universe within an infinite singularity or point-particle that we call a Psyche. A Psyche is a real quantum computer. It exists at the quantum level, not the physical level. It's made from energy or light, and NOT physical matter. Physical matter is a highly inefficient form of data storage in comparison.

Remember, there is psyche or life force even in inanimate matter that has yet to manifest. The psyches that are controlling the physical matter and the elementary particles have chosen to submit themselves to physical restrictions or physical laws. Physical atoms seem very much inert or inanimate; however, their innards are alive with motion and activity. Energy is motion. Energy is life. The quarks and the electrons that make up atoms are very much alive and are in constant perpetual motion. Energy is the ultimate perpetual motion machine. Energy or psyche is always conserved. It never wears out; and, it never runs out.

Now, within this essay or chapter, David Bohm is ready to start talking about "thought" or "consciousness" or "intelligence". He never uses the word "psyche", because back in 1980 it wasn't public knowledge that Psyche has been repeatedly experienced and observed. Psyche is experienced first-hand as an immaterial viewpoint in space. Psyche is seen as a pinpoint of light, a spark, or a point-particle of light – even at the quantum level. Psyche seems to be an infinite singularity, or the ultimate point-particle. Like everything else, it's made from energy; and, every psyche has a certain amount of energy that's under its control. Psyche and the energy under its control form a symbiosis or a life force. This is the way things really work at the quantum level.

Remember, when it comes to physics, the smaller dwells within and controls the larger. What is smaller than quantum fields? What is smaller than string theory's strings? Below the quantum level or the implicate order, we have the thought level or the psyche level. We have the level of consciousness or intelligence or light.

> **To obtain an understanding of the relationship of matter and consciousness has, however, thus far proved to be extremely difficult, and this difficulty has its root in the very great difference in their basic qualities as they present themselves in our experience. This difference has been expressed with particularly great clarity by Descartes, who described matter as "extended substance" and consciousness as "thinking substance." Evidently, by "extended substance" Descartes meant something made up of distinct forms existing in space, in an order of extension and separation basically similar to the one that we have been calling explicate. By using the term "thinking substance" in such sharp contrast to "extended substance" he was clearly implying that the various distinct forms**

appearing in thought do not have their existence in such an order of extension and separation (i.e., some kind of space), but rather in a different order, in which extension and separations have no fundamental significance. The implicate order has just this latter quality, so in a certain sense Descartes was perhaps anticipating that consciousness has to be understood in terms of an order that is closer to the implicate than it is to the explicate.

However, when we start, as Descartes did, with extension and separation in space as primary for matter, then we can see nothing in this notion that can serve as a basis for a relationship between matter and consciousness, whose orders are so different. Descartes clearly understood this difficulty and indeed proposed to resolve it by means of the idea that such a relationship is made possible by God, who being outside of and beyond matter and consciousness (both of which He has indeed created) is able to give the latter "clear and distinct notions" that are currently applicable to the former. Since then, the idea that God takes care of this requirement has generally been abandoned, but it has not commonly been noticed that thereby the possibility of comprehending the relationship between matter and consciousness has collapsed.

In this chapter, we have, however, shown in some detail that matter as a whole can be understood in terms of the notion that the implicate order is the immediate and primary actuality (while the explicate order can be derived as a particular, distinguished case of the implicate order). The question that arises here, then, is that of whether or not (as was in a certain sense anticipated by Descartes) the actual "substance" of consciousness can be understood in terms of the notion that the implicate order is also its primary and immediate actuality. If matter and consciousness could in this way be understood together, in terms of the same general notion of order, the way would be opened to comprehending their relationship on the basis of some common ground. Thus, we could come to the germ of a new notion of unbroken wholeness, in which consciousness is no longer to be fundamentally separated from matter.

Let us now consider what justification there is for the notion that matter and consciousness have the implicate order in common. First, we note that matter in general is, in the first instance, the object of our consciousness. However, as we have seen throughout this chapter, various energies such as light, sound, etc., are continually enfolding information in principle concerning the entire universe of matter into each region of space. Through this process, such information may of course enter our sense organs, to go on through the nervous system to the brain. More deeply, all the matter in our bodies, from the very first, enfolds the universe in some way. Is this enfolded structure, both of information and of matter (e.g., in

601

the brain and nervous system), that which primarily enters consciousness?

Let us first consider the question of whether information is actually enfolded in the brain cells. Some light on this question is afforded by certain work on brain structure, notably that of Pribram. Pribram has given evidence backing up his suggestion that memories are generally recorded all over the brain in such a way that information concerning a given object or quality is not stored in a particular cell or localized part of the brain, but rather that all the information is enfolded over the whole. This storage resembles a hologram in its function, but its actual structure is much more complex. We can then suggest that when the "holographic" record in the brain is suitably activated, the response is to create a pattern of nervous energy constituting a partial experience similar to that which produced the "hologram" in the first place. But it is also different in that it is less detailed, in that memories from many different times may merge together, and in that memories may be connected by association and by logical thought to give a certain further order to the whole pattern. In addition, if sensory data is also being attended to at the same time, the whole of this response from memory will, in general, fuse with the nervous excitation coming from the senses to give rise to an overall experience in which memory, logic, and sensory activity combine into a single unanalyzable whole.

Of course, consciousness is more than what has been described above. It also involves awareness, attention, perception, acts of understanding, and perhaps yet more. We have suggested in the first chapter [of *Wholeness and Implicate Order*] that these must go beyond a mechanistic response (such as that which the holographic model of brain function would by itself imply). So, in studying them we may be coming closer to the essence of actual conscious experience than is possible merely by discussing patterns of excitation of the sensory nerves and how they may be recorded in memory. (pp. 105-106.)

This is the primary discovery that I made while developing Quantum Neuroscience. The memories that survive the death of your physical brain and show up in your after-death Life Review cannot be stored within your physical brain. Your memories are NOT stored at the physical level within your physical brain. It would require at least an exabyte of data storage capacity to store all of your memories at the physical level within your physical brain. That's physically impossible. Your memories or thoughts have to be stored as quantum waves at the quantum level within something psyche. Psyche produces, broadcasts, receives, and stores thoughts or quantum waves. This is the way consciousness really works.

Information or thought or memory is constantly enfolding into the light, and then when it's needed it's being unfolded or decoded from the light. This isn't taking place within your physical brain. There are NO wires within a physical brain. There is no RAM within a physical brain. There are NO memory engrams within a physical brain. This information is being produced by Psyche, broadcast as quantum waves, received by other psyches, and then stored within those psyches as thoughts, memories, or quantum waves. The information or thoughts or quantum waves are enfolded and broadcast by Psyche, and then later received, unfolded or decoded, and then stored by other psyches. It's all based upon energy; and, it's completely non-physical and pre-physical. God's Psyche had to make the quantum fields from energy before physical matter could be made and sustained.

Information is energy; and, all it takes is information or energy to eliminate entropy or disorder completely. Psyche or energy or information can eliminate entropy at will. In other words, Psyche can order disorder at will, because the energy under its control is always conserved. Entropy is hidden information. That information is no longer accessible at the physical level. That information is typically hidden within black holes; but, that information or energy also becomes inaccessible at the physical level whenever a region of space reaches thermal equilibrium or maximum entropy or heat death. Entropy is hidden information; but, that information is still there and it's still fully accessible and fully usable at the quantum level by Someone Psyche. There's no such thing as entropy or thermodynamics at the Quantum Level, because energy or psyche is always conserved and always available at the quantum level. This is the Quantum Law of Thermodynamics, which states that entropy does not exist at the quantum level.

What's the common ground between Psyche and Physical Matter? Energy is that common ground. They both are made from energy. Psyche makes physical matter from energy. Both psyche and physical matter are made of light or the implicate order. What is this thinking substance that Descartes was talking about? It's energy or psyche. It's obvious that this has to be true, because there's no other way to explain the order, organization, and intelligence that's obviously there within the quantum fields and within the physical matter that emerges from the quantum fields. A physical atom is a marvelous miracle of order and organization, and the same blueprint for making a physical atom has already been used trillions of trillions of times thereby constantly and reliably producing physical atoms that are compatible with each other and can actually be formed into molecules by Someone Psyche or Someone Intelligent. Energy or psyche is "thinking substance". Physical matter was thought into existence by Someone Psyche.

Of course, consciousness is more than what has been described above. It also involves awareness, attention, perception, acts of understanding, and perhaps yet more. We have suggested in the first chapter [of *Wholeness and Implicate Order*] that these must go beyond a mechanistic response (such as that which the holographic model of brain function would by itself imply). So, in studying them we may be coming closer to the essence of actual conscious experience than is possible merely by discussing patterns of

603

excitation of the sensory nerves and how they may be recorded in memory. (p. 106.)

This is the start of Karl Pribram and David Bohm's Holonomic Brain Theory.

https://en.wikipedia.org/wiki/Holonomic_brain_theory

The Materialists, Mechanists, Naturalists, and Atheists that write for the Wikipedia declare the holonomic brain theory to be pseudoscience, because they don't have what it takes to understand it or explain it. Everything in our physical universe, except for physical matter, proves that the Materialists, Naturalists, Darwinists, Nihilists, and Atheists are wrong. The proven and verified existence of anything non-physical FALSIFIES Materialism, Naturalism, and their derivatives such as Darwinism, Nihilism, and Atheism. The false is falsified by the truth; and, the truth is repeatedly experienced and observe. Psyche, energy, consciousness, intelligence, life force, or light is being constantly experienced and observed. Is it not? Only the Materialists and Naturalists deny the existence of their own intelligence, consciousness, or life force. To the rest of us, it's obvious that it exists.

By 2016, scientists had successfully proven the existence of quantum holograms, and they had effectively proven that holonomic brain theory is true. It took thirty years, but it has been verified. The point I want to make is that over and over again, David Bohm has been vindicated by science and scientific experimentations, while at the same times the claims of the Skeptics, Atheists, Materialists, and Naturalists have constantly been falsified or proven false.

Quantum Hologram ESP Edgar Mitchell, Apollo Astronaut

https://www.youtube.com/watch?v=EQuFtyruewo

https://www.experiencer.org/the-quantum-hologram-and-the-nature-of-consciousness/

David Bohm was right; and, the Materialists, Naturalists, Darwinists, Nihilists, and Atheists have constantly been proven wrong. This quantum stuff, or psychic stuff, or spiritual stuff is real and truly exists. Consciousness or Psyche is real, and it truly exists at the quantum level. We KNOW this is true because energy is real and truly exists at the quantum level, as well as the psyche level. Energy or light is the common substance from which everything is made; and, every Psyche has a certain amount of energy that's under its direct control. A Psyche can form the energy under its control into anything that it wants that energy to be, because energy is infinitely malleable or infinitely transformable. Energy can be formed into anything that Someone Psyche or Someone Intelligent wants it to be.

David Bohm started to figure this out back in 1980, and since then he has been vindicated by Science, Quantum Mechanics, Holography, Near-Death Experiences, Out-of-body Experiences, and Quantum Field Theory. Meanwhile, Materialism and Naturalism have constantly been falsified.

As we have seen, however, according to the quantum theory, movement is not fundamentally continuous. So even as an algorithm its current field of application is limited to theories expressed in terms of classical concepts (i.e., in the explicate order) in which it provides a good approximation for the purpose of calculating the movements of material objects.

When we think of movement in terms of the implicate order, however, these problems do not arise. In this order, movement is comprehended in terms of a series of interpenetrating and intermingling elements in different degrees of enfoldment *all present together*.

We see, then, that through thinking in terms of the implicate order, we come to a notion of movement that is logically coherent and that properly represents our immediate experience of movement. Thus, the sharp break between abstract logical thought and concrete immediate experience, that has pervaded our culture for so long, need no longer be maintained. Rather, the possibility is created for an unbroken flowing movement from immediate experience to logical thought and back, and thus for an ending to this kind of fragmentation.

Moreover, we are now able to understand in a new and more consistent way our proposed notion concerning the general nature of reality, that *"what is"* is movement. Actually, what tends to make it difficult for us to work in terms of this notion is that we usually think of movement in the traditional way as an active relationship of what is to what is not. Our traditional notion concerning the general nature of reality would then amount to saying that *"what is"* is an active relationship of what is to what is not. To say this is, at the very least, confused.

In terms of the implicate order, however, movement is a relationship of certain phases of what is to other phases of what is, that are in different stages of enfoldment. This notion implies that the essence of reality as a whole is the above relationship among the various phases in different stages of enfoldment (rather than, for example, a relationship between various particles and fields that are all explicate and manifest).

Of course, actual movement involves more than the mere immediate intuitive sense of unbroken flow, which is our mode of directly experiencing the implicate order. The presence of such a sense of flow generally implies further that, in the next moment, the state of affairs will actually change – i.e., it will be different. How are we to understand this fact of experience in terms of the implicate order?

A valuable clue is provided by reflecting on and giving careful attention to what happens when, in our thinking, we say that one set of ideas implies an entirely different set. Of course, the word "imply" has the same root as the word "implicate" and thus also involves the notion of enfoldment. Indeed, by saying that something is implicit we generally mean more than merely to say that this thing is an inference following from something else through the rules of logic. Rather, we usually mean that from many different ideas and notions (of some of which we are explicitly conscious) a new notion emerges that somehow brings all these together in a concrete and undivided whole.

We see, then, that each moment of consciousness has a certain explicit content, which is a foreground, and an implicit content, which is a corresponding background. We now propose that not only is immediate experience best understood in terms of the implicate order, but that thought also is basically to be comprehended in this order. Here we mean not just the content of thought for which we have already begun to use the implicate order. Rather, we also mean that the actual structure, function and activity of thought is in the implicate order. The distinction between implicit and explicit in thought is thus being taken here to be essentially equivalent to the distinction between implicate and explicate in matter in general. (*The Essential David Bohm*, pp. 110-111.)

Bohm lays the foundation for all of this when it comes to physics, but then he talks himself out of it whenever he switches over to biology and the theory of evolution. The two ideas are incompatible. Quantum Field Theory and Quantum Mechanics falsify Materialism, Naturalism, and their derivatives such as Darwinism, the Theory of Evolution, Spontaneous Generation, and Creation Ex Nihilo. Atheism is Creation Ex Nihilo, or creation from nothing by nothing.

When Bohm talks about an "end to fragmentation", he is in fact talking about the end of Materialism, Naturalism, Mechanism, Nihilism, Darwinism, and even Atheism – although he doesn't seem to know it yet. He's talking about wholeness and the implicate order. This "wholeness" consists of a Matrix of Quantum Fields, which comprises Bohm's implicate order. Quantum fields have been proven to exist. They have to exist, or physical matter couldn't exist. The Gods had to make the quantum fields BEFORE physical matter could be made and sustained. That's just the way it is. It's the truth. The movement or motion of energy is what actually exists. Energy is psychic, which means that energy is alive. It moves. It lives. It's eternal and everlasting. Energy, life force, or psyche has always existed, and it will always exist.

Bohm is right. The actual structure (pinpoint-of-light or psyche), psychic function or action at a distance, and activity of thought (quantum waves) are all taking place in the implicate order, the quantum realm, or the spirit world. It's there. It's real. It truly exists. The explicate order or manifest physical level emerges from the quantum level or the implicate order. In other words, physical matter is made from energy by a bunch of different psyches who control the energy from which physical matter is made. This is obviously true, because energy and consciousness and intelligence have been experienced and observed trillions of trillions of trillions of times.

The truth has always been there and will always be there, because it has always existed, even though David Bohm wasn't able to see it nor understand it in its entirety back in 1980. Back then, while discussing the Science behind psyche or energy and the obvious existence of information or intelligence, Bohm was simultaneously denying the existence of Psyche. He couldn't see nor understand back then that they are in fact the same thing. Psyche is energy; and, energy is psychic, intelligent, conscious, perceptive, and alive. We've made a little progress during the past forty years; but, we are not there yet, because we still have Naturalists, Materialists, and Mechanists in this world who don't get it and can't get it because they have put up mental blocks that prevent them from finding, recognizing, and knowing the truth.

All of this makes a great deal of sense to me now (2019), but it didn't while I was a materialist, naturalist, nihilist, and atheist. Back then (2012), I didn't have a clue. I knew what Atheism was and knew that I was an atheist; but, I didn't even know what these other things meant or that they even existed. It was only after I began to study, and think, and look it up that I finally figured out what a physicalist, materialist, nihilist, behaviorist, and naturalist is.

I've come a long way in a very short time, once I finally saw through all the deceptions and lies that are associated with Materialism, Naturalism, Darwinism, Nihilism, Behaviorism, Determinism, Physical Reductionism, Atomism, and Atheism. It took me over fifty years, but I finally got there in the end. My eyes have been opened, and now I can see. I'm no longer hiding from the evidence like I used to do when I was a materialist, naturalist, nihilist, and atheist. I have changed. The fear is gone. God took it all away, and I'm finally at peace within my body and my soul. The Atonement of Christ can do that for you, if you let it.

Mark My Words

Quantum Field Theory

What are quanta or particles made from?

Energy!

What are quantum fields made from?

Energy!

What is physical matter made from?

Energy!

What is Psyche or Intelligence made from?

Energy!

What are quantum waves or thoughts made from?

Energy!

What is always conserved? What has always existed and will always exist?

Energy!

Are you starting to see a pattern here? Energy can be used to make anything that you can possibly imagine. Energy is infinitely malleable. God is literally in the light.

The idea that Quantum Fields were designed, made, and organized from energy by Someone Intelligent or Someone Psyche makes infinitely more logical sense to me than the materialistic and atheistic idea that Quantum Fields sprang into existence whole from nothing. Science itself, including the Law of the Conservation of Energy, proves that there is no such thing as creation ex nihilo or spontaneous generation. These laws and truths are greatly enhanced by the Law of Psyche which effectively states that Someone Psyche is needed to organize raw chaotic energy into useful forms, including quantum fields, physical matter, and genomes. The Gods had to make quantum fields BEFORE physical matter could be made. Quantum fields are non-physical and pre-physical.

Every Psyche or Intelligence has a certain amount of energy that's under its control. This Psyche can form that energy into anything that it wants that energy to be. Therefore, design and organization by Someone Intelligent are infinitely more realistic and plausible than design and creation from nothing by nothing. Spontaneous generation and the theory of evolution were proven false in 1859 by Louis Pasteur. Genomes and quantum fields are intelligently designed, made, and deployed. According to science, there's no other way they could have come into existence, because there is no such thing as spontaneous generation or creation ex nihilo. The false is falsified by the truth; and, the truth is repeatedly experienced and observed.

Random chance, blind luck, spontaneous generation, and creation nihilo violate the fundamental principle of Science, in that they are not reliably replicable. These principles of random chance do not permit and predict the reliable replication of compatible physical atoms over and over again for trillions of years of time. In order for each atom to function reliably and predictably with every other atom in our physical universe, each atom must be deliberately designed and made by Someone Intelligent or Someone Psyche to fulfill that specific purpose. There has to be quality control; and, when it comes to energy, only Psyche or Intelligence has been known to be capable of fulfilling the role of Quality Controller or Reliable Replicator. Physical matter is the end-product, and NOT the Producer or Manufacturer, because there's no such thing as creation ex nihilo or spontaneous generation from nothing. Design and creation require an intelligent Designer and Creator. That's the way it has always been.

Quantum Field Theory falsifies the philosophies of Materialism, Naturalism, Nihilism, Physicalism, Atomism, Physical Reductionism, and even Atheism and Darwinism.

David Bohm viewed quantum theory and relativity as contradictory, which implied a more fundamental level in the universe. He claimed both quantum theory and relativity pointed towards this deeper theory, which he formulated as a quantum field theory. This more fundamental level was proposed to represent an undivided wholeness and an implicate order [non-physical order or matrix of quantum fields]**, from which arises the explicate order** [physical order] **of the universe as we experience it.**

Bohm's proposed implicate order applies both to matter and consciousness. He suggested that it could explain the relationship between them. He saw mind and matter as projections into our explicate order from the underlying implicate order. Bohm claimed that when we look at matter, we see nothing that helps us to understand consciousness.

https://en.wikipedia.org/wiki/Quantum_mind#Bohm

When it comes to physics, one must define what one is talking about; otherwise, the whole exercise becomes pointless, fruitless, and meaningless. When it comes to science, Quantum Field Theory has become my one of my most favorite areas of study because when it's properly understood it ends up explaining what everything is and how it works.

Who needs quantum field theory?

Quantum field theory arose out of our need to describe the ephemeral nature of life.

No, seriously, quantum field theory is needed when we confront simultaneously the two great physics innovations of the last century of the previous millennium: special relativity and quantum mechanics. Consider a fast-moving rocket ship close to light speed.

You need special relativity but not quantum mechanics to study its motion. On the other hand, to study a slow-moving electron scattering on a proton, you must invoke quantum mechanics, but you don't have to know a thing about special relativity.

It is in the peculiar confluence of special relativity and quantum mechanics that a new set of phenomena arises: Particles can be born, and particles can die. It is this matter of birth, life, and death that requires the development of a new subject in physics, that of quantum field theory.

Let me give a heuristic discussion. In quantum mechanics the uncertainty principle tells us that the energy can fluctuate wildly over a small interval of time. According to special relativity, energy can be converted into mass and vice versa. With quantum mechanics and special relativity, the wildly fluctuating energy can metamorphose into mass, that is, into new particles not previously present.

Special relativity tells us that energy can be converted to matter.

Zee, A. (2010). *Quantum Field Theory in a Nutshell* (2nd ed.). Princeton, NJ: Princeton University Press.

https://epdf.tips/queue/quantum-field-theory-in-a-nutshellf65ca91cbc5fdf00a51fd98ad1007dd18493.html

https://syntropy.site/Quantum-Field-Theory-in-a-Nutshell

Quantum field theory is needed to describe the ephemeral.

What does the word "ephemeral" mean? I originally thought that the word meant "ethereal" – which is defined as unworldly, spiritual, lacking material substance, immaterial, intangible, non-local, and non-physical. My scientific research revealed to me that Quantum Mechanics is absolutely necessary for describing the ethereal or the non-physical. However, Zee used the word "ephemeral" to define what quantum field theory was designed to explain. Ephemeral means "lasting for a very short time". So, what lasts for a very short time? Particles! "Particles are born, and particles can die." Particles are made from energy! There's the sum-total of particle physics and quantum field theory there in a nutshell.

"Particles are born, and particles can die." This statement implies that particles of any kind are alive. Does it not? Each particle is made from energy, and every form of energy has some type of Psyche, Intelligence, or Life Force within it that is choosing what form the energy under its control will assume. Energy is infinitely malleable, which means that Someone Psyche is choosing whether that energy will function as quantum fields, photons, electrons, quarks, dark matter (spirit matter), or entropic physical matter. It requires some type of Intelligence in order for the energy to know what each of these different forms does and how they

610

work with other forms of energy. There has to be some kind of universal law or universal blueprint that these Psyches or Intelligences are choosing to follow and obey; otherwise, chaos would continue to reign supreme.

Think about it. If particles are born and particles can die, then it's physically impossible for particles or physical matter to be the fundamental unit of reality as the Naturalists and Atheists claim. Rather instead, while searching for the fundamental stuff of reality and existence, we find ourselves looking for something that cannot be born and that cannot die – something like energy, and the psyche or intelligence that controls that energy. Particles are made from energy, and when those particles cease to exist, they are converted back into raw usable energy, because energy or psyche is always conserved. Energy is the fundamental substance of reality, and the Psyche or the Intelligence who controls that energy is the thing that decides which form the energy under its control is going to take. There's the sum-total of energy mechanics, spiritual mechanics, or quantum mechanics in a nutshell.

This to me is one of the most convincing Scientific Proofs of God's Existence, or Scientific Proofs of God's Psyche. Someone Psyche or Someone Universal had to make and has to enforce these universal laws or universal blueprints, into which the psyches within the energy are forming themselves or conforming themselves. Laws obviously require some type of intelligent Law-Giver, Law-Enforcer, and Law-Obeyer. Anything that is obviously made obviously has a Maker or Creator who designed it and made it. The universal blueprints for compatible physical matter or compatible atoms were obviously designed and made, which means that they obviously have some kind of Designer and Creator. Anything that is obviously organized obviously has an Organizer who organized it. Quantum fields are non-physical and pre-physical. The Gods had to make the quantum fields BEFORE physical matter could be made and sustained. The quantum fields were obviously organized from energy by Someone Psyche or Someone Intelligent; otherwise, they would not exist and would not be uniform at a universal scale. There's intelligence and consciousness behind all of this, or it wouldn't exist in the first place. Everything would be random chaos instead.

This is where science starts to get interesting. In contrast, we have the Materialists, Naturalists, Darwinists, Nihilists, and Atheists assuring us that psyche or intelligence or life force does not exist. They are wrong. They are obviously wrong. Psyche or intelligence is the ONLY reasonable and logical cause of the order and organization that we see all around us. If there were NO universal laws and NO universal blueprints, then everything around us would be nothing but random chaos, and we wouldn't exist. Existence is based upon psyche, intelligence, consciousness, or life force. Is it not? It's obviously true; whereas, the materialistic and naturalistic claim that the non-physical or the non-local does not exist is obviously false.

The proven and verified existence of quantum fields, quantum mechanics, magnetism, light, microwaves, radio waves, quantum waves (thoughts), dark matter or spirit matter, dark energy, gravity, space, time, mind, consciousness, intelligence, psyche, universal laws, universal blueprints, and everything else that is non-physical in nature and origin FALSIFIES Materialism, Naturalism, Darwinism, Nihilism, Behaviorism, Determinism, Atheism, and their derivatives such as Spontaneous Generation, Design and Creation by Natural Selection, Chemical Evolution, and Creation Ex Nihilo. The false is falsified by the truth; and, the truth is repeatedly experienced and observed.

Unlike energy, psyche, or the ethereal which seems to be eternal and everlasting without a beginning of days or an end of years, physical particles are ephemeral which means that they are born, and later they die or are converted back into usable energy again. Physical particles are made, and then at some point they cease to exist. According to Quantum Mechanics, physical particles have a "source" where they start their existence, and they have a "sink" where they end their existence and are converted back into raw usable energy once again. Physical particles are made from energy, and they eventually return back into that energy sink from whence they came. Only the ethereal energy or ethereal psyche is eternal and everlasting. In contrast, physical matter is born or made, and physical matter can die or cease to exist as physical matter.

Only the spiritual, the quantum, the psychic, or the ethereal is eternal and everlasting and indestructible. The psyche or the energy is the part of us that's truly real. In contrast, physical matter is ephemeral – it is made, and it can cease to exist at any time. These realities and truths are exactly opposite of what the Materialists, Naturalists, Darwinists, Nihilists, Behaviorists, Physical Reductionists, and Atheists teach and believe. These people teach that the quantum, the psychic, the spiritual, and the ethereal do not exist; and, these people erroneously teach and believe that physical matter is eternal and always conserved. They don't understand how things really work. They don't understand the Science because they formally reject the Science.

Because the Materialists, Naturalists, Nihilists, and Atheists have chosen to believe that physical matter is the only thing that exists, they are forced to erroneously believe that entropic physical matter is a conserved entity in complete violation of Quantum Field Theory which teaches that particles of any kind are ephemeral or temporary entities and therefore NOT conserved. Particles are born, and particles die. What happens when particles die? They become raw usable energy once again. They return to the state from whence they came. It's obvious. Particles can be born, and that particles can die. The process has been experienced and observed. That is the fundamental law of Quantum Field Theory, which means that there is nothing fundamental, necessary, essential, or foundational about entropic physical matter because it's ephemeral or temporary. Entropic physical matter and non-physical particles like photons have expiration dates built into them at the physical level, even though the energy and psyche within them is always conserved. This is what the Science is trying to tell us, although apparently the majority of our scientists aren't listening because they don't want it to be true.

According to a number of self-serving polls, it has been claimed that the majority of scientists formally reject this Quantum Science or Psyche Science, because they are in fact materialistic, atheistic, and naturalistic philosophers rather than true scientists who are willing to follow the evidence wherever it leads them.

The Materialists, Naturalists, Nihilists, and Atheists don't understand the Science because they formally reject the Science and choose to embrace their philosophical wishful-thinking, groupthink, and brain-washing propaganda instead. Their a priori commitment to physical matter as the ONLY reality prevents them from seeing and understanding anything else.

Furthermore, these people don't realize that entropy does not exist in any usable form. Entropy is NOT a person, place, or thing. There's nothing tangible or physical about entropy. Entropy is an emergent property associated with the passage of time. If there is no passage of time, then there is no entropy. That means that there is no such thing as entropy at the timeless level of existence, or the quantum level, or the psyche level where the energy or light is always being conserved. Conservation of energy falsifies and eliminates entropy. There can be no entropy in the timeless realm where energy, psyche, or light is being conserved. Entropy is only possible at the physical level where everything has been deliberately slowed down and the passage of time or the measurement of time has been introduced into the system.

Entropy only applies at the physical level. Entropy is non-existent at the quantum level and the psyche level. Entropy is a measurement of thermal equilibrium at the physical level. When a physical system has achieved thermal equilibrium at the physical level, then the system has achieved maximum entropy and then no further work can be done within that system at the physical level; however, all of the energy within the system is being conserved and is still fully usable at the quantum level and the psyche level. The internal energy is always conserved.

The Materialists, Naturalists, Darwinists, Nihilists, and Atheists are incapable of understanding this Science because they have chosen to believe that only physical matter exists and therefore are forced to belief that physical matter is eternal and is therefore always being conserved. They are wrong. Quantum Field Theory proves to us that they are wrong. Quantum Field Theory states that particles are ephemeral or temporary – particles are born, and particles die – and that reality and truth includes entropic physical matter most of all. Physical matter is born, and it can die or be turned back into raw usable energy by Someone Psyche. When entropic physical matter has no further use, then it's time for it to be recycled or recharged. It's time for further energy and heat to be introduced into the system.

The theory of Dark Energy implies that the Gods are currently infusing or transferring massive amounts of energy into our physical universe at an accelerating rate. In other words, the Gods are building quantum fields in the dark regions of our universe at an accelerating rate; and, the Gods can birth or begin physical matter within those dark regions anytime they choose to do so. Entropic physical matter at maximum entropy can be dissolved by the Gods, turned back into raw usable energy, quantum tunneled anywhere in the universe, and then used as Dark Energy and Quantum Fields for the expansion of our universe. When our solar system has reached maximum entropy, and has gone completely dark and cold, the Gods can simply dissolve it and turn it into Dark Energy instead. It won't be missed if it were to go missing.

By turning or transforming entropic physical matter back into raw energy, a huge amount of usable energy and heat can be generated as a result – think of a nuclear explosion for example. Trigger a nuclear explosion within entropic physical matter that has reached maximum entropy or thermo-equilibrium, and a ton energy and heat is automatically generated as a result. This is possible because the energy is always conserved; whereas, the entropic physical matter is born, and it dies or is turned back into raw energy whenever the Gods or Psyches or Intelligences see fit to do so. This is a whole other way of looking at Science, Entropy, Energy, Psyche, and Physical Matter which the Materialists, Naturalists, and Atheists absolutely refuse to consider or think about.

Psyche or God is literally within the energy or the light. A Psyche is a God over the energy that's under its control. That Psyche can convert that energy into anything that it wants it to be at any time it decides to do so. Psyche, and therefore the Gods, can take entropic physical matter that has reached maximum entropy or maximum thermo-equilibrium, and convert that physical matter into Dark Energy, Spirit Matter (Dark Matter), or Quantum Fields any time that Psyche chooses to do so. There's nothing sacred or eternal, or necessary and essential, about physical matter that has reached maximum entropy. Take the iron in the dead core of a dead star, transmute that iron into uranium, and then detonate it turning the whole thing back into energy and heat. Or simply take the iron in a dead core of a dead star and transform it into Dark Energy, Spirit Matter, or Usable Heat. Energy is infinitely malleable; and, the Psyche or God who controls it can transform it into anything instantaneously any time that this Psyche or God chooses to do so. This is a whole other way of looking at Science that the Materialists, Naturalists, and Atheists completely refuse to consider because it falsifies their personally chosen beliefs.

From any angle that we choose to look, the Science tells us that the Materialists, Naturalists, Darwinists, Nihilists, Behaviorists, Determinists, Physical Reductionists, Atomists, and Atheists are WRONG. These falsified philosophies are falsified by the truth; and, the truth is repeatedly experienced and observed. That's the way Science works.

I find this whole thing fascinating because I used to be a materialist, naturalist, nihilist, and atheist until the Science convinced me that I was wrong. It is obvious that Quantum Fields are made out of energy and not physical matter. It's obvious from the Science that Quantum Fields are non-physical and pre-physical. The Gods had to make the Quantum Fields out of energy before they could make physical matter out of energy. It's obvious that physical matter is made out of energy or light. Yet, all of these obvious truths are categorically denied and rejected by the Atheists, Naturalists, and Physicalists.

In spite of all the evidence to the contrary, these people successfully convince themselves that physical matter is the only thing that exists, and that physical matter is conserved. Self-deception works, and it works every time.

These people don't want the non-physical or the spiritual or the quantum to exist, so they define it out of existence and interpret everything else accordingly. I've been there and done that. We atheists don't want God to exist because we don't want to be held responsible for our choices, actions, vices, addictions, and sins. That's ironic in a way because those kinds of things can have fatal consequences here in mortality. We don't have to wait until the next life in order to suffer the consequences of our addictions, sins, and bad choices. We don't have to wait until the next life to go to hell – we can do so right here, right now, if we want to. I know, because I have been there and done that already.

Anyway, the Materialists, Naturalists, Nihilists, and Atheists are always wrong when it comes to Science. Physical matter is NOT the only thing that exists. In a very real sense, physical matter is made from the interaction of Quantum Fields; and, Quantum Fields are clearly made out of energy or light, and NOT physical matter. Physical matter is made from energy or light; and, the Gods or Controlling Psyches can turn physical matter back into raw usable energy or light anytime they choose to do so. Entropy is an ephemeral ethereal illusion, or a temporary non-physical illusion, because any physical matter that has reached maximum entropy or complete thermal equilibrium can be converted back into raw usable energy any time its controlling Psyche chooses to do so. Energy is the ultimate perpetual motion machine, and Psyche is its driver or controller. That's the true Science according to Quantum Field Theory, and it FALSIFIES what the Materialists, Naturalists, Nihilists, and Atheists have chosen to preach, teach, and believe.

Energy or light is in motion. It seems to be purposeful, conscious, and alive. It fluctuates. Energy is infinitely malleable. Energy can be converted into mass, and mass or physical matter can be converted back into energy. Converted by whom? Who or what decides to do this conversion or transformation? Who or what has the power, and the ability, the intelligence, and the knowledge necessary to convert unorganized energy into mass or physical matter? According to special relativity, energy can be converted into matter. Converted by whom? Who makes this decision to convert a specific batch of energy into physical matter, or spirit matter, or dark matter? Who makes the decision to dissolve this matter or mass back into the quantum fields from which it came? Somewhere along the line, the choice has to be made. Who or what is making this CHOICE?

Remember, when it comes to physics, the smaller dwells within and controls the larger. What's smaller than a quantum fluctuation or a particle? What's smaller than the string within string theory? Who determines the shape a string will take, or the type of quantum fluctuation of energy that a particle will be? If there is such a thing as a point particle, then psyche, intelligence, consciousness, or life force would be it. Out-of-Body Travelers have observed, while at the psyche level within the quantum realm itself, that individual Psyches or Intelligences are seen as being pin-points of light. Even at the quantum level or the level of strings, Psyches or Intelligences are observed as being point-particles or infinite singularities. That tells me that Psyche or Intelligence is the smallest thing that exists; and, when it comes to physics, the smaller always dwells within and controls the larger. The observational experiences of the Out-of-Body Travelers tell us who is converting energy into mass or matter. The smaller always dwells within and controls the larger.

With quantum mechanics and special relativity, the wildly fluctuating spiritual energy or quantum energy can metamorphose into mass or physical matter – that is, into new physical particles that were not previously present. Physical matter is made from energy or light by Someone Psyche or Someone Intelligent. Someone Intelligent or Someone Psyche within the energy itself DECIDES or CHOOSES to transform that energy into new physical particles that were not previously present. This same Psyche or Intelligence can also CHOOSE to transform that energy into non-physical particles that have NO mass – particles or wave packets of energy such as photons. It's all there within the physics; and, it's NOT good news for the Materialists, Naturalists, Darwinists, Nihilists, and Atheists who have chosen not to believe in the non-physical. The very existence of Quantum Field Theory and Energy FALSIFIES Materialism, Naturalism, Nihilism, and Atheism.

Energy or psyche is infinitely malleable or infinitely transformable. Energy can be converted into anything, including quantum fields, psyche or the spark of life, spirit matter (dark matter), dark energy, radio waves, microwaves, magnetic fields, gravity, entropic physical matter, and even immortal eternal resurrected physical matter. There are no limits to what energy or psyche can be transformed into.

The false is falsified by the truth; and, the truth is repeatedly experienced and observed. This is the way that science should work.

As far as I know, I am the first person on the planet to propose a Psyche Ontology wherein Psyche or Intelligence is the fundamental unit of reality.

Why did I do this?

It's because Psyche has been experienced and observed functioning completely separate from its assigned spirit body and its assigned physical body. Psyche or Intelligence is a completely separate entity. It has been observed that whenever a Psyche has separated from its spirit body and its physical body, and that Psyche is looking at its spirit body, it is the Psyche who is having experiences and remembering those experiences and NOT its spirit body NOR its physical body. When it comes to human beings, the Life Force or Intelligence is contained within the Human Psyche, and NOT our spirit body NOR our physical body. The energy within our spirit body and the energy within the atoms of our physical body have a Psyche or an Intelligence all their own, which I have chosen to call Nature's Psyche within many of my books.

https://psyche-ontology.com/

https://psyche-ontology.com/buhlman/

https://psyche-ontology.com/psyche-observed/

https://psyche-ontology.com/psyche-experienced-and-observed/

At the physical level, the Intelligences within each atom have chosen to follow and obey physical restraints or physical laws. Consequently, those atoms seem inert, inanimate, and inactive to us. However, all of the different particles within an individual atom are alive with motion, purpose, and coordinated intelligence. The energy within an atom is in constant motion; and, the Psyches or Intelligences controlling that energy KNOW precisely what they are doing and why. Nature's Psyche has direct access to infinitely more knowledge and information than the Human Psyche does. We human beings have been dumbed-down, made ignorant and innocent, so that we can be tested and tried – so that we can prove to ourselves the type of Psyche or Personality that we really are and really want to become. Psyche really is the fundamental unit of life, existence, and reality; and, energy really is the stuff of which everything is made. Contrary to what the Materialists, Naturalists, Nihilists, and Atheists have chosen to believe and teach, physical matter is NOT the fundamental unit of reality, nor is physical matter the fundamental stuff from which everything is made. Physical matter is not the only thing that exists.

Technically, there's no such thing as physical particles. What the Atheists, Materialists, and Naturalists call "empty space" is actually a Matrix made from Quantum Fields. Quantum Fields are made from energy or light.

If you get a chance, look at the table of contents within Zee's book *Quantum Field Theory in a Nutshell* (2nd ed.). Notice all the dozens of different quantum phenomena that are made out of energy by Someone Psyche or Nature's Psyche; and, these are just some of the quantum phenomena that are associated with Quantum Field Theory, not all of them. There are a lot more associated with Quantum Mechanics. These are quantum phenomena or quantum functionality that's being produced by energy under the control of Someone Psyche or Nature's Psyche. Energy really is infinitely malleable; and, Psyche really can form energy into anything that it can possibly imagine. All of these different quantum phenomena or non-physical phenomena explain how things really are; and, the whole of it FALSIFIES Materialism, Naturalism, Darwinism, Nihilism, Behaviorism, Determinism, Physical Reductionism, Atomism, and even Atheism. Psyche is indeed the God over the energy that's under its control.

https://epdf.tips/queue/quantum-field-theory-in-a-nutshellf65ca91cbc5fdf00a51fd98ad1007dd18493.html

https://syntropy.site/Quantum-Field-Theory-in-a-Nutshell

Energy is infinitely malleable or infinitely transformable. Energy or light can take on an infinite number of different forms – including psyche, spirit matter, quantum fields, and physical matter. Psyche can transform the energy under its control into anything that psyche wants it to be! That's the true nature of energy – it's capable of constantly changing its form. Energy under the control of psyche can take on an infinite number of different forms. This truth is essential and fundamental if we want to know how things really work.

For a small sampling of all the different things that energy can be made to do and can be formed into, watch the documentary, *Inner Worlds, Outer Worlds*.

This documentary focuses primarily on the truths about energy or light which the various different world religions have discovered. When it comes to energy or light, these different world religions are quite consistent with each other; and, they match extremely well with the Energy Mechanics or Light Mechanics that David Bohm talked about in his books and that I'm talking about within these essays. These books, essays, and movies are talking about psychic phenomena, energy phenomena, and the capabilities of light which the Materialists, Naturalists, and Atheists formally reject and deny. These people have taught themselves to HATE spiritual phenomenon, psychic phenomenon, energy phenomena, and the quantum phenomena that falsify their personal beliefs.

There's another reason why the Materialists, Naturalists, Nihilists, Behaviorists, Determinists, and Atheists formally reject Psyche, Quantum Mechanics, Quantum Field Theory, the spiritual, the psychic, and the non-physical. It's because these things cannot be controlled from the physical level. These people can't make money out of these things. They can't bank on it. Quantum Mechanics and the quantum level can only be controlled and manipulated by Someone Psyche at the quantum level and the psyche level. Quantum tunneling or teleportation can't be triggered nor controlled at the physical level.

I watched a movie called *Anti Matter* in which they found a way to trigger quantum tunneling or teleportation from the physical level. They got all the science completely wrong, because they didn't understand the science. They used the jargon, but they used it incorrectly. The creators of the movie were Materialists, Naturalists, and Atheists trying to explain the psyche or the soul from the physical level. They called the jumping process a wormhole, instead of quantum tunneling or teleportation. Furthermore, they made the energy-being duplicate of the heroin into the ephemeral or temporary being constantly losing her memories, whereas it was the physical being with the brain who was the permanent, stable, or real being with the memories. There's nothing permanent or stable about a physical body and physical brain. It was all completely backwards from the true reality of the Science. The memories that survive the death of our physical body and show up in our after-death Life Review cannot be stored within our physical brain at the physical level. Memories are stored within the Human Psyche. In this movie, they got it all wrong.

Finally, I'm pretty much sure that there's no way to trigger quantum tunneling or teleportation from the physical level. Quantum phenomena can only be triggered and controlled from the psyche level by Someone Psyche or Someone Intelligent. We are not going to be accidentally triggering quantum tunneling from the physical level with physical processes and physical instruments. When it comes to physics, the smaller always dwells within and controls the larger. The Psyche dwells within and controls the energy; and, the controlling Psyche chooses what form it wants that energy to take. Quantum fields and physical matter are made from energy by Someone Psyche. That's the way things really work. The false is falsified by the truth; and, the truth is repeatedly experienced and observed.

Quantum Fields by Art Hobson

Some say everything is made of atoms, but this is far from true. Light, radio, and other radiations aren't made of atoms. Protons, neutrons, and electrons aren't made of atoms, although atoms are made of them. Most importantly, 95% of the universe's energy comes in the form of dark matter and dark energy, and these aren't made of atoms.

The central message of our most fundamental physical theory, namely quantum physics, is that everything is made of quantized fields. To see what this means, we need to understand two things: fields, and quantization.

https://blog.oup.com/2017/02/quantum-fields/

https://quantum-neuroscience.com/Quantum-fields-_-OUPblog

https://blog.oup.com/2017/01/universe-fields-particles/

https://quantum-neuroscience.com/Fields-or-Particles-_-OUPblog

Quantum Fields are made from energy or light, and energy is infinitely malleable. The Psyche or Intelligence who is in control of that energy can form that energy into anything that that Psyche wants it to be, including quantum fields, spirit matter (dark matter), spiritual landscapes, dark energy, and physical matter.

https://en.wikipedia.org/wiki/Quantum_field_theory

https://quantum-neuroscience.com/Quantum-Field-Theory

Quantum fields exist at the transition between thought and physical matter. Quantum fields exist in David Bohm's Implicate Order. Quantum fields are organized energy or light. Physical particles or physical matter are part of Bohm's Explicate Order or Manifest Order. Quantum Fields comprise Bohm's Implicate Order, the Quantum Realm, or what some people call the Spirit World. Energy or light make up Bohm's Super-Implicate Order thereby making energy or light more fundamental than quantum fields. Finally, Bohm talks about "significance" or "meaning" or "consciousness" and states that it's more fundamental and essential than energy and matter. For Bohm, consciousness, psyche, intelligence, or thought are sort of a Super-Super Implicate Order. That pretty much covers everything that we human beings have ever experienced or observed.

https://quantum-neuroscience.com/The-Essential-David-Bohm

Quantum Fields and the Implicate Order are where physics and quantum mechanics start to get interesting. Quantum Fields are non-physical and pre-physical. Energy or light is interesting too, because it seems to be more fundamental or primal than quantum fields. Of course, everything begins with psyche or thought.

https://quantum-neuroscience.com/The-Effect-of-Mind-upon-Brain

According to the Quantum Zeno Effect, the energy within atoms is psychic, conscious, intelligent, sentient, perceptive, aware, telepathic, and alive. Our understanding of reality and quantum mechanics is never complete until after we have successfully included psyche, consciousness, intelligence, or thought into our theoretical models.

Scientific Premise: Anything that is obviously organized obviously has an Organizer who organized it.

Scientific Observation: Quantum fields were obviously organized.

Scientific Conclusion: Therefore, it is obvious that quantum fields have an Organizer who designed them, organized them, and made them from energy or light. Quantum fields are organized energy or light.

The false is falsified by the truth; and, the truth is repeatedly experienced and observed. Psyche, intelligence, consciousness, or thought is repeatedly experienced and observed, which is why the Materialists, Naturalists, Darwinists, Nihilists, Behaviorists, and Atheists constantly reject these things. Their ultimate goal in life is to turn Psyche and Quantum Mechanics into non-sense.

> **Between a stimulus and the responses that follow it, there must be an unbroken network of causally related events potentially observable by an outside beholder, no matter whether or not some of the neural events have correlates in the consciousness of the subject. Thus, an image of which I am not conscious now will appear in my imagination if appropriate stimuli activate certain of my neurons. It can have no independent existence apart from the stimulation of those neurons. Within my nervous system, a *potentiality* which includes the evocation of this image may be said to exist, but *only in the same sense as it may be said that my nervous system harbors the potentiality of a knee jerk given the stimulus of a patellar tap*. If the relationship of images to the nervous system is so conceived, all talk of "mind structure" becomes nonsensical. – Joseph Wolpe**

From: Rychlak, J. (1981). *A Philosophy of Science for Personality Theory*. 2nd ed. (pp. 225-226.)

Notice how Wolpe defines consciousness, psyche, and mind out of existence BEFORE doing any actual science or scientific experimentation. Wolpe admits that the goal of the Behaviorists, Mechanists, Naturalists, and Atheists is to turn Consciousness, Quantum Mechanics, Intelligence, and Psyche into nonsense. They can't allow psyche to exist because it falsifies Behaviorism, Naturalism, Materialism, and Mechanism. Quantum Mechanics, Mind Structure, Consciousness, Intelligence, and Psyche are indeed non-sense from a materialistic, atheistic, and naturalistic perspective. That's precisely what they intend to happen, and they want you to fall for the ruse. Most scientists do. One of our goals in life should be to be smarter than the average scientist. Materialism, Naturalism, Behaviorism, and Atheism were designed to dumb us down, make us ignorant, and keep us stupid. It works. I know because I lived it and experienced it first-hand for myself. I used to be a materialist, naturalist, nihilist, and atheist.

In truth, though, the outsider observer is another Psyche because only Psyche can do observation and actually remember having done so. Furthermore, only Psyche or Mind can do direct stimulation of neurons from the inside of those neurons. When it comes to physics, the smaller dwells within and controls the larger. When I decide to lift my finger, Someone Psyche finds and triggers the appropriate neuron to get the job done. That doesn't come from any outside stimulus. That comes from me – my Human Psyche. It's a conscious choice. Only psyche can do conscious choice.

We KNOW this is true because Psyche has been experienced and observed. The Human Psyche continues to exist and continues to function long after its physical brain is dead and offline.

https://psyche-ontology.com/psyche-observed/

https://psyche-ontology.com/psyche-experienced-and-observed/

https://psyche-ontology.com/buhlman/

Mind over matter, as well as mind over energy, is one of the fundamental aspects of physics that we have to wrap our minds around if we want to know how things truly work. I found all of this extremely interesting when I first discovered it.

https://quantum-neuroscience.com/Mind-Brain-Interaction

https://quantum-neuroscience.com/on-nature-of-things/

https://quantum-neuroscience.com/physicalism-and-quantum-mechanics/

https://quantum-neuroscience.com/neuroscience-and-the-human-person/

https://quantum-neuroscience.com/Mindful-Universe

https://quantum-neuroscience.com/Mind-Matter-and-Quantum-Mechanics

https://quantum-neuroscience.com/Consciousness-Beyond-Life

https://quantum-neuroscience.com/The-Spiritual-Brain

These scientists and books are thousands of years ahead of the Materialists, Naturalists, Nihilists, and Atheists. Whether these people realize it or not, the very existence of quantum fields is Proof of Psyche or Proof of Intelligence. During the past forty or fifty years, Bohm's crazy and unorthodox ideas have been proven true through scientific experimentation. The Placebo Effect and the Quantum Zeno Effect are scientific proof of mind-over-matter. It's real. It truly exists. Psyche or mind has been proven to exist. The proven and verified existence of Psyche

FALSIFIES Materialism, Naturalism, Darwinism, Nihilism, Atheism, and their derivatives.

Of course, most scientists reject this evidence, science, and information because they don't want it to be true. They won't buy it. Most of them won't even accept it when it's given away to them for free. They don't want it because it successfully falsifies Materialism, Naturalism, Darwinism, Nihilism, Behaviorism, Determinism, Mechanism, Physical Reductionism, Atomism, and Atheism.

I know how it goes because I used to be a materialist, naturalist, nihilist, and atheist. One day, though, I chose to go with the science – with the observations, experiences, and experiments – rather than the wishful thinking, dogmatic propaganda, and blind desires of the Materialists, Naturalists, Darwinists, Nihilists, and Atheists. One day, I finally saw the light.

Mark My Words

The Super-Implicate Order (1986)

I have a great interest in updating Bohm's ideas, and I want to bring them into the 21st Century. Science has made a great deal of progress during the past forty or fifty years proving that Bohm was right. Bohm has been vindicated by Science, while at the same time Materialism, Naturalism, Darwinism, Mechanism, Physicalism, Nihilism, Atheism, and their derivatives have been falsified by Science and scientific experimentation.

Bohm's Super-Implicate Order is made from energy or light by meaning, purpose, active information, consciousness, intelligence, or psyche. Bohm's Super-Implicate Order consists of the psyche, energy, and light from which his Implicate Order or the Quantum Fields are made.

This is one of my favorite chapters in his book, *The Essential David Bohm*.

https://philosophy-of-science.com/The-Essential-David-Bohm

https://epdf.tips/the-essential-david-bohm.html

https://ultimate-model-of-reality.com/The-Super-Implicate-Order

This particular link seemed to bring up the whole chapter on Google Chrome:

https://books.google.com/books?id=KqnHo5RlWp4C&pg=PA155&lpg=PA155&dq=%22the+null+ray%22+quantum+mechanics&source=bl&ots=NTtkiFba5H&sig=pDc6UChAmPZi0v-ps4H7UWTJYj8&hl=en&sa=X&ved=2ahUKEwiep8ClzNTfAhXnCTQIHVlFCugQ6AEwBnoECAgQAQ#v=onepage&q=%22the%20null%20ray%22%20quantum%20mechanics&f=false

THIS IS THE ONE YOU WANT TO HAVE, READ, KNOW, AND UNDERSTAND. Bohm's Philosophy of Science based upon his Super-Implicate Order is a successful and useful replacement for the falsified Philosophy of Materialism or the falsified Philosophy of Mechanism upon which classical physics or Newtonian physics is based. Bohm's implicate order and super-implicate order made possible the Holonomic Brain Theory.

Holonomic brain theory is a branch of neuroscience investigating the idea that human consciousness is formed by quantum effects in or between brain cells. This is opposed by traditional neuroscience, which investigates the brain's behavior by looking at patterns of neurons and the surrounding chemistry, and which assumes that any quantum effects will not be significant at this scale. The entire field of quantum consciousness is often criticized as pseudoscience, as detailed on the main article thereof.

This specific theory of quantum consciousness was developed by neuroscientist Karl Pribram initially in collaboration with physicist David Bohm. It describes human cognition by modeling the brain as a

holographic storage network. Pribram suggests these processes involve electric oscillations in the brain's fine-fibered dendritic webs, which are different from the more commonly known action potentials involving axons and synapses. These oscillations are waves and create wave interference patterns in which memory is encoded naturally, and the waves may be analyzed by a Fourier transform. Gabor, Pribram and others noted the similarities between these brain processes and the storage of information in a hologram, which can also be analyzed with a Fourier transform. In a hologram, any part of the hologram with sufficient size contains the whole of the stored information. In this theory, a piece of a long-term memory is similarly distributed over a dendritic arbor so that each part of the dendritic network contains all the information stored over the entire network. This model allows for important aspects of human consciousness, including the fast associative memory that allows for connections between different pieces of stored information and the non-locality of memory storage (a specific memory is not stored in a specific location, i.e. a certain cluster of neurons).

https://en.wikipedia.org/wiki/Holonomic_brain_theory

I found this quote after writing my book, *Quantum Neuroscience: The Answer to Life, the Universe, and Everything*; and, holonomic brain theory does indeed provide a realistic quantum mechanism to explain the truths that I discovered while researching Quantum Neuroscience.

https://www.amazon.com/gp/product/B079Z6QQQB/

As I have done within many of my books, within this essay and conversation about the Super-Implicate Order or the Psyche Level, David Bohm marries the theory of relativity with quantum mechanics by targeting their common denominator which is energy or light.

Here we get to the true reality of our situation; but first, we have to introduce and define Bohm's Super Implicate Order.

Here Bohm refines his original model of the implicate order, addressing how the unfoldment of an implicate order results in manifest order and structure. Why is the multidimensional medium of space (the vacuum, or plenum) capable of unfolding the forms we experience in the sensual, three-dimensional world? Why does the plenum of space correspond more to the analogies of ink enfolded into glycerin, or to light enfolded in a hologram, than to the diffuse and random order of eggs enfolded into cake batter?

It is the activity of a super-implicate order, says Bohm, that accounts for this ordered manifestation. Bohm derives the super-implicate order by applying his causal interpretation to quantum field theory. This super-implicate order infuses the implicate order of space with active information, which generates various levels of

organization, structure, and meaning. To illustrate the relationship of active information to these levels of organization, Bohm introduces the principle of *soma-significance*, whereby meaning and form are transmitted throughout a hierarchical continuum of matter and consciousness. – Lee Nichol (*The Essential David Bohm*, p. 139.)

https://syntropy.site/The-Essential-David-Bohm

Bohm uses words that nobody understands to talk about and explain this stuff; but, it all makes perfect sense once we translate this material into English.

The Implicate Order is the Quantum Realm of Energy, Quantum Fields, Quantum Mechanics, Dark Energy, or the Spirit World.

The Manifest Order or Explicate Order is our physical reality. Our physical reality unfolds or manifests from its underlying spiritual reality or quantum reality.

Multi-dimensional means multiple levels of existence or reality. We have the psyche level, the quantum level, and the physical level. Bohm suggests that there might be other levels as well that we currently know nothing about. Transdimensional is translated as being non-physical. Quantum Mechanics is Transdimensional Physics.

Multi-dimensional space is capable of unfolding or manifesting the forms that we experience in our sensual reality or physical reality, because space is comprised of various different quantum fields, and energy is infinitely malleable, which means that energy and the psyche that controls it can form itself into anything it wants to be, including spirit matter and physical matter.

The activity or choices of a super-implicate order (psyche) accounts for this ordered physical manifestation that we currently see all around us. Psyche is the ultimate causal agent. Psyche is supernatural or super-implicate.

Energy is infinitely malleable or infinitely transformable. Psyche can transform the energy under its control into anything that psyche wants it to be! That's the true nature of energy – it's capable of constantly changing its form. Energy under the control of psyche can take on an infinite number of different forms. That's a spiritual phenomenon, and not a physical phenomenon.

The Gods created our physical reality so that we would finally have a platform that is stable and can't change at will. The energy within physical matter has covenanted with God to get God's permission before changing its form. The energy within physical matter stays true-to-form unless God tells it to do otherwise. Physical matter is a whole other way of doing business. It brings order, stability, predictability, and dependability to our universe. "Energy constantly changing its form at will" is the very definition of Chaos. The Gods brought order to Chaos by organizing energy into quantum fields and physical matter. The Gods gave energy a purpose for being.

The Gods organized energy into quantum fields; and then, the Gods used those quantum fields to produce some physical matter. Physical matter is less than 5% of our physical universe. The Gods didn't make a whole lot of the stuff; and, the Gods can make more anytime they choose to do so. In fact, from what we can see through our telescopes, it really IS as if galaxies spring into existence whole as if from nowhere or nothing. It's as if galaxies are born all at once instead of coalescing into existence from huge clouds of gas. The only way that's possible is if the Gods take a huge chunk of energy and transform it in-place all at once into physical matter. After all, clouds of hydrogen and helium gas DO NOT coalesce into planets and stars naturally. They have to be forced to do so, or it isn't going to happen.

Whether we are talking about quantum fields, dark energy, dark matter (spirit matter), or physical matter, it was all designed to be stable, predictable, dependable, and semi-permanent. Entropic physical matter even has clocks, time, an aging process, or expiration dates built into it. The energy within the physical matter is always conserved; but, entropic physical matter is not conserved. The quantity of physical matter is changing all the time according to Quantum Mechanics and Science. We human beings create new physical matter from energy within our particle accelerators; and, nuclear explosions destroy physical matter by converting some of it back into energy.

Bohm isn't afraid to falsify the Philosophy of Mechanism, because he found something infinitely better – Quantum Mechanics, Quantum Field Theory, the Implicate Order, and the Super-Implicate Order.

Bohm doesn't talk about psyche or mind until he gets to his Super-Implicate Order; but, once he develops his Super-Implicate Order, then he finally has the philosophical foundation that he needs to delve into psyche, mind, and consciousness. Psyche or Intelligence is the foundation of the Supernatural Order or the Super-Implicate Order.

Psyche, Intelligence, or Consciousness is the primary source of active information, organization, structure, meaning, form, quantum fields, physical matter, and the different levels or dimensions of reality that we have discovered so far. Soma is the physical body; and, significance or meaning or purpose is produced by the Psyche or the Mind. That's how things really work contrary to what our atheistic college teachers are trying to make us believe.

Mark My Words

Psyche and Light Are the Super-Implicate Order

In this chapter, David Bohm effectively defines the Super-Implicate Order as raw energy or light. In the next chapter of *The Essential David Bohm*, around the same point in time, David Bohm effectively calls Psyche (significance or meaning or consciousness) the Super-Super Implicate Order. David Bohm states that "meaning" is more fundamental than energy and matter.

https://ultimate-model-of-reality.com/The-Super-Implicate-Order

When I made my own conceptual breakthroughs in physics, they ended up being directly in-line with what David Bohm has to say here within *The Essential David Bohm*:

Weber: Speaking of mysticism, there is one important idea that I would like to discuss and understand and that is the idea of light. That is especially important to me because you are a physicist. Light has been used as the privileged metaphor in the language of mysticism and experimental religions, going back to the Greeks and the east. In all these, light is the symbol of our union with the divine. They talk about a light without shadow, an all-suffusing light, and it comes up as the central metaphor in near-death experiences. Do you have any hypothesis as to why light has been singled out as the privileged metaphor?

Bohm: If you want to relate it to modern physics (light and more generally anything moving at the speed of light, which is called the null-velocity, meaning null distance), the connection might be as follows. As an object approaches the speed of light, according to relativity, its internal space and time change so that the clocks slow down relative to other speeds, and the distance is shortened. You would find that the two ends of the light ray would have no time between them and no distance, so they would represent immediate contact. (This was pointed out by G. N. Lewis, a physical chemist, in the 1920s.) You could also say that from the point of view of present field theory, the fundamental fields are those of very high energy in which mass can be neglected, which would be essentially moving at the speed of light. Mass is a phenomenon of connecting light rays which go back and forth, sort of freezing them into a pattern.

So, matter, as it were, is condensed or frozen light. Light is not merely electromagnetic waves but, in a sense, other kinds of waves that go at that speed. Therefore, all matter is a condensation of light into patterns moving back and forth at average speeds which are less than the speed of light. Even Einstein had some hint of that idea. You could say that when we come to light we are coming to the fundamental activity in which existence has its ground, or at least coming close to it.

Weber: Why is speed the determinant?

Bohm: Well, let's turn it around. If you look at Piaget and young children, movement is primary in perception. They see movement first and its unfoldment as time, and only perceive distance later. They have a tendency to say that if something went further it must have been going faster. They only learn later how to do it right. They are carrying some deeper perception into the ordinary explicate level, where it is inappropriate. In the deeper perception, movement is the primary reality in perception. The thing that is not moving is the result of the cancellation of movement. We say that there is no speed at all at light. To call it speed is merely using ordinary language. In itself, when it is self-referential, there's no time, no space, no speed [at the speed of light].

Weber: What is it?

Bohm: It's just a primary conception. As you move faster and faster according to relativity your time rates slow down and the distance gets smaller, so as you approach very high speeds your own internal time and distance become less, and therefore if you were at the speed of light you could reach from one end of the universe to the other without changing your age at all.

Weber: Isn't that saying that it's approaching a timeless state?

Bohm: That's right. We're saying that existentially speaking or logically speaking, time originates out of the timeless.

Weber: This is primary, and time is derivative of it, cutting it down, freezing it, and arresting it.

Bohm: Yes, arresting it to a certain extent, not absolutely, but to a large extent.

Weber: When mystics use the visualization of light they don't use it only as a metaphor, to them it seems to be a reality. Have they tapped into matter and energy at a level where time is absent?

Bohm: It may well be. That's one way of looking at it. As I've suggested the mind has two-dimensional and three-dimensional modes of operation. It may be able to operate directly in the depths of the implicate order where this [timeless state] is the primary actuality. Then we could see the ordinary actuality as a secondary structure that emerges as an overtone on the primary structure. It's again the business of what is emphasized and what is secondary – the two kinds of music. The ordinary consciousness is one kind of music, and the other kind of consciousness is the other kind of music.

Weber: For the mystics there is always light. The primary clear light in the *Tibetan Book of the Dead* is the first thing the dying person is aware of. If he doesn't move towards it or away from it or feel awe or fear or manipulate it in any way as if it were outside himself, then he merges with it and is liberated, enlightened. Christ says: "I am the light," and so on. I've always asked myself, why light? You're saying that from the point of view of a physicist, it has to do with the absence of speed and the closeness of contact.

Bohm: Light is what enfolds all the universe as well. For example, if you're looking at this room, the whole room is enfolded into the light which enters the pupil of your eye and unfolds into the image and into your brain. Light in its generalized sense (not just ordinary light) is the means by which the entire universe unfolds into itself.

Weber: Is this a metaphor for you or an actual state?

Bohm: It's an actuality. At least as far as physics is concerned.

Weber: Light is energy, of course.

Bohm: It's energy and it's also information – content, form and structure. It's the potential of everything.

Weber: Physicists are not satisfied that they have understood light up to now because of the particle-wave paradox, right?

Bohm: Yes, I think that to understand light we'll have to understand the structure underlying time and space more deeply. You can see that these issues are related in the sense that light transcends the present structure of time and space and we will never understand it properly in that present structure [of physical space-time].

Weber: How would the implicate order philosophy handle light?

Bohm: It could handle it more naturally, mathematically speaking, because it doesn't commit itself to the idea of separate points in space; but it may say that the underlying reality is something which is not localized, and light is also something which is not localized. [Non-local means non-physical.] One view [linear physics or mechanical classical physics] says that light moves from one place to another through a series of positions, and the other view [quantum mechanics] says it doesn't do that at all. Rather, light exists, it just simply is. [Light is omnipresent at the speed of light and time ceases to exist according to quantum mechanics and the theory of relativity.]

Weber: It is at all points?

Bohm: Points are defined by the intersections of different rays of light. That's the way we actually do it in perception. We infer a point from the fact that many light rays are coming from it, say a star or any point. In this view, points would be understood as the intersection of many light rays. The light is fundamental, the null ray. That's a technical term that shows the recognition of this fact in ordinary physics.

Weber: It's where every particle of matter is in contact without the slightest gap between them.

Bohm: Yes, it's possible to have that contact without a gap.

Weber: So, light is one continuous, unbroken, undivided whole?

Bohm: You would have to look at it that way, yes; especially if you consider the quantum theory of it which says the action in it is undivided as well. What G. N. Lewis had in mind was to explain the quantum in that way. It was very mysterious to say that light is a wave which spreads continuously through space and yet that a single quantum of energy goes from one point to another. How could that happen? G. N. Lewis said this wave was some sort of an abstraction, and he said what actually happened in each ray was that there was an immediate contact from the source to the absorber. [There's distance between start and finish at the speed of light because there is no passage of time at the speed of light.] One understood the quantum in that way, that there was no spreading out of energy.

Weber: It therefore takes no time, no transmission, no distance. There isn't any, is what you're saying.

Bohm: That is the view I'm proposing. The ordinary view is another map of it. You can take many maps of the world; one of them is Mercator's projection, which is quite good near the equator, but it says near the poles that the space is infinite. So, maps can have the wrong structure. We can say that the ordinary space-time is a map which holds fairly well for ordinary speeds, but when you get to the speed of light it's as wrong in structure as Mercator's projection is at the pole.

Weber: We say light is clarity, light illumines, light is energy, some mystics have said light is love, compassion, understanding, light can make whole or heal. If light is the background of everything, what would be its relationship to the foreground?

Bohm: Light is this background which is all one, but its information content has the capacity for immense diversity. Light can carry information about the entire universe. The other point is that light, by interactions of different rays (as field theory in physics is investigating today), can produce particles and all the diverse structures of matter.

Weber: You've stressed information and that has to do with knowing the universe.

Bohm: A kind of knowing.

Weber: The other aspect would have to do with its being. Maybe there's an undifferentiated realm of light and when it radiates itself as being, as particles, those might be its "shadows" or finite expression.

Bohm: They are expressions, but they are ripples on this vast ocean of light. This ocean of energy could be thought of as an ocean of light. But the information-content may be such as to predispose certain light rays to combine so that they move back and forth rather than moving straight ahead, and thus forming particles.

Weber: Are those ripples, those particles, the silhouette of that light?

Bohm: Implicit in the information-content of the light – you could say that. About silhouette, I don't know. Something would have to throw the shadow. What is going to do that? The light, as it were, determines itself to make particles.

Weber: In order to do what?

Bohm: I don't know. But we're proposing that this allows for a richer universe.

Weber: To be consistent one might have to say that the light transforms aspects of itself into particles in order that those particles will reveal the light.

Bohm: That's right, they will reveal the potential of the light in a new way. So, the light and the particles together make a higher unity. Most physicists subtract off this infinity and say it doesn't count and what's left over are the particles, and they claim that these are all that count.

Weber: But you're claiming that's incorrect and shallow because it's subtracting off the very thing in which these particles have their roots and being.

Bohm: That's why I say present physics doesn't understand it, it's merely a system of computing and getting empirical results.

Weber: We've given light a cosmological, a physical, and a metaphysical interpretation. What about the psychological and spiritual interpretation? Why do people who tap into that realm of light feel a rare peace and happiness even though light is considered neutral and value-free by physics?

Bohm: The mind may have a structure similar to the universe, and in the underlying movement we call empty space there is actually a tremendous energy, a movement. The particular forms which appear in the mind may be analogous to the particles, and getting to the ground of the mind might be felt as light. The essential point is not that it's light but rather this free, penetrating movement of the whole.

Weber: Somehow the energy it triggers in the experiencer is an integrated whole and that perhaps is what accounts for this profound sense of peace.

Bohm: Yes. The analogy has often been made that even though the ocean is all stirred up and quite stormy on the surface, if you get to the bottom it is peaceful.

https://philosophy-of-science.com/The-Essential-David-Bohm

https://epdf.tips/the-essential-david-bohm.html

https://www.amazon.com/Essential-David-Bohm-Lee-Nichol/dp/0415261740/

https://ultimate-model-of-reality.com/The-Essential-David-Bohm-PDF-Free-Download

THIS!

Light is energy and information. In other words, light is energy and psyche. Information is worthless without Someone Psyche or Someone Intelligent to understand it and know how to use it.

Bohm and Weber repeatedly state that light contains information. Light knows the universe. Light is omnipresent, omniscient, and omnipotent. The light is fundamental.

All types of energy or light have some kind of psyche or intelligence within them controlling them, including physical matter. This must be so; otherwise, the energy within physical matter wouldn't be able to understand and obey physical laws, and therefore wouldn't know how to act like physical matter rather than acting randomly or chaotically. Once I finally saw this and understood it, I realized that it is obviously true. Laws need some kind of Law-Giver; and, laws also need Someone Psyche or Someone Intelligent who is smart enough to understand them and obey them.

Since psyche is seen as a pin-prick of light even at the quantum level within the spirit world, I have hypothesized that psyche in its raw original format is an infinite singularity. If there is such a thing as a point-particle, then psyche or intelligence would be it. The smaller always dwells within and controls the larger. There is psyche, intelligence, or usable information within all types of light; and, there are many different types of light or forms of light besides the visible light that we can see with our physical eyes.

Energy is infinitely malleable or infinitely transformable. Energy or light can take on an infinite number of different forms – including psyche, spirit matter, quantum fields, and physical matter. Psyche can transform the energy under its control into anything that psyche wants it to be! That's the true nature of energy – it's capable of constantly changing its form. Energy under the control of psyche can take on an infinite number of different forms. This truth is essential and fundamental if we want to know how things really work.

At the speed of light, from light's perspective, there is no time and there is no distance, which means that light is omnipresent or that light has an infinite velocity at the speed of light, from its perspective. At the speed of light, a photon experiences no passage of time from the start of its journey to the end of its journey. From its timeless perspective, light simply teleports or quantum tunnels to its destination. In fact, it has been hypothesized that a photon knows where it is going to land before it launches, otherwise it doesn't launch.

At the quantum level and at the speed of light, light is psychic, and light is infinite. It is omniscient, omnipresent, and omnipotent. The only way that's realistically possible is if light has some kind of psyche or intelligence within it. According to Bohm, light contains active information, or usable information, or information that is being used. Light chooses and acts, which means that light is intelligent, conscious, psychic, aware, and alive. What David Bohm calls "active information" is in fact psyche or intelligence; and, all types of light contain psyche or intelligence within it.

Bohm's Super-Implicate Order is psyche, or light, or intelligence. Bohm and Weber call it an "ordering principle". Order and organization require information and intelligence both on the part of the Organizer and within the thing being organized. It needs a Law-Giver and it needs Someone Psyche who is intelligent enough to be able to understand and follow those Laws. It's obvious that Someone Psyche or Someone Intelligent organized raw chaotic energy into usable quantum fields – in the beginning. These quantum fields were absolutely necessary before physical matter could be made or before physical matter could manifest. Quantum fields are non-physical and pre-physical. These quantum fields contain a great deal of information, law, structure, organization, purpose, consciousness, and intelligence; otherwise, Nature's Psyche or the Gods wouldn't be able to use these quantum fields to make and sustain physical matter.

Time is an emergent property. Time originates or manifests when God or Psyche slows things down putting distance or time between them. At the speed of light, at the quantum level, there is no distance and there is no time between objects. That's what Quantum Mechanics and the Theory of Relativity are trying to tell us.

Here we see the whole of physics as it really is at all levels of existence. It's powerful information, or powerful knowledge. With this information, we know how everything works.

Here Bohm refines his original model of the implicate order, addressing how the unfoldment of an implicate order results in manifest order and structure. Why is the multidimensional medium of space (the vacuum, or plenum) capable of unfolding the forms we experience in the sensual, three-dimensional world? Why does the plenum of space correspond more to the analogies of ink enfolded into glycerin, or to light enfolded in a hologram, than to the diffuse and random order of eggs enfolded into cake batter?

It is the activity of a super-implicate order, says Bohm, that accounts for this ordered manifestation. Bohm derives the super-implicate order by applying his causal interpretation to quantum field theory. This super-implicate order infuses the implicate order of space with active information, which generates various levels of organization, structure, and meaning. To illustrate the relationship of active information to these levels of organization, Bohm introduces the principle of *soma-significance*, whereby meaning and form are transmitted throughout a hierarchical continuum of matter and consciousness.

The final portion of the conversation is concerned with the nature of light. Bohm suggests an intimate relation, even an identity, between light and the "ocean" of energy that emanates from multi-dimensional space, or the vacuum. The movement and activity of light can be understood as a first phase in explicate manifestation, foundationally necessary for the subsequent emergence of all material structures. Indeed, Bohm proposes that all matter is a form of "frozen" light, resulting from the oscillation of intersecting light rays.

By way of explaining the concept of a super-implicate order, Bohm gives a brief sketch of Louis de Broglie's model of the electron, first put forward in 1923. This model is particularly significant as it is the basis for what eventually becomes the causal-ontological interpretation, Bohm and Hiley's quantum formalism for the implicate order. – Lee Nichol (*The Essential David Bohm*, pp. 139-140.)

One time I made the mistake of mentioning this Quantum Sea of Light on Amazon within the comments, and the Materialists, Naturalists, Darwinists, and Atheists laughed me to scorn. These people can't wrap their minds around such a concept; and, they don't want to because it falsifies their belief system. Their only recourse is to laugh at it and mock it because their uncontrollable compulsion is to deny it. This Ocean of Energy, or Zero-Point Field of Light, or Quantum Sea of Light falsifies their chosen beliefs. The necessary and proven existence of Quantum Fields and Quantum Field Theory falsifies Materialism, Naturalism, Mechanism, Physicalism, and their derivatives. Psyche, light, and quantum fields are what's REAL, and not the philosophical musings of the Materialists, Naturalists, Darwinists, and Atheists.

https://books.google.com/books?id=5aYpmixdF7gC&pg=PA79&lpg=PA79#v=onepage&q&f=false

https://ultimate-model-of-reality.com/The-Quantum-Sea-of-Light

Psyche or light is fundamental. This Quantum Sea of Light is comprised of Quantum Fields. It's non-physical and pre-physical. When it comes to the Darwinists and Atheists, anything non-physical falsifies their Philosophy of Life, World-View, or Model of Reality. Quantum Mechanics and Quantum Field Theory FALSIFY Materialism, Naturalism, Physicalism, Mechanism, Atomism, Physical Reductionism, and their derivatives such as Atheism, Behaviorism, Determinism, and Darwinism. These people take a great deal of pride in their ignorance, and they will fight tooth and nail to maintain it. I know because I used to be a materialist, naturalist, nihilist, and atheist until Science itself taught me that I was wrong.

When it comes to the Ultimate Model of Reality and Existence, the Materialists, Naturalists, Darwinists, Nihilists, and Atheists are always wrong where it matters most. You have to know and understand Quantum Field Theory and something like Bohm's Super-Implicate Order or my Ultimate Model of Reality, or your science and scientific knowledge comes to a dead-end under the sway of Scientific Naturalism. You can never find and know the truth through a falsified philosophy or based upon a falsified philosophy such as Materialism, Naturalism, Darwinism, Nihilism, or Atheism. These false and falsified philosophies were designed to hide the truth from us; and, they work as advertised if you choose to subscribe to them and follow them religiously.

Mark My Words

Experiencing Quantum Phenomena

When I was eighteen, in 1979, I experience the quantum phenomenon that scientists call "quantum tunneling" and that others call "teleportation".

At the time, I had no scientific explanation for what I experienced because the science hadn't been fully discovered yet. From my perspective, it was supernatural, miraculous, unexplainable, and at times unbelievable.

A large portion of my life has been an effort to find a scientific explanation for what happened to me. Now, it's 2019, and I finally have a logical scientific explanation for what was done to me and the car I was driving.

I was driving north at 45 mph approaching a free-way overpass when another car suddenly pulled out in front of me. I remember hitting the brakes; but, I didn't have time for anything else. I was right on the side and the bumper of that car with nowhere to go.

Time stopped. It felt like the whole of eternity folded into me, and time literally came to a standstill. I was frozen in time and could seem to sense everything all around me. I could also sense where I wanted to be, but there was no way to get there except through that other car.

When time resumed, I was in the middle of a lawn, with the car completely stopped, and the car turned off. It was as if I had passed out. I had no memory of how I got into the middle of that lawn or how I was able to get the car stopped and turned off. I was just there instantly.

A man came running out of the store to see if I was okay. He said that that was the coolest thing he had ever seen in his life. He was checking my car for signs of damage, and he said that there's no way I could have missed hitting that other car but somehow I did. From his perspective, I had apparently phase-shifted and passed through that other car unscathed. From my perspective, I had simply teleported to my destination. I have no memory of doing anything besides hitting the brakes. I have no idea how I got from point A to point B unscathed. From my perspective, I was on the side of that car, and then I was instantly someplace else, with my car stopped, and my car turned off.

I didn't have a scientific explanation for any of it back in 1979. Now I do. According to the Theory of Relativity, time stops at the speed-of-light. From the perspective of a photon traveling at the speed-of-light, time stops, and that photon simply teleports or quantum tunnels to its destination. According to David Bohm, at the speed-of-light time stops, and the distance between start and finish becomes zero. At the speed-of-light, the photon or physical object simply quantum tunnels or teleports to its destination. That's precisely what I experienced. Time stopped, and then I was simply someplace else. I have no memory of what happened between start and finish.

I went instantly from 45 mph northward, to 0 mph northeast. It's good that God stopped my car and turned it off; otherwise, I would have gone over the embankment and down onto the free-way into oncoming traffic if I would have kept going at 45 mph in my new direction. I and others would have probably been killed. My falling onto and into the freeway at 45 mph wouldn't have been good for any of us.

I have no idea why God chose to save me and any of the other people I could have killed. I didn't even fully understand what had happened to me until decades later.

Whenever God stops the passage of time for a photon or a physical object, the distance goes to zero; and, that photon or quantum object or physical object simply quantum tunnels or teleports to its destination. This is precisely the way that David Bohm explained the speed-of-light, the theory of relativity, and quantum mechanics. At the speed-of-light, times stops, and the distance between start and finish goes to zero. At the speed-of-light, you can literally quantum tunnel anywhere in the universe instantaneously and experience no passage of time or no aging process while doing so.

Furthermore, quantum phenomena are additive – a quantum phenomenon called Quantum Superposition. Many different quantum phenomena can transpire in the same space at the same time simultaneously. That's how quantum fields work. I suppose that some sort of Quantum Phase-Shifting was also involved in what happened to me, although I have no memory of it. My best friend has experienced phase-shifting, and he remembers it. Quantum Phase-Shifting takes place when two physical objects pass through each other unphased and unscathed. A huge elk passed through my friend's truck unphased and unscathed. In his case, he actually saw the elk pass through his truck. Apparently God wanted that elk to live, or God wanted my friend to live. Phase-shifting is when two physical objects occupy the same space at the same time without crashing into each other, or damaging each other, or merging into each other. They simply pass through each other unphased and unscathed. The car I was driving apparently passed through that other car unscathed and unphased. There was no damage to my mother's car, when I ended up on the middle of that lawn, with the car stopped, and the car turned off.

I was supernaturally calm. There hadn't been enough time for my adrenaline to start pumping. That other car was simply in front of me, and then I was instantly someplace else. That's it. The only way to explain it scientifically is through quantum tunneling or traveling at the speed-of-light, which are the same thing. Time stops, the distance between start and finish goes to zero, and you simply quantum tunnel to your destination at the speed-of-light which is an infinite speed from a timeless perspective. It makes sense to me now because I have the science to explain it. It seemed quite miraculous and supernatural to me back in 1979 when I had no scientific explanation for what was done for me and done to me. It was nothing that I did. It was something that God did for me to save me and others from being hurt and killed.

Mark My Words

Soma-Significance and the Activity of Meaning (1985)

Lee Nicole wrote:

> **In this chapter Bohm outlines the nature of soma-significant and signa-somatic activity. Here "soma" refers to the body, and by extension to any material structure or process, while "significance" refers to mind or meaning. These terms are meant to suggest complementary aspects of one indivisible process, rather than two qualitatively distinct domains. With this model Bohm furthers his argument that there is no essential difference between reciprocal processes in the objective world (Chapter 1) and reciprocal processes in the perception and cognition of human beings (Chapter 2), suggesting that active meaning is enfolded and unfolded throughout the whole of existence. Soma-significant and signa-somatic processes are thus seen as aspects of the dynamics of implicate and explicate orders.** (*The Essential David Bohm*, p. 158.)

Rather than calling it "psyche" or "intelligence", Bohm calls it "meaning" and "significance". It's the same difference. But for Bohm and the scientific community at that point in time (1985), Psyche is a four-letter word. If they can help it, they are not going to have anything to do with it. It wasn't until 2010 that scientists started putting Psyche back into science and psychology. This development was dependent upon the successful falsification of Darwinism and the Theory of Evolution, which scientists started to do in earnest by the year 2000. The false has to be falsified by the truth; and, the truth has to be discovered, experienced, and observed before it can be known.

Most of this essay is thirty years ahead of its time. This essay is extremely fascinating because Bohm finally separates our physical universe into the three different levels or dimensions from which it is comprised – soma, significance, and energy – or matter, meaning, and energy. It has been observed that our physical universe is comprised of physical matter, quantum fields made out of energy, and information or intelligence or psyche or thought.

Bohm's soma and matter are talking about our physical body and physical objects, which are part of Bohm's explicate order. I call this level the physical level. It is obvious that everything within our physical universe is made out of energy or light, including quantum waves, quantum fields, dark energy, dark matter (spirit matter), and physical matter. Everything is made from energy or light. This is Bohm's implicate order and what I like to call the quantum level of existence and reality.

What isn't so obvious is that every type of energy contains information, or psyche, or intelligence within it that is capable of forming the energy under its control into anything that that psyche or intelligence wants that energy to be. This is Bohm's super-implicate order and what I typically find myself calling the psyche level of existence and reality. This is where science starts to get interesting – when we finally leave the Materialists, Naturalists, Darwinists, Nihilists, Behaviorists, Determinists, and Atheists behind in the dust where they want to stay and be and belong.

Most of what Bohm has written matches perfectly with what has been subsequently experienced and observed; however, David Bohm and Lee Nichol do indeed confound or confuse the physical brain with psyche from time to time, because they still have some of that Materialism or Mechanism ingrained within them. We all do.

For example, Lee Nichol wrote:

The brain produces an intention to stop, which works its way out through increasingly "manifest" levels of soma.

This is wrong. The brain doesn't produce intentions. Intentions are a product of Psyche. The Super-Implicate Psyche produces the intention to stop, an intention that it has to send through the brain in order to accomplish the task. The way it really works according to Henry P. Stapp and his Orthodox Interpretation of Quantum Mechanics is that the Human Psyche makes a request to move its finger. Then Nature's Psyche analyzes the requests and determines if it's physically possible. If possible, then Nature's Psyche (who designed, made, and mapped the physical brain) fires the appropriate neuron and your finger rises off the table. According to the Orthodox Interpretation, it is Nature's Psyche who collapses the wave function and fires the neurons, NOT the Human Psyche and definitely not the physical brain. The physical brain is simply a machine. Someone Psyche is driving that machine. In the case of the brain, lots of different Psyches or Minds are driving that machine. The Human Psyche is the overlord; but, Nature's Psyche is found within every physical atom in the universe, and Nature's Psyche makes choices and collapses wave functions. The Human Psyche doesn't collapse wave functions. The Human Psyche isn't consciously aware of these quantum mechanical phenomena. The Human Psyche simply chooses what it wants to do with its physical body; and, that's it. Nature's Psyche (or God's Psyche) takes care of everything else.

For a much more detailed explanation of these scientific truths, study my book, *Quantum Neuroscience*. Henry P. Stapp's "Orthodox Interpretation of Quantum Mechanics" was absolutely essential for the successfully creation and development of Quantum Neuroscience.

Quantum Neuroscience: The Answer to Life, the Universe, and Everything (2018)

https://www.amazon.com/gp/product/B079Z6QQQB/

Within *The Essential David Bohm* and Bohm's other books, Bohm doesn't always get this wrong.

Even the electron is informed with a certain level of mind. – David Bohm

How's that possible?

How is it possible that an electron has a Psyche, Intelligence, or Mind? It's possible because an electron is made from energy; and, the Psyches or Intelligences controlling that energy have decided that that energy is going to take the form of an electron, rather than a proton or a photon or a quantum field.

When I was creating my Ultimate Model of Reality, I had to start from scratch because NONE of this is taught nor understood in our public colleges that are run by the Materialists, Naturalists, and Atheists.

Bohm's "Soma-Significance" and "Activity of Meaning" are the same exact thing as my Ultimate Model of Reality. We scientists are constantly re-inventing the same wheel because the Materialists, Naturalists, and Atheists who run our public schools won't allow this information or "non-physical science" into our public schools. None of the non-physical terminology has been standardized within our public schools, so each one of us who delves into the meaning of Quantum Mechanics ends up developing his or her own unique vocabulary to explain what's really going on.

While creating my Ultimate Model of Reality, I used names that were self-explanatory. Bohm used Latin and Greek for his names, so Bohm's theories and ideas have to be translated into English before they can be understood and used.

The Ultimate Model of Reality: Psyche Is the Ultimate Cause

https://www.amazon.com/gp/product/B071NC9JK6/

The Ultimate Model of Reality is a Psyche Ontology wherein psyche is the fundamental unit of reality. In this essay, Bohm states that "significance" or "meaning" or "consciousness" is the fundamental unit of reality. That's the same difference. Meaning, significance, intention, purpose, desires, consciousness, intelligence, and so forth are all a function or a product of Psyche. Bohm's Super-Super Implicate Order is the psyche level of existence that I talk about in my different books, wherein Psyche or Thought becomes the fundamental unit of reality. It's all got to start somewhere.

Scientific Axioms: By definition, in principle, quantum fields are non-physical and pre-physical. Quantum fields are made out of energy or light. Someone Psyche or Someone Intelligent had to design and make the quantum fields before physical matter became possible. Chaos of any kind needs an Organizer in order to become organized. It's obvious that each quantum field was designed to fulfil a specific purpose.

Scientific Premise: Anything that is obviously made obviously has a Maker who designed it and made it. This premise is obviously true.

Scientific Observation: Quantum fields were obviously designed and made from raw chaotic energy to serve the specific purpose of being a substrate for physical matter.

Scientific Conclusion: Therefore, it is logical to conclude that quantum fields obviously had a Designer and Maker who designed them and made them. Some people call this psyche or this person God.

This part of *The Essential David Bohm* introduces these concepts into Bohm's Model of Reality.

Lee Nicole writes:

At the quantum level of matter, says Bohm, soma-significant processes also occur. In Bohm's version of quantum theory (Chapters 4 and 6), a "pilot wave" reads the somatic form of the environment and conveys this form to its accompanying particle (a soma-significant flow). The subtler somatic structure of the particle – which Bohm suggests is at least as complex as a radio receiver – develops a cumulative "orientational" significance from this information. When this significance is fully developed it also becomes outwardly active, giving rise to specific movements on the part of the particle (a signa-somatic flow).

For Bohm, the pilot wave model is not merely analogous to human soma-significance. He sees each of these examples as abstracted nodes in a continuum that includes the quantum level, the human domain, and the largescale evolution of the cosmos. As a magnet divided into multiple parts will always exhibit positive and negative poles in each part, so also will any aspect of reality we select for examination show somatic and significant aspects. It is not possible to find an independent somatic phenomenon, or an independent significant phenomenon. Anywhere we make a cut in the fabric of reality, we will find this mutual interpenetration of soma and significance.

The implication of this perspective is central to Bohm's overall world view: meaning is not an exclusively human activity. (*The Essential David Bohm*, p. 159.)

Here, Bohm once again introduces us to the quantum level, or his implicate order. Lee Nicol calls it the quantum level, as do I. "A pilot wave reads." Reading of any kind implies Psyche or Intelligence. Our physical brain is a complex radio receiver. This pilot wave, or quantum wave, or psyche reads the information from the universal Akashic Field or from the mind of God, and then it conveys this information or this "form" to the physical particle telling the physical particle how to behave and act. The sub-atomic structure or energy of a particle is informed and controlled by the Psyche that has been assigned to it. Within a physical atom, there is enough space for something like a radio receiver and a radio transmitter to exist. There's enough "empty space" within an atom to store a quantum computer within an atom. When the information or significance or psyche within the atom becomes active, it forms the energy into the different types of particles that a physical atom needs.

Quantum waves, pilot waves, or thoughts imply some kind of Intelligence or Psyche that's capable of reading and understanding all of this information or meaning that Bohm is talking about.

Both the quantum level and the human domain at the physical level are controlled by Nature's Psyche, or this significance and meaning which permeates the whole universe. Here Bohm also introduces the holographic principle, which is the idea that each physical atom in the universe contains ALL of the knowledge and information of the universe within it. The individual parts contain the whole. Psyche, energy, and matter merge together into one because they are all comprised of energy or light. God is in the light, and it has been observed that whenever the Human Psyche enters into a non-consensus reality in the spirit world, the Human Psyche is God within that domain. Meaning or psyche is not an exclusively human activity. All of the energy in the universe contains within it meaning or psyche, and Nature's Psyche has been programmed to treat the Human Psyche as if it were an Overlord or a God.

I find it interesting to observe that the magnet is physical; but, the magnetic fields are non-physical. Whether we are talking about magnetic fields or quantum fields, their verified and proven existence falsifies Materialism, Naturalism, Nihilism, Mechanism, and Atheism which claim that these psychic fields or quantum fields DO NOT EXIST. To claim that psychic fields or quantum fields do not exist is rather worthless is it not? The proposed non-existence of these quantum fields or psychic fields doesn't explain what they are and how they work, now does it? I want to know what things are and how they work, not be told that they don't exist, especially when it's obvious that they do exist.

While reading Bohm's essays and books, half of your time is spent translating his ideas into English. Yet, it all makes perfect sense and dovetails perfectly with my Ultimate Model of Reality once you get Bohm translated into English. Let me demonstrate.

David Bohm wrote:

> **I want to introduce a new notion of meaning which I call soma-significance, and also a notion of the relationship between the physical and the mental. This relationship has been widely considered under the name psycho-somatic. "Psyche" comes from a Greek word meaning mind or soul and "soma" means the body. If we generalize soma to mean the physical, the term psycho-somatic suggests two different kinds of entities, each existent in itself – but both in mutual interaction. In my view such a notion introduces a split, a fragmentation, between the physical and the mental that doesn't properly correspond to the actual state of affairs. Instead I want to suggest the introduction of a new term which I call "soma-significance." This emphasizes the unity of the two, and more generally, with meaning in all its implications and aspects. That is, "significance" goes on to "meaning," which is a more general word.**

In this approach meaning is clearly being given a key role in the whole of existence. (*The Essential David Bohm*, pp. 159-169.)

So, we have soma, which is the physical level of existence. We also have significance or meaning, which is the psyche level of existence. They are each self-existent in themselves, but both are in mutual interaction. They are different aspects of the whole of existence.

This quote was interesting, because it's one of the few times that Bohm uses the word "psyche", and he uses it to falsify it and eliminate it. Bohm wants to merge psyche or mind whole into physical matter and then treat it as one united whole. This is another thing that Bohm got wrong. He had no way of knowing that Psyche would be experienced and observed functioning completely separate from its assigned spirit body and assigned physical body. Each psyche is observed as an individual pin-prick of light, a spark, which physicists would call a point-particle or an infinite singularity.

Still, Bohm repeatedly makes allowances for the possibility that these things can exist and function separately from each other. He really has no choice but to do so. He has to define them separately in order to talk about them separately in order to then try to find a way to unite them into one great whole.

Bohm wants to take the whole thing one step further and unite soma and psyche into one through the Implicate Order or the Quantum Realm of energy and light. Bohm is pursuing a holistic approach to science. Although Bohm doesn't always seem to realize it, energy is the common denominator in all of this. But, every now and then Bohm sees the light. Thereby, your psyche, your physical body, and the quantum fields end up being different ASPECTS of the same universal reality.

It all makes sense to me once I finally understood what he was talking about.

The total physical response of the human being is profoundly affected by what physical forms mean to him. A change of meaning can totally change your response. This meaning will vary according to all sorts of things, such as your ability or background, conditioning, and so on.

This is different from psycho-somatic, because with psycho-somatic you say that mind affects matter as if they were two different substances – mind substance affects material substance. Now I am saying there is only one flow, and a change of meaning is a change in that flow. Therefore, any change of meaning is a change of soma, and any change of soma is a change of meaning. So, we don't have this distinction. (*The Essential David Bohm*, p. 163.)

Bohm wants to unite psyche and physical matter. The unity of these two takes place through energy or light. Psyche and matter of any kind are both made from energy or light.

Psyche and matter are connected together, integrated, and entangled through quantum waves, thoughts, energy, or light. However, David Bohm didn't realize at the time back in 1985 that Psyche has been experienced and observed existing and functioning separately from its spirit body and physical body. As we approach 2020, the world is now aware that Psyche or Intelligence can function and exist separate from its spirit body and physical body, which is why I separate the three from each other within my books.

Bohm in constantly trying to unite them into one, which is also a legitimate observation, because they can and do unite and function as one. You see, when it comes to physics, the smaller dwells within and controls the larger. That's how psyche is able to dwell within a spirit body, an atom, or a physical body and at the same time control them. It's all based upon energy or light. Energy or light is the common denominator. David Bohm is on the right track. Within this chapter Bohm emphasizes the fact that there is a two-way interaction or two-way communication between Psyche (significance) and physical matter (soma). He calls the whole process soma-significant and signa-somatic.

There are many different types of light or forms of energy in addition to the visible light that we can see with our physical eyes. Quantum fields are non-physical, and they are made out of energy or light. Over 95% of our physical universe is in fact non-physical. This is common knowledge in the whole of science, and only the Materialists, Naturalists, and Atheists reject it. But, we never learn about any of this in our public classrooms because the Atheists, Naturalists, and Materialists are currently running our public schools and selecting their curriculum.

According to David Bohm, there are three different levels or aspects of reality and existence.

So, it is in principle possible in this view to encompass both the outward universe of matter and the inward universe of mind.

In this approach, the three basic aspects arise:

Soma

Significance

Energy

(*The Essential David Bohm*, p. 172.)

The soma aspect includes Bohm's Explicate Order, and it consists of the physical level including your physical body and physical brain. It's obvious that physical matter exists. It has been experienced and observed. There's no sense denying it. Matter of any kind belongs to the soma aspect.

The energy aspect includes Quantum Mechanics, Quantum Fields, and Quantum Field Theory. These fundamental constructs are made from energy and NOT physical matter. This basic aspect of reality is non-physical in nature and origin. Quantum Mechanics is synonymous with Bohm's Implicate Order. I call this level of existence and reality the quantum level. Others call it the spirit world.

Quantum fields and quantum mechanisms have been proven to exist through science experiments. Energy of any kind belongs to the energy aspect.

The significance aspect deals with conscious meaning. Consciousness, meaning, purpose, intention, attention, interpretation, design, creation, and choice are all functions of the Human Psyche. The supernatural psyche is synonymous with Bohm's Super-Implicate Order. It's the organizing principle or the organizing aspect within his model of reality. I tend to call this level of existence and reality the psyche level because psyche has been experienced and observed. Meaning or consciousness of any kind belongs to Bohm's significance aspect.

Each one of these ASPECTS of reality is made out of energy or light. Energy or light is the common denominator in all of this. That would make energy or light the fundamental unit of reality and existence as we know it.

However, David Bohm goes on to state that meaning (psyche or consciousness) is more fundamental than energy or matter. That's not going to sit well with the Mechanists, Materialists, and Naturalists – and David Bohm took a lot of flak because of it.

> **So quite generally, energy enfolds matter and meaning, while matter enfolds energy and meaning. But also meaning enfolds both matter and energy. The way we find out about matter and energy is by seeing what they mean. So, each of these basic notions enfolds the other two. It is through this mutual enfoldment that the whole notion obtains unity. So, we can put all these relationships together.**
>
> **However, in some sense the enfoldment by meaning seems to be more fundamental than the enfoldment of the other types, because we can discuss the meanings of meaning. In some sense meanings enfold meanings. But we cannot have the matter of matter, or the energy of energy. There seems to be no intrinsic enfoldment relation in matter-energy. Matter enfolds energy, and energy enfolds matter, according to this view, by way of significance. But meaning refers to itself directly, and this is in fact the basis of the possibility of that intelligence which can comprehend the whole, including itself. On the other hand, matter and energy obtain their self-reference only indirectly, firstly through meaning. That is, we can refer matter back to itself by first seeing what it means to us, and then going back. Or we can refer matter to energy, or energy to matter, by seeing what they mean. We refer them to each other reflexively, but only through their meaning.** (*The Essential David Bohm*, p. 175.)

Meaning or psyche is more fundamental or essential than energy or matter.

The Ultimate Model of Reality: Psyche Is the Ultimate Cause

https://www.amazon.com/gp/product/B071NC9JK6/

Within this and my other Psyche books, I introduce a Psyche Ontology in which Psyche is the fundamental unit of reality. David Bohm seems to agree with this assessment, even though he doesn't use the word "psyche" to describe meaning, consciousness, or mind.

I use the word "psyche", because I know that Psyche has been experienced and observed.

This type of information wasn't available back in the 1980's when I first went looking for it; but, it's available in great abundance now thanks to YouTube and the internet.

https://psyche-ontology.com/psyche-observed/

https://psyche-ontology.com/psyche-experienced-and-observed/

Buhlman, William L. (1996). Adventures Beyond the Body: How to Experience Out-of-Body Travel. New York: HarperCollins.

https://psyche-ontology.com/buhlman/

I also found a revealed religion that is in complete harmony with all of this. The Biblical God Jesus Christ revealed that Psyche is Intelligence, or light and truth.

Doctrine and Covenants 93: 29-30, 36:

May 6, 1833

29 Man was also in the beginning with God. Intelligence [psyche], or the light of truth, was not created or made, neither indeed can be.

30 All truth is independent in that sphere in which God has placed it, to act for itself, as all intelligence also; otherwise there is no existence.

36 The glory of God is intelligence, or, in other words, light and truth.

Within *The Essential David Bohm*, he periodically tries to make the case that there is no such thing as absolute truth. He periodically states the claim that science cannot be used to find and know the truth. He's also constantly trying to define the truth. He hasn't arrived yet. He hasn't found the truth yet.

Only that which is eternal and everlasting can in fact be absolutely true in every situation and every case. So, what part of us is absolutely true? What part of us is eternal and everlasting? When it comes to a human being, I have found only two parts of us that are actually eternal and everlasting, and therefore indestructible absolute truth. The energy that is used to construct our spirit bodies and the atoms within our physical bodies is eternal and everlasting. That energy or light cannot be made, and it cannot be destroyed; therefore, it is absolute true. It has always existed, and it will always exist; therefore, it is absolutely true. The other part of us that is eternal and everlasting, and therefore absolutely true, is our

645

Psyche or Intelligence or Consciousness or Primal Life Force. It's absolutely true, because it cannot be made, and it cannot be destroyed. It has always existed, and it will always exist. Energy, psyche, intelligence, or light is absolutely true. It is always conserved. It is eternal and everlasting. Nature's Psyche can take the energy under its control and form that energy into anything that it wants it to be, including spirit matter (dark matter), dark energy, quantum fields, quantum particles, and even physical matter. It's all made from energy by Someone Psyche. This is the absolute truth of our existence and reality and surroundings.

Truth verifies and confirms the truth. The false is falsified by the truth; and, the truth is repeatedly experienced and observed. There is NO existence or significance without Intelligence, or Psyche, or Consciousness, or Meaning.

According to Bohm, his meaning aspect of reality or significance aspect of reality is what makes consciousness and intelligence possible. Bohm explains how things really are, and so does God.

Physical matter is made, and it can be destroyed. Physical matter is temporary or temporal.

In contrast, Psyche and Energy are conserved, which means that they are eternal and everlasting. They can neither be made, nor destroyed. Intelligence or Psyche was not created nor made, neither can be. Without Intelligence or Psyche, there is NO existence or life force. Your Intelligence or Psyche was also in the beginning with God. Your Intelligence and God's Psyche are co-eternal. They cannot be made, and they cannot be destroyed. This reality and truth is in complete harmony with the First Law of Thermodynamics or the Conservation of Energy and Psyche.

It all fits together seamlessly. It works. It's perfect. It's complete. It's true. We know that it is true because it falsifies Materialism, Mechanism, Naturalism, Nihilism, Darwinism, Behaviorism, Determinism, Physical Reductionism, Atomism, and Atheism. Once you successfully eliminate everything that is false and everything that has been falsified, then ONLY the truth will remain. It's elementary.

I wish I would have found this and understood this back in 1980 when I was first in college; but, I got there in the end, so I have to be satisfied with that. Imagine the billions of people who have died without any clue whatsoever that any of these truths are known and have been revealed to mankind.

The truth is out there. You just have to know where to find it and then recognize it when you see it. I didn't have a clue when I was a materialist, naturalist, nihilist, and atheist; but, I'm seeing it now. We each have to start where we currently are, and then go on from there.

Mark My Words

The Causal-Ontological Interpretation and Implicate Orders (1987)

An Ontological Basis for the Quantum Theory

https://psyche-ontology.com/bohm_hiley_kaloyerou_1986

http://www.tcm.phy.cam.ac.uk/~mdt26/local_papers/bohm_hiley_kaloyerou_1986.pdf

https://psyche-ontology.com/Bohm-HV-I-Phys-Rev-1952

https://psyche-ontology.com/Bohm-HV-II-Phys-Rev-1952

https://psyche-ontology.com/Quantum-Potential

I used to be a materialist, naturalist, nihilist, and atheist before I learned better and Science convinced me that I was wrong. Quantum Mechanics never made any sense to me from a materialistic or mechanistic perspective. My car and my computer NEVER showed any signs of being conscious, self-willed, or alive. Perception, intention, psyche, consciousness, and mind never made any sense from a materialistic or mechanistic perspective either. The Materialists, Naturalists, Mechanists, and Physicalists define "consciousness" as an emergent property of a physical brain – it's magic! It's creation ex nihilo. It's Atheism – it's the creation of something by nothing out of nothing. It's magic! There's no science there. It's magic. These atheistic philosophies are creation ex nihilo – creation from nothing by nothing. Creation ex nihilo or atheism is patently absurd. It makes no logical sense whatsoever because it is magic.

What interests me is the observed and proven fact that the Human Psyche continues to think, observe, experience, perceive, and learn after its physical brain has ceased to exist. Near-Death Experiences and Out-of-Body Experiences prove that we continue to perceive, think, and live long after our physical brain is dead and gone. We need a Science that actually explains the science behind these kinds of phenomena, experiences, and observations.

We need a quantum mechanical explanation for how psyche or mind works because it's definitely not a physical phenomenon. The Physicalists and Naturalists have no way to define or explain something non-physical like psyche or mind using classical mechanics or Newtonian mechanics. It can't be done, which means it isn't done. The Materialists, Naturalists, Darwinists, Nihilists, Behaviorists, Determinists, Mechanists, and Atheists are NEVER going to be able to accept and understand a psyche-based interpretation of Quantum Mechanics because they have ruled it out a priori.

In the early 1950s, Bohm's causal quantum theory program was mostly negatively received, with a widespread tendency among physicists to systematically ignore both Bohm personally and his ideas. There was a significant revival of interest in Bohm's ideas in the late 1950s and the early 1960s; the Ninth Symposium of the

Colston Research Society in Bristol in 1957 was a key turning point toward greater tolerance of his ideas.

https://en.wikipedia.org/wiki/David_Bohm

The Materialists, Naturalists, Nihilists, and Atheists didn't like David Bohm and his ideas because David Bohm was thousands of years ahead of them when it came to philosophy and science. The Scientific Naturalists and Atheists are jealous, vengeful, hateful, and retaliatory. Their only goal is to shut down any scientific research into anything that deals with psyche, consciousness, intelligence, mind, and quantum mechanics. The Atheists have a program in place to censor and delete any evidence that falsifies their chosen beliefs. That's how they do science.

When it comes to Quantum Mechanics, it's all about the different interpretations that have been given to Quantum Mechanics. There are as many different interpretations of Quantum Mechanics as there are scientists and philosophers on this planet. Each person creates his own unique world-view or philosophy of life; and, I am constantly adjusting mine so that it matches more closely with what has actually been experienced and observed.

On the Wikipedia, the official interpretations or the most-favored interpretations of Quantum Mechanics are materialistic and naturalistic in nature. The most interesting and useful Interpretations of Quantum Mechanics are not listed on the Wikipedia, nor given any kind of significant coverage. The Materialists, Naturalists, Nihilists, and Atheists formally reject the Interpretations of Quantum Mechanics that have the most explanatory power and the most experiential and observational evidence supporting them. Nevertheless, look at all the different things that can be made out of energy or light by Someone Psyche at the quantum level. The Naturalists and Atheists can't suppress it all, because energy really exists, and energy is infinitely malleable or transformable under the control of Someone Psyche or Someone Intelligent. This is what's actually being experienced and observed; and, the collective whole of it FALSIFIES Materialism, Naturalism, Nihilism, and Atheism.

https://en.wikipedia.org/wiki/Interpretations_of_quantum_mechanics

https://en.wikipedia.org/wiki/Glossary_of_quantum_philosophy

https://en.wikipedia.org/wiki/Quantum_mechanics

https://en.wikipedia.org/wiki/Atomic_orbital

An atomic orbital is essentially a force field surrounding an atomic nucleus, thereby giving that atom substance, presence, and shape. Look at all the different shapes that that energy can take. You can see the intelligent organization, design, and control within all these different quantum phenomena or energy phenomena. It's not chaos that we are witnessing. It is intelligent order and design and purpose and control. In contrast, the spiritual or non-physical interpretations of Quantum Mechanics are typically deleted, suppressed, ridiculed, and censored away by the Materialists, Naturalists, and Atheists who control our public schools. That's they only way these people can make their case – by deleting the evidence and suppressing the evidence – because the evidence FALSIFIES their chosen beliefs. Quantum Fields are non-physical and pre-physical. That's just the way it is. First things first.

A materialistic and naturalistic interpretation of Quantum Mechanics never made any sense to me; and, it doesn't make sense to the Naturalists and Atheists either. We have to expand our levels or dimensions and go deep if we want to understand Quantum Mechanics. We have to go down the rabbit hole and try to see how far down it goes. Physical matter is left floating on the surface.

Pim van Lommel's "Non-Local Consciousness Interpretation of Quantum Mechanics" was the first time in my life that I actually understood Quantum Mechanics in a way that made logical sense to me.

Quantum Mechanics: From a Non-Physical Spiritual Perspective (2016)

https://www.amazon.com/gp/product/B01J023TGU/

https://www.amazon.com/author/science

--

Henry P. Stapp's "Orthodox Interpretation of Quantum Mechanics" was absolutely essential for the successfully creation and development of Quantum Neuroscience.

Quantum Neuroscience: The Answer to Life, the Universe, and Everything (2018)

https://www.amazon.com/gp/product/B079Z6QQQB/

--

In 2019, David Bohm's "Causal-Ontological Interpretation of Quantum Mechanics" was the third time in my life that Quantum Mechanics actually made logical sense to me, but first I had to translate Bohm's Interpretation into English so that I could understand it.

https://origin-science.org/The-Essential-David-Bohm/

--

These different interpretations of Quantum Mechanics match well with the Ultimate Model of Reality that I developed (2017).

The Ultimate Model of Reality: Psyche Is the Ultimate Cause (2017)

https://www.amazon.com/gp/product/B071NC9JK6/

https://ultimate-model-of-reality.com/

Within this and my other Psyche books, I introduce a Psyche Ontology in which Psyche is the fundamental unit of reality.

--

When it comes to Science, it's all about observation and experience. The false is falsified by the truth; and, the truth is repeatedly experienced and observed.

2 Corinthians 13: 1: **In the mouth of two or three witnesses shall every truth be established.**

When it comes to Quantum Mechanics, I now have three witnesses who are telling me the same thing – namely that the Philosophy of Materialism or the Philosophy of Mechanism is inadequate, limited, and false. That's precisely what I discovered all on my own after studying science for fifty-five years. It's nice to know, though, that some people actually agree with me.

The Materialists, Naturalists, Darwinists, Nihilists, and Atheists hate this kind of science and reject it categorically, which tells me that I'm finally on the right track after all. If you want to find and know the truth, then find and study the things that the Materialists, Naturalists, Darwinists, and Atheists are trying to hide from you. It works.

--

Not only am I trying to present some Science that has some teeth to it; but, I'm also trying to present a Science that could actually dissolve your teeth and convert them back into raw energy if it were to choose to do so. I'm trying to present a Psyche Ontology wherein psyche is the fundamental unit and controller of existence and reality. A Psyche really is a God over the energy that's under its control. That Psyche can turn that energy into anything that that Psyche wants that energy to be. This is what has been experienced and observed.

When it comes to Quantum Mechanics and Quantum Field Theory, all of the math has been successfully developed. There's math supporting everything that I have written in these essays. But, what they have never been able to agree upon is what the math actually means. There seem to be as many different interpretations of Quantum Mechanics as there are philosophers, mathematicians, and scientists on this planet. They can't agree was to what the math means. The math is there; but, it has to be interpreted; and, that's where science, the scientific method, and math all fall down – they require interpretation in order to be meaningful and useful. Unfortunately, scientific evidence, scientific methods, math, observations, experiences, and experiments are never self-explanatory. They all require some type of interpretation from fallible human beings.

In my research, I have found three holistic, all-inclusive Interpretations of Quantum Mechanics; and, David Bohm's Causal-Ontological Interpretation is now one of them.

--

Lee Nichol's introduction to this chapter does an excellent job of describing what's at stake when it comes to Quantum Mechanics and our on-going attempt to achieve a correct interpretation of Quantum Mechanics with the scientific community.

Bohm's Causal-Ontological Interpretation

In the following selections, Bohm gives concise descriptions of three aspects of his alternative to the standard interpretation of quantum mechanics, referred to here as the "causal interpretation" (later to become Bohm and Hiley's "ontological interpretation"). He goes on to explain in some detail the manner in which this interpretation is related to the implicate order.

To grasp the significance of the causal interpretation, it is useful to understand two aspects of the standard (Copenhagen) interpretation. The first of these is that particles, such as an electron in a laboratory, have only potential existence until they are observed. Once observed (e.g. with a measuring device), this potentiality "collapses" down into the concrete manifestation of the actual particle. Second, when not manifesting as a particle, the state of potentiality is represented by a mathematical wave form known as the "wave function." This wave function is understood to be mathematical only – there is no "real" wave there behind the numbers.

In the early causal interpretation, Bohm proposes that the particle is an objectively existing entity that does not depend on observation to bring it into existence. Further, the wave phenomenon also is understood to have objective existence – every objective particle has an objective wave which accompanies it. From this perspective there is no collapse from a (mathematical) wave state into a manifest particle. There is always the simultaneous objective existence of wave and particle, with no observer required for their actualization. – Lee Nichol (*The Essential David Bohm*, p. 183.)

https://ultimate-model-of-reality.com/The-Essential-David-Bohm-PDF-Free-Download

I like this here because Bohm's Causal Interpretation gets rid of the Copenhagen idea that the moon doesn't exist when nobody is looking at it. I always thought that was a stupid idea. Einstein thought it was stupid too. According to the Copenhagen Interpretation, the quantum wave doesn't exist, and nothing exists until the whole thing magically becomes a physical particle. It only exists while it's a physical particle. That's a stupid idea that we get from the Materialists, Naturalists, Nihilists, and Atheists – creation ex nihilo. Of course, the energy exists. Of course, the quantum waves exist. Of course, the quantum fields exist. It all exists, and it all existed BEFORE the first physical particle was made and came into existence. ALL of the stupid ideas, such as creation ex nihilo, come from the Atheists and their materialistic philosophies. Creation ex nihilo IS atheism

651

– creation from nothing by nothing – magic. The atheists are sorcerers and magicians in their philosophy and orientation, and they don't even know it. The existence of "non-existence" is a stupid idea that we get from the Materialists, Naturalists, Nihilists, and Atheists.

Of central importance in the causal interpretation is the relationship between the particle and the wave. The wave is understood to carry complex but passive information about the form of the environment that surrounds the particle. Through the quantum potential – a feature unique to Bohm and Hiley's interpretation – the information contained in the wave is transmitted to the particle. This information – now active rather than passive – is "processed" by the particle and subsequently directs its movement. This relationship between wave and particle is analogous to radar waves directing a ship on automatic pilot – in both cases, a self-sufficient energy (the particle; the ship) is "informed" and directed by a wave whose form, rather than its intensity, is significant. This process can be understood as yet another variation of the "two-way" reciprocal relations discussed in Chapters 1 and 2, and of the soma-significant relations of Chapter 5. – Lee Nichol (*The Essential David Bohm*, p. 184.)

Now, wrap your mind around this because it's important. The Energy or Psyche is REAL. The energy is really there whether one is looking at it or not. The quantum waves and quantum fields are really there, whether a human is looking at them or not. The energy or psyche has always existed, and it will always exist because it is conserved. The energy or psyche carries within it complex but passive information. ALL energy has a psyche, is intelligent, and carries within it passive information. According to Henry P. Stapp and Paul Dirac, it is Nature's Psyche who collapses the wave function thereby producing a physical particle – NOT the human psyche and NOT the human observer. Nature's Psyche (or God's Psyche) collapses the wave functions producing the physical particles from energy, quantum waves, quantum information, and quantum fields – NOT the human psyche. It all fits together perfectly and makes logical sense once you get the correct interpretation of Quantum Mechanics.

The psyche within the energy collapses the wave function. In other words, the psyche decides what form the energy under its control will take. Energy is infinitely malleable. Furthermore, there is an infinite amount of information within this universal Energy Field or universal Akashic Field. Each psyche uses this universal information whenever it decides what form the energy under its control will take. The quantum information or quantum potential contained within the wave or the energy is used to make the particle. The information within the energy becomes active when the psyche forms the particle by collapsing the wave function; and then, the particle knows how to behave or how to move and interact with its environment. Energy is psychic. Energy is perceptive, sentient, intelligent, and alive. This is how things really work. It's NOT magic nor creation ex nihilo as the Materialists, Naturalists, Nihilists, and Atheists claim. It's all being controlled by Psyche or Intelligence. It's deliberate, not magical.

The Gods or Intelligent Psyches, who designed and made this information field or Akashic Field, can do transmutation at will. Energy is infinitely malleable. The Psyches in control of that energy can form the energy under their control into anything they want it to be. The Gods can contact the Psyches within nature directly – the Psyches controlling the different parts of a physical atom – and encourage those Psyches to transmute the energy under their control from carbon to nitrogen, or from lead to gold. The question is whether we humans can find ways to do the same thing at the physical level from the physical level.

The above features are characteristic of the wave-particle relation in an isolated, "one particle" system. Extended to a "many particle" system, the causal interpretation provides a framework for considering non-local connections between distant particles when they are in an "entangled" state; in such a state the whole system has primacy over the behavior of its parts. When the interpretation is applied to quantum field theory, further novel features emerge, particularly the super-quantum potential, which has the capacity to inform the sub-structures of the entire universe. Of equal significance, the super-quantum potential (equivalent to the super-implicate order as discussed in Chapter 4) coordinates the movement of quantum waves in a manner that allows for the very creation of "particle-like manifestations" in the first instance. – Lee Nichol (p. 184.)

All of Bohm's talk about a super-quantum potential, super-quantum information, and a super-implicate order is directly synonymous with the Supernatural Psyche or Supernatural Intelligence that resides within all the different forms of energy. This Supernatural Psyche or Super-Quantum Potential coordinates the movements of the quantum waves and produces the physical particles in the first place. That's the way it really works. Only Psyche or Intelligence or Life Force can process and understand information. Psyche or that Spark of Life is the ultimate causal agent within this physical universe that we observe. This reality and truth applies just as much at the physical level as it does the quantum level.

The math for all of this has already been developed. The problem is that nobody can agree what the math means in principle or practice. We have to find a way to get the true interpretation of Quantum Mechanics, or we will never know what all that math means.

In this latter version of the causal interpretation there is a clear convergence with the perspective of the implicate order. The particle is no longer seen as a purely objective entity, but as a relatively autonomous and stable unfoldment from the implicate order, governed in its activity and creation by the super-quantum potential. Bohm thus outlines a coherent model in which the mechanically limited analogies of the glycerin and the hologram (Chapter 3) are given a more complete theoretical underpinning, while at the same time accounting for the organizational principles of whole and sub-whole in the universe at large. – Lee Nichol (p. 184.)

The "particle" is made from energy by Someone Psyche or by Bohm's Super-Quantum Potential. It's the same difference. Bohm took so much heat from the materialistic and naturalistic censors around him, who accused Bohm of being a mystic, that Bohm learned to create and use terms that sound scientific rather than spiritual. Bohm didn't use the word "psyche" or "soul" because the Materialists, Naturalists, and Atheists would ridicule him, mock him, and call him names if he did. Instead, Bohm called the Supernatural Psyche the "Super-Quantum Potential".

Bohm created lots of different terms to make his scientific theories sound scientific rather than mystical because the Atheists and Naturalists aren't capable of understanding the mystical, the spiritual, or the non-local. Therefore, Bohm's implicate order is in fact comprised of the couple dozen non-physical quantum fields that are matrixed together into one great whole throughout our physical universe. I tend to call this the quantum level within my writings, and it's the basis of Quantum Field Theory. Bohm's super-implicate order is comprised of the energy, psyche, and information from which the quantum fields and physical particles were made. I tend to call this the psyche level within my writings. Bohm has called the physical level the "explicate order" or the "manifest order". The physical level is something that we can get our hands on and see with our physical eyes.

My Psyche Ontology states that Psyche or Intelligence is present in all the different types and forms of energy or light – whether we are talking about quantum waves, quantum fields, spirit matter (dark matter), dark energy, magnetism, gravity, radio waves, or physical matter. All of this stuff is made from energy by Someone Psyche or Someone Intelligent. Psyche is the ultimate causal agent within our physical universe. Bohm was right. This is what has been experienced and observed. Unlike Bohm, I'm no longer afraid to talk about the Psyche as a unique entity because Psyche has been experienced and observed by Out-of-Body Explorers and Near-Death Experiencers.

https://psyche-ontology.com/buhlman/

https://psyche-ontology.com/psyche-observed/

https://psyche-ontology.com/psyche-experienced-and-observed/

Bohm didn't have access to this information back in the 1980's; but, this evidence for psyche is in great abundance on YouTube as we approach 2020.

Science as originally conceived was supposed to be about observation and experience – knowledge – and not the wishful thinking, restrictive extremism, and dogmatic religion of the Atheists, Materialists, Naturalists, Darwinists, and Mechanists. Science is supposed to allow all of the evidence into evidence, and then pursue a preponderance of that evidence. As of 2020, Bohm has been vindicated, while his materialistic and naturalistic opposition are currently going down in flames, even though they don't know it yet.

I used to be a materialist, naturalist, nihilist, and atheist; but, I have learned to love Quantum Mechanics and Quantum Field Theory because they falsify Materialism, Naturalism, Nihilism, Darwinism, Behaviorism, Determinism, and Atheism replacing these falsified philosophies with the truth instead.

Mark My Words

Quantum Potential

Energy is infinitely malleable. Each psyche controls a certain amount of energy. The psyche in control of energy or light can form that energy into anything that it wants that energy to be. This is the way things really work at the quantum level and the psyche level. Energy contains information, psyche, consciousness, life force, and intelligence. This intelligence, information, psyche, and knowledge is what gives energy the ability or the **potential** to become anything that it wants to be. Bohm talks about this "quantum potential" within his essays.

David Bohm wrote:

> **There are several reasons for including a discussion of this theory within this chapter. To begin with, it provides a relatively intelligible and intuitively graspable account of how an actual quantum process may take place. Moreover, it does not require a conceptual or formal separation between the quantum system and its surrounding "classical" apparatus. In other words, there is no fundamental "incommensurability" between classical and quantum concepts and, therefore, a greater unity between the formal and informal languages used in its exposition. In addition, this theory has never before been presented in a non-technical way and it may be of interest to the reader to learn of a quite novel approach to the quantum theory.**

> [This is an interpretation of quantum mechanics that actually explains both quantum physics and classical Newtonian physics. Bohm's interpretation takes the psyche level, the quantum level, and the physical level into consideration.]

> **Although the interpretation is termed causal, this should not be taken as implying a form of complete determinism. Indeed, it will be shown that this interpretation opens the door for the creative operation of underlying, and yet subtler, levels of reality. The theory begins, in its initial form, by supposing the electron, or any other elementary particle, to be a certain kind of particle which follows a causally determined trajectory. (In the later, second quantized form of the theory, this direct particle picture is abandoned.) Unlike the familiar particles of Newtonian physics, the electron is never separated from a certain quantum field which fundamentally affects it and exhibits certain novel features. This quantum field satisfies Schrödinger's equation, just as the electromagnetic field satisfies Maxwell's equation. It, too, is therefore causally determined.**

> ["Creative operation" signifies Psyche or Intelligence. Creation and choice are a function of Psyche or Intelligence, and therefore sure signs of Psyche or Intelligence. The electron in enmeshed or enfolded within a quantum field; and, the quantum fields are organized energy. Anything that

is obviously organized obviously has an Intelligent Organizer who designed it and organized it. Psyche or Intelligence is the thing that determines what novel features an electron and a quantum field will manifest. Here Bohm is suggesting Psyche Determinism or Psyche Causality; and, that's appropriate because Psyche or Intelligence is the ultimate causal agent within our physical universe both at the quantum level and the physical level.]

Within Newtonian physics, a classical particle moves according to Newton's laws of motion and the forces that act on the particle are derived from a classical potential V. The basic proposal of the causal interpretation is that, in addition to this classical potential, there also acts a new potential, called the quantum potential Q. Indeed, all the new features of the quantum world are contained within the special features of this quantum potential. The essential difference between classical and quantum behavior, therefore, is the operation of this quantum potential. Indeed, the classical limit of behavior is precisely that for which the effects of Q become negligible.

[Potential of any kind is a product of Psyche or Intelligence, especially when that potential is actualized or made real. Classical mechanics or Newtonian physics is comprised of different Physical Laws. Laws of any kind require an intelligent Law Maker to make and enforce those Laws. Furthermore, Laws are absolutely worthless without Someone Psyche or Someone Intelligent who is capable of understanding, following, and choosing to obey those laws. There has to be something intelligent with the energy or light itself that gives it the ability to understand and follow the Physical Laws that are associated with classical physics or Newtonian mechanics.]

[Potential is synonymous with Psyche. They are the same thing. The psyche who controls energy can potentially turn that energy into anything that it wants that energy to be. Energy is infinitely malleable, which means that it has infinite potential. The only way that's possible is if that energy is psychic, intelligent, conscious, sentient, perceptive, and self-aware. In order to have infinite potential, energy has to be conscious, intelligent, and alive. "Behavior" of any kind is chosen into existence by Someone Psyche. Behavior is a product of Psyche. If quantum potential or quantum behavior goes to zero, then all we are left with is classical behavior or classical potential. The essential difference between classical and quantum behavior, therefore, is the operation of this quantum potential; and, it's obviously Someone Psyche or Someone Intelligent who chooses, operates, and controls quantum potentials. Behavior of any kind is chosen into existence by Someone Psyche; otherwise, chaos and randomness would reign supreme.]

[Later on, Hiley started calling "quantum potential" by the name "information potential". It absolutely requires Someone Psyche or Someone Intelligent in order to develop, access, understand, and implement information. Information is produced by Psyche or Intelligence; and, it requires some kind of Psyche or Intelligence in order to make intelligent use of that information.]

https://en.wikipedia.org/wiki/Quantum_potential

(For the mathematically minded, the quantum potential is given by:

$$Q = (-h^2 / 2m) * (\nabla^2 |\psi|^2 / |\psi|^2)$$

where ψ is the quantum field or "wave function" derived from Schrödinger's equation, h is Planck's constant, and m is the mass of the electron or other particle. Clearly the quantum potential is determined by the quantum wave field, or wave function. But what is mathematically significant in the above equation is that this wave function is found in both the numerator and the denominator. The curious effects that spring from this relationship will be pointed out in the following paragraphs.)

[Clearly, the quantum potential, the quantum wave field, and quantum wave function are chosen into existence by the Controlling Psyche who has taken control of those things. Mass is made from energy. Mass is just a different form of energy. Psyche can form energy into anything that it wants that energy to be, including physical matter or mass. The physical constants were made by Someone Psyche or Someone Intelligent. Furthermore, there must be some type of intelligence or psyche within the system that's actually capable of understanding what the Planck constant is and how it's supposed to be enacted or performed. The quantum fields were NOT made by physical matter. In fact, quantum fields are non-physical and pre-physical. Quantum fields were made by Someone Psyche. When it comes to physics, the smaller dwells within and controls the larger. The Psyche who dwells within the energy decides what form that that energy will take.]

[What is mathematically significant is that Psyche or Intelligence is an essential component of each part of the quantum potential equation. It actually requires Someone Psyche or Someone Intelligent to figure out what that equation means. Furthermore, changes in the wave function or the quantum field are produced by Someone Psyche, and NOT physical matter. Physical matter is just one form of energy. Physical matter is an emergent property of the many different choices that Someone Psyche has made. Physical matter doesn't decide anything, nor does it choose anything. Instead, physical matter is the decision and the choice of Someone Psyche. These truths are obvious for anyone who is willing to look and see. When it comes to physics, the smaller dwells within and controls the larger. The Psyche dwells within and controls the energy; and, the Psyche determines what form that energy will take. Psyche is the ultimate causal agent in this physical universe, and Psyche determines what curious effects will spring forth from the energy under its control.]

[There wasn't enough evidence back in the 1950's or even the 1980's to make a convincing case for the existence of Psyche. I know, because I was looking for it periodically, and I didn't find it until 2015. I can fully understand why people believe that Psyche does not exist, because I couldn't find any convincing evidence for its existence until 2015 when I finally started finding tons of evidence for its existence on YouTube. Evidence for Psyche was there in the public record as early as the mid-1990's; but by then, I wasn't looking for it because I had convinced myself that it didn't exist. That's why it took me nearly 55 years to find it. Even though people were finally writing about the phenomenon, it still took me 20 years to find and read their books. However, as we now approach 2020, there's tons of evidence for the existence of Psyche on YouTube, and it's obvious now that Psyche has been experienced and observed. I no longer have any doubts about Psyche's existence because it has been proven to exist. Psyche is experienced first-hand as an immaterial viewpoint in space. Whenever Out-of-Body Explorers are separated from their spirit body looking at their spirit body, there is nothing to be found when they go looking for their Psyche, or Intelligence, or Spark, or Life Force. The reason why is that Psyche is observed or seen as being a pin-point of light, a spark of light, or a pin-prick of light. If there is such a thing as a point-particle or an infinite singularity, Psyche is it. Psyche would be smaller than the strings within string theory, which means that Psyche would be the thing that determines what shape or form a string will assume. Psyche dwells within and controls energy. Psyche decides what form or shape the energy under its control will take. That's the way it really works at the quantum level and the psyche level. I didn't know that until 2015; but, it's obvious to me now. This is very exciting, if you think about it. As a race, we have finally reached an era or a point in time when we can actually figure out how things really work both at the quantum level and the psyche level. Bohm was guessing or theorizing back in the 1980's; but now, we KNOW that he was on the right track because we now have the evidence necessary to prove that he was right.]

[The only thing I can do at this point in time is to tell you what I'm currently getting out of Bohm's theories and writings, and then hope that somebody else benefits from the process. It wasn't until 2015 that I personally discovered the existence of Psyche; but, everything changed once I did. I consider myself to be an open-minded scientist, which means that I'm willing to follow the evidence wherever it leads me. That might in fact be my only saving grace. Eventually it became obvious to me that Psyche is real and truly exists. Psyche has to exist or order to choose what form the various types of energy will take or assume. There's intelligence behind that process; otherwise, everything would be nothing but random chaos.]

https://psyche-ontology.com/psyche-observed/

https://psyche-ontology.com/psyche-experienced-and-observed/

https://psyche-ontology.com/buhlman/

At first sight, it may appear that to consider the electron as some kind of particle, causally effected by a quantum field, is to return to older, classical ideas which have clearly proved inadequate for understanding the quantum world. However, as the theory develops, this electron turns out not to be a simple, structureless particle but a highly complex entity that is affected by the quantum potential in an extremely subtle way. Indeed, the quantum potential is responsible for some novel and highly striking features which imply qualitative new properties of matter that are not contained within the conventional quantum theory.

[Electrons are made from energy. Someone Psyche within that energy decided to form that energy into an electron rather than a photon or a neutron. Someone Psyche has formed a master plan; and, some other psyche within the energy of the electron has chosen that method or means by which to follow that master plan. That energy could have chosen to become a photon or a neutron, but it chose to be an electron instead. A choice was made somewhere along the way; and, only Psyche can do choice. Causality or choice is violated by chance; and, chance is overridden or violated by causality or choice. We are talking about psychic causality here, or Psyche Determinism, where it's the Psyche who chooses what form the energy under its control will take. Quantum potential or psyche potential is responsible for some novel and highly striking features which imply qualitatively new properties of matter and existence that cannot be found within materialistic and naturalistic interpretations of Quantum Mechanics.]

The fact that ψ is contained both in the numerator and the denominator for Q means that Q is unchanged when ψ is multiplied by an arbitrary constant. In other words, the quantum potential Q is independent of the strength, or intensity, of the quantum field but depends only on its form. This is a particularly surprising result. In the Newtonian world of pushes and pulls on, for example, a floating object, any effect is always more or less proportional to the strength or size of the wave. But with the quantum potential, the effect is the same for a very large or a very small wave and depends only on its overall shape.

[Remember, form or shape is always determined by the controlling psyche, when it comes to Quantum Mechanics and the pre-physical. Something intelligent and psychic within the energy itself has to be deciding the overall shape and form that that energy will assume; otherwise, nothing would ever happen. The quantum potential is determined by the overall shape or the form that the Psyche within it chooses for it. That's what the math is trying to tell us. Something intelligent within the energy itself is choosing what form or shape that energy will take. At the quantum level, the distance between quantum objects or particles ceases to be a factor. Therefore, at the quantum level, particles and the energy within those particles are omnipresent in range and scope, and omniscient and telepathic when it comes to their access to information. The Intelligence or Psyche

660

within a physical atom is infinitely more knowledgeable and intelligent than the Human Psyche because Nature's Psyche has direct access to all of the information in the universe. Nature's Psyche has access to the mind of God. We are talking about Psyche Determinism or Psyche Causality here, and NOT the creation ex nihilo of the Atheists and Naturalists.]

[The Materialists, Mechanists, Darwinists, Naturalists, Nihilists, Behaviorists, and Atheists formally reject everything non-local, non-physical, supernatural, quantum, transdimensional, psychic, and spiritual that has ever been experienced and observed. These people formally reject every non-physical truth that has ever been discovered, proven through experimentation, experienced, or observed. If it's true and has been truly experienced and observed, they reject it anyway if it is non-physical in nature. Materialism, Naturalism, Darwinism, and their derivatives were designed to prevent us from finding and knowing the truth. I have come to realize that if it successfully falsifies Materialism, Naturalism, Atheism, and their derivatives, then that greatly increases the probability that what you are looking at or studying contains an element of truth within it. Even though demons or evil spirits will lie to you, the fact that they have been experienced and observed provides evidence that falsifies Materialism, Naturalism, and their derivatives such as Darwinism and Atheism. The false is falsified by the truth; and, the truth is repeatedly experienced and observed by Someone Psyche or Someone Intelligent. That's the way science, observation, or knowledge works. If it has been seen, experienced, discovered through experimentation, or observed, that greatly increases the probability that it is in fact real and true – infinitely more real and true than the creation ex nihilo of the Atheists and Naturalists. Creation ex nihilo is one of the dumbest ideas that we have ever gotten from the Materialists, Naturalists, Nihilists, and Atheists. It's nothing but magic, or smoke and mirrors. Creation ex nihilo or Atheism is obfuscation – pseudoscience, philosophy, and wishful thinking masquerading as science.]

By way of an illustration, think of a ship that sails on automatic pilot, guided by radio waves. The overall effect of the radio waves is independent of their strength and depends only on their form. The essential point is that the ship moves with its own energy but that the information within the radio waves is taken up and used to direct the much greater energy of the ship. In the causal interpretation, the electron moves under its own energy, but the information in the form of the quantum wave directs the energy of the electron. Clearly the term causal is now being used in a very new way from its more familiar sense.

[I too have used the radio analogy a lot within my own books and essays. Telepathy is Wi-Fi at the quantum level. Thoughts and memories are quantum waves. Quantum waves are radio waves at the quantum level. The information within radio waves and quantum waves depends upon the form that Someone Psyche assigns to those radio waves and quantum waves. Again, when it comes to physics of any kind, the smaller dwells within and controls the larger. There's information within the quantum waves, or thoughts, or memories just as there's information within the radio waves. Psyche produces, transmits, receives, and stores quantum waves. The information or psyche in the form of quantum waves directs the energy within the electron. Clearly, we are no longer dealing with efficient causality (the billiard-ball causality of classical physics) but instead we are dealing with formal causality (information blueprints) and final causality (purpose, teleology, and intelligence). Clearly, Intelligence or Psyche is starting to come into play when we get to the quantum level of existence and reality.]

[It is obvious that information of any type has to be produced by Someone Psyche or Someone Intelligent. Information doesn't just spring into existence whole from nothing as the Materialists, Naturalists, Darwinists, Atheists, and Mechanists claim. Information is obviously made. Anything that is obviously made obviously has a Maker who thought of it and made it. Therefore, it is logical and rational to conclude that any type of information has Someone Psyche or Someone Intelligent who thought of it and made it. The source of any type of information is Someone Psyche. Remember, information doesn't come into existence ex nihilo as the Atheists and Naturalists claim. The Materialists, Naturalists, Nihilists, and Atheists are wrong most of the time when it comes to science, because these people preach and teach creation from nothing or "magic", which is not science but dogmatic philosophical wishful thinking. These people have a talent for identifying every truth and then formally rejecting that truth. It's what they do; and, they are proud of it. They greatly admire and cherish their ignorance and wear it as a badge of honor.]

[According to David Bohm, the information or the Psyche in the form of a quantum wave (thought) directs the energy of the electron.]

The result is to introduce several new features into the movement of particles. First, it means that a particle that moves in empty space, with no classical forces acting on it whatsoever, still experiences the quantum potential and therefore need not travel uniformly in a straight line. This is a radical departure from Newtonian theory. The quantum potential itself is determined from the quantum wave ψ, which contains contributions from all other objects in the particle's environment. Since Q does not necessarily fall off with the intensity of the wave, this means that even distant features of the environment can affect the movement in a profound way.

[At the quantum level, particles can teleport or quantum tunnel whenever the psyche controlling them chooses to do so. The quantum potential is determined from the quantum wave, and the shape and form of the quantum wave is chosen by Someone Psyche. At the quantum level, each physical particle is interacting telepathically with every other particle within this physical universe instantaneously and simultaneously through all the various different quantum fields that have been created and made to exist. At the quantum level, the effects of distance cease to exist, because everything at the quantum level exists at the speed-of-light which means that it experiences NO passage of time, and the distance between object therefore moves to zero. At the speed-of-light, there is no distance between objects because there is no passage of time. Time stops at the speed-of-light, which means that the distance between quantum objects becomes zero. Quantum mechanics or Psyche mechanics is a whole other way of doing physics.]

[When it comes to the Quantum Potential, someone has to choose the frequency of the wave, or the vibration of the string, of the form that the energy is to assume. This choice is deliberate and intelligent. It can't be done randomly, or all we would have is chaos.]

The explanation of the quantum properties of the electron given above emphasized how the form of the quantum potential can dominate behavior. In other words, information contained within the quantum potential will determine the outcome of a quantum process. Indeed, it is useful to extend this idea to what could be called active information. The basic idea of active information is that a form, having very little energy, enters into and directs a much greater energy. This notion of an original energy form acting to "inform," or put form into, a much larger energy has significant applications in many areas beyond quantum theory.

[It's all about information or intelligence; and, Psyche is the thing that produces, transmits, receives, and stores information. Psyche is Intelligence. Psyche is the part of you that knows. The form of the quantum potential dominates and determines the quantum behavior of a particle; and, Someone Psyche chooses the form that the energy under its control will take. The information, or the intelligence, or the psyche contained within a quantum potential will determine the outcome of a quantum process. Active information is Psyche in action. Active information is a psyche form, or a little energy form, that enters into and directs and controls a much greater or much larger energy. Psyche is an "original energy form" or a primal energy form that acts to "inform" or to put into form the larger energies that are under its control. Back in 1987, David Bohm defines Psyche and predicts Psyche at least a decade before convincing evidence for Psyche starts to show up on the internet and in the public record. This is Science at its very best – making predictions that are subsequently confirmed by evidence and observation. The existence of Psyche has significant applications that go far beyond quantum theory – to infinity and beyond. I keep getting drawn back to Quantum Mechanics and Quantum Field Theory because it is infinite in potential. Its explanatory power is through the roof! It's infinitely more significant, meaningful, and powerful than anything the Materialists, Naturalists, Darwinists, Nihilists, and Atheists have been able to produce.]

[Remember, ALL order, organization, purpose, and creation are done by Someone Psyche. It doesn't spring into existence whole from nothing as

663

the Materialists, Naturalists, Darwinists, Mechanists, and Atheists claim. When it comes to Science, these people don't know what they are talking about. Creation is always done by Someone Psyche or Someone Intelligent. It's obvious. According to Quantum Mechanics, the form that a subset of energy takes is always determined by the controlling psyche. We are talking about Psyche Determinism or Psyche Causality here, and NOT creation ex nihilo.]

[In all of Bohm's talk about "active information", he is in fact talking about Psyche in action – Psyche doing what Psyche does – processing and using information or intelligence. All of the individual particles within a physical atom are intelligence, active, purposeful, and in motion. Energy moves, which is what makes it alive. You are going to have to decide for yourself if I know what I'm talking about or not; but, I'm not worried about it because I have finally seen the light. What we have here is infinitely superior to anything that I have ever gotten from the Materialists, Naturalists, Darwinists, Mechanists, and Atheists. All I have ever gotten from them is their claim that Psyche does not exist. The difference is obviously night and day.]

Consider a radio wave, whose form carries a signal – the voice of an announcer, for example. The energy of the sound that is heard from the radio does not in fact come from this wave but from the batteries or power plug. This latter energy is essentially "unformed," but takes up its form from the information within the radio wave. This information is potentially active everywhere but only actually active when its form enters the electrical energy of the radio.

[Energy takes up its form from the "information" or the intelligence or the Psyche that is controlling it. This information is potentially active everywhere in the universe; but, it only becomes actually active when the Psyche within the energy decides to use that information to accomplish a specific purpose. The information is activated when the Psyche within the energy chooses to use that information to form the energy under its control into a specific type of particle or wave.]

The analogy with the causal interpretation is clear. The quantum wave carries "information" and is therefore potentially active everywhere, but it is actually active only when and where this energy enters into the energy of the particle. But this implies that an electron, or any other elementary particle, has a complex and subtle inner structure that is at least comparable with that of a radio. Clearly this notion goes against the whole tradition of modern physics, which assumes that as matter is analyzed into smaller and smaller parts, its behavior grows more elementary. By contrast, the causal interpretation suggests that nature may be far more subtle and strange than was previously thought.

[Quantum waves are thoughts and memories. The quantum wave carries "information" or intelligence, and it is therefore potentially active everywhere; but, it is actually active only when and where this "information" or intelligence is being used to form energy into a specific type particle, quantum wave, or quantum field. Atoms, electrons, protons, and neutrons have a complex inner structure that is informed and controlled by this energy or this psyche within the atom. Clearly, the Materialists, Naturalists, Mechanists, and Atheists are going to throw a fit when presented with this type of information and truth because it falsifies their pre-chosen philosophical beliefs.]

[Information, intelligence, or psyche becomes the ultimate causal control within Bohm's causal interpretation of Quantum Mechanics. Bohm's analogy works and makes logical sense to me. It's infinitely superior to the "magic" or creation ex nihilo that we get from the Atheists and Naturalists. Quantum waves carry information or thoughts. That's what quantum waves are – thoughts and memories. Quantum waves carry information just like radio waves carry information. Every elementary particle has this inner network of information in the form of quantum waves or thoughts. The Quantum Zeno Effect has also proven beyond a shadow of a doubt that the Psyche or Intelligence within atoms is telepathic and can read your mind. When it comes to an atom or a photon, the information within the atom is actually active when the controlling psyches use that energy or information to form and maintain the particle itself.]

But this inner complexity of elementary matter is not as implausible as it may appear at first sight. For example, a large crowd of people can be treated by simple statistical laws, whereas individually their behavior is immensely subtler and more complex. Similarly, large masses of matter reduce to simple Newtonian behavior whereas atoms and molecules have a more complex inner structure. And what of the subatomic particles themselves? It is interesting to note that between the shortest distance now measurable in physics (10^{-16} cm) and the shortest distance in which current notions of space-time probably have meaning (10^{-33} cm), there is a vast range of scale in which an immense amount of yet undiscovered structure could be contained. Indeed, this range is roughly equal to that which exists between our own size and that of the elementary particles.

[There's plenty of room within an atom to contain and store trillions of point-particles that we call Psyche or Intelligence. In fact, there is plenty of space within an atom to store a quantum computer and exabytes of data storage in holographic form. When it comes to the Materialists, Naturalists, Mechanists, and Atheists, there is more in heaven and earth and within an atom than is ever dreamt of within their philosophy. These people deliberately choose to remain ignorant of the possibilities and the potential associated with energy, quantum fields, quantum waves, and psyche. I know because I have been there and done that. You can't find the truth by deliberately embracing the lie.]

A further feature of the causal interpretation is its account of what Bohr called the wholeness of the experimental situation. In, for example, the double slit experiment, each particle responds to information that comes from the entire environment. For while each particle goes through only one of the slits, its motion is fundamentally affected by information coming from both slits. More generally, distant events and structures can strongly affect a particle's trajectory so that any experiment must be considered as a whole. This [causal interpretation] gives a simple and tangible account of Bohr's wholeness, for since the effects of structures may not fall off with distance, all aspects of the experimental situation must be taken into account. (*The Essential David Bohm*, pp. 184 – 187.)

[Each individual particle that is brought into existence by Someone Psyche responds to all of the information that comes from the entire environment. The Psyche or Intelligence within quantum particles, quantum fields, and quantum waves is omnipresent, prescient, and omniscient. At the quantum level – at the speed-of-light – time stops, and distance becomes zero. There is no distance between objects at the quantum level according to the theory of relativity. It's all one integrated timeless whole at the quantum level where Psyche reigns supreme.]

[We definitely need a quantum mechanical explanation for how psyche or mind works because it's definitely not a physical phenomenon. Quantum fields are non-physical and pre-physical; and, it's obvious that quantum fields had to be designed and made by Someone Psyche because physical matter didn't exist yet. Psyche is the ultimate cause of all the different quantum fields.]

This is cool stuff because it actually explains how everything really works. You will never get anything like this from the Materialists, Naturalists, Darwinists, Mechanists, Behaviorists, Determinists, Physical Reductionists, and Atheists because these people have nothing like this to give you. Instead, all they will ever tell you is that all of this does not exist. That's sloppy science or Bad Science. The Atheists preach the existence of "non-existence", or creation ex nihilo, or magic; and, that's all they have to give you. Atheism is creation ex nihilo or magic – the creation of something from nothing by nothing. All of the most stupid, illogical, and irrational ideas come from the Materialists, Naturalists, Darwinists, Nihilists, Mechanists, and Atheists; and, they don't even know it. I didn't have a clue when I was a materialist, naturalist, nihilist, and atheist. That's the way it works. Materialism, Naturalism, Darwinism, Nihilism, and Atheism were designed to keep us stupid, ignorant, uneducated, and dumb. They work as intended.

For me personally, the Naturalists and Atheists had successfully kept Bohm's theories and true beliefs from me. They associated Bohm with the problematic materialistic and naturalistic pilot-wave interpretation of Quantum Mechanics that they themselves had falsified, so the Materialists and Mechanists had successfully convinced me that Bohm was just another nutter like them. I had no idea the Bohm had developed a useful and realistic interpretation of Quantum Mechanics until after I bought and started reading The Essential David Bohm. Then my eyes were opened to the truths that the Materialists and Naturalists had successfully hidden from me.

For the past few years, I hadn't been able to subscribe to nor accept everything that Bohm has produced because the Naturalists and Atheists had successfully convinced me that Bohm doesn't really include Psyche or Consciousness or Intelligence, as such, into any of his theories. The Scientific Naturalists fooled me and tricked me. It's what they do. I used to be a materialist, naturalist, nihilist, and atheist; but, when I finally realized that they are lying to me and trying to trick me and deceive me, that's when I started to separate myself from them and went in search of the truth instead. The false is always falsified by the truth; and, the truth is repeatedly experienced and observed.

The Materialists, Naturalists, Darwinists, Nihilists, Behaviorists, Mechanists, Physical Reductionists, and Atheists deliberately lie to us and suppress evidence in order to convince us that Psyche and God do not exist. Bohm ended up being yet another case where the Materialists and Naturalists had successfully taken his real message and true beliefs, hidden them from me, changed them into something false and falsified, and destroyed them so that they could no longer be found nor understood. That's what these people do. They are the masters of the art of deception, and they spend all of their time ridiculing, hiding, changing, and destroying the truth. Bohm's own words, theories, and ideas are nothing like the Materialists' presentation of his ideas. Bohm's theories, ideas, and beliefs FALSIFY Materialism, Naturalism, Mechanism, and their derivatives which is why these people never tell us the truth about what Bohm truly believed.

When one looked at the many-particle system, this new kind of wholeness became much more evident, for the quantum potential was now a function of the positions of all the particles which (as in the one-particle case) did not necessarily fall off with the distance. Thus, one could at least in principle have a strong and direct (non-local) connection between particles that are quite distant from each other. This sort of non-locality would, for example, give a simple and direct explanation of the paradox of Einstein, Podolsky, and Rosen, because in measuring some property of one of a pair of particles with correlated wave functions, one will alter the "non-local" quantum potential so that the other particle responds in a corresponding way.

[At the quantum level, everything is one united whole – it's tied together by a Universal Matrix of Quantum Fields. Particles, including physical matter, emerge from this Matrix of Quantum Fields. Influence, quantum potential, telepathy, or action at a distance does not fall off with distance at the non-local level or the quantum level or the spiritual level. Non-locality is the only way to explain Quantum Entanglement. A few years ago, I realized that "non-local" means non-physical; and, "local" means classical physics and physical matter. Once I made that connection, then everything made logical sense to me. Quantum Mechanics is Energy Mechanics, Psyche Mechanics, or Spiritual Mechanics. It explains how the non-physical or the non-local works at the quantum level.]

Because the above response is instantaneous, however, it would seem at first sight to contradict the theory of relativity, which requires that no signals be transmitted faster than the speed of light. At the time of proposing these notions I regarded this as a serious difficulty, but I hoped that the problem would ultimately be resolved with the aid of further new orders. This indeed did happen later in connection with the application of the causal interpretation to the quantum mechanical field theory, but as this question is not relevant to the subject of the present paper, I shall not discuss it further here.

[The Theory of Relativity is the pinnacle of Classical Physics. It explains how physical matter works. It explains physical limitations. At the quantum level, response is instantaneous in violation of the theory of relativity because at the quantum level everything is non-local or non-physical. At the quantum level, nothing has any physical limitations. Physical limitations, such as entropy and time and distance and the theory of relativity only show up at the physical level. They are non-existent at the quantum level. There is no entropy, no distance, no time, and no theory of relativity at the quantum level. There are NO physical limitations at the quantum level. Quantum fields and quantum field theory have NO physical limitations. This reality and truth has monumental explanatory power. It explains how everything works. Physical matter is subjected to physical limitations or "locality"; whereas, quantum matter or spirit matter is NOT. Non-locality has NO physical limitations. There is NO theory of relativity at the quantum level, or the psyche level, or non-local level. Did the lights just go on? They did for me. God is in the light. God is in the Quantum Fields.]

Meanwhile, however, I felt that the causal interpretation was affording valuable insight into a key difference between classical and quantum properties of matter. Classically, all forces are assumed to fall off eventually to zero, as particles separate, whereas in the quantum theory the quantum potential may still strongly connect particles that are even at macroscopic orders of distance from each other. In fact, it was just this feature of the quantum theory, as brought out in the causal interpretation, that later led Bell to develop his theorem, demonstrating quite precisely and generally how quantum non-locality contrasts with classical notions of locality.

[Psyche, Intelligence, Consciousness, or Life Force is the Ultimate Cause or the ultimate causal agent at the quantum level, as well as at the physical level. The verified and proven existence of Action at a Distance in 1982 successfully falsified Materialism, Naturalism, Darwinism, Nihilism, and Atheism which claim that there is no such thing as Psyche, Telepathy, Telekinesis, or Action at a Distance.]

["Specifically, Bell demonstrated an upper limit, seen in Bell's inequality, regarding the strength of correlations that can be produced in any theory obeying local realism, and he showed that quantum theory predicts violations of this limit for certain entangled systems. His inequality is experimentally testable, and there have been numerous relevant experiments, starting with the pioneering work of Stuart Freedman and John Clauser in 1972 and Alain Aspect's experiments in 1982, all of which have shown agreement with quantum mechanics rather than the principle of local realism." https://en.wikipedia.org/wiki/Quantum_entanglement]

[Quantum entanglement, or quantum non-locality, or quantum non-physicality has been repeatedly verified through scientific experiments. This means that Local Realism, Materialism, Naturalism, Mechanism, Physicalism, and their derivatives have been simultaneously falsified. The false is falsified by the truth; and, the truth is repeatedly experienced and observed.]

As important as this new feature of non-local connection is, however, the quantum potential implies a further move away from classical concepts that is yet more radical and striking. This is that the very form of the connection between particles depends on the wave function for the state of the whole. This wave function is determined by solving Schrödinger's equation for the entire system, and thus does not depend on the state of the parts. Such a behavior is in contrast to that shown in classical physics, for which the interaction between the parts is a predetermined function, independent of the state of the whole. Thus, classically, the whole is merely the result of the parts and their preassigned interactions, so that the primary reality is the set of parts while the behavior of the whole is derived entirely from those parts and their interactions.

[The quantum whole has NO physical limitations. The Theory of Relativity doesn't apply to Quantum Mechanics. Entropy, time, and distance do NOT apply to Quantum Mechanics and Quantum Field Theory. There is no thermodynamics at the quantum level or the psyche level. Non-locality has NO thermodynamics, and therefore no entropy. The non-local level has NO time and NO distance, so there is no such thing as a light-speed limit at the quantum level, and therefore, the theory of relativity does not exist at the quantum level. These things are local phenomena or physical phenomena, and they don't apply at the quantum level. There are NO physical restrictions and NO physical limitations at the quantum level. Determination or classical physics or conventional cause-and-effect only apply at the physical level.]

With the quantum potential, however, the whole has an independent and prior significance such that, indeed, the whole may be said to organize the activities of the parts. For example, in a superconducting state it may be seen that electrons are not scattered because, through the action of the quantum potential, the whole system is undergoing a coordinated movement more like a ballet

dance than like a crowd of unorganized people. Clearly, such quantum wholeness of activity is closer to the organized unity of functioning of the parts of a living being than it is to the kind of unity that is obtained by putting together the parts of a machine. (pp. 191-192.)

[There's wholeness or unity at the non-local level or the quantum level, because it is all meshed together by a Matrix of Quantum Fields. There is no distance, time, entropy, or physical limitations at the quantum level. Everything at the non-local level is coordinated and instantaneous. Time, distance, space, entropy, and speed-limits (physical limitations) are introduced into the system at the physical level; whereas, they are non-existent at the quantum level. The physical limitations and physical laws are real at the physical level; and, they had to be designed and brought into existence by Someone Psyche at the quantum level or the non-local level. Life and Life Force functions on unity or quantum principles, whereas physical machines are subjected to physical limitations.]

[Neurons communicate with each other telepathically (through Wi-Fi) and NOT through chemical neurotransmitters. Neurotransmitters open ion gates. They don't transmit messages at the physical level. There are no physical wires at the physical level within a physical brain. There are no memory engrams or no RAM at the physical level within a physical brain. ALL of the thinking, communication, intention, analysis, interaction, and memory storage is taking place at the quantum level when it comes to a physical brain.]

The implicate order thus plays a primary role, while the explicate order is secondary, in the sense that its main qualities and properties are ultimately derived in its relationship with the implicate order, of which it is indeed a special and distinguished case. (p. 193.)

[The implicate order, psyche order, quantum order, energy order, non-local order, or spiritual order is PRIMARY; whereas, the explicate order or physical order is derived from it by Someone Psyche. The verified and proven existence of Action at a Distance falsifies Materialism, Naturalism, Darwinism, Nihilism, and Atheism. Atheism is Creation Ex Nihilo, and Atheism states that there is no such thing as psyche or intelligence. That's obviously false. Particles are communicating instantaneously with each other psyche-to-psyche telepathically at a distance, whether we are talking about neurons within a physical brain, or entangled photons that are separated by galaxies.]

All throughout *The Essential David Bohm*, Bohm verifies Energy Mechanics, Psyche Mechanics, Spiritual Mechanics, or Quantum Mechanics while at the same time successfully falsifying the philosophies of Materialism, Naturalism, and Mechanism. Remember, every time Action at a Distance, Quantum Field Theory, Quantum Waves, Psychic Entanglement, or Non-Locality is verified, Materialism, Naturalism, and their derivatives are automatically falsified. These two separate philosophies or worldviews are mutually exclusive. They falsify each other. If one of them can be demonstrated to be true, then the other has been successfully falsified. Quantum Mechanics or Psyche Mechanics is repeatedly verified and proven true, which means that Materialism, Naturalism, and their derivatives are repeatedly falsified.

My reason for quoting Bohm so extensively in these essays is to demonstrate that Bohm's "unorthodox" theories have subsequently been vindicated, while at the same, the materialistic, naturalistic, and atheistic philosophies have been successfully falsified.

Back in the 1950's, the Materialists, Naturalists, and Mechanists hammered on Bohm mercilessly; and, seventy years later, Science (observation and experience) and Quantum Mechanics (scientific experimentation) are now in a position to repay the favor by falsifying Materialism, Naturalism, Physicalism, and Mechanism. There is no logical, rational reason to believe in Materialism, Naturalism, Darwinism, Nihilism, Atheism, and their derivatives because they have been successfully falsified by Science itself. Quantum Field Theory, Quantum Mechanics, Near-Death Experiences, After-Death Life Reviews, and Out-of-Body Experiences are the Science (observations and experiences) that falsify Materialism, Naturalism, and their derivatives. Science is now thousands of years ahead of the Materialists, Naturalists, and Atheists. I used to be a materialist, naturalist, nihilist, and atheist. There was only one way to bring my Science up-to-date, and that was by falsifying and abandoning My Materialism, My Naturalism, My Nihilism, and My Atheism. If you successfully eliminate everything that is false, then only the truth will remain.

Fascinating, is it not? We are now able to verify everything that David Bohm wrote, except for the occasions where he slips back into Materialism, Naturalism, Darwinism, Mechanism, and Atheism. We can't verify Materialism, Naturalism, and their derivatives because they are unverifiable. The only thing we can do is to falsify them, which we have now done trillions of times in thousands of different ways.

Now, as we approach 2020, the only time Darwinism, Naturalism, and Atheism are intellectually stimulating is when we are successfully debunking them and falsifying them. The false is falsified by the truth; and, the truth is repeatedly experienced and observed.

Mark My Words

Super-Quantum Potential

On PBS Spacetime, Matt does an episode entitled, "Why Quantum Information Is Never Destroyed". Conservation of information is a fundamental and essential aspect of Quantum Mechanics.

https://www.youtube.com/watch?v=HF-9Dy6iB_4

Quantum Information is always conserved. The only way that that is realistically possible is if there is something at the quantum level within the implicate order that is capable of storing information and that is always conserved. We KNOW of two such things – Psyche and Energy. Energy is information, thoughts, or quantum waves – while Psyche has the innate ability to manufacture, transmit, receive, process, and store quantum waves or thoughts. Psyche, quantum waves, and thoughts are made from energy. Every Psyche has a certain amount of energy that's under its control. Information is stored within energy, which means that information is stored within Psyche. It is a proven fact of physics that Energy or Psyche is always conserved, which means that Quantum Information can never be destroyed once it has been produced.

Being a Darwinism, Materialist, Naturalist, and Atheist, Matt comes to the same truths and conclusions from a completely different direction; but, the truth will always be true no matter how it is derived. My model for the Conservation of Quantum Information is more parsimonious and easier to understand; but, Matt's model – a model based upon particle physics – works just fine as well.

It's unavoidable. Eventually, Bohm is forced to talk about and consider a Super-Implicate Order and a Super-Quantum Potential. The implicate order consists of a Matrix of Quantum Fields that were designed and made by Someone Psyche to support and sustain the existence of physical matter and other quantum particles.

Bohm's Super-Implicate Order ends up being "command and control", or the psyche level of existence. It's obvious that Psyche or Intelligence is made from energy – energy which is in a non-physical, immaterial, non-local, quantum, or spiritual form. Bohm's Super-Quantum Potential ends up being "an inner principle of organization within the implicate order that determines" what the energy under its control will form into. Bohm's Super-Quantum Potential is obviously Psyche or Intelligence. What is always missing from the materialistic, naturalistic, mechanistic, and atheistic models and analogies is the Psyche, or the Chooser, or the Determiner. Both the Super-Implicate Order and the Super-Quantum Potential are supernatural, non-local, non-physical, quantum, or spiritual in nature and origin.

Bohm could see the necessity for some type of non-physical, supernatural Determiner, Chooser, or Command and Control, even though the Materialists and Naturalists can't and won't. Even the Atheists and Naturalists recognize the need for some type of determination, and these people assign chance, blind luck, creation ex nihilo, chemical evolution, natural selection, spontaneous generation, or some other type of magic to get the job done. These people rely upon the most irrational and illogical of designers and creators to produce order, organization, and life – anything but God – anything but Intelligence or Psyche. These people prefer to deny the obvious existence of Intelligence, Life Force, Choice, or Psyche rather than accepting the abundant and obvious evidence proving that such a thing actually exists and functions within our universe.

Why?

Well, the proven and verified existence of Intelligence (Psyche) falsifies Materialism, Naturalism, Darwinism, Nihilism, and Atheism. Likewise, the proven and verified existence of anything non-physical such as energy, radio waves, magnetic fields, quantum fields, and quantum mechanisms (spiritual mechanisms) FALSIFIES Materialism, Naturalism, and their derivatives.

Even physical objects such as cars, computers, airplanes, rockets, televisions, refrigerators, and cell phones require Someone Intelligent or Someone Psyche to make them and to make good use of them. It's obvious that only Intelligence or Psyche has the innate ability to design and create. The Gods or God's Psyche had to design and make the quantum fields from raw unorganized chaotic energy BEFORE they could make and sustain physical matter.

Anything that is obviously made obviously has a Maker who made it. Quantum Fields were obviously designed and made; otherwise, there would be nothing but random chaos right now. What would be capable of designing and making Quantum Fields from raw chaotic energy? The only thing we know of is some type of non-physical and pre-physical Psyche or Intelligence – or as Bohm would say, some type of Super-Quantum Potential.

Try to follow Bohm's train of thought:

> **I shall not go into great detail about the implicate order here; I shall assume that the reader is somewhat familiar with this. What I want to emphasize is only that the implicate order provided an image, a kind of metaphor, for intuitively understanding the implication of wholeness which is the most important new feature of the quantum theory. Nevertheless, it must be pointed out that the specific analogies of the ink drop and the hologram are limited, and do not fully convey all that is meant by the implicate order. What is missing in these analogies is an inner principle of organization in the implicate order that determines which sub-wholes shall become actual and what will be their relatively independent and stable forms.**

[It has been observed that this Quantum Wholeness that Bohm used to talk about is in fact comprised of a Matrix of Quantum Fields that fill the immensity of space and permeate each and every particle and physical atom. The Quantum Fields that were designed and made by Someone Psyche are the most important new feature of quantum theory. Everything in this universe is made from Quantum Fields, and the Quantum Fields were made from energy by Someone Psyche or Someone Intelligent. Holograms are

673

organized energy; and, anything that is obviously organized obviously has an Organizer who organized it. Someone had to design and make the holograms and the quantum fields. They didn't just spring into existence whole from nothing as the Materialists, Naturalists, and Atheists claim. There's no such thing as Creation Ex Nihilo or Spontaneous Generation.

What is missing within Materialism, Mechanism, Darwinism, Naturalism, and Atheism is an inner organizing force or an inner organizing principle such as Psyche or Intelligence. Psyche, Intelligence, Consciousness, or Life Force is the organizing principle that Bohm keeps talking about. Psyche determines what forms the energy under its control will take. It's obvious that Someone Psyche or Someone Intelligent has to make that choice. There has to be something within the energy itself that determines which "sub-whole" or which form shall become actual, or physical, or particalized. Remember, the energy can literally become anything that its Controlling Psyche wants it to be. There has to be Someone Psyche or Someone Intelligent within the energy who determines which form that energy will assume. Once that decision or determination has been made by the Controlling Psyche, then the form becomes actual and relatively independent and stable, at least until the Controlling Psyche decides to change its form into something else besides what it was before. In other words, at every vertex or convergence on a Feynman Diagram, there is some type of Psyche who decides the outcome of each particle interaction.

https://en.wikipedia.org/wiki/Feynman_diagram

These quantum particles are obviously following specific rules. Rules or laws of any kind obviously require some kind of intelligent Law-Maker. However, even more important is the obvious fact that obedience to specific rules or laws requires some kind of intelligence within the particles themselves. Energy can become anything it wants to become, so why is that energy limiting itself to becoming only certain types of particles during each type of particle interaction? Who made the rules? And more importantly, what is there within energy itself that gives it the ability to understand, follow, and obey the rules? The only logical answer to these questions is some type of Psyche, or some type of Intelligence, or some type of Living Force. It can't be random blind luck that causes the energy to follow specific rules of behavior each and every time there is some type of particle interaction at the quantum level.

Bohm calls the psyche level the "super-implicate order"; and, Bohm calls the psyche by the name "super-quantum potential". All this super stuff is obviously supernatural; however, Bohm hides this fact in an attempt to keep the Materialists, Naturalists, Darwinists, Nihilists, and Atheists engaged in the science. These people have trained themselves to turn their brains off whenever they encounter the words "psyche", "spirit", "soul", "immaterial", "non-physical", and "spiritual". They are deathly afraid of these concepts – "Ain't gonna look at it! Not gonna think about it! It ain't real! It can't be true!" That is their mantra when it comes to quantum phenomena, or spiritual phenomena, or energy phenomena, or non-physical phenomena. It's the head-in-the-sand approach to doing quantum mechanics, energy mechanics, or spiritual mechanics.

These people don't realize that energy is non-physical and pre-physical in nature and origin. Physical matter is made from energy. They don't realize that matter is deliberately made. They don't realize that energy is spiritual, non-physical, immaterial, and pre-physical in nature and origin. They don't realize that every form of energy is intelligent or has some type of Psyche within it. They don't realize that something intelligent within the physical atoms gives those atoms the ability to understand, follow, and obey physical laws. They don't realize that energy is alive and has some type of Life Force or Psyche within it. They don't realize that their ignorance is getting in the way of their Science. I KNOW, because I used to be a materialist, naturalist, nihilist, and atheist. I had no idea how ignorant I really was until I started to study and understand the Psyche-based or Consciousness-based interpretations of Quantum Mechanics. Once I started to look at it and study it, then the lights finally went on, and I could see.]

Gradually, throughout the 1970s, I became more aware of the limitations of the hologram and ink droplet analogies to the implicate order. Meanwhile, I noticed that both the implicate order and the causal interpretations had emphasized this wholeness signified by quantum laws, though in apparently very different ways. So, I wondered if these two rather different approaches were not related in some deep sense – especially because I had come at least to the essence of both notions at almost the same time. At first sight, the causal interpretation seemed to be a step backwards toward mechanism, since it introduced the notion of a particle acted on by a potential. Nevertheless, as I have already pointed out, its implication that the whole both determines its sub-wholes and organizes their activity clearly goes far beyond what appeared to be the original mechanical point of departure. Would it not be possible to drop this mechanical starting point altogether? I saw that this could indeed be done by going on from the quantum mechanical particle theory to the quantum mechanical field theory.

[In order to get at the truth, we are eventually forced to drop and abandon Materialism, Naturalism, Mechanism, Darwinism, Nihilism, and Atheism. Atheism is creation ex nihilo or spontaneous generation or chemical evolution, all of which are physically impossible. Creation ex nihilo can't be done, which means that it wasn't done. Psyche is the Ultimate Cause of everything, especially at the quantum level within Bohm's implicate order. Even holograms, with their massive explanatory power, are non-existent and impossible without Someone Psyche or Someone Intelligent to design them and make them. Whenever we go deep, we always find Psyche, Intelligence, or Consciousness staring back at us.

Remember, in order for a particle to be acted upon by a potential at the quantum level that potential has to be some type of intelligent Actor or Psyche. Psyches are not only point-particles; but, they are also omnipresent and omniscient because they are quantum in nature and origin. The "whole" who is determining and organizing the sub-wholes has to be some type of Intelligence or Psyche. There's no other logical explanation for what we are observing and experiencing. It's obvious that Intelligence exists; and, it's time for scientists to figure out what it is and how it works.

Bohm is right. In order to get at the truth, we have to eliminate Materialism, Naturalism, Darwinism, Nihilism, and Atheism (physical particles) as the fundamental unit of reality and switch over to the non-local and non-physical Quantum Field Theory which consists of a Matrix of Quantum Fields that were made from raw chaotic energy by Someone Psyche. In other words, we have to abandon classical physics and switch over to Psychic Field Theory or Quantum Field Theory instead. It takes a while, but eventually the truth prevails. God is in the details, which means that God is in the energy or God is in the light.]

This is accomplished by starting with the classical notion of a continuous field (e.g., the electromagnetic) that is spread out through all space. One then applies the rules of the quantum theory to this field. The result is that the field will have discrete "quantized" values for certain properties, such as energy, momentum, and angular momentum. Such a field will act in many ways like a collection of particles, while at the same time it still has wave-like manifestations such as interference, diffraction, etc.

[Quantum fields fill the immensity of space, and they permeate to the very core of every particle that is brought into existence. In order for energy fields to "act", they have to have some type Actor within them who is capable of making decisions or choices as to how it wants the energy under its control to act. The primary axiom of Quantum Field Theory is that particles are born, and particles die. Someone Psyche or Someone Intelligent has to be giving birth to them; and, that same Someone Psyche has to decide when it is time for them to die and be absorbed back into the quantum fields from whence they came. This all requires some type of intelligence.

Remember, the choosing of discrete quantized values and forms requires the intervention of some type of Chooser at the quantum level. Psyche has been experienced and observed, so it is safe to conclude that some type of Psyche is in fact this Quantum Chooser who is choosing what form the energy will assume at the quantum level. Energy can be formed into anything, including spirit matter (dark matter), physical matter, and quantum fields. Who is making this choice? What is small enough to dwell within the energy and control the energy? It has to be some type of psyche. Physical matter is an emergent phenomenon – not a controlling phenomenon. Physical matter was designed and made by Someone Psyche. Everything requires Command and Control, including energy.

Once I realized that Psyche or Intelligence has been experienced and observed, then all my confusion in regard to Quantum Mechanics and Science in general just simply disappeared. Knowledge trumps philosophical speculation and wishful thinking every time. Psyche or Intelligence has way too much explanatory power to simply ignore it. With knowledge of Psyche and Quantum Mechanics or Energy Mechanics, you can literally explain everything that comes your way. We no longer have to be ignorant, like we were when we were Materialists, Naturalists, Darwinists, Nihilists, and Atheists. The best thing I ever did for my Science was to falsify and abandon My Materialism, My Naturalism, My Nihilism, and My Atheism.]

Of course, in the usual interpretation of the theory, there is no way to understand how this comes about. One can only use the mathematical formalism to calculate statistically the distribution of phenomena through which such a field reveals itself in our observations and experiments. But now one can extend this causal interpretation to the quantum field theory. Here, the actuality will be the entire field over the whole universe. Classically, this is determined as a continuous solution of some kind of field equation (e.g., Maxwell's equations for the electromagnetic field). But when we extend the notion of the causal interpretation to the field theory, we find that these equations are modified by the action of what I called a super-quantum potential. This is related to the activity of the entire field as the original quantum potential was to that of the particles. As a result, the field equations are modified in a way that makes them, in technical language, non-local and non-linear.

[Of course, the usual interpretations of quantum theory found on the Wikipedia are materialistic, naturalistic, and atheistic. Bohm is right. There's NO way to use Materialism and Naturalism to understand how energy mechanics, non-physical mechanics, spiritual mechanics, or quantum mechanics comes about or works. Energy and psyche are non-physical, immaterial, and pre-physical. God's Psyche had to make the quantum fields from raw chaotic energy BEFORE physical matter could be made. The usual materialistic and atheistic interpretations of quantum theory cannot be used to understand how this comes about. Materialism, Naturalism, Mechanism, and their derivatives completely lack explanatory power when it comes to the non-physical and the pre-physical realm of psyche and energy.

Bohm states that activity within the quantum fields or the entire field is modified, caused, or activated by a "super-quantum potential" or some type of Intelligent Psyche. Bohm never uses the word "psyche", but that's what he is talking about whenever he starts talking about the super-implicate order or the psyche level of existence and reality. Bohm emphasizes that Quantum Mechanics and Quantum Field Theory are non-local (non-physical) and non-linear (capable of quantum tunneling or teleporting).]

What this implies for the present context can be seen by considering that, classically, solutions of the field equations

represent waves that spread out and diffuse independently. Thus, as I indicated earlier in connection with the hologram, there is no way to explain the origination of the waves that converge to a region where a particle-like manifestation is actually detected, nor is there any factor that could explain the stability and sustained existence of such a particle-like manifestation.

[What this implies is that Classical Physics, Materialism, Naturalism, Mechanism, Atheism, and their derivatives are seriously lacking in explanatory power. There is no way to explain the origin, stability, and maintenance of quantum waves, quantum fields, and quantum particles using Classical Physics, Materialism, Naturalism, Atheism, and their derivatives.

Likewise, it's obvious that with a hologram Someone Psyche or Someone Intelligent is the origin, cause, and source of the whole system. Anything that was obviously organized and made – such as a hologram – obviously has an Organizer and Maker who organized it and made it. This is Logic 101. Holograms that contain usable information required an Intelligent Psyche of some sort to make them and require an Intelligent Psyche of some sort to access that information and make use of it. This is obviously true. Holograms don't just spring into existence whole from nothing, as the Materialists, Naturalists, Darwinists, Nihilists, and Atheists claim.]

However, this lack is just what is supplied by the super-quantum potential. Indeed, as can be shown by a detailed analysis, the non-local features of this latter will introduce the required tendency of waves to converge at appropriate places, while the non-linearity will provide for the stability of recurrence of the whole process. And thus, we come to a theory in which not only the activity of particle-like manifestations, but even their actualization, e.g. their creation, sustenance, and annihilation, is organized by the super-quantum potential.

[The actualization, creation, sustenance, and annihilation of particles is organized and done by the "super-quantum potential" or what we have come to recognized as the Supernatural Psyche, Life Force, Consciousness, or Intelligence.

Naturalism's lack of explanatory power is rectified or supplied by the "super-quantum potential or Intelligent Psyche. Notice how Bohm states that quantum particles are organized by the super-quantum potential or the Supernatural Psyche. Bohm couldn't call it the "psyche" back in the 1980's because he wanted the Naturalists and Mechanists to take him seriously. Now as we approach 2020 and Naturalism has been successfully falsified by Science itself, we are free to come out of the closet and tell it how it truly is. The non-local, non-physical, supernatural Psyche is the Ultimate Organizing Force behind everything that is made from energy. It has to exist; otherwise, all we would have would be random chaos or static. There has to be an Organizing Force or Life Force at the quantum level in order for there to be order, organization, and quantum fields at the quantum level, the non-local level, or the non-physical level of existence and reality.

Non-locality is non-physicality. The non-linear features of Quantum Mechanics such as quantum tunneling, quantum superposition, omniscience, entanglement, action at a distance, and omnipresence cannot be explained by Classical Physics, Materialism, Naturalism, and their derivatives.

Bohm is right. The actualization, creation, sustenance, and annihilation of particles is handled by the "super-quantum potential" or the Supernatural Psyche. Remember, the primary axiom of Quantum Field Theory states that particles are born, and particles die. The "super-quantum potential" creates those particles, and the same "super-quantum potential" is the thing that dissolves those particles back into the quantum fields from whence they came. Bohm is telling us how things really are.]

The general picture that emerges out of this is of a wave that spreads out and converges again and again to show a kind of average particle-like behavior, while the interference and diffraction properties are, of course, still maintained. All this flows out of the super-quantum potential, which depends in principle on the state of the whole universe. But if the "wave function of the universe" falls into a set of independent factors, at least approximately, a corresponding set of relatively autonomous and independent sub-units of field function will emerge.

[The general picture that emerges is that quantum waves spread out through quantum fields, and they show a particle-like behavior because they have Something Psyche within them who is choosing how they should behave. Chosen behaviors imply Psyche or Intelligence. Choice is a function of Psyche at all levels of existence. In order for quantum particles to maintain specific properties and to follow specific rules or laws reliably and dependably requires some kind of Command and Control within the energy itself. All of this flows out of the Super-Quantum Potential, the Intelligence, or the Psyche. The Psyche within energy has the innate ability to form various different types of particles, some of which seem to become relatively autonomous and independent. If laws and rules are being followed and obeyed, then the quantum particles doing the following and obeying obviously have some kind of Psyche or Intelligence within them that gives them the ability to follow and obey rules and laws. Physical atoms are made from some type of universal blueprint, which ends up making each physical atom compatible with every other physical atom. If atom production were a random process as the Materialists, Naturalists, and Atheists claim, then no physical atom would be compatible with any other physical atom because they would no longer be following a universal blueprint or universal law.

Remember, laws or rules of any type require some type of intelligent Law-Giver in order to design them, implement them, and enforce them.

Likewise, laws and rules are worthless, unless the energy has within it some type of intelligent Law-Follower who is capable of understanding, following, and obeying those rules and laws. Laws are intelligently designed and created and intelligently enforced and obeyed; otherwise, they won't work. Energy is infinitely malleable or infinitely transformable, which means that Something Psyche within the energy has to be there to determine which form the energy will take and which blueprint or set of laws the energy will follow and obey. This is obvious to anyone who is willing to look, see, and think.

Bohm is dancing around these concepts; but, as we approach 2020, now that Materialism and Naturalism have been falsified by Science itself, there's no reason to dance around it. We can simply tell it as it is. It's obvious that there is Something Intelligent or Something Psyche within the energy that is giving that energy the ability to choose what form it will assume and what form it will continue to maintain.]

And, in fact, as in the case in the particle theory, the wave function will under normal conditions tend to factorize at the large-scale level in an entirely objective way that is not basically dependent on our knowledge or on our observations and measurements. So now we see quite generally that the whole universe not only determines and organizes its sub-wholes, but also that it gives form to what has until now been called the elementary particles out of which everything is supposed to be constituted. What we have here is a kind of universal process of constant creation and annihilation, determined through the super-quantum potential so as to give rise to a world of form and structure in which all manifest features are only relatively constant, recurrent and stable aspects of this whole.

[What we have here is a universal blueprint or universal law that is used by different psyches to constantly create and annihilate quantum particles. The form and structure of these quantum particles, including physical particles, is DETERMINED by the super-quantum potential or the Supernatural Psyches who are controlling the energy from which the quantum particles are made. This is Psyche Determinism, wherein Psyche decides what form the energy under its control will assume. In this paragraph, Bohm also states that "the universe" determines what particles will be made and when they will be made. This is the same thing as saying that God's Psyche determines what particles will be made and when they will be made. What we have here is an explanation of how things really work at the quantum level and the psyche level. It's not controlled by physical matter. It's obviously controlled by Psyche or Intelligence, or by what Bohm calls the "super-quantum potential" or the "universe".

Remember, Psyche or Intelligence IS potential. Psyche is Bohm's Super-Quantum Potential. Psyche is also Quantum Potential. Bohm's Super-Implicate Order is in fact the psyche level of existence. Psyche is potential because Psyche is Life Force and Psyche is intelligent. It's obvious that Life

Force exists. It's obvious that intelligence exists. It's obvious, therefore, that Psyche exists. Our job as scientists is to try to explain what it is and how it works.]

To see how this is connected with the implicate order, we have only to note that the original holographic model was one in which the whole was constantly enfolded into and unfolded from each region of an electromagnetic field, through dynamical movement and development of the field according to the laws of classical field theory. But now, this whole field is no longer a self-contained totality; it depends crucially on the super-quantum potential. As we have seen, however, this in turn depends on the "wave function of the universe" in a way that is a generalization of how the quantum potential for particles depends on the wave function of a system of particles.

[God actually told us that the "wave function of the universe" actually exists. God called it the Light of Christ. God would know because God's Psyche made the thing. It's a Matrix of Quantum Fields. The whole is contained within each individual Psyche. That's how holograms work. Bohm actually states that Quantum Fields are NOT a self-contained or self-made totality. Quantum Fields actually depend upon the super-quantum potential (Supernatural Psyche) for their origin and continued existence. Psyche is potential, whether we are talking about quantum potential, or super-quantum potential.

You can look up the Light of Christ for yourself on the internet:

https://www.google.com/search?q=%22Light+of+Christ%22

It's there to be found for those who are willing to look and see. Christ himself calls it the Light of Life. It's Psyche or Life Force. It's also what made the Quantum Fields. God is in the light.]

But all such wave functions are forms of the implicate order (whether they refer to particles or to fields). Thus, the super-quantum potential expresses the activity of a new kind of implicate order. This implicate order is immensely more subtle than that of the original field, as well as more inclusive, in the sense that not only is the actual activity of the whole field enfolded in it, but also all its potentialities, along with the principles determining which of these shall become actual.

[There it is. There's how things really work. Quantum fields are the implicate order, and psyches use those quantum fields to make particles. The super-quantum potential is the Activating Force, Life Force, Psyche, or Intelligence who produces the activity necessary to form quantum fields, quantum particles, and physical matter. The explicate order is a substrate or sub-domain of the implicate order or emerges from the implicate order; and, the implicate order is a small part of the super-implicate order. Each new level introduces new laws and new restrictions into the system. The explicate order or physical matter comprises or uses only 4% of the energy within our physical universe. The other 96% remains as energy that we call dark matter (spirit matter) and dark energy. The Materialists and Naturalists are wrong about 96% of our physical universe. Most of our physical universe is in fact non-physical or immaterial in nature and origin.

The verified and proven existence of energy FALSIFIES Materialism, Naturalism, Darwinism, and their derivatives. Remember, Psyches are determining principles, or potentialities, or life force, or intelligence, or super-quantum potential. Bohm is right. There has to be an Organizing Force or an Organizing Principle in order for there to be organization. Psyche is the only thing we know of that can overcome and override chaos and entropy at will. Whether he realized it or not, Bohm was trying to point scientists to Psyche Determinism or Psyche Causality, wherein Psyche or Intelligence is the Ultimate Cause of everything that was made.]

I was in this way led to call the original field the first implicate order, while the super-quantum potential was called the second implicate order (or the super-implicate order). In principle, of course, there could be a third, fourth, fifth implicate order, going on to infinity, and these would correspond to extensions of the laws of physics going beyond those of the current quantum theory, in a fundamental way. But for the present I want to consider only the second implicate order, and to emphasize that this stands in relationship to the first as a source of formative, organizing, and creative activity.

[The implicate order consists of a Matrix of Quantum Fields that was made by Someone Psyche who exists at the supernatural super-implicate order. The original quantum field formed the basis for the quantum order or the implicate order. Remember, Quantum Potential is stored within Something Psyche, and then accessed or used by that Something Psyche. Potential, possibilities, and choice are a function of Psyche. Psyche is the thing that designed and made the quantum fields. Psyche or Intelligence is Bohm's super-implicate order, or second implicate order, or supernatural implicate order.

The super-implicate order or Supernatural Psyche is in fact the ultimate source of formative, organizing, and creative activity. Only Psyche or Intelligence can design and create. At times, I have hypothesized that Psyche is an infinite singularity, or the ultimate point-particle. I could in fact be wrong. It's just as likely and possible that Psyche is a Quantum Computer and actually has size and volume. In fact, there's evidence that a Psyche has CPUs (processing power) and RAM (memory storage) within it, as well as some type of system that transmits and receives quantum waves or thoughts. If so, then these CPUs, RAM, transceivers, and thoughts would in fact represent the third, fourth, and fifth implicate orders that Bohm was talking about. Psyche becomes the second implicate order or the super-implicate order; and, Psyche ends up being the thing that designed and made the implicate order or the Quantum Fields. Of course, the Gods or God's Psyche had to design and make the quantum fields BEFORE physical matter could be made and sustained.]

[My best friend has had a few out-of-body experiences. He's familiar with my Psyche Ontology, wherein Psyche is the ultimate point-particle. While in the spirit, his Psyche or Intelligence has looked out from the third-eye position at the hands, arms, and chest of his spirit body. He has told me that my Ultimate Model of Reality or Psyche Ontology is correct, as far as it goes. But he told me that when I get to the other side, I will discover that it goes much further or much deeper than Psyche, or that point-particle of light that we call Life Force, Consciousness, or Intelligence.

https://psyche-ontology.com/

https://ultimate-model-of-reality.com/

Bohm's implicate order is there, and it is comprised of a Matrix of Quantum Fields. Quantum fields can support or sustain spirit matter (dark matter) just as easily as they can support or sustain physical matter. It's the same basic stuff, just a different phase of existence. Two physical bodies can occupy the same space at the same time as long as they are out-of-phase with each other.

My friend told me that it is matter or organization all the way down. Technically, as the ultimate point-particle, Psyche, Life Force, Intelligence, or that Pin-Prick of Light is also an organized form of energy, and therefore an organized form of matter. Matter of any kind is organized energy. Psyche or Life Force is a type of matter, or a type of organized energy. Psyche resides within Bohm's super-implicate order, second implicate order, or what I call the psyche level of existence and reality. Bohm also calls the super-implicate order an "organizing principle" or an "organizing power". That's Psyche or Life Force. That's what it does. It chooses.

Are Virtual Particles A New Layer of Reality?

https://www.youtube.com/watch?v=ztFovwCaOik

Matt's answer is, "NO. Virtual particles do not exist. They are simply a mathematical convenience." That's the typical materialistic, naturalistic, and atheistic answer. He's wrong. The Materialists, Naturalists, and Atheists are always wrong. Once you understand and accept the existence of Psyche and Bohm's super-implicate order, then you understand why Matt is wrong.

You see, whenever particles of any kind interact at the quantum level, there is a psyche-to-psyche communication that takes place between the particles, while they decide what to do with their interaction. Psyche, the ultimate point-particle at the super-implicate level produces quantum waves (thoughts), transmits quantum waves, receives quantum waves, processes quantum waves, and stores quantum waves. These quantum waves are the virtual particles that show up in the math. They really do exist. Thoughts or quantum waves are real.

Within that pin-point of light that we call Psyche or Intelligence, there's plenty of room for dozens of quantum computers and exabytes of RAM. These virtual quantum computers and virtual RAM made of light would end up being Bohm's third implicate order. They, too, would be comprised of matter, or organized energy. What do these third-level processors do? They process quantum waves or thoughts or ideas. These quantum waves, or thoughts, or ideas would then be Bohm's fourth implicate order. Behind all of that or below all of that sustaining all of that would most likely be some type of Super-Psyche, or Super-Intelligence, or Super Life Force who is running and using the quantum processors and producing the thoughts or quantum waves. This would be Bohm's fifth implicate order. Bohm suggests that this nesting process could go all the way to infinity for all we know.

The existence of virtual particles or the mathematical necessity of virtual particles suggests that this nesting of implicate orders does NOT stop at the second level that we call the psyche level. Remember, matter is organized energy. In this part of the multiverse, it has been organized all the down, and all the way up. It's matter, or organized energy. Even the quantum fields would technically be matter, or organized energy. This organization of energy shows up as particles within the science and the math, when it shows up. But, there are levels of organization or "organizing principles" that run much deeper than the point-particle of light that we call Psyche, Intelligence, or Life Force. Out-of-Body Travelers and Near-Death Experiencers have SEEN the pinpricks of light or the psyches – they are different colors and contain different amounts of energy depending upon how advanced they are. But, the fact that they can be SEEN at the quantum level or the implicate level suggests that there are much deeper levels functioning that are permitting these psyches or pin-points of light to be seen, felt, and communicated with.

This is where science gets interesting – when we realize that it is matter or organized energy all the way down as far as it goes. There's something within that pin-point of light that gives it the ability to produce quantum waves, transmit quantum waves, receive quantum waves, process quantum waves, and store quantum waves. There's something within the Psyche or Life Force that gives it the ability to produce quantum waves or thoughts; and, these quantum waves or organized energy packets are going to be smaller than and more refined than the pinpoints of light that we call Psyche. Remember, psyche has been experienced and observed; however, these quantum waves or thoughts have only been experienced. We don't know if they have actually ever been observed or seen with spiritual eyes.

With Bohm, we have an infinite number of levels of organization. Remember, in this part of the multiverse, it is matter or organized energy all the way down. Whenever you organize energy, it becomes particles of energy or matter, no matter how far down the rabbit hole you chose to go.

We always organize energy into particles or matter. At the psyche construction level, Bohm's third implicate order, it could in fact be quantum computers and RAM within that pinpoint of light that we call Psyche. At the quantum wave level, or Bohm's fourth implicate order, it could be qubits – distinct bits and bytes of information – distinct particles that form quantum waves or thoughts. At the Super-Psyche Level, or Bohm's fifth implicate order, there would be again organized energy or particles of energy, and therefore an even different type of matter. This organized matter or organized energy could go all the way down to infinity for all we know.

Doctrine and Covenants 131: 7-8:

7 There is no such thing as immaterial matter. All spirit is matter, but it is more fine or pure, and can only be discerned by purer eyes;

8 We cannot see it; but when our bodies are purified we shall see that it is all matter.

My friend is right. There is more being experienced there in the spirit world or the implicate order than just the pin-points of light that we call Psyche, Intelligence, or Life Force. It goes much deeper than that. It's organized energy, or matter, or particles all the way down. The virtual particles that show up in the math are just the beginning of it. It goes way further than that.]

It should be clear that this notion now incorporates both of my earlier perceptions – the implicate order as a movement of outgoing and incoming waves, and of the causal interpretation of the quantum theory.

[Bohm is talking about Psyche Causality or Psyche Determinism within this essay.

It should be clear now that Bohm tried to take everything into consideration – the explicate order or manifest order (physical matter), the implicate order (quantum fields and quantum particles), the super-implicate order (psyche or intelligence), and the third, fourth, and fifth implicate orders (the quantum computers and RAM within the pinprick of light that we call a Psyche or a Soul). Psyche becomes the Ultimate Cause and the ultimate causal agent when it comes to Bohm's causal interpretation of quantum theory. It's all made out of energy or light by Someone Psyche or Someone Intelligent. Quantum Fields were obviously made, each to fulfill a specific purpose. Anything that is obviously made obviously has a Maker who made it. What would be capable of designing and making quantum fields at the quantum level? It certainly wouldn't be physical matter, because physical matter didn't exist yet. The Gods had to make the quantum fields BEFORE they could make the physical matter. This is logic 101.

Bohm tried to figure out and explain how everything works at every level of existence; and, he did an excellent job in my opinion. In contrast, we Materialists, Naturalists, Nihilists, and Atheists are idiots because we deliberately stop at the physical level of existence (the explicate order) and refuse to go any further. I know how it goes. I've been there and done that because I used to be a materialist, naturalist, nihilist, and atheist until I learned better. Materialism or Naturalism is the grand illusion, and these people don't even know it.]

So, although these two ideas seemed initially very different, they proved to be two aspects of one more comprehensive notion. This can be described as an overall implicate order, which may extend to an infinite number of levels and which objectively and self-actively differentiates and organizes itself into independent sub-wholes, while determining how these are interrelated to make up the whole.

Moreover, the principles of organization of such an implicate order can even define a unique explicate order, as a particular and distinguished sub-order, in which all the elements are relatively independent and externally related. To put it differently, the explicate order itself may be obtainable from the implicate order as a special and determinate sub-order that is contained within it.

All that has been discussed here opens up the possibility of considering the cosmos as an unbroken whole through an overall implicate order. Of course, this possibility has been studied thus far in only a preliminary way, and a great deal more work is required to clarify and extend the notions that have been discussed in this paper. (pp. 193-197.)

[That's what I'm trying to do here – clarify and extend the Science.

Although quantum mechanics and classical physics seem to be very different notions, they are in fact two aspects of a more comprehensive notion – namely, the organization of energy into quantum fields and physical particles under the command and control of something psyche or something intelligent.

The implicate order or quantum realm or spirit world may extend to an infinite number of levels. Within that pinprick of light that we call Psyche or Soul, there may in fact be rows of quantum computers made from light and exabytes of RAM made from light. There may even be a mini-psyche in there somewhere who is running those quantum computers and using those exabytes of RAM in order to get things done.

Psyche or Intelligence objectively and self-actively differentiates and organizes itself into independent sub-wholes, while determining how these are interrelated to make up the whole. Bohm's super-quantum potential is in fact Psyche Determinism, wherein the Psyche determines what form the energy under its control will take. Elsewhere, Bohm calls the super-implicate order and super-quantum potential a "generative order". In this essay,

Bohm calls it an "organizing principle". Remember, Psyche is potential, no matter what level of existence you might be talking about.

These principles of organization, which form the explicate order or physical level from the implicate order or the spiritual energy level, are indeed Psyche or Intelligence. Only Psyche can do design, creation, order, organization, and choice. Bohm is right, the explicate order (physical matter) is made from the energy that exists at the implicate order (quantum level or spiritual level).

Obviously, I and others have done a great deal more study, observation, and work to clarify and extend the notions that have been discussed in Bohm's paper.]

The only logical way to consider the cosmos as some kind of unbroken whole is to conclude that the immensity of space is filled with the Light of Christ or a Matrix of Quantum Fields. Physical matter obviously is NOT filling the whole of space in some kind of unbroken whole. Physical matter is the emergent explicate order, and it is made by Someone Psyche from immaterial and non-physical energy and quantum fields that exist at the quantum level within Bohm's implicate order.

You are going to have to decide for yourself if you believe any of this to be true, or not. The only thing I can tell you for sure is that it has infinitely more explanatory power than My Materialism, My Naturalism, My Nihilism, and My Atheism ever had.

Mark My Words

Signs of Psyche within Origins

Ever since I first discovered that Psyche has actually been experienced and observed, towards the end of 2015, I have been looking for signs of psyche within everything that I read and watch.

These are situations where psyche, mind, intelligence, intention, perception, consciousness, meaning, purpose, and choice are the best explanation for the Science that is currently being studied and observed. Even the Materialists, Naturalists, Mechanists, and Atheists will use the words "consciousness" or "mind" while interpreting and describing their scientific discoveries. These are situations where psyche or mind is obviously the best explanation for what these people are trying to explain. "Signs of Psyche" demonstrate that psyche or intelligence is necessary to explain what's going on and how things really work.

https://psyche-ontology.com/

https://psyche-ontology.com/buhlman/

https://psyche-ontology.com/psyche-observed/

https://psyche-ontology.com/psyche-experienced-and-observed/

Within *The Essential David Bohm*, there are dozens if not hundreds of times where Bohm exhibits "signs of psyche" even though he doesn't actually use the word "psyche" while doing so.

Any time the scientists mention energy or light, they are in fact talking about psyche, because every type of energy in existence has some type of psyche or intelligence within it who is capable of studying and understanding information and then using that information to change the form or the shape of the energy that's under its control. Energy is infinitely malleable, and the psyche controlling that energy determines what form or shape that energy will take. Someone Psyche or Someone Intelligent is making this choice. It doesn't happen spontaneously or magically out of thin air ex nihilo as the Materialists, Naturalists, Darwinists, Nihilists, Mechanics, Behaviorists, and Atheists claim.

Anything that changes its form or transforms has Someone Psyche driving it and choosing what new form that that energy will take. There are obvious Laws and Rules behind transformation and transmutation. It's observable, predictable, and at times even controllable. Someone Psyche is always involved whenever it comes time to choose what new form a collection of energy will take. Someone Psyche or Someone Intelligent is always needed if a collection of energy is going to successfully understand, follow, and obey specific Laws and Rules that have been established by God's Psyche. When it comes to the formation of energy or light into objects and things, it always comes down to some sort of psyche who is choosing the form that that energy will take. This is obviously true to anyone who is willing to look at it with an open mind.

While the qualitative infinity of nature is consistent with an infinity of levels, it does not necessarily imply such an infinity.

If such relatively and approximately autonomous things [such as psyche] **did not exist, then laws would lose their essential significance**

It is clear that the results of scientific research to date strongly support the notion that nature is inexhaustible in the qualities and properties that it can have or develop. If the laws of nature are to be expressible in any kind of terms at all, however, it is necessary that the things into which it can be analyzed shall have at least some degree of approximate and relative autonomy in their modes of being, which is maintained over some range of variation of the conditions in which they exist. – David Bohm (*The Essential David Bohm*, p. 20.)

It is clear from Science itself that Nature's Psyche is inexhaustible in the qualities and properties and forms that it can have and develop from the energy under its control. Whenever Bohm or any other scientist starts talking about "autonomy" and "choice", they are in fact talking about Psyche.

Chance violates causality; and, causality of any kind is ultimate the result of Someone Psyche. Psyche can override chance and start a causal chain instead. Whenever you decide to lift your finger off the table, Someone Psyche literally has to fire or trigger the specific neuron that lifts that specific finger off the table. The Human Psyche isn't consciously aware of the quantum mechanical processes; therefore, it has to be Nature's Psyche who is in fact collapsing the wave functions and firing the specific neuron that lifts your finger off the table. There are NO physical wires within your physical brain; therefore, your finger-lifting neuron seems to fire out-of-the-blue as if someone is telling it telepathically or wirelessly to fire. That's precisely what's happening. Telepathy is in fact Wi-Fi at the quantum level; and, Psyche is the thing that produces, transmits, and receives quantum waves, thoughts, or telepathy at the psyche level and the quantum level.

David Bohm does a lot of talk about different levels of existence and reality. All of these different levels or dimensions of existence are made out of energy by Someone Psyche. These different energy levels and energy fields didn't just magically spring into existence from nothing as the Materialists, Naturalists, Darwinists, Nihilists, and Atheists claim. Quantum fields were deliberately made to exist by Someone Psyche at the quantum level.

Physical Laws are pointless and worthless without Someone Autonomous or Someone Psyche who is capable of understanding them and capable of choosing to obey them. When it comes to causality and origins of any kind, Psyche or Choice is ultimately the best explanation. There's actually intelligence there that's capable of understanding and choosing to obey the Laws that have been established. That's something that you will never find in the "mindless physical matter" that the Behaviorists and Naturalists choose to preach and worship.

All throughout *The Essential David Bohm*, within practically every chapter, Bohm talks about Psyche or Intelligence without actually mentioning the word "psyche".

On the other hand, in terms of the notion of the qualitative infinity of nature, one is led, as we have seen in previous sections, to the conclusion that every entity, however fundamental it may seem, is dependent for its existence on the maintenance of appropriate conditions in its infinite background and substructure. The conditions in the background and substructure, however, must themselves evidently be affected by their mutual interconnections with the entities under consideration. Indeed, as we have shown in many examples, this interconnection can, under appropriate conditions, grow so strong that it brings about qualitative changes in the modes of being of every kind of entity known thus far. This type of interconnection we shall denote by the name of *reciprocal relationship*, to distinguish it from mere interaction.

The question now follows quite naturally, "If everything is in this very fundamental kind of reciprocal relationship with everything else, a relationship in which even the basic qualities and modes of being can be transformed, then how can we disentangle these relationships in such a way as to obtain an intelligible treatment of the laws governing the universe, or any part of it?" The answer is that all effects of reciprocal connections are not in general of equal importance. (*The Essential David Bohm*, p. 24.)

Relationships are formed by Psyches. This infinite background and substructure of quantum fields is pre-physical, which means that it was designed and made by Someone Psyche or Someone Intelligent before physical matter was designed and made. Mutual interconnections within the background or quantum fields takes place telepathically, psyche-to-psyche, through quantum waves or thoughts. These psychic interconnections can become so strong, that physical matter is produced as a result and physical matter can be transmuted or transformed into something else as a result. Psyche controls these interconnections and transformations. Psyche controls energy or quanta.

Energy of any kind can be transformed into anything that its controlling Psyche wants it to be. The reciprocal connections between the different types of energy or the different forms of energy are controlled by different Psyches. Some of these reciprocal connections are dominate or controlling, and others are simply responsive or reactive.

The thing which has the major effect on the other is the dominant and controlling factor in the relationship.

A fundamental problem in scientific research is then to find what are the things that in a given context, and in a given set of conditions, are able to influence other things without themselves being significantly changed in their basic qualities, properties, and

laws. These are, then, the things that are, within the domain under consideration, autonomous.

When we have found such things, then we can make use of them for the prediction and control of the other things whose modes of being and basic characteristics are dependent on them. (*The Essential David Bohm*, pp. 24-25.)

This is 1957 when David Bohm is writing this, and he is talking about Psyche and laying the foundation for the introduction of Psyche back into science and psychology as a part of Quantum Mechanics.

The Psyche is the thing that has the major effect on energy and is therefore the dominant and controlling factor in the relationship between psyche and energy. Psyche is the thing that decides what form the energy under its control will take. Psyche or Intelligence is the thing that decides what energy will do. Psyche is the dominant and controlling factor in any physical relationship whether we are talking about quantum physics or classical physics.

Psyche is able to influence energy, quanta, spirit matter, and physical matter without being significantly changed in its basic qualities, properties, capabilities, intelligence, and laws. Psyches are autonomous. Your Psyche is eternal and everlasting. It has always existed and it will always exist. It doesn't change. Your Psyche has remained the same and it will always remain the same – a pin-point of light that is capable of dominating and commanding the energy that's under its control, all the way up to the physical level.

Remember, energy is infinitely malleable by Someone Psyche. When it comes to physics, the smaller dwells within and controls the larger. At the quantum level and the psyche level, Psyche is observed as being a pin-point of light, a point-particle, a spark of light, or an infinite singularity. Psyche is smaller than everything else which is why it is capable of dwelling within and controlling everything else.

Psyche is the thing that does prediction. Psyche also controls the other things whose modes of being and basic characteristics are dependent on them.

David Bohm does this all throughout *The Essential David Bohm* – Bohm lays the foundation for the introduction of Psyche into science and quantum mechanics without actually introducing Psyche into science and quantum mechanics.

THE PROCESS OF BECOMING

Thus far, we have been discussing the properties and qualities of things mainly in so far as they may be abstracted from the processes in which these things are always changing their properties and qualities and becoming other things. We shall now consider in more detail the characteristics of these processes which may be denoted by the general term of "motion." By "motion" we mean to include not only displacements of bodies through space, but also all possible changes and transformations of matter, internal and

external, qualitative and quantitative, etc. Both the existence and the necessity for the process of motion described above have now been demonstrated in innumerable ways in all the sciences. (p. 26.)

Energy is constantly in motion. Energy is infinitely malleable. The Psyche in control of energy can transform that energy into anything that that Psyche wants it to be. That's the true nature of energy – it's capable of constantly changing its form. Both the existence and the necessity of Psyche for the process of motion and transformation has been described and demonstrated in innumerable ways within all the different sciences. Energy under the control of Someone Psyche is constantly changing its form. Psyche can cause the energy under its control to become something other than what it was before. This scientific truth becomes obvious once we start studying Feynman diagrams and Quantum Field Theory. The false is falsified by the truth; and, the truth is repeatedly experienced and observed.

https://en.wikipedia.org/wiki/Feynman_diagram

https://en.wikipedia.org/wiki/Quantum_field_theory

According to Quantum Field Theory, these particles of energy are constantly in motion and constantly interacting with each other at the quantum level. They are following specific rules that were designed for them to follow.

In sum, then, no feature of anything has as yet been found which does not undergo necessary and characteristic motions. In other words, such motions are not inessential disturbances superimposed from outside on an otherwise statically existing kind of matter. Rather, they are inherent and indispensable to what matter is, so that it would in general not even make sense to discuss matter apart from the motions which are necessary to define its mode of existence.

We conclude, then, that opposing and contradictory motions are the rule throughout the universe, and this is an essential aspect of the very mode of things. (pp. 27-28.)

Physical matter consists of energy in motion. Think about it. All of the component parts of a physical atom are made from energy, and that energy is constantly in motion. The atom itself may be stationary, but all of its innards are constantly in motion. Psyche or Intelligence is the thing that determines the form and the motion of energy. An atom is a choreographed dance, which means that each atom requires of necessity some type of Choreographer. The dance isn't going to work right either, without a bunch of intelligent dancers. Each component within a physical atom is one of these intelligent psychic dancers. This reality is obviously true. It's self-evident.

In other words, we have to use our intelligence if we want to figure out how things really work.

There must be opposition in all things. Opposing forces under the control of a bunch of different Psyches are what's needed to produce a physical atom. There are forces holding an atom together; and, there are other counter-forces preventing the electrons and protons from merging together and annihilating each other. All of these different forces are controlled by Someone Psyche. When it comes to a physical atom, the whole thing is intelligent and coordinated, rather than random and chaotic. It's been choreographed! It's obvious for anyone who is willing to look and see.

It takes a while for a person to overcome his or her Materialism and Naturalism. I know, because I have experienced it firsthand. While he was alive, Bohm started to do so where Quantum Mechanics is concerned; but, he continued to fall for the materialistic and atheistic lies within the other sciences that predominated at that point in time. Back in the 1950's and even in the 1980's it was possible in good conscience to be a Darwinist, Materialist, and Naturalist because we simply didn't know better at the time. There are times within *The Essential David Bohm* when Materialism, Naturalism, and Mechanism creep into Bohm's theories and science. Here's yet another example for your consideration.

> **Over periods of time of the order of billions of years, new stars, new planets, new nebulae, new galaxies, new galaxies of galaxies, etc., can come into existence, while the older organization of things passes out of existence. – David Bohm** (1957, p. 27.)

Order and organization of any type is the product of Someone Psyche or Someone Intelligent. Anything that is obviously organized obviously has an Organizer who organized it. Organization and order don't just spring into existence whole from nothing as the Materialists, Naturalists, Nihilists, Mechanists, and Atheists claim. Order and organization are designed, created, and made by Someone Psyche.

Furthermore, it doesn't require billions of years of time for a galaxy to come into existence. In fact, when we look out into the universe, we see that galaxies are born whole and complete all at once. The galaxies come into existence instantaneously as a united whole. They don't coalesce out of hydrogen and helium over billions of years of time, because hydrogen and helium never coalesce in the cold dark of empty space. Hydrogen molecules and helium atoms naturally repel each other; and according to the Big Bang Theory and Dark Energy Theory, all of the hydrogen molecules and helium atoms are supposed to be expanding away from each other rather than coalescing together. It would take an act of God to get expanding clouds of hydrogen and helium gas to coalesce into a planet or a star or a galaxy. When we compress hydrogen or helium into a canister, does it ever coalesce and become a star when we are forcing all of that hydrogen and helium into the same place? No! If it can't and won't coalesce within a pressurized canister, then it certainly won't do it in the vacuum of expanding space.

https://en.wikipedia.org/wiki/Solid_hydrogen

It would require tons of external pressure from God in order to form hydrogen and helium into a planet or a star. Furthermore, the stuff is going to turn into a solid block of ice long before it ignites and becomes a star. You will never get a solid block of ice or a solid block of hydrogen out of expanding clouds of hydrogen. It can't be done. It's physically impossible. Hydrogen molecules in a hydrogen balloon tend to repel each other rather than coalesce. Force hydrogen molecules into a tank, and they repel each other rather than coalescing. Think about it! The only logical conclusion, given the physics of the situation, is that planets and stars and galaxies are made by Someone Psyche.

https://en.wikipedia.org/wiki/Helium#Liquid_helium

Furthermore, even under tons of external pressure from God, helium is never going to coalesce into a solid. The best that can be done with helium is to turn it into a liquid. We can't even begin to imagine the amounts of external pressure and external coordination and external energy that would be required to force expanding clouds of hydrogen and helium to coalesce into something the size of a star. Think about it! The only logical conclusion, given the physics of the situation, is that planets and stars and galaxies are made by Someone Psyche.

Planets, stars, and galaxies are born. They are quantum tunneled into existence or phased into existence whole from the quantum realm or the spirit realm by Someone Psyche. This is what it would take to make a star out of an expanding cloud of hydrogen gas. It would require some kind of intervention from God to get the whole process started in the first place. God is Psyche. God is Life Force. Psyche is the ultimate cause. God is in the light. God also has a spirit body and a resurrected physical body. Everything is made from energy or light by Someone Intelligent or Someone Psyche.

Like Bohm, my goal is to take Science to its ultimate conclusion. Materialism and Naturalism are dead. It's time to embrace Psyche and Quantum Mechanics instead.

The cool thing about Bohm is that he lays the foundation for the introduction of Psyche into science, psychology, and quantum mechanics; but to survive and pacify the Naturalists and Mechanists and other Extremists, Bohm doesn't actually introduce Psyche into science, psychology, and quantum mechanics. Bohm allows these people to maintain their illusions, self-deception, and lies if they want to – while at the same providing the rest of us with something that we can use sixty years later. Bohm has been vindicated by Science and History, while the Materialists, Naturalists, and Mechanists have recently gone down in flames. As we approach 2020, the evidence falsifying Materialism, Naturalism, Darwinism, Nihilism, Mechanism, Physical Reductionism, Atomism, and Atheism has achieved critical mass. There's no longer any logical or rational reason to be a Materialist or Naturalist or Darwinist because these philosophies or dogmatic religions have been falsified by Science itself. It's all happened in the past ten years or so; and, it was made possible by people like Bohm who started to pursue the truth back in the 1950's over seventy years ago.

In the different books that I have written, I try to explain why I am no longer a materialist, naturalist, nihilist, and atheist. I have logical reasons and scientific reasons why I abandoned my former philosophical beliefs. Materialism, Mechanism,

Darwinism, and Naturalism were not able to give me the kinds of information, understanding, knowledge, science, and truth that I was seeking. I wanted to find out how things really work rather than be told that they don't exist. Being told that Psyche does not exist does not explain what it is or how it works.

https://www.amazon.com/author/science

Ultimately, I didn't find the "creation ex nihilo" or the "magic" of materialistic and atheistic philosophies satisfying in the end. Non-causality, atheism, or the creation of something from nothing by nothing just didn't make sense to me. Electrons DO NOT POP IN AND OUT OF EXISTENCE as the Atheists, Mechanists, and Materialists claim. Instead, electrons are half-in and half-out of the quantum world. They timeshare between the physical level and the quantum level. The Atheists preach the existence of "non-existence", and I just don't buy it. The energy that went into making an electron has always existed, and it will always exist. It doesn't just magically cease to exist when it's not in physical mode, as the Materialists, Naturalists, and Atheists claim. Whenever the electron is non-physical, it's very much there, and it's all there – just at the quantum level rather than the physical level – because energy is always conserved. The conservation of energy says that it's always there all the time – not popping in and out of existence willy-nilly as the Mechanists, Naturalists, and Atheists claim. These people don't understand Psyche, Conservation of Energy, Quantum Mechanics, and Quantum Field Theory. If they did, they would no longer be Materialists, Naturalists, Darwinists, Nihilists, Mechanists, Behaviorists, Determinists, Physical Reductionists, Atomists, or Atheists. The truth always falsifies the lies; and, the truth is repeatedly experienced and observed.

In the opening chapter of The Essential David Bohm, Bohm talks about an obligatory topic for the subject that we call physics – the origin or beginning of our physical universe. After mentioning the Big Bang Theory, Bohm states: **"Now, it is very important to emphasize how speculative and provisional large parts of this theory are."** Ironically, sixty years later, large parts of the Big Bang Theory are still speculative, provisional, and even falsified. The only thing we can be sure of is that if the Big Bang ever happened, then then the Big Bang Singularity was composed of nothing but pure raw condensed energy. The other thing that I'm sure of is that it would take Someone Psyche or Someone Intelligent to transform all of that raw chaos into the order and organization that we see around us today. It's not going to happen automatically ex nihilo as the Materialists, Naturalists, Darwinists, Nihilists, and Atheists claim.

> **In any case, whatever may have been the at present practically unknown earlier phase of the process of evolution of this particular part of the universe, there exists by now a considerable amount of evidence suggesting that the galaxies, the stars, and the earth come from some quite different previously existing state of things.** (The Essential David Bohm, p. 31.)

If the Big Bang happened, then it was ALL chaotic unorganized energy or material in the beginning. Then everything was standardized and organized.

Anything that is obviously standardize obviously has a Standardizer who standardized it. It's obvious that planets, stars, and galaxies have been made from raw unorganized energy and matter. Anything that is obviously made obviously has a Maker who made it. It's obvious. It seems to be obvious to David Bohm as well.

In conclusion, the notion of the qualitative infinity of nature leads us to regard the eternal but ever-changing process of motion and development described above as an inherent and essential aspect of what matter is. In this process there is no limit to the new kinds of things that can come into being, and no limit to the number of kinds of transformations, both qualitative and quantitative, that can occur. This process, in which exist infinitely varied types of natural laws, is just the process of becoming. (p. 32.)

Energy is infinitely malleable. Under the control of Psyche, there's no limit to the number and types of things that can come into being from energy. Energy is eternal but ever-changing. It is obvious that physical atoms are comprised of different forms of energy. It's obvious that atoms were designed and made. Anything that is obviously designed and made obviously has Someone Psyche or Someone Intelligent who designed it and made it. It doesn't just spring into existence whole from nothing as the Materialist, Naturalists, Darwinists, Mechanists, and Atheists claim. These people don't know what they are talking about whenever they make such a claim. The Atheists are all sitting around waiting for nothing to do something, and it ain't never gonna happen.

Indeed, the very fact that a thing is able to undergo a qualitative change is itself a property that is an essential part of the mode of being of the thing, and yet a property that is not contained in the original concept of it. (p. 33.)

When it comes to something like a physical atom, it's obvious that it required Someone Psyche or Someone Intelligent to produce the original concept of it. Not only did the original concept of it have to be originated; but, then that concept had to be reproduced or replicated over and over again trillions of trillions of times. If the Psyche that makes the atom doesn't get it right, then it isn't going to work. It's not going to be able to interface with the other atoms that were made right. Think about it. It's obvious. Somewhere there is a blueprint at the psyche level for the proper construction of a physical atom. There's a way to do it right, and there is an infinity of different ways for doing it wrong.

But if all things eventually undergo qualitative transformations, then the process described above will never end. Thus, we conclude that the notion that all things can become other kinds of things implies that a complete and eternally applicable definition of any given thing is not possible in terms of any finite number of qualities and properties.

If, however, we now start from the opposite side, viz., from the notion of the qualitative infinity of nature, we are then immediately

able to arrive at a type of definition of the mode of being of any given kind of thing that does not contradict the possibility of its becoming something else. (p. 33.)

When it comes to energy, it requires Someone Psyche to do the qualitative transformations that are needed to transform energy from chaos into something organized like a physical atom. The raw chaotic energy has to learn and know how to transform into a physical atom, or the process is never going to happen. Remember, Psyche can transform the energy under its control into anything that it wants that energy to be, and then that same Psyche can set that newly created particle into motion in telepathic coordination with other types of particles and energy. An atom is a choreographed dance between different intelligently produced dancers or particles of energy. Energy is the stuff from which everything is made, and Psyche is the thing that makes it. Physical matter is just one of the things that can be made.

Transformation from one form of energy into a different form of energy is the KEY to physics as a whole. It's at the very heart of Quantum Field Theory and Feynman diagrams. Even back in 1957, David Bohm was aware of this reality and truth.

> **We may illustrate the above conclusions by returning to a more detailed discussion of the transformations between steam, liquid water, and ice. Thus, the macroscopic concept of a certain state of matter (e.g., gaseous, liquid, or solid) leaves out of account an enormous number of kinds of factors that are not and cannot be defined in the macroscopic** [mechanistic] **domain alone. Among these are the motions of the molecules constituting the fluid quantum fluctuations, field fluctuations, nuclear motions, mesonic motions, motions in a possible sub-quantum mechanical level, and so on. In short, we may say that the real fluid is enormously richer in qualities and properties than is our macroscopic** [classical physics] **concept of it. It is richer, however, in just such a way that these additional characteristics may, in a wide variety of applications, be ignored in the macroscopic** [physical] **domain. Nevertheless, when we come to the problem of understanding why transformations between gas, liquid, and solid are possible, we can no longer completely ignore the additional properties of the real fluid.** (*The Essential David Bohm*, p. 34.)

We can't ignore Psyche and Quantum Mechanics if we really want to know how things work at the quantum level and the psyche level. Obviously, we can and do ignore Psyche and Quantum Mechanics at the macroscopic mechanistic physical level because the whole machine is purring along just fine at the physical level thanks to all the quantum fine-tuning that Someone Psyche has done at the quantum level.

> **According to the Pauli Exclusion Principle, any two electrons are said to be "identical." This conclusion follows from the fact that within the framework of the current quantum theory there can be no**

property by which they could be distinguished. On the other hand, the conclusion that they are completely identical in all respects follows only if we accept the assumption of the usual interpretation of the quantum theory that the present general form of the theory will persist in every domain that will ever be investigated. If we do not make this assumption, then it is evidently always possible to suppose that distinctions between electrons can arise at deeper levels. (p. 38.)

At the formal level, electrons are all the same because they all have the same blueprint. However, if we go deeper towards the level of ultimate cause, we quickly discover that each electron has a different psyche or personality at the psyche level or super-implicate level of existence. Go deep! Quantum Mechanics and Quantum Field Theory simply don't make sense without including Someone Psyche to design, make, and implement the quantum fields and the quantum particles in the first place. It's illogical and irrational to assume that all of this order and organization sprang into existence from nothing as the Materialists, Naturalists, Darwinists, Mechanists, and Atheists assume. Assumptions are not science. Science is based upon evidence; and, there's now plenty of evidence that falsifies the claims of the Materialists, Naturalists, and Atheists. When properly understood, Quantum Field Theory falsifies Materialism, Naturalism, and their derivatives such as Atheism and Darwinism.

In both theory and experiment this mechanistic order [physical order] **has been undermined by a series of developments in twentieth-century physics. Bohm points to four features of the "new" physics that, taken together, require the formulation of a larger, more encompassing order. – Lee Nicol** (p. 3).

Bohm is laying the foundation for a Science of Psyche, but he doesn't introduce a Science of Psyche. He leaves it up to someone like me to come along sixty years later and complete the process.

Psyche is in the energy or the light. In other words, God is in the light. You may in fact be one of the Psyches that helped to design, create, and make this physical universe that we see around us. Our spirit bodies and physical bodies are descended from the Gods. We are Gods. Our surname is God. Adam was a son of God – both his spirit body and his physical body. The Human Psyche has been elevated to the level of a God. It has been observed that whenever a Human Psyche enters into a non-consensus reality within the quantum realm or spirit world, the controlling psyches within that reality reorganize the energy in that reality according to the demands and expectations of the Human Psyche. Within a non-consensus reality, the Human Psyche is God. That's why you are able to understand the words that I write here, and that's also why you are able to choose to believe that they are false and worthless. The Human Psyche is capable of creating realities; and, the Human Psyche is also capable of denying reality.

It is clear from the preceding section that the empirical evidence available thus far shows that nothing has yet been

discovered which has a mode of being that remains eternally defined in any given way. Rather, every element, however fundamental it may seem to be, has always been found under suitable conditions to change even in its basic qualities, and to become something else. Moreover, as we have also seen, the notion of the qualitative infinity of nature implies that every kind of thing not only can change in this very fundamental way but that, given enough time, conditions in its infinite background and substructure will alter so much that it must do so. – **David Bohm** (p. 32.)

As we approach 2020, there's abundant empirical evidence available showing that Psyche and Energy have a mode of being that remains eternally defined. Anything, that is eternal and everlasting, is absolutely true and remains true in every case and every situation. Energy, intelligence, psyche, life force, or light is still the same; and it will always be the same. Psyche or energy is always conserved! Psyche is still seen as a pin-point of light or a point-particle, even at the quantum level and the psyche level. Psyche is still experienced first-hand as an immaterial viewpoint in space. Energy is still infinitely malleable. Energy is still the fundamental stuff from which everything is made by Someone Psyche. Energy under the control of Psyche can still take on an infinite number of different forms. Energy is infinitely and eternally transformable. We have now found a couple of things that are absolutely true and are eternally defined in very specific ways.

Mark My Words

Signs of Psyche within Quantum Mechanics

Back in 1962, David Bohm was demonstrating the need for Psyche as one of the levels of Quantum Mechanics.

> **The fact is that there is an individual who is carrying out an ego process. In this process, the whole mind produces signals that excite the reflexes of conflict and confusion.**

> **It is very important then to get below the usual level of thoughts, words, and feelings, coming to a direct awareness of the process-structure of the ego.**

> **Our words function as a sort of "living card index system," such that when the right "button" is pressed, the memory projects pictures, sounds, smells, and feelings in such a profusion that we easily confuse them with real perceptions.**

> **So the important point is to notice that there is a level below words and thoughts (a structure-process level) in which action, thought, and feeling are determined. In fact action is always determined at the structure-process level.**

> **The cortex is always actually a totality, and indeed thalamus and cortex together are always a totality. Now, the whole thought, feeling, and action of the individual responds to this whole awareness.**

> **In the cortex, the idea is produced that the signals that one feels are the operation of one's "very self," something that it would be senseless and meaningless to deny or question. And the reason for this is clear.** (pp. 221-224.)

Though Bohm couldn't see it at the time, it is clear that the source or cause of this wholeness and structure-process level is in fact the Human Psyche.

Bohm is talking about Psychology's Binding Problem. The observation that the cortex, thalamus, brain, and body function as a unity or a totality goes contrary to contemporary neuroscience, and it is only possible thanks to Psyche. The individual is the Psyche. Bohm's "structure-process level" is in fact the psyche level; and, the wholeness or whole awareness that Bohm talks about is produced and maintained by the Human Psyche.

The Binding Problem or Combination Problem is the problem of how sensory objects, background and abstract brain functions, and emotional features are combined into a single experience. That's not physically possible according to classical physics and neuroscience. The only logical solution to the Binding Problem is the observation that all of these sensations, observations, perceptions, and experiences feed into a Single Processor or a Single Psyche.

All of the unsolvable problems in Psychology are caused by Materialism, Naturalism, Darwinism, Nihilism, Behaviorism, Determinism, Atheism, and their derivatives. They are all SOLVED instantly once we introduce Psyche into the system. Once I understood that Psyche has been experienced and observed, then all of my confusion simply disappeared. Psyche has infinitely more explanatory power than the materialistic and atheistic claim that psyche does not exist.

As we approach 2020, the existence and experience of Psyche is a proven reality. The Human Psyche is "one's very self". The Human Psyche is indeed experiencing and remembering physical sensations. It's interesting to observe that when you fall asleep, your Human Psyche disconnects from your physical sensations or stops paying attention to them. You actually have to train yourself to recognize the need to urinate while you are asleep, because the Human Psyche disconnects from the physical body during the sleep process. If you are in a lot of pain, as I once was, there is a moment of peace when you first wake up, and then all of the pain comes flooding back into your awareness when your Human Psyche reacquires its connection with its physical body. If all we were was physical matter, then it would be physically impossible to disconnect our consciousness or psyche from our physical body; and, all physical sensations would be present all the time. We would never be able to sleep.

The memories that survive the death of our physical brain, and show up in our after-death Life Reviews, are definitely stored someplace else besides our physical brain. This is obviously true. Psyche or Intelligence or Life Force has been experienced and observed. See the links below for a sampling.

Bohm repeatedly talks about that unitary wholeness that takes place between the Human Psyche and its physical body. The physical matter and physical sensations makes it real for us. The physical matter also places a veil over the Human Psyche so that it can no longer remember what it used to know. This veil of forgetfulness and innocence is a quantum phenomenon. In this case, it's matter-over-mind.

Quanta are packets of energy. Energy or psyche is always conserved. Quantum Mechanics is the scientific study of energy – namely, all the different things that can be formed from energy by Someone Psyche.

Energy or Light is the stuff from which our physical universe was made. Energy is infinitely malleable, which means that energy can literally be formed into anything by the Psyche or the Intelligence who controls it. Transformation of energy from one form into a different form at the quantum level is one of the surest signs of Psyche, because Psyche is the thing which decides which form the energy under its control will take. That energy can be transformed into anything at the quantum level; and, the controlling Psyche is the personality or the life force who makes that decision or choice. Choice is always a function or a product of Someone Psyche, both at the physical level and at the quantum level. That's the only thing that ever made logical sense.

The idea that choice, deliberation, purpose, and organization magically spring forth from nothing is one of the stupidest ideas that we ever gotten from the

701

Mechanists, Determinists, Physical Reductionists, Naturalists, Materialists, Nihilists, Behaviorists, and Atheists. Creation ex nihilo is Atheism – the creation of something from nothing by nothing – or magic. Creation from nothing by nothing or Atheism is patently absurd and one of the most irrational and illogical ideas that has ever been created by the mind of man.

We don't know what the self is, nobody has ever managed to look at the self. – David Bohm (1987).

Forty years ago, this was probably true. Nobody had ever seen Psyche from what I could tell. During the past decade or two, though, as we approach 2020, dozens of people have looked at Psyche or seen Psyche. It's seen as a pinprick of light or a point-particle at the quantum level.

I've been searching for Signs of Psyche.

https://psyche-ontology.com/

https://psyche-ontology.com/buhlman/

https://psyche-ontology.com/psyche-observed/

https://psyche-ontology.com/psyche-experienced-and-observed/

The direct observation and experience of Psyche, functioning separately from its assigned spirit body and physical body while looking at its spirit body, had a profound influence on my interpretation of Science and Quantum Mechanics. After uncovering dozens of examples of a person's Psyche separating from his or her spirit body and looking at his or her spirit body or looking at other Psyches, I realized and accepted the fact that the phenomenon is real. Then I started looking for Signs of Psyche within Science, Quantum Mechanics, and Scientific Theories because I knew that they should be there to be found since Psyche had been experienced and observed.

Seek, and ye shall find. Knock, and it shall be opened unto you.

I quickly discovered that there are Signs of Psyche – where Psyche is the best and most parsimonious explanation – all throughout Science, especially in the sciences associated with Quantum Mechanics, Quantum Field Theory, Quantum Neuroscience, Syntropy, Energy, Light, and Psychology. These sciences don't make any real sense without including psyche, intelligence, intention, teleology, purpose, perception, attention, consciousness, intelligently designed blueprints, or information somewhere in the system. For example, Quantum Fields are non-physical, non-local, and pre-physical. The Gods or the Controlling Psyches had to design and make the quantum fields first BEFORE they could design and make physical atoms and physical objects. Physical matter isn't possible without these quantum fields to sustain it and transmit it.

There's really no such thing as particles or physical matter. The universe is made of quanta, and each quantum is made from energy by Someone Psyche. Only psyche is capable of forming raw chaotic energy into specific different types of quanta. Physical matter can't do that because physical matter or a physical atom is

702

one of these quanta that that was made from energy by a bunch of different Psyches working together to produce the choreographed dance that we see and experience as a physical atom.

Psyche has explanatory power when it comes to Quantum Mechanics, Quantum Field Theory, Feynman Diagrams, and the transformation of Energy from one thing into another. In contrast, the materialistic and atheistic claim that Psyche DOES NOT EXIST has no explanatory power whatsoever. Stating that Psyche doesn't exist doesn't explain what it is and how it works. I want to know how things really work at the quantum level and the psyche level rather than being told that they don't exist. Obviously, Psyches exist because there has to be something intelligent or something psychic within each type or form of energy who decides what form the energy under its control will take. Energy is infinitely malleable. Psyche can transform the energy under its control into anything that that Psyche wants it to be; and, God's Psyche is the overlord or the mastermind who is choreographing this whole thing at a universal and multi-dimensional scale.

It's obvious that intelligence or information or carefully designed blueprints are necessary to make the whole of this physical universe work. It's obvious that raw chaotic energy needed and needs some type of intelligent consciousness or psyche to form that energy into specific quanta, and then to organize all those different quanta and quantum fields into the object that we call a physical atom. Each time a quantum is made from raw chaotic unorganized energy, a choice has to be made to determine which type of quantum that energy will become. Psyche is the only thing that we are aware of who is capable of choosing at the psyche level and the quantum level what form the energy under its control will take. Even within a physical atom, there are Psyches who are deciding which form that atom will take and how that atom will interface with and work with the rest of this physical universe.

Throughout the rest of *The Essential David Bohm*, I saw many Signs of Psyche within the words, theories, and concepts that were use. Here are a few of them.

> **Whereas the reciprocal relations outlined in Chapter 1 primarily describe the "external" world, the thread of this chapter goes a step further, linking the "internal" process of perception to analogous processes in the external world. In positing this linkage, Bohm is anticipating his later formulation of soma-significance (Chapter 5), in which mind and matter are understood as reciprocal [interactive], evolving aspects of a more fundamental implicate order.** (*The Essential David Bohm*, p. 40.)

Relationships of any kind are a function or a product of Psyches. Perception is a function of Psyche. Anytime a scientist talks about internal processes, perception, linkages, relationships, significance, meaning, purpose, intelligence, fundamentals, and mind, he or she is in fact talking about Psyche. He's definitely NOT talking about the external world or physical matter, because physical matter is generally seen and observed at the physical level as being inert and inanimate. It's only within the atom at the quantum level that the atom is alive with intentional and deliberate motion, information exchange, purpose, intelligence, planning, perception, telepathic communication, interaction, and sentience.

From our external perspective at the physical level, a single atom can seem to be perfectly still, inanimate, inert, and motionless; but, at the quantum level thanks to all the different Psyches within that atom who are controlling the quanta or the components within that atom, that atom is literally alive with motion, purpose, intelligence, stability, form, and sentient consciousness.

The Quantum Zeno Effect has been repeatedly verified; and, the Quantum Zeno Effect proves that atoms can read your mind and respond accordingly. Mind-to-mind or psyche-to-psyche telepathic communication is only possible in terms of Psyche at the psyche level through the quantum level. The Materialists, Naturalists, Nihilists, and Atheists assure us that there is no such thing as telepathy or mind-reading at the physical level. The fact that mind-reading or telepathy has been proven to exist through scientific experimentation does in fact prove that some sort of Psyche or Mind-Reader does exist at the psyche level, because it's pretty much obvious that such a thing can't exist and doesn't exist at the physical level. There's no way to develop a physical device that can read your thoughts and record your thoughts. All the different brain imaging devices that we have created can't record your thoughts. It requires a quantum device, such as a Psyche, to be able to read and understand and record thoughts or quantum waves. This is all logical common sense; and, it's precisely the type of rational common sense that the Materialists, Naturalists, Darwinists, Mechanists, and Atheist formally reject and deny.

For David Bohm, his essay entitled "Physics and Perception" was seen as a significant step in restoring the primacy of perception. Perception is always a function of Psyche. In contrast, sensations are produced by our physical body. The five physical senses provide us with our physical sensations. Perception is a different animal, because it always involves quantum waves or thoughts. Thanks to the Human Psyche, we continue to perceive and continue to think and continue to remember long after our physical brain is dead and gone. Do you see how that works? It's important, if we want to understand how everything really works.

Throughout this chapter, Bohm repeatedly mentions the "inner show". This inner show is perception, and it's handled by the Human Psyche. The Human Psyche perceives or thinks. The Human Psyche remembers. Then the Human Psyche decides what it's going to do with its assigned spirit body and its assigned physical body. Bohm mentions "the ability to abstract". The ability to abstract is a function of Psyche. The Materialists, Mechanists, and Naturalists assure us that physical brains and physical matter cannot do consciousness, intelligence, abstraction, intention, choice, or psyche. Therefore, if they are right in this regard, then it's obvious that all of these "inner show" phenomena are taking place at the psyche level within the Human Psyche, as well as at the quantum level between the Human Psyche and Nature's Psyche.

Anytime we encounter mind-over-matter or the placebo effect, we have in fact demonstrated the necessity and existence of Psyche.

> **The over-all or general structure of our total perceptual process can be regarded not only from the standpoint of its development from infancy but can also be investigated directly in the adult. Such studies have been made by Hebb and his group, by isolating individuals in environments in which there was little or nothing to be perceived. The extreme cases of such isolation involved putting people in tanks of water at a comfortable temperature, with nothing to be seen or heard, and with hands covered in such a way that nothing could be felt. Those individuals who were hardy enough to volunteer for such treatment found that after a while the structure of the perceptual field began to change. Hallucinations and other self-induced perceptions, as well as distortions of awareness of time, became more and more frequent. Finally, when these people emerged from isolation, it was found that they had undergone a considerable degree of general disorientation, not only in their emotions but also in their ability to perceive. For example, they often found themselves unable to see the shapes of objects clearly, or even to see their forms as fixed. They saw changing colors which were not there, etc. (In time, normal perception was, of course, regained.)**
> (*The Essential David Bohm*, p. 60.)

The neurologists and neuroscientists erroneously teach that whenever we start hallucinating, it's the physical brain that's misfiring and causing the phenomenon. However, that technically can't be the case when we are dealing with sensory deprivation, because there are NO physical sensations entering into the physical brain. In such a situation, the Human Psyche takes over and starts to generate its own reality and existence separate from the physical brain. Whenever our physical brain goes offline, as it does while we are sleeping and dreaming or while we are exploring the astral plane, the Human Psyche takes over. Millions of Near-Death Experiencers and Out-of-Body Explorers have observed first-hand that the Human Psyche takes over and takes control whenever their physical brain dies or goes offline.

Now, if we are talking about substance-induced psychosis and withdrawal from substance abuse (a phenomenon which I have personally experienced thanks to my addiction to a variety of different prescription drugs), then the physical brain is indeed receiving stimuli or sensations from the physical environment, but the brain is also misfiring and thrashing thereby generating the false stimuli, false sensations, hallucinations, delusions, and false beliefs that are typically associated with substance-induced psychosis and withdrawal from addictive prescription drugs. The fear, pain, and confusion were through the roof until I stopped caring whether I live or die. Once I decided to commit suicide, I had had enough with all of the pain and confusion and insanity and fear. I wanted to be anywhere but here; and, the people on the television and the radio knew my name and were telling me that I had to kill myself. By that point in time, after two years of this, they were telling me precisely what I wanted to hear. I wanted out, and I took bottles of sleeping pills in an attempt to get out. It didn't work.

They got me addicted to everything all over again in the looney bin; and, when I got out of the psych ward a month later, I realized that I was never going to get well if I stayed on all the drugs that they were giving me. I went cold turkey, and I went completely insane. The withdrawal took six months of hellish confusion, delusions, and hallucinations before I finally started thinking rationally again and finally realized how insane I had been. You don't realize that you are insane and psychotic while you are in the middle of it. It seems real at the time.

My decision to go cold-turkey, and go insane, and go through the withdrawal process was purely a decision of my Human Psyche. My society, my doctors, my psychiatrists, and the behaviorists and mechanists and materialists in control of my psychotherapy wanted me on the addictive drugs. My physical body and physical brain wanted me on the addictive drugs. When it comes to addictive prescription drugs, you will lie, kill, and steal to get your next fix. It was purely my Human Psyche who decided to quit taking the drugs and to suffer the consequences. It was mind-over-matter. That was one of my first solid indications that the mind matters. According to the BioPsychoSocial Model of reality and abnormal behavior, there are three different aspects that influence our behavior. Our biology or addictions can influence our behavior. Our society and psychiatrists and drug dealers can influence our behavior. And, our Human Psyche can influence and change our behavior. I eventually learned and observed that the Human Psyche can trump its nature and nurture at will. I made a choice to disobey my society and my physical body by making a choice to get sober instead. It worked. It took six months, but it worked.

My blind faith in the Materialists, Naturalists, Psychiatrists, Pharmacologists, and Medical Doctors almost got me killed. Substance-Induced Psychosis is an official mental illness; and in order to achieve sanity and sobriety once again, one must stop taking the addictive substances that your Psychiatrist is giving you. Ironically, Psychiatrists often treat your mental illness by giving you additional mental illnesses. The process has proven fatal. These people are Materialists, Naturalists, and Behaviorists; and, they have no idea whatsoever as to what's actually good for the psyche or the soul. They think they can cure spiritual and psychological issues by using a sledgehammer on the physical brain. I no longer trust the Materialists, Naturalists, Behaviorists, Pharmacologists, and Psychiatrists, because they make their money by pushing drugs and doing therapies that were never designed to cure but only designed to make you dependent on them. Behaviorists exempt themselves from their treatment; and, psychiatrists should be forced to take the drugs they prescribe so that they can know first-hand the kind of hell that they are putting people through. Anytime I would complain about the side effects of the psych drugs, then they would give me another one. They had me nested twelve deep, and it took six months for my brain to normalize after I stopped taking all their crap.

Speaking about Science, the Scientific Methods, and Perception, David Bohm wrote:

**There is a continual process of "trial and error" in which what
is shown to be false is continually being set aside, while new
structures are continually being put forth for "criticism."** (p. 63.)

The Materialists, Naturalists, Darwinists, Nihilists, Mechanists, and Atheists
demand and grant themselves an exemption or special pleading where their own
philosophies are concerned. They declare themselves to be exempt from
falsification; and, whenever their theories and beliefs are falsified, then they cry
foul and state that it's not fair to falsify their theories because their ideas are
axiomatically true by definition. The Materialists, Naturalists, Darwinists, and
Atheists seldom put aside anything that is false or anything that is falsified.
Instead, these people are continually making falsehoods and then embracing them.

Speaking indirectly about Science and the Scientific Method, Bohm explains
how it should work:

> **Each abstraction constitutes, as it were, a kind of "hypothesis,"
> put forth to explain what has been found to be invariant in such
> earlier experiences. Only the abstractions which stand up to further
> tests and probings will be retained.**

> **This abstraction, expressed in terms of the notion that a
> circular object accounts for all the changing views of it, is the basis
> of the "construction" of it that we perceive in the "inner show." The
> "hypothesis" that this object is really a circle is then further probed
> and tested in subsequent ways of coming in contact with it
> perceptually, and it is retained as long as it stands up to such
> probing and testing. But the realization that the perceived object is a
> circle depends also on knowledge going beyond the level of
> immediate perception.** (*The Essential David Bohm*, pp. 67 and 68.)

Except for Materialism, Naturalism, Darwinism, Nihilism, and Atheism. The
more they are falsified, the more they are dogmatically retained. The Materialists,
Naturalists, Darwinists, Mechanists, Nihilists, Behaviorists, and Atheists always
make themselves exempt from the rules of Science. It's fascinating how they lie,
cheat, and deceive themselves in order to make their case. This is one of the most
fascinating psychological phenomena to study and observed. Under materialistic
and atheistic philosophies, any mechanistic hypothesis that doesn't stand up to
further tests and probings is always retained and is instead declared to be
axiomatically true – God does not exist, Psyche does not exist, the mind or "inner
show" does not exist, there's no such thing as quantum consciousness, there is no
universal field, physical matter is conserved, physical matter has always existed
and will always exist, the physical atom is the elementary particle from which the
universe was made, and intelligently designed physical matter, galaxies, and
genomes spring into existence from nothing. These falsified ideas are declared to
be axiomatically true, and therefore they can't be falsified by science because they
are declared to be absolutely true by Materialism, Naturalism, and their derivatives.

The Materialists, Naturalists, Darwinists, Nihilists, and Atheists take things
that have been experienced and observed, and then these people declare these

707

things to be absolutely false instead. That's how these people do Science. The false is enshrined and made sacrosanct, while at the same time the experienced and the observed is formally rejected and declared to be false or non-existent. These falsified philosophies violate the rules of Science and the rule of Logic; and therefore, they are philosophically flawed.

In contrast, while doing real science rather than dogmatic philosophy or dogmatic religion, the false is eventually falsified by the truth, and the truth is repeatedly experienced and observed.

The objective content of our perceptions is then implicit in the process of falsification and confirmation described above. Indeed, the very fact that our vision of the world can be falsified as a result of further movement, observation, probing, etc., implies that there is more in the world than what we have perceived and known. That is to say, we do not actually create the world. In fact, we only create an "inner show" of the world in response to our movements and sensations. It is, however, the possibility of confirmation of the "inner show" which demonstrates that there is more in it than merely a summary of past experiences. For this "inner show" is based on the abstraction of the general structure of these past experiences, the structure having predictive inferences for later experiences.

A little reflection shows that there is an enormous number of cases in which the above-described kinds of predictive inferences based on the general structure of our perceptions have turned out to be correct. That is to say, the "world" that we see in immediate perception has, at a given moment, a general structure, which has withstood a long series of tests, in the observations that have led up to the moment in question. And as a rule, it happens that the natural projection of this structure in accordance with the known state of movement of the observer and of what is in the field of perception will continue to be more or less in accord with later observations in a great many respects. This means that the general structure of our perceptions has a certain similarity to the general structure of what is actually in our environment. (p. 64.)

Unlike Materialism, Darwinism, and Naturalism which can never be questioned and therefore can never be falsified, the "inner show" of Psyche is constantly subjected to trial and error, and it is therefore either falsified or verified as the Psyche has further experiences and makes further observations. The Psyche is the scientist who does all of the hypothesizing, abstracting, theorizing, speculating, experiencing, testing, and experimentation BOTH at the quantum level and at the physical level. Out-of-Body Travelers and Near-Death Experiencers continue to explore their environment, even while they are separated from their physical body. This "inner show" or psyche is intelligent, perceptive, and capable of abstraction, prediction, and inference. It's as if the Psyche is alive or some type of Life Force. It learns through trial and error. It learns through its mistakes. If it's truly smart, then it abandons everything that has been falsified such as Materialism, Naturalism, Nihilism, and Atheism and moves on instead in search for what is true, which ends up being everything that is repeatedly experienced and observed.

Psyche and energy are inherent and necessary features of all that exists, in every possible domain and field of experiencing and investigation. Psyche or energy is eternal and everlasting.

Every type of wave carries specific information within it – most obviously quantum waves or thoughts, as well as radio waves and other types of electromagnetic waves. Information is always the product of Someone Psyche or Something Intelligent. Usable information doesn't just spring into existence from nothing as the Materialists, Naturalists, Nihilists, and Atheists claim. Information requires intentional purpose on the part of Someone Psyche in order for that information to be usable and reusable on a universal scale. It's hard to imagine and quantify the amount of information that had to be designed, organized, and then implemented just to reliably reproduce physical atoms at will. The whole of Quantum Mechanics, Quantum Field Theory, Relativity, Classical Physics, and String Theory probably just scratches the surface of the amount of information and knowledge that was necessary to acquire and use, just to produce the first physical atom. That kind and amount of information isn't just going to spring into existence from nothing as the Atheists and Naturalists and Darwinists claim so that multiple compatible atoms can be reproduced at will. When it comes to physical matter, Someone Psyche planned it all and designed it all before the first physical atom was made. There's too much intelligence and information within the thing and behind the thing for it to have sprung into existence from nothing as the Atheists repeatedly claim that it did.

Remember, information and organization are always the product of Someone Psyche or Someone Intelligent. Information, organization, intelligence, and quantum fields were necessary BEFORE the first physical particle could be made and sustained. The quantum foundation and quantum fields had to be designed and made BEFORE the physical particles could then be produced and sustained. This is obviously true for anyone who understands the Science.

So there is always finally a stage where an *essentially perceptual process* is needed in scientific research – a process taking place within the scientist himself. (p. 73.)

This is the fatal flaw in Science and the Scientific Method – the reason why the Scientific Method can't be used to prove the truth – *interpretation of the evidence* by a flawed, limited, and inadequate Human Psyche. Interpretation of the evidence is a perceptual process that's done by the Human Psyche. Interpretation and perception are functions of psyche or products of psyche. For every piece of scientific evidence that is discovered or produced, there are literally an infinite number of possible interpretations that can be given to that evidence, and there's no guarantee that any of those interpretations are actually correct and true. Science, Quantum Mechanics, Psyche or Energy or Light, and Quantum Field Theory

FALSIFY Materialism, Naturalism, Atheism, and their derivatives, which means that a materialistic, naturalistic, and atheistic interpretation is always going to be flawed, incomplete, deceptive, and false. Garbage in, then garbage out. It's unavoidable. The truth doesn't magically spring forth from theories and philosophies that are false and have been falsified.

Like many scientists, David Bohm had convinced himself that there is no such thing as absolute truth and that consequently the Scientific Method can't be used to find and know the truth. Technically, they are right in that "affirming the consequent" prevents us from using the Scientific Method to prove the truth. However, as a master of the Philosophy of Science, I developed the work-around and therefore the solution to this conundrum.

Further developments in modern physics, including quantum theory and the studies of the transformations of the so-called "elementary" particles suggest that the notion of permanent entities constituted of substances with unchanging qualitative and quantitative properties may have to be dropped altogether, and that physics will be left with nothing but the study of what is relatively invariant in as wide as possible a variety of movements, transformations of coordinates, changes of perspective, etc.

Moreover, it seems that the notion that science is collecting absolute truths about nature, or even approaching such truths in a convergent fashion, is not in good accord with the facts concerning the actual development of scientific theories thus far, and has indeed also been a major source of confusion in scientific research. Rather, as Professor Popper has emphasized, science actually progresses through the putting forth of falsifiable hypotheses, which are confirmed up to a certain point and thereafter, as a rule, eventually falsified. New hypotheses are then put forth, which are criticized and tested by a process of "trial and error" very similar to that to which our immediate perceptions are continually being subjected.

In science this process takes place at a very high level of abstraction, on a scale of time involving years. In immediate perception it occurs on a lower level of abstraction, and it is very rapid. In science the process depends strongly on collective work, involving contributions of many people, and in immediate perception it is largely individual. But fundamentally both can be regarded as limiting cases of one over-all process, of a generalized kind of perception, in which no absolute knowledge is to be encountered. (*The Essential David Bohm*, pp. 66 and 76.)

Bohm has convinced himself that NO absolute truths are obtainable from nature and science. However, I eventually observed, thanks to Sherlock Holmes, that if you successfully eliminate everything that is false, then only the truth will remain. It works. The Scientific Method is the perfect tool for eliminating everything that is false, starting with Materialism, Naturalism, Darwinism, Nihilism, Behaviorism, and Atheism.

I love to study psychology. I find it fascinating to observe that when it comes to the Materialists, Naturalists, Darwinists, Nihilists, Mechanists, Behaviorists, Determinists, and Atheists, these people are constantly employing a special pleading and thereby making their theories and science exempt from falsification. There's no way to falsify the non-existence of something because there's no way to verify the non-existence of something. Instead, these people simply make their theories, ideas, and science exempt from falsification by definition in principle by turning them into axioms or absolute truths. They define Science axiomatically as Materialism, Naturalism, Nihilism, and Atheism so that their science will always be true. In other words, they cheat in order to make their case. Self-deception works, and it works every time. Nobody is immune.

These people stack the deck so that they can't lose. They turn the "non-existence of Psyche" and the "non-existence of God" into axioms or absolute truths which permit no falsification to take place. That's not science. It's religious dogma, sophistry, and philosophical extremism instead. Nevertheless, that's how these people operate. It's tricky and deceptive, but few people are willing to call them on it, so they get away with it at will.

Anyway, I falsified their materialistic and atheistic theories, philosophies, and ideas back in 2016, and then I went in pursuit of the truth instead. The Scientific Methods can indeed be used to falsify Materialism, Naturalism, Darwinism, and Atheism but only if you permit yourself to do so. I falsify these dogmatic religions in a few of my books. Once I eliminate everything that is obviously false from science, then I was finally free to go in pursuit of the truth; and, I quickly realized that the truth is constantly experienced and observed. The truth is always being verified. The truth is obviously true. There's really no question that it's true.

The Theory of Evolution Proved to Me that God Exists: Why I Am No Longer an Atheist and Why I No Longer Believe in the Theory of Evolution

https://www.amazon.com/gp/product/B01HZYBZ7K/

Summary Of: The Theory of Evolution Proves that God Exists

https://www.amazon.com/gp/product/B01GQCWED6/

The Scientific Method: Proves That the Theory of Evolution Is False

https://www.amazon.com/gp/product/B01IAAIRT2/

Using the Scientific Method: To Eliminate the Usual Suspects and to Prove the Truth

https://www.amazon.com/gp/product/B01J6STHP0/

While using the Scientific Methods, if you can successfully identify and eliminate everything that is false, then only the truth will remain. The false is falsified by the truth; and, the truth is repeatedly experienced and observed.

Psyche, Consciousness, Intelligence, or Thought is constantly being experienced and observed. It's unavoidable, because it's true. Your psyche is made from energy, and energy is always conserved. Your psyche has always existed, and it will always exist. It cannot be created from nothing, and it cannot be destroyed. Your psyche is eternal and everlasting. Your Psyche or thought processes are being constantly experienced and verified every day of your life. It's inevitable. The fact that you can read this and understand it proves that your Psyche, Consciousness, Intelligence, or Life Force actually exists. The only task that remains is to figure out what it is and how it works at the quantum level and the psyche level, since most people are pretty much certain that it doesn't exist at the physical level.

The false is falsified by the truth; and, the truth is repeatedly experienced and observed. The non-existence of God and the non-existence of Psyche cannot be experienced and observed; therefore, we know that these philosophical ideas are false, especially since both the Biblical God Jesus Christ and the Psyche have been abundantly experienced and observed. The false is falsified by the truth; and, the truth is repeatedly experienced and observed. That's Science. That's the way science should work. Science is observation and experience, the law of witnesses, and the falsification and elimination of everything that is false such as Materialism, Naturalism, Darwinism, Nihilism, Physical Reductionism, Behaviorism, Determinism, and Atheism.

Mark My Words

Psyche Is the Best Explanation

I'm constantly looking for Science where psyche or mind or intelligence ends up being the best explanation for what we are experiencing, seeing, and observing.

What we would like is a view in which thought itself is part of the reality. – David Bohm (p. 176.)

What we need and have always needed is a Model of Reality that includes Psyche, Intelligence, Consciousness, or Thought as an integral part of that Model of Reality. As scientists, we need to define what Psyche or Thought is and how it works, both at the quantum level and the physical level. This is precisely what Bohm tries to do in his 1985 chapter entitled, "Soma-Significance and the Activity of Meaning".

That's what I tried to do when I created my Ultimate Model of Reality back in 2017.

https://ultimate-model-of-reality.com/

https://www.amazon.com/gp/product/B071NC9JK6/

You see, because the Materialists, Naturalists, and Atheists who control our public schools won't permit us to explore and study the evidence that falsifies their beliefs, each scientist has to develop his or her own Model of Reality, that includes psyche or thought, from scratch – or go without. We can't standardize the terms and standardize the Science because the Materialists, Naturalists, and Atheists who hand out the doctorate degrees in Science won't permit us to create a Model of Reality that falsifies Materialism, Naturalism, Darwinism, Nihilism, and Atheism within our public schools. These people won't permit us to find and know the truth. They won't permit us to teach the truth. Therefore, we have to find it on our own if we want to have it. Each scientist ends up re-inventing the very same wheel, if he or she wants to find and know the truth. Bohm created his version of this Psyche Ontology or Mind Ontology during the 1980's; and, here I am thirty or forty years later doing the same exact thing. We each develop our own model, our own ideas, and our own unique vocabulary for quantum phenomena or energy phenomena or spiritual phenomena because the Materialists and Atheists who control our public schools won't allow us to do so at the public level within our university courses and degrees.

Psyche or Intelligence is the BEST EXPLANATION for what we are observing at the physical level; but, the Materialists, Naturalists, Darwinists, and Atheists won't allow us to enter that information into evidence because it falsifies their beliefs.

The modern mechanistic approach says that this area covers everything: but what I am saying is that it is a small area within a much vaster field. (p. 176.)

The Mechanists, Materialists, Naturalists, and Atheists say that physical matter covers everything. What scientists and quantum mechanists have learned is that physical matter is a very small area within a much vaster field of existence and reality.

ONLY 4% of our physical universe's cosmic density or energy is in the form of physical matter. The other 96% is comprised of dark matter and dark energy which are in fact non-physical in nature and origin. The Naturalists and Mechanists claim that only physical matter exists. These people are in fact WRONG about 96% of our physical universe. And, once you realize that physical matter is frozen energy or frozen light, then you realize that even physical matter is non-physical in nature and origin. There really is no such thing as physical matter. What we call physical matter is made up of force fields that are made out of energy or light. Physical matter is technically an illusion. It's not solid at all. So, we could in fact state that the Materialists and Naturalists and Mechanists are WRONG about 100% of the universe, because the whole thing is non-physical and immaterial in nature and origin.

In the past fifty years or so, Science has slowly but successfully FALSIFIED Materialism, Naturalism, Darwinism, Mechanism, Nihilism, Behaviorism, Determinism, Atomism, Physical Reductionism, and Atheism. Remember, the false is falsified by the truth; and, the truth is repeatedly experienced and observed by Someone Psyche both at the quantum level and at the physical level.

Now we KNOW that all energy and matter contains psyche, thought, or intelligence within it. Every Psyche has a certain amount of energy that's under its control; and, physical matter is made from different forms of energy by a bunch of different psyches who are collectively known as Nature's Psyche or God's Psyche. Furthermore, the Matrix of Quantum Fields was designed and made by God's Psyche in order to make the existence of physical matter possible. Quantum Fields are non-physical and pre-physical. The Gods had to make the Quantum Fields BEFORE physical matter became possible to make.

When it comes to the organization of Quantum Fields from the raw, chaotic, unformatted energy that pervades the whole of reality and existence, Psyche is the BEST EXPLANATION. Psyche is the only thing we know of that can actually form the energy under its control into anything that it wants that energy to be.

David Bohm wrote:

> **At first a person is able to take in only various bits of knowledge, the relationship of which is not yet clear. But at a certain stage, in a very rapid process often described as "click" or as a "flash," he understands what is being explained. When this happens he says "I see," indicating the basically perceptual character of such a process. (Of course, he does not see with optical vision but rather, as it were, with the "mind's eye.") But what is it that he sees? What he perceives is a new total structure in terms of which the older items of knowledge all fall into their proper places, naturally related, while many new and unsuspected relationships suddenly come into view. Later, to preserve this understanding, to communicate it to other people, to apply it, or to test its validity, he may translate it into words, formulas, diagrams, etc. But initially it seems to be a single act, in which older structures are set aside and a new structure comes into being in the mind.**
>
> **There seems to be no limit to the possibility of the human mind for developing new structures in the way described above. And it is this possibility that seems to be behind our ability to put forth new theories and concepts, which lead to knowledge that goes beyond the facts that are accessible at the time when the theories are first developed. It should be recalled that this possibility exists as much in immediate perception as in scientific research, since very often what is constructed in the "inner show" leads, as we have seen earlier, to many correct predictive inferences for future perceptions. It is evident that such an ability cannot be due merely to some sort of mechanism that randomly puts forth "hypotheses" until one of them is confirmed.** (*The Essential David Bohm*, p. 75.)

Bohm is always slow or reticent to talk about Psyche or Mind, but he tends to get there towards the end of every one of his essays, because Psyche or Mind or Intelligence is always the best way to explain what we have experienced and observed, especially when it comes to something intelligent like Science.

Psyche and mind and consciousness are the same thing.

How do we know?

We KNOW because the Psyche has been experienced and observed functioning separately from its assigned spirit body and physical body. Whenever a Psyche is outside of its spirit body looking at its spirit body, it's the Psyche who is having the experience and forming the memories, and not the spirit body. The spirit body and physical body don't have experiences and don't form memories while their Psyche is separated from them. This is what has been experienced and observed. You don't feel the pain while you are unconscious and your Psyche is separated from your physical body. Only when you wake up, when your Psyche engages your physical body, does the pain flood back into you. When you are sick and damaged, it hurts to wake up. You much prefer to remain unconscious and asleep. The spirit body and physical body don't have experiences and don't form memories while their Psyche is separated from them. This is what has been experienced and observed. The false is falsified by the truth; and, the truth is repeatedly experienced and observed.

If the Psyche or Consciousness were in fact nothing but the physical brain, if the Psyche couldn't separate or disconnect from its physical body, then some of us would be in pain all the time and would never be able to fall asleep. It's a God-send that we can separate our Psyche or Consciousness from our physical body so that we can fall unconscious and be relieved from all that pain. If our Consciousness or Psyche were indeed produced by our physical brain, then there would be no way to disconnect our Consciousness or Psyche from all of our physical pain. Our consciousness would be online all the time, and we would never be able to fall asleep. When we are in a great deal of pain, we need to be able to disconnect our Psyche from our physical body in order to fall asleep.

For two years, in my twenties, it felt like my skin was on fire. I would wake up and for a split second be just fine, and then I would feel the fire and the pain flood over me from head to toe as my Psyche re-engaged my physical body once again. The physical pain was always there; but, my Psyche or Spirit had managed to disconnect itself from it. This, once again, is yet another scientific and experiential example where Psyche is in fact the best explanation for what is being experienced and observed. I didn't realize it at the time, of course; but, I understand it now, because now I'm allowing myself to understand it. If our Psyche or Consciousness was in fact produced by our physical brain, then there would be no way for our Psyche or Consciousness to disconnect from our physical pain. This truth is obvious for anyone who has experienced it and is willing to understand it and accept it. Each one of us is born with an in-built quantum mechanism that permits us to temporarily disconnect our Psyche or Spirit from our physical pain. It happens automatically when needed.

In my fifties, I was again in so much pain that I wanted to die and cease to exist. They had me on so many different drugs that I lost the ability to fall asleep, so my body couldn't repair itself and I could never seem to escape from all the pain. That's when I decided to kill myself by overdosing on the sleeping pills that they had given me. Apparently, I came close to succeeding, but I'm still here. It took six months of withdrawal to get off all the pills they had me addicted to; and, once I got sober, my level of pain decreased noticeably, and I finally gained the ability to fall asleep naturally once again. Substance-induced psychosis, or matter-over-mind, is the very definition of hell. I know, because I lived it and experienced it. I didn't understand it at the time, but I understand it now. Death-by-medicine is one of the leading causes of death; but, that fact has been hidden from us by the Materialists, Naturalists, Pharmacologists, Doctors, and Psychiatrists who produced death-by-medicine in the first place.

Matter-over-mind is the absolute worst way to do psyche-therapy. It makes things worse rather than making things better. The Psychiatrists treat your mental illness by giving you half a dozen additional mental illnesses, each of which is its own unique substance-induced psychosis. There's no way to win, except by killing yourself or getting off all the drugs. Any time that I would complain about the side-effects, they would give me another addictive drug to supposedly counteract those side effects. They had me cycling through at least a dozen different drugs, including four different types of sleeping pills. I lost all sense of reality, and I eventually got to the point where I didn't even know how to turn on a television. I'd just sit there in pain and stare at the wall all day, until I couldn't take it any longer and felt like carving off all of my skin in order to get rid of all the bugs that were crawling under my skin. Matter-over-mind is the very definition of hell. You don't have to die to go to hell. You can do so right here, right now, if you want to. Addictive prescription drugs can help you get to hell even faster if you let them.

https://psyche-ontology.com/truths-about-psychiatry/

We discover the truth through observation and experience, not philosophical speculation and wishful thinking. You can wish all you want that Psyche or God would cease to exist; but, that will never change the fact that the Biblical God Jesus Christ and Psyche have been repeatedly experienced and observed.

After two years of addiction to prescription drugs, I realized that matter-over-mind is the very definition of prison or hell. Your physical body and the atoms within your physical body have a mind of their own; and, when your body is in control of your Human Psyche, then Satan has you in his grasp. He can pull your strings just like a puppeteer can pull the strings of a puppet. You have no control when you are enslaved to matter-over-mind.

One of the reasons we are here on this earth in this physical format is to learn mind-over-matter. It's an important lesson to learn. It can set you free on many different levels.

David Bohm laid the scientific foundation for all of this back in the 1980's. Bohm was a maverick when it came to Quantum Mechanics; but, he seemed to be a conformist when it came to everything else. When the Materialists, Naturalists, Mechanists, and Atheists erroneously told him that Psyche does not exist, Bohm conformed and concluded that Psyche, Spirit, or Soul does not exist. When the Materialists, Naturalists, Darwinists, and Atheists told Bohm that Creation Ex Nihilo, Spontaneous Generation, or the Theory of Evolution is true, Bohm conformed and concluded that it is true. That's what most of us did back in the 1980's. My college professor in 1980, told me that I have to believe in the Theory of Evolution because the scientists have no other explanation for the origin of life on this planet. Now, as we approach 2020, there's no logical reason to believe in the Theory of Evolution because Science has FALSIFIED Darwinism, Macro-Evolution, Creation by Natural Selection, Creation by Chance, Chemical Evolution, Spontaneous Generation, and Creation Ex Nihilo trillions of different times in thousands of different ways.

It's physically impossible for a sexually reproducing species to produce a genetically compatible Mr. and Mrs. Mutant over and over again, in the same place at the same time, for millions of years of time so that chimp-like ancestors can evolve naturally into chimpanzees and humans over millions of years of time. It can't be done, which means that it wasn't done. It's physically impossible. The Theory of Evolution is physically impossible. Macro-evolution is prevented from happening both by entropy and genetics. Q.E.D.

Only Psyche or Intelligence can do design and creation, both at the quantum level and at the physical level. When it comes to design, engineering, manufacturing, field-testing, fine-tuning, and creation, Psyche or Intelligence is the BEST EXPLANATION. There are many other situations where psyche is the best explanation. David Bohm listed some of them in his chapter entitled, "Soma-Significance and the Activity of Meaning (1985)".

> **In these higher levels this soma-significant and signa-somatic activity shows up most directly. In fact the word "meaning" indicates not only the significance of something to us, but also our intention toward it. Thus "I mean to do something" means "I intend to do it." This double meaning of the word "meaning" is not just an accident of our language, but rather it implicitly contains an important insight into the structure of meaning.** (*The Essential David Bohm*, p. 166.)

Intention and meaning are always a function and a produce of Psyche. When it comes to significance, meaning, purpose, language, information processing, and intention, Psyche is the best explanation for these phenomena, because you (your psyche or intelligence) continue to do these things long after your physical brain is dead and gone.

https://www.youtube.com/results?search_query=nde

Psyche, Intelligence, Consciousness, or Life Force is always the best explanation whenever it comes to the things that LIFE DOES. It's an infinitely better explanation than to say that Psyche, Intelligence, Consciousness, or Life Force DOES NOT EXIST, as the Materialists, Naturalists, Darwinists, Nihilists, Mechanists, and Atheists claim. It's obvious that consciousness and intelligence exist. It's ludicrous and absurd to say that they don't exist. Since we KNOW for a fact that consciousness and intelligence exist, the job of a scientist is to try to explain what they are and how they work. Consciousness or intelligence is Psyche or Life Force. It's a quantum phenomenon. Psyche is experienced first-hand at the quantum level as an immaterial viewpoint in space. Psyche is seen or observed at the quantum level as a pin-point of light, a point-particle, or a spark of life.

Whenever the Human Psyche is separated from its spirit body looking at its spirit body, it is always the Human Psyche who is having the experience and remembering the experience – NOT the spirit body, and definitely NOT the physical body. Your physical body doesn't have and remember experiences while your Psyche or Intelligence is offline or separated from it. This is what has been experienced and observed. You don't feel the pain until after your Psyche syncs up or links up with your physical body. Then all the pain comes rushing in all at once, where there was nothing before. That's what I have experienced and observed.

https://www.youtube.com/results?search_query=nde+spark

https://www.youtube.com/results?search_query=nde+psyche

Psyche is seen as a pin-point of light, and it seems to be an infinite singularity. Psyche is the ultimate point-particle, even at the quantum level where it is still seen to be a pin-prick of light. It's a quantum phenomenon – not a physical phenomenon. Everything at the quantum level is a non-physical phenomenon, until the different psyches get together and decide to organize themselves into a physical atom. Physical matter definitely is NOT the fundamental unit of reality. Physical matter is made from different forms of energy by the different psyches who control that energy. Physical matter is a cooperative effort.

Here David Bohm seems to provide a scientific explanation for the powerful life-changing effects that Cognitive Therapy can have in our lives:

> **Meaning and intention are therefore inseparably related as two sides or aspects of one activity. This is the same as we discussed with soma and significance, and the subtle and the manifest. We are saying that there is one whole of activity abstracted at a certain point conceptually – we make a cut in it – and we say it always has two sides.**
>
> **Intentions are commonly thought to be conscious and deliberate. But you really have very little ability to choose your intentions. Deeper intentions generally arise out of the total significance in ways of which one is not aware, and over which one has little or no control. So you usually discover your intentions by observing your actions. These in fact often contain what are felt to be unintended consequences leading one to say, "I didn't mean to do that. I missed the mark." In action, what is actually implicit in what one means is thus more fully revealed. That is the importance of giving attention.**
>
> **To learn the full meaning of our intentions in this way can very often be costly and destructive. What we can do instead is to display the intention along with its expected consequences through imagination, and in other ways. The word "display" means "unfold," but for the sake of revealing something other than the display itself. As such a display is perceived one can then find out whether or not one still intends going on with the original intention. If not, the**

intention is modified, and the modification is in turn displayed in a similar way. Thus to a certain extent, by means of trying it out in the imagination, you can avoid having to carry it out in reality and having to suffer the consequences, although that is rather limited.

So intention constantly changes in the act of perception of the fuller meaning. Even perception is included within this over-all activity.

What one perceives is not the thing in itself, which is unknown or unknowable, but however deep or shallow one's perceptions, all one perceives is what it means at that moment, and then intention and action develop in accordance with this meaning.

The point is that as you act according to your intention, and as the perception comes in, there can arise an indefinite extension of inward signa-somatic and soma-significant activity. That is, you go to more and more subtle levels and the thing is, as it were, looking at itself at different levels ever more deeply.

Such activity is roughly what is meant by the mental side of experience. When something is going on that is not strongly coupled with the outer physical manifestation of some soma-significant and signa-somatic activity in which it is looking at itself, then we call that the mental side of experience. Now this is only a side. Once again I want to repeat that there is no separation between the mental and the physical. When it gets to the other side where it is primarily concerned with actions it just gets more physical. (*The Essential David Bohm*, pp. 166-167.)

The mental controls the physical. It's mind-over-matter, or the placebo effect. The placebo effect, or mind-over-matter, or psyche-over-matter is a scientifically verified phenomenon. It's real. It truly exists. The best way to explain it is through Psyche and Quantum Mechanics or Spiritual Mechanics. The verified and proven existence of the placebo effect FALSIFIES Materialism, Naturalism, Mechanism, and their derivatives.

The physical realm is a playground where the psyche can act-out its intentions and desires.

Cognitive therapy teaches us that we can learn to school or control our choices, thoughts, desires, feelings, and intentions. The "unconscious" is in fact the Human Psyche. Psyche is the best explanation for our unconscious thoughts. The goal of cognitive therapy is to make the unconscious conscious. You learn to analyze your thoughts, and then you replace your self-defeating thoughts with more productive and useful thoughts. With cognitive therapy, you learn to take control of your automatic thoughts, and you learn to shut down the thoughts that are working against you. Again, it's mind-over-matter, rather than having the matter control the mind.

Matter has a mind of its own, too. You can learn to discern what your physical body needs. If you learn to listen, your physical body will actually tell you what it needs or what it wants you to eat. You, the Human Psyche, has to decide to go find food and has to decide what to eat. You can't eat everything that is available, so if you learn to listen, your body will actually tell you what to eat when it next comes time to eat.

Bohm talks about thought, intention, or meaning "getting to the other side". What is he talking about? Well, you have the physical side, and you have the quantum side. Psyche or intentions or meaning reside on the quantum side (implicate order); but, they manifest on the physical side (explicate order), when Someone Psyche decides to move the physical body in order to act upon that psyche's intentions in the physical realm. It's really simple to understand once a person chooses to do so.

It's all made from energy. Energy is constantly in motion. Energy acts. In contrast, physical matter was deliberately designed to react to the commands that its controlling psyche sends to it. There are the things that act – psyche – both at the quantum level and the physical level; and then, there are the things that react, which we call physical matter or physical atoms. Physical matter was designed to react to the commands of psyche. That's the way things really work. It has been experienced and observed.

Now we can look at this in terms of the implicate or enfolded order, for all these levels of meaning enfold each other and may have a significant bearing on each other. Within this context, meaning is a constantly extending and actualizing structure – it is never complete and fixed. (p. 168.)

Energy is infinitely malleable or infinitely transformable. Each psyche controls a certain amount of energy, and that psyche has complete control over the energy that's under its control. Psyche can form the energy under its control into anything that it wants that energy to be. Energy is never complete or fixed. Energy can be transformed any time the controlling psyche chooses to transform it.

Within this essay or chapter, Bohm tries to explain the relationship between the physical level (explicate order) and the quantum level (implicate order). Bohm also tries to figure out if there is a common denominator or a "bottom level" to all of this. Bohm doesn't seem to see it, because he has chosen to believe the mechanistic and naturalistic claim that Psyche does not exist. However, it's obvious to most out-of-body travelers and near-death experiencers that the common denominator in all of this is energy or light, and that the "bottom level" or "controlling level" is in fact Psyche, Intelligence, Consciousness, Thought, the Spark of Life, or Life Force. Thoughts and memories are quantum waves. Psyche produces quantum waves, transmits quantum waves, receives quantum waves, and stores quantum waves. Thoughts or quantum waves are how Psyches interact with each other at the quantum level. Quantum waves are Wi-Fi at the quantum level. Information is transferred psyche-to-psyche at the quantum level through quantum waves or thoughts. It's simple to understand, once a person chooses to do so.

In this chapter, Bohm starts to get into this and starts to explain it. Bohm lays the foundation for it, even though he doesn't completely arrive. Bohm died in 1992 before the existence of psyche became common knowledge on YouTube on the internet. As we approach 2020, Proof of Psyche has proliferated all across the internet; and, it's impossible to deny Psyche's existence because it has been experienced zillions of times by billions of different people, and because it has been observed thousands of times at the quantum level by out-of-body explorers and near-death experiencers. We KNOW that psyche exists, because it has been experienced and observed. However that wasn't common knowledge back in the 1980's when Bohm was first formulating his theories and ideas about the implicate order, or the psyche order, or the quantum order. Knowledge should trump philosophical speculation and wishful thinking; but, millions of Naturalists, Mechanists, Materialists, and Atheists aren't quite there yet. I KNOW, because I used to be a materialist, naturalist, nihilist, and atheist until I learned better. The false is falsified by the truth (observation and experience); and, the truth is repeatedly experienced and observed.

As we approach 2020, we KNOW that the existence of the psyche is real and true, because the thing has been experienced and observed. Psyche is now a well-documented phenomenon on YouTube and the internet. We now KNOW what it is and how it works.

Back in 1985, Bohm wrote: **These processes have access to an, in principle, unlimited depth in the implicate order.** (p. 169.)

Even though he doesn't know it, he is talking about Psyche here. Psyche is an infinite singularity. Psyche appears to have unlimited depth. Psyche is the ultimate point-particle. The reason why the physicists converted to string theory is because the strings have size. Without size or length, the math goes to infinity. When dealing with a point-particle like Psyche, ALL of the math goes to infinity, because Psyche is an infinite singularity. You can't do math by multiplying or adding infinities; and, Psyche is infinite and eternal. Psyche or energy has always existed and it will always exist. Psyche or energy is always conserved. Psyche or energy is infinite, just as the math suggests. In principle, Psyche has unlimited depth in the implicate order.

Remember, the implicate order relies upon the conservation of energy or the conservation of psyche in order to achieve unlimited depth, unlimited memories, or infinite depth. This is only realistically possible if Psyche is some type of infinite singularity or point-particle.

Now, let's look at some of the information that Bohm provides us around this particular quote. Notice how psyche ends up being the best explanation.

At any stage the perception of new meanings may dissolve these discrepancies, but there will still continue to be a limit, so that the resulting knowledge is still incomplete.

What this implies is that meaning is capable of an indefinite extension to ever greater levels of subtlety as well as of

comprehensiveness – in which there is a movement from the explicate toward the implicate. This can only take place however when new meanings are being perceived freshly from moment to moment. But if significance comes solely from memory and not from fresh perceptions it will be limited to some finite depth of subtlety and inwardness.

Memory, being some kind of recording, necessarily has a certain stable quality which cannot transform its structure in any fundamental way, and has only a limited capacity to adapt to new situations – for example, by forming new combinations of known principles, either through chance or through rules already established in memory. Memory is thus necessarily bounded both in scope and in the subtlety of its content. Any structure arising solely out of memory will be finite, and will be able to deal with some finite limited domain; but of course, to go beyond this, a fresh perception of new meanings is needed. And in fact, when you have a fresh perception you may also see new meanings of your memories. In other words, memory may cease to be so limited when there is fresh perception. To go on in this way to new meanings that are not arbitrarily limited requires a potentially infinite degree of inwardness and subtlety in our mental processes. And I am suggesting that these processes have access to an, in principle, unlimited depth in the implicate order. (*The Essential David Bohm*, pp. 168-169.)

Memories are bounded or fixed, yet they can also be modified or enhanced by the Psyche who is storing those memories. Psyche produces thoughts or quantum waves, can broadcast those thoughts or quantum waves, can receive quantum waves from other psyche, can process quantum waves, and can then store quantum waves or thoughts. Psyche is the best explanation for thoughts and memories, especially at the quantum level BEFORE physical matter and physical brains were designed and made. Even at the physical level, Psyche is the best explanation for all the different memories that survive the death of our physical brain and show up in our after-death Life Reviews.

https://www.youtube.com/results?search_query=nde+life+review

It was my attempt to find a logical explanation for all the different memories that show up in our after-death Life Reviews that led to my desire to create Quantum Neuroscience as a possible explanation for what we are witnessing and experiencing as a race when it comes to memories – especially the memories and perceptions that we have after our physical brain is offline or dead.

Quantum Neuroscience: The Answer to Life, the Universe, and Everything

https://www.amazon.com/gp/product/B079Z6QQQB/

Bohm laid the scientific foundation for all of this and provided a scientific explanation for all of this back in 1985; but, it took us a couple of decades to prove that he was right. The implicate order is non-physical or spiritual or quantum in nature and origin. The quantum realm, the spirit realm, or the energy realm in non-physical and pre-physical. The Gods or the Psyches had to design, create, and organize the quantum fields from raw chaotic energy BEFORE physical matter could be made and sustained. This is obviously true, or what logicians call "self-evident". The quantum fields had to be made by Someone Psyche or Someone Intelligent BEFORE physical matter could be made.

Psyche appears to be an infinite singularity whenever it is viewed at the quantum level. However, one would suppose that it would have some type of size or existence. Why? Well, Psyches produce quantum waves (thoughts), transmit quantum waves, receive quantum waves, and store quantum waves. One would guess that there might be some type of mechanism within a Psyche that makes this possible.

These processes have access to an, in principle, unlimited depth in the implicate order. (p. 169.)

If one computes the amount of energy that would be in one cubic centimeter of space, with this shortest possible wavelength [10^{-33} cm], **it turns out to be very far beyond the total energy of all the matter in the known universe.** (p. 98.)

You might say that everything has a kind of mental side, rather like the magnetic poles. In inanimate matter the mental side is very small, but as we go deeper into things the mental side becomes more and more significant. (*The Essential David Bohm*, p. 171.)

Each physical atom has some type of Psyche or Consciousness or Mental Side.

Even within string theory, there has to be some type of intelligence who is making the strings and choosing what frequency they should vibrate at. This would be some type of Psyche. Psyche controls and organizes energy. Remember, when it comes to physics, the smaller dwells within and controls the larger. Psyche would be smaller than the strings of string theory. Psyche would make the strings of string theory. Psyche designs and creates.

If any of this true, and Psyche has some type of measurable size, then it's theoretically possible to store all of the knowledge or information in the universe within each and every Psyche with plenty of room to spare. An "unlimited depth" is the same thing as saying "an infinite smallness". In principle, these psychic processes or quantum processes would have access to an unlimited depth or an infinite depth with the implicate order, the spirit world, or the quantum realm.

I find this fascinating, and it's worth repeating. Bohm writes: **I am suggesting that these processes have access to an, in principle, unlimited depth in the implicate order.**

What he is suggesting is that there is an unlimited depth when it comes to something like a Psyche. In other words, there's something within the point-

723

particle that we call Psyche that gives that Psyche the ability to control energy, form thoughts, transmit thoughts, receive thoughts, store thoughts, and communicate with other psyches through quantum waves. In principle, Psyche is theoretically capable of functioning at $10^{-\infty}$. Something has to be able to do this if what the math is telling us is actually true. Psyche deals with the realm of the infinite – being an infinite singularity that has infinite range or scope.

If any of this true, and Psyche has some type of measurable size, then it's theoretically possible to store all of the knowledge or information in the universe within each and every Psyche with plenty of room to spare. You could have a quantum computer within a Psyche with exabytes of RAM for memory storage, if the implicate order has unlimited depth or infinite depth as Bohm hypothesizes.

Remember, one of the obvious laws of physics is the fact that the smaller always dwells within and controls the larger. So, what's dwelling within the Psyche and controlling a Psyche? There may be no end to it, for all we know.

Once a person starts to wrap his mind around all of this, suddenly Materialism, Naturalism, Mechanism, and even Atheism become ludicrous and absurd because they are so extremely limited and restricted as to be practically worthless. They only have any meaning or value at the physical level. When we get to the quantum level, where the total energy of all the matter in the known universe can be stored within a cubic centimeter of space, Physicalism, Classical Physics, Mechanism, Materialism, Naturalism, and even Atheism cease to have any value or meaning. Physical matter is the absolute worst way to store information, because physical matter was purposefully designed to take up space. If you want to store an infinite amount of information, then you must store it in something that has an infinite amount of depth – something like Psyche. You'll never get an infinite amount of memory storage at the physical level, because that's physically impossible, because it would take up more space than there is. Physical matter was designed to take up space; and, the quantum fields seemed to have been designed to produce space-time.

In the next couple of paragraphs, Bohm actually tries to speculate as to what might be at the bottom of the physical level or the bottom of the explicate order.

Thus far I have suggested reasons why meaning is capable of infinite extension to ever greater levels of subtlety and refinement. However, it might seem at first sight that in the other direction – of the manifest and the somatic – there is a clear possibility of a limit in the sense that one might arrive at a "bottom level" of reality. This could be, for example, some set of elementary particles out of which everything would be constituted such as quarks, or perhaps yet smaller particles. Or in accordance with currently accepted views of modern physics it might be a fundamental field, or set of fields, that was the "bottom level."

What is of crucial importance is that its meaning would be in principle unambiguous. In contrast, all higher order forms in this supposedly basic structure of matter are ambiguous – that is, their

meaning is incomplete. There is an inherent ambiguity in any concrete meaning. That is to say, how the meanings arise and what they signify depends to a large extent on what a given situation means to us, and this may vary according to our interests and motivations, our background of knowledge, and so on. But if for example, there were a "bottom level" of reality, these meanings would be exactly what they were, and anybody who looked correctly could find them. They would be a reality that was just simply there, independent of what it meant to us.

Of course you also have to keep in mind that all scientific knowledge is limited and provisional so that we cannot be certain that what we think is the "bottom level" is actually so. For example, possibly something other than the present theories will come to reveal a "bottom level." But this uncertainty of knowledge cannot of itself prevent us from believing in the existence of some kind of "bottom level" if we wish to do so. It is not commonly realized however that the quantum theory implies that no such "bottom level" of unambiguous reality is possible. (*The Essential David Bohm*, p. 169.)

The physical level is a level of reality that is "simply there". It exists whether the Human Psyche is thinking about it or not. The physical level is sustained in its existence by all the trillions of different psyches that control the energy and the quantum fields from which the physical matter is organized or made. Psyche is at the bottom of it all controlling it all. However, it's impossible to tell for sure when it switches over from being spiritual or quantum to being "physical" or what the Materialists and Naturalists all "real". It's all made from energy, and it's all real, even before it becomes physical. It was all real and it all existed long before the first particle of physical matter was designed and made.

The quantum level or spiritual level is highly ambiguous because energy is infinitely malleable or infinitely transformable. The Gods or the First Psyches designed and created physical matter in order to greatly reduce the ambiguity. Physical matter is highly controllable and highly predictable. That was the ultimate intent of its creation – to have something that is unambiguous or something that is predictable, reliable, dependable, and controllable. We needed something that reacts to Psyche rather than acting chaotically, randomly, and independently all on its own. Physical matter is it. Physical matter is the coordination of energies by a bunch of different psyches all acting together for a common, predictable, reliable, and dependable purpose.

Bohm ends his "soma-significance" essay by mentioning and discussing the three levels of existence that we humans have experienced and observed. On page 172, he calls them "soma" (physical matter), "significance" (psyche, intelligence, or consciousness), and "energy" (the matrix of quantum fields). We KNOW that these things exist because they have been experienced and observed.

On page 175, Bohm lists these three aspects of reality and existence as "matter", "energy", and "meaning". Physical matter obviously exists. The solipsists who state that physical matter does not exist (that only mind exists) are just as stupid and ignorant as the Materialists and Naturalists who claim that only physical matter exists. Physical matter obviously exists, and mortal beings have to deal with it every day of their lives. Likewise, the quantum fields or the spirit world or the energy realm exists. Energy is always conserved, which means that the spirit world or the quantum world or the energy realm has always existed and will always exist. It's eternal and everlasting. It cannot be destroyed. Psyche, Consciousness, or Intelligence is the third obvious aspect of our existence. If you can read and understand this, then obviously you are conscious and intelligent, sentient and aware. Only Psyche or Life Force can do intelligence, purpose, meaning, planning, design, creation, and significance. It's obvious that intelligence exists. Our job as scientists is to try to figure out what it is and how it works.

So, there are three levels of reality or existence that we have to take into consideration if we want to know how things really work. Bohm was talking about all of this back in 1985, and it has taken us twenty to thirty years to start to catch up with him. As we approach 2020, there's now NO logical rational reason to believe in Materialism, Naturalism, Mechanism, Darwinism, Nihilism, and Atheism because Science itself has FALSIFIED these false beliefs trillions of times in thousands of different ways. The false is falsified by the truth; and, the truth is repeatedly experienced and observed by Someone Psyche or Someone Intelligent both at the quantum level and the physical level. Science is observation and experience; and, both observation and experience have successfully FALSIFIED Materialism, Naturalism, and their derivatives. I had to get used to this reality and truth because I used to be a materialist, naturalist, nihilist, and atheist; but, Science itself, particularly Quantum Mechanics and Quantum Field Theory, convinced me that I was wrong. Quantum fields are non-physical and pre-physical in nature and origin. The proven and verified existence of quantum fields FALSIFIES Materialism, Naturalism, and their derivatives such as Darwinism, Nihilism, Physicalism, and Atheism.

The existence of Quantum Fields doesn't make any logical sense from a materialistic or naturalistic perspective. It's obvious that organization of any kind requires intelligence, whether we are talking about the quantum level or the physical level. One of the definitions for entropy is "disorder". The opposite of entropy is syntropy, or what we typically call intelligence, organization, information, or psyche. It's obvious that some type of Psyche or Intelligence was required to design and organize the Quantum Fields; otherwise, chaos would have continued to reign supreme. Likewise, something like Quantum Neuroscience – neuroscience at the quantum level or the non-physical level – doesn't make any sense without knowledge of energy or psyche. Psyche or Intelligence is the only thing that can organize energy at the quantum level or the spiritual level.

Quantum Consciousness or Quantum Mind doesn't make any sense from a physical, or mechanistic, or materialistic perspective.

The quantum mind or quantum consciousness group of hypotheses propose that classical mechanics cannot explain consciousness. It posits that quantum mechanical phenomena, such as quantum entanglement and superposition, may play an important part in the brain's function and could contribute to form the basis of an explanation of consciousness.

https://en.wikipedia.org/wiki/Quantum_mind

On page 173, Bohm talks about "intentional activity", "deeper intelligence", and "intelligent action". Experimentation is intentional activity or intelligent action. Only Psyche or Life Force has the innate capability of doing experimentation, intentional activity, or intelligent action. Only Psyche can do Science. Only Psyche can do conscious meaning, both at the quantum level and at the physical level.

This is a very important development of intentional activity which makes possible an unending movement of learning and discovering what has not been known before. So we want to say that this soma-significant and signa-somatic activity, constantly going back and forth, is what is involved in learning. And we can say that this is going on, not only in regard to outward objects, but inwardly – that is, for example, with regard to thought. And there may be another level which picks up the meaning of the thought and takes an action toward that thought while thinking another thought to see if it is consistent. If it is not, then the intention changes until we get a consistent relationship between the thought which arises from the deeper intention and the thought that was first being looked at. You see, you may have a thought that you want to look at, and there may be a deeper intelligence which is able to grasp the meaning of that thought in a broader context and take an action toward it by, as it were, thinking again and seeing whether the thought which comes out is coherent with the thought with which you started. And if it's not, then you can start to change that action until it is. Or you can change the thought. Change can occur at various levels.

So all of these levels of meaning enfold each other and have a certain bearing on each other. This whole process is always soma-significant and signa-somatic, going to ever deeper levels. When I talk of these processes I don't only mean going outward into the manifest world, but also the deeper mental processes being explored by still more subtle mental processes. So you could say that the mind has available in principle an unlimited depth of subtlety, and learning can take place at all these levels.

Meaning and matter may not have the same sort of consciousness that we have, but there is still a mental pole at every level of matter, and there is some kind of soma-significance. And eventually, if you go to infinite depths of matter, we may reach something very close to what you reach in the depths of mind. So if

727

you consider it, we no longer have this division between mind and matter.

Now we have in this whole process these three aspects: soma and significance and an energy which carries the significance of soma to a subtler level and gives rise to a backward movement in which the significance acts on the soma. Modern physics has already shown that matter and energy are two aspects of one reality. Energy acts within matter, and even further, energy and matter can be converted into each other, as we all know.

From the point of view of the implicate order, energy and matter are imbued with a certain kind of significance which gives form to their over-all activity and to the matter which arises in that activity. The energy of mind and of the material substance of the brain are also imbued with a kind of significance which gives form to their over-all activity. So quite generally, energy enfolds matter and meaning, while matter enfolds energy and meaning. (*The Essential David Bohm*, pp. 173-174.)

He got it. He understood it back in 1985. Mind or Psyche controls both energy and matter, both at the quantum level and at the physical level. Psyche is the control factor in all of this. One of the reasons we are here in this physical format is to learn mind-over-matter control. When you are suffering from addictions or going through withdrawal, the matter is actually controlling your psyche or your mind, and you have NO control. Matter-over-mind is the very definition of prison or hell. Trust me, you never want to go there if you can prevent it.

You see, your physical body, or the atoms within your physical body, have a psyche or a mind of their own. If you get addicted to crap, your physical body takes over and takes control, and the Human Psyche can't stand up against it effectively or efficiently. My medical doctors and psychiatrists wanted me taking the crap – they wanted me addicted to the drugs. Whenever I was going through withdrawal, I was an animal. You would lie, cheat, steal, and kill to get your next fix. My physical body definitely wanted me to keep taking the drugs. Now, enter the Human Psyche. One day I finally realized that if I stayed on the drugs, I would never get better. So, I chose to quit. I went cold-turkey, and I went insane. I was hallucinating and delusional for six solid months before I achieved sobriety and started thinking rationally once again. Six months is a long time to go through withdrawal.

My society wanted me taking the drugs, and my physical body wanted me taking the drugs. ONLY my Psyche had the power to make the choice to go cold-turkey, and quit the drugs, and suffer the consequences. I experienced substance-induced psychosis, a documented and official mental illness, and it only went away after I got sober. It took six months for my physical brain and physical body to normalize, once I decided to stop taking all the different drugs that they were feeding me. Mind-over-matter was the cure to my substance-induced psychosis;

and, while I was addicted and experiencing matter-over-mind, then I was trapped and in hell.

Psyche or Life Force is the only thing we know of that is capable of going to infinite depth. Here, Bohm suggests a Psyche Ontology or a Mind Ontology wherein psyche becomes the fundamental unit of reality. If you go deep enough, you will find your mind or your psyche staring back at you. In fact, it will be you.

Every Psyche has a certain amount of energy or information that's under its control. Some psyches have more intelligence or light than other psyches do. There are different types of psyches in different stages of progress or development. All of this was revealed to Abraham thousands of years ago; and, we are just starting to catch up with him thousands of years later.

Abraham 3: 21-23:

21 I dwell in the midst of them all; I now, therefore, have come down unto thee to declare unto thee the works which my hands have made, wherein my wisdom excelleth them all, for I rule in the heavens above, and in the earth beneath, in all wisdom and prudence, over all the intelligences thine eyes have seen from the beginning; I came down in the beginning in the midst of all the intelligences [psyches] thou hast seen.

22 Now the Lord had shown unto me, Abraham, the intelligences that were organized before the world was; and among all these there were many of the noble and great ones;

23 And God saw these souls that they were good, and he stood in the midst of them, and he said: These I will make my rulers; for he stood among those that were spirits, and he saw that they were good; and he said unto me: Abraham, thou art one of them; thou wast chosen before thou wast born.

It has been estimated that Abraham was born 2052 BC.

https://www.johnpratt.com/items/docs/lds/dates.html

So, we know that God was teaching us about Psyches or Intelligences over 4,000 years ago. Psyches exist, and God knows how to interact with them. God knows how to advance psyches or how to help them to progress, change, and grow. Psyches or Intelligences were there from the beginning.

If you were a God, how would you organize a Psyche or an Intelligence? I've given it some thought. You would first give that Psyche or Intelligence a spirit body, so that it has some spirit matter and energy to work with at the quantum level in the spirit world. Later, you would give that Psyche or Intelligence a physical body, so that it has some physical matter to work with and learn from at the physical level in the physical realm.

Abraham 3: 17:

17 Now, if there be two things, one above the other, and the moon be above the earth, then it may be that a planet or a star may exist above it; and there is nothing that the Lord thy God shall take in his heart to do but what he will do it.

18 Howbeit that he made the greater star; as, also, if there be two spirits, and one shall be more intelligent than the other, yet these two spirits, notwithstanding one is more intelligent than the other, have no beginning; they existed before, they shall have no end, they shall exist after, for they are gnolaum, or eternal.

19 And the Lord said unto me: These two facts do exist, that there are two spirits, one being more intelligent than the other; there shall be another more intelligent than they; I am the Lord thy God, I am more intelligent than they all.

Here, God calls psyches "spirits". Though one psyche is more intelligent than another, they all have NO beginning, for they have always existed and they will always exist. The same thing can be said for energy. The energy has always existed and it will always exist. Psyche and spirit are different types of energy or different forms of energy. Intelligence, thought, or quantum waves are a unique form of energy. It's all made from energy, and all energy seems to be intelligent and psychic to a certain degree. God is God because he is more intelligent – has more truth and light – than all the rest of us combined. God is jacked into the Matrix of Quantum Fields. He's got access to it all. God is in the light.

Four thousand years later, we are finally starting to figure out that God and Abraham were right; and, they actually knew what they were talking about. All of this has been verified during the past few decades. It's real. It truly exists. And, there are different levels of existence. The intelligence level or the psyche level or the consciousness level truly exists. It has been both experienced and observed. The spirit world, the energy level, or the matrix of quantum fields truly exists. It has been both experienced and observed. Quantum Mechanics is Spiritual Mechanics or Energy Mechanics – the way psyche and energy really work at the quantum level. The physical realm is real and truly exists. It has been experienced and observed. It's stupid to deny the existence of these things because they have actually been experienced and observed.

Remember, every psyche has a certain amount of energy, intelligence, or information that's under its control. There's no limit to what's possible at the energy level, or the quantum level, or the psyche level because energy is infinitely malleable or infinitely transformable. The controlling psyche can form the energy under its control into anything that it wants that energy to be. In order to eliminate the ambiguity associated with psyche, thought, and the quantum realm or spirit world, we need a physical realm that is dependable, predictable, replicable, reliable, and controllable. The Gods provided us with such a thing by designing and organizing the Matrix of Quantum Fields upon which physical matter is based and sustained.

This is what Science and Quantum Mechanics are trying to tell us; and, this is precisely what the Materialists, Naturalists, Darwinists, Nihilists, Behaviorists, Mechanists, and Atheists formally deny and reject. These people deny the truth (the experienced and the observed) so that they can have the fiction or the lie instead.

Are you able to see and understand and accept all of this? Most scientists can't because they have chosen not to. They have chosen to deny it and reject it instead. One of our goals in life is to become smarter than your average scientist. These people limit themselves and restrict themselves so as to prevent themselves from finding and knowing the truth. You could say that these people have a knack for finding and then rejecting the truth. I KNOW how it goes, because I used to be a materialist, naturalist, nihilist, and atheists until I started searching for and finding the truth. The false is falsified by the truth; and, the truth is repeatedly experienced and observed by Someone Psyche either at the quantum level or at the physical level.

Remember, Quantum Mechanics is the science of probabilities or the science of possibilities. Psyche or Intelligence is the BEST EXPLANATION we have for the origin of infinite possibilities or infinite choices, because energy is infinitely malleable and because Psyche is the thing that decides what form the energy under its control will take. Decision or choice is always a function and a product of Someone Psyche. Psyche acts, and physical matter reacts. We KNOW this is so because the Materialists, Naturalists, and Atheists assure us that physical matter isn't making choices – physical matter is simply following the physical laws that Someone Psyche has designed and made for that physical matter to follow or obey.

The Materialists, Naturalists, Darwinists, Nihilists, and Atheists don't have a scientific explanation for CHOICE. Instead, these people simply claim that choice does not exist. These people claim that psyche or choice does not exist. They are essentially right when it comes to the physical level. The psyches, who are controlling the energies from which the physical atoms are made, have collectively chosen to follow or obey Physical Laws. That choice has already been made, and there are no further choices to be made when it comes to physical matter. Only God can override those choices and set physical matter on a different "supernatural" path, because God made the physical matter and the physical laws in the first place and convinced the psyches within the physical matter to obey those laws until He commands otherwise.

It all makes perfect sense once we include Psyche, Consciousness, Life Force, or Intelligence in the mix. Without Psyche, though, we have NO scientific explanation for choice and NO scientific explanation for the order, organization, and origin of quantum fields and physical matter. Without Psyche or Intelligence, all there would be is entropy, disorder, and random chaos. Psyche or Intelligence is obviously needed to produce the order, organization, purpose, and syntropy that we see around us today. We can't exist without it.

Mark My Words

Bohm on Psychology and Sociology (1958 – 1989)

Even though he knows that it is never going to happen in actuality, David Bohm proposes an open-minded and open-ended approach to science.

Science is predicated on the concept that science is arriving at truth – at a *unique* truth. The idea of dialogue is thereby in some way foreign to the current structure of science, as it is with religion. In a way, science has become the religion of the modern age. It plays the role which religion used to play of giving us truth; hence different scientists cannot come together any more than different religions can, once they have different notions of truth.

As one scientist, Max Planck, said, "New ideas don't win, really. What happens is that the old scientists die and new ones come along with new ideas." But clearly that's not the right way to do it. This is not to say that science couldn't work another way. If scientists could engage in a dialogue, that would be a radical revolution in science, in the very nature of science. Actually, scientists are in principle committed to the concepts involved in dialogue. They say, "We must listen. We shouldn't exclude anything."

However, they find that they can't do that. This is not only because scientists share what everybody else shares – assumptions and opinions – but also because the very notion which has been defining science today is that we are going to *get* truth. Few scientists question the assumption that thought is capable of coming to know "everything." But that may not be a valid assumption, because thought is abstraction, which inherently implies limitation. The whole is too much. There is no way by which thought can get hold of the whole, because thought only abstracts; it limits and defines. And the past from which thought draws contains only a certain limited amount. The present is not contained in thought; thus, an analysis cannot actually cover the moment of analysis.

There are also the relativists, who say that we are never going to get at an absolute truth. But they are caught in a paradox of their own. They are assuming that relativism is the absolute truth. So it is clear that people who believe that they are arriving at any kind of absolute truth can't make a dialogue, not even among themselves. Even different relativists don't agree.

So we can see that there is no "road" to truth. What we are trying to say is that in this dialogue we share all the roads and we finally see that none of them matters. We see the meaning of all the roads, and therefore we come to the "no road." Underneath, all the roads are the same because of the very fact that they are "roads" – they are rigid. (*The Essential David Bohm*, pp. 330-331.)

Even here, even though Bohm falsifies relativism, we do see signs of Bohm's relativistic assumption that absolute truth cannot be found. If you choose to believe that, then you are right, and you'll never be able to find it, because you won't even go looking for it.

The flaw in the Scientific Method is the interpretation phase. There are literally an infinite number of interpretations possible for each piece of scientific evidences. We KNOW that Quantum Mechanics and Quantum Field Theory are true; but, how many interpretations of Quantum Mechanics are there? Thousands! They can't all be right because they contradict each other. I found three interpretations for Quantum Mechanics that actually explain what Psyche is doing at the quantum level in order to get things done for us at the physical level. Most interpretations of Quantum Mechanics are worthless because they exclude Psyche, or Consciousness, or Intelligence, or Information a priori, before they actually start to make any type of explanation for anything. Choosing to believe that Psyche does not exist doesn't do anything to explain to us what it is and how it works. I want to know how things work.

Intelligence or Psyche is constantly experienced, observed, verified, and caught in the act. It takes intelligence to recognize intelligence. David Bohm seems to have convinced himself that there's no such thing as Absolute Truth. However, one of my greatest scientific discoveries came when I first realized that anything that is conserved – anything that is eternal and everlasting – is Absolutely True. Psyche or Intelligence and the energy associated with it ARE eternal and everlasting. They are conserved. They are without a beginning of days or an end of years. Energy cannot be made, and it cannot be destroyed. Therefore, energy and the psyche who controls it are eternal, everlasting, and Absolute Truth. I made this scientific discovery by listening and by refusing to exclude anything. I chose to admit all of the evidence into evidence, and then I chose to pursue a preponderance of that evidence. In the process, I found Absolute Truth.

Psyche or Intelligence or Life Force or Consciousness has always been true, and it will always be true because it is conserved. Psyche and energy are the absolute truths upon which everything else is based. Psyche is comprised of energy. Every psyche controls a certain amount of energy; and, both the energy and the psyche are always conserved. The Conservation of Energy and the Conservation of Psyche seem to be rigid Eternal Laws. They can't be falsified because they are always true. They are always verified. Anything that is always verified starts to rise to the level of LAW or Absolute Truth. The false is falsified by the truth; and, the truth is repeatedly experienced and observed and verified by Someone Psyche or Someone Intelligent.

By definition, our senses come from our physical body; however, conscious perception is a function or a product of the Psyche. We continue to perceive and think and analyze long after our physical brain is dead and gone. It has been observed by Near-Death Experiencers that the Human Psyche continues to live and continues to make choices long after its physical body is dead.

Sensitivity is being able to sense that something is happening, to sense the way you respond, the way other people respond, to sense the subtle differences and similarities. To sense all this is the foundation of perception. The senses provide you with information, but you have to be sensitive to it or you won't see it. If you know a person very well, you may pass him on the street and say, "I saw him." If you are asked what the person was wearing, however, you may not know, because you didn't really look. You were not sensitive to all that, because you saw that person through the screen of thought. And that was not sensitivity.

So sensitivity involves the senses, and also something beyond. The senses are sensitive to certain things to which they respond, but that's not enough. The senses will tell you what is happening, and then the consciousness must build a form, or create some sense of what it means, which holds it together. Therefore, meaning is part of it. You are sensitive to the meaning, or to the lack of meaning. It's perception of meaning, if you want to put it that way. In other words, it is a more subtle perception. The meaning is what holds it together. As I said, it is the "cement." Meaning is not static – it is flowing. (*The Essential David Bohm*, p. 332.)

This information enters into your eyes and your brain while you read it; but, you (your Human Psyche) is the thing that had to decide to pay attention to it, to think about it, and to decide whether it has any meaning or value to you or not. We are bombarded constantly by physical senses while we are awake; but, it's the Human Psyche who decides which of all those sensations we are going to pay attention to. Paying attention is a function of the Human Psyche. If we didn't have a Psyche to decide what it wants to pay attention to, then all of our physical senses would hit us simultaneously and constantly; and, there would be no way to filter them out and decide which ones have value at the current moment and which ones can be ignored.

Bohm suggests that as a society we choose **"the opening up of the mind, and looking at all opinions."** Then he concludes, **"If there is some sort of spread of that attitude, I think it can slow down the destruction."** Science is supposed to be an open-minded project; but, it never is in actuality thanks to the Materialists, Naturalists, Darwinists, Nihilists, and Atheists who define "science" as Materialism, Naturalism, Darwinism, Nihilism, and Atheism. We will never find and know the truth until someone opens their mind and chooses to look. Atheism of any kind is based upon a refusal to look at evidence. These people refuse to look at the evidence that falsifies their pre-chosen conclusions.

Meaning flows just like energy flows. Energy is constantly in motion because it is alive. As I see it, energy is the common denominator for everything, and Psyche or some kind of intelligence is the thing that decides what form that energy will take. Energy is infinitely malleable. It can take on any form that it desires, including spirit matter and physical matter and quantum fields. Whenever Psyche has been seen or observed by Near-Death Experiencers and Out-of-Body Explorers, it is always seen as a pinpoint of light or a spark of life. It's obvious that Psyche or Intelligence is comprised of energy or light. Everything is made from energy or light, including Psyche or Intelligence. The Psyche or Intelligence within the energy decides what form that energy will take. Someone has to make that decision, because energy can be anything that it wants to be.

The last half of *The Essential David Bohm* contains chapters, essays, letters, and conversations from Bohm dealing with Psychology and Sociology, especially in the last third of this book. Within these essays, David Bohm explores the Individual Level or the Psyche Level. David Bohm never uses the word "psyche"; but, it's clear that that is what he is talking about whenever he starts discussing consciousness, thought, mind, freedom, choice, and the super-implicate order. Physical matter was designed to be mechanistically determined; whereas, Psyche or Consciousness or Energy is eternal and everlasting which means that it cannot be made and it cannot be destroyed.

https://origin-science.org/The-Essential-David-Bohm/

David Bohm is thousands of years ahead of the Materialists, Naturalists, Nihilists, and Atheists when it comes to the non-physical sciences or Quantum Mechanics. In contrast, modern (2010-2020) psychology and sociology have advanced quite a bit since the time when Bohm was writing about these subjects in the 1960's to the 1980's. I didn't find Bohm's pop psychology to be as useful nor as interesting as his Quantum Mechanics.

However, one has to be willing to allow the Science of Psyche into science if one is to make any real progress with the non-physical sciences such as Quantum Mechanics, Psyche-Study (Psychology), Consciousness, Non-Locality, and Quantum Field Theory. The non-physical sciences don't make any sense when interpreted by and explained by Materialism, Naturalism, Nihilism, Darwinism, Behaviorism, Determinism, and Atheism. That's one of the first things that I observed when I finally started to study Quantum Mechanics with an intent to actually understand what I was studying.

Quantum Mechanics never made any sense to me while I was a materialist, naturalist, nihilist, and atheist because Quantum Mechanics explains how spirit matter, quantum fields, and psyche work. Quantum Mechanics only makes sense from a spiritual or non-physical perspective.

Quantum Mechanics: From a Non-Physical Spiritual Perspective

https://www.amazon.com/gp/product/B01J023TGU/

Even though David Bohm never seems to use the word "psyche", the last half of *The Essential David Bohm* contains essays and chapters wherein David Bohm discusses mind, consciousness, thought, and the super-implicate order which are all involved with what I like to call the Psyche Level of existence and reality.

David Bohm buries his discovered truths under confusing titles and words; but, once I was able to translate his terms into English, I quickly discovered that his theories and ideas match extremely well with the truths that I have discovered and tried to explore and expose in the books that I have written. I'm a scientist, mathematician, philosopher, logician, theoretician, and psychologist, so it didn't take me too long to figure out that David Bohm discusses the very same things that have always been of interest to me.

In his chapter entitled "Soma-Significance and the Activity of Meaning", David Bohm lays a quantum mechanical foundation for meaning, significance, consciousness, thought, and psyche.

These meanings change as human beings live, work, communicate and interact. These changes are based for the most part on adaptation of existent meanings. But it has also been possible from time to time for new meanings to be perceived and realized – in other words, made real. Perceptions of this kind have generally occurred when someone became aware that certain sets of older meanings no longer made any sense. This may be understood as a vast extension of what happens in the development of intelligence in young children. That is, as they see something about which they are puzzled, they have to see its meaning in a new way.

It is only when one's purpose or intention changes that a new meaning can be realized. Then, often in a flash that seems to take no time at all, a coherent new whole of meaning is formed, within which the older meanings may be comprehended as having a limited validity within their proper context.

Now if meaning is an intrinsic part of not only our reality but reality in general, then I would say that a perception of a new meaning constitutes a creative act. As their implications are unfolded, when people take them up, work with them, and so on, the new meanings that have been created make their corresponding contributions to this reality. And these are not only in the aspect of significance but also in the aspect of soma. That is, the situation changes physically as well as mentally. [At least while we have a physical body. When we no longer have a physical body, then the Psyche takes control and continues to handle "meaning" at the quantum level.]

Therefore each perception of a new meaning by human beings actually changes the over-all reality in which we live and have our existence – sometimes in a far-reaching way. This implies that this reality is never complete. In the older view, however, meaning and reality were sharply separated. Reality was not supposed to be changed directly by perception of a new meaning. Rather it was thought that to do this was merely to obtain a better "view" of reality that was independent of what it meant to us, and then to do something about it. But once you actually see the new meaning and take hold of your intention, reality has changed. No further act is needed.

Seeing something intellectually or abstractly, though, will not change your intention. You may say that you need an act of will to change it, but I think that when you really see something deeply with great energy, no further act of will is needed. If you really see a new meaning to be true, then your intention will change – unless there is something blocking it, such as your conditioning, or the "program." And if something is blocking it, then the will is not going to help, because you don't know what the block is. Therefore you have to see the meaning of the block. So choice and will are of limited significance – valid in certain areas. But I think something deeper is needed if you are discussing the transformation of mind or consciousness or matter – they really all change together. (*The Essential David Bohm*, pp. 177-178.)

Notice that it's all based upon energy, not physical matter!

A concept called Quantum Superposition states that quantum waves (quantum particles) can be added together so as to become a unified whole, but they can also be separated from each other into different individual parts. Quantum Superposition states that it's possible for consciousness, or psyche, or a spirit body to be in two separate locations at the same time. But typically, the different parts of the psyche, spirit body, or consciousness are united together into one united whole.

Thoughts, memories, or quantum waves can and do function independently at the quantum level; however, thanks to Quantum Superposition, all those thoughts or memories can be added together and stored within the Human Psyche where they can later be retrieved and used again. Thoughts or the "presence of psyche" can in fact be omnipresent or in more places than just one, thanks to Quantum Superposition; yet, at the same time, thanks to Quantum Superposition, all of your thoughts or memories or quantum waves can be added together and stored within your Psyche.

In order to make sense of Psychology from a quantum perspective, we actually have to put Psyche back into Psychology where it belongs. Psychology is the scientific study of the Human Psyche. Likewise, Sociology is the scientific study of all those other psyches. Both Psychology and Sociology are valid sciences, whether you have a physical body or not.

Within this chapter, Bohm actually lays the foundation for Quantum Psychology and Quantum Sociology. It's all there to be found once a person chooses to go looking for it.

You see, the deep change of meaning is a change in the deep material structure of the brain as well, and this unfolds into further changes. Every time you think, the blood distribution all over the brain changes; every emotion changes it. Between thinking and the somatic activity there is also a tremendous connection with the heartbeat and the chemical constitution of the blood, and so on. The

new meaning will produce different thought and therefore possibly an entirely different functioning of the brain.

We already know that certain meanings can greatly disturb the brain, but other meanings may organize it in new ways. And when the brain comes to a new state, new ideas become possible. But the new meaning is what organizes the new state. If the brain holds the old meanings, then it cannot change its state. The mental and the physical are one. A change in the mental is a change in the physical, and a change in the physical is a change in the mental. In fact, there has been some discussion of what is called subtle brain damage in animals in which no physical abnormality can be found; but some disturbance of function takes place when the animals are put under stress. So you see, we could say that living as we do, we probably have a great deal of subtle brain damage. In other words, the brain is damaged at a subtle level that might not show up at the cellular level, but deep in the implicate order. Eventually of course, it shows up in the cellular level too. So instead of saying that when we see a new meaning we make a choice and then act, we say that the perception and realization of the new meaning in our intention is already the change.

This point is crucially significant for understanding psychological and social change. For if meaning is something separate from human reality, then any change must be produced by an act of will or choice, guided perhaps by our new perception of meaning. But if meaning itself is a key part of reality, then once society, the individual and their relationships are seen to mean something different from what they did before, a fundamental change has already taken place. So social change requires a different, socially accepted meaning, such as in the change from feudalism to the forms that followed it, or from autocracy to democracy, or to communism, and so on. According to the meanings accepted, the entire society went.

These meanings may have been correct or incorrect. But once the meanings become fixed, the whole thing must gradually go wrong. Or to put it differently, what man does is an inevitable signa-somatic consequence of what the whole of his experience, inward and outward, means to him. For example, once the world came to mean a set of disjointed mechanical fragments, one of which was himself, people could not do other than begin to act accordingly and engage in the kind of ceaseless conflict that this meaning implies. The meaning of fragmentation includes conflict and self-centeredness – in other words, not creative tension but meaningless conflict. (*The Essential David Bohm*, pp. 178-179.)

Materialism, Naturalism, Darwinism, Nihilism, Atheism and their derivatives produce "meaningless conflict" because these falsified philosophies make the claim that intelligence, energy, quantum waves, consciousness, psyche, and other non-physical things DO NOT EXIST. Claiming that Psyche or Intelligence does not exist doesn't explain what it is and how it works. Claiming that God's Psyche does not exist doesn't explain what it is and how it works. Instead, it's completely meaningless. The purpose of Materialism, Naturalism, and their derivatives is to eliminate "meaning" from Science. Materialism, Naturalism, and Atheism were designed to make Psyche, Science, and Quantum Mechanics meaningless. They were designed to destroy the science of Quantum Mechanics.

All of this has indeed been verified – new meaning does indeed produce different thoughts and therefore entirely different functioning within the physical brain. It's mind-over-matter, or the placebo effect, which has been verified and proven to exist. It's Cognitive Therapy. Cognitive Therapy is mind-over-matter. It really works. Change your thoughts, and you change the way your physical brain functions. We've known about this for decades; but, this is precisely the type of Science that the Materialists and Naturalists are determined to eliminate, suppress, and destroy.

Feeling Good: The New Mood Therapy

https://www.amazon.com/dp/B009UW5X4C/

When the Human Psyche changes its thoughts and changes the meaning of things, the function of the physical brain actually changes. It's mind-over-matter. The physical brain was designed by the Gods to be controlled by the Human Psyche. Change has to be produced by Psyche or Intelligence, because Psyche or Life Force is the only thing that can do choice, decision, or will. Change the meaning or the intentions, and you change the functioning of the physical brain and the actions of the physical body. It's mind-over-matter or psyche-over-matter, which is a quantum phenomenon and not a physical phenomenon.

What I intend by "meaningless" therefore is that there is a meaning, but that it is inadequate because it is mechanical and constraining and is hence of little value and not creative. A change in this is possible only if new meaning is perceived that is not mechanical. Such a new meaning, sensed to have a high value, will arouse the energy needed to bring a whole new way of life into being. You see, only meaning can arouse energy.

At present people don't seem to have the energy to face this sea of troubles that threatens to overwhelm us, generally speaking. If we take a mechanical meaning, it tends to deaden the energy so that people remain indefinitely as they have been, or at best allows change in limited directions, such as the continuation of the development of technology, and so on. So I am saying that meaning is fundamental to what life actually is.

Now you can extend this to the cosmos as a whole. We can say that human meanings make a contribution to the cosmos, but we can also say that the cosmos may be ordered according to a kind of

"objective" meaning. New meanings may emerge in this over-all order. That is, we may say that meaning penetrates the cosmos, or even what is beyond the cosmos. For example, there are current theories in physics and cosmology that imply that the universe emerged from the "big bang." In the earliest phase there were no electrons, protons, neutrons, or other basic structures. None of the laws that we know would have had any meaning. Even space and time in their present, well-defined forms would have had no meaning. All of this emerged from a very different state of affairs. The proposal is that, as happens with human beings, this emergence included a creative unfoldment of generalized meaning. (*The Essential David Bohm*, pp. 179-180.)

Bohm is spot on. The Big Bang was physically impossible, because physical matter and physical laws didn't exist yet. Physical matter and physical laws had to be thought into existence by Someone Psyche or Someone Intelligent. Only Psyche can do meaning, intention, purpose, and intelligence. Think about it. The Big Bang was physically impossible because physical matter and physical laws didn't exist yet.

The Materialists, Mechanists, Naturalists, and Atheists don't even realize it; but, even the Big Bang Theory falsifies Materialism, Naturalism, and Creation Ex Nihilo or Atheism. Amazing, is it not? The false is eventually falsified by the truth; and, the truth is repeatedly experienced and observed. Intelligence or Psyche is repeatedly experienced or repeatedly caught in the act. We KNOW it exists because it has been experienced and observed.

Extended meaning as "intention" [PSYCHE] is the ultimate source of cause and effect, and more generally, of necessity – that which cannot be otherwise. (*The Essential David Bohm*, p. 181.)

There you have it, the answer to life, the universe, and everything. Psyche, meaning, intention, thought, or intelligence is the ULTIMATE SOURCE of cause and effect. It cannot be otherwise. God's Psyche must of necessity exist, or we wouldn't exist. Someone Psyche had to think the first particles of physical matter into existence, or we wouldn't exist as physical beings. That's just the way it is. Physical atoms are repeatedly built from the very same non-physical quantum blueprints by something psyche, so that every physical atom ends up being reliably compatible with every other physical atom.

Similarly there is no point in asking the meaning of life, as life too is its meaning, which is self-referential and capable of changing, basically, when this meaning changes through a creative perception of a new and more encompassing meaning.

You could also ask another question: What is the meaning of creativity itself? But as with all other fundamental questions we cannot give a final answer, but we have to constantly see afresh. For the present we can say that creativity is not only the fresh perception of new meanings, and the ultimate unfoldment of this perception

within the manifest and the somatic, but I would say that it is ultimately the action of the infinite in the sphere of the finite – that is, this meaning goes to infinite depths.

What is finite is, of course, limited. These limits may be extended in any number of ways, but however far you go, they are still limited. What is limited in this way is not true creativity. At most it leads to a kind of mechanical rearrangement of the kinds of elements and constituents that are possible within those limits. One may think of anything finite as being suspended in a kind of deeper infinite context or background. Therefore the finite must ultimately be dependent on the infinite. And if it is open to the infinite then creativity can take place within it. So the infinite does not exclude the finite, but enfolds within it and includes and overlaps it. Every finite form is somewhat ambiguous because it depends on its context. This context goes on beyond all limits, and that is why creativity is possible. Things are never exactly what they mean; there is always some ambiguity. (*The Essential David Bohm*, pp. 181-182.)

That's the problem with Materialism, Naturalism, Darwinism, Nihilism, Atheism, and their derivatives. It's too finite and too limited to be meaningful, and useful, and true. It lacks explanatory power. It's Bad Science. Materialism, Mechanism, and Naturalism are NOT creative! Physical matter can't design and create. Only Psyche can design and create, both at the quantum level and at the physical level.

The "background" is made from quantum fields. The background is a Matrix of Quantum Fields. Quantum fields are non-physical and pre-physical. The Gods had to design and make the quantum fields BEFORE physical matter could be made and sustained. The verified and proven existence of quantum fields FALSIFIES Materialism, Naturalism, Mechanism, and their derivatives. The false is falsified by the truth; and, the truth is repeatedly experienced, observed, and verified.

David Bohm figured out all of this back in the 1980's, and now as we approach 2020, we scientists are finally starting to verify and prove that he was right. We are no longer apologizing for Materialism, Darwinism, Naturalism, Mechanism, and Atheism. Instead, we are falsifying them, as should have happened centuries ago. If you successfully eliminate everything that is false, then only the truth will remain.

In physics, we can see that from the general laws of physics we have come to the generative source of energy, and so on. – David Bohm.

The generative order is in fact the Psyche Order. Psyche is command and control. Only Psyche can do design and creation at the quantum level. Energy is the generative source. Everything is made from energy, including Psyche; and, every Psyche has a certain amount of energy that's under its control. The Controlling Psyche can form the energy under its control into anything that it wants that energy to be. This is how things really work at the quantum level and the psyche level.

741

Knowledge [or thought] literally disrupts the brain physically. – David Bohm.

Bohm is constantly talking about mind-over-matter. Change your thoughts and you literally change the way your brain functions. This reality has been experienced and observed. It's mind-over-matter or psyche-over-matter. It's the placebo effect. It works. It's real.

The physical brain doesn't think and make choices.

How do we KNOW?

When the brain is dead and gone, it's the Human Psyche who controls its spirit body and environment. It's the Human Psyche who is thinking and making choices. Whenever the Human Psyche is separated from its spirit body looking at its spirit body, it is the Human Psyche who is having the experience and remembering the experience – NOT the spirit body, and definitely not the physical body. Your physical body doesn't have and remember experiences while your Human Psyche is separated from it.

It has been observed that the physical brain has no memory engrams, no memory storage, and no physical RAM. Thoughts and memories are processed and stored at the quantum level by the Human Psyche and Nature's Psyche.

Attention, learning, observation, interpretation, attribution, thought, perception, and experience are all functions of Psyche or Mind. We KNOW this is so, because we continue to do these things long after our physical brain is dead and gone. In order to make real progress in Science and Quantum Mechanics, we have to put Psyche back into Psychology where it belongs. There has to be a generative order or a creative order; otherwise, everything would be random chaos and we wouldn't exist.

Towards the end of his book, David Bohm talks about a generative order or the order that does design and creation. This interesting and useful essay takes place in the section of *The Essential David Bohm* which is entitled "Dialogue as a New Creative Order (1987)".

A generative order may also be an implicate order, particularly if the whole is relevant to the creation of the parts, or if processes of enfoldment and unfoldment are present. Such an example is Bohm's causal interpretation as applied to quantum field theory (Chapters 4 and 6). In this model, particle-like structures (an explicate order) unfold from a latent field (a first-level implicate order). Yet active information is required for this unfoldment to occur, which comes from a super-implicate, or generative order (a second-level implicate order).

Bohm proposes that a similar generative order can be discerned in the consciousness of society. (Lee Nicol, *The Essential David Bohm*, p. 289.)

According to Quantum Field Theory, particles are made from energy or light – including physical particles or physical matter – by Someone Psyche or Someone Intelligent who is following specific rules or laws. Particle-like structures (an explicate order) unfold from latent quantum fields (the first-level implicate order). Active information or intelligence or the word of command is required for this unfoldment to occur, and this active information comes from a super-implicate order, or a generative order, or a psyche order which is a second-level implicate order or a psyche level of existence.

Right there is the truth of everything, and it explains everything that Science has discovered so far. By 1987, David Bohm had found the truth. All the pieces of the puzzle were now in place. This is how things really work; and, it scales up to the macro-level or the physical level or the social level.

All of a society's overt activities, artifacts, and individuals are its explicate order. Its first-level implicate order is the latent, relatively passive content of the entire culture – the pool of knowledge it has accumulated for millennia, as well as the somatic correlates of this knowledge.

Its second-level implicate order – its generative order – is the values and meanings that inform the pool of knowledge with specificity and order, giving rise to dispositions, intentions, and actions that unfold into the explicate social order. This model, while not exhaustive of Bohm's view of the implicate nature of consciousness – which goes on to include the holomovement as a generative order – nonetheless provides a reference point for exploring Bohm's view of dialogue.

To effectively resolve the conflicts and contradictions in the structure of human experience, Bohm claims that values and meanings must be transformed at their concrete generative level. (Lee Nicol, *The Essential David Bohm*, pp 289-290.)

The generative level is the psyche level. Psyche is the thing that produces order, dispositions, intentions, purpose, teleology, values, meaning, actions, choices, information, and intelligence which unfolds into the explicate order or the mechanistic order or the physical order. Values and meaning are produced and transformed at the generative level or the psyche level. This is the way things really work, according to Quantum Field Theory.

A generative order is a conscious order or a psyche order – a super-implicate order. Bohm doesn't call it Psyche or Soul because at that point in time (1987) few people knew that Psyche has been experienced and observed. Thirty years later, on YouTube, it has become common knowledge that Psyche has been experienced and observed.

Psyche is experienced as an immaterial view-point in space. Whenever a Psyche is separated from its spirit body looking at its spirit body, it is the Psyche who his having the experience and forming the memories of that experience.

Psyche is the intelligence or the life force that has experiences and remembers having done so. While the Psyche is separated from its physical body and spirit body, its spirit body and physical body don't have experiences and don't form memories.

Psyche is seen at the quantum level as a point-particle, a pinpoint of light, a spark, or a pinprick of light. Psyche is the ultimate point-particle. Every Psyche has a certain amount of energy that it controls. When it comes to physics, the smaller always dwells within and controls the larger. Within every type of energy, there is a Psyche who dwells within it and controls it. That's the way it works.

Psyches can coordinate their efforts telepathically through thoughts or quantum waves. Psyche or Intelligence is creative. It decides and controls the form which the energy under its control will assume. Each Psyche controls a certain amount of energy; and, that psyche decides what it wants to do with the energy under its control. This is the absolute truth; and, everything else was designed to prevent us from finding and knowing this truth.

When it comes to our pursuit of the truth, we each have to start where we currently are; and for most of us, that starts with Materialism, Naturalism, and their derivatives. Speaking about the assumptions of Materialism and Naturalism and Mechanism, without mentioning them by name, David Bohm writes:

> **Creativity, in almost every area of life, is blocked by a wide range of rigidly held assumptions that are taken for granted by society as a whole. Some of these have already been discussed, but in addition, every society holds additional assumptions that are of such a shaky nature that they are not even admitted into discussion. There is therefore an unspoken requirement that everyone must subscribe to these assumptions, but that no one should ever mention that any such assumptions indeed exist. They are tacitly denied as operating within society, and even this denial is denied. The overall effect is to lead people to collude in "playing false" so they constantly distort all sorts of additional thoughts in order to protect these assumptions. Such bad faith enters deep into the overall generative order of society.**

> **These rigidities and fixed assumptions, many of which must not be mentioned but must nevertheless be defended, may be compared with a kind of pollution that is constantly being poured into the stream of the generative order of society. It makes no sense to attempt to "clean up" parts of this pollution farther downstream while continuing to pollute the source itself. What is needed is either to stop the pollution at its source, or to introduce some factor into the stream that naturally "cleans up" pollution.** (*The Essential David Bohm*, pp. 290-291.)

Materialism, Naturalism, Darwinism, Mechanism, and Atheism are pollutions that are destroying our social consciousness or collective consciousness. Within our society, we are not allowed to question them or falsify them, because they are declared to be axiomatically true; and thus, the rot continues from one generation to the next corrupting and making false everything that they touch.

Of course, there is no way to tell that Bohm is talking about Mechanism and Materialism. He could just as well be calling Christianity and Psyche and God "pollution" and "misinformation". It's only at the end of the essay that we begin to find out that he's talking about the Philosophy of Mechanism, as being the thing that is meaningless and worthless. Bohm is subtle at times, and he's trying to trick these people into a dialogue, where hopefully the truth will someday be revealed.

Meanwhile, now that Creativity is a function or a product of Psyche or Intelligence or Mind; and, this creativity or intelligence is deliberately rejected and destroyed by Materialism, Naturalism, and their derivatives.

Remember, according to Bohm and according to Quantum Field Theory, the generative level is in fact the psyche level, or the level of consciousness and intelligence. Materialism, Naturalism, and their derivatives were designed to destroy these truths. Naturalism and Quantum Field Theory are mutually exclusive, because they falsify each other. If one of them can be demonstrated to be true, then the other has been successfully falsified, because the false is always falsified by the truth. So, which one has been demonstrated to be true – Materialism or Quantum Field Theory – the non-existence of Intelligence and Quantum Fields, or the existence of Psyche and Quantum Fields? Which one has actually been experienced and observed? Has the non-existence of Intelligence and Quantum Fields been experienced and observed? Have Materialism and Naturalism been experienced and observed? Or is it Psyche, Intelligence, Quantum Fields, Energy, Consciousness, and Quantum Mechanics that has in fact been repeatedly experienced and observed? Quantum Field Theory FALSIFIES Materialism, Naturalism, Creation Ex Nihilo (Atheism), and their derivatives. The false is falsified by the truth; and, the truth is repeatedly experienced and observed. The truth naturally cleans up the pollution that is being produced by Materialism, Naturalism, and their derivatives.

The immune system itself is particularly complex and contains a very subtle kind of information that can respond to the whole "meaning" of what is happening to the order of the body. In this way it is able to distinguish misinformation from information needed for the body's healthy operation. It can be compared to a kind of "intelligence" that works within the body. Moreover there is evidence that this sort of "intelligence" can respond to the higher levels that are usually associated with thought and feeling. It is well known that depressing thoughts can inhibit the activity of the immune system, with the result that a person becomes more susceptible to infections. Indeed there is much evidence that a vigorous, creative state of mind and a strong "will to live" are conducive to general health and even to recovery from dangerous illnesses. More generally, it could be said that good health is basically a manifestation of the overall creative intelligence, working in concert with the body, through various means that include exercise, diet, relaxation, and so on.

Returning to a consideration of society, clearly there is also a vast amount of misinformation in circulation which acts toward society's degeneration. The media and various modern means of communication have the effect of rapidly disseminating and magnifying this misinformation, just as they do with valid information. It should be clear that by "misinformation" is meant a form of generative information that is inappropriate, rather than simply incorrect statements of fact. In a similar way a small "mistake" in DNA can have disastrous consequences because it forms part of the generative order of the organism and may set the whole process in the wrong direction. (*The Essential David Bohm*, pp. 291-292.)

What is it that we see when we study and observe the immune system? We see active information being used and implement. We see intelligence and intelligent intervention. We see mind-over-matter! We see Psyche. Psyche is the best explanation for how intelligence works at the quantum level or the generative level. Psyche is the best explanation for all the different mind-over-matter and placebo-effect phenomena that we experience and observe. Psyche is the best explanation for the source and for the cause of thoughts or quantum waves at the quantum level. Psyche or Intelligence is the best explanation for how all these various different quantum waves or thoughts are intercepted, deciphered, understood, and then implemented. The proven and verified existence of Quantum Mechanics and Quantum Field Theory FALSIFIES Materialism, Naturalism, Mechanism, Creation Ex Nihilo (Atheism), and their derivatives.

Materialism, Naturalism, Darwinism, Mechanism, Nihilism, and Creation Ex Nihilo (Atheism) are part of the misinformation that acts towards society's degeneration. They set the whole process of Science in the wrong direction.

Still, we are all influenced by the principles of Materialism and Naturalism. Even David Bohm would erroneously promote relativism and try to find ways to support it and sustain it. David Bohm had erroneously chosen to believe that there is no such thing as absolute truth. Anything that cannot be made and that cannot be destroyed is in fact absolutely true. It is eternal and everlasting. It has always existed, and it will always exist. The Conservation of Energy is absolutely true. Energy or psyche has always existed, and it will always exist. The Conservation of Psyche is absolutely true. Psyche or Intelligence cannot be

When a given principle is regarded as universally valid, it means that it is taken as absolutely necessary. In other words, things cannot be otherwise, under any circumstances whatsoever. Absolute necessity means "never to yield." To have something in the generative order that can never give way, no matter what happens, is to put an absolute restriction on free play of the mind, and thus to introduce a corresponding block to creativity that is very difficult to move.

Of course, both the individual and society require a certain stability, and for this, thought must be able to hold itself fixed within

certain appropriate limits and with a certain kind of *relative* necessity. Over a limited period of time, certain values, assumptions, and principles may usefully be regarded as necessary. They are relatively constant, although they should always be open to change when evidence for the necessity of the latter is perceived. The major problem arises, however, when it is assumed, usually tacitly and without awareness and attention, that these values, assumptions, and principles have to be absolutely fixed, because they are taken as necessary for the survival and health of the society and for all that its members hold to be dear.

We have argued elsewhere that science, which is in principle dedicated to the truth, tends to be caught up in necessity, which then leads to false play and a serious blockage of creativity. It is now clear that the assumptions of absolute necessity, with their predispositions to unyielding rigidity, are only part of a much broader spectrum of similar responses that pervade society as a whole. General principles, values, and assumptions, which are taken in this way to have absolute necessity, are thus seen as a major source of the destructive misinformation that is polluting the generative order of society. (*The Essential David Bohm*, pp. 292-293.)

Whenever Bohm starts talking about the dangers of absolute truth, we have no idea what the antecedent is. He could be talking about Christianity and religion and Jesus Christ and Psyche, just as much as he could be talking about Materialism, Naturalism, Darwinism, and Mechanism. All of these things are set up as absolute truths within our society, even though they are in fact mutually exclusive and falsify each other.

Whenever Materialism and Naturalism and Atheism are declared to be axiomatically true or absolutely true, it leads to false play and a serious blockage of creativity and intelligence. Of course, the Materialists, Naturalists, and Atheists say the same exact thing about Psyche, Intelligence, Christianity, Religion, and God. Which group is right? Which group is holding us back and preventing us from making scientific discoveries? I know which one has set me free and which one was holding me back; but, this is something that each person has to figure out for himself, or go without.

Bohm talks about the "free play of mind". I have learned that Psyche or Mind is eternal and everlasting. It cannot be made, and it cannot be destroyed; therefore, it is absolutely true. Not even God can create and destroy energy or the Human Psyche. Intelligence or Psyche is Light, and the absolute truth of our existence.

Again, supposedly speaking about Materialism, Mechanism, Naturalism, and their derivatives, Bohm writes:

A particularly important piece of misinformation is the key assumption that creativity is necessary only in specialized fields. This assumption pervades the whole culture, but most people are generally not aware of it; there is always a tendency for misinformation to defend itself by leading people to collude in playing false, whenever such an assumption is questioned. Assuming the restricted nature of creativity is obviously of serious consequence for it clearly predetermines any program that is designed to clear up the misinformation within society and suggests that it cannot be creative.

All that seems to be left is to ask whether society contains some kind of "immune system" that could spontaneously and naturally clear up misinformation. If such a system exists, then it is certainly not obvious, nor does it appear to be in common operation within our society today. (*The Essential David Bohm*, pp. 293-294.)

Again, Bohm doesn't identify the source or the cause of all this misinformation or pollution, so it is left up to the reader to decide for himself whether Bohm is talking about Mechanism and Materialism, or Psyche and God.

It is obvious, though, that the assumption of Materialism and Naturalism pervades the whole culture, and there is always a tendency for these falsified philosophies to defend themselves by colluding, playing false, axiomizing falsehoods, and spreading misinformation and lies whenever Materialism and Naturalism are questioned. One of the main reasons that I abandoned My Materialism, My Naturalism, My Nihilism, and My Atheism was that I finally realized that these people were lying to me and trying to deceive me. They lied to me one too many times, and I caught them in the act. Something like the Theory of Evolution is based upon hundreds of different lies, or a whole wall of misinformation and deception. The same can be said for Creation Ex Nihilo or Atheism. I stopped falling for their lies, and they lost me. I couldn't reconcile all the contradictions and falsehoods.

The false is falsified by the truth; and, the truth is repeatedly experienced and observed. Bohm assumes that society has no immune system against all the misinformation and falsehoods that permeate our whole culture; however, the absolute truth of Psyche or Intelligence, namely the proven fact that Psyche or Intelligence is eternal and everlasting, becomes that immune system that falsifies Materialism, Naturalism, Darwinism, Nihilism, and Creation Ex Nihilo or Atheism. We have in fact discovered this "societal immune system" that Bohm was searching for. It's found in Absolute Truth, an absolute truth that Bohm repeatedly says does not exist. However, it is obvious that anything that is eternal and everlasting – such as Psyche or Energy – is in fact absolutely true. It has always existed, and it will always exist. It cannot be made, and it cannot be destroyed. Psyche or Energy is always conserved. The conservation of energy or psyche is the very definition of absolute truth. It has always been true, and it will always be true.

It is obvious that physical matter exists. Therefore, I try to use discussion of Psyche and Quantum Field Theory to explain why it exists and how it was chosen into existence. As I see it, energy or psyche is the common denominator for

748

everything; and, Psyche or some kind of intelligence is the thing that decides what form that the energy under its control will take.

The eternal nature or the absolute truth of Psyche or Energy is the very thing that the Materialists, Naturalists, Darwinists, Nihilists, and Atheists formally deny and reject. Yet, it's obvious that Someone Psyche had to design and make the quantum fields from raw chaotic unorganized energy; otherwise, they would not exist, and chaos or randomness would reign supreme.

This brings us to an important root feature of science, which is also present in dialogue: to be ready to acknowledge any fact and any point of view as it actually is, whether one likes it or not. In many areas of life, people are, on the contrary, disposed to collude in order to avoid acknowledging facts and points of view that they find unpleasant or unduly disturbing. Science is, however, at least in principle, dedicated to seeing any fact as it is, and to being open to free communication with regard not only to the fact itself, but also to the point of view from which it is interpreted. Nevertheless, in practice, this is not often achieved. What happens in many cases is that there is a blockage of communication.

For example, a person does not acknowledge the point of view of the other as being a reasonable one to hold, although perhaps not correct. Generally this failure arises when the other's point of view poses a serious threat to all that a person holds dear and precious in life as a whole. (*The Essential David Bohm*, p. 295.)

It's obvious that physical matter exists. There's no sense denying it. What I'm trying to do is to find a quantum mechanical and psychic explanation for why it exists. I'm trying to explain the existence of physical matter, because the stuff shouldn't exist if Materialism, Naturalism, Darwinism, Mechanism, and Nihilism were in fact true. Materialism, Naturalism, and their derivatives deny the existence of the non-physical, which means that they deny the existence of Psyche and Quantum Fields, which are by definition in principle non-physical and pre-physical. If Quantum Fields did not exist, then physical matter would not and could not exist, because physical matter is based upon and dependent up these non-physical Quantum Fields. This is obviously true, and this is the very thing that the Materialists and Naturalists reject and deny.

Bohm wasn't able to find this "societal immune system" or "resistance to falsehoods and lies" that he was looking for because he too denied the existence of Absolute Truth. However, it is obvious that anything that is eternal has always been true and will always be true. The existence of Psyche, Energy, Intelligence, Light, or Life Force is eternal; and therefore, it is absolutely true because it is always conserved. Psyche or Energy cannot be made and it cannot be destroyed, therefore it is Absolutely True and will always be true.

In dialogue it is necessary that people be able to face their disagreements without confrontation and be willing to explore points of view to which they do not personally subscribe. If they are able to

749

engage in such a dialogue without evasion or anger, they will find that no fixed position is so important that it is worth holding at the expense of destroying the dialogue itself. This tends to give rise to a unity in plurality of the kind we have discussed elsewhere. This is, of course, quite different from introducing a large number of compartmentalized positions that never dialogue with each other. Rather, a plurality of points of view corresponds to the earlier suggestion that science and society should consist not of monolithic structures but rather of a dynamic unity within plurality. (*The Essential David Bohm*, p. 296.)

Here we find Bohm promoting "no fixed position", and that's why his program repeatedly fails. He's not looking for absolute truth, and therefore, he is unable to find it or see it. In contrast, I finally realized that there has to be one fixed position to serve as the basis of everything else, if we are indeed successfully to find and know the truth. There has to be an absolute truth somewhere, or we will never find and know the truth, just as Bohm repeatedly states. According to Science, observation, and experience, the Eternal Psyche and the energy it controls seems to be that eternal fixed position or absolute truth that Bohm and others are trying to avoid so that they can maintain their philosophical dialogue that goes nowhere in the end.

Still, Bohm doesn't completely exclude Psyche or Mind from his theories or ideas; and, that's why Bohm's ideas can be adjusted to serve as the basis for a Model of Absolute Truth upon which our universe can be designed, made, and sustained both at the quantum level and the physical level.

The importance of the principle of dialogue should now be clear. It implies a very deep change in how the mind works. What is essential is that each participant is, as it were, suspending his or her point of view, while also holding other points of view in a suspended form and giving full attention to what they mean. In doing this, each participant has also to suspend the corresponding activity, not only of his or her own tacit infrastructure of ideas, but also of those of the others who are participating in the dialogue. Such a thoroughgoing suspension of tacit individual and cultural infrastructures, in the context of full attention to their contents, frees the mind to move in quite new ways. The tendency toward false play that is characteristic of the rigid infrastructures begins to die away. The mind is then able to respond to creative new perceptions going beyond the particular points of view that have been suspended. In this way, something can happen in the dialogue that is analogous to the dissolution of barriers in the "stream" of the generative order, as discussed previously. In the dialogue, these blockages, in the form of rigid but largely tacit cultural assumptions, can be brought out and examined by all who take part. (pp. *The Essential David Bohm*, 296-297.)

Bohm was trying to start a purposeless and pointless dialogue, while also trying to promote open-minded scientific research. My goal has always been a bit different – to find and know the truth. We each have our thing. It is my observation that the existence of Psyche or Mind is absolutely necessary; otherwise, we wouldn't have Quantum Fields, and as a result, we wouldn't have physical matter. Someone Psyche had to design and make the quantum fields at the quantum level BEFORE physical matter became possible to make at the physical level. This is obviously true to me, or Absolute Truth. I found what I was looking for, because I went looking for it. I found the Absolute Truth, which consists of the things that have always existed and will always exist – namely Psyche or Energy or Light. The Absolute Truth is found in everything that is being conserved. The Absolute Truth is found in the Law of the Conservation of Psyche or the Law of the Conservation of Energy. The stuff has always existed, and it will always exist; therefore, it is Absolutely True.

Remember, anything that is eternal ends up being absolutely true, because it has always been true, and it will always be true.

Every time that you think Bohm has gone over the deep end into Materialism, Mechanism, Nihilism, Darwinism, and Atheism, then Bohm ends his essay by talking about consciousness or mind.

Within this generally fragmentary order of consciousness, the social order of language is largely for the sake of communicating information. This is aimed, ultimately, at producing results that are envisaged as necessary, either to society or to the individual, or perhaps to both. Meaning plays a secondary part in such usage, in the sense, for example, that what are put first are the problems that are to be solved, while meaning is arranged so as to facilitate the solution of these problems. Of course, a society may try to find a common primary meaning in myths, such as that of the invincibility of the nation or its glorious destiny. But these lead to illusions, which are in the long run unsatisfactory, as well as dangerous and destructive. The individual is thus generally left with a desperate search for something that would give life real meaning. But this can seldom be found either in the rather crude mechanical, uncaring society, or in the isolated and consequently lonely life of the individual. For if there is not common meaning to be shared, a person can be lonely even in a crowd.

What is especially relevant to this whole conflict is a proper understanding of the nature of culture. It seems clear that in essence culture is meaning, as shared in society. And here "meaning" is not only significance but also intention, purpose, and value. It is clear, for example, that art, literature, science, and other such activities of a culture are all parts of the common heritage of shared meaning, in the sense described above. Such cultural meaning is evidently not primarily aimed at utility. Indeed, any society that restricts its knowledge merely to information that it regards as useful would hardly be said to have a culture, and within it, life would have very little meaning. Even in our present society, culture, when considered

in this way, appears to have a rather small significance in comparison to other issues that are taken to be of vital importance by many sectors of the population.

The gulf between individual consciousness and social consciousness is similar to a number of other gulfs that have already been described, for example, between descriptive and constitutive orders, between simple regular orders of low degree and chaotic orders of infinite degree, and, of course, between the timeless and time orders. But in all these cases, broad and rich new areas for creativity can be found by going to new orders that lie between such extremes. In the present case, therefore, what is needed is to find a broad domain of creative orders between the social and individual extremes. Dialogue therefore appears to be a key to the exploration of these new orders.

If, however, the dialogue is sustained sufficiently, then all who participate will sooner or later be able to see, in actual fact, how a creative movement can take place in a new order between these extremes. This movement is present both externally and publicly, as well as inwardly, where it can be felt by all. As with alert attention to a flowing stream, the mind can then go into an analogous order. In this order, attention is no longer restricted to the two extreme forms of individual and social. Rather, attention is transformed so that it, along with the whole generative order of the mind, is in the rich creative domain "between" these two extremes.

The mind is then capable of new degrees of subtlety, moving from emphasis on the whole group of participants to emphasis on individuals, as the occasion demands. (*The Essential David Bohm*, pp. 298-299.)

Bohm is suggesting that we take on a project to enhance social consciousness or the collective consciousness or the Akashic Field. This is very much in line with what Carl Gustav Jung and the Essentialists have been suggesting that we do. Here Bohm finally suggests that we reject the crude life of Mechanism, Materialism, Naturalism, and Utilitarianism; and go in search instead for a collective consciousness that has intentional significance, meaning, purpose, and value. Intention, meaning, purpose, value, and significance are all functions or products of Psyche or Intelligence. They are all produced by the Psyche or the Mind.

Energy is constantly in motion. It's alive. A physical particle may seem to be stationary and held within a fixed position; but, it innards – its internal energy – is constantly in motion. Every Intelligence or Psyche is capable of enlargement or growth. It can always take on more energy, information, light, knowledge, responsibility, and intelligence.

Remember, anything that is eternal ends up being absolutely true, because it has always been true, and it will always be true. Psyche, Consciousness, Intelligence, or Mind has always been true, and it will always be true because it is made from energy or light, and energy or psyche is always conserved. Anything that is always conserved ends up being Absolutely True.

> **In the ordinary situation, consensus can lead to collusion and to playing false, but in a true dialogue there is the possibility that a new form of consensual mind, which involves a rich creative order between the individual and the social, may be a more powerful instrument than is the individual mind. Such consensus does not involve the pressure of authority or conformity, for it arises out of a spirit of friendship dedicated to clarity and the ultimate perception of what is true.** (*The Essential David Bohm*, p. 300.)

While denying the existence of absolute truth, Bohm does allow for the possibility that collective consciousness might somehow reach a collective consensus, or form a new "consensual mind". While dancing around the truth and suggesting that the "spirit of friendship" might be dedicated to the "ultimate perception of what is true", Bohm also tries to hold onto the idea that everything is relative and there is no such thing as absolute truth. He's trying to have it both ways in order to draw in the Materialists and Naturalists into a dialogue; but, as we approach 2020, more and more people KNOW that Materialism, Naturalism, Darwinism, and their derivatives have been FALSIFIED by Science itself including Psyche, Intelligence, Quantum Mechanics, and Quantum Field Theory.

Back in 2016, I had a dialogue (a debate) with these people on their turf, where they tried to convince me that they are right and that I am wrong. It continued until they banned me from participating on their websites. I no longer have an active presence online, because these people don't like having their philosophies, theories, and ideas falsified or proven wrong. A refusal to look at evidence that falsifies their theories is how they maintain the illusion that they are right and that everyone else is wrong. Once you start providing evidence that proves that they are wrong – that falsifies their theories – then the debate is over and they disconnect you from their websites and no longer allow you to post. That's the way the operate; and, soon you realize that there's no logical reason to have a debate or a dialogue with these people because they have already made up their mind to reject anything that you might happen to say.

Furthermore, there's no need to have a dialogue with these people to help them find the truth, because the Absolute Truth has been found in Quantum Consciousness, Psyche, Quantum Mechanics, Quantum Field Theory, Out-of-Body Experiences, and the Conservation of Psyche or Energy. The dialogue we need to have now is to help these benighted people to find and know the truth, as many of them who are willing to do so. Materialism, Naturalism, Darwinism (spontaneous generation), and Creation Ex Nihilo (Atheism) have been falsified by Science; and, it's time to send out the memo so that people know that it is so. Then the ball is in their court, and they can decide what they want to do with the information that they have received.

Many of them will continue to preach and teach Materialism, Naturalism, Darwinism, Nihilism, and Atheism because that's how they make their money.

There's a lot of money to be made telling the Atheists what they want to hear; but, for those of us who aren't making any money out of the process (people won't actually pay to hear the truth), there's no reason to continue to promote the lies. Therefore, we are free to tell people how it really is and how it really works at the quantum level and the psyche level. That's where things start to get interesting; and consequently, that's what I have chosen to study and write about, because it's obvious that physical matter wouldn't exist if Someone Psyche hadn't designed the information and created the Quantum Fields from energy in the first place. The false is falsified by the truth; and, the truth is repeatedly experienced and observed.

By the end of his life (1992), David Bohm has started to point us to all the scientific truths that are soon to be discovered, verified, and proven to exist. As Materialism, Naturalism, Mechanism, Darwinism (spontaneous generation), and Atheism (creation ex nihilo) decline, the truth ascends. As we approach 2020, the truth is now available to be found and known for anyone who wants to find it and know it. The Quantum Fields were made from raw unorganized chaotic energy by Someone Psyche – not by physical matter as the Materialists, Naturalists, Darwinists, Nihilists, and Atheists claim. The Quantum Fields had to be designed and made from energy by Someone Psyche BEFORE physical matter could be made. Psyche or Energy is Absolute Truth because Psyche or Energy is always conserved. Psyche or Energy cannot be made, and it cannot be destroyed. It has always existed, and it will always exist; therefore, it is Absolute Truth. Here we finally have the truth that explains everything that has ever been experienced and observed.

Propositions and theories – including those in this book – are essential for exposing us to new possibilities, but they can easily ensnare us and provide a new framework for delusion and false comfort.

At the end of the day, it may be that Bohm's vision, and his challenge, were deceptively simple. When books and speculations have served their purpose, we are left with the unvarnished truths of who we are and how we live in this world. Perhaps what matters then is that we take seriously our capacity, as Bohm suggests, to "soften up," to "open up the mind" – and to allow for the possibility of that strange energy we call love. (Lee Nichol, *The Essential David Bohm*, p. 302.)

It's all made from energy. Energy is intangible, immaterial, and non-physical in nature and origin. The verified and proven existence of energy FALSIFIES Materialism, Naturalism, Darwinism, Nihilism, Atheism, and their derivatives.

Everything around us has been determined by thought – all the buildings, factories, farms, roads, schools, nations, science, technology, religion – whatever you care to mention. (David Bohm, *The Essential David Bohm*, p. 352.)

All of these things were chosen into existence by Someone Psyche or Someone Intelligent. Thoughts are quantum waves or energy. Psyche is Life Force or energy; and, every Psyche has a certain amount of energy that's under its control. Thought is a function or a product of Psyche. Psyche produces quantum waves, transmits quantum waves, receives quantum waves, and stores quantum waves. Thoughts don't just spring into existence ex nihilo as the Materialists, Naturalists, and Atheists claim. Something Psyche produces them, transmits them, receives them, analyzes them, and stores them.

The point is that thought produces results, but thought says it didn't do it. And that is a problem. The trouble is that some of those results that thought produces are considered to be very important and valuable. Thought produced the nation, and it says that the nation has an extremely high value, a supreme value, which overrides almost everything else. The same may be said about religion. Therefore, freedom of thought is interfered with, because if the nation has high value it is necessary to continue to think that the nation has high value. Therefore you've got to create a pressure to think that way. You've got to have an impulse, and make sure everybody has got the impulse, to go on thinking that way about his nation, his religion, his family, or whatever it is that he gives high value. He's got to defend it.

You cannot defend something without first thinking the defense. There are those thoughts which might question the thing you want to defend, and you've got to push them aside. That may readily involve self-deception – you will simply push aside a lot of things you would rather not accept by saying they are wrong, by distorting the issue, and so on. Thought defends its basic assumptions against evidence that they may be wrong.

As long as we have this defensive attitude – blocking and holding assumptions, sticking to them and saying, "I've got to be right," and that sort of thing – then intelligence is very limited, because intelligence requires that you don't defend an assumption. There is no reason to hold to an assumption if there is evidence that it is not right. The proper structure of an assumption or of an opinion is that it is open to evidence that it may not be right. (*The Essential David Bohm*, pp. 306-327.)

This is scary stuff for some people because it is true. Materialism, Naturalism, Darwinism, Mechanism, Atheism and their derivatives involve a massive amount of self-deception. When it comes to these people, their Psyche thinks of ways to defend their pre-chosen belief that psyche does not exist. Self-deception works, and it works every time; and, it's highly unscientific. Atheism of any type is based upon a refusal to look at evidence. These people reject all the evidence that falsifies their beliefs. That's NOT science. That's dogmatism or religion instead. Materialism, Naturalism, and Atheism is a religion – it's a form of dogmatic extremism. It's every bit as bad as every other fanatical, dogmatic, closed-minded, and extremist religion on this planet. When it comes to *The Essential David Bohm*, I'm able to verify or conform everything that Bohm writes, except when he starts slipping back into Darwinism, Naturalism, Mechanism, Materialism, and Atheism. There's no way to verify those things because they have been proven false or falsified.

PSYCHE IS THOUGHT; and, it is obvious that thought or intelligence exists. It has been observed that we continue to think long after our physical brain is dead and gone. Whenever a Psyche is separated from its spirit body looking at its spirit body, it is the Psyche who is having the experiencing and forming memories of the experience, and not the spirit body. Your physical body doesn't have experiences and doesn't form memories while your Psyche is separated from it. I had to get used to all of this, because I used to be a materialist, naturalist, nihilist, and atheist until I learned better and saw the light. The verified existence of thought or psyche or intelligence FALSIFIES Materialism, Naturalism, and their derivatives because thought is non-physical, quantum, spiritual, or energy-based in nature and origin. It's all made from energy; and, the verified and proven existence of energy falsifies Materialism, Naturalism, and their derivatives. Think about it. That's what I chose to do. The false is falsified by the truth; and, the truth is repeatedly experienced and observed. Thoughts are repeatedly experienced and observed, are they not? To claim that Psyche does not exist – as the Materialists, Naturalists, and Atheists do – doesn't help us to determine what Psyche is and how it works. Now does it?

PSYCHE IS THOUGHT.

Mark My Words

Freedom and the Value of the Individual (1986)

The individual or one's personality is in fact one's Human Psyche. It remains individual, unique, conscious, and functional even while it is completely separate from its assigned spirit body and assigned physical body. Within Greek Philosophy, Socrates, Plato, and Aristotle defined the Psyche as the Soul, and vice versa. They understood that the Psyche or the Soul is indeed something completely different than our physical body. They were thousands of years ahead of our modern-day Materialists, Naturalists, Mechanists, Darwinists, Nihilists, and Atheists.

The following is found within *The Essential David Bohm*, page 254.

By loosening these conservative restrictions, we may begin to discern our inherent link to the holomovement – the unknown timeless present, the qualitative infinity of nature at the experiential level. In this way we embody true individuality, manifesting our creative potential through participation in – rather than objectification of – a total field of experience.

Freedom has been commonly identified (especially in the West) with free will or with the closely associated notion of freedom of choice. In these terms, the basic question is: Is will actually free, or are our actions determined by something else (such as our hereditary constitution, our conditioning, our culture, our dependence on the opinions of other people, etc.)? Alternatively, can we or can we not choose freely among whatever courses of action may be possible?

Such a way of putting the question presupposes that the mind is always able to know what are the various alternative possibilities, and which of these is the best. Evidently, however, if one does not have correct knowledge of the consequences of one's actions, freedom of will and choice have little or no meaning. It must be admitted that, in most of human life, lack of knowledge of what will actually flow out of one's choice prevails.

In order to deal with this, we try constantly to improve our knowledge. But as we have seen, reality is infinite both in its depth and in its extension. (*The Essential David Bohm*, p. 254.)

According to Jesus Christ, knowledge of the truth is the only thing that can make us truly free. John 8: 32.

I came across a piece of sophistry that actually made sense to me. We humans often talk about free will. According to this point of view, the human will is not free because you always have to make a choice whenever you are presented with options and opinions. Whenever you are given options, you are forced to choose between them. Whenever you encounter an opinion, you are forced to decide for yourself whether you believe it to be true or believe it to be false.

According to this perspective, the human will is not free because we are always forced to choose between options and opinions whenever we encounter them.

Out-of-Body Explorers have observed that whenever they are separated from their spirit body (and physical body) looking at their spirit body from an immaterial viewpoint in space, it is in fact their Psyche or Spark who is having and remembering the experience, and NOT their spirit body and definitely NOT their physical body. Our psyche, intelligence, or life force is the part of us that is actually alive and self-aware. Our psyche is the part of us that makes decisions and choices and then has to live with those decisions and choices.

https://psyche-ontology.com/buhlman/

https://psyche-ontology.com/psyche-observed/

https://psyche-ontology.com/psyche-experienced-and-observed/

Psyche or Intelligence is the fundamental unit of our personal existence and reality. Our spirit body and physical body are secondary. Our spirit body allows us to interact with quantum realms or spirit worlds; and, our physical body allows our psyche to interact with our physical reality at the physical level.

A specific Psyche is identified by the choices that it makes. Your psyche is the thing that decides what things mean to you. Your psyche functions just fine while separated from its spirit body and its physical body. Your psyche is the part of you that has experiences and remembers them. This is what has been experienced and observed.

In this chapter, David Bohm had some interesting things to say about freedom and individuality – freedom and psyche. I don't know if David Bohm ever uses the word "psyche", but that's what he is talking about whenever he talks about individuality, thought, or the mind.

David Bohm wrote:

No person can be said to be free who is for reasons of internal confusion unable consistently to carry out his or her chosen aims and purposes, for evidently such a person is driven by inner compulsions of which he or she is unaware. This inner lack of freedom is far more serious than a lack arising from external constraints or a lack of adequate knowledge of external circumstances.

If something is not considered a real possibility, there is no chance at all that it will appear among one's choices. Is there any meaning to freedom of will when the content of this will is thus determined by false knowledge of what is possible, false knowledge that we do not even know we possess (or, more accurately, that possesses us)? (*The Essential David Bohm*, pp. 255-256.)

Bohm talks a lot about the "unconscious" or the inner compulsions of which we are not aware. Psychologists have proven that the "unconscious" does in fact exist. The "unconscious" is purely a psyche phenomenon. It's a part of the Human

Psyche and has nothing to do with one's physical brain. Our physical brain only comes into play when we are conscious and awake.

The Materialists, Naturalists, Nihilists, and Atheists have developed "false knowledge" and they purposefully and deliberately embrace "falsified philosophies". These people are possessed by "false knowledge"; and therefore, they are not free to find and understand the truth. There's no chance that they can find the truth because they have formally rejected everything that is true. That's the power of choice – it can shut-down free will. It can enslave us.

David Bohm suggests as much within his writings, especially while indirectly talking about the Materialists, Naturalists, Nihilists, and Atheists.

The problem is seen to be even sharper if one considers that the question of choosing what is right and good so often arises in circumstances in which one's desire is in some other direction. Indeed, if something is clearly seen to be right and good, and if one has no desire to do otherwise, it hardly seems that any particular act of choice is ever needed. Will one then not spontaneously have the urge to act according to what one has perceived to be right and good? But often, as has been mentioned above, one finds that one has an irresistible desire in some other direction. One has by no means chosen this desire. Rather, it also arises from the totality of remembrances, reactions, and "knowledge" accumulated over the past, which responds to present "needs" as overwhelmingly urgent in ways of which one is not aware.

The attempt of will to struggle against such desire has no meaning, for this sort of desire contains in it a movement of self-deception, along with a further movement aiming to conceal this self-deception and to conceal the fact that concealment is taking place. Thus, one will often accept as true any false thought that makes one feel better (or more secure) or that makes one believe that the object of one's desire can be realized. This is, for example, the basis of the activity of the confidence trickster, who paints a false picture of satisfying greed that the victim cannot resist accepting as true. As long as one is ignorant of how this sort of self-deceptive desire operates, what can it mean to talk of freedom of any kind?

It appears, then, that the principal barrier to freedom is ignorance, mainly of "oneself" and secondarily of the "external world." This ignorance is also the main barrier to true individuality. For any human being who is governed by opinions and models unconsciously picked up from the society is not really an individual.

It is important to note that the main kind of ignorance that destroys freedom and prevents true individuality is ignorance of the activity of the past. As has been brought out earlier, although the past is gone, it nevertheless continues to exist and to be active in the

759

present, as a nested structure of enfoldments, going into even the distant past, which are carried along (with modifications) from one moment to the next. (I have treated these as projections of various kinds.) Here I have been indicating how this activity of the past can interfere with freedom and individuality as long as one is not aware of this past. (*The Essential David Bohm*, pp. 256-257.)

Self-deception works, and it works every time. I know, because I used to be a materialist, naturalist, nihilist, and atheist. As David Bohm suggests, we aren't free whenever we are forced to accept as true any false thought or falsified philosophy because it makes us feel better to do so. These people are enslaved by their past. Their rejection of evidence and suppression of evidence makes them ignorant of the truth; and therefore, the Materialists, Naturalists, Darwinists, Nihilists, and Atheists are no longer free to go and find the truth. It's a prison of their own making, and as David Bohm says, they are not even aware that they are enslaved.

David Bohm emphasizes the importance of learning how to rise above our inbuilt Mechanistic Philosophy.

How, then, is it possible for there to be the self-awareness that is required for true freedom? Along the lines of what has been said in this [book], I propose that self-awareness requires that consciousness sink into its implicate (and now mainly unconscious) order. It may then be possible to be directly aware, in the present, of the actual activity of past knowledge, and especially of that knowledge which is not only false but which also reacts in such a way as to resist exposure of its falsity. Then the mind may be free of its bondage to the active confusion that is enfolded in its past. Without freedom of this kind, there is little meaning even in raising the question as to whether human beings are free, in the deeper sense of being capable of a creative act that is not determined mechanically by unknown conditions in the untraceably complex interconnections and unplumbable depths of the overall reality in which we are embedded. (*The Essential David Bohm*, pp. 258-259.)

Here David Bohm defines the implicate order as the "unconscious order", which is in fact the Psyche Level of our existence. I have indeed observed that Materialism, Naturalism, Darwinism, Nihilism, Behaviorism, Determinism, and Atheism act in such a way as to resist exposure of their falsity. David Bohm repeated states that through conscious self-awareness, we can indeed free our minds or psyches from the mechanical prisons that we have created for ourselves in the past. Without freedom of the psyche or the mind, we are never truly free.

By this point in David Bohm's development, he has created the Science and the Philosophy necessary to understand and explain the Human Psyche and the Quantum Level of existence.

If we are the unknown, which is the present, then time can be seen in its proper meaning only in the context of that which is

beyond time (i.e., the holomovement or eternity). Any attempt to treat the whole meaning of existence in terms of time alone will lead to arbitrary and chaotic limitation of this existence, which then takes on the quality of being rather mechanical. If we are to be creative rather than mechanical, our consciousness has to be primarily in the movement beyond time. Implicitly, this is well known to us.

No one will be creative who does not have an intense interest in what one is doing. With such an interest, one can see that one will be at most only dimly conscious of the passage of time. That is to say, though physical time still goes on, consciousness is not organized mainly in the order of psychological time; rather, it acts from the holomovement.

On the other hand, if the mind is constantly seeking the goal of finishing its task and reaching its aim (so that it is organized in terms of psychological time) it will lack the real interest needed for true creativity.

A human being may be creative when his or her consciousness arises directly from the "timeless" holomovement. (*The Essential David Bohm*, p. 259.)

The holomovement is a part of David Bohm's implicate order, and they both are non-physical, timeless, spiritual, psychic, and quantum in nature and origin. In contrast, the Mechanical Order is in fact Bohm's explicate order, which is in fact physical and material in nature.

David Bohm often confuses or confounds his terms whenever he is discussing psychology, because he doesn't have a strong theological background and because there was little convincing evidence of Psyche's true nature and existence back in 1986 when David Bohm was writing about these things. It wasn't until the past decade (2010-2020), when Near-Death Experiences and Out-of-Body Experiences achieved critical mass on YouTube and became truly convincing.

https://www.youtube.com/results?search_query=nde

https://www.youtube.com/results?search_query=nde+atheist

Some of these are more convincing than others; but, they got through to me in the end. They led me to upgrade my science to Science 2.0. Science 2.0 allows all of the evidence into evidence and then pursues a preponderance of that evidence.

Science 2.0: I Upgraded My Science

https://www.amazon.com/gp/product/B0771K6WTX/

Back in 1986, David Bohm was finally starting to explore and take seriously these higher orders of existence and reality. He was no longer a materialist, naturalist, and nihilist even though he still seemed to be a Darwinist and atheist.

David Bohm was starting to approach the complete truth, but he hadn't yet arrived by the time that he died in 1992.

But once these higher levels are possible, why are they not always fully and harmoniously realized? I have proposed that, at least in part, this is because of ignorance. Such ignorance leads the mind to continue its past, mechanically, through identification rather as if it were a form of matter at a grosser level. The mind is trying in a confused way to realize the kind of creativity appropriate to such grosser levels of matter. In doing so, it is clearly unable to realize the kind of creativity appropriate to its own level.

Ending this state of ignorance may then open a new possibility for the mind to be creative at its own level. When it does this, it is still participating in the universal creativity, but now it is realizing its proper potential.

I suggest that this is the essence of freedom, to realize one's true potential, whatever the source of the potential may be.

So I propose that the question of freedom has to be looked at in a different way. This new way flows out of giving sustained and serious attention to how unfreedom arises basically from identification with the past, in which the mind commits itself to act as if it were determined mechanically in the ways in which grosser levels of matter are determined. We have to use the past, but to determine what we are from it is the mistake. To do this implies that such grosser mechanical existence in time has supreme value and that the main function of the mind is to sustain this sort of existence by continuing the past with modifications.

The clear perception that we are the unknown, which is beyond time, allows the mind to give time its proper value, which is limited and not supreme. This is what makes freedom possible, in the sense of realizing our true potential for participating harmoniously in universal creativity, a creativity that also includes the past and future in their proper roles. (*The Essential David Bohm*, p. 260.)

Here David Bohm is talking about mind over matter. Bohm encourages us to free ourselves from the grosser levels of mechanism or materialism, so that we are then free to be creative and to reach our full potential.

A couple of years ago, I came across a definition for "sin" that I found useful. This definition is completely in line with what David Bohm is saying here. **Sin is a failure to reach one's full potential.** This definition for sin is useful for everyone, whether they are religious or not. Failure to reach one's full potential is a sin.

For me personally, I was failing to reach my full potential while I was a materialist, naturalist, nihilist, and atheist. That was a sin. It was holding me back and keeping me ignorant. I was enslaved to it.

I have a dozen different friends who are homosexual and suffer from same-sex attraction, and each one of them has chosen a different way to deal with her or her misfortune.

Homosexual activity is a failure to achieve one's full potential. From a social perspective, it has been our observation that homosexual activity is unfruitful and unproductive; therefore, I am able to understand why some of my homosexual friends refuse to give into it and choose instead to try to do something better with their lives. A few of my gay friends have chosen not to be enslaved to a gay life-style nor to engage in homosexual activity, because they realize that it's pointless and fruitless to do so. I have a great deal of admiration for their courage, strength, and example. They made some hard choices. They chose to rise above their physical prison.

Of course, I have many gay friends who chose to give into it, one way or the other; and, I have a great deal of compassion for their misery and misfortune, because I too gave into Scientism and Scientific Naturalism when I was a materialist, naturalist, nihilist, and atheist and surrendered myself to that particular prison.

In this particular essay, David Bohm encourages to free ourselves from our mechanistic prisons or naturalistic prisons so that we can achieve our full potential. It's a lesson and invitation that has personal meaning to me on many different levels.

Mark My Words

Proof of Concept

A documentary entitled "Mind Control: HAARP Conspiracy" provides scientific proof that Quantum Neuroscience is correct and that Materialism, Naturalism, and contemporary Neuroscience are false.

https://www.youtube.com/results?search_query=Mind+Control+HAARP+Conspiracy

https://www.amazon.com/Mind-Control-Dr-Nick-Begich/dp/B079KD1N42/

https://en.wikipedia.org/wiki/High_Frequency_Active_Auroral_Research_Program

The government has known since the 1980's at least that we can use electromagnetic waves to beam actual words or voices into the auditory system of the human brain from a distance. It's also possible to project feelings of fear, anxiety, stress, and heat into the physical body from a distance using electromagnetic waves. This is Action at a Distance or Quantum Mechanics – NOT classical physics.

Contemporary Neuroscience teaches that the neurons communicate with each other through chemicals called neurotransmitters. That is FALSE. We've known for decades that this is false; but, this knowledge is being suppressed.

Remember, it's physically impossible to send a message through neurotransmitters from one neuron to the next. You see, if you tried to send a message using neurotransmitters, that message would be scrambled or randomized every time. Remember, neurotransmitters are dumped randomly into the synapse. The only thing that can be transmitted with a neurotransmitter is one single physical bit – OPEN the gate. One bit does not a message make. One bit simply acts as a key to OPEN a gate. It's physically impossible to store memories within synapses, and it's physically impossible to transmit messages or thoughts through neurotransmitters from one neuron to the next. Classical Neuroscience or contemporary Neuroscience has been falsified. Materialism, Naturalism, Mechanism, Darwinism, Classical Physics, and the Chemical Model for Neuroscience have all been FALSIFIED.

It's time for a better model. That's why I produced Quantum Neuroscience. It's a better model that attempts to explain what's really going on within a physical brain.

Quantum Neuroscience: The Answer to Life, the Universe, and Everything

https://www.amazon.com/gp/product/B079Z6QQQB/

Mind control actually requires access to the Psyche or the Mind. That's not possible at the physical level. It requires Action at a Distance at the quantum level in order to beam messages into the mind.

https://en.wikipedia.org/wiki/Alpha_wave#Brain_waves

The brain generates delta waves, alpha waves, and beta waves. That's how neurons truly communicate with each other. Neurons are communicating non-physically through various different quantum waves. If you get the right frequencies, you can actually beam words and sentences into the physical brain from a distance. There's no way to tell the difference between words beamed into the brain and words that come into the brain through the ears.

We send messages all the time through radio waves, and radio waves are just one type of quantum wave. It's all based upon Action at a Distance or Quantum Mechanics. The physical brain was designed by the Gods to be a receiver of quantum waves of various different types. Information can be communicated directly to the Human Psyche through the right kind of quantum waves. You can send messages or voices into a physical brain with the right frequency of electromagnetic waves; and, you can also communicate psyche-to-psyche with the right frequency of quantum waves. Thoughts are quantum waves; and, quantum waves are in fact Wi-Fi at the quantum level. It's really easy to understand, once a person chooses to do so.

The verified and proven existence of Action at a Distance FALSIFIES Materialism, Naturalism, Darwinism, Nihilism, and even Atheism. The false is falsified by the truth; and, the truth is repeatedly experienced and observed.

Even though it is science fiction, the "Skyline Series" is on the right track when it comes to quantum messaging. In those movies, the invading aliens use light to control the physical body and physical brain. That's Quantum Mechanics and not classical physics. Anytime you use light (quantum waves) to send messages at a distance, you are in fact using Quantum Mechanics and not classical physics. Mind control is theoretically possible through quantum waves, even though it's physically impossible at the physical level through physical manipulation.

https://www.amazon.com/Skyline-Eric-Balfour/dp/B004UTA5BG/

https://www.amazon.com/Beyond-Skyline-Frank-Grillo/dp/B077TVGKG5/

Study the Science Fiction and Fantasy. What do you observe? You soon observe that the Radically Advanced Aliens have mastered some sort of mind-over-matter technique or have developed some sort of matter-over-mind machine. The Super Advanced Aliens have learned how to control and manipulate quantum fields; whereas, we stupid humans have convinced ourselves that only physical matter exists. Our technology is physically-based; whereas, their technology or power is quantum-based or spiritually-based. In *Skyline*, whenever we destroy the alien ships, the aliens use quantum fields or "action at a distance" to reassemble their ships. That's quantum mechanics. In all of these alien movies, we are going up against Quantum Mechanics with Materialism, Naturalism, and Classical Physics. The aliens always win because they have mastered Quantum Mechanics, and we have not. The aliens use quantum fields as shields, and if we nuke or crash one of their ships, the aliens use quantum fields and action at a distance to reassemble their ships. It's definitely mind-over-matter, or quantum fields, or psyche, or intelligence that the aliens are using to conquer the humans who only know about Materialism or Naturalism and only know how to use Classical Physics.

Look at the New Testament and study what Jesus was able to do. He was a radically advanced alien. He was a God. He could do mind-over-matter and control quantum fields with his mind. He could also teleport or quantum tunnel his physical body. He could do transmutation. He could levitate and walk on water. He could convert energy into physical matter. He could control the weather. He could use his faith or knowledge to heal another physical body at a distance. Those are all quantum mechanical phenomena that have been proven to be real and truly exist. Every one of us could do the very same things that Jesus did if we had mastery of mind-over-matter or mastery of Quantum Mechanics. Psyche is the only thing we know of that can control quantum fields at every level of existence and reality. It is mind-over-matter at the physical level, and it is mind-over-energy at the quantum level. It works. It's proven technology. It's real. The Quantum Zeno Effect has been proven to be real and true.

You can study *The Expanse* television series, and once again, we see that the super advanced aliens have somehow learned to control and manipulate quantum fields. They have mastered Quantum Mechanics, can do action at a distance, can control quantum fields, can change the laws of physics, can do quantum tunneling, and are telepathic. Telepathy is Wi-Fi at the quantum level – psyche-to-psyche communication at a distance. The alien's technology is only interesting to us when they start demonstrating their mastery of Quantum Mechanics and Quantum Fields. To us, it looks like magic. To them, it's simply mind-over-matter.

https://www.amazon.com/gp/video/detail/B018BZ3UWU/

At the physical level, we humans have learned how to make science fiction into science reality. Our next step is to learn how to make the mind-over-matter science fiction into quantum reality or science reality.

Remember, the Human Psyche can break its conditioning at will. Within the *Skyline* movies, there are human beings (Human Psyches) who learn to break their conditioning, brainwashing, or mind control. The invading aliens use quantum waves within the light to control the brains of the humans; and, then some of the humans learn to use mind-over-matter (their Human Psyche) to regain control of their brains and the physical machine that their brains are connected to. The light changes color or frequency when the Human Psyche regains control of its physical brain. For the most part, the Science is accurate and correct. The only time they get it wrong is when they slip up and re-introduce Materialism and Naturalism into the mix. Mind control is purely a quantum phenomenon – not a physical phenomenon. If your physical brain is being controlled remotely by quantum waves, the only solution is for your Human Psyche to retake control and override the external mind control that's being projected into your physical brain. It's mind-over-matter or quantum mechanics that these Human Psyches are using to regain control of their physical brain. It's NOT their physical brain that they are using to regain control of their physical brain, because they have lost control of their physical brain and only their Human Psyche can regain control of their physical brain. It's obvious, once a person chooses to think about all of this logically and rationally.

The same thing applies to us whenever we lose control of our physical body and physical brain through addiction. The only way to regain control is mind-over-matter. Society will keep you addicted to the drugs. Your physical body and physical brain will keep you addicted to the drugs. Only your Human Psyche can decide to stop taking the drugs and suffer the consequences. It's mind-over-matter. Obviously, I have become interested in the Science where psyche is the best explanation or mind-over-matter is the best explanation for what we are observing and experiencing. It's a lot more interesting than Classical Physics.

Remember, the proven and verified existence of mind-over-matter – the placebo effect – FALSIFIES Materialism, Naturalism, Mechanism, and their derivatives such as Nihilism, Determinism, Darwinism, and Atheism. Contemporary materialistic and naturalistic Neuroscientists deny the existence of quantum waves, quantum consciousness, and mind control calling it nothing but pseudoscience. These people deny the existence of the psyche or the mind. They have to do so, because the verified and proven existence of Action at a Distance FALSIFIES

Materialism, Naturalism, and their derivatives. These people have to lie to themselves in order to maintain their false beliefs. Self-deception works, and it works every time. That's the power of mind-over-matter. The psyche can actually convince itself that psyche does not exist.

These people believe God out of existence and choose God out of existence, and for them personally, God really does not exist. They are without God in the world because they want to be. I KNOW, because I have been there and done that. I used to be a materialist, naturalist, nihilist, and atheist. God is in the light, and I chose to be ignorant and in the dark. Being without God in the world is another definition for hell. We don't have to die in order to go to hell. We can do so right here, right now, if we so choose. All we have to do is push the light out of our lives.

Mark My Words

Quantum Field Model for Origins

Origin Science is the most interesting type of Science to study and learn about. We modern scientists are in desperate need of a Model for Origins that actually matches with reality. We need a Model for Origins that actually matches with the observed reality which states that everything is made from energy or light, and that energy or light is always conserved. A Model for Origins based upon energy or light is the best type of model that we could possibly have because energy and light have actually been experienced and observed. They are as real as anything possibly can be.

Ever since I falsified and eliminated My Materialism, My Naturalism, My Nihilism, and My Atheism, I have found myself creating a New Model for Origins every year as each year goes by. This Quantum Field Model for Origins is my most recent version (2019). At first, I was trying to salvage the Big Bang Theory, but then I stopped trying when I finally started to find and develop something better.

Back in 2012, I was a materialist, naturalist, nihilist, and atheist. I don't know when exactly I overcame My Materialism and My Atheism because it took a couple of years to do so, once I started studying Science and Theology in earnest looking for the truth. It was Science – especially Quantum Mechanics – that finally convinced me that Materialism, Naturalism, and Atheism are false. By 2016, I started writing about what I had discovered.

We each have to overcome the False and the Falsified – Materialism, Naturalism, Darwinism, Nihilism, Behaviorism, Determinism, Mechanism, Physical Reductionism, and Atheism – BEFORE we can start to find and know the truth. The truth cannot be built upon falsehoods and falsified philosophies. The false and the falsified always produce abortions of the truth or pseudo-sciences masquerading as Science. The false and the falsified never produce the truth from nothing. It can't be done. The truth is discovered by studying what has actually been experienced and observed.

The problem with the Big Bang Theory and Darwinism is that they are Creation Ex Nihilo, or Atheism. There is no such thing as Creation Ex Nihilo, and the Spontaneous Generation that became the "creative cause" behind the Theory of Evolution was FALSIFIED in 1859 by Louis Pasteur. It's obvious that "nothing" can't design and create anything. There is no such thing as Spontaneous Generation or Creation Ex Nihilo, which means that there is no such thing as Atheism or that Atheism has been FALSIFIED by Science itself. The reason that the Big Bang Theory and the Theory of Evolution are false is that they were designed to eliminate Psyche or God or Intelligence from the equation. They were designed to be Creation Ex Nihilo, which is impossible. There's no such thing as Creation Ex Nihilo. Even God can't do creation ex nihilo or creation from nothing. It's totally illogical and absurd to have "nothing" designing and creating everything from nothing. Atheism or Creation Ex Nihilo is patently absurd.

The creation of a Big Bang Singularity from nothing doesn't make logical sense, especially since the Big Bang Singularity is by definition the collection of a nearly infinite amount of energy into a single point or singularity. If you choose to believe that the Big Bang happened for real – and most scientists do – then you must believe in the God who gathered and organized all of that energy into the Big Bang Singularity in the first place. Big Bang Singularities don't just spontaneously generate out of thin air from nothing as the Materialists, Naturalists, Nihilists, and Atheists claim. They have to be carefully designed and made and implemented. If the Big Bang happened for real, then Someone Psyche or Someone Intelligent at the psyche level had to design and create the Big Bang Singularity at the quantum level.

However, there are more plausible and more realistic explanations for the origin of our physical universe besides the Big Bang Theory, which has been falsified on many different levels and from many different aspects. The Quantum Field Model for Origins is one such model. Here's how it works.

The original primal construct was the Chaos Realm, or what some people call "matter unorganized" or "energy unorganized". Energy has always existed, and it will always exist.

Out-of-Body Travelers have observed that it is in fact matter all the way down the rabbit hole.

How is that possible?

It's because Psyche has been seen or observed to be the Ultimate Point-Particle. When it comes to physics, the smaller dwells within and controls the larger. Psyche is a pin-prick of light. Psyche is a point-particle or an infinite singularity, even at the quantum level. Every Psyche or Intelligence has a certain amount of energy that's under its control. That's just the way it is, and that's the way it has always been.

In the Chaos Realm or the Primal Construct, all of these Psyches and energy were unorganized. They needed Someone Intelligent or Someone Psyche with sufficient knowledge and intelligence to organize them into something valuable and useful.

What a newly-minted Head God does is pick a vast stretch of unorganized Psyches (matter) and unorganized energy within the Chaos Realm; and then that Head God establishes his throne and home in the middle of all that chaos; and then, He starts to organize it.

What's the first thing that the Gods organize?

It isn't the Psyches, per se, because Psyches are eternal and everlasting. They cannot be made, and they cannot be destroyed. What the Gods are organizing is the energy that's under the control of these Psyches or Intelligences. The Gods teach these fledgling Psyches how to organize the energy under their control into useful structures. The Gods teach these Psyches how to organize. The Gods establish a union among them. It's all about organization – not creation ex nihilo. There's no such thing as creation ex nihilo, or Atheism.

Whenever a God enters into a new stretch of Chaos, the first thing He organizes is the quantum fields. Quantum fields are made out of energy by Someone Psyche. The creation of quantum fields is a joint effort that takes place between the Head God and all the different Psyches that are controlling the energy from which the quantum fields are made.

As this organization or formation of quantum fields spreads out from the central starting point, it's as if a physical universe is born from a Big Bang Singularity. It will leave a mark on the universe as a whole. We can actually see that our physical universe was organized from some kind of central starting point expanding outwards in all directions at the speed of light or faster. The organization of pre-existing energy is a lot more plausible and believable than the creation of a Big Bang Singularity from nothing, whether we are talking about the Big Bang Theory or the Quantum Field Model for Origins.

The Gods have to make the quantum fields first BEFORE physical matter becomes possible to make, because physical matter is made from energy or fields and because physical matter requires quantum fields to sustain it and propagate it.

The Quantum Field Model for Origins is a much more feasible, logical, rational, and parsimonious explanation for the physical universe that we see around us, because it employs Intelligence in order to get the job done in a timely and efficient manner. Processes of any kind are no longer based upon chance whenever Psyche or Intelligence is involved. Organization becomes infinitely more realistic, plausible, possible, and guaranteed when Psyche or Intelligence is involved.

Look at our own earth. We've got mountains made out of rock and beaches made out of sand; and, that's the way it would always be, until the Human Psyche or Human Intelligence gets involved. Once Psyche or Intelligence starts to act on that rock and sand, then computers, houses, televisions, cars, factories, airplanes, rocket ships, bridges, roads, skyscrapers, microwave ovens, the internet, and the electric grid are the result. None of these things would exist without some kind of Psyche or Intelligence to organize them from the raw materials that we call physical matter. This same truth applies to genomes. Genomes have to be designed, created, engineered, field-tested, manufactured, and then deployed. That requires intelligence. Functional genomes don't just spontaneously generate out of thin air ex nihilo as the Materialists, Naturalists, Darwinists, Nihilists, and Atheists claim. This same truth also applies to quantum fields. The quantum fields were obviously designed and made to fulfill a specific purpose. Without the quantum fields, physical matter could not be made. Someone Psyche or Someone Intelligent had to design and then make the quantum fields, which permeate everything today, from the raw chaotic unorganized energy that was once a part of the original Chaos Construct.

This is how it was really done.

Someone Psyche had to design and then organize the quantum fields from raw energy BEFORE physical matter became possible. It wasn't some kind of physical brain that designed and organized these quantum fields because physical matter didn't exist yet. Quantum fields had to be designed and made by Someone Psyche or Someone Intelligent. There is no other way it could have been done, because there is no such thing as spontaneous generation or creation ex nihilo, even at the quantum level and the psyche level.

Quantum fields are non-physical and pre-physical. The Gods had to make the quantum fields first BEFORE physical matter became possible to make and sustain. The proven and verified existence of quantum fields FALSIFIES Materialism, Naturalism, and their derivatives such as Darwinism, Nihilism, and Atheism. The absolute and necessary existence of quantum fields for the production of physical matter FALSIFIES Materialism, Naturalism, and their derivatives. The false is falsified by the truth; and, the truth is repeatedly experienced and observed. The effects of these invisible, non-physical quantum fields are constantly being experienced and observed.

The effects of Intelligence or Psyche are also constantly being experienced and observed. Whereas, it's impossible to experience and observe the non-existence of Psyche or the non-existence of God, because non-existence cannot be experienced and observed because by definition in principle it does not exist, so there's nothing to experience and observe. Instead, we go with the truth and accept the fact that the resurrected Jesus Christ, Psyches or pin-points of light, the conservation of energy, and quantum fields (or their effects) are constantly being experienced and observed both at the physical level and also at the quantum level by out-of-body explorers and near-death experiencers. Psyche, Intelligence, Quantum Fields, and Light have been experienced and observed on both sides of the veil. They are real and truly exist; and, they are infinitely better than "nothing".

Once I realized and accepted the fact that Science, experience, and common sense have proven that there is no such thing as spontaneous generation or creation ex nihilo, then I realized that it's pointless to try to create a Model for

Origins that excludes Intelligence, Psyche, or God from the process because order and organization are impossible without the direct intervention of Someone Psyche or Someone Intelligent. Science has proven that disorder and entropy cannot be overridden, reversed, or overcome by anything other than Intelligence or Psyche. Science has also proven that entropy or disorder can be eliminated at will by information, intelligence, or psyche. It's obvious that this is true; otherwise, this physical universe would not exist. Someone Psyche had to provide some syntropy or organization at the quantum level, or this physical universe would not exist. Someone Psyche had to make quantum fields from disorganized chaotic energy, or physical matter could not exist. Order, organization and syntropy can only be produced by Someone Psyche at the quantum level because everything tends toward disorder and entropy and thermal equilibrium at the physical level. Our very existence, as physical beings, tells us that this is true.

Only Psyche or Intelligence can do order, organization, syntropy, and creation at the quantum level or the energy level; and, Someone Psyche had to design and make the quantum fields from energy before particles and physical matter could be made. This is obviously true. First things first!

Our existence comes from the quantum level and the psyche level, and not the physical level. Our life force and existence come from the energy level or the quantum level where everything is being conserved and where everything is eternal and everlasting. In contrast, the primary axiom of quantum field theory is the idea that particles are born, and particles die, including physical particles. There's nothing eternal and everlasting about particles and physical matter, except for the energy from which they are made. Furthermore, it is obvious that Someone Psyche or Someone Intelligent has to organize energy into quantum fields and particles; otherwise, that level of order and organization would never exist because there is no such thing as creation ex nihilo or the spontaneous generation of order and organization from nothing. Creation from nothing by nothing can't happen, which means that it didn't happen. Only Psyche or Intelligence or Information can reverse and eliminate disorder and entropy. Only Psyche or Intelligence can do creation.

It has been observed that Psyche or Intelligence can overcome and negate entropy at will, both from the quantum level and at the physical level. Information or intelligence is all that's needed to override entropy, reverse entropy, or eliminate entropy. Entropy can be eliminated at will by Someone Psyche or Someone Intelligent. This is what has been experienced and observed. This is what Science and History are trying to tell us.

Based upon the Second Law of Thermodynamics, the Materialists, Naturalists, Darwinists, Nihilists, Mechanists, and Atheists erroneously teach and believe that our physical universe is running out of energy as it approaches thermal equilibrium or heat death. They are wrong. The energy is always there, and the energy is always conserved, and the energy is always available at the quantum level, even if that energy achieves thermal equilibrium at the physical level and becomes unavailable for doing work at the physical level. Energy is the ultimate perpetual motion machine – like Psyche, energy cannot be made and it cannot be destroyed. It is always conserved. It is always available for use.

Remember, there is no such thing as the Second Law of Thermodynamics at the quantum level, because there is no thermodynamics or thermal equilibrium at the quantum level. At the quantum level, the Psyches and the energy are always conserved and therefore always available for use. Entropy is non-existent at the quantum level and the psyche level. This is what has been experienced and observed. The bane of Classical Physics, Materialism, Naturalism, and their derivatives – entropy – ceases to exist at the quantum level and the psyche level.

Remember, each Psyche has a certain amount of energy that's under its control. The Psyches that control that energy at the quantum level can turn that energy into anything that they want it to be, any time they choose to do so. Energy is infinitely malleable by Someone Psyche or Someone Intelligent. Once the energy reaches thermal equilibrium within a certain part of a physical system and that physical system achieves maximum entropy, the Psyches that control that energy can transform all of that energy into a new planet, star, or galaxy whenever they choose to do so. The energy is always there and always usable and reusable at the quantum level. Our physical universe is not going to end in heat death. At some point, it will transform into something else. That's the way energy works. Energy is always conserved, and the Controlling Psyche can form the energy under its control into anything that that Psyche wants its energy to be. If it wants that energy to become part of a new sun and a heat-producing factory, it can choose to do so at any time.

There really is no such thing as entropy. Entropy is NOT a person, place, or thing that's going to someday reign supreme. Energy doesn't work that way. Under the control of Someone Psyche, energy is infinitely malleable and infinitely transformable; and, energy or psyche is always conserved. Furthermore, Psyche or Intelligence can override and eliminate entropy at will.

If the Second Law of Thermodynamics and Materialism were in fact the controlling laws of our universe as the Naturalists and Atheists claim, then there could be NO big bang singularity and no big bang and no physical beginning, because everything would have reached thermal equilibrium an eternity ago and we wouldn't exist today. Thank goodness that Materialism, Naturalism, and the Second Law of Thermodynamics are FALSE, because physical matter, and this physical universe, and our physical earth as we now see them and experience them would not exist and could not exist without Someone Psyche or Someone Intelligent to overcome the effects of entropy or thermal equilibrium, and establish these planets and stars in the first place.

This Quantum Field Model for Origins is infinitely more plausible, believable, and possible than the big bang theory, creation ex nihilo, or the spontaneous generation associated with the theory of evolution because there is actually some Intelligence behind this Quantum Field Model of Origins; and, that makes all the difference in the world.

The primary advantage of this Quantum Field Model for Origins is that every part of it has actually been experienced and observed; and therefore, every part of it has been proven to be real and true. None of it is theory. It's all Law, because it has been experienced and observed. According to the Law of the Conservation of Energy, psyche or energy is always conserved. It cannot be made, and it cannot be destroyed. It has always existed, and it will always exist. It has no beginning, and it cannot die.

In contrast, according to Quantum Field Theory, particles are born, and particles die. When particles die, their energy is absorbed back into the quantum fields from whence they came. This means that planets, stars, and galaxies are born; and, they die or cease to exist when their energy is absorbed back into the quantum fields from whence they came. If protons were to decay, they would in fact turn back into energy where they could be reused again and made into something else, including another fully recharged proton. According to Quantum Field Theory, particles are popping into existence from the quantum fields, and then annihilating or ceasing to exist by being absorbed back into the quantum fields, all the time. Particles are born, and particles die. Entropic physical matter is born, and entropic physical matter can die or have its energy absorbed back into the quantum fields.

The Big Bang Theory doesn't work because it is Creation Ex Nihilo, and therefore violates the Law of the Conservation of Energy. In contrast, the Quantum Field Model for Origins is based exclusively on the Conservation of Energy which has been proven to be real and true. Particles are born, and particles die. Planets, stars, and galaxies are born whole; and, when they die, they are dissolved, and their energy is returned back into the quantum fields for reuse. The cosmic background radiation is the result of all these different galaxies being born whole all at once. Stars produce heat and light.

One of the main reasons why the Big Bang Theory is false and has been falsified is because it's physically impossible for expanding clouds of hydrogen and helium gas to coalesce into planets, stars, and galaxies. It can't be done, which means that it wasn't done. When we start pumping hydrogen and helium into a balloon or a canister, does it coalesce into a pea-sized lump or a planetoid? NO! Hydrogen molecules and helium atoms repel each other, even while under pressure. That's why hydrogen and helium balloons float, rather than coalescing into a brick or a planetoid. Even under the pressure of gravity and an atmosphere, hydrogen and helium repel each other. They never coalesce.

We have observed through our telescopes that planets, stars, and galaxies are born whole all at once. According to Quantum Field Theory, they pop into existence as a unit all at once. That's the only way to form them and make them work, because expanding clouds of hydrogen and helium gas are NEVER going to coalesce into planetoids, planets, stars, and galaxies. It's physically impossible, which means it can't be done. However, quantum fields have been proven to be real and true; and furthermore, the fact that particles are born, and that particles can die by having their energy absorbed back into the quantum fields, has also been proven to be real and true. Every part of this Quantum Field Model for Origins has been proven to be real and therefore truly exists.

The Big Bang Theory is based upon ideas that are physically impossible and that have been falsified. The Big Bang Theory was designed to eliminate psyche, intelligence, or God from the equation. In contrast, everything about the Quantum Field Model for Origins has been experienced and observed, which means that it has been proven to be true and continues to be proven that it's true. The one has been proven to work, and the other one has been falsified in multiple different ways. The false is falsified by the truth; and, the truth is repeatedly experienced and observed. That's the way we are supposed to do Science.

Even Intelligence, or Psyche, has been experienced and observed. Whenever a particle pops into existence, some type of intelligence or psyche has to determine what type of particle it will be. It could become anything. Energy is infinitely malleable. Someone has to decide that that energy is going to be a quark rather than a photon. If it were all completely random as the Materialists, Naturalists, and Atheists claim, then it would be nothing but chaos. We KNOW that intelligence or psyche or consciousness is involved in Quantum Field Theory because according to Feynman diagrams, the creation of particles, the transformation of particles, and the absorption of particles back into the quantum fields follows specific repeatable and predictable rules or quantum laws. There's nothing random about it. It's been intelligently designed, and the rules or quantum laws are intelligently followed and obeyed. Any type of Law requires in intelligent Law-Maker to make and enforce the Law; and, any type of Law also requires an intelligent Law-Follower to understand, follow, and obey the Law. This is obviously true. At the quantum level, the intelligent Law-Maker and the intelligent Law-Follower have to be some kind of Psyche, because when it comes to physics, the smaller always dwells within and controls the larger.

So, even the intelligent aspects of this Quantum Field Model for Origins have actually been experienced and observed; and, we KNOW for a fact that intelligence exists and has an influence when it comes to physics. It's obvious. The Quantum Field Model of Origins is based upon a Psyche Ontology, wherein the point-particle that we call Psyche is in fact the fundamental unit of all reality and existence. The existence of Psyche is obviously true because these point-particles or pinpricks of light have actually been experienced and observed at the quantum level. This Quantum Field Model for Origins is based upon principles and laws that are obviously true because they have actually been experienced and observed; and, they continue to be experienced and observed. It works because every part of it is real and it is true. There's really no theory here when it comes to the Quantum Field Model for Origins because every part of has been experienced and observed, and therefore proven to be real and true.

Mark My Words

PART VII — NIRVANA, PSYCHE, OR QUANTUM CONSCIOUSNESS

Every eastern religion defines Nirvana as some type of non-physical existence – or non-physical non-existence. In Hinduism's version of Nirvana, the Atman (the individual psyche, soul, life force, intelligence, or consciousness) unites with Brahman (God's psyche or soul, or universal cosmic consciousness); and, they become one. The psyche or soul returns to the God who gave it life.

Buddhism erroneously teaches that Atman, Psyche, or Soul ceases to exist when we die. Buddhism is spirituality for Atheists, Materialists, and Naturalists. Buddhism erroneously teaches that psyche or energy is NOT conserved. They are wrong. Energy, psyche, or light is always conserved.

There is a video entitled, "Inner Worlds, Outer Worlds", that talks about this Universal Quantum Consciousness, Akashic Field, Matrix of Quantum Fields, Universal Psyche, or Universal Intelligence.

https://www.amazon.com/gp/video/detail/B07K5HBZFV/

https://www.scripts.com/script-pdf/1356

Here are some of the paraphrased highlights:

There is one vibratory field that connects all things. It has been called Akasha, Logos, the primordial OM, the music of the spheres, the Higgs field, dark energy, and a thousand other names such as the Matrix of Quantum Fields, Quantum Field Theory, the Big Bang Singularity, the Universal Sound, the Cosmic Background Radiation, the Zero-Point Field of Light, the Quantum Sea of Light, Inflation or Expansion, the Fabric of the Universe, Cosmic Consciousness, the Universal Broadcast, Quantum Consciousness, the Universal Blueprint, the Mind of God, Nirvana, Enlightenment, Vitality, the Word of God, God's Word of Command, God's Priesthood Power, the Holy Ghost, the Spirit, or the Light of Christ.

Every religion and every science has some version of this thing. It exists. Quantum Field Theory is verified science. It's true.

Here are some quotes from this movie:

In the beginning was the Logos, the Big Bang, the primordial Om. Big Bang theory says that the physical universe spiraled out of an unimaginably hot and dense single point called a singularity – billions of times smaller than the head of a pin. It does not say why or how. The more mysterious something is, the more we take for granted that we understand it.

Images from the Hubble space telescope show that the universe's expansion seems to be actually accelerating. Expanding faster and faster as it grows out of the Big Bang. Somehow, there is more mass in the universe than physics predicted. To account for the missing mass, physicists now say that the universe consists of only 4% atomic matter or what we consider normal matter. 23% of the universe is dark matter and 73% is dark energy – what we previously thought of as empty space. It is like an invisible nervous system that runs throughout the universe connecting all things.

The ancient Vedic teachers taught Nada Brahma – the universe is vibration. The vibratory field is at the root of all true spiritual experience and scientific investigation. It is the same field of energy that saints, Buddhas, yogis, mystics, priests, shamans and seers have observed by looking within themselves. It has been called Akasha, the Primordial Om, Indra's net of jewels, the music of the spheres, and a thousand other names throughout history. It is the common root of all religions, and the link between our inner worlds and our outer worlds.

Indra's net of jewels is a metaphor used to describe a much older Vedic teaching which illustrates the way the fabric of the universe is woven together. Indra, the king of the gods, gave birth to the sun and moves the winds and the waters. Imagine a spider web that extends into all dimensions. The web is made up of dew drops and every drop contains the reflection of all the other water drops, and in each reflected dew drop you will find the reflections of all the other droplets. The entire web, in that reflection and so on, to infinity. Indra's web could be described as a holographic universe, where even the smallest stream of light contains the complete pattern of the whole.

This Holographic Universe is the idea that each individual physical atom contains within it the sum-total of the universe, or all the knowledge and information in the universe. God is in the energy, or the quantum field, or the light.

Tesla used the term Akasha to describe the etheric feel that extends throughout all things.

In the Vedic teachings, Akasha is space itself; the space that the other elements fill, which exists simultaneously with vibration. The two are inseparable. Akasha is yin to prana's yang. A modern concept that can help us to conceptualize Akasha, or the primary substance, is the idea of fractals.

Just like with holograms, when it comes to fractals, each individual part contains within it the sum-total of the whole. A fractal has infinite depth. It never ends.

They are limited, but at the same time, infinite. A fractal is a rough geometric shape that can be split into parts, each of which is approximately a reduced sized copy of the whole pattern – a property called self-similarity. Mandelbrot's fractals have been called the thumbprint of God.

Humans have long associated beauty and the sacred with fractal patterns. Infinitely complex, yet every part contains the seed to recreate the whole. Fractals have changed mathematicians' views of the universe and how it operates. With each new level of magnification, there are differences from the original. Constant change and transformation occurs as we traverse from one level of fractal detail to another. This transformation is the cosmic spiral. The embedded intelligence of the matrix of time space.

At the speed-of-light, everything transforms or transmutes from the physical to the quantum or the spiritual. Quantum Mechanics is Energy Mechanics – the scientific study of how energy and psyche really work at the quantum level. Every Psyche or Intelligence has a certain amount of energy that's under its control. Energy is infinitely malleable, or infinitely transformable. Energy or psyche is always conserved, which means that it is always preserved. According to the Law of the Conservation of Energy, energy can neither be created nor destroyed; rather, it can only be transformed or transferred from one form to another. In other words, Psyche, Intelligence, or Consciousness can transform the energy under its control into anything that it wants that energy to be, any time that it chooses to do so.

All energy in the universe is neutral, timeless, dimensionless. Our own creativity and capacity for pattern recognition is the link between the microcosm and macrocosm. The timeless world of waves and the solid world of things. Observation is an act of creation through limitations inherent in thinking.

The Theory of Relativity correctly teaches that at the speed-of-light, time stops, distance goes to zero, length contracts to zero, and velocity goes to infinity. The Quantum Realm or the Energy Realm in its original native format is indeed timeless, dimensionless, and without physical limitations. Because time stops a speeds faster than light, there is NO distance in the Quantum Realm or the Energy Realm. If something chooses to move within the Quantum Realm or the Spirit World, it can simply quantum tunnel or teleport to its destination instantaneously at an infinite velocity because there is NO distance within the Quantum Realm or the Matrix of Quantum Fields. A photon traveling at the speed-of-light experiences nothing, because from its perspective it simply quantum tunnels to its destination.

Creativity is our highest nature. With the creation of things comes time, which is what creates the illusion of solidity. Einstein was the first scientist to realize that what we think of as empty space is not nothing. It has properties, and intrinsic to the nature of space is nearly unfathomable amounts of energy. The renowned physicist Richard Feynman once said, "There is enough energy in a single cubic meter of space to boil all the oceans in the world."

And, that's at the physical level! At the quantum level, there is enough energy within a cubic centimeter of space to create ALL of the physical matter within our observable physical universe. Feynman was a Materialist, Naturalist, and Atheist. He only thought in terms of physical matter. He didn't allow himself to think in terms of the non-physical.

Advanced meditators know that in the stillness lies the greatest power. The Buddha had yet another term for the primary substance; what he termed kalapas, which are like tiny particles or wavelets that are arising and passing away trillions of times per second. Reality is, in this sense, like a series of frames in a holographic film camera moving quickly as to create the illusion of continuity. When consciousness becomes perfectly still, the illusion is understood because it is consciousness itself that drives the illusion.

As a scientist, who has studied Quantum Mechanics and Quantum Field Theory, it soon became obvious that Quantum Fields are non-physical and pre-physical. Quantum Fields were made from Energy by some type of Cosmic Consciousness or Quantum Consciousness. It's obvious that the Gods, or God's Psyche, or Brahma had to design and make the Quantum Fields BEFORE they could make and sustain physical matter. Your Psyche is eternal and everlasting. It's very much possible that your Psyche was involved in the design and creation of the Quantum Fields, and later involved in the design and creation of physical matter within those quantum fields. These universal fields or quantum fields are indeed the foundation of every science and world religion. Within this book, I have defined "Nirvana" as Cosmic Consciousness, Quantum Conscious, God's Mind, or God's Psyche. I have also defined "Nirvana" as a Matrix of Quantum Fields. Every science and every religion on the planet does so, whether we realize it or not. The only exception are the falsified philosophies such as Materialism, Naturalism, Darwinism, Nihilism, Behaviorism, Determinism, Physical Reductionism, Atomism, and Atheism or Creation Ex Nihilo. These falsified philosophies are falsified by Quantum Mechanics, Quantum Field Theory, and Quantum Consciousness. The false is falsified by the truth; and, the truth is repeatedly experienced and observed.

In the ancient traditions of the East, it has been understood for thousands of years that all is vibration. "Nada Brahma" – the universe is sound. The word "nada" means sound or vibration and "Brahma" is the name for God. Brahma, simultaneously IS the universe and IS the creator. The artist and the art are inseparable. In the Upanishads, one of the oldest humans records in ancient India, it is said "Brahma the creator, sitting on a lotus, opens his eyes and a world comes into being. Brahma closes his eyes, and a world goes out of being." Ancient mystics, yogis and seers have maintained that there is a field at the root level of consciousness.

There is an Energy Field or a Matrix of Quantum Fields at the root of everything. Every religion has this at its foundation or core, including the revealed religions. The resurrected Biblical God Jesus Christ has revealed to us that this is true.

Doctrine and Covenants 88: 6-13:

6 He that ascended up on high, as also he descended below all things, in that he comprehended all things, that he might be in all and through all things, the light of truth;

7 Which truth shineth. This is the light of Christ. As also he is in the sun, and the light of the sun, and the power thereof by which it was made.

8 As also he is in the moon, and is the light of the moon, and the power thereof by which it was made;

9 As also the light of the stars, and the power thereof by which they were made;

10 And the earth also, and the power thereof, even the earth upon which you stand.

11 And the light which shineth, which giveth you light, is through him who enlighteneth your eyes, which is the same light that quickeneth your understandings;

12 Which light proceedeth forth from the presence of God to fill the immensity of space —

13 The light which is in all things, which giveth life to all things, which is the law by which all things are governed, even the power of God who sitteth upon his throne, who is in the bosom of eternity, who is in the midst of all things.

The Light of Christ, or this Akashic Field, or the Matrix of Quantum Fields proceeds from the presence of God or Brahma to fill the immensity of space. They are all talking about the same thing. This Light of Christ is the power by which everything was made. It ends up being the Universal Matrix of Quantum Fields that fill the immensity of space and penetrate to the very heart of every physical atom. It all makes sense. It's real. It's true. Every science and world religion has discovered the truthfulness of it, except for the Materialists and Naturalists who ignorantly deny the existence of it. Been there and done that! I used to be a Materialist, Naturalist, Nihilist, and Atheist until I learned better. We all have to start where we are and go on from there.

The Akashic field or the Akashic records where all information, all experience past, present and future, exists now and always. It is this field or matrix from which all things arise. From sub-atomic particles, to galaxies, stars, planets and all life. You never see anything in its totality because it is made up of layer upon layer of vibration and it is constantly changing, exchanging information with Akasha.

This Universal Matrix of Quantum Fields or Light of Christ is intelligent, which means that it is conscious, psychic, and alive. Remember, every Psyche has a certain amount of energy that's under its control; and together, they form some type of Cosmic Consciousness or Quantum Consciousness. It's all linked together into one united whole at the quantum level; yet, zillions of Psyches or Intelligences exist together simultaneously at the same time. Remember, energy fields contain information. They are conscious, sentient, perceptive, and alive. Every form of energy has some type of controlling Psyche within it.

In the Native American and other indigenous traditions it is said that everything has spirit which is simply another way of saying everything is connected to the one vibratory source. There is one consciousness, one field, one force that moves through all. This field is not happening around you, it is happening THROUGH you and happening AS you. You are the "U" (you) in universe. You are the eyes through which creation sees itself. When you wake from a dream you realize that everything in the dream was you. You were creating it. So called real life is no different. Everyone and everything is you. The one consciousness looking out of every eye, under every rock, within every particle.

There is One Hologram, One Universe, or One Head God; but, just like a hologram, each individual part of the hologram contains within it the whole hologram. You can take a hologram and tear it apart into Individual Psyches, and yet every Individual Psyche or hologram contains within it the overall Universal Hologram. This is the Many within One; and, the One within Many that Jesus Christ and other spiritual teachers have taught.

International researchers at CERN, the European laboratory for particle physics, are searching for this field that extends throughout all things. But instead of looking within, they look to the outer physical world.

The only ones who can't get this, understand it, and accept it are the Materialists, Naturalists, Darwinists, Nihilists, Behaviorists, and Atheists because the proven and verified existence of Quantum Fields falsifies Materialism, Naturalism, and their derivatives such as Darwinism, Nihilism, and Creation Ex Nihilo or Atheism.

The Higgs Boson experiments prove scientifically that an invisible energy field fills the vacuum of space.

ALL of the science experiments in Quantum Mechanics, Quantum Field Theory, Out-of-Body Exploration, and Near-Death Experiments prove the Conservation of Psyche or the Conservation of Energy, and prove that invisible non-physical energy fields fill the immensity of space. Each individual physical atom is chock full of these invisible non-physical quantum fields.

The standard model cannot account for how particles get their mass.

Mass is resistance to acceleration, and it only applies at the physical level or the space-time level. Remember, at the speed-of-light, mass ceases to exist, time stops, length contracts to zero, distance goes to zero, and velocity goes to infinity. Mass is simply another form of energy. $E = mc^2$. Mass is just another form of energy. Energy can assume an infinite number of different forms. Every Psyche has a certain amount of energy that's under its control; and, that controlling Psyche can form the energy under its control into anything that it wants that energy to be, including mass or physical matter. Mass functions as physical brakes on the infinite velocity of the quantum realm. Mass is synonymous with physical limitations. Remember, there are NO physical limitations at the quantum level, only at the physical level. It's really easy to understand once a person chooses to do so.

Everything appears to be made of vibration but there is no 'thing' being vibrated. It is as if there has been an invisible dancer, a shadow dancing hidden in the ballet of the universe. All the other dancers have always danced around this hidden dancer. We have observed the choreography of the dance, but until now we could not see that dancer. The so-called "God Particle", the properties of the base material of the universe, the heart of all matter which would account for the unexplained mass and energy that drives the universe's expansion. But far from explaining the nature of the universe, the discovery of the Higgs Boson simply presents an even greater mystery, revealing a universe that is even more mysterious than we ever imagined. Science is approaching the threshold between consciousness and matter. The eye with which we look at the primordial field and the eye with which the field looks at us are one and the same.

Energy is constantly in motion. Energy is conscious and alive. A physical atom does indeed have locality in space-time; but, the energy and psyche within that physical atom are constantly in motion because they are conscious, psychic, and alive. The Quantum Zeno Effect has proven that the Psyche or Intelligence within a physical atom can read your mind and knows when you are looking at it or not. The false is falsified by the truth; and, the truth is repeatedly experienced and observed. It's all about Psyche, Consciousness, Intelligence, or Life Force. Life wouldn't exist without it. Every science and every religion contains this truth within it. It's unavoidable. We wouldn't exist without some kind of Intelligence, Psyche, or Life Force within us.

<u>Abraham 3: 18-23</u>**:**

18 Howbeit that he made the greater star; as, also, if there be two spirits, and one shall be more intelligent than the other, yet these two spirits, notwithstanding one is more intelligent than the other, have no beginning; they existed before, they shall have no end, they shall exist after, for they are gnolaum, or eternal.

19 And the Lord said unto me: These two facts do exist, that there are two spirits, one being more intelligent than the other; there shall be another more intelligent than they; I am the Lord thy God, I am more intelligent than they all.

21 I dwell in the midst of them all; I now, therefore, have come down unto thee to declare unto thee the works which my hands have made, wherein my wisdom excelleth them all, for I rule in the heavens above, and in the earth beneath, in all wisdom and prudence, over all the intelligences thine eyes have seen from the beginning; I came down in the beginning in the midst of all the intelligences thou hast seen.

22 Now the Lord had shown unto me, Abraham, the intelligences that were organized before the world was; and among all these there were many of the noble and great ones;

23 And God saw these souls that they were good, and he stood in the midst of them, and he said: These I will make my rulers; for he stood among those that were spirits, and he saw that they were good; and he said unto me: Abraham, thou art one of them; thou wast chosen before thou wast born.

If you run simple sine waves through a dish of water, you can see patterns in the water. Depending on the frequency of the wave, different ripple patterns will appear. The higher the frequency, the more complex the pattern. These forms are repeatable, not random. The more you observe, the more you start to see how vibration arranges matter into complex forms from simple repeating waves. This water vibration has a pattern similar to a sunflower. Simply by changing the sound frequency, we get a different pattern.

Life itself exists within these Psychic Vibrations. Life Force is energy, and it's also psychic vibration. Remember, the controlling Psyche can form the energy under its control into anything that it wants it to be. The fact that these vibrational patters are repeatable and not random is scientific proof that there is some kind of Universal Blueprint or Akashic Field that the psyches within physical matter have access to. The Psyches or Intelligences within our physical universe are deliberately and knowingly following some kind of Universal Law or Universal Blueprint; and, they are coordinating their actions. This is the very definition of intelligence or life.

Water is a very mysterious substance. It is highly impressionable. That is, it can receive and hold onto vibration. Because of its high resonance capacity and sensitivity and an inner readiness to resonate, the water responds instantaneously to all types of sonic waves. Vibrating water and earth make up the majority of mass in plants and animals. It is easy to observe how simple vibrations in water can create recognizable natural patterns but as we add solids and increase the amplitude, things get even more interesting. Adding cornstarch to water, we get more complex phenomena. Perhaps the principles of life itself can be observed as vibrations move the cornstarch blob into what appears to be a moving organism.

Every atom within the cornstarch and the water is connected with some kind of Universal Intelligence or Universal Blueprint – what David Bohm called an "Organizing Principle" or a Super-Implicate Order. Life Force or Psyche or Intelligence is based upon vibrations or quantum waves. Quantum waves are in fact thoughts. Each Psyche or Pin-Prick of Light has the innate ability to form quantum waves, transmit quantum waves, receive quantum waves, process quantum waves, and store quantum waves. Quantum waves are thoughts. Thoughts are transmitted through vibrations or quantum waves. That's the way things really work at the quantum level. This is what has actually been experienced and observed. Every religion, every science, every observation, and every experience comes down to this truth.

The animating principle of the universe is described in every major religion using words that reflect the understanding of that time in history. In the language of the Incas, the largest empire in pre-Columbian America, the word for "human body" is "alpa camasca" which means literally, "animated earth". In Kaballah, or Jewish Mysticism, they talk about the divine name of God. The name that cannot be spoken. It cannot be spoken because it is a vibration that is everywhere. It is all words, all matter. Everything is the sacred word. The tetrahedron is the simplest shape that can exist in three dimensions. Something must have at least four points to have physical reality. The triangle structure is nature's only self-stabilizing pattern. In the Old Testament the word "tetragrammaton" was often used to represent a certain manifestation of God. It was used when talking about the word of God or the special name of God, Logos or primordial word. The ancient civilizations knew that at the root structure of the universe was the tetrahedral shape. Out of this shape, nature exhibits a fundamental drive toward equilibrium; Shiva. While it also has a fundamental drive towards change; Shakti.

Psyche or Life Force is the animating principle and the organizing principle in this universe. Psyche or Intelligence is described and mentioned in every science and religion, except for the naturalistic and atheistic religions that formally reject it. Just because they reject it and deny it doesn't mean that they are right. The whole of Science falsifies their materialistic, naturalistic, and atheistic beliefs. Quantum Mechanics and Quantum Field Theory falsify their beliefs. The false is falsified by the truth; and, the truth is repeatedly experienced and observed. Notice how these non-physical quantum truths are being repeatedly experienced and observed and verified.

In the Bible, the gospel of John usually reads, "In the beginning was the word" but in the original text the term used was "Logos". The Greek philosopher Heraclitus, who lived around 500 years before Christ, referred to the Logos as something fundamentally unknowable. The origin of all repetition, pattern and form. The Stoic philosophers who followed the teachings of Heraclitus identified the term with the divine animating principle pervading the universe. In Sufism the Logos is everywhere and in all things. It is THAT out of which the unmanifest becomes manifest.

Through God's Word of Command, the invisible unorganized non-physical energies and psyches are organized into quantum fields, quantum waves, quantum mechanisms, spirit matter or dark matter, mass, and physical matter. The unmanifest or non-physical becomes manifest or physical. That's the way things really work. It's obvious that the Gods, or Brahma, or Primal Psyches had to make the quantum fields BEFORE they could make and sustain the existence of mass or physical matter. Every world religion talks about these things, including Christianity. Only the Materialists, Naturalists, Darwinists, Nihilists, and Atheists deny the reality of these truths.

In the Hindu tradition Shiva Nataraja literally means "lord of the dance". The whole cosmos dances to Shiva's drum. All is imbued or ensouled with the pulsation. Only as long as Shiva is dancing can the world continue to evolve and change, otherwise it collapses back into nothingness (chaos). While Shiva is representative of our witnessing consciousness, Shakti is the substance or stuff of the world. While Shiva lies in meditation, Shakti tries to move him, to bring him into the dance.

Without God's ongoing Word of Command, our physical universe would dissolve and sink back into the chaos from whence it came. God's Word of Command or His Physical Laws or Physical Restrictions is what's keeping the atoms within your physical body from quantum tunneling away from you at will. God is literally holding your physical body together. Without these physical restrictions that God has created, the atoms in your physical body could quantum tunnel anywhere in the universe that they want to go. You would literally dissolve into nothing, and only your Psyche or Intelligence would remain, because psyche or energy is always conserved which means that it cannot be made and it cannot be destroyed. Every religion contains these truths hidden somewhere within it, except for the falsified religions that deny the existence of the non-physical, or the spiritual, or the quantum.

There is NO existence without some type of Organizing Force or Life Force, which we typically call Psyche, Soul, or God. Every religion contains this truth within it. It's unavoidable, because without psyche or life force or consciousness, we would not exist.

Doctrine and Covenants 93: 27-34:

26 The Spirit of truth is of God. I am the Spirit of truth, and John bore record of me, saying: He received a fulness of truth, yea, even of all truth;

27 And no man receiveth a fulness unless he keepeth his commandments.

28 He that keepeth his commandments receiveth truth and light, until he is glorified in truth and knoweth all things.

29 Man was also in the beginning with God. Intelligence [Psyche], or the light of truth, was not created or made, neither indeed can be.

30 All truth is independent in that sphere in which God has placed it, to act for itself, as all intelligence also; otherwise there is no existence.

31 Behold, here is the agency of man, and here is the condemnation of man; because that which was from the beginning is plainly manifest unto them, and they receive not the light.

32 And every man whose spirit receiveth not the light is under condemnation.

33 For man is spirit. The elements are eternal, and spirit and element, inseparably connected, receive a fulness of joy;

34 And when separated, man cannot receive a fulness of joy.

Without Intelligence, Psyche, or Consciousness there is NO existence! Without some kind of Organizing Force or Life Force, there would be NO quantum fields, NO physical matter, and NO physical life. It's obvious that the Gods had to design and make the quantum fields BEFORE they could design and make physical matter and physical life. It's obvious that quantum fields are made from energy. It's obvious that quantum fields are non-physical and pre-physical. The verified and proven existence of quantum fields, quantum mechanisms, dark matter, dark energy, intelligence, consciousness, life force, psyche, and non-physical types of energy FALSIFY Materialism, Naturalism, and their derivatives. The false is falsified by the truth; and, the truth is repeatedly experienced and observed. Every religion contains these truths, except for the falsified religions that we call Materialism, Naturalism, Darwinism, Nihilism, and Atheism.

Intelligence or Psyche or Energy is always conserved. It has always existed, and it will always exist. It cannot be made, and it cannot be destroyed.

The law of conservation of energy states that the total energy of an isolated system is constant; energy can be transformed from one form to another, but can be neither created nor destroyed.

The controlling Psyche can form the energy under its control into anything that it wants that energy to be. Energy is infinitely malleable or infinitely transformable. Energy can be transformed into anything, including quantum fields, gravity, dark matter or spirit matter, and mass or physical matter.

The fundamental axiom of Quantum Field Theory states that particles are born, and particles can die. This implies that particles are alive. You see, the controlling psyche determines what type of particle the energy under its control will be. It can be anything it wants to be anytime it wants to. The Gods created Universal Blueprints or Universal Laws that the psyches can choose to follow if they want to organize the energy under their control into physical matter or quantum fields or anything else for that matter. The Gods or Controlling Psyches provide the order and the organization for the rest of the universe; otherwise, everything would still be nothing more than random chaos.

This is the way things really work at the quantum level or the psyche level. There has to be some type of Organizing Force, Intelligence, Quantum Consciousness, Life Force, or Universal Blueprint; otherwise, we wouldn't exist, and you wouldn't be reading this.

Like yin and yang, the dancer and the dance exist as one. Logos also means unconcealed truth. He who knows the Logos, knows the truth. Many layers of concealment exist in the human world as Akasha has been swirled into complex structures concealing the source from itself. Like a divine game of hide and seek, we have been hiding for thousands of years, eventually forgetting about the game completely. We somehow forgot that there is anything to find.

Materialism, Naturalism, Darwinism, Nihilism, Atheism and their derivatives teach us that there is nothing to find at the quantum level or the psyche level. These falsified philosophies teach that psyche and non-physical quantum fields DO NOT EXIST. How does claiming that psyche and quantum fields do not exist explain what they are and how they work? It doesn't, does it? Materialism, Naturalism, Physicalism, and their derivatives completely lack explanatory power when it comes to the non-physical quantum fields that have been experienced and that must exist in order for physical matter to exist.

Once we realize there is one vibratory field that is the common root of all religions, how can we say "my religion" or "this is my primordial Om", "my quantum field"? The true crisis in our world is not social, political or economic. Our crisis is a crisis of consciousness, an inability to directly experience our true nature. An inability to recognize this nature in everyone and in all things.

https://www.amazon.com/gp/video/detail/B07K5HBZFV/

https://www.scripts.com/script-pdf/1356

Each individual Psyche or Intelligence has its own unique Quantum Field! Every verified and proven science and religion on this planet presents us with this very same truth.

Materialism, Naturalism, Darwinism, Nihilism, Atheism, and Creation Ex Nihilo are indeed "The Crisis" in our modern-world. We have indeed forgotten who we are and what we are.

John 8: 32:

32 And ye shall know the truth, and the truth shall make you free.

Doctrine and Covenants 76: 19-23:

19 And while we meditated upon these things, the Lord touched the eyes of our understandings and they were opened, and the glory of the Lord shone round about.

20 And we beheld the glory of the Son, on the right hand of the Father, and received of his fulness;

21 And saw the holy angels, and them who are sanctified before his throne, worshiping God, and the Lamb, who worship him forever and ever.

22 And now, after the many testimonies which have been given of him, this is the testimony, last of all, which we give of him: That he lives!

23 For we saw him, even on the right hand of God; and we heard the voice bearing record that he is the Only Begotten of the Father —

24 That by him, and through him, and of him, the worlds are and were created, and the inhabitants thereof are begotten sons and daughters unto God.

Conclusion

If you can read this and understand it, you are literally a son or daughter of God. Both your spirit body and your physical body are descended from the Gods. Adam was a son of God, both physically and spiritually. All of this chaotic energy was organized by the Power of God's Word or God's Word of Command.

This script only covers the first half-hour of this two-hour movie, or about one-fourth of the movie; but, that's good enough for my purposes here. You will have to buy or rent the movie in order to see and experience the rest of it. The music within the movie is good for meditation, and can actually put you to sleep if you are tired. That's a bit different than what we typically get from Hollywood or Movie Land.

The Theory of Relativity and the Speed-of-Light are the pinnacle of Classical Physics and Physical Reality. Once an object achieves the speed-of-light, then it is no longer subject to physical limitations, and it switches over to Quantum Mechanics and Quantum Field Theory instead. Remember, Quantum Mechanics is Energy Mechanics, or the scientific study of how psyche or energy really works at the quantum level.

According to both the Theory of Relativity and Quantum Mechanics, when an object achieves the speed-of-light, it enters the quantum realm and transmutes into something else. Time stops; mass, entropy, and resistance to acceleration cease to exist; distance and length contract to zero; space-time, locality, and physical limitations cease to be a factor; and, the quantum object becomes capable of infinite velocity or can quantum tunnel at will to any destination of its choosing. At the speed-of-light, objects enter into a whole other dimension, a whole other reality, and a whole other set of set of physical laws that we call Quantum Mechanics. The rules change. Reality and existence change. Physical limitations literally cease to exist. The quantum object effectively becomes omnipresent and omniscient. Conservation of Psyche or Conservation of Energy takes over instead, which means that psyche or energy is eternal and everlasting without a beginning of days or an end of years. According to the Quantum Law of Thermodynamics, entropy or thermodynamics ceases to exist at the quantum level, and everything becomes syntropic in the Spirit World or the Quantum realm.

Mark My Words

The Law of Psyche

Web Page:

Psyche has been experienced and observed by Out-of-Body Explorers and Near-Death Experiencers. The First Law of Thermodynamics tells us that Psyche or Energy is conserved, which means that it is eternal and everlasting. It cannot be made, and it cannot be destroyed. Psyche or Energy is syntropic which means that it is always conserved both at the quantum level and the physical level.

I prefer the scientific approach. I define science as observation and experience. Psyche has been experienced and observed; therefore, we KNOW that it exists. Quod erat demonstrandum!

Law of Psyche: The Law of Psyche states that every massless quantum particle is some type of Psyche or Intelligence. Psyche is the innate intelligence within all the different forms of energy which gives that energy the inherent ability to understand, follow, and obey God's Laws and God's Commands. Every massless elementary particle is some type of Psyche because it is psychic, quantum, telepathic, makes choices, can self-generate or self-propagate, can think, can communicate, can transmute, can move, can phase-shift, can collapse its own wave function, is perceptive, is sentient, is omniscient, is omnipresent, is intelligent, is conscious, is alive, must travel at the speed-of-light from our perspective at the physical level, is prescient, and without any physical limitations imposed upon it can quantum tunnel at will in its native original environment at the quantum level. Remember, every massless quantum particle is some type of Psyche, or Intelligence, or Life Force. Psyche is seen by Out-of-Body Travelers as a pinpoint of light or a Photon. Each Photon is a type of Psyche or Intelligence. A Psyche is a Photon, a massless elementary particle. If it looks like a Photon and acts like a Photon, then it is some type of Psyche. Every massless quantum is a Psyche. Psyche USES mass, energy, or matter to get things done. This is what has been experienced and observed.

Psyche has been hiding in plain sight all the time where nobody can see it nor find it because they aren't looking for it; but, there's no great mystery here. Psyche is seen as a Photon BOTH at the quantum level and the physical level because it is a Photon. Psyche or Intelligence IS what it appears to be, a Photon of Light or a Pinpoint of Light. No surprise there. Psyche is what it has been experienced and observed to be – a massless quantum of energy that looks like and acts like a Photon. Psyche, whenever it has been seen both at the quantum level and the physical level, looks like a Photon or a Pinpoint of Light.

I know that it's hard to believe because we have been trained, brainwashed, and conditioned all of our lives not to believe it; but, it is true, nonetheless. Psyche has been experienced and observed. Every massless quantum or every massless elementary particle looks like a photon and acts like a photon because is a Controlling Psyche who USES energy, mass, or matter to get things done both at the quantum level and the physical level.

Mark My Words

Experiential and Observational Proof of Psyche

Web Page: https://psyche-ontology.com/psyche-experienced-and-observed/

The Materialists, Naturalists, Darwinists, Nihilists, Behaviorists, Determinists, Physical Reductionists, and Atheists teach and believe that there is NOTHING fundamental about energy or consciousness. I have even heard them say the words, "There is nothing fundamental about energy." These people teach us and tell us that Psyche or Non-Local Consciousness DOES NOT EXIST. These people erroneously teach and believe that ONLY entropic physical matter exists; consequently, these people teach and believe that energy or consciousness is an emergent property of entropic physical matter. Whether they realize it or not, these people also teach and believe that entropic physical matter and entropy are conserved. They got their priorities and science upside-down.

Psyche is the innate intelligence within all the different FORMS of energy which gives that energy the inherent ability to understand, follow, and obey God's Laws and God's Commands. Energy is psychic and intelligent. Energy is sentient, conscious, aware, and alive. Psyche is Energy, and Energy is Psychic. Energy or Psyche is the fundamental unit of reality because Psyche, Intelligence, Life Force, or Energy is always conserved. In contrast, physical matter, particles, and entropy are simply different FORMS of energy. The FORM is never conserved. This is what has been experienced and observed. Particles or quanta are popping in and out of existence all the time. However, they don't really cease to exist as the Materialists and Naturalists claim. They simply change state, form, phase, or dimension while their underlying Intelligence or Energy is always conserved. Particles or quanta are made and then destroyed. They are NOT conserved. Only the underlying energy, psyche, or intelligence is conserved. This is what has been experienced and observed. Entropic physical matter is NOT the fundamental unit of reality after all. Entropic physical matter is MADE from many different FORMS of energy; and, the FORM is never conserved. This is the Ultimate Law of Thermodynamics.

I learn best through comparison and contrast by comparing what has been experienced and observed with the philosophical speculation and wishful thinking of the Materialists, Naturalists, Darwinists, Nihilists, and Atheists.

Psyche has been experienced and observed. Psyche is experienced as an immaterial, non-physical, intangible viewpoint in space. Psyche is seen or observed by out-of-body travelers as a pinpoint of light, a point particle of light, or a spark of light.

As scientists, we have to expand, enhance, and upgrade Theoretical Physics and make it explain what has already been experienced and observed. Psyche is like a photon, or a particle of light, or a quantum of light, except for the fact that Psyche actually emits light, or emits quanta, or emits and stores quantum waves. Thoughts and memories are quantum waves. Psyche is like a permanent, infinitely stable, perfectly conserved particle of energy or quantum of energy. Psyche is NOT ephemeral – coming and going – transforming and being transformed – like the other elementary particles or quanta that have been discovered and observed. Psyche, and the energy within psyche, is conserved. It's stable. It's eternal and everlasting. Because Psyche is conserved, its FORM never changes. Psyche is always a pinprick of light, or a point particle of light, or a spark of light. Psyche is like a Starbase that instantiates, or starts, or makes the other quanta or quantum waves. The traveling quanta have to come from someplace, and they come from the Starbases that we call Psyche. Psyche makes and stores quantum waves. That's how Psyche is able to communicate with other Psyches – through quantum waves. Psyche ends up being the fundamental unit of reality, or the conserved unit of reality. Psyche is Energy, and Energy is Psychic.

The vacuum of space is NOT empty. It is comprised of many different fields of energy which are omnipresent in time and space. The various different Quantum Fields were MADE from energy. "Particles" or quanta are excitations or "packets of energy" within Quantum Fields. Psyche is the innate intelligence within ALL the different FORMS of energy which gives that energy the inherent ability to understand, follow, and obey God's Laws and God's Commands.

The **gluons** binding quarks together comprise 99% of the energy or mass in a proton and neutron. The quarks themselves comprise only 1% of the energy or mass in a proton and neutron. The massive energy of the **gluons** within a proton, as compared to the rest energy of the quarks alone in the QCD vacuum, accounts for almost 99% of the mass within a proton. It's ALL made from energy, not entropic physical matter! The gluons or energy are what give the neutrons and protons stability and make them last essentially for forever. Proton decay has NOT been observed. Why? It's because the underlying energy is always conserved. Gluons and quarks are just different FORMS of energy. **Gluons** are viciously strong attractive forces within the Quantum Chromodynamic Field. Who is making these quarks and gluons ACT this way rather than some other way? It has to be Someone Psyche. Therefore, Psyche or Consciousness ends up being the fundamental unit of reality – NOT quarks and gluons.

Protons and neutrons can be transformed into each other which means that they are NOT conserved; but, the underlying energy is always conserved. Remember, the FORM of the energy is NEVER conserved; but, the underlying energy is always conserved. If the Gods were to pull all of the **gluons** out of this physical universe into a different dimension or universe, the whole thing would dissolve back into the chaos or anarchy from whence it came. The gluons, quarks, forces, fields, laws, restrictions, order, and organization were MADE, which means that entropy and physical matter were also made. Anything that is made or organized is NOT being conserved. In contrast, the underlying Energy or Psyche or

Intelligence was NOT made and it cannot be destroyed because it is always conserved. This is what has been experienced and observed.

I define Science as observation and experience; and, I have observed that the observations and experiences of the human race as a whole FALSIFY Materialism, Naturalism, Darwinism, Nihilism, Atheism, and their derivatives. Contrary to what these people claim, entropic physical matter is NOT the only thing that exists. Entropic physical matter is NOT the fundamental unit of reality. Materialism, Naturalism, Darwinism, Nihilism, and even Atheism are FALSIFIED by Quantum Field Theory and the Ultimate Law of Thermodynamics. Psyche or Quantum Non-Local Consciousness ends up being the fundamental unit of reality. A Psyche Ontology ends up being the Ultimate Model of Reality.

We have to expand, enhance, and upgrade our Theories of Physics to include Psyche or the Conserved; otherwise, we will NEVER be able to explain what has been experienced and observed.

Quantum Field Theory supports and predicts the Ultimate Law of Thermodynamics which states that physical matter and the other FORMS of energy such as entropy are NOT conserved. It's all made from energy, and entropic physical matter consists of different FORMS of energy. Quanta or "particles", including physical matter, are packets of energy. Quanta or particles are changing FORM all the time, which means that they are NOT conserved. Quantum Field Theory supports these ideas. The energy in the quantum fields is conserved; whereas, the "particles", or physical matter, or quanta come and go which means that they are NEVER conserved. You can see the truthfulness of these assertions with your own eyes by studying the Feynman Diagrams. There is NOTHING fundamental about physical matter. Everything is made from different FORMS of energy, including physical matter. Made by whom? Organized by whom? By Someone Psyche! Who else would be able to organize energy into its different FORMS at the quantum level or the psyche level?

Packets of energy, fields of energy, quarks made from energy, gluons made from energy, and the Psyche or Intelligence within all that energy ends up being the fundamental basis of reality. Psyche or Intelligence ends up being the fundamental unit of reality or the causal unit of reality – the conserved unit of reality. In contrast, entropic physical matter is MADE from many different quanta or packets of energy that we often call "particles"; and, when it comes to the electrons and the virtual photons in the Feynman Diagrams, it is obvious that they are NOT conserved. They are being transformed all the time! ONLY the underlying Energy, or Psyche, or Intelligence is conserved. This is what has been experienced and observed!

The Secrets of Feynman Diagrams

https://www.youtube.com/watch?v=fG52mXN-uWI

https://syntropy.website/wp-content/uploads/2018/08/Feynman-Diagrams.zip

http://hyperphysics.phy-astr.gsu.edu/hbase/Forces/feyns.html

Remember, Psyche is the person who makes the particles or the quanta and starts them moving in the first place. Psyche or Intelligence is the causal force or organizing force behind the quantum fields, quantum packets of energy, "particles", and the choices being made during the particle interactions that are diagrammed on the Feynman Diagrams. Psyche or Intelligence is the reason why the quarks, gluons, bosons, and electrons ACT the way they ACT rather than acting some other way or not acting at all. Psyche or Intelligence explains why there is something rather than nothing. Psyche is the thing that instantiates quanta, makes quanta, or causes the "particles" to begin in the first place. Psyche is the thing that chooses or decides what type of quantum a newly generated energy packet will be. Psyche is the Organizing Force behind everything else. Psyche is like a permanent Starbase – it generates, transmits, receives, and stores quantum waves or packets of energy. Psyche is conserved. Psyche is the fundamental unit of reality and existence. By allowing Psyche and Quantum Mechanics in to play, we can literally explain everything that comes our way. Try it, you might like it!

Due to the observations and the experiences of the human race as a whole, I upgraded my science to Science 2.0. According to Science 2.0, the BEST and FASTEST way to find and know the truth is to live it and experience it first-hand for yourself, or to choose to trust someone who has. The Human Psyche is KNOWN by living it and experiencing it, especially while your physical brain and physical body are dead and gone. Science 2.0 allows all of the evidence into evidence and then pursues a preponderance of that evidence. That's how I was able to identify and discover the Ultimate Law of Thermodynamics. That's how I was able to FALSIFY Materialism, Naturalism, Darwinism, Nihilism, and Atheism.

Phenomenology or the Lived Experiences of the human race had a very powerful influence in changing and altering my life, my science, my philosophy of life, and my worldview. The Phenomenologists, Near-Death Experiencers, and Out-of-Body Travelers changed the way that I look at the world. Thanks to these people, I'm no longer the same individual that I used to be. I'm no longer a Materialist, Naturalist, Nihilist, and Atheist.

In order to fully comprehend Quantum Mechanics and Syntropy, you must understand and accept the impact and influence of Quantum Non-Local Consciousness (or Psyche) on quantum mechanisms. Without that knowledge and understanding, your comprehension of Quantum Mechanics will always be incomplete. You must find an interpretation of Quantum Mechanics and Syntropy that explains what the Human Psyche and Nature's Psyche are doing at the quantum level or psyche level in order to get things done for us at the physical level. Without this bridge between the quantum and the physical, your knowledge and understanding of science will always be incomplete and ineffective. You must know how Psyche or CHOICE fits into the picture, if you want to understand how science really works at every level of existence.

Here's a list of some of the observational experiences of Psyche or Intelligence that changed my life.

Psyche or Consciousness without a Spirit Body

Nowadays, I define Science as observation and experience; and, I treat experience as scientific evidence. After reading William Buhlman's account of how he left his physical body, asked to see his spirit body, and found that he was an immaterial viewpoint in space while looking at his spirit body, I started looking for corroborating evidence to verify what William Buhlman reported. I began looking for the people who had the experience of being an immaterial pinpoint of consciousness, a pinprick of light, or a spark on the ceiling instead of looking for themselves and seeing parts of their spirit body. I have been looking for the experience of Psyche and the observation of Psyche.

William Buhlman wrote the following:

Journal Entry, October 2, 1982

I hear the buzzing, engine-like sounds and will myself out-of-body. [He left his physical body behind.] **I step to the bedroom door and automatically request "Clarity now!" My vision improves, and I step through the door, into the living room. Still feeling a little out of sync, I verbally repeat my request with more emphasis, "Clarity now!" I feel my awareness and vision snap into place.**

My thoughts are clear, and I make a verbal demand, "I need to see the form I'm in now!" Instantly I feel an intense sensation of being drawn within myself. I'm suddenly different, weightless as though I'm floating in space. As I look forward I see a sparkling, bluish white form. For some reason, I seem to know that I'm looking at my nonphysical body from a different perspective. I stare in amazement at this form before me that shines and flows with energy and light. It looks like an energy mold created from a million tiny points of light; it radiates a bluish glow but appears to have a defined outer structure. The body of light before me is naked and is identical to my physical form. Even though my body looks firm, there is a noticeable energy motion and radiation present. I can see what appears to be an ocean of blue stars throughout my body. It's difficult to describe because the stars are stable yet moving at the same time; the light and energy of my [spirit] **body appear to change and flow almost like the waves of an ocean.**

As I stare at the body of light, it hits me that I must be in another body. Yet I can't perceive any form or substance; I'm like a viewpoint in space without shape or form of any kind. [He, his immaterial viewpoint in space, is pure psyche or intelligence or consciousness.] **As I reflect upon my new state of being, I feel a sensation of rapid motion and I'm instantly back within my physical body.**

Lying still and reviewing my experience, I'm struck by an inescapable conclusion: I must possess multiple energy-bodies. The form I just experienced was noticeably lighter (less dense) than even my second nonphysical body. I realize that the traditional view of our possessing two bodies — a physical body and a spiritual body — is far too simplistic; we are much more complex than this. Just as there are multiple nonphysical energy dimensions within the universe, each of us must consist of multiple energy-bodies or vehicles of expression.

Now I seriously wonder just how many nonphysical bodies or forms this involves. I suspect that there must be one within each dimension of the universe and that all of these are interrelated and connected, just as the physical body is connected to its first nonphysical (spiritual) body.

https://psyche-ontology.com/buhlman/

[See: Buhlman, W. L. (1996). *Adventures Beyond the Body: How to Experience Out-of-Body Travel.* New York: HarperCollins.]

For me personally, this was the turning point. This was the straw that broke the camel's back. After reading this book, and this journal entry in particular, I've KNOWN that Psyche or Intelligence is something completely different than one's spirit body. Since that moment, I've KNOWN that Psyche or Intelligence is an immaterial viewpoint in space – the part of us that has experiences and forms memories of those experiences. Ever since then, I've KNOWN that our spirit body and physical body don't have experiences and don't form memories while our Psyche is separated from them.

I treated this account as Scientific Evidence; and, I went searching for confirming evidence. Seek and ye shall find. I found what I was looking for, which tells me that it is true. Psyche is experienced as an immaterial viewpoint in space. Psyche is seen or observed as a point particle of light, a pinprick of light, or a spark of light. Psyche has been experienced and observed! That REALITY makes it Science because Science is observation and experience.

Science 2.0 allows ALL of the evidence into evidence, and then it pursues a preponderance of the evidence. The BEST and FASTEST way to find and know the truth is to live and experience it for yourself, or to choose to trust someone who has. If a phenomenon is real, then you should be able to find corroborating evidence; and I do, when it comes to Psyche or Quantum Non-Local Consciousness.

When people are separate from their spirit body and physical body, and they are viewing their physical body while they are an immaterial spark on the ceiling, they are experiencing their Psyche first-hand rather than their spirit body. It happens to some out-of-body travelers, but not all of them.

A more common out-of-body experience is to see parts of their spirit body – hands, arms, legs, and feet – but not their skull or eyes. If you close your eyes right now and try to sense where you are when your physical eyes are closed, you will sense that you (your psyche) is in the third eye position just behind your skull.

That's the position from which you perceive things as well while inside your spirit body exploring the astral plane.

Psyche has been experienced and observed. Our job as scientists is to figure out what it is and how it works rather than trying to explain it away, dismissing it, and pretending that it doesn't exist. Denial of evidence and rejection of evidence makes for bad science. Feynman Diagrams map or portray the interaction of "particles" or quanta within Quantum Fields. Psyche is the thing that makes those particles and starts those particles moving in the first place.

One of my most interesting and useful Scientific Discoveries came to me when I first realized that Intelligences or Psyches, whenever they are seen or observed, manifest as a Pinpoint of Light, a Spark of Light, a Pinprick of Light, or a Photon of Light BOTH at the quantum spiritual level and at the macro physical level.

Direct experiential evidence of Psyche is rather rare; but, I have found a few experiential accounts of the nature and functionality of Psyche or Non-Local Consciousness while it is separated from its spirit body. Most of them were found on YouTube. Here are a few of them:

A Point of Consciousness in a Wonderful Black Void and Pinpricks of Light

https://www.youtube.com/watch?v=quU1xPeOtWs

She went looking for her body, hands, and feet and couldn't find them. "I was a point of consciousness," she said. She also saw other pinpricks of light that she recognized as souls that had passed on.

This NDE and OBE is probably the closest to what I experienced when I died. I remember floating in a crystal sharp, bright, black, obsidian void; and, for the first time in memory, I was completely and totally at peace. A part of me died. My five decades of fear and anxiety died. I experienced some kind of spiritual rebirth.

I didn't see my spirit body, didn't look for my spirit body, and didn't see anything else either that I remember. It was bright, black, and I was totally at peace. I had finally found rest and peace for my troubled soul.

Based upon my experience, the Atheists are right, there's nothing there when we die; however, the Atheists are wrong, because I was there when I died.

Each person who dies has a different experience and comes back with a different perspective, which is why it's important to allow ALL of the evidence into evidence, and then pursue a preponderance of the evidence.

I seemed to experience many of the benefits of a Near-Death Experience and none of the side-effects or bad-effects – without having any kind of astounding or amazing experience worth reporting. I was finally at peace; and, it stayed with me to this very day. The fear and anxiety were gone. I'm finally at peace. I see it as a gift from God.

A Unit of Consciousness

She talks about not being in a spirit body anymore but being a unit of consciousness. She went on a tour of the universe with the Being of Light at the speed of thought. The Human Psyche or Human Intelligence can leave its spirit body behind and separate from its spirit body.

https://www.youtube.com/watch?v=WUjP9kKoXfU

Lightning Strike Near-Death Experience – Dr. Anthony Cicoria

When Dr. Anthony Cicoria was struck by lightning, he died. He, his spirit body, was outside his physical body. As he walked up the stairs, his spirit body dissolved beneath him and transformed into an orb of light. He, his Psyche, was looking at his spirit body and later that orb of light from an immaterial third-person perspective or that Third Eye Perspective. This NDE is Scientific Evidence or Empirical Proof that Psyche, or Intelligence, is a different entity than our spirit body and our physical body.

https://www.youtube.com/watch?v=WUXzj0Tczz4

Consciousness without a Body Attached to It

https://www.youtube.com/watch?v=DmBnCTuQUOc

She Described Herself as Pure Consciousness While Out-of-Body

https://www.youtube.com/watch?v=V7xWffB2nH0

No Body – Just Pure Consciousness

https://www.youtube.com/watch?v=1kql9eD9qO4

Near Death Experience of Barbara Wilcox

In the next NDE, Barbara Wilcox explains not having a body, or spirit, or a soul but being Total Intelligence.

https://www.youtube.com/watch?v=JQzap_jFXT8

The Human Psyche or Human Intelligence is something completely different than our spirit body; and, our spirit body is something completely different than our physical body.

Intelligence or Psyche pre-dated the birth and organization of our Spirit Body. Our Psyche is described as an immaterial viewpoint in space, a point of consciousness, a spark on the ceiling, a pinprick of light, a speck of dust, a mote, or total intelligence.

Our spirit body seems to be malleable and can take on different shapes at will. In non-consensus realities, your spirit body can take on any shape that you want it to be. An orb of light is a common shape used for describing a spirit body. For the spirit children of God, our human shape is the most common shape that we see whenever our spirit body is viewed; however, Satan, the devils, and the evil spirits have denied and rejected their heritage as the spirit children of God, and these evil spirits can and do take on any shape that they want – serpents, dragons, alien grays, spacecraft, Nordic aliens, purple monstrosities, forked tail, pitchfork, red skin, and horns.

Our physical body is a part of the Ultimate Consensus Reality, which means that there's not a lot that we can do to change our physical body at will, because our physical body was designed to stay the same.

Both our physical body and our spirit body have NO memorable experiences and form NO new memories while the Human Psyche is separated from them. It's the Human Psyche who forms and stores new memories and has experiences worth remembering. Our spirit body is simply a convenient way to interface with the spirit realm; and, our physical body is an essential necessity for interfacing with this physical reality, which is the Ultimate Consensus Reality.

Susan Volt Talks about the Divine Spark within Each of Us

https://www.youtube.com/watch?v=-Gig4lwKNfQ

The word "spark" is used quite often by NDErs and others to describe the Human Psyche or the Human Intelligence. It's the spark of life. Psyche is life. Psyche is the Life Force.

Even in science fiction – something like the *Transformer* movies – they talk about the Spark or the AllSpark, which is the Life Force within us and within them.

The series *Earth: Final Conflict* did a great deal of discussion about Psyches and our spirit or ghost, often portraying them as two different things. There's a lot of talk about Quantum Mechanics within that series, because Psyche and Quantum Mechanics are integrated together and inseparable. In that series, Psyche or Consciousness has NO physical limitations, which means that it isn't limited by the speed-of-light. Entropy and the speed-of-light limitation applies only to physical matter. They don't apply to quantum mechanisms, spirit matter, and psyche. Syntropy is the natural state in the quantum realm, or the psyche realm, or the spirit world. There are NO physical limitations at the quantum level or the psyche level.

The Materialists and Naturalists have a hard time accepting and understanding Quantum Mechanics and Psyche, because Quantum Mechanics or Transdimensional Physics is supernatural in nature and origin. Quantum Tunneling or Teleportation is a supernatural phenomenon that happens automatically at the quantum level or the psyche level but is greatly limited and restricted at the physical level by the physical limitations that God has imposed on physical matter. You don't want your physical atoms within your physical body quantum tunneling away on you one at a time until you dissolve into thin air. God has imposed physical limitations on physical matter to prevent that from happening to you.

It's interesting to observe that whenever the Human Psyche is separated from its spirit body, it's the Human Psyche or Human Intelligence who is having the experiences and forming the memories, and NOT the spirit body. The Life or the

Spark is something completely different than the spirit body and the physical body. It's always the Psyche or the Spark who has the experiences and forms the memories, not the spirit body and not the physical body.

Nicole Swann Is an Awareness Looking Out but No Body

https://www.youtube.com/watch?v=Bo9IAGeYG7A

It's our psyche, or intelligence, or non-local consciousness who has and experiences awareness. It's not our spirit body, and it's definitely not our physical body. Our physical body has no experiences and forms no memories while our psyche is separated from our physical body. The same can be said of our spirit body. It has no experiences and forms no memories while our psyche is separated from our spirit body.

Jessica Near-Death Experience of a Different Kind

In the following NDE, Jessica describes being a Point of Light, a Small Spec, and a Pinpoint of Light during her near-death experience. She, too, is describing Psyche or Non-Local Consciousness. Jessica also described the fact that we don't see everything and don't understand everything which is there to be seen and understood while out-of-body. Instead, there is ever-growing awareness and understanding, with still more to go that is never reached. Jessica could sense the presence of others, but she didn't engage with them during her NDE as much as others seem to do. Hers is the most "self-centered" NDE that I have encountered so far. Jessica's Psyche is very much the center of her universe.

https://www.youtube.com/watch?v=Ve6RG9K3qrA

https://www.youtube.com/watch?v=k_dKVndThVg

https://www.youtube.com/watch?v=VD0Pd7twFBY

There are many things about Jessica's NDE that are atypical, abnormal, weird, unusual, and strange – making some people believe that she's faking the whole thing. I include her in this short list, because she talks about being that immaterial spark during her NDE. I've never heard Jessica talk about her spirit body. She always describes herself as a spark; and because of my research, I define that "immaterial spark" or "viewpoint in space" as Psyche, Intelligence, or Non-Local Consciousness within my books.

It's interesting how they each describe something different, yet something similar as well.

Due to all the empirical evidence or observational evidence from NDEs and OBEs, I believe that I have finally found the Correct Model of Reality or the Ultimate Model of Reality.

"The Ultimate Model of Reality: Psyche Is the Ultimate Cause"

https://www.amazon.com/dp/B071NC9JK6

Psyche or Intelligence or Non-Local Consciousness is something completely different than a spirit body. The psyche and spirit body exist on different frequencies in different dimensions. The Psyche or Intelligence is eternal and everlasting without beginning of days or an end of years – PURE SYNTROPY.

In contrast, our spirit body had a point in time when it was organized by the Gods – a birth date. If you can read this, then your spirit body is a child of God, a child of your Heavenly Father and Heavenly Mother. Raw spirit matter is also PURE SYNTROPY, without beginning of days or an end of years. But, the birthdate of our spirit body or the organization of our spirit body did indeed have a beginning.

It's ALL matter and energy, all the way down to the most elementary of particles. If there is such a thing as a point particle, then Psyche or Intelligence is it. Psyche is some kind of smart dust or intelligent energy. Psyche or Intelligence is smaller than a string. Psyche is the thing that sets the frequency at which the string vibrates. Thoughts and memories are like quantum waves or vibrations within a string. The smaller can dwell within and control the larger. Psyche produces and controls the quantum waves or the vibrations within the strings. Psyche is like an infinite singularity.

Two particles of matter can occupy the same space at the same time if they are out of phase with each other, existing a different frequency or dimension from each other. Consequently, a spirit body and its assigned physical body as well as their Psyche can all occupy the same space at the same time because they are out of phase with each other. The quantum functionality known as phase-shifting allows the Psyche, the Spirit Body, and the Physical Body to occupy the same location or the same space at the same time. It works due to a feature known as Quantum Superposition. Different quantum waves can occupy the same space at the same time in a cumulative or additive fashion, forming a united whole, but also at the same time remaining completely separate from each other as well.

Due to the same Quantum Superposition feature, it's possible for your Psyche and your Spirit Body to be in two different places on opposite sides of our earth at the same time. Due to the fact that our physical body is part of the Ultimate Consensus Reality, this dual-presence feature doesn't seem to apply to our physical body. Although dual-presence is theoretically possible for a physical body thanks to Quantum Superposition, God seems to prevent that from happening where the physical body is concerned, because physical matter and physical bodies have been subjected to physical limitations and entropy by God in order to make our physical reality dependable, reliable, predictable, and controllable. There are NO physical limitations at the quantum level or the psyche level, so infinitely more is possible at the quantum level or the syntropy level.

Quantum Mechanics is infinitely more versatile and infinitely more powerful than Materialism, Naturalism, Nihilism, and Classical Physics. Quantum Mechanics has infinitely more explanatory power than Materialism and Naturalism do. Quantum Mechanics makes for better science and better philosophy than Materialism, Naturalism, and Classical Physics do.

WE KNOW that Psyche or Intelligence exists and that it is REAL because it has been observed as pinpricks of light and experienced first-hand from a first-person perspective. The existence of Psyche or Intelligence or Non-Local Consciousness has been VERIFIED by observation and experience.

In contrast, something like the Multi-Me Theory or the Parallel Worlds Theory has been FALSIFIED due to a complete lack of observation. Nobody has observed this phenomenon nor experienced it, outside of science fiction. Likewise, chemical evolution and the associated theory of evolution have been FALSIFIED due to a complete lack of observational evidence because chemical evolution of genes and proteins from atoms is physically impossible thanks to entropy or the second law of thermodynamics.

Do you see how that works?

Truths are repeatedly VERIFIED by being observed and experienced. In contrast, falsehoods and lies and fiction – like chemical evolution or the evolution of completely new and different life forms from random mutations and natural selection – are repeatedly FALSIFIED by a complete lack of verification or a complete lack of observational evidence.

According to Science 2.0, the BEST and FASTEST way to find and know the truth is to observe it, live it, experience it, and VERIFY it. The second-best way to find and know the truth is to use the *negating the consequent* version of the Scientific Methods to falsify and eliminate everything that is false so that only the truth remains. If you successfully eliminate everything that is false, then only the truth will remain. This is Logic 101.

Science 2.0: I Upgraded My Science

https://www.amazon.com/dp/B0771K6WTX

Materialism, Naturalism, Darwinism, Nihilism, and Atheism are FALSIFIED due to a complete lack of observational evidence, or a complete lack of supporting evidence, or a complete lack of verification. That's the same way that we have falsified the Multi-Me Theory or the Parallel Worlds Theory. Chemical evolution, spontaneous generation, abiogenesis, and macro-evolution have ALL been FALSIFIED due to a complete lack of observation. These things are physically impossible and prevented from happening by entropy or the second law of thermodynamics.

Every aspect of the Theory of Evolution has been FALSIFIED due to a complete lack of observation, or a complete lack of empirical evidence, or a complete lack of verification. Random mutations and natural selection have NEVER been caught in the act of designing and creating completely new and different life forms from the pre-existing life forms that God designed and created; and, evolution (genetic drift), random mutations, and natural selection did NOT exist until AFTER God designed and created the genes, genomes, proteins, eyes, brains, and life forms in the first place. These truths are so obvious that I sometimes wonder how I was able to overlook them for fifty years of my life.

Science is all about observation and experience. Go with the experienced and the observed; and then, get rid of all the rest – the philosophical speculation and wishful thinking.

Remember, Psyche or Quantum Non-Local Consciousness has been experienced and observed. Psyche or Intelligence is a massless quantum or a massless elementary particle; and, Psyche looks like and acts like a Photon whenever it is seen or observed, both at the quantum spiritual level and at the macro physical level. At every level of existence, Psyche or Intelligence uses mass, energy, or matter in order to get things done. Photons and other massless quantum particles have been experienced and observed which means that different types of Psyches or Intelligences have also been experienced and observed. The massless elementary particles within fermions, leptons, quarks, electrons, and neutrinos are Controlling Psyches or Controlling Particles who command and control the mass, energy, forces, and fields that surround them. This is what has been experienced and observed.

Mark My Words

My Time in the Void

I don't consider my experience to be an NDE. I don't think it happened when I was near-death during my suicide attempt, but sometime later during the withdrawal phase of my addiction when I was near death or wishing I could die. For me, withdrawal lasted six months after I decided to go cold turkey from all the different prescription drugs that they had me on and had gotten me addicted to. During this six-month withdrawal period, my sense of time was gone, my sense of reality was distorted, I was psychotic and delusional, my sense of right and wrong was gone, I was apathetic, I wanted to die, I couldn't remember how to turn on a computer or television, and I was in a lot of pain and confusion.

I was tripping the light fantastic. I found out later that one of the psychiatrists had diagnosed me as being schizophrenic. That was a misdiagnosis, because I was actually suffering from substance-induced psychosis and withdrawal symptoms; but, I didn't figure out any of this until years later. I didn't know what the hell was going on, but it was hell.

I didn't believe in God anymore. I was an Atheist. But, at a couple of the worst times during the withdrawal, I remember crying out, "God help me. I can't take this anymore." And apparently, He did. When the pain is severe enough, you will say and do anything to get it to stop.

At some point along the way, a part of me did die; and, for the first time in my existence, I finally achieved peace. I don't think it was a gift from my suicide attempt, but more of a gift from finally getting sober. Like I said, I lived for over six months without any real sense of time; and, I was very much in an atheistic frame of mind. I was delusional and hallucinating the whole time. I'm one of those who convinced himself that my NDE didn't happen – that I simply imagined it all – but the profound change in psyche, personality, psychology, and perspective was definitely real and life-changing.

After dreaming of the void or floating in the void completely at peace, my psychology completely changed from anxiety, dread, fear, paranoia, and anger to one of peace and compassion. It was a spiritual rebirth of some sort, and the peace has continued to last ever since.

I consider October 31, 2012, to be my birthday. By then I was back to the land of the living, and I fully realized how insane, psychotic, delusional, and paranoid had really been. By that day, I had finally started thinking rationally once again. All the addictive medications were out of my system. My brain had rebalanced or normalized, and the withdrawal period was coming to an end.

They wouldn't let me quit the Zoloft, but by October 31, 2012, they had had me tapering down the Zoloft to a maintenance dose for a month or more; and, I felt an infusion of happiness, peace, and joy. I was back, and better than before. Good enough.

The Blind Can See While Out-of-Body

Spiritual consciousness is like the internet. It continues to exist when the computer or the brain is turned off. Spiritual vision continues to exist, even though the physical brain was never able to see.

https://www.youtube.com/watch?v=gKyQJDZuMHE

https://www.youtube.com/watch?v=azIh8gsXVRg

Powerful Life Reviews

https://www.youtube.com/watch?v=gF_Dj6EduLY

Merging and Becoming One or Omniscient

Renee Pasarow - Near Death Experience

https://www.youtube.com/watch?v=qlFTanblpvq

https://www.youtube.com/watch?v=rSrHE8zkwYg

https://www.youtube.com/watch?v=xB-T78qgfHM

Eben Alexander: A Neurosurgeon's Journey through the Afterlife

https://www.youtube.com/watch?v=qbkgj5J91hE

Confirmed Out-of-Body Experiences

Saw and Described the Operating Theater

https://www.youtube.com/watch?v=J5_x8U7SR0I

Maria Sees a Tennis Shoe on a Ledge While Dead and Out-of-Body

https://www.youtube.com/watch?v=3gGqpxa32og

The Dead Person Describes the Operating Theater

https://www.youtube.com/watch?v=JL1oDuvQR08

A Shared-Death Experience (SDE)

https://www.youtube.com/watch?v=a1qGJfSZ_LQ

The Bubble of Protection, Quantum Tunneling, and Teleportation

https://www.youtube.com/watch?v=DmBnCTuQUOc

Near-Death Experience: Conversations with God

https://www.youtube.com/watch?v=Zrx8C2lxhJI

Rudolf Smit - The Self Does Not Die: 104 Cases of Verified NDEs

https://www.youtube.com/watch?v=2qJ5UGbBGhg

NDE Research Proves Afterlife Exists

https://www.youtube.com/watch?v=mcb2cQMTPRg

Dr. Jeffrey Long - God and the Afterlife - Science & Spirituality Collide

https://www.youtube.com/watch?v=SyhZV-LGtJ8

Dr. Gary Habermas - Near Death Experiences

https://www.youtube.com/watch?v=ac9pF32gRxU

Penny Satori – Documented Cases Near-Death Experiences

https://www.youtube.com/watch?v=yS2ITQzSPLk

https://www.youtube.com/watch?v=F6TRsTUj8WM

https://www.youtube.com/watch?v=MwXQaRwqUpk

https://www.youtube.com/watch?v=T_KJNQPPoZk

https://www.youtube.com/watch?v=n1JAPxxFkkA

https://www.youtube.com/watch?v=YkW0ikd8i7U

Shared-Death Experiences

https://www.youtube.com/watch?v=Z31cI73DI7M

https://www.youtube.com/watch?v=-0-R3nz0cdg

https://www.youtube.com/watch?v=IWjYjsh8i0w

https://www.youtube.com/watch?v=N8_2P8s77lc

William Buhlman Out-of-Body Experiences

William Buhlman has turned out-of-body experiences into a business, a science, and a way of life. There is no end to what he has available for free on the internet. What you won't get from William Buhlman is any contact with God or religion because he seems to have had no experience with either.

https://psyche-ontology.com/buhlman/

I used to be a Materialist, Naturalism, Nihilist, and Atheist; so, William Buhlman was the first person to get through to me and convince me that there's more to existence than this physical reality, because he treated his OBEs as a science and actually experimented with the phenomenon.

https://www.youtube.com/results?search_query=William+Buhlman

https://www.youtube.com/watch?v=v-Oa2lWrKOg

https://www.youtube.com/watch?v=FjbwXI2-0n8

https://www.youtube.com/watch?v=HtHEtWntLiw

https://www.youtube.com/watch?v=6HWGPSRLBTo

https://www.youtube.com/watch?v=4OR6Kiwlohw

https://www.youtube.com/watch?v=ZlZNmwCD1pA

https://www.youtube.com/watch?v=JEimkt3Wl98

https://www.youtube.com/watch?v=NaaUkJF2JMc

https://www.youtube.com/watch?v=Apw2WpuW60o

https://www.youtube.com/watch?v=Zy1-0wbRJpY

https://www.youtube.com/watch?v=IoQ9T7H4OrE

https://www.youtube.com/watch?v=gewT3DtsRWM

After Effects of NDEs

Alicia Fagan on some of the after effects of an NDE.

https://www.youtube.com/watch?v=71zXemHbHaE

Nancy Rynes on NDE after effects and adjustments.

https://www.youtube.com/watch?v=ii1UDGWi6Gk

The After Effects of My Personal Experiences

After my brush with death, spiritual death, and spiritual rebirth, I have experienced some of the positive effects of NDEs. I had fifty years of anxiety and fear. The fear is gone. For the first time in my remembered existence, I'm at peace. I'm not afraid of death.

I worked out a system with God to get answers to my questions. I want to know how things really work; and, whenever I ask God a question, He either tells me in my mind what book to read next where I find my answer, or He reveals to me the next morning while waking up the answer to my question if the answer is not easily available in one of the books that I own.

Sometimes, I'll spend a month or two writing a book based upon a hypnopompic flash of insight that came to me in answer to prayer while waking up the next morning. It can take months to put into words that single flash of insight, which I had months before in answer to my prayer.

I don't see dead people, and don't have visions or any of that. That doesn't seem to be my calling or gift. But, I do seem to get answers to my prayers whenever it comes to science, reality, and how things really work at all levels of existence or in every dimension of existence. Sometimes the answers have been quite surprising, and far outside the box of the natural sciences or the physical science.

Once I got rid of My Materialism, My Nihilism, and My Atheism, it opened me up to infinitely more. Materialism, Naturalism, and their derivatives are the greatest bane and hindrance to scientific discovery that mankind has ever created. Quantum Mechanics and Psyche are supernatural in nature and origin. The very existence of Psyche and Quantum Mechanics as well as verified proof of quantum mechanisms FALSIFIES Materialism, Naturalism, Darwinism, Nihilism, Behaviorism, Scientism, Determinism, and Atheism.

Quantum Mechanics and Naturalism are mutually exclusive because Quantum Mechanics is supernatural and Naturalism states that the supernatural does not exist. They both can't be true at the same time. If one of them is true, then the other has to be false. Quantum Mechanics has observational evidence, experimental evidence, and verified evidence supporting it; whereas, there is NO evidence and can NEVER be any evidence to support the claim of Naturalism which states that the supernatural does not exist. It's logically impossible to provide evidence of something's non-existence. It can't be done. It's physically impossible.

Quantum Mechanics is supernatural. Its very existence falsifies Naturalism, Materialism, and their derivatives such as Darwinism and Atheism. Quantum Mechanics and Psyche are true and verified; whereas, Materialism, Naturalism, and their derivatives are false and have been falsified. If you choose to follow the evidence and choose to allow ALL of the evidence into evidence, it is immediately clear that Psyche and Quantum Mechanics are true because they have been observed and experienced, and it's equally clear that Materialism and Naturalism are false because their claims are impossible to observe and experience. Observation trumps philosophical speculation every time, or at least it should.

Science is observation. Materialism and Naturalism are unscientific, because their hidden assumptions or major premises cannot be observed nor experienced. Psyche and Quantum Mechanics have been observed and experienced. Materialism and Naturalism have not been observed nor experienced because they can't be. You can't experience nor observe the non-existence of something, such as the non-existence of the supernatural or the non-existence of Quantum Mechanics. You cannot observe the non-existence of God and Psyche, either; but, you can definitely experience these things and observe these things for yourself or choose to trust someone who has.

Can you see the difference between the observed and the verified in comparison to the unobservable, unverified, and falsified? The one is true and the other is false. Quantum Mechanics and Naturalism are mutually exclusive. If one of them is true, then the other one is automatically false. Psyche, Intelligence, Life, Syntropy, the Supernatural, and Quantum Mechanics are true, which means that Materialism, Naturalism, Darwinism, and their derivatives are false. It's really simple to understand once a person chooses to do so.

How Quantum Physics and Psychology Affirm NDEs

https://www.youtube.com/watch?v=rBshmwf-iaw

https://www.youtube.com/watch?v=V9KnrVlpqoM

Since I'm a scientist by nature, I also turn to Quantum Mechanics in order to explain Quantum Non-Local Consciousness or Psyche, Out-of-Body Experiences, and Near-Death Experiences. Quantum Mechanics is vastly superior to classical physics when it comes time to try to explain what's happening to us at the quantum level or the psyche level. Quantum Mechanics is a proven and verified science, and it's time that we as a race start using Quantum Mechanics to explain our observational experiences while we are out-of-body exploring the Astral Plane.

Your consciousness, the Human Psyche, is what makes your reality in any plane of existence that we can possibly imagine. Wherever you go, there you are. You don't cease to exist when your physical body and physical brain die or go offline for an extended period of time.

Scientists Who Have Had NDEs

Life-long scientists have interesting and unique perspectives on NDEs, especially after they have had one. By best friend is a patent holding scientist. He is also a seer and has had NDEs and visions. He has been in the presence of Jesus Christ our Savior, and he has had a lot of other amazing and interesting experiences while out-of-body or in-the-spirit.

I have gone out of my way to collect NDEs from scientists who have had them; and, TED talks have proven to be a good source.

https://www.youtube.com/watch?v=EtbiUsX1klk

https://www.youtube.com/watch?v=Y8WIdDz4RxI

https://www.youtube.com/watch?v=MyaBeHeRK6M

https://www.youtube.com/watch?v=rbnBe-vXGQM

https://www.youtube.com/watch?v=mMYhgTgE6MU

There are others:

https://www.youtube.com/watch?v=PMICW2aaplA

https://www.youtube.com/results?search_query=Eben+Alexander

I'm a scientist. I define Science as observation and experience. As I see it, we scientists have to take these observations and experiences seriously if we want to find out what things really are and how they truly work. The BEST and FASTEST way to find and know the truth is to live it, experienced it, and observe it for yourself, or to choose to trust someone who has.

Encounters with God or Jesus Christ

Whenever the being of light and love identifies himself, it's always Jesus Christ whom NDErs encounter and experience when they have separated from their physical body. These various NDEs convinced me that Jesus Christ does in fact exist. Should you find yourself in hell, Jesus Christ will come to you and get you out of there just for the asking. That's probably the most valuable lesson that I have learned from all of this.

Howard Storm Saved from Hell by Jesus

https://www.youtube.com/watch?v=UPj4wci_bcI

https://www.youtube.com/watch?v=Y9AjcfM75gI

Taught by Jesus Christ: Ralph Jensen Shares NDE (Near Death Experience)

https://www.youtube.com/watch?v=uWshfNnyEQA

Lee Stoneking Addresses UN General Assembly

https://www.youtube.com/watch?v=FYt8sv4vzQs

George Ritchie

https://www.youtube.com/watch?v=DsQIU2dNY44

Richard Met God

https://www.youtube.com/watch?v=HAR5MpYNnRE

Erica Had Two Different Types of Life Review during Her Encounter with God

https://www.youtube.com/watch?v=xG_hEi8E4U8

Ian McCormack - NDE - former atheist - near death experience

https://www.youtube.com/watch?v=sTU7MfOgDKM

Dr. Mary Neal - Raised from the Dead

https://www.youtube.com/watch?v=DX473dF7ChY

https://www.youtube.com/watch?v=ULsl92H-Noc

https://www.youtube.com/watch?v=63wY2fylJD0

A Wide Variety of NDE Encounters with Jesus Christ

https://www.youtube.com/results?search_query=nde+jesus

A preponderance of the evidence tells us that Jesus Christ is the being of light and the being of love whom we encounter after we die. A preponderance of the evidence tells us that Jesus Christ exists and that He truly rose from the dead. Science is supposed to be about observation and experience, not philosophical speculation. Science is supposed to be phenomenological, not just hypothetical wishful thinking. The Materialists, Naturalists, Darwinists, Behaviorists, Determinists, Nihilists, and Atheists don't want God to exist, so these people deliberately reject, censor, block, ban, and destroy any evidence to the contrary.

That's NOT science. That's the actions of religious dogmatism, fanatical extremism, wishful thinking, and blind faith. It's a perversion of rationality, logic, and science to block, ban, censor, ridicule, and destroy evidence. Materialism, Naturalism, Atheism, and their derivatives are based upon a refusal to look at evidence. *Refusing to look at evidence* is a logic fallacy.

Truth Is Known through a Preponderance of the Observational Evidence

According to Science 2.0, the BEST and FASTEST way to find and know the truth is to observe it, live it, and experience it for yourself, or to choose to trust someone who has. Through phenomenology or lived experiences, we can go directly to knowing the truth without having to run expensive and time-consuming science experiments. Furthermore, our physical science experiments definitely infer that the quantum level, quantum realm, and quantum mechanisms are real and truly exist; but, our physical experiments cannot provide direct evidence that the quantum level or psyche level is there nor explain what these things are truly like. When it comes to the quantum level, psyche level, or syntropy level, the BEST and FASTEST way to know that it is real and truly exists is to observe it, live it, and experience it for yourself or to choose to trust someone who has.

The preponderance of the evidence tells me that the NDE and OBE phenomenon is real, and that there really is an afterlife and a spirit world.

In contrast, we have the Materialists, Naturalists, Nihilists, and Atheists telling us that there is no such thing as psyche, God, non-locality, spirit, non-local consciousness, an afterlife, NDEs, OBEs, SDEs, Life Reviews, quantum mechanisms, supernatural mechanisms, and spiritual experiences.

How do they know?

They don't know. They can't know. It's impossible to know that something does not exist.

How do they prove that they are right?

The Materialists and Naturalists can't prove that they are right. There's NO way to prove that something doesn't exist. It's physically impossible, philosophically impossible, and logically impossible.

The major premises or primary assumptions of the Materialists, Naturalists, Nihilists, and Atheists are FALSIFIED by a complete lack of observational evidence or a complete lack of verification. You can't observe or verify that something doesn't exist.

ALL of the observational evidence, experiential evidence, empirical evidence, and eye-witness evidence tells us that the Materialist, Naturalists, Darwinists, Nihilists, Behaviorists, Determinists, and Atheists are wrong.

Materialism, Naturalism, and their derivatives are FALSIFIED by observational evidence and a preponderance of that evidence.

People choose to trust and believe the Materialists, Naturalists, Darwinists, Nihilists, Behaviorists, Determinists, and Atheists; BUT, these people haven't even observed, nor experienced, nor seen anything at all. It's impossible to observe that God does not exist. These people are flying on blind-faith and wishful thinking, and nothing more. When it comes to science, truth, and reality, you are much better served by choosing to trust and believe someone who has actually observed and experienced something.

Science 2.0 elevates observation and experience over the philosophical speculation of the Materialists, Naturalists, and Atheists. Science 2.0 allows all of the evidence into evidence and then pursues a preponderance of the evidence. Science 2.0 is the way that science should have always been done but wasn't. We should NEVER have let science be hijacked by the Materialists, Naturalists, Darwinists, Nihilists, and Atheists, because these people don't KNOW anything because they have never experienced anything and never observed anything.

Science 2.0: I Upgraded My Science

https://www.amazon.com/dp/B0771K6WTX

Remember, the BEST and FASTEST way to find and know the truth is to observe it, live it, and experience it for yourself, or to choose to trust someone who has. In contrast, know that Materialism, Naturalism, Nihilism, and Atheism are FALSIFIED due to a lack of observational evidence supporting their primary assumptions or major premises.

The truth is repeatedly observed, repeatedly experienced, and therefore repeatedly VERIFIED. In contrast, falsehoods like Materialism, Naturalism, and their derivatives are repeatedly FALSIFIED by a complete lack of observational evidence or a complete lack of verification. All you want is the truth, unless of course you are a Materialist, Naturalist, Darwinist, Nihilist, Behaviorist, or Atheist – then any old lie will do.

Mark My Words

—

Source

Putting Psyche Back into Psychology: Restoring Science to Consciousness

https://www.amazon.com/dp/B071NC987S

References

The Ultimate Model of Reality: Psyche Is the Ultimate Cause

https://www.amazon.com/dp/B071NC9JK6

Science 2.0: I Upgraded My Science

https://www.amazon.com/dp/B0771K6WTX

Quantum Neuroscience: The Answer to Life, the Universe, and **_Everything_**

https://www.amazon.com/dp/B079Z6QQQB

The Associated Websites

This book is copyrighted. Please do not hand it out for free to your friends.

Instead, please send them to one of our websites where they can acquire a free booklet related to our message by subscribing to our newsletter. This process will help us to spread our message to those who are the most interested and receptive.

http://markme.website/free-book/

http://markme.us/free-book/

—

The Associated Website: http://markme.website/

The Associated Forum: http://markme.us/forums/

The Associated Website: http://markme.website/nirvana/

The Associated Forum: http://markme.us/forums/forum/nirvana/

—

We will be hosting the occasional Free Giveaway or Promotional Offer for one of our books on our Facebook Page and our Twitter Following.

The Facebook Page for Mark My Words:
  https://www.facebook.com/MarkMyScience/

Twitter for Mark My Words:
  https://twitter.com/Mark_Me_Words

—

Have a nice day, and a wonderful life!

Mark My Words!

Other Books by This Author, Mark My Words!

My Author Page on Amazon:

https://www.amazon.com/-/e/B01IAEF2Y6

https://amazon.com/author/science

My Facebook Page:

https://www.facebook.com/MarkMyScience/

My Twitter Page:

https://twitter.com/Mark_Me_Words

—

1. "Summary Of: The Theory of Evolution Proves that God Exists"

https://www.amazon.com/dp/B01GQCWED6

https://www.amazon.com/dp/1521130485

07JUN2016

—

2. "The Theory of Evolution Proved to Me that God Exists: Why I Am No Longer an Atheist and Why I No Longer Believe in the Theory of Evolution"

https://www.amazon.com/dp/B01HZYBZ7K

https://www.amazon.com/dp/1521131228

04JUL2016

—

3. "The Scientific Method Proves That the Theory of Evolution Is False"

https://www.amazon.com/dp/B01IAAIRT2

https://www.amazon.com/dp/1521133611

11JUL2016

—

4. "The Second Comforter: Supping with Our Resurrected Lord Jesus Christ"

   https://www.amazon.com/dp/B01IAKHTY6

   https://www.amazon.com/dp/152113281X

11JUL2016

—

5. "Quantum Mechanics from a Non-Physical Spiritual Perspective"

   https://www.amazon.com/dp/B01J023TGU

   https://www.amazon.com/dp/1521132380

24JUL2016

—

6. "Using the Scientific Method to Eliminate the Usual Suspects and to Prove the Truth"

   https://www.amazon.com/dp/B01J6STHP0

   https://www.amazon.com/dp/1521133581

27JUL2016

—

7. "NATURE vs. NURTURE vs. NIRVANA: An Introduction to Reality"

   https://www.amazon.com/dp/B01JWRCSVA

   https://www.amazon.com/dp/1521132615

07AUG2016

—

8. "I Am Not a Creationist: So What Am I?"

   https://www.amazon.com/dp/B071XTM8XY

18APR2017

—

9. "The Ultimate Model of Reality: Psyche Is the Ultimate Cause"

https://www.amazon.com/dp/B071NC9JK6

22APR2017

—

10. "Putting Psyche Back into Psychology: Restoring Science to Consciousness"

https://www.amazon.com/dp/B071NC987S

22APR2017

—

11. "BioPsychoSocial: Including Psyche or Light into our Theoretical Models"

https://www.amazon.com/dp/B0713NDHVW

22APR2017

—

12. "God Is in the Light: God is light, and in Him is no darkness at all."

https://www.amazon.com/dp/B07168S37N

23APR2017

—

13. "Tripping the Light Fantastic: How Prescription Drugs Almost Killed Me"

https://www.amazon.com/dp/B071RJP9T8

23APR2017

—

14. "Scientific Proof of God's Existence: A Primer"

https://www.amazon.com/dp/B071713NNL

https://www.amazon.com/dp/1521325170

18MAY2017

15. "Science 2.0: I Upgraded My Science"

https://www.amazon.com/dp/B0771K6WTX

31OCT2017

—

16. "Scientific Proof of God's Existence: Finding God Where the Atheists Refuse to Look for Him"

  https://www.amazon.com/dp/B07B26CRHX

25FEB2018

—

17. "Quantum Neuroscience: The Answer to Life, the Universe, and Everything"

  https://www.amazon.com/dp/B079Z6QQQB

25FEB2018

—

18. "Syntropy in Defense of Quantum Mechanics: The Answer to Life, the Universe, and Everything"

https://www.amazon.com/dp/B07BPT3W8R/

25MAR2018

—

Scientific Observation: It is obvious that at the speed-of-light, photons are massless or non-physical. Quantum waves or photons are purely quantum objects or massless non-physical objects while traveling at the speed-of-light. Photons have to slow down to zero or splash-down, in order to manifest in the physical realm. This reality is obviously true because it has been experienced and observed. It's obvious that the massless or the non-physical exists. At the speed-of-light, photons are quantum phenomena, supernatural phenomena, spiritual phenomena, or non-physical phenomena. At the speed-of-light, we can't get our hands on them.

Science Needs to Prove Stuff to You

If your Science isn't proving stuff to you, then you ain't doing it right or your theories are false. Remember, a false theory can be used to point us to the truth, because the truth is often the opposite of any theory which has been demonstrated to be false. Therefore, false theories can be used to point us to the truth and thereby prove the truth.

THE TRUTH, whenever I find it and wherever I find it, IS parsimonious and makes logical sense. THE TRUTH tastes good and feels good. There is NO confusion or doubt associated with THE TRUTH. You just know that it is TRUE.

The primary goal of my books is to set people free from the chains of Materialism, as I have been set free. I hope to do for my readers what other authors have so kindly done for me. To accomplish that goal, I try to explain in every way possible why I am no longer an Atheist and why I no longer believe in Darwinism or Materialism. I have changed a great deal in the past few of years, and I am no longer the person that I used to be. My blinders have been taken off, and I have been set free to pursue my full potential. Come join me!

Have a nice day, and a wonderful life!

Mark My Words!

—

Original Release: 07AUG2016
Revised Edition: 07JUN2019
Current Edition: 07JUN2019